# PATTY'S INDUSTRIAL HYGIENE AND TOXICOLOGY

**Fourth Edition**

Volume I, Parts A and B
GENERAL PRINCIPLES

Volume II, Parts A, B, C, D, E, and F
TOXICOLOGY

**Third Edition**

Volume III, Parts A and B
THEORY AND RATIONALE
OF INDUSTRIAL HYGIENE
PRACTICE

# PATTY'S INDUSTRIAL HYGIENE AND TOXICOLOGY

## Fourth Edition
Volume II, Part B
Toxicology

**GEORGE D. CLAYTON**
**FLORENCE E. CLAYTON**
Editors

**CONTRIBUTORS**
R. E. Allan
D. M. Aviado
T. J. Benya
F. Cavender
H. H. Cornish
R. D. Harbison
A. P. Leber

A Wiley-Interscience Publication
**JOHN WILEY & SONS, INC.**
New York / Chichester / Brisbane / Toronto / Singapore

This text is printed on acid-free paper.

Copyright © 1994 by John Wiley & Sons, Inc.

All rights reserved. Published simultaneously in Canada.

Reproduction or translation of any part of this work beyond that permitted by Section 107 or 108 of the 1976 United States Copyright Act without the permission of the copyright owner is unlawful. Requests for permission or further information should be addressed to the Permissions Department, John Wiley & Sons, Inc., 605 Third Avenue, New York, NY 10158-0012.

*Library of Congress Cataloging in Publication Data:*
Patty's industrial hygiene and toxicology.

"A Wiley-Interscience publication."
Rev. ed. of: Industrial hygiene and toxicology / F. A. Patty.
Includes bibliographical references and index.
Contents: v. 1. General principles (2 v.)—v. 2 Toxicology pts. A, B.
1. Industrial hygiene.  2. Industrial toxicology.
I. Clayton, George D.  II. Clayton, Florence E.
III. Allan, R. E. (Ralph E.)  IV. Patty, F. A. (Frank Arthur), 1897–1981 Industrial hygiene and toxicology.
RC967.P37 1991     613.6'2          90-13080
ISBN 0-471-54724-7 (v. 2, pt. A)
ISBN 0-471-54725-5 (v. 2, pt. B)

Printed in the United States of America

10  9  8  7  6  5  4  3  2  1

Volume II of this series is dedicated to Paul D. Halley, past president of the American Industrial Hygiene Association, former Director of Environmental Health Services for AMOCO, (Industrial Hygiene, Safety and Toxicology), and Editor-in-Chief for eight years of the *AIHA Journal*, who, after retirement from the latter position, initiated contacts for the fourth revision of this volume, but was forced to relinquish its editorship because of illness.

# Contributors

**Ralph E. Allan**, J.D., C.I.H.,
University of California, Irvine,
California

**Domingo M. Aviado**, M.D.,
Atmospheric Health Sciences, Short
Hills, New Jersey

**Theordore J. Benya**, Ph.D.,
D.A.B.T., Ethyl Corporation, Baton
Rouge, Louisiana

**Finis Cavender**, Ph.D., D.A.B.T.,
C.I.H., Abilene Christian University,
Abilene, Texas

**Herbert H. Cornish**, Ph.D.,
D.A.B.T., Consultant, Ypsilanti,
Michigan

**Raymond D. Harbison**, Ph.D.,
Director, Center for Environment &
Human Toxicology, Alachua, Florida

**A. Philip Leber**, Ph.D., D.A.B.T.,
Chem-Tox Consulting, Akron, Ohio

# Preface

This book is the second of the fourth revision of Volume II of the Patty series. There will be a total of six books, twice the number in the third revision, all of which are now being processed by the publisher. Without exception, this expanded revision is the most comprehensive one on industrial toxicology produced to date.

In the preface of Volume IIA the uniqueness of the field of toxicology and the toxicologist were stressed. We also pointed out that it would be desirable that each book be characterized by class of chemistry within its contents; however, such an ideal arrangement was not practical because of the diverse flow of material to the editors. It is always in the best interest of authors that their manuscripts be published with dispatch.

Since beginning the revision of Volume II, three years of extensive work by a myriad of professionals has elapsed. The contributors to the revision of Volume II are to be commended for undertaking a monumental task. Considering the time constraints placed on all scientists, especially toxicologists, in addition to their normal workload, that they undertook this assignment reflects their deep and abiding interest in the profession and in the welfare of humanity.

GEORGE D. CLAYTON<br>FLORENCE E. CLAYTON

*San Luis Rey, California*
*November 1993*

# Contents

| | | |
|---|---|---|
| Useful Equivalents and Conversion Factors | | xiii |
| Conversion Table for Gases and Vapors | | xiv |

**16 Aromatic Nitro and Amino Compounds** — 947
Theordore J. Benya, Ph.D., D.A.B.T., and Herbert H. Cornish, Ph.D., D.A.B.T.

**17 Aliphatic and Alicyclic Amines** — 1087
Theordore J. Benya, Ph.D., D.A.B.T., and Raymond D. Harbison, Ph.D., D.A.B.T.

**18 Fluorine-Containing Organic Compounds** — 1177
Domingo M. Aviado, M.D.

**19 Aliphatic Hydrocarbons** — 1221
Finis Cavender, Ph.D., D.A.B.T., C.I.H.

**20 Alicyclic Hydrocarbons** — 1267
Finis Cavender, Ph.D., D.A.B.T., C.I.H.

**21 Aromatic Hydrocarbons** — 1301
Finis Cavender, Ph.D., D.A.B.T., C.I.H.

| | | |
|---|---|---|
| **22** | **Halogenated Benzenes** | **1443** |
| | A. Philip Leber, Ph.D., D.A.B.T., and Theordore J. Benya, Ph.D., D.A.B.T. | |
| **23** | **Chlorinated Hydrocarbon Insecticides** | **1503** |
| | A. Philip Leber, Ph.D., D.A.B.T., and Theordore J. Benya, Ph.D., D.A.B.T. | |
| **24** | **Phenols and Phenolic Compounds** | **1567** |
| | Ralph E. Allan, J.D., C.I.H. | |
| **Subject Index** | | **1631** |
| **Chemical Index** | | **1661** |

## USEFUL EQUIVALENTS AND CONVERSION FACTORS

1 kilometer = 0.6214 mile
1 meter = 3.281 feet
1 centimeter = 0.3937 inch
1 micrometer = 1/25,4000 inch = 40 microinches = 10,000 Angstrom units
1 foot = 30.48 centimeters
1 inch = 25.40 millimeters
1 square kilometer = 0.3861 square mile (U.S.)
1 square foot = 0.0929 square meter
1 square inch = 6.452 square centimeters
1 square mile (U.S.) = 2,589,998 square meters = 640 acres
1 acre = 43,560 square feet = 4047 square meters
1 cubic meter = 35.315 cubic feet
1 cubic centimeter = 0.0610 cubic inch
1 cubic foot = 28.32 liters = 0.0283 cubic meter = 7.481 gallons (U.S.)
1 cubic inch = 16.39 cubic centimeters
1 U.S. gallon = 3.7853 liters = 231 cubic inches = 0.13368 cubic foot
1 liter = 0.9081 quart (dry), 1.057 quarts (U.S., liquid)
1 cubic foot of water = 62.43 pounds (4°C)
1 U.S. gallon of water = 8.345 pounds (4°C)
1 kilogram = 2.205 pounds

1 gram = 15.43 grains
1 pound = 453.59 grams
1 ounce (avoir.) = 28.35 grams
1 gram mole of a perfect gas ≎ 24.45 liters (at 25°C and 760 mm Hg barometric pressure)
1 atmosphere = 14.7 pounds per square inch
1 foot of water pressure = 0.4335 pound per square inch
1 inch of mercury pressure = 0.4912 pound per square inch
1 dyne per square centimeter = 0.0021 pound per square foot
1 gram-calorie = 0.00397 Btu
1 Btu = 778 foot-pounds
1 Btu per minute = 12.96 foot-pounds per second
1 hp = 0.707 Btu per second = 550 foot-pounds per second
1 centimeter per second = 1.97 feet per minute = 0.0224 mile per hour
1 footcandle = 1 lumen incident per square foot = 10.764 lumens incident per square meter
1 grain per cubic foot = 2.29 grams per cubic meter
1 milligram per cubic meter = 0.000437 grain per cubic foot

To convert degrees Celsius to degrees Fahrenheit: °C (9/5) + 32 = °F
To convert degrees Fahrenheit to degrees Celsius: (5/9) (°F − 32) = °C
For solutes in water: 1 mg/liter ≎ 1 ppm (by weight)
Atmospheric contamination: 1 mg/liter ≎ 1 oz/1000 cu ft (approx)
For gases or vapors in air at 25°C and 760 mm Hg pressure:
    To convert mg/liter to ppm (by volume): mg/liter (24,450/mol. wt.) = ppm
    To convert ppm to mg/liter: ppm (mol. wt./24,450) = mg/liter

## CONVERSION TABLE FOR GASES AND VAPORS[a]

*(Milligrams per liter to parts per million, and vice versa; 25°C and 760 mm Hg barometric pressure)*

| Molecular Weight | 1 mg/liter ppm | 1 ppm mg/liter | Molecular Weight | 1 mg/liter ppm | 1 ppm mg/liter | Molecular Weight | 1 mg/liter ppm | 1 ppm mg/liter |
|---|---|---|---|---|---|---|---|---|
| 1  | 24,450 | 0.0000409 | 39 | 627 | 0.001595 | 77  | 318   | 0.00315 |
| 2  | 12,230 | 0.0000818 | 40 | 611 | 0.001636 | 78  | 313   | 0.00319 |
| 3  | 8,150  | 0.0001227 | 41 | 596 | 0.001677 | 79  | 309   | 0.00323 |
| 4  | 6,113  | 0.0001636 | 42 | 582 | 0.001718 | 80  | 306   | 0.00327 |
| 5  | 4,890  | 0.0002045 | 43 | 569 | 0.001759 | 81  | 302   | 0.00331 |
| 6  | 4,075  | 0.0002454 | 44 | 556 | 0.001800 | 82  | 298   | 0.00335 |
| 7  | 3,493  | 0.0002863 | 45 | 543 | 0.001840 | 83  | 295   | 0.00339 |
| 8  | 3,056  | 0.000327  | 46 | 532 | 0.001881 | 84  | 291   | 0.00344 |
| 9  | 2,717  | 0.000368  | 47 | 520 | 0.001922 | 85  | 288   | 0.00348 |
| 10 | 2,445  | 0.000409  | 48 | 509 | 0.001963 | 86  | 284   | 0.00352 |
| 11 | 2,223  | 0.000450  | 49 | 499 | 0.002004 | 87  | 281   | 0.00356 |
| 12 | 2,038  | 0.000491  | 50 | 489 | 0.002045 | 88  | 278   | 0.00360 |
| 13 | 1,881  | 0.000532  | 51 | 479 | 0.002086 | 89  | 275   | 0.00364 |
| 14 | 1,746  | 0.000573  | 52 | 470 | 0.002127 | 90  | 272   | 0.00368 |
| 15 | 1,630  | 0.000614  | 53 | 461 | 0.002168 | 91  | 269   | 0.00372 |
| 16 | 1,528  | 0.000654  | 54 | 453 | 0.002209 | 92  | 266   | 0.00376 |
| 17 | 1,438  | 0.000695  | 55 | 445 | 0.002250 | 93  | 263   | 0.00380 |
| 18 | 1,358  | 0.000736  | 56 | 437 | 0.002290 | 94  | 260   | 0.00384 |
| 19 | 1,287  | 0.000777  | 57 | 429 | 0.002331 | 95  | 257   | 0.00389 |
| 20 | 1,223  | 0.000818  | 58 | 422 | 0.002372 | 96  | 255   | 0.00393 |
| 21 | 1,164  | 0.000859  | 59 | 414 | 0.002413 | 97  | 252   | 0.00397 |
| 22 | 1,111  | 0.000900  | 60 | 408 | 0.002554 | 98  | 249.5 | 0.00401 |
| 23 | 1,063  | 0.000941  | 61 | 401 | 0.002495 | 99  | 247.0 | 0.00405 |
| 24 | 1,019  | 0.000982  | 62 | 394 | 0.00254  | 100 | 244.5 | 0.00409 |
| 25 | 978    | 0.001022  | 63 | 388 | 0.00258  | 101 | 242.1 | 0.00413 |
| 26 | 940    | 0.001063  | 64 | 382 | 0.00262  | 102 | 239.7 | 0.00417 |
| 27 | 906    | 0.001104  | 65 | 376 | 0.00266  | 103 | 237.4 | 0.00421 |
| 28 | 873    | 0.001145  | 66 | 370 | 0.00270  | 104 | 235.1 | 0.00425 |
| 29 | 843    | 0.001186  | 67 | 365 | 0.00274  | 105 | 232.9 | 0.00429 |
| 30 | 815    | 0.001227  | 68 | 360 | 0.00278  | 106 | 230.7 | 0.00434 |
| 31 | 789    | 0.001268  | 69 | 354 | 0.00282  | 107 | 228.5 | 0.00438 |
| 32 | 764    | 0.001309  | 70 | 349 | 0.00286  | 108 | 226.4 | 0.00442 |
| 33 | 741    | 0.001350  | 71 | 344 | 0.00290  | 109 | 224.3 | 0.00446 |
| 34 | 719    | 0.001391  | 72 | 340 | 0.00294  | 110 | 222.3 | 0.00450 |
| 35 | 699    | 0.001432  | 73 | 335 | 0.00299  | 111 | 220.3 | 0.00454 |
| 36 | 679    | 0.001472  | 74 | 330 | 0.00303  | 112 | 218.3 | 0.00458 |
| 37 | 661    | 0.001513  | 75 | 326 | 0.00307  | 113 | 216.4 | 0.00462 |
| 38 | 643    | 0.001554  | 76 | 322 | 0.00311  | 114 | 214.5 | 0.00466 |

## CONVERSION TABLE FOR GASES AND VAPORS (*Continued*)
(*Milligrams per liter to parts per million, and vice versa; 25°C and 760 mm Hg barometric pressure*)

| Molecular Weight | 1 mg/liter ppm | 1 ppm mg/liter | Molecular Weight | 1 mg/liter ppm | 1 ppm mg/liter | Molecular Weight | 1 mg/liter ppm | 1 ppm mg/liter |
|---|---|---|---|---|---|---|---|---|
| 115 | 212.6 | 0.00470 | 153 | 159.8 | 0.00626 | 191 | 128.0 | 0.00781 |
| 116 | 210.8 | 0.00474 | 154 | 158.8 | 0.00630 | 192 | 127.3 | 0.00785 |
| 117 | 209.0 | 0.00479 | 155 | 157.7 | 0.00634 | 193 | 126.7 | 0.00789 |
| 118 | 207.2 | 0.00483 | 156 | 156.7 | 0.00638 | 194 | 126.0 | 0.00793 |
| 119 | 205.5 | 0.00487 | 157 | 155.7 | 0.00642 | 195 | 125.4 | 0.00798 |
| 120 | 203.8 | 0.00491 | 158 | 154.7 | 0.00646 | 196 | 124.7 | 0.00802 |
| 121 | 202.1 | 0.00495 | 159 | 153.7 | 0.00650 | 197 | 124.1 | 0.00806 |
| 122 | 200.4 | 0.00499 | 160 | 152.8 | 0.00654 | 198 | 123.5 | 0.00810 |
| 123 | 198.8 | 0.00503 | 161 | 151.9 | 0.00658 | 199 | 122.9 | 0.00814 |
| 124 | 197.2 | 0.00507 | 162 | 150.9 | 0.00663 | 200 | 122.3 | 0.00818 |
| 125 | 195.6 | 0.00511 | 163 | 150.0 | 0.00667 | 201 | 121.6 | 0.00822 |
| 126 | 194.0 | 0.00515 | 164 | 149.1 | 0.00671 | 202 | 121.0 | 0.00826 |
| 127 | 192.5 | 0.00519 | 165 | 148.2 | 0.00675 | 203 | 120.4 | 0.00830 |
| 128 | 191.0 | 0.00524 | 166 | 147.3 | 0.00679 | 204 | 119.9 | 0.00834 |
| 129 | 189.5 | 0.00528 | 167 | 146.4 | 0.00683 | 205 | 119.3 | 0.00838 |
| 130 | 188.1 | 0.00532 | 168 | 145.5 | 0.00687 | 206 | 118.7 | 0.00843 |
| 131 | 186.6 | 0.00536 | 169 | 144.7 | 0.00691 | 207 | 118.1 | 0.00847 |
| 132 | 185.2 | 0.00540 | 170 | 143.8 | 0.00695 | 208 | 117.5 | 0.00851 |
| 133 | 183.8 | 0.00544 | 171 | 143.0 | 0.00699 | 209 | 117.0 | 0.00855 |
| 134 | 182.5 | 0.00548 | 172 | 142.2 | 0.00703 | 210 | 116.4 | 0.00859 |
| 135 | 181.1 | 0.00552 | 173 | 141.3 | 0.00708 | 211 | 115.9 | 0.00863 |
| 136 | 179.8 | 0.00556 | 174 | 140.5 | 0.00712 | 212 | 115.3 | 0.00867 |
| 137 | 178.5 | 0.00560 | 175 | 139.7 | 0.00716 | 213 | 114.8 | 0.00871 |
| 138 | 177.2 | 0.00564 | 176 | 138.9 | 0.00720 | 214 | 114.3 | 0.00875 |
| 139 | 175.9 | 0.00569 | 177 | 138.1 | 0.00724 | 215 | 113.7 | 0.00879 |
| 140 | 174.6 | 0.00573 | 178 | 137.4 | 0.00728 | 216 | 113.2 | 0.00883 |
| 141 | 173.4 | 0.00577 | 179 | 136.6 | 0.00732 | 217 | 112.7 | 0.00888 |
| 142 | 172.2 | 0.00581 | 180 | 135.8 | 0.00736 | 218 | 112.2 | 0.00892 |
| 143 | 171.0 | 0.00585 | 181 | 135.1 | 0.00740 | 219 | 111.6 | 0.00896 |
| 144 | 169.8 | 0.00589 | 182 | 134.3 | 0.00744 | 220 | 111.1 | 0.00900 |
| 145 | 168.6 | 0.00593 | 183 | 133.6 | 0.00748 | 221 | 110.6 | 0.00904 |
| 146 | 167.5 | 0.00597 | 184 | 132.9 | 0.00753 | 222 | 110.1 | 0.00908 |
| 147 | 166.3 | 0.00601 | 185 | 132.2 | 0.00757 | 223 | 109.6 | 0.00912 |
| 148 | 165.2 | 0.00605 | 186 | 131.5 | 0.00761 | 224 | 109.2 | 0.00916 |
| 149 | 164.1 | 0.00609 | 187 | 130.7 | 0.00765 | 225 | 108.7 | 0.00920 |
| 150 | 163.0 | 0.00613 | 188 | 130.1 | 0.00769 | 226 | 108.2 | 0.00924 |
| 151 | 161.9 | 0.00618 | 189 | 129.4 | 0.00773 | 227 | 107.7 | 0.00928 |
| 152 | 160.9 | 0.00622 | 190 | 128.7 | 0.00777 | 228 | 107.2 | 0.00933 |

## CONVERSION TABLE FOR GASES AND VAPORS (*Continued*)
(*Milligrams per liter to parts per million, and vice versa; 25°C and 760 mm Hg barometric pressure*)

| Molecular Weight | 1 mg/liter ppm | 1 ppm mg/liter | Molecular Weight | 1 mg/liter ppm | 1 ppm mg/liter | Molecular Weight | 1 mg/liter ppm | 1 ppm mg/liter |
|---|---|---|---|---|---|---|---|---|
| 229 | 106.8 | 0.00937 | 253 | 96.6 | 0.01035 | 277 | 88.3 | 0.01133 |
| 230 | 106.3 | 0.00941 | 254 | 96.3 | 0.01039 | 278 | 87.9 | 0.01137 |
| 231 | 105.8 | 0.00945 | 255 | 95.9 | 0.01043 | 279 | 87.6 | 0.01141 |
| 232 | 105.4 | 0.00949 | 256 | 95.5 | 0.01047 | 280 | 87.3 | 0.01145 |
| 233 | 104.9 | 0.00953 | 257 | 95.1 | 0.01051 | 281 | 87.0 | 0.01149 |
| 234 | 104.5 | 0.00957 | 258 | 94.8 | 0.01055 | 282 | 86.7 | 0.01153 |
| 235 | 104.0 | 0.00961 | 259 | 94.4 | 0.01059 | 283 | 86.4 | 0.01157 |
| 236 | 103.6 | 0.00965 | 260 | 94.0 | 0.01063 | 284 | 86.1 | 0.01162 |
| 237 | 103.2 | 0.00969 | 261 | 93.7 | 0.01067 | 285 | 85.8 | 0.01166 |
| 238 | 102.7 | 0.00973 | 262 | 93.3 | 0.01072 | 286 | 85.5 | 0.01170 |
| 239 | 102.3 | 0.00978 | 263 | 93.0 | 0.01076 | 287 | 85.2 | 0.01174 |
| 240 | 101.9 | 0.00982 | 264 | 92.6 | 0.01080 | 288 | 84.9 | 0.01178 |
| 241 | 101.5 | 0.00986 | 265 | 92.3 | 0.01084 | 289 | 84.6 | 0.01182 |
| 242 | 101.0 | 0.00990 | 266 | 91.9 | 0.01088 | 290 | 84.3 | 0.01186 |
| 243 | 100.6 | 0.00994 | 267 | 91.6 | 0.01092 | 291 | 84.0 | 0.01190 |
| 244 | 100.2 | 0.00998 | 268 | 91.2 | 0.01096 | 292 | 83.7 | 0.01194 |
| 245 | 99.8 | 0.01002 | 269 | 90.9 | 0.01100 | 293 | 83.4 | 0.01198 |
| 246 | 99.4 | 0.01006 | 270 | 90.6 | 0.01104 | 294 | 83.2 | 0.01202 |
| 247 | 99.0 | 0.01010 | 271 | 90.2 | 0.01108 | 295 | 82.9 | 0.01207 |
| 248 | 98.6 | 0.01014 | 272 | 89.9 | 0.01112 | 296 | 82.6 | 0.01211 |
| 249 | 98.2 | 0.01018 | 273 | 89.6 | 0.01117 | 297 | 82.3 | 0.01215 |
| 250 | 97.8 | 0.01022 | 274 | 89.2 | 0.01121 | 298 | 82.0 | 0.01219 |
| 251 | 97.4 | 0.01027 | 275 | 88.9 | 0.01125 | 299 | 81.8 | 0.01223 |
| 252 | 97.0 | 0.01031 | 276 | 88.6 | 0.01129 | 300 | 81.5 | 0.01227 |

[a] A. C. Fieldner, S. H. Katz, and S. P. Kinney, "Gas Masks for Gases Met in Fighting Fires," U.S. Bureau of Mines, Technical Paper No. 248, 1921.

# PATTY'S INDUSTRIAL HYGIENE AND TOXICOLOGY

Fourth Edition

Volume II, Part B
TOXICOLOGY

CHAPTER SIXTEEN

# Aromatic Nitro and Amino Compounds

Theodore J. Benya, Ph.D., D.A.B.T., and Herbert H. Cornish, Ph.D., D.A.B.T.

## 1 GENERAL CONSIDERATIONS

The aromatic nitro and amino compounds* are reasonably considered together because their acute toxic responses are often remarkably similar. The response, however, may vary markedly in degree and intensity. In addition, the most common metabolic fate of the nitroaromatic compounds is a reduction to corresponding amines. The toxic response anticipated from these compounds is tempered by the nature of other substitutions on the ring and on the excretory rates of the various metabolites, which may include phenols and conjugated products such as glucuronides and sulfates. These more water-soluble metabolites tend to be more readily excreted than their less soluble parent compounds. Many aromatic nitro and amino compounds are methemoglobin formers, and this is of immediate concern in acute human exposures. Some of these compounds are sensitizers and others are known animal and/or human carcinogens. In animal studies certain compounds of this group may produce liver, kidney, spleen, and/or testicular damage. Generally, liquid and vapor forms in both groups are readily absorbed through the intact skin.

---

*A literature search on individual chemicals was conducted in the following computer data bases maintained by the National Library of Medicine: Toxline, 1981–1992; Toxlit, 1981–1992; and Medline Files, 1980–1992.

---

*Patty's Industrial Hygiene and Toxicology, Fourth Edition, Volume 2, Part B*, Edited by George D. Clayton and Florence E. Clayton.
ISBN 0-471-54725-5   © 1994 John Wiley & Sons, Inc.

An extensive report by Bronaugh and Maibach (1a) demonstrated this in animals, and for selected compounds it was also demonstrated on human volunteers.

Major end uses include production of polymers, including flexible and rigid foams, and in dyestuffs, pesticides, pharmaceuticals, rubber products, explosives, pesticides, herbicides, fungicides, optical brighteners, photographic chemicals, coatings, adhesives, and paints. Some compounds may have limited use as solvents.

## 1.1 Aromatic Nitro Compounds

The aromatic nitro compounds are manufactured by the direct nitration of benzene, naphthalene, or other aromatic precursors with a hot mixture of nitric and sulfuric acids. Controlled nitration can produce mono- or disubstituted products. Nitrophenols may be produced by the hydrolysis of chloronitrobenzenes. Approximately 1.8 billion lb of nitroaromatics was produced in 1985. Approximately 5 percent of the benzene, toluene, and xylene production is used in the synthesis of nitroaromatics and essentially all commercial $N$-substituted aromatic chemicals are derived from nitroaromatics (1b).

Nitroaromatics are used to produce consumer products that accounted for nearly 10 percent of chemical industry sales in 1985. The polyurethane industry consumes almost half of the nitrobenzene production and essentially all of the dinitrobenzene production. This accounts for approximately half of the total derivative nitroaromatic volume (1b). A number of pesticides such as methyl- and ethylparathion are organophosphate derivatives of $p$-nitrophenol. The toxic response to aromatic nitro compounds may vary considerably; however, most of them cause methemoglobin formation on an acute basis along with anemia on more prolonged exposures. Other effects reported for some of these compounds include corneal opacity, altered metabolic rate, skin irritation, kidney, liver, and spleen damage, and bladder tumors. In experimental animals effects on reproduction have also been reported.

## 1.2 Aromatic Amino Compounds

The aromatic (or aryl) amino compounds are aromatic hydrocarbons in which at least one ring hydrogen has been replaced by an amino group. The aromatic amines are generally synthesized by nitration of the aromatic ring with subsequent reduction to the amine. They may also be formed by the reaction of ammonia with a chloro or hydroxy aromatic hydrocarbon. The most common toxic effects are methemoglobin formation, anemia, hematuria, and skin sensitization. A number of these compounds are known human carcinogens. Bladder tumors were recognized as resulting from occupational exposure to certain aromatic amines as early as 1895. Cancers of the bladder are most common but tumors have been reported at several other sites. Inhalation and skin absorption are the common routes of exposure.

The aromatic amines have industrial use as intermediates in the production of various dyes and pigments, and in the textile, rubber, polymer, pharmaceutical, paper, and other industries.

## 1.3 Physical Properties and Toxicity

The physical nature of a compound determines to a large extent its method of entry in the body. Compounds that have a degree of volatility or that may be present in the atmosphere as an aerosol of small particle size may readily penetrate to the alveoli of the lung and be absorbed into the systemic circulation. The absorption of aerosols from the lung depends primarily on the solubility of the particulate. If lipid soluble, these aerosols may readily pass through the alveolar membrane into the bloodstream. The absorption of vapor is variable and depends to a large extent on solubility of the compound in the bloodstream. Nonpolar compounds that are soluble in most organic solvents are also fat-soluble and readily penetrate the intact skin. Such compounds pass through the major skin barrier, the stratum corneum, by dissolving in and diffusing through the nonaqueous lipid matrix between the protein filaments (2). Many of the aromatic nitro and amine compounds are in this category. Control methods for safe handling in industry are more difficult because of the necessity of preventing skin absorption. Gastrointestinal absorption also occurs and, although it may not be an obvious industrial route of exposure, it is an important route of entry because of the swallowing of inhaled materials trapped in the upper respiratory tract.

The possible formation of carcinogenic nitrosamines in the stomach by reaction of amines with nitrite, dependent upon specific acidic pH ranges required for individual reactions to occur, presents a specialized problem in toxicology.

## 1.4 Toxicology

### 1.4.1 Methemoglobinemia

A common major effect of many of the aromatic nitro and amino compounds is their ability to convert hemoglobin to the oxidized ferric form, methemoglobin, sometimes referred to as ferrihemoglobin in the literature. Methemoglobin is unable to transport oxygen and symptoms of oxygen deficiency may begin to appear, the severity depending upon the amount of conversion to methemoglobin. Cyanosis may occur in humans at levels of 15 to 25 percent methemoglobin. Cyanosis in humans is a bluish color beginning with the lips and ears. Early symptoms of methemoglobinemia may include a feeling of euphoria and headache. At approximately a 40 percent level symptoms of lightheadedness, ataxia, and weakness may occur. Further elevations may cause tachycardia, dyspnea, and severe cyanosis. Death can ensue if treatment is not instituted.

Table 16.1 illustrates a variety of compounds that produce methemoglobin in the dog (3a).

In addition to aromatic nitro and amino compounds, nitrites, nitrates, chlorates, and quinones can also produce methemoglobinemia, as can a number of drugs. In the oxidized state the oxygen-carrying capacity of the hemoglobin is lost.

Some chemicals, such as sodium nitrite and inorganic hydroxylamine, can directly oxidize hemoglobin to methemoglobin. Considerable information suggests that the aromatic nitro and amino compounds are generally active methemoglobin pro-

Table 16.1. Methemoglobin-Forming Compounds—Dog (3a)

| Compound | Single Dose (mg/kg) | Route[a] |
|---|---|---|
| Acetanilide | 200 | T |
| o-Aminophenol | 20 | V |
| p-Aminophenol | 20 | V |
| Aniline | 50 | V, T |
| Dimethylaniline | 50 | T |
| Hydroxylamine | 5 | V |
| α- or β-Naphthylamine | 200 | T |
| p-Nitroaniline | 15 | V |
| Nitrobenzene | 200 | T |
| Nitroglycerin | 10 | V |
| Sodium dichromate | 60 | V |
| Sodium nitrite | 30 | V |

[a] T = stomach tube administration; V = intravenous administration.

ducers only after metabolism to a more reactive form. The following evidence supports such a conversion:

1. Dimethylaniline, a methemoglobin former, does not form methemoglobin when mixed with blood in in vitro studies.
2. Marked species differences in response to methemoglobin formers suggest they may not react immediately upon the red cell.
3. There is a delay in methemoglobin formation after administration of these compounds to experimental animals. For example, the maximum effect from nitrobenzene may not occur until 12 to 15 hr after dosing.

The conversion of inactive methemoglobin formers to an active form appears to be primarily a function of the liver mixed-function oxidase system, although metabolism to a reactive form also occurs via intestinal microflora (4).

The metabolites specifically responsible for methemoglobin generation have not been identified with certainty, although aromatic hydroxylamine and/or nitroso interactions appear to be involved. Phenylhydroxylamine has a very high potency for methemoglobin formation, evidently due to an erythrocyte recycling mechanism (5). Phenylhydroxylamine reacts with hemoglobin to form methemoglobin and nitrosobenzene. In the red cell, mechanisms exist for the reduction of nitrosobenzene to regenerate phenylhydroxylamine; thus a cyclic mechanism exists, resulting in the formation of many moles of methemoglobin per mole of phenylhydroxylamine. Nicotinamide adenine dinucleotide phosphate reduced (NADPH) is involved in this complex recycling mechanism, as is the enzyme glutathione reductase.

Nicotinamide adenine dinucleotide reduced (NADH)–methemoglobin reductase is the principal enzyme system involved in methemoglobin reduction in mammalian red cells. It is also known as cytochrome $b_5$ reductase, diaphorase, ferrihemoglobin reductase, NADH methemoglobin reductase, or NADH–cytochrome $b_5$ reductase. Genetic NADH–methemoglobin reductase deficiency can occur, which results in a distinct "blue" color of the skin. The enzyme deficiency usually occurs in red cells only; however, white cells, platelets, or all cells can be involved. Cellular deficiency can produce severe neurological disorders because the enzyme plays an important role in lipid metabolism by brain microsomes. An excellent review article by Charache, which includes a discussion on sulfhemoglobinemia, is recommended (6).

A secondary pathway for reduction of ferrihemoglobin is by reduced glutathione (GSH). In erythrocytes, glucose-6-phosphate dehydrogenase (G6PD) is responsible for maintaining low levels of reduced glutathione. Reduced glutathione accounts for only approximately 12 percent of the total erythrocytes' reductive capacity (7). Individuals with hereditary G6PD deficiency are at higher risk for developing methemoglobinemia by oxidative chemicals or their metabolites (8). Although this is a more common congenital deficiency than NADH reductase deficiency, individuals are not normally methemoglobinemic (9). Congenital abnormality of the globin portion of the heme molecule, owing to abnormal amino acid substitutions, can also occur, referred to as hemoglobin M disease. Such individuals are also more sensitive to methemoglobin-forming chemicals.

The presence of a so-called "dormant NADPH reductase" system has also been identified in erythrocytes; this system requires NADPH as a cofactor and appears to be activated only by the administration of electron carriers such as methylene blue. An extremely rare congenital deficiency of this enzyme has been reported (10).

Many investigators have reported molecular ratios (molar ratio of methemoglobin formed to dose of test compound) to illustrate the methemoglobin-forming capacity of compounds.

The molar ratio is considerably greater than 1 for many compounds, indicating that 1 mol of compound forms more than 1 mol of methemoglobin. Table 16.2 illustrates the methemoglobin-forming capacity of a number of aromatic nitro and amine compounds (3b).

There are marked species differences in response to methemoglobin formers; thus the experimental animal must be considered when evaluating the study and its relevance to humans. If a value of 100 is assigned to the sensitivity of the cat, then the sensitivity of other species is as follows for acetanilide: human, 56; dog, 29; rat, 5; rabbit, 0; and monkey, 0. For acetophenetidin, the sensitivities are as follows: cat, 100; human, 63; dog, 35; and rat, 5 (3c).

Although the metabolism of certain aromatic amines and nitroaromatics appears to be involved in mechanisms of methemoglobin formation, metabolism by alternate pathways may also be involved in decreasing the capacity of some compounds of this group to form methemoglobin. Compounds that are hydroxylated to phenols and then rapidly excreted as glucuronides or sulfates would be expected to have less opportunity to form methemoglobin than those with slower excretory rates. A comparison indicates that $p$-dinitrobenzene is a powerful methemoglobin former, whereas the dinitrophenols and dinitrocresols are relatively poor methemoglobin formers. See Section 2 for other effects of these compounds.

**Table 16.2.** Methemoglobin-Forming Capacity of Aryl Amino and Nitro Compounds in the Cat (3b)

| Compound | Molecular Ratio |
|---|---|
| Aniline | 2.5 |
|  | 0002.7 |
| Acetanilide | 1.0 |
| m-Phenylenediamine | 1.4 |
| Acetophenetidin | 0.14 |
| o-Aminophenol | 6.8 |
| Nitrosobenzene | 8.6 |
| p-Aminophenol | 3.6 |
| p-Aminophenol | 1.3 |
| Phenylhydroxylamine | 34.0 |
| Nitrobenzene | 0.86 |
| o-Dinitrobenzene | 1.9 |
|  | 3.7 |
| m-Dinitrobenzene | 7.1 |
|  | 7.8 |
|  | 6.4 |
| p-Dinitrobenzene | 55 |
|  | 198 |
| Trinitrobenzene | 4.8 |
| o-Nitrotoluol | 0.05 |
| m-Nitrotoluol | 0.04 |
| p-Nitrotoluol | Very slight |
| 2,4-Dinitrotoluol | 1.4 |
| 2,6-Dinitrotoluol | 0.55 |
| 2,4,6-Trinitrotoluol | 1.7 |
| m-Chloronitrobenzene | 2.3 |
| m-Aminonitrobenzene | 3.0 |
| 2,4-Dinitrochlorobenzene | 0.6 |
| p-Nitro-o-toluidine | 3.7 |

2,4-Dinitrophenol    p-Dinitrobenzene    2-Methyl-4,6-dinitrophenol

Because phenolic compounds tend to be rather rapidly excreted it may be that excretory rates may be partially responsible for the various capacities of these compounds to form methemoglobin.

Small amounts of methemoglobin that are normally formed within the red blood cell are usually reduced to hemoglobin by red cell methemoglobin reductase. This capacity, however, is overwhelmed in cases where methemoglobin levels are significantly increased. Certain rare individuals who are deficient in methemoglobin reductase may normally have as much as 50 percent of their normal hemoglobin present as methemoglobin. These individuals are at unusual risk from exposure to methemoglobin-forming chemicals. Newborn infants normally have a low level of NADH–methemoglobin reductase and consequently are also unusually susceptible to development of methemoglobinemia. Historically, "aniline dyes" used for indelible laundry ink have been associated with infant outbreaks of methemoglobinemia due to laundry marking of diapers, in some instances with fatalities of 5 to 10 percent. Such episodes of poisoning have been reported repeatedly since 1886 (11).

A recent incident of methemoglobinemia due to occupational exposure to di-nitrobenzene was provided by the Morbidity and Mortality Report Centers for Disease Control (12): On April 23, 1986 five steam-press operators at an Ohio rubber plant became ill with symptoms including discoloration of the hands, blue discoloration of the lips and nail beds, headache, nausea, chest pain, dizziness, confusion, and difficulty in concentrating. One worker suffered a seizure. Medical examination showed blood methemoglobin (MetHb) levels in workers as high as 41.2 percent. The workers had been using an adhesive to bond metal studs into rubber strips to be attached to car bumpers. Plant officials had stopped the steam-press operation and asked the Occupational Safety and Health Administration (OSHA) at the Ohio Industrial Commission to investigate. Five days later, a plant supervisor operated the steam press for about 2 hr in order for an industrial hygienist to take air samples. After the 2-hr simulation the supervisor's MetHb level was 12.5 percent. Because the cause remained obscure, technical assistance was requested from the National Institute for Occupational Safety and Health (NIOSH). The product being used was composed of carbon black (less than 5 percent), a proprietary curative system, and xylene as a solvent.

$p$-Dinitrobenzene was identified as a contaminant (1 percent) in the adhesive being used. A new lot of adhesive contained only a trace (0.03 percent) of 1,4-dinitrobenzene. The NIOSH investigation revealed that 1,4-dinitrobenzene was inadvertently formed during the manufacturing of one of the proprietary substances used in the adhesive. The manufacturer recalled all lots thought to be contaminated with significant quantities of 1,4-dinitrobenzene. The material safety data sheet was revised to indicate that trace quantities of 1,4-dinitrobenzene, which can cause cyanosis, may be present. NIOSH recommended that workers use butyl rubber gloves to avoid skin contact and that periodic medical monitoring be instituted. Monitoring, after resumption of operations, indicated no further problems.

This case history illustrates the rather insidious nature of this group of methemoglobin formers and the difficulties they may cause in an industrial setting.

### 1.4.2 Metabolism

As a rule, the aromatic nitro compounds are reduced to the amino derivative. Nitrobenzene, for example, undergoes both oxidation and reduction to form amino and phenol derivatives. Aromatic compounds containing more than one nitro group, such as dinitrophenol, usually have only one nitro group reduced.

When nitrobenzene was fed to rabbits all of the metabolites identified in the urine were free phenolic compounds of their conjugated products (13, 14). The major urinary metabolites identified in the urine of rats dosed with nitrobenzene included *p*-nitrophenyl sulfate (19.9 percent), *p*-hydroxyacetanilide sulfate (10 percent), and *m*-nitrophenyl sulfate (10.2 percent). One strain of rats also excreted small quantities of free and glucuronide-conjugated *m*-nitrophenol, *p*-nitrophenol, and *p*-hydroxyacetanilide. In the same study mice, in addition to the above, also excreted free and conjugated *p*-aminophenol. If the metabolites of nitrobenzene, the nitrophenols, are fed directly to rabbits or rats they are excreted in the urine within 24 hr. Specific data with 4-nitrophenol injected intravenously into rats showed that the urinary metabolites 4-nitrophenyl sulfate and 4-nitrophenyl glucuronide constituted approximately 75 percent of the radioactive dose. The presence of the phenolic group in 4-nitrophenol apparently facilitates the rapid formation and excretion of the corresponding sulfate and glucuronide conjugates. These studies also provided data that indicated approximately 20 percent of the glucuronide and sulfate conjugation occurred in the kidney with secretion directly into the urine (15). No *N*-acetylated urinary metabolites of the nitrophenols were reported in these studies of the metabolism of the nitrophenols, although they were reported as urinary metabolites of nitrobenzene. Thus it appears that phenolic conjugation takes precedence over nitro reduction and acetylation in the metabolism of 4-nitrophenol.

In a study of radioactive carbon-labeled 1,2-, 1,3-, and 1,4-dinitrobenzenes, Nystrom and Rickert (16a) reported the excretion of 85, 60, and 70 percent of the oral dose, respectively. This was largely in the urine, although significant biliary excretion was also reported. The major urinary metabolites of 1,2-nitrobenzene were *S*-(2-nitrophenyl)-*N*-acetylcysteine (42 percent of dose), 2-nitroaniline-*N*-glucuronide (4 percent), 2-amino-3-nitrophenyl sulfate (1.5 percent), and 2-(*N*-hydroxylamine)-nitrobenzene (1 to 2 percent). The major metabolites of 1,3-dinitrobenzene were 3-aminoacetanilide (22 percent), 4-acetamidophenyl sulfate (6 percent), 1,3-diacetamidobenzene (7 percent), and 3-nitroaniline-*N*-glucuronide (4 percent). The major metabolites of 1,4-dinitrobenzene were 2-amino-5-nitrophenyl sulfate (35 percent), *S*-(4-nitrophenyl)-*N*-acetylcysteine (13 percent), and 1,4-diacetamidobenzene (7 percent). The authors point out that 1,3-dinitrobenzene, which produces testicular toxicity in rats, is the only one of the three dinitro compounds that was metabolized exclusively by reduction. It has also been demonstrated in vitro that nitroreductase activity is located mainly in the microsomes and aromatic nitro reductions are thought to be mainly mediated by cytochrome P450 and/or the flavoprotein NADPH-cytochrome *c* reductase; gastrointestinal microflora also possess a large reservoir of nitroreductase activity (16b).

The in vivo and in vitro metabolism of nitroaromatic compounds was reviewed in 1987 (16c) and is recommended for additional background.

A study in humans occupationally exposed during the manufacture of technical-grade dinitrotoluene (DNT) (76 percent 2,4-DNT, 18.8 percent 2,6-DNT, 4.8 percent other isomers) demonstrated that the major urinary metabolites were 2,4-dinitrobenzoic acid and 2-amino-4-nitrobenzoic acid (17). These metabolites are qualitatively similar to those found in rats. Guest et al. (18) reported that technical-grade 2,4-DNT, which is a hepatocarcinogen in rodents, consisted of 76 percent of the 2,4-isomer. The 2,4-DNT isomer was metabolized under anaerobic conditions by human feces and ileal contents and by rat and mice cecal microflora to diaminotoluene. Under these anaerobic conditions a reductive sequence was observed involving nitroso intermediates that were identified by gas chromatography–mass spectroscopy. The authors suggest the reduction of the nitroso intermediates to the amine compound would involve a hydroxylamino intermediate that could not be isolated. It was concluded that the intestinal microflora of rodents represent a major site of reductive metabolism of 2,4-dinitrotoluene and may play an important role in the carcinogenic action of this compound in the rodent. This study emphasizes the need to consider the anaerobic site of the gastrointestinal tract in the metabolism of ingested chemicals and chemicals that may be absorbed by other routes and that then undergo biliary excretion into the gut.

Metabolic products of aniline excreted in the urine of rats and sheep accounted for approximately 80 percent of an oral dose within 24 hr. N-Acetylated metabolites accounted for most of this excretion. N-Acetyl-p-glucuronide accounted for 63 percent of the N-acetyl metabolites in sheep and pigs but in the rat N-acetyl-p-phenyl sulfate accounted for 56 percent of these metabolites (19). Extensive structure–activity relationship (SAR) research has been conducted on alkyl and halo derivatives of aniline to determine substituent effects upon methemoglobinemia activity (20).

Aromatic amines with free para or ortho positions are usually oxidized to the corresponding aminophenol. These metabolites can undergo further dehydrogenation to the more reactive quinoid derivatives or can conjugate with sulfuric or glucuronic acid.

Demethylation can also occur, as in the case of dimethylaniline in the dog and rabbit. A relatively large number of aromatic amines have been shown to undergo acetylation in the body, the rate of which may vary depending on whether genetically controlled slow or fast acetylation is involved. Biotransformation to reactive electrophilic species is discussed in Section 1.4.3.

Few metabolic data are available for o-, m-, and p-nitroanilines. Mate et al. (21) injected carbon labeled p-nitroaniline intravenously into rats and measured 76.5 percent excretion of the dose within 24 hr.

p-Amino-5-nitrophenol was the major metabolite (43.1 percent), with smaller quantities of p-phenylenediamine (26.1 percent) and p-nitroaniline (14.1 percent). These data indicate that considerable ring hydroxylation occurred during the metabolism of these aniline compounds, suggesting a major role of the hepatic microsomal enzyme system in their metabolism.

Benzidine, a well established mutagenic and carcinogenic amine, undergoes extensive biotransformation in the rat (22–26). An extensive study reported the separation of a minimum of 17 urinary or biliary metabolites in the rat after oral dosing with [$^{14}$C]benzidine. Some metabolites were acetylated, others were glucuronide or glutathione conjugates. The major urinary metabolite at all doses studied was characterized as 3-hydroxy-N,N'-diacetylbenzidine glucuronide. N-

Hydroxydiacetylbenzidine glucuronide, which was a minor urinary and biliary metabolite at low doses, became a major metabolite at high doses (50 mg/kg). Although a number of metabolites were mutagenic in the *Salmonella typhimurium* (strain TA98) assay containing liver S-9 fraction and β-glucuronidase, this latter compound exhibited a potency 10-fold greater than any other compound. Only 3 percent of the radioactive oral dose appeared in the feces of bile duct-cannulated rats compared with 63 percent of the dose found in the feces of intact rats, indicating that most of the normal fecal excretion of metabolites originated in the bile, presumably via liver metabolism.

Perhaps one of the best-known aromatic amines is β-naphthylamine (2-naphthylamine) because of its long history as a known carcinogen in humans. Metabolism of 2-naphthylamine may proceed by a variety of pathways, depending to some extent upon the species studied: *N*-hydroxylation followed by oxidation to 2-nitrosonaphthalene; rearrangement of the nitroso derivative to form 2-amino-1-naphthol; conjugation with sulfate to form 2-amino-1-naphthylmercapturic acid; oxidation to form 5,6-dihydroxydihydronaphthalene with conjugation to form 5-hydroxy-6-mercapturic acid; acetylation of the amino group; or conjugation at the amino group with sulfuric or glucuronic acid (27, 28). Further studies by others have generally confirmed these early reports and greatly expanded the metabolic findings. *N*-Hydroxy derivatives, which are relatively unstable, have been isolated in small quantities in the urine of treated dogs (28, 29) and in monkeys (30). In vitro studies have indicated that canine bladder microsomes and hepatic microsomes from humans, monkeys, rats, dogs, and pigs bring about *N*-hydroxylation of 2-naphthylamine. Dogs, which do not acetylate aromatic amines, excrete primarily conjugated sulfate or glucuronic acid derivatives. On the basis of animal studies and human epidemiology the International Agency for Cancer Research (IARC) (31) concluded that 2-naphthylamine is carcinogenic to humans, dogs, monkeys, hamsters, and mice but has little, if any, carcinogenic effect in the rat. Studies have not clearly demonstrated metabolic differences in the rat that might account for its apparent resistance to the carcinogenic effects of this compound.

### 1.4.3 Carcinogenicity

In 1984, more than 450,000 Americans died of cancer (32). It is speculated that one-third to two-thirds of all cancers may be accounted for by our total environment, which includes everything ingested, including alcohol and foods, anything smoked, sunshine, drugs, natural and contaminant materials in air, water, and soil, the workplace, and the home and social environment. For example, grilling a hamburger brings about profound changes in its chemical composition. Fat dripping onto the coals produces smoke which is highly irritating and contains a large number of complex chemicals, including carcinogens, which may become part of the hamburger.

Carcinogenicity induced by nitro and amino compounds is believed to occur as a result of metabolic activation to a reactive electrophilic species that is capable of reaching and alkylating genetic material as the ultimate target. In addition to the possibility of aromatic epoxidation, it is believed that *N*-hydroxylation can lead to a reactive species. This reaction appears to be followed by degradation to a reactive electrophilic arylnitrenium ion (Aryl-$N^+$-H). (33a–33c). $N$-Arylhydroxylamines are apparently not reactive at physiological pH and are rapidly conjugated with glucuronic acid and excreted in the urine. The conjugate may hydrolyze in the bladder at urinary pH, liberating the hydroxylamine and generating a reactive nitrenium ion (34). Arylhydroxylamines can also be activated by cytosolic acetyltransferase in the presence of acetyl coenzyme A to form $N$-acetoxyarylamine, which then generates the ultimate reactive arylnitrenium ion (35).

Enzymatic reduction of aromatic nitro compounds has also been recognized as a common metabolic pathway both in vivo and in vitro, as previously discussed. The reduction of aromatic nitro compounds to their corresponding hydroxylamino derivatives may also explain the carcinogenic action of such compounds as 2-nitrofluorine, 4-nitroguinoline $N$-oxide, certain heterocyclic nitro compounds, 4-nitrobiphenyl, and 4,4'-dinitrobiphenyl (36).

The potential for formation of $N$-arylnitrosamines or $N$-nitroso compounds (Aryl—NR—N=O) from nitrite reaction with secondary amines or diazo compounds following nitrite and primary aromatic amine (or heterocyclic amine) reactions is discussed in Chapter 11 in Part A. Virtually all nitrosamines are carcinogenic.

In the case of diphenylamine, the nitrosamine or $N$-nitroso derivative was not expected to be carcinogenic because the substitution of two bulky phenyl groups would prevent oxidative formation of a reactive electrophilic intermediate, as is true for the corresponding $t$-butyl derivative. Yet diphenylnitrosamine appears to be a very weak bladder carcinogen. The mechanism is uncertain (37).

It is difficult to predict carcinogenicity on the basis of chemical structure, although sometimes it is successful. Stula et al. (38) investigated the carcinogenic potential of 4,4'-methylenebis(2-chloroaniline) because of its structural similarity to 3,3'-dichlorobenzidine and 4,4'-methylenebis(2-methylaniline), known rat carcinogens:

3,3'-Dichlorobenzidine
(DCB)

4,4'-Methylenebis(2-methylaniline)
(Me-MDA)

4,4'-Methylenebis(2-chloroaniline)
(Cl-MDA)

They demonstrated that all three agents were carcinogenic when high dose levels were included in the diet of rats.

Heredity also appears to play a role in the development of tumors, and thus individuals may vary markedly in their response to various types of carcinogens. Specific patterns of heredity and susceptibility are undergoing considerable study at present. For example, it is known that acetylation of a chemical compound occurs at different rates in different strains of rats and also at different rates in the human population. One step in the metabolism of some carcinogens is acetylation, which may alter tissue distribution patterns and rates of excretion of the compound. This in turn may affect the response of the exposed animal or person. Many more complex patterns of inherited metabolic differences are known to exist.

One's overall view of carcinogens in our environment is imprinted to a large degree by the definition of the term "carcinogen" or "carcinogenic compound." When a carcinogenic response can be directly related to human exposure to a well-defined chemical, as may happen in a work population, the definition of the causative agent may be simplified. With most work populations, however, the situation is more complex. Worker turnover may be high; workers move from one job to another within the plant; chemical exposure is often to a mixture of compounds; alcohol ingestion, smoking, and the home environment may complicate the response; and finally, a carcinogenic response may develop only after many years of exposure. In spite of these difficulties there are a number of human carcinogens that have been identified by epidemiologic studies of worker populations. IARC (32a) and the U.S. Department of Health and Human Services (USPHS) (32b) define two categories anticipated to be carcinogens as follows:

1. Known to be carcinogens:
   There is "sufficient evidence of carcinogenicity" from studies in humans "which indicate a causal relationship between the agent and human cancer."
2. Reasonably anticipated to be carcinogens:
   A. There is "limited evidence of carcinogenicity" from studies in humans, "which indicates that causal interpretation is credible, but that alternative explanations such as chance bias or confounding could not adequately be excluded."
   B. There is "sufficient evidence of carcinogenicity" from studies in experimental animals "which indicate there is an increased incidence of malignant tumors: (*a*) in multiple species or strains, or (*b*) in multiple experiments (preferably with different routes of administration or using different dose levels), or (*c*) to an unusual degree with regard to incidence, site or type or tumor, or age at onset. Additional evidence may be provided by data concerning dose-response effects, as well as information on mutagenicity or chemical structure."

A total of only 23 "substances or groups of substances, and medical treatments" are listed in the first category, "known to be carcinogens." Three of these substances are aromatic amines, well-known carcinogens in the history of chemical exposures, and three are more complex drugs that could be considered substituted amines.

The three aromatic amines are:

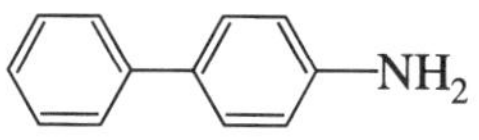

4-AMINODIPHENYL

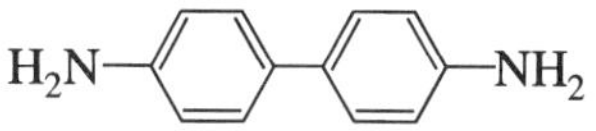

BENZIDINE

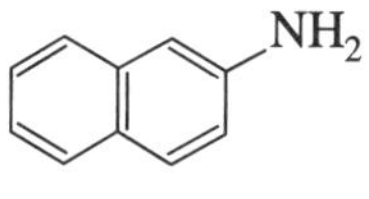

β-NAPHTHYLAMINE

Included in the category "reasonably anticipated to be carcinogenic" are approximately 200 other substances that are considered carcinogens. Approximately 10 percent of these are aromatic amines.

The above USPHS categories are rather stringent requirements for scientific evidence of carcinogenicity or anticipated carcinogenicity in humans. For example, a chemical with a weak positive carcinogenicity response in only one species of animals might not be included in either category given above, unless there were considerable additional supporting evidence.

IARC has definitions quite comparable to those of USPHS (32a). Additional categories would include:

3. The agent does not fall into any group, and is not classifiable as to its human carcinogenicity.
4. The agent is probably not carcinogenic to humans.

NIOSH and OSHA publish "Occupational Guidelines for Chemical Hazards," and in 1975 NIOSH developed a "Current Intelligence Bulletin" series which provides information on health and safety hazards that have gone unrecognized or are greater hazards than generally known. NIOSH also issues Criteria Documents with recommended occupational exposure standards. These agencies have a rather specific assignment of protecting the health of occupationally exposed workers. Thus one will find a number of aromatic nitro and amino compounds under NIOSH guidelines that should be regarded as "potential human carcinogens in the workplace" that are not included in the *Sixth Annual Report on Carcinogens* (32b). This is also true for the guidelines provided by the American Conference of Governmental and Industrial Hygienists (ACGIH) (39).

The NIOSH/OSHA, *Pocket Guide to Chemical Hazards*, June 1990 (40) states in Appendix B, "Thirteen OSHA Regulated Carcinogens":

Without establishing PELS, OSHA promulgated standards in 1974 to regulate the industrial use of 13 chemicals identified as occupational carcinogens:

(2-acetylaminofluorene, 4-aminodiphenyl, benzidine, bis-chloromethyl ether, 3,3'-dichlorobenzidine, 4-dimethylaminoazobenzene, ethyleneimine, methyl chloromethyl ether, alpha-naphthylamine, beta-naphthylamine, 4-nitrobiphenyl, N-nitrosomethylamine, and beta-propriolactone).

Exposures of workers to these 13 chemicals are to be controlled through the required engineering controls, work practices, and personal protective equipment, including respirators. See 29 CFR (Code of Federal Regulations) 1910.1003–1910.1016 for specific details of these requirements."

Eight of these 13 designated occupational carcinogens were aromatic amines and two were aliphatic amines. This indicates the carcinogenic potential within this group of compounds.

NIOSH (40) further points out in Appendix A, "NIOSH Occupational Carcinogens":

NIOSH has identified numerous chemicals that should be treated as occupational carcinogens even though OSHA has not identified them as such. In determining their carcinogenicity NIOSH uses a classification outlined in 29 CFR 1990.103 which states in part:

Potential occupational carcinogens means any substance, or combination or mixture of substances, which causes an increased incidence of benign and/or malignant neoplasms, or a substantial decrease in the latency period between exposure and onset of neoplasms in humans or in one or more experimental mammalian species as the result of any oral respiratory or dermal exposure, or any other exposure which results in the induction of tumors at a site other than the site of administration. This definition also includes any substance which is metabolized into one or more potential occupational carcinogens by mammals.

Interesting additions to the NIOSH workplace carcinogen list were phenyl-β-naphthylamine and 2-nitronaphthalene (41), which were included under the guidelines because they are metabolized to β-naphthylamine, a known human carcinogen.

This "occupational classification" accounts for many of the chemicals considered carcinogens in the occupational setting. For example, there are no dinitrotoluenes in either the 1991 U.S. Dept. of Health and Human Services *Sixth Annual Report* (32b) or in the 1974 OSHA list of 13 carcinogens (40). Commercial-grade dinitrotoluene (DNT) is approximately 76.5 percent 2,4-DNT and 18.8 percent, 2,6-DNT. A long-term study of commercial-grade trinitrotoluene showed a significant increase in the incidence of liver tumors in male and female rats. Studies with purified 2,4- and 2,6-dinitrotoluenes have provided equivocal data in several instances. Other related data are available on these compounds. Thus the carcinogenicity data are sufficient for NIOSH to recommend in Current Intelligence Bulletin No. 44 that technical-grade dinitrotoluene and the 2,6 isomer be "regarded as potential human carcinogens in the workplace" (42). The Environmental Protection Agency (EPA) has also reviewed the data on the dinitrotoluenes and has provided a cancer risk assessment for these compounds (43). In addition to OSHA and NIOSH, the Food and Drug Administration (FDA) and the EPA [which includes administration of FIFRA, TSCA, Clean Air Act, Federal Water Pollution Control Act, Safe Drinking Water Act, hazardous wastes or Resource Conservation and Recovery Act, uncontrolled waste sites or Comprehensive Environmental Response, Compensation and Liability Act ("Superfund"), and several other areas including ionizing radiation] have authority to regulate food additives, drugs, pesticides, air, water, and waste or dump sites. Other regulatory agencies also charged with regulating carcinogens include the Consumer Product Safety Commission and the U.S. Department of Agriculture.

The publications of Imbus (44a), Paustenbach (44b), and Cothern (44c) provide additional background information concerning governmental regulations of carcinogens. For additional information of specific aromatic nitro and amine compounds, see Section 2.

Epidemiologic studies have played an important role in detecting carcinogenic compounds and in determining their potential effect on exposed workers.

Case et al. (45) examined the data on 4622 men from 21 companies who were employed in the manufacturing of 1-naphthylamine, 2-naphthylamine, benzidine, and aniline. Workers had been employed for more than 6 months between 1920 and 1950 in England and Wales. Of the 209 men who were involved in the manufacture of 2-naphthylamine, 55 were diagnosed as having bladder cancer and 27 had already died. A comparison with the general population indicated that less than approximately one death would have been expected in this size population.

Goldwater et al. (46) studied the occurrence of bladder tumors in workers exposed to coal tar dyes and their intermediates. Of 366 men, 48 were presumably exposed only to 2-naphthylamine. Twelve of these individuals had bladder tumors. Tumors were first diagnosed within 6 to 32 years after the first exposure.

Zavon et al. (47) reported 13 cases of bladder cancer in 25 workers exposed to workroom levels of benzidine of 0.005 to 17.6 mg/m$^3$ for a mean exposure period of 11.46 years.

Mancuso and El-Attar (48) studied workers exposed to benzidine and/or 2-naphthylamine. Seventy-nine white males exposed to either of these compounds had died prior to 1965; 14 of these had died of bladder tumors. These data provided an annual mortality rate of 78/100,000 compared with an annual mortality rate in the state of 4.4/100,000 due to bladder cancer. A number of additional epidemiologic studies have confirmed these findings with 2-naphthylamine and benzidine.

4,4'-Methylenebis(2-chloroaniline) (MBOCA), a structural analogue of benzidine, has been shown to be carcinogenic in rats, mice, and dogs. Tumors of the liver, lung, mammary gland, and bladder have been reported. Ward et al. (49) reported two cases of bladder cancer in 540 workers occupationally exposed to MBOCA. These workers were under the age of 30 and had never smoked. A note added "in proof" indicated that a third worker with bladder cancer had been detected in this group. The authors indicate that the incidence of apparent bladder tumors in U.S. males age 25–29 is only 1 per 100,000. The compound is included in the *Sixth Annual Report on Carcinogens* (32b) as a compound that may reasonably be anticipated to be a carcinogen. OSHA and ACGIH have proposed an 8-hr time-weighted average (TWA) of 0.02 ppm with a skin notation for occupational exposure to this compound.

Because epidemiologic studies can provide evidence of carcinogenicity only after tumors have developed in humans, animal studies are routinely utilized to predict human carcinogenicity. The complexity of long-term animal studies utilized as a basis for human carcinogenicity risk assessment must be understood in order to appreciate fully the inherent difficulties in the process.

Because the carcinogenic response may be species specific, the animal species selected may affect the results. The choice is rather arbitrary and, unless metabolic and pharmacokinetic data suggest otherwise, males and females of the rat and/or the mouse are normally used. The purpose of the long-term study is to determine the "potential" for a chemical substance to produce benign and/or malignant tumors; however, data on other effects are always noted. At least three dose levels are used, including a dose level that is as high as possible without excess mortality, even though such a dose may be greatly above normal or anticipated exposure levels. There has been some concern over the interpretation of data from animals given massive doses of compounds because they may completely overwhelm normal metabolic pathways, thus altering metabolic patterns and the ensuing toxic response. High chemical doses may also alter normal physiological processes such as hormonal balance or immunologic response, producing an animal more prone to a tumorigenic response. The animals are maintained essentially for a lifetime, 2 years if possible. Animal deaths, however, often shorten the study period. Usually 100 animals of each sex are utilized along with comparable groups of control animals. A number of factors may affect the results of such studies. Many control or test animals may die from disease or other causes unrelated to the test compound. The dose levels may be too low to be effective or so high that animals die too soon. The number of test animals with tumors may be greater in the test group than in the control animals, but the difference must be statistically significant to be meaningful. At times the number of tumors in the control group may be unusually high, thus confounding the interpretation of the results. Numerous other problems may arise during 2 years of animal study, which includes periodic weighing and examination of animals to determine early adverse effects. It is not at all unusual for tumors to show up in the mouse and not in the rat, but the reverse may also happen. Rather complex statistical procedures are necessary for appropriate analysis of the data. Federal agencies have attempted to provide guidelines for such long-term studies, for appropriate statistical analysis, and for the inclusion of other relevant data in the overall risk assessment.

As a result of the many difficulties in animal studies, some compounds are clouded with uncertainty with respect to their carcinogenic potential even after a number of long-term studies. In such cases reputable scientists may disagree on the interpretation of the data. Consequently, regulatory agency staff, toxicologists, and industrial hygienists must use their "best scientific judgment" in developing reasonable guidelines to protect the public and occupationally exposed workers.

The aromatic amines and nitroaromatics include compounds that are known human carcinogens and many more that, on the basis of accumulated data, may reasonably be considered to be carcinogens.

As a group, they require extensive surveillance and good industrial hygiene control in the industrial setting. Depending upon the specific compound and extent of exposure, medical supervision is often recommended.

### 1.4.4 Reproductive Effects

Several animal studies have indicated that nitrobenzene can produce reproductive effects in experimental animals. Bond et al. (50) reported that, in the rat, single oral doses of 300 mg/kg of nitrobenzene produced necrosis of the primary and secondary spermatocytes in the seminiferous tubules. Necrotic debris and decreased numbers of spermatozoa were seen in the epididymis as early as 3 days after nitrobenzene administration. The authors point out that the accompanying methemoglobinemia did not appear to be responsible for the testicular effects because similar levels of methemoglobinemia produced by sodium nitrite do not result in this type of toxicity. Dodd et al. (51) conducted a two-generation study by inhalation exposure of rats to 1.0, 10, or 40 ppm of nitrobenzene vapor. No reproductive effects were noted at 1.0 or 10 ppm. At 40 ppm, a decrease in the fertility index of the $F_0$ and $F_1$ generations occurred. In male rats, weight of testes and epididymides was reduced and seminiferous tubule atrophy, spermatocyte degeneration, and the presence of giant syncytial cells were observed. The only significant finding in the litters of rats exposed to the 40-ppm level was approximately a 12 percent decrease in mean body weight of $F_1$ pups on postnatal day 21. Survival indexes were unaltered. The $F_1$ males from the 40-ppm exposure were allowed a 9-week recovery period, then mated to unexposed females. At this time there was a fivefold increase in the fertility index, indicating at least a partial recovery of function. The number of giant syncytial spermatocytes and degenerated spermatocytes were also greatly reduced. The authors suggest 10 ppm as a no observable effect level for reproduction and fertility in the rat. Levin et al. (52) demonstrated that after a single oral dose of 300 mg/kg to male rats degenerative changes in the seminiferous tubules occurred as early as 3 days after dosing. At 32 days after treatment all sperm production ceased and aspermia continued until approximately day 48. By days 76 to 100 after dosing the rate of sperm output had reached 78 percent of the control group. These studies indicate the dramatic effect of the simplest of the aromatic nitro compounds, nitrobenzene, on the testes in experimental animals.

A number of papers have demonstrated that 1,3-dinitrobenzene also produces severe testicular effects in the rat and mouse (53–56). In general these findings were similar to those reported for nitrobenzene. Sperm production in rats was decreased at a dose level as low as 1.5 mg/kg/day, 5 days/week for 10 weeks. Altered sperm morphology and decreased motility at single oral doses of 16 and 24 mg/kg were reported. A no-effect level for a single oral dose was reported as

8 mg/kg in one study, and no effects were noted at 5 or 10 mg/kg in another study. Decreased reproductive ability in male rats also occurred with a single oral dose of 16 mg/kg. A no-effect single oral reproductive dose was 8 mg/kg. Overall the testicular toxicity of 1,3-dinitrobenzene appears to be considerably greater than that of nitrobenzene. In a study of the three dinitrobenzene isomers Blackburn (57) found that after a single oral dose of 50 mg/kg only the rats receiving 1,3-dinitrobenzene showed testicular damage 12 hr after dosing. By 24 hr widespread Sertoli cell damage was evident associated with widespread degeneration of primary spermatocytes. The no-effect single oral dose in rats was 5 or 10 mg/kg. In contrast 1,2- and 1,4-dinitrobenzene were without effect on the testes.

These findings have resulted in considerable interest in the mechanisms responsible for the marked differences in testicular toxicity among these three isomers. Many of the metabolic studies presented have attempted to define metabolic pathways that might provide a basis for these effects.

Species differences have been noted in the testicular effects of 1,3-dinitrobenzene (58). The testicular toxicity and methemoglobinemia induced by 1,3-dinitrobenzene were compared in the Sprague–Dawley rat and in the Syrian golden hamster. At single oral dose levels up to 50 mg/kg the hamster showed no testicular lesions whereas marked damage was apparent in rat testicles following the same dosage. Methemoglobin formation was also much lower (15 percent) in the hamster than in the rat (80 percent). McEuen and Miller (59) reported that when both the rat and the hamster were given intraperitoneal doses of 25 mg/kg, peak blood levels of 1,3-dinitrobenzene were 46 nmol/ml in the hamster and 99 nmol/kg in the rat. The rats excreted more unconjugated metabolites and fewer phenolic metabolites than the hamster. Phenolic compounds are rapidly excreted; this may account for the difference in blood levels of the compound found in these animals. Such differences in blood levels would be expected to relate to the toxic response. The authors suggest that the rat's ability to form reductive metabolites may play a role in its increased susceptibility to testicular toxicity.

1,3-Dinitrobenzene is metabolized, both in vivo and in vitro, to *m*-nitroaniline and *m*-nitroacetanilide. Cave and Foster (60) pointed out that because neither of these metabolites produced testicular toxicity, a metabolite intermediate such as *m*-nitrosonitrobenzene might be involved. When tested in rat Sertoli–germ cell cultures at equimolar doses, *m*-nitrosonitrobenzene produced similar morphological changes to those formed by 1,3-dinitrobenzene. The authors suggest that an intermediate, such as *m*-nitrosonitrobenzene or a further metabolite of this compound, may be the species responsible for testicular toxicity.

Topham (61) reported that rats receiving intraperitoneal doses of *p*-aminophenol had abnormally shaped sperm heads. These abnormalities persisted for at least 26 weeks after treatment.

Four-week oral studies demonstrated that 2,4- and 2,6-dinitrotoluene (62, 63) produced depression of spermatogenesis and atrophy of the testes in mice, rats, and dogs. A critical review of the acute and chronic toxicity of these two compounds has been recently published (43). Reader and Foster (64) found that the four dinitrotoluene isomers produced toxic effects comparable to that of 1,3-dinitrobenzene when tested in a Sertoli cell culture system.

Bloch et al. (65) reported that in rats treated with 2,4-dinitrotoluene electron microscopy of the Sertoli cells showed extensive damage. Much of the cytoplasm was occupied by varying size vesicles associated with swollen mitochondria and distended endoplasmic reticulum. Sperm counts were reduced by 62 percent in rats whose diets contained 0.2 percent 2,4-dinitrotoluene for 3 weeks. Serum levels of luteinizing hormone and follicle-stimulating hormone were significantly elevated, indicating an effect on pituitary function, either direct or indirect.

Several aromatic amines have been tested for teratological effects on the fetus. Aniline hydrochloride was found not to be teratogenic in rats even at maternally toxic dose levels (66). Ortho, meta, and para aminophenols were tested for teratogenicity in the Syrian golden hamster. Ortho and para aminophenols were teratogenic at dose levels that did not compromise maternal health. *m*-Aminophenol could not be conclusively considered as a teratogen in this study. Technical-grade *o*-toluenediamine (2,3-isomers approximately 40 percent, 3,4-isomers approximately 60 percent) was studied in rats and rabbits. Missing and incompletely ossified sternebra were noted in the rat at the high (300 mg/kg/day) and mid-dose (100 mg/kg/day), respectively. No effects were noted at the 30 or 10 mg/kg/day dose levels. These delayed fetal development effects, noted only at dosages that were toxic to the dam, are not generally considered to be evidence of a teratogenic effect. *p*-Nitroaniline and *p*-nitrobenzene have also been reported to produce effects on the fetus only at dose levels producing maternal toxicity. Other compounds reported as having no teratogenic effects in animals include *m*-aminophenol, *o*-nitroaniline, *m*- and *p*-phenylenediamine, and 4-dimethylaminoazobenzene.

Reader et al. (68) evaluated the potential use of plasma enzymes of testicular origin and plasma hormones as indicators of testicular damage. Lactic dehydrogenase-C4 (LDH-C4) was reduced in testis and increased in plasma after dosing of rats with 1,3-dinitrobenzene. The plasma increases in LDH-C4 preceded noticeable histological change and may be of diagnostic value in monitoring testicular toxicity produced by these compounds.

To date there are no definitive human data that have demonstrated testicular effects in exposed workers. The extensive data on a number of aromatic nitro compounds that produce a degenerative effect on the testes in several animal species does suggest that these compounds must be closely monitored in the workplace. Appropriate periodic medical examination of exposed workers should be encouraged.

### 1.4.5 Environmental Concerns

In recent years the increasing concern about environmental contaminants and the proliferation of guidelines and regulations have resulted in a greater effort by industry to control not only releases into the atmosphere, but releases into water and soil as well. This expanded activity at times requires toxicologic information not only on humans, birds, and wild mammalian species, but also on aquatic species that may be in the area of concern. This responsibility may often be passed on to industrial hygienists because of their familiarity with hazardous chemicals.

Although the industrial release of chemicals into the environment is of concern, other sources may also contribute to environmental "contamination." The environment has never been free of contamination, if by that term we mean chemicals that may be harmful to human health. Lightning produces ozone; bacterial degradation releases a host of products to soil and water, including methane and hydrogen sulfide; volcanic activity releases not only particulate matter but enormous quantities of toxic gases; and forest fires release not only carbon dioxide but also hundreds of known products of combustion, including irritating aldehydes and complex aromatic hydrocarbons. In addition to these natural sources of contamination we have superimposed environmental releases from our life-styles: automobiles, air conditioners, spray cans, barbeque grills, and so on. As a result it is not always easy to pinpoint the source of widespread air, water, or soil contamination.

Often, however, chemical releases from industrial plants can be readily traced from the use or synthesis of a specific compound in the plant to plant effluents and further into the environment.

A few examples of the types of information that are available for aromatic amines and nitro aromatics in the environment are given below.

***1.4.5.1 Biodegradation.*** It is important to know whether or not a chemical decomposes under environmental conditions; thus there are a large number of biodegradation studies in the literature.

Alexander and Lustigman (69) studied the biodegradation of a relatively large number of aromatic nitro and amine compounds by soil microorganisms. Incubation continued for a 64-day period. Metabolism of the compound was measured by ultraviolet absorption near the absorption maximum for each compound. Under the conditions of this study aniline decomposed in 4 days, whereas nitrobenzene was not completely decomposed after 64 days. This is consistent with other reports that aniline but not nitrobenzene was readily degraded by sewage sludge. *o*-Nitrobenzoic and *p*-nitrobenzoic acids were degraded within 4 to 8 days, whereas *m*-nitrobenzoic acid was not degraded in 64 days. The authors indicated that the stability of *m*-nitrobenzoic acid has been reported in other studies. With nitrophenols stability was apparently conferred by the ortho position, for *o*-nitrophenol was not degraded in 64 days whereas *m*- and *p*-nitrophenol were degraded in 4 to 16 days. The dinitrobenzene isomers and the diaminobenzenes were not degraded by these soil microorganisms. Studies of this nature provide some comparative persistence data, but one needs to recall that environmental conditions are so diverse that what occurred with these soil microorganisms under aerobic conditions may not occur in sediment, sewage sludge, or in soils from other sources where the bacterial composition may be quite different, or in studies under anaerobic conditions.

Hallos and Alexander (70) reported that a 60-day incubation of *o*-nitroaniline with activated sludge resulted in less than 15 percent decomposition under aerobic conditions and about 60 percent decomposition under anaerobic conditions. Tabak et al. (71) studied the microbial degradation of a large number of substituted phenolic compounds by various phenol-adapted bacterial cultures from soil, compost, river mud, and sediment from a waste lagoon. Phenol was degraded within 1 or 2 days, $o$-, $m$-, and $p$-nitrophenols were metabolized more slowly (3 to 6 days), and 2,4-dinitrophenol was metabolized in 7 to 10 days. In this study $o$-, $m$-, and $p$-aminophenols were degraded even more rapidly than phenol itself. In general, the addition of a nitro group to a phenolic compound markedly increased its resistance to microbial degradation.

4,4'-Methylene-bis-2-chloroaniline is degraded slowly, if at all, by wastewater sludge under aerobic conditions. This compound has a strong affinity for soil and should adsorb to aquatic sediments.

McCormick et al. (72) reported on the reduction of trinitrotoluene (TNT) by hydrogen in the presence of enzyme preparations from the anaerobic microorganism *Veillonella alkalescens*. With approximately 40 nitro compounds 3 mol of hydrogen was required to convert each nitro group to the corresponding amine. The authors indicate that this is consistent with the reduction of nitro to amino compounds through the nitroso and hydroxylamine pathway according to the following equations:

$$RNO_2 \xrightarrow{H_2} RNO + H_2O \qquad (1)$$

$$RNO \xrightarrow{H_2} RNHOH \qquad (2)$$

$$RNHOH \xrightarrow{H_2} RNH_2 + H_2O \qquad (3)$$

By this metabolic pathway reduction of one nitro group requires 3 mol of hydrogen. Metabolism of dinitrobenzenes under these reductive conditions utilized approximately 6 mol of hydrogen per mole of compound, consistent with the reduction of two nitro groups. Trinitrotoluene approached the required 9 mol of hydrogen per mole of compound, indicating the conversion of all three nitro groups to the amine. The proposed hydroxylamine intermediate was identified in the incubation mixture. The proposed nitroso compound could not be detected. These data suggest that the 4-nitro group of TNT is reduced first to form 4-amino-2-nitrotoluene, which is in turn reduced to 2,4-diaminotoluene. Further reduction forms triaminotoluene. The only nitro compounds not converted to the corresponding amines under these reductive conditions were $o$-, $m$-, and $p$-nitrotoluenes, where hydrogen uptake leveled off much below the anticipated value of 3 mol of hydrogen per mole of compound. This suggests incomplete reduction of the nitro group of these compounds under the conditions of this study.

*1.4.5.2 Aquatic Toxicity.* Considerable toxicity information is available on aquatic species but it is widespread in the literature and often not as readily available to industrial hygienists as the human and laboratory animal data. In general, the data are acute in nature because of difficulties in the maintenance and dosing regimens for these species. Examples of the types of studies available are presented below.

Several studies of the effects of 1,3-dinitrobenzene on aquatic flora and fauna have been reported. Bailey and Spanggord (73) reported static 96-hr $LC_{50}$ values for 1,2-, 1,3-, and 1,4-dinitrobenzenes of 0.6, 7.0, and 1.7 mg/l, respectively, in fathead minnows. Hermens and Leeuwangh (74) determined a 14-day $LC_{50}$ of 9.4 mg/l for 1,3-dinitrobenzene in the guppy *(Poecilia reticulata)*. Fifty percent of the blue-green algae in an exponentially growing culture were killed by a concentration of 5 mg/l of 1,3-dinitrobenzene (75). A 96-hr $LC_{50}$ of 9.8 mg/l of *o*-nitroaniline was reported for the guppy (76).

1,3,5-Trinitrobenzene has been found as a component of the condensate wastewater effluent from a TNT production plant (77). The compound was thought to be formed by sequential oxidation and decarboxylation of TNT during the manufacturing process. There is no evidence of general environmental distribution of this compound. The 96-hr static $LC_{50}$ for 1,3,5-trinitrobenzene for fathead minnows is reported to range from 0.49 to 1.1 mg/l (73, 78). In flowing water the 96-hr $LC_{50}$ for bluegills was 0.57 mg/l. Over 6 days the lowest observable adverse effect level (LOAEL) was 0.128 mg/l, and the no-observable effect level (NOEL) was 0.061 mg/l (79). A calculated bioconcentration factor of 1.36 suggests that trinitrobenzene does not accumulate in aquatic species (78).

Studies on "environmental species" such as those described above serve as general guidelines on contaminant levels that may have serious consequences on environmental flora and fauna.

## 2 SPECIFIC COMPOUNDS

This section is intended to summarize the accumulated information from the literature regarding animal and human toxicology, physical and chemical properties, synonyms, uses, and hygienic standards. In addition, structural activity relationships and mechanisms are also included for a few selected compounds. The order of presentation is based upon parent structure whenever possible in order to facilitate congeneric comparisons.

### 2.1 2-Acetylaminofluorene [CAS # 53-96-3]

The formula is $CCH_2C_6H_4C_6H_3NHCOCH_3$; other names for this compound are *N*-2-fluorenylacetamide, 2-acetaminofluorene, *N*-acetylaminophenathrene, and 2-AAF.

#### 2.1.1 Uses

2-Acetylaminofluorene is used for research purposes; its use as a pesticide has been abandoned.

## 2.1.2 Physical and Chemical Properties (100)

Molecular weight   223.28
Melting point      194°C
Solubility         Insoluble in water; soluble in alcohol, ether, acetic acid

## 2.1.3 Toxicity

There is sufficient evidence of carcinogenicity in animals but no data are available to evaluate its carcinogenic potential in humans; however, EPA regulates 2-AAF under the Comprehensive Environmental Response, Compensation, and Liability Act (CERCLA); Resource Conservation and Recovery Act (RCRA); and the Superfund Amendments and Reauthorization Act (SARA), and has designated it to be a hazardous constituent of waste and a potential human carcinogen under RCRA (81). The compound has caused cancer in the bladder, kidney, pelvis, liver, and pancreas of animals; no human toxicity has been reported.

In 1973 Oyasu reported formation of bile duct carcinoma in Syrian golden hamsters (82). Cholangiolar carcinoma rarely occurs with AAF carcinogenesis; and in a more recent study in male F344 rats, consisting of feeding 2-AAF in a choline-deficient diet to enhance 2-AAF hepatocellular carcinogenicity activity previously demonstrated in Sprague–Dawley rats (83), metaplastic ductlike structures that formed appeared to differentiate into bile-ductlike structures. Although the rat has been shown to excrete 2-AAF in bile, the authors concluded that these ductular cells could also be a non-tumor-related adaptive change to carcinogens that exhibit hepatocellular but not cholangiolar properties (84, 85). This illustrates the danger of extrapolation on the basis of a single species.

The metabolism of 2-AAF in mammals occurs by two routes: $N$-hydroxylation, leading to the carcinogenic $N$-hydroxy derivative; and aromatic hydroxylation, leading to the noncarcinogenic 7-hydroxy derivative. The guinea pig does not produce the $N$-hydroxy derivative and in this species 2-AAF is not carcinogenic. P450 oxidases catalyze the $N$-oxidation of AAF to give $N$-hydroxy-2-AAF; a soluble sulfotransferase then catalyzes the formation of a highly reactive, mutagenic, sulfuric acid ester electrophile. $N$-Hydroxy-2-AAF can also undergo a peroxidase-catalyzed one-electron oxidation to give a free nitroxide radical. 2-AAF is a "redox cycler," which is thereby capable of undergoing single-electron oxidation-reduction reactions to yield free-radical species. 2-AAF can accept electrons from a variety of biologic reducing agents (flavoproteins, NADPH, NADP, GSH, other thiol compounds, and ascorbate). In the presence of oxygen, the reduced 2-AAF can be reoxidized to the original parent compound by donating a single electron to oxygen (86).

This compound has been extensively studied as a carcinogen and continues to serve as a model compound in many studies of interest including developmental studies, in which 2-AAF has been shown to be teratogenic in rats, mice, and chicks, as well as studies of immunotoxicity.

### 2.1.4 Hygienic Standards of Permissible Exposure

OSHA regulates 2-AAF under the Hazard Communication Standard and as a chemical hazard in laboratories (81).

## 2.2 o-Aminophenol [CAS # 95-55-6]

The formula is $NH_2C_6H_4OH$; o-aminophenol is also known as o-hydroxyaniline and 2-aminophenol.

### 2.2.1 Uses

o-Aminophenol is used as an azo and sulfur dye intermediate and for dying fur and hair; it is widely used in the cosmetics, dye, and drug industries.

### 2.2.2 Physical and Chemical Properties (100)

| | |
|---|---|
| Physical state | Colorless rhombic needles or plates |
| Molecular weight | 109.12 |
| Melting point | 170-174°C |
| Boiling point | Sublimes |
| Solubility | 1.7 in water at 0°C; 4.3 in alcohol at 0°C; very soluble in ether |
| Odor and warning properties | None |

### 2.2.3 Toxicity

o-Aminophenol is not readily absorbed through intact skin but may prove to be a sensitizing agent with resultant contact dermatitis. Inhalation of dust should be avoided because if it is inhaled in excessive amounts it may cause methemoglobinemia. In rare instances, o- and p-aminophenol have caused a bronchial asthma.

Intraperitoneal administration (100 to 200 mg/kg) to Syrian golden hamsters on day 8 of gestation produced a significant teratogenic response similar to that of p-aminophenol (see below), including neural tube defects (exencephaly, encephalocele, and spina bifida), eye defects, and skeletal defects (87). However, no teratogenicity study by the oral route of administration was found, and it must be remembered that effects can be inconsistent within or among species depending on the route of administration, as illustrated below under p-aminophenol.

### 2.2.4 Hygienic Standards of Permissible Exposure

Standards have not been assigned.

## 2.3 m-Aminophenol [CAS # 591-27-5]

The formula for m-aminophenol is $NH_2C_6H_4OH$; it is also known as 3-amino-1-hydroxybenzene, m-hydroxyaniline, 3-aminophenol, and 3-hydroxyaniline.

### 2.3.1 Uses

*m*-Aminophenol is used chiefly in the synthesis of dyes and occasionally as a hair dye [red-brown color obtained with *p*-phenylenediamine or light orange with *p*-aminophenol (88)] and in the manufacture of *p*-aminosalicylic acid.

### 2.3.2 Physical and Chemical Properties (100)

| | |
|---|---|
| Physical state | Colorless prisms |
| Molecular weight | 109.12 |
| Melting point | 122-123°C |
| Solubility | 2.6 in water at 0°C; very soluble in alcohol and ether |
| Odor and warning properties | None |

### 2.3.3 Toxicity

Intraperitoneal administration (100 to 200 mg/kg) of *m*-aminophenol to Syrian golden hamsters on day 8 of gestation produced inconsistent results. A teratogenic response was demonstrated, expressed as a percentage of the total number of litters with one or more live fetuses, at a dose of 150 mg/kg in only one (six malformed fetuses) of six litters, and teratogenicity was not evident at a dose of 200 mg/kg (87).

In a follow-up teratology study in which Sprague–Dawley rats were fed a diet of 0.1, 0.25, and 1.0 percent for 90 days prior to mating, maternal toxicity was demonstrated at the highest dose level, and a significant reduction in body weight was noted in the 0.25 percent group, but there was no evidence of teratogenic or embryo–fetal toxicity at any dose level tested. Accumulation of iron-positive pigment within the liver, kidneys, and spleen was observed in dams fed a 1 percent diet, together with significant reduction in red blood cell count and hemoglobin level, as well as an increase in mean corpuscular volume, indicating a hemolytic effect; histomorphologic appearance of the thyroid indicated hyperactive activity (at 0.25 and 1.0 percent diet) (89a). In contrast to *o*- and *p*-aminophenol and their glucuronides, neither *m*-aminophenol nor its conjugate with glucuronic acid has been shown to form methemoblobin in vitro (89b).

### 2.3.4 Hygienic Standards of Permissible Exposure

Standards have not been assigned.

## 2.4 *p*-Aminophenol [CAS # 123-30-8]

The formula for *p*-aminophenol is $NH_2C_6H_4OH$; it is also known as *p*-hydroxyaniline, 4-aminophenol, 4-amino-1-hydroxybenzene, 4-Hydroxybenzanamine, and PAP.

### 2.4.1 Uses

*p*-Aminophenol is used in the manufacture of sulfur and azo dyes and in dying furs. The hydrochloride salt is used as a photographic developer in conjunction with sodium or potassium carbonates. *p*-Aminophenol was used medicinally as an analgesic in the late 1800s.

The introduction of aniline derivatives as analgesics was based on the accidental discovery by Cahn and Hepp in 1886 that aniline and acetanilid both had powerful antipyretic properties. They introduced acetanilid into medicine under the name of antifebrin (90). This led to this group of aniline analgesics being called "coal tar analgesics." Acetanilid was introduced by these workers because of the known toxicity of aniline. The acyl derivatives of aniline were thought to exert their analgesic and antipyretic effects by first being hydrolyzed to aniline and the corresponding acid, following which the aniline was oxidized to *p*-aminophenol. This was then found to be excreted in combination with glucuronic or sulfuric acid. Acetanilid (acetanilide, phenylacetamid or acetanilid U.S. circa 1860, antifebrin) was made by heating a mixture of aniline and glacial acetic acid to the boiling point, cooled, congealed, and purified by sublimation or recrystallization to yield the medicinal monoacetyl derivative of aniline. In 1907 the recommended dose as an antipyretic was from 300 to 600 mg (91). By 1955 the usual dose was 200 mg (92). The compound caused methemoglobinemia, skin reactions, and jaundice.

*p*-Aminophenol was tried as an analgesic because of the belief that acetanilid was ultimately oxidized to *p*-aminophenol; however, it was found to be more toxic than its predecessor, acetanilid (90).

Phenacetin, the ethyl ether of *N*-acetyl-*p*-aminophenol, also known as acetophenetidin, was widely prescribed as an analgesic; however, there were reports of possible nephrotoxicity when used in large amounts for long periods of time (93a). In normal individuals 75 to 80 percent is metabolized by deethylation to yield acetaminophen, or *N*-acetyl-*p*-aminophenol (90, 93a). Phenacetin has been employed worldwide for more than a century. The signs of its chronic toxicity in humans did not become apparent until several decades after the introduction of the drug into medical practice (93b). Mixtures were often prescribed consisting of phenacetin, aspirin, and caffeine, known as aspirin compound or APC compound tablets, or PAC compound capsules. Chronic hemolytic anemia, methemoglobinemia, renal and liver necrosis, hypoglycemic coma, and uroepithelial cancer of the renal pelvis and lower urinary tract are associated with the ingestion of 1 g of phenacetin per day for at least 1 year or smaller daily intakes for many years (Dacie, 1967; Radomski, 1979; Carro-Ciampi, 1978) (93b). Lethal doses of phenacetin are not associated with hepatic damage, but cyanosis, respiratory depression, and cardiac arrest occur. Currently phenacetin is considered a carcinogenic drug, although no mutagenic activity was found in *Salmonella typhimurium* Ames et al., 1975) or by the micronucleus test (Hoover et al., 1981); in the latter test phenacetin was not assayed at low doses (93b).

Acetaminophen was first utilized by von Mering in 1893 as a medicinal analgesic. However, it has gained popularity only since 1949, after it was recognized as the major active metabolite of both acetanilid and phenacetin (90). Chronic toxicity of acetaminophen, caused by overdoses, has been associated with hepatotoxicity (Mitchell, et al., 1973) and nephrotoxicity (McMurtry et al., 1978; Breen et al., 1982), both requiring metabolic activation (93b).

## 2.4.2 Physical and Chemical Properties (100)

| | |
|---|---|
| Physical state | Colorless prisms |
| Molecular weight | 109.12 |
| Melting point | 189.6 to 190.2°C |

## 2.4.3 Toxicity

*p*-Aminophenol is a significantly toxic chemical and one mechanism associated with its cytotoxicity has been attributed to its activity as a tissue respiratory (oxidative phosphorylation) inhibitor (94).

Intraperitoneal administration (100 to 200 mg/kg) on day 8 of gestation of Syrian golden hamsters produced a significant teratogenic response including encephalocele and limb, tail, and eye defects; rare malformations observed included ectopic heart, cleft palate, occult cranioschisis, and abnormal genitalia. It was proposed that the mechanism may be related to the formation of a reactive quinone. Oral administration of 100 or 200 mg/kg was not teratogenic [$p > .1$] (87).

However, *p*-aminophenol is nonteratogenic in Sprague–Dawley rats fed a diet containing 0.07, 0.2, or 0.7 percent for up to 6 months. After 13 weeks, 25 females/group were mated to untreated males in a teratology study; after 20 weeks, 20 males/group were mated to untreated virgin females in a dominant lethal mutagenicity study. Dose-related nephrosis was seen in both sexes after 13 and 27 weeks and in the high-dose males that were removed from the test diet for a 7-week recovery period. The authors noted an increase in developmental variations associated with maternal toxicity at the mid- and high-dose levels. The dominant lethal study was equivocal (95).

*p*-Aminophenol is considered a minor nephrotoxic metabolite of acetaminophen in humans. Long-term use of acetaminophen can result in an increased lipofuscin deposition in kidneys. In vitro studies have demonstrated that *p*-aminophenol can undergo oxidative polymerization to form melanin, a component of soluble lipofuscin. Hemolysis accompanies this process in whole blood. Long-term excessive use of phenacetin or acetaminophen has been associated with chronic renal disease, hemolytic anemia, and increased solid lipofuscin deposition in tissues (96).

*p*-Aminophenol has also been demonstrated to be nephrotoxic to rats. Administration of 25 to 100 mg/kg to male F344 rats resulted in a dose-related proximal nephropathy. The observed increased excretion of enzymes, glucose, and urine total protein, resulting in glycosuria and amino aciduria, indicated functional defects in the proximal tubule and reduced solute reabsorption efficiency (97). The necrosis is apparently restricted to the straight segment of the proximal tubule of the Fischer 344 rat; Sprague–Dawley rats, on the other hand, are more resistant to the nephrotoxicity of acetaminophen and its nephrotoxic metabolite *p*-aminophenol. The authors postulated that the strain differences in *p*-aminophenol-induced nephrotoxicity may be related to differences in the intrarenal activation of *p*-aminophenol (98).

*p*-Aminophenol has not been demonstrated to be carcinogenic (99). It may cause skin sensitization, contact dermatitis, bronchial asthma, and methemoglobinemia (100a, 101).

### 2.4.4 Hygienic Standards of Permissible Exposures

Standards have not been assigned.

### 2.5 o-Nitrophenol [CAS # 88-75-5]

The formula for o-nitrophenol is $NO_2C_6H_4OH$; it is also known as 2-nitrophenol.

#### 2.5.1 Uses

It is used in the synthesis of dyestuffs and other intermediates and also as a chemical indicator.

#### 2.5.2 Physical and Chemical Properties (100)

| | |
|---|---|
| Physical state | Light yellow crystals |
| Molecular weight | 139.11 |
| Density | 1.657 (20°C) |
| Melting point | 44 to 46°C |
| Boiling point | 214 to 216°C |
| Solubility | 0.21 in water at 20°C, 1.08 at 100°C; 46.0 in alcohol at 25°C; soluble in ether, acetone, benzene, alkali |
| Odor and warning properties | Aromatic odor and sweet taste properties |

#### 2.5.3 Toxicity

o-Nitrophenol causes central and peripheral vagus stimulation, central nervous system (CNS) depression, methemoglobinemia, and dyspnea in animal experimentation. The oral $LD_{50}$ in mice is 1.297 g/kg; in rats, 2.828 g/kg (100a).

o-Nitrophenol is reportedly negative in the Ames mutagenicity test (102). An additional in vivo study, which also suggests relatively minimal biologic activity, demonstrated that o-nitrophenol is metabolically reduced less readily than the meta or para isomers to the corresponding amino derivatives (103).

#### 2.5.4 Hygienic Standards of Permissible Exposure

Standards have not been assigned.

### 2.6 m-Nitrophenol [CAS # 554-84-7]

The formula for m-nitrophenol is $NO_2C_6H_4OH$; it is also known as 3-nitrophenol.

#### 2.6.1 Uses

It is used in the synthesis of dyestuffs and as a chemical indicator.

## 2.6.2 Physical and Chemical Properties (100)

| | |
|---|---|
| Physical state | Colorless to yellowish crystals |
| Molecular weight | 139.11 |
| Density | 1.485 (20°C) |
| Melting point | 97°C |
| Boiling point | 194°C (70 mm Hg) |
| Solubility | 1.35 in water at 25°C; very soluble in ether; soluble in benzene, alkali, ether, alcohol, acetone |

## 2.6.3 Toxicity

In contrast to the ortho isomer, *m*-nitrophenol is more readily biotransformed to its corresponding amino derivative; however, it is reportedly nonmutagenic in the Ames bacterial mutagenicity test (103, 104).

## 2.6.4 Hygienic Standards of Permissible Exposure

Standards have not been assigned.

## 2.7 *p*-Nitrophenol [CAS # 100-02-7]

The formula for *p*-nitrophenol $NO_2C_6H_4OH$; it is also known as 4-nitrophenol.

### 2.7.1 Uses

It is used in the synthesis of dyestuffs and as a chemical indicator.

### 2.7.2 Physical and Chemical Properties (100)

| | |
|---|---|
| Physical state | Colorless to yellowish monoclinic prisms |
| Molecular weight | 139.11 |
| Density | 1.479 (20°C) |
| Melting point | 113-116°C |
| Boiling point | 279°C (sublimable) |
| Solubility | 1.6 in water at 25°C, 26.9 at 90°C; 189.5 in alcohol at 25°C; very soluble in ether; soluble in benzene, acetone, alkali |

### 2.7.3 Toxicity

*p*-Nitrophenol has been demonstrated to undergo glutathione (105) and glucuronide conjugation (106). Isolation of a rat liver glucuronosyltransferase isozyme has also been reported; this enzyme is responsible for the glucuronide formation (107). In addition, *p*-nitrophenol can readily undergo reduction to its amino derivative. However, in vivo nitro reduction conditions require bacterial enzymes in the natural anaerobic environment of the gut (possibly in localized cellular ischemic conditions); in vitro, mammalian enzymes are required under artificial anaerobic conditions (103).

$p$-Nitrophenol is nonmutagenic in the Ames assay (104). The oral $LD_{50}$ in mice is 467 mg/kg, and in rats, 616 mg/kg (100a). An NTP carcinogenicity study was completed by Litton Bionetics in 1991; results were not indicated (108).

### 2.7.4 Hygienic Standards of Permissible Exposure

Standards have not been assigned.

## 2.8 Dinitrophenol Isomers

The isomers of dinitrophenol are usually not separated for commercial use, but are prepared as mixtures. They are used in the synthesis of dyestuffs, picric acid, and picramic acid. They are also used in the preservation of timber and in the manufacture of the photographic developers.

### 2.8.1 2,3-Dinitrophenol [CAS # 66-56-8]

The formula for 2,3-dinitrophenol is $C_6H_3OH(NO_2)_2$.

*2.8.1.1 Uses.* Specific applications for 2,3-dinitrophenol were not located.

*2.8.1.2 Physical and Chemical Properties (100).* 2,3-Dinitrophenol has the following properties:

| | |
|---|---|
| Physical state | Yellow monoclinic prisms, flammable |
| Molecular weight | 184.11 |
| Melting point | 144 to 145°C |
| Solubility | Slightly soluble in water; very soluble in hot alcohol; very soluble in ether; soluble in benzene, alcohol |

*2.8.1.3 Toxicity.* Dermal exposure to 2,3-dinitrophenol causes yellow staining of skin and it may also cause dermatitis or allergic sensitivity. Systematically, it disrupts oxidative phosphorylation causing increased metabolism, oxygen consumption, and heat production. Chronic exposure may result in kidney and liver damage and cataract formation (109).

*2.8.1.4 Hygienic Standards of Permissible Exposure.* Standards have not been assigned.

### 2.8.2 2,4-Dinitrophenol [CAS # 51-28-5]

2,4-Dinotrophenol, $C_6H_3OH(NO_2)_2$, is also known as DNP. It is a flammable solid.

*2.8.2.1 Uses.* DNP finds use as a herbicide, insecticide, and wood preservative, and in the manufacture of dyes.

*2.8.2.2 Physical and Chemical Properties (100).* 2,4-Dinitrophenol has the following properties:

| | |
|---|---|
| Physical state | Pale yellow rhombic crystals or needles, flammable |
| Molecular weight | 184.11 |
| Melting point | 115 to 116°C |
| Solubility | 0.56 in water at 18°C, 4.3 at 100°C; 3.9 in alcohol at 19°C; 3.065 in ether at 15°C; soluble in chloroform, benzene |
| Warning properties | bitter taste |

*2.8.2.3 Toxicity.* The most characteristic toxicologic effect of DNP is an increase in metabolic rate, which is a consequence of its activity as an oxidative phosphorylation uncoupler, preventing phosphorylation of ADP to ATP, thereby inhibiting energy transport. An early sign of exposure may be yellow staining of the skin (pseudojaundice) and hair. The skin also becomes hot and flushed, and profuse perspiration may occur. Delayed hearing impairment and delayed cataract formation can ultimately occur. The increased oxidative metabolism, or oxygen consumption, will lead to hyperpyrexia, tachycardia, increased respiratory rate, acidosis dehydration, and the ultimate depletion of fat stores. The individual can also experience intense thirst, severe headache, nausea, vomiting, abdominal pain, restlessness, anxiety, delirium, and generalized weakness. Hepatic dysfunction, blood dyscrasias, circulatory or respiratory collapse, and death may occur (109–111).

During the mid-1930s DNP was introduced as a nonprescription "over-the-counter" antiobesity or slimming agent, owing to its ability to stimulate cellular respiration by several hundred percent (112, 113). Several hundred human cataracts were reported as a consequence of ingestion (112).

Cataract formation has been suggested to be due to tissue anoxia via inhibition of mitochondrial respiration (oxidative phosphorylation uncoupling), which damages the lens epithelium. Experimental animals are insensitive to the cataractogenic activity of DNP, with the exception of young fowl and rabbits (114).

Using an in vitro lens culture system, Lebeau and Gehring in 1971 demonstrated that DNP at a concentration of $2 \times 10^{-5}$ $M$ caused cataracts to develop in lenses obtained from rabbits. Known metabolites of DNP did not produce this effect (109).

Experimental studies indicate that lens metabolism is essentially anaerobic, with ATP synthesis depending on glycolysis and therefore insensitive to DNP; however, mitochondrial oxidative phosphorylation in the epithelial cells may play a greater role in ATP synthesis in human lenses (114). The exact mechanism(s) involved have yet to be completely elucidated. See Chapter 21, Aromatic Hydrocarbons, for further details on mechanisms of uncouplers.

In 1973, England et al. reported that DNP also increased the production of thyroxine, and suggested that DNP may compete for sites on T4 plasma binding proteins, resulting in a decline in PBI (protein-bound iodine), resulting in increased secretion of thyroxine, which is excreted in the bile largely as thyroxine glucuronide (109). More recently it has been suggested that in cases of DNP toxicity, the β-adrenergic system may also be brought into play be promoting elevated peripheral glycogenolysis, as a consequence of DNP inhibition of oxidative phosphorylation; it apparently helps to compensate for the loss of mitochrondrial synthesis of ATP by greatly accelerating its cytoplasmic synthesis (115).

This special ability to affect metabolism has been found useful to biochemists. A review of this phenomenon by Parascandola in 1974 reported the following: "Dinitrophenol had proved to be a useful tool for biochemists in the study of metabolism. For example, it has been used to determine whether or not a metabolic process, such as peptide-bond formation, is dependent upon energy rich phosphate compounds" (109).

Dinitrophenol neither promotes nor causes tumor formation (109).

*2.8.2.4 Hygienic Standards of Permissible Exposure.* Standards have not been assigned.

### 2.8.3  2,5-Dinitrophenol [CAS # 329-71-5]

The formula for 2,5-dinitrophenol is $C_6H_3OH(NO_2)_2$.

*2.8.3.1 Uses.* Specific applications for 2,5-dinitrophenol were not located.

*2.8.3.2 Physical and Chemical Properties (100).* The properties of 2,5-dinitrophenol are as follows:

| | |
|---|---|
| Physical state | Yellow needles, flammable |
| Molecular weight | 184.11 |
| Melting point | 108°C |
| Solubility | Slightly soluble in water, alcohol; very soluble in hot water, alcohol; soluble in ether, benzene |

*2.8.3.3 Toxicity.* No specific studies were located. Toxicity is assumed to be identical with that of 2,3-dinitrophenol (109).

*2.8.3.4 Hygienic Standards of Permissible Exposure.* Standards have not been assigned.

### 2.8.4  2,6-Dinitrophenol [CAS # 573-56-8]

The formula for 2,6-dinitrophenol is $C_6H_3OH(NO_2)_2$.

*2.8.4.1 Uses.* Specific applications for 2,6-dinitrophenol were not located.

*2.8.4.2 Physical and Chemical Properties (100).* The properties of 2,6-dinitrophenol are as follows:

| | |
|---|---|
| Physical state | Pale yellow rhombic crystals flammable solid |
| Molecular weight | 184.11 |
| Melting point | 63 to 64°C |

AROMATIC NITRO AND AMINO COMPOUNDS 979

Solubility              Insoluble in cold, very soluble in hot water; readily soluble in hot alcohol, ether; soluble in benzene, acetone

*2.8.4.3 Toxicity.* No specific studies were located. Toxicity has been assumed to be identical with that of 2,3-dinitrophenol (109).

*2.8.4.4 Hygienic Standards of Permissible Exposure.* Standards have not been assigned.

### 2.8.5  3,4-Dinitrophenol [CAS # 577-71-9]

The formula is $C_6H_3OH(NO_2)_2$.

*2.8.5.1 Uses.* Specific applications for 3,4-dinitrophenol were not located.

*2.8.5.2 Physical and Chemical Properties (100).* 3-4-Dinitrophenol has the following properties:

| | |
|---|---|
| Physical state | Colorless needles, flammable |
| Molecular weight | 184.11 |
| Melting point | 134°C |
| Solubility | Very soluble in alcohol, ether |

*2.8.5.3 Toxicity.* No specific studies were located. Toxicity has been assumed to be identical with that of 2,3-dinitrophenol (109).

*2.8.5.4 Hygienic Standards of Permissible Exposure.* Standards have not been assigned.

### 2.8.6  3,5-Dinitrophenol [CAS # 586-11-8]

The formula is $C_6H_3OH(NO_2)_2$.

*2.8.6.1 Uses.* Specific applications for 3,5-dinitrophenol have not been located.

*2.8.6.2 Physical and Chemical Properties (100).* 3,5-Dinitrophenol has the following properties:

| | |
|---|---|
| Physical state | Colorless, monoclinic prisms, flammable |
| Molecular weight | 184.11 |
| Melting point | 134°C |
| Solubility | Very soluble in alcohol, ether; soluble in chloroform, benzene |

*2.8.6.3 Toxicity.* No specific studies were located. Toxicity has been assumed to be identical with that of 2,3-dinitrophenol (109).

*2.8.6.4 Hygienic Standards of Permissible Exposure.* Standards have not been assigned.

## 2.9 Picric Acid [CAS # 88-89-1]

The formula for picric acid is $HOC_6H_2(NO_2)_3$; it is also known as 2,4,6-trinitrophenol, carbazotic acid, picronitric acid, phenol trinitrate, trinitrophenol, nitroxanthic acid, and TNP.

### 2.9.1 Uses

Picric acid is used in the manufacture of explosives, as a burster in projectiles, rocket fuels, fireworks, colored glass, matches, electric batteries, and disinfectants. It is also used in the pharmaceutical and leather industries, as a fast dye for silk and wool, in copper and steel etching, forensic chemistry, and photographic emulsions.

At one time it was used in medicinal formulations in the treatment of malaria, trichinosis, herpes, smallpox, and antiseptics (116). The rationale was apparently based upon picric acid antiseptic activity, together with its ability to form insoluble picrates with proteins and nitrogenous bases. A 1 percent solution was also used in the treatment of burns (117).

### 2.9.2 Physical and Chemical Properties (100)

| | |
|---|---|
| Physical state | White to yellowish needles, flammable |
| Molecular weight | 229.11 |
| Melting point | 122 to 123°C |
| Boiling point | >300°C (sublimes, explodes) |
| Solubility | Slightly soluble in water; soluble in alcohol, ether, benzene; very soluble in acetone |

NOTE: Picric acid explodes when heated rapidly or when subjected to percussion; it is very explosive when dry, and for safety in transportation it is usually mixed with from 10 to 20 percent of water (118).

### 2.9.3 Toxicity

Picric acid causes sensitization dermatitis (119). Allergic hepatitis has been induced in guinea pigs by picric acid (120). Dust or fumes cause eye irritation, which can be aggravated by sensitization (121). It will dye animal fibers yellow (including skin) upon contact (117).

Systemic absorption can cause weakness, myalgia, anuria, polyuria, headache, fever, hyperthermia, vertigo, nausea, vomiting, diarrhea, and coma. High doses may cause destruction of erythrocytes, hemorrhagic nephritis and hepatitis, yellow coloration of the skin ("pseudojaundice") including conjunctiva and aqueous humor, and yellow vision (122, 110).

The strange visual effects where objects appear yellow may be related to the fact that systemic toxicity results in coloring all tissues yellow, including the conjunctiva and aqueous humor, so that yellow vision appearing under these circumstances seems to be explained on the basis of optical effects (121).

# AROMATIC NITRO AND AMINO COMPOUNDS

It is reportedly nonmutagenic in the Ames assay (123). Picric acid is metabolized largely to picramic acid (103).

### 2.9.4 Hygienic Standards of Permissible Exposure

A TLV–TWA value of 0.1 mg/m$^3$ has been adopted (ACGIH 5th Documentation). A caution against cutaneous or mucous membrane exposure is noted (40).

## 2.10  4,6-Dinitro-o-cresol [CAS # 534-52-1]

The formula for 4,6-dinitro-o-cresol is $C_6H_2(CH_3)OH(NO_2)_2$; it is also called 2-methyl-4,6-dinitrophenol, 4,6-dinitro-2-methylphenol, 3,5-dinitro-2-hydroxytoluene, dinitrocresol, 2,4-dinitro-6-methylphenol, 4,6-dinitro-o-cresol, 3,5-dinitro-o-cresol, DNOC, DNC, and DN.

### 2.10.1  Uses

The compound exists in nine isomers, all similar; however, 4,6-dinitro-o-cresol and 3,5-dinitro-o-cresol are the most common commercially. It is used as a herbicide and as an insecticide.

### 2.10.2  Physical and Chemical Properties (100)

| | |
|---|---|
| Physical state | Yellow prisms |
| Molecular weight | 198.14 |
| Melting point | 86.5°C |
| Solubility | Slightly soluble in water; very soluble in ether; soluble in acetone |

### 2.10.3  Toxicity

The oral $LD_{50}$ in rats is 31 mg/kg (124). Methemoglobinemia has been demonstrated in sheep (125), but not in humans.

The compound is an oxidative phosphorylation uncoupler (109), increasing body temperature and metabolism rate, which results in weight loss. The most severe toxicity occurs when workers are concurrently exposed to hot, humid environments. Nail damage has been reported in workers progressing to a painless inflammation, edema of the proximal nail folds accompanied by a white exudate (126). Bilateral cataracts have been reported following long-term ingestion (109, 127).

When human blood levels of DNOC exceed 15 to 20 µg/g, symptoms of poisoning appear (headache, fever, excessive sweating, unusual thirst, shortness of breath, cough, emesis, tachypnea, restlessness, weakness, yellow pigmentation of the skin, hair, sclera, and conjunctivae, pulmonary edema, narcosis, coma, and death) (128).

Neurotoxic effects have included cerebral edema, clinically evidenced by toxic psychosis and, at times, convulsions; peripheral neuropathy, asthenia or fatigue, and autonomic dystonia have also been reported (129).

Degenerative changes can occur in the liver parenchyma and renal tubules, resulting in clinical jaundice, albuminuria, hematuria, pyuria, and increased blood urea nitrogen (130).

Levels below 10 ppm are of trivial importance; levels greater than 20 ppm are toxic. Illness runs a rapid course with death or recovery within 24 to 48 hr. Rapid rigor mortis after death is characteristic of compounds that uncouple oxidative phosphorylation (131).

*2.10.4 Hygienic Standards of Permissible Exposure*

The adopted TLV–TWA value is 0.2 mg/m$^3$. Caution against cutaneous or mucous membrane absorption is noted (40).

### 2.11  2,6-Dinitro-*p*-cresol [CAS # 609-93-8]

The formula for 2,6-dinitro-*p*-cresol is $C_6H_2(CH_3)OH(NO_2)_2$; the compound is also called 4-methyl-2,6-dinitrophenol.

*2.11.1 Uses*

Its uses are reportedly identical with those of 4,6-dinitro-*o*-cresol (109).

*2.11.2 Physical and Chemical Properties (100)*

| | |
|---|---|
| Physical state | Long yellow prisms |
| Molecular weight | 198.14 |
| Melting point | 85°C |
| Solubility | Insoluble in water; soluble in alcohol benzene; very soluble in ether |

*2.11.3 Toxicity*

The toxicity of 2,6-dinitro-*p*-cresol is reportedly identical with that of the ortho isomer (109). Peripheral neuropathy (dying back) has apparently been observed following exposure (132).

*2.11.4 Hygienic Standards of Permissible Exposure*

Standards have not been assigned.

### 2.12  Aniline [CAS # 62-53-3]

The formula for aniline is $C_6H_5NH_2$; it is also known as aminobenzene, phenylamine, aniline oil, benzenamine, aminophen, and arylamine.

It has been produced commercially for decades by catalytic vapor phase hydrogenation of nitrobenzene. Production of aniline amounted to 180,000 metric tons in 1975 and was expected to reach 334,000 tons by 1980. Current production figures were not available (133).

### 2.12.1 Uses

Aniline is used in the manufacture of dyestuffs, dyestuff intermediates, rubber accelerators, and antioxidants; it is also an intermediate in the manufacture of pharmaceuticals, photographic developers, plastics, isocyanates, hydroquinones, herbicides, fungicides, and ion-exchange resins.

### 2.12.2 Physical and Chemical Properties (100)

| | |
|---|---|
| Physical state | Oily liquid |
| Molecular weight | 93.13 |
| Density | 1.002 (20/4°C) |
| Melting point | −6.3°C |
| Boiling point | 184.4 to 186°C |
| Vapor density | 3.22 (air = 1) |
| Vapor pressure | 15 mm Hg (77°C) |
| Refractive index | 1.5863 (20°C) |
| Solubility | 3.4 in water at 20°C; soluble in alcohol, ether, benzene, chloroform, carbon tetrachloride, acetone |
| Flash point | 76°F (closed cup) |
| Autoignition | 1418°F |
| Odor and warning properties | Characteristic, peculiar odor and burning taste |

### 2.12.3 Toxicity

Aniline exposure in humans can cause headaches, methemoglobinemia, paresthesias, tremor, colicky pain, narcosis/coma, cardiac arrhythmia, and possibly death (109, 110). An acute oral $LD_{50}$ for rats has been reported as 0.44 g/kg (134).

It has been demonstrated in rats (Toth, 1971) to inhibit corticosteroidogenesis, lowering plasma corticosterone levels and causing adrenocortical hyperplasia. The responsiveness of the adrenals to insulin, vasopressin, or adrenocorticotrophic hormones was also diminished (109).

Aniline hydrochloride was not teratogenic to F344 rats, even at maternally toxic doses via gavage of 10, 30, or 100 mg/kg. However, methemoglobinemia, increased relative spleen weight, and decreased red blood cell count were noted in 20 dams (135).

Historically, bladder tumors have been associated with aromatic amines and the dye industry. These cancers were referred to as "aniline tumors" because the cases were from workers in the aniline dye industries. The term "aniline tumor" was eventually considered to be a misnomer, however, because convincing evidence could not be established primarily as a result of mixed exposures to a multiplicity of compounds and dyestuff intermediates within the same work area.

Investigators began to believe that β-naphthylamine, benzidine, and other analogues and homologues of aniline were more likely the etiologic agents rather than aniline. Hamilton and Hardy reported in 1974 that the so-called aniline tumor of the bladder was currently believed to be related to absorption of any one of the following four aromatic compounds: β-naphthylamine, 4-aminodiphenyl, 4-nitrodiphenyl, or 4,4'-diaminodiphenyl (109).

IARC has classified aniline as a Group 3 carcinogen, that is, not classifiable as to its carcinogenicity (32a). However, NIOSH has determined that there is sufficient evidence to recommend that OSHA require labeling this substance a potential occupational carcinogen. This position followed an evaluation of a high-dose feeding study of aniline hydrochloride in F344 rats and B6C3F1 mice (3000 or 6000 ppm and 6000 or 12,000 ppm, respectively). The test was negative in both sexes of mice; hemangiosarcomas of the spleen and combined incidence of fibrosarcomas and sarcoma of the spleen were statistically significant in the male rats; the number of female rats having fibrosarcomas of the spleen was also significant. At high dose levels the resulting methemoglobinemia has been postulated to be the effect leading to chronic splenic congestion, fibrosis, and sarcoma formation (136).

In a recent epidemiology study, excess bladder cancers were reported in workers exposed to o-toluidine and aniline; the authors stated that o-toluidine is an animal carcinogen more potent than aniline and is most likely responsible, although aniline may have played a role (137).

### 2.12.4 Hygienic Standards of Permissible Exposure

A TLV–TWA value of 2 ppm (7.6 mg/m$^3$) has been adopted, with caution against cutaneous and mucous membrane exposures. Biologic monitoring of total p-aminophenol in urine has been suggested (138). NIOSH considers aniline (and its homologues) occupational carcinogens (40). Neither OSHA nor ACGIH has classified aniline as an occupational carcinogen (40, 138).

### 2.13 N-Methylaniline [CAS # 100-61-8]

The formula for N-methylaniline is $C_6H_5NH(CH_3)$; it is also known as methylaniline, methylphenylamine, monomethylaniline, or N-methylbenzenamine.

### 2.13.1 Uses

N-Methylaniline is used as a solvent and in organic synthesis.

### 2.13.2 Physical and Chemical Properties (100).

| | |
|---|---|
| Physical state | Colorless or slightly yellow liquid, turns brown on exposure to air |
| Molecular weight | 107.15 |
| Specific gravity | 0.989 at 20°C |
| Melting point | −57°C |

| | |
|---|---|
| Boiling point | 194.6 to 196°C |
| Flash point | 79.4°C (closed cup) |
| Solubility | Slightly soluble in water; soluble in alcohol, ether |

### 2.13.3 Toxicity

The oral $LD_{50}$ (rabbit) is reportedly 280 mg/kg (100a).

### 2.13.4 Hygienic Standards of Permissible Exposure

A TLV–TWA of 0.5 ppm is recommended by the ACGIH (139).

## 2.14 N,N-Dimethylaniline [CAS # 121-69-7]

The formula for N,N-dimethylaniline is $C_6H_5N(CH_3)_2$; it is also known as n-phenyldimethylamine, dimethylaniline, DMA, dimethylaminobenzene, N,N-dimethylbenzenamine, and N,N-dimethylphenylamine.

### 2.14.1 Uses

N,N-Dimethylamine is used in the synthesis of dyestuffs and dyestuff intermediates and as a solvent, reagent in methylation reactions, and catalytic hardener in fiber glass resins.

### 2.14.2 Physical and Chemical Properties (100)

| | |
|---|---|
| Physical state | Yellow liquid |
| Molecular weight | 121.18 |
| Density of liquid | 0.9557 (20/4°C) |
| Melting point | 2.45°C |
| Boiling point | 192.5°C |
| Density of vapor | 4.17 (air = 1) |
| Refractive index | 1.55819 (20°C) |
| Solubility | Slightly soluble in water; soluble in alcohol and ether |
| Flash point | 145°F (closed cup) |
| | 170°F (open cup) |
| Autoignition | 700°F |

### 2.14.3 Toxicity

The oral $LD_{50}$ in rats is reported to be 1410 mg/kg; the dermal $LD_{50}$ for rabbits is reported as 1770 mg/kg (140).

Metabolism studies have demonstrated that N,N-dimethylaniline can undergo N-oxidation and N-demethylation. In male Sprague–Dawley rats, following enzyme induction with either phenobarbital or 3-methylcholanthrene (MCA), two distinct pathways were observed. N-Oxide formation was seen in both phenobarbital- and MCA-induced systems; the rate of formation was higher with MCA. The second reaction involved the *N*-dealkylation of the *N*-oxide of DMA; the authors concluded that DMA is converted to a monomethylaniline by two separate pathways within the P450 system (141a). DMA has also been reported to be completely demethylated and oxidized to *o*-aminophenol, then conjugated with sulfuric and glucuronic acid in the dog (141b).

Developmental toxicity was not observed in a preliminary Chemoff and Kavlock screening test in mice (142).

Subchronic toxicity investigation by gavage at doses of 31.25, 62.5, 125, 250, and 500 mg/kg, 5 days/week for 13 weeks was conducted in F344 rats and B6C3F1 mice. Clinical signs of toxicity including cyanosis and decreased motor activity were observed in both species and sexes in a dose-dependent manner. Splenomegaly was observed in all treated animals. Hemosiderin was observed in the spleen, liver, testes, and kidneys. Bone marrow hyperplasia and increased hematopoiesis in the spleen were observed in rats; hematopoiesis was increased in the spleen and liver of treated mice. It was concluded that the cyanosis observed in the rats indicated erythrocyte destruction and reduced blood oxygenation, possibly as a result of methemoglobin formation. A no observable effect level was not reached in rats; the NOEL for mice was estimated at 31.25 mg/kg DMA (143).

A 2-year corn oil-gavage bioassay conducted by NTP in F344/N rats (0.3 or 30 mg/kg) and B6C3F1 mice (0, 15, or 30 mg/kg) for 103 weeks concluded that there was some evidence of carcinogenic activity for male F344/N rats as indicated by the increased incidences of sarcomas or osteosarcomas in the spleen; there was no evidence of carcinogenic activity in the female rats or male mice; there was equivocal evidence of carcinogenic activity for female mice as indicated by an increased incidence of squamous cell papillomas of the forestomach. Both rats and mice could have tolerated doses higher than those used in these studies (144).

### 2.14.4 Hygienic Standards of Permissible Exposure

The cutaneous TWA values adopted are 5 ppm, 25 mg/m$^3$. The tentative STEL values are 10 ppm, 50 mg/m$^3$; a skin notation was indicated (138).

## 2.15 *N*-Ethylaniline [CAS # 103-69-5]

The formula for *N*-ethylaniline is $C_6H_5NHC_2H_5$; it is also known as ethylphenylamine.

### 2.15.1 Uses

*N*-Ethylaniline is used as an explosive stabilizer and in dyestuff manufacturing as a cyclic intermediate.

### 2.15.2 Physical and Chemical Properties (100)

| | |
|---|---|
| Physical state | Colorless liquid, darkens in air |
| Molecular weight | 121.18 |
| Boiling point | 204.5°C |

Melting point            −63.5°C
Solubility               Insoluble in water; miscible with alcohol, ether; soluble in acetone, benzene

### 2.15.3 Toxicity

The oral $LD_{50}$ (rat) is reportedly 1.1 g/kg (100a).

### 2.15.4 Hygienic Standards of Permissible Exposure

Standards have not been assigned.

## 2.16 N,N-Diethylaniline [CAS # 91-66-7]

The formula for N,N-diethylaniline is $C_6H_5N(C_2H_5)_2$; it is also called n-phenyldiethylamine, DEA, and diethylaniline.

### 2.16.1 Uses

Diethylaniline is used in dyestuffs and in the synthesis of other intermediates and pharmaceuticals.

### 2.16.2 Physical and Chemical Properties (100)

Physical state       Colorless or yellow or brown flammable oil
Molecular weight    149.23
Density               0.93507 (20/4°C)
Melting point        −38.8°C
Boiling point         215.5 to 216°C
Refractive index     1.54105 (22°C)
Solubility            1.44 in water at 12°C; soluble in alcohol, ether

### 2.16.3 Toxicity

Diethylaniline is reportedly less toxic than aniline, but very similar in its effects. It is readily absorbed through the intact skin and precautions must be taken to avoid inhalation of its vapors (109).

### 2.16.4 Hygienic Standards of Permissible Exposure

Standards have not been assigned.

## 2.17 o-Nitroaniline [CAS # 88-74-4]

The formula for o-nitroaniline is $NO_2C_6H_4NH_2$; it is also called 1-amino-2-nitrobenzene, 2-nitroaniline, 1-nitro-2-aminobenzene, 2-nitrobenzenamine, and o-nitraniline.

### 2.17.1 Uses

*o*-Nitroaniline is used as a dyestuff intermediate.

### 2.17.2 Physical and Chemical Properties (100)

| | |
|---|---|
| Physical state | Golden yellow to orange rhombic needles |
| Molecular weight | 138.13 |
| Density | 1.442 (20/4°C) |
| Melting point | 69 to 71.5°C |
| Boiling point | 284.11°C |
| Solubility | 0.126 in water at 25°C; 15.8 in alcohol at 15°C, 27.87 at 25°C; very soluble in ether, acetone, benzene |

### 2.17.3 Toxicity

In contrast to *m*-nitroaniline (mutagenic) and *p*-nitroaniline (weakly or not mutagenic), *o*-nitroaniline is nonmutagenic (145).

The compound is nonteratogenic in Charles River CD rats administered 0, 100, 300, and 600 mg/kg by gavage on days 6 to 15 of gestation. Maternal toxicity was noted at 300 and 600 mg/kg, but there were no compound-related changes in pregnancy rates, fetal resorptions, fetal viability, postimplantation losses, total implantations, mean litter weights, or signs of teratological changes (146).

### 2.17.4 Hygienic Standards of Permissible Exposure

Standards have not been assigned.

## 2.18 *m*-Nitroaniline (CAS # 99-09-2)

The formula for *m*-nitroaniline is $NO_2C_6H_4NH_2$; it is also called 1-amino-3-nitrobenzene, 3-nitroaniline, 3-nitrobenzenamine, and *m*-nitraniline.

### 2.18.1 Uses

*m*-Nitroaniline is used in the synthesis of dyestuffs and other intermediates.

### 2.18.2 Physical and Chemical Properties (100)

| | |
|---|---|
| Physical state | Yellow rhombic needles |
| Molecular weight | 138.13 |
| Density | 1.430 (20/4°C) |
| Melting point | 114°C |
| Boiling point | 305 to 307°C |

| | |
|---|---|
| Solubility | Slightly soluble in water; soluble in alcohol, ether |
| Warning property | Burning sweet taste |

### 2.18.3 Toxicity

$m$-Nitroaniline is a powerful methemoglobin former with attendant hemolytic effect. On prolonged and excessive exposures it may also cause liver damage. It is readily absorbed through the intact skin and irritates skin, eyes, and mucous membranes, and its vapors are highly toxic as well (109).

In contrast to $o$-nitroaniline and $p$-nitroaniline, $m$-nitroaniline is a positive mutagen in the Ames assay (145b).

### 2.18.4 Hygienic Standards of Permissible Exposure

Standards have not been assigned.

## 2.19 p-Nitroaniline [CAS # 100-01-6]

The formula is $NO_2C_6H_4NH_2$; $p$-nitroaniline is also called 1-amino-4-nitrobenzene, 4-nitroaniline, $p$-aminonitrobenzene, 4-nitrobenzenamine, $p$-nitrophenylamine, $p$-nitraniline, or PNA.

### 2.19.1 Uses

$p$-Nitroaniline is used in the synthesis of dyes, such as $p$-nitroaniline red, as a corrosion inhibitor, and in the synthesis of antioxidants.

### 2.19.2 Physical and Chemical Properties (100)

| | |
|---|---|
| Physical state | Pale yellow crystals |
| Molecular weight | 138.13 |
| Specific gravity | 1.442 |
| Melting point | 146 to 149°C |
| Boiling point | 331.7°C |
| Vapor pressure | <1 torr at 20°C |
| Flash point | 198.9°C (closed cup) |

### 2.19.3 Toxicity

Human exposure can result in severe headache, methemoglobinemia, narcosis, or coma (110, 147). This compound is a powerful methemoglobin former that can result in hemolytic anoxia and anemia. Other clinical symptoms include sleepiness, weakness, respiratory distress, and jaundice.

Chronic administration of $p$-nitroaniline for 2 years in Sprague–Dawley rats (0, 0.25, 1.5, or 9.0 mg/kg) by gavage in corn oil produced elevated methemoglobin in the mid- and high-dosage groups; slight anemia was also observed in the high dosage group, and spleen weights were significantly increased, but no treatment-related increase in tumors was observed (148a).

In a parallel reproductive study, groups of male and female Sprague–Dawley rats, designated as the $F_0$ generation, were administered $p$-nitroaniline at the same dosage levels as the above bioassay for 14 weeks prior to mating and during mating, gestation, and lactation. Selected groups of male and female rats of the $F_1$ generation received the same dose of $p$-nitroaniline for 18 weeks prior to mating and during mating, gestation, and lactation. No consistent pattern of effect from the treatment between the $F_0$ and $F_1$ generation was seen in mating, pregnancy, or fertility indexes (148a).

Definitive teratogenicity studies in rats and rabbits indicated that $p$-nitroaniline was not teratogenic. Groups of 24 mated Sprague–Dawley rats were administered 25, 85, or 250 mg/kg/day by gavage on days 6 to 19 of gestation; New Zealand mated rabbits were dosed with 15, 75, or 125 mg/kg/day on gestation days 7 to 19 via a stomach tube. In the rat study, evidence of embryo toxicity as measured by increased resorptions was observed at the 250 mg/kg/day dosage level, and at this level the incidence of external soft-tissue and skeletal malformations was significantly higher than for controls on both a per fetus and per litter basis. In the rabbit study, $p$-nitroaniline elicited a higher degree of maternal toxicity than in the rats; however, no significant differences in the number of external, soft-tissue, or skeletal malformations were observed, either on a per fetus or on a per litter basis. The authors concluded that $p$-nitroaniline should not be labeled as a teratogen because the fetal effects could not be dissociated from the maternal effects; agents producing fetal effects only at dosages overtly toxic to the dam are not classifiable as teratogenic (148b).

$p$-Nitroaniline has also been reported to be weakly or not mutagenic in the Ames assay (145).

$p$-Nitroaniline carcinogenicity bioassay was initiated in 1991 under the auspices of the National Toxicology Program at Southern Research Institute (108).

### 2.19.4 Hygienic Standards of Permissible Exposure

The adopted TLV–TWA value is 3 mg/m$^3$. A caution against cutaneous or mucous membrane exposure is noted (139).

## 2.20 *o*-Chloroaniline [CAS # 95-51-2]

The formula for *o*-chloroaniline is $ClC_6H_4NH_2$; it is also known as 2-chlorophenylamine or 2-chloroaniline.

### 2.20.1 Uses

The uses of *o*-chloroaniline are reportedly the same as those of the para isomer (109).

## 2.20.2 Physical and Chemical Properties (100)

| | |
|---|---|
| Physical state | Colorless liquid |
| Molecular weight | 127.57 |
| Density of liquid | 1.2125 (20/4°C) |
| Melting point | α, −14°C; β, −1.9°C |
| Boiling point | 208.8°C |
| Refractive index | 1.5895 (20°C) |
| Solubility | Insoluble in water; miscible with alcohol; soluble in ether, acetone |
| Odor and warning properties | Characteristic sweet odor |

## 2.20.3 Toxicity

The compound is absorbed through the intact skin. It has a relatively low vapor pressure, but precaution should be taken to avoid inhalation of vapors (109).

In a comparative nephrotoxicity study of aniline with its monochlorophenyl derivatives, o-, m-, and p-chloroaniline, it was observed that chloro substituents on the phenyl ring of aniline increased nephrotoxic potential of aniline, the ortho substitution producing the greatest enhancement. The in vivo experiments were conducted by administering a single intraperitoneal injection of the chloro derivatives at 0.4, 1.0, or 1.5 mmol/kg and renal function monitored. o-Chloroaniline was the only compound tested that decreased urine volume, elevated blood area nitrogen concentration, and depressed both basal and lactate-stimulated p-aminohippurate (PAH) accumulation by renal cortical slices at the 1.0 mmol/kg dose. Similar results were produced following m- and p-chloroaniline administration, but these compounds required a dose of 1.5 mmol/kg. Aniline had little effect on renal function at the doses used in this study (149).

The National Toxicology Program scheduled o-chloroaniline for the prechronic phase of toxicity studies, by gavage in rats and mice, beginning in fiscal year 1991 (150).

## 2.20.4 Hygienic Standards of Permissible Exposure

Standards have not been assigned.

## 2.21 m-Chloroaniline [CAS # 108-42-9]

The formula is $ClC_6H_4NH_2$; it is also known as 3-chloroaniline or 3-chlorophenylamine.

### 2.21.1 Uses

Its uses are reportedly the same as those of p-chloroaniline (109).

### 2.21.2 Physical and Chemical Properties (100)

| | |
|---|---|
| Physical state | Colorless liquid |
| Molecular weight | 127.57 |
| Melting point | $-10.3°C$ |
| Boiling point | 229.8 to 230.5°C |
| Refractive index | 1.59424 (20°C) |
| Solubility | Insoluble in water; miscible with alcohol; soluble in ether, acetone, benzene |
| Odor and warning properties | Characteristic sweet odor |

### 2.21.3 Toxicity

*m*-Chloroaniline is readily absorbed through the intact skin. It may cause methemoglobinemia, anemia, and liver and kidney damage. It has a relatively low vapor pressure, but precaution should be taken to avoid inhalation of its vapors (109, 110).

The National Toxicology Program (NTP) scheduled *o*-chloroaniline for the prechronic phase of toxicity studies, by gavage in rats and mice, beginning in fiscal year 1991 (150).

### 2.21.4 Hygienic Standards of Permissible Exposure

Standards have not been assigned.

## 2.22 *p*-Chloroaniline [CAS # 106-47-8]

The formula for *p*-chloroaniline is $ClC_6H_4NH_2$; it is also known as 4-chlorophenylamine, *p*-chloroaniline, or *p*-aminochlorobenzene.

### 2.22.1 Uses

*p*-Chloroaniline is used as a dye intermediate in the synthesis of azo dyes and other dyestuff intermediates (109).

### 2.22.2 Physical and Chemical Properties (100)

| | |
|---|---|
| Physical state | Colorless solid, rhombic crystals |
| Molecular weight | 127.57 |
| Density | 1.427 (19/4°C) |
| Melting point | 72.5°C |
| Boiling point | 231 to 232°C |

| Solubility | Insoluble in water; miscible with alcohol; soluble in ether |
| Odor and warning properties | Characteristic sweet odor |

### 2.22.3 Toxicity

*p*-Chloroaniline is absorbed through the intact skin and may cause methemoglobinemia. It is also less hazardous than aniline or mononitrobenzene in industrial exposures. It has a relatively low vapor pressure, but precautions should be taken to avoid inhalation of vapors (109). The oral $LD_{50}$ (rats) is reportedly 0.31 g/kg (100a).

*p*-Chloroaniline is a microsomal P450 inhibitor (151), postulated to be the result of *N*-oxidation by microsomal P450 to *N*-hydroxylamine, which then complexes with P450 via a nitroxide radical intermediate generated during conversion to the corresponding nitroso metabolite (152). In addition to microsomal enzyme inhibition, *N*-hydroxylamine also oxidizes $Fe^{2+}$ of heme to $Fe^{3+}$, resulting in ferri- or methemoglobin. Methemoglobin per se does not lead to oxidative hemolysis and there is poor correlation between the ability of a chemical to produce methemoglobin and its ability to produce Heinz bodies (dark-staining granules found in red cells associated with oxidative hemolytic anemia and thought to consist of denatured hemoglobin) (153). However, it has been reported that *p*-chloroaniline has been associated with oxidative hemolytic anemia in humans as a result of oxidative hemolysis of the red blood cells. The authors point out that this particular risk is greater in genetic glucose-6-phosphate dehydrogenase (G6PD) deficiency (154).

In a more recent metabolism study in which [$^{14}$C]*p*-chloroaniline was administered to male F344 rats, female C3H mice, and a male rhesus monkey, greater than 90 percent of the radiocarbon was eliminated through the urine in all species; a major route of metabolism was identified as ortho hydroxylation, whereby 2-amino-5-chlorphenyl sulfate is the major excretion product. In addition, *p*-chloroacetanilide was also found to be a major circulating metabolite (155). This metabolite has also been demonstrated to cause hemolysis (154).

An initial NTP bioassay reportedly demonstrated equivocal results and the compound was assigned a Class E carcinogen rating by EPA (156).

In a second study to reevaluate the carcinogenicity potential of *p*-chloroaniline (as the hydrochloride), groups of 50 F344/N rats of each sex were given by gavage in water doses of 0, 2, 6, or 18 mg/kg, 5 days/week for 103 weeks; groups of 50 male and female B6C3F1 mice of each sex were given 0, 3, 10, or 30 mg/kg, 5 days/week for 103 weeks. Rats developed mild hemolytic anemia and slight increases in methemoglobin at the 18 mg/kg level, and male rats at all dose levels showed increased fibrosis of the spleen and sarcomas, the ratios of incidence of sarcomas in control to high-dose groups being 0/49, 1/50, 3/50, and 38/50, respectively; female rats at 18 mg/kg showed fibrosis of the spleen. There was a slightly increased incidence of pheochromocytomas of the adrenal gland occurring in male and female rats. Male mice had increased incidences of hepatocellular adenomas or carcinomas (control to high-dose group ratios of 11/50, 21/49, 20/50, and 21/50, respectively). Hemangiosarcomas of the liver or spleen were also increased in the high-dose group of male mice (control to high-dose group ratios of 4/50, 4/49, 1/50 and 10/50, respectively). The authors concluded that *p*-chloroaniline was carcinogenic in male rats and male mice (157).

### 2.22.4 Hygienic Standards of Permissible Exposure

Standards have not been assigned.

## 2.23 Dichloroaniline Isomers

### 2.23.1 3,4-Dichloroaniline [CAS # 95-76-1]

The formula is $C_6H_5Cl_2NH_2$; the compound is also known as 3,4-dichlorobenzenamine.

*2.23.1.1 Uses.* Specific applications were not located.

*2.23.1.2 Physical and Chemical Properties (100).* 3,4-Dichloroaniline has the following properties:

| | |
|---|---|
| Molecular weight | 162.02 |
| Melting point | 71 to 72°C |
| Boiling point | 272°C |
| Solubility | Practically insoluble in water; very soluble in alcohol; slightly soluble in benzene |

*2.23.1.3 Toxicity.* 3,4-Dichloroaniline has been demonstrated to cause methemoglobin in mice. When mice were administered toxic doses of the amide herbicide propanil (3',4'-dichloropropionanilide), the resulting cyanosis was found to be due to methemoglobin formation following hydrolysis to 3,4-dichloroaniline (158). Interestingly, rice plants were found to be selectively resistant to propanil owing to hydrolysis to 3,4-dichloroaniline and propionic acid (159).

Acute dermal $LD_{50}$ in albino rabbits was reported as approximately 300 mg/kg; clinical observations included cyanosis (160a). The acute oral $LD_{50}$ value in Chr-CD male rats (5 percent in acetone–corn oil 15–85) was estimated to be 545 mg/kg; clinical observations included lethargy, cyanosis, and chromodacryorrhea in all dose groups (160b).

Human exposure reportedly caused chloracne (161).

*2.23.1.4 Hygienic Standards of Permissible Exposure.* Standards have not been assigned.

### 2.23.2 2,3-Dichloroaniline [CAS # 608-27-5]

The formula is $C_6H_3Cl_2NH_2$.

*2.23.2.1 Uses.* No specific applications were located in the literature.

### AROMATIC NITRO AND AMINO COMPOUNDS

*2.23.2.2 Physical and Chemical Properties (100).* The properties of 2,3-dichloroaniline are as follows:

| | |
|---|---|
| Molecular weight | 162.02 |
| Boiling point | 252°C |
| Melting point | 24°C |
| Solubility | Soluble in ether, alcohol, acetone |

*2.23.2.3 Toxicity.* No toxicology information was located in the literature.

*2.23.2.4 Hygienic Standards of Permissible Exposure.* Standards have not been assigned.

### 2.23.3  2,4-Dichloroaniline [CAS #554-00-7]

The formula is $C_6H_3Cl_2NH_2$.

*2.23.3.1 Uses.* Specific applications were not located in the literature.

*2.23.3.2 Physical and Chemical Properties (100).* 2,4-Dichloroaniline has the following properties:

| | |
|---|---|
| Molecular weight | 162.02 |
| Boiling point | 245°C |
| Melting point | 63 to 64°C |
| Solubility | Soluble in alcohol, ether |

*2.23.3.3 Toxicity.* No toxicology information was located in the literature.

*2.23.3.4 Hygienic Standards of Permissible Exposure.* Standards have not been assigned.

### 2.23.4  2,5-Dichloroaniline [CAS # 95-82-9]

The formula is $C_6H_3Cl_2NH_2$.

*2.23.4.1 Uses.* No specific applications were found in the literature.

*2.23.4.2 Physical and Chemical Properties (100).* 2,5-Dichloroaniline has the following properties:

| | |
|---|---|
| Molecular weight | 162.02 |
| Boiling point | 251°C |
| Melting point | 50°C |
| Solubility | Soluble in alcohol, ether, benzene |

*2.23.4.4 Toxicity.* No toxicology information was found in the literature.

*2.23.4.5 Hygienic Standards of Permissible Exposure.* Standards have not been assigned.

### 2.23.5  2,6-Dichloroaniline [CAS # 608-31-1]

The formula is $C_6H_3Cl_2NH_2$.

*2.23.5.1 Uses.* No specific applications were located in the literature.

*2.23.5.2 Physical and Chemical Properties (100).* The properties of 2,6-dichloroaniline are as follows:

| | |
|---|---|
| Molecular weight | 162.02 |
| Melting point | 39°C |
| Solubility | Soluble in ether |

*2.23.5.3 Toxicity.* No toxicology information was found in the literature.

*2.23.5.4 Hygienic Standards of Permissible Exposure.* Standards have not been assigned.

### 2.23.6  3,5-Dichloroaniline [CAS # 626-43-7]

The formula is $C_6H_3Cl_2NH_2$.

*2.23.6.1 Uses.* No specific applications were found in the literature.

*2.23.6.2 Physical and Chemical Properties (100).* 3,5-Dichloroaniline has the following properties:

| | |
|---|---|
| Molecular weight | 162.02 |
| Boiling point | 260°C |
| Melting point | 51 to 53°C |
| Solubility | Soluble in alcohol, benzene, ether |

*2.23.6.3 Toxicity.* No toxicology information was found in the literature.

*2.23.6.4 Hygienic Standards of Permissible Exposure.* Standards have not been assigned.

## 2.24  2,6-Dichloro-4-nitroaniline [CAS # 99-30-9]

The formula is $C_6H_2NH_2(NO_2)Cl_2$; and the compound is also known as Dichloran and DCNA.

### 2.24.1  Uses

DCNA is used as a fungicide.

### 2.24.2 Physical and Chemical Properties (100)

| | |
|---|---|
| Molecular weight | 206.07 |
| Melting point | 46°C |
| Boiling point | 275°C |
| Solubility | Soluble in alcohol, ether, benzene |

### 2.24.3 Toxicity

Bernstein (1970) reported the development of irreversible corneal and lens opacities in dogs following administration of high doses of DCNA (25 to 50 mg/kg). In the anterior cornea, there were discrete areas of degeneration of the anterior corneal lamella associated with histiocytes containing lipid granules. The predominant effect upon the lens was edema around the anterior Y suture. These corneal and lens abnormalities could be produced only when the animal was exposed to outdoor or natural sunlight illumination, indicating a photoreactive mechanism (109). Dichloran and one of its metabolites (3,5-dichloro-4-aminophenol) have also been demonstrated to be potent uncouplers of oxidative phosphorylation in vitro (159).

2,6-Dichloro-4-nitroaniline is included in a tabulated list of 465 suspected or known carcinogens published by Rinkus and Legator in 1979 (162). No toxicology citations were found for its homologue 2,4-dichloro-6-nitroaniline (CAS # 2683-43-4).

### 2.24.4 Hygienic Standards of Permissible Exposure

Standards have not been assigned.

## 2.25 4-Chloro-2-Nitroaniline [CAS # 09-63-4]

The formula is $C_6H_3NH_2(NO_2)Cl$.

### 2.25.1 Uses.

No specific applications were found in the literature.

### 2.25.2 Physical and Chemical Properties (100)

| | |
|---|---|
| Physical state | Dark orange crystals |
| Molecular weight | 172.57 |
| Melting point | 116–117°C |
| Solubility | Soluble in alcohol, ether |

### 2.25.3 Toxicity

NTP has scheduled this compound as a candidate for future testing (108). Other homologues include 2-chloro-4-nitro-, 2-chloro-5-nitro-, 4-chloro-3-nitro-, and 5-chloro-2-nitroaniline.

## 2.25.4 Hygienic Standards of Permissible Exposure

Standards have not been assigned.

## 2.26 Tetryl [CAS # 479-45-8]

The formula for tetryl is $(NO_2)_3C_6H_2N(NO_2)CH_3$; it is also known as trinitrophenylmethylnitramine, nitramine, tetranitromethylaniline, pyrenite, picrylmethylnitramine, picrylnitromethylamine, N-methyl-N, 2,4,6-tetranitroaniline, tetralite, and 2,4,6-tetryl.

### 2.26.1 Uses

Tetryl is used as an explosive and in explosives as an intermediary charge in detonators and blasting caps for less sensitive explosives, as a booster charge for military devices, and as a chemical indicator.

### 2.26.2 Physical and Chemical Properties (100)

| | |
|---|---|
| Physical state | Yellow monoclinic crystals |
| Molecular weight | 287.15 |
| Density | 1.57 at 19°C |
| Melting point | 130°C |
| Boiling point | 187°C, explodes |
| Solubility | Insoluble in water; slightly soluble in alcohol; very soluble in ether |

### 2.26.3 Toxicity

Tetryl is a potent sensitizer and causes allergic dermatitis. It is also highly irritating to the skin and mucous membranes of the respiratory tract and eyes. Anemia has also been found among exposed workers (109).

It also stains the skin and hair yellow to orange and causes bone marrow depression, hepatotoxicity, and central nervous system effects including weakness, headache, malaise, lassitude, and sleeplessness (129b, 147, 163, 164).

### 2.26.4 Hygienic Standards of Permissible Exposure

The adopted TLV–TWA value is 1.5 mg/m$^3$ (139). A caution against cutaneous or mucous membrane exposure is noted.

## 2.27 4,4'-Methylenebis(2-chloroaniline) [CAS # 101-14-4]

The formula is $CH_2(C_6H_4ClNH_2)_2$ or $C_{13}H_{14}Cl_2N_2$; the compound is also known as methylenebis(o-chloroaniline), 4,4'-diamino-3,3'-dichlorodiphenylmethane, 4,4'-methylene-2,2-dichloroaniline, MBOCA, DACPM, Cl-MDA, and MOCA.

## 2.27.1 Uses

MOCA is currently used as a curing agent for isocyanate-containing polymers and in the manufacture of polyurethane foams, epoxy resins, gun mounts, jet engine turbine blades, radar systems, and components in home appliances. It is also a model compound for studying carcinogens. MOCA has not been produced in the United States since 1979, but it is imported, largely from Japan (165).

## 2.27.2 Physical and Chemical Properties (100)

| | |
|---|---|
| Physical state | Tan colored solid |
| Molecular weight | 267.15 |
| Specific gravity | 1.44 at 4°C |
| Melting range | 99 to 110°C |
| Solubility | Soluble in hot methyl ethyl ketone, dimethylformamide, dimethyl sulfoxide, acetone, esters, aromatic hydrocarbons |

## 2.27.3 Toxicity

Russfield et al. (1975) investigated the carcinogenicity of MOCA in male and female random-bred albino mice and male Charles River CD rats. The maximally tolerated dose of the chemical in the diet was 1000 mg/kg in rats and 2000 mg/kg in mice. The compound was administered in the diet (500 or 1000 mg/kg rats; 1000 or 2000 mg/kg mice) for 19 months, followed by control diet for another 6 months. MOCA was carcinogenic in female mice but not in male rodents; a statistically significant incidence of hepatoma at both dose levels was observed. In addition, a higher incidence of tumors was observed in treated animals than in controls as follows: hepatomas and lung adenomatosis in rats, hemangiosarcomas, and hemangiomas in mice (109).

In a comparative study, Stula et al. (1975) investigated the carcinogenic potential of MOCA because of its molecular similarity to two known rat carcinogens, 3,3'-dichlorobenzidine (DCB) and 4,4'-methylenebis(2-methylaniline) (Me-MDA). They demonstrated that all three agents were carcinogenic to rats (ChR-CD, Charles River cesarean derived, Sprague–Dawley rats). DCB was carcinogenic at a level of 1000 ppm in a standard diet; it produced tumors of the mammary gland in rats of both sexes, together with Zymbal gland and hematopoietic tumors in males. (Zymbal gland tumors have been reported in rats following exposure to benzidine, o-toluidine, 2-acetylaminofluorine, 4-aminostilbene, 4-dimethylaminostilbene, 4-acetaminostilbene, 3,3'-dimethylbenzidine, vinyl chloride, and vinyl bromide.) The same level of MOCA in a standard diet produced tumors of the lung in both sexes; in a protein-deficient diet, MOCA caused lung tumors in both sexes, liver tumors in males, and malignant mammary tumor in females. MOCA in a low-protein diet resulted in a shorter average life-span with termination after 16 months. Me-MDA at a level of 200 ppm in a standard diet produced tumors in the livers of rats of both sexes, together with skin and mammary tumors in males. From previous studies

(Munn, 1967), it was found that 4,4'-diaminodiphenylmethane was considerably less carcinogenic than Me-MDA in rats, and the authors concluded that methyl or chlorine substitution in the 3,3' positions brings on additional carcinogenic risk for the rat, the methyl substitution resulting in the greater carcinogenic activity (38, 109). When administered by gavage, MOCA induced transitional cell carcinogen of the urinary bladder in dogs (81). MOCA is a bacterial mutagen and positive in a number of genotoxicity assays (166).

The mechanism of carcinogenesis is thought to involve the formation of chemical adducts in genetic material through covalent binding of nucleophilic sites by electrophilic metabolites (167). MOCA is a structural analogue of benzidine, an aromatic amine known to cause bladder cancer in humans. MOCA and benzidine have similar potency to induce bladder tumors in beagle dogs, the species considered to be the best animal model for humans. Two case reports have appeared that identified noninvasive papillary tumors of the bladder from a screening study of 540 workers exposed to MOCA during its production at a Michigan chemical plant from 1968 to 1979. Both tumors occurred in men under 30 years of age who had never smoked. The incidence of clinically apparent bladder tumors in U.S. males aged 25 to 29 is only 1 per 100,000 per year (SEER, 1977). This case report was considered to be consistent with the hypothesis that MOCA induces bladder neoplasms in humans (165).

HPLC urine analysis from two MOCA-exposed workers revealed four peaks, corresponding to the β-N-glucuronide, and the authors suggested that urine monitoring should include determination of the β-N-glucuronide (168). Following the accidental, massive exposure of a worker to MOCA, serial urinary MOCA samples from the worker over a 2-week period allowed calculation of biologic half-life, based on a one-compartment model, of approximately 23 hr (169).

There are no reports indicating that MOCA affects the thyroid gland. It has been suggested that the large ortho chlorine groups, in contrast to the thyroid activity of MDA discussed below, may have obstructed the methylene group or prevented the chemical from binding to the thyroid hormone receptor sites (170).

### 2.27.4 Hygienic Standards of Permissible Exposure

MOCA is classified as an A2 carcinogen and is considered a suspect human bladder or liver carcinogen (139). A TLV–TWA of 0.02 ppm (approximately 0.22 mg/m$^3$) has been adopted with caution against cutaneous or mucous membrane exposure.

## 2.28  4,4'-Methylenedianiline [CAS # 101-77-9]

The formula for 4,4'-methylenedianiline is $CH_2(C_6H_4NH_2)_2$; it is also known as 4,4'-diaminodiphenylmethane, bis(p-aminophenyl)methane, 4,4'-methylenedianiline, bis(4-aminophenyl)methane, methylene dianiline, DADPM, DAPM, or MDA.

### 2.28.1  Uses

Approximately 98 percent of the MDA produced domestically, usually by the reaction of aniline and formaldehyde, is used as a chemical intermediate in the closed-system production of isocyanates and polyisocyanates. It is used extensively in the manufacture of rigid semiflexible polyurethane foams for automobile safety cushioning, and also in the production of wire coatings. MDA is also widely used as a hardener or curing agent for epoxy resins (providing cross-linkages). It is also used in the manufacture of various natural and synthetic rubbers as an antioxidant, in the production of polyurethane elastomers, as a curing agent for neoprene, as an antioxidant in rubber, in the preparation of polyamide–imide resins, as a metal deactivator fuel additive (neutralizes the catalytic effect of copper associated with fuel oxidation), and in the production of polymers for synthetic fibers in the textile industry.

### 2.28.2 Physical and Chemical Properties (100)

| | |
|---|---|
| Physical state | Pale yellow crystals that darken on exposure to air |
| Molecular weight | 198.26 |
| Melting point | 91.5 to 92°C |
| Boiling point | 263°C at 25 torr |
| Solubility | Soluble in alcohol, benzene, ether, acetone, methanol; slightly soluble in water, carbon tetrachloride |
| Flash point | 220°C (open cup) |

### 2.28.3 Toxicity

The oral $LD_{50}$ in Wistar rats has been reported to be 0.83 g/kg (139). Dermal absorption of MDA can be as high as 80 percent of the applied dose (171). Cats exposed to DMA, in contrast to dogs, developed methemoglobinemia (172).

MDA also reportedly caused retinal and liver damage in cats administered 25 to 50 mg/kg orally. The changes consisted of degeneration of the rods and cones and the pigmented epithelial cells of the retina, with no damage to the neuronal structures beyond the pigmented area; in the same report the authors indicated that there was no occular damage to albino species (rabbit, guinea pig, and rat) (173). In contrast to the above study, both male guinea pigs of albino (Hartley) and pigmented strains (mixed variety), exposed nose-only to average concentration of 440 mg/m³ MDA in polyethylene glycol 4 hr/day, 5 days/week for 2 weeks developed similar retinal degenerative changes that were not related to the presence or absence of melanin in the retinal epithelium. In addition, MDA did not induce pulmonary distress or sensitization. The authors concluded that although the prevention of acute and chronic hepatoxicity and carcinogenesis is important for safe handling, fast developing retinopathy is an additional risk that should be emphasized (173).

Dermal exposure can result in allergic contact dermatitis (174). In addition, cross-sensitization with *p*-phenylenediamine has also been reported (175). Photosensitization has also been observed (176), as well as yellow staining of the skin (177).

MDA has reportedly demonstrated activity as an anti-thyroid drug; however, it is not as potent as derivatives of thiourea (178). Exposure may therefore constitute a potential reproductive hazard as a consequence of anti-thyroid activity. Rodent tumorigenicity studies have demonstrated thyroid effects as discussed below. Although anti-thyroid drugs can be administered during pregnancy for the treatment of hyperthyroidism, careful monitoring is required, and very often thyroid hormone is administered to prevent hypothyroidism in the mother and possibly the fetus. A chicken yolk sac screening test demonstrated beak deformity (short mandible) and leg damage following 5 mg (10% in ethanol) injected through the air cell into the yolk sac of fertilized (randomly selected to avoid series of infertile eggs) eggs, followed by incubation, 10 per dosage level. The use of yolk sacs of fertile chicken eggs for teratogenicity investigations was developed initially by Fere (1893) (179).

Human exposure to MDA, which can occur by dermal, inhalation, or oral routes, can result in fever, hyperthermia, discolored urine, skin rash, and bile duct epithelial damage resulting in jaundice, hepatitis, cholestasis, cholangitis, and bile duct proliferation (180). Acute myocardiopathy has also been reported as a consequence of exposure (181). Impaired visual acuity has been reported following ingestion of MDA; however, the liquid also contained $\gamma$-butyrolactone (182). Animal studies indicate a potential adverse effect upon the retinal epithelium (173). In addition, sensitization and yellow skin staining may also occur as indicated above. MDA does not have any warning properties with the exception of a faint amine odor and therefore renders industrial handling hazardous.

Between 1966 and 1972 hepatitis developed in 12 young male workers exposed to MDA. This compound caused the so-called Epping jaundice when 84 persons ate bread contaminated with MDA. The clinical pattern of the cases reported here resembled that in the earlier report: a short prodome, severe right-upper-quadrant pain, high fever, and chills, with subsequent jaundice. All patients recovered by 7 weeks. When reexamined 9 months to 3½ years later, all were in excellent health, without clinical or biochemical evidence of chronic liver disease. A 13th worker in another factory also had hepatitis; after working with the compound he may have inhaled toxic amounts, but circumstantial evidence in the 12 cases suggested that the skin was the major portal of entry (109).

Chronic studies in animals have demonstrated necrotic changes in liver, kidneys, and lungs. In female beagle dogs fed 4.0 to 6.3 g/kg (70 mg/dog dissolved in corn oil) three times a week for 3 to 7 years, there was no evidence of neoplasia of the bladder (183). In a study in which rats were fed MDA for 90 days at 30 or 100 mg/kg/day for 33 doses, bile duct proliferation and reduced liver weights occurred at all doses in the dog, and only at the 100 mg/kg/day dose in the rat. Hemoglobin levels were low in both species; no methemoglobinemia occurred in either species (184).

In order to investigate the carcinogenicity of MDA, the dihydrochloride was administered in drinking water (0.015 or 0.03 percent) to F344 rats and B6C3F1 mice of both sexes for 103 weeks. Increased incidence of neoplastic nodules occurred, but a significant increase in hepatocellular carcinomas was not observed in the rats (1 male in both low and high dose group). However, hepatocellular carcinomas were increased significantly in female and male mice. In addition, a significant incidence of thyroid hyperplasia and adenomas occurred in both male and female mice treated at the higher dose level; only two female mice developed follicular cell carcinoma, which was not considered to be statistically significant. Significant thyroid hyperplasia was observed in female rats at the high dose level only, and statistically nonsignificant thyroid carcinoma was observed at both dosage levels (two female rats at each dose level). Male rats did not exhibit statistically significant thyroid hyperplasia; however, an increased incidence was observed at the high dose level only (7/50; $p = 012$). Nephropathy and mineralization of renal papilla in the kidneys of mice were the outstanding treatment-related nonneoplastic lesions. In contrast to previous reports, bile duct proliferation was not noted in the rodents (which may be a dose-related effect at higher dosage levels). The authors pointed out that a survey of aromatic amines tested in the National Cancer Institute (NCI)–NTP bioassay program indicated that thyroid tumors usually were caused by compounds with an ether or thioether function. The ether function may simulate part of the thyroxine molecule; and MDA contains the methylene group (—$CH_2$—), which approximates the oxygen group (—O—) in size. It is not unusual for thyroxine analogues with —S— or —$CH_2$— groups in place of oxygen to show hormonal activity. A previous acute study indicated that MDA had a goitrogenic effect and produced adrenal hypertrophy in the ovariectomized female rat; however, in the chronic study under discussion the adrenal glands of rats were not affected, although male mice did show some adrenal pheochromocytomas. In contrast to 4,4′-oxydianiline, MDA did not induce harderian gland tumors in mice. It is believed that although the harderian gland is not an endocrine gland, development is affected by hormonal influences (170). MDA is mutagenic in the Ames test (185).

Epidemiologic evidence suggests an association between MDA and bladder cancer, colon cancer, lymphosarcoma, and reticulosarcoma in workers exposed to MDA and other chemical agents. Although airborne exposures to MDA can occur, dermal contact is considered the major route of occupational exposure (186).

### 2.28.4 Hygienic Standards of Permissible Exposure

Workplace exposure levels reportedly range from 0.041 mg/m$^3$ (the suggested nonhepatotoxic concentration for safe worker exposures) to 3.1 mg/m$^3$ (187). The current TLV–TWA is 0.1 ppm (approximately 0.8 mg/m$^3$), and MDA is classified as a suspected human carcinogen (139).

## 2.29  2-Amino-5-Nitrophenol [CAS # 121-88-0]

The formula for 2-amino-5-nitrophenol is $H_2NC_6H_3(NO_2)OH$; it is also known as 2-hydroxy-4-nitroaniline.

### 2.29.1  Uses

Specific applications were not located in the literature.

### 2.29.2 Physical and Chemical Properties (100)

Molecular weight    154.13
Melting point    207–208°C
Solubility    Soluble in alcohol, benzene

### 2.29.3 Toxicity

2-Amino-5-nitrophenol administered in corn oil to F344/N rats and B6C3F1 mice by gavage for 2 years demonstrated some evidence of carcinogenic activity for male F344 rats receiving 100 mg/kg, indicated by increased incidence of acinar cell adenomas of the pancreas. No evidence of carcinogenic activity was found among female F344 rats. Marginal increases were noted in the incidence of clitoral gland adenomas or carcinomas in male and female rats given 200 mg/kg. There was no evidence of carcinogenic activity in the mice receiving 400 mg/kg (188).

### 2.29.4 Hygienic Standards of Permissible Exposure

Standards have not been assigned.

## 2.30 4-Amino-2-nitrophenol [CAS # 119-34-6]

The formula for 4-amino-2-nitrophenol is $H_2NC_6H_3(NO_2)OH$; it is also known as 4-hydroxy-3-nitroaniline.

### 2.30.1 Uses

This compound is used in dying human hair and animal fur (109).

### 2.30.2 Physical and Chemical Properties (100)

Molecular weight    154.13
Melting point    125 to 127°C

### 2.30.3 Toxicity

4-Amino-2-nitrophenol is reportedly classified as a cancer suspect agent (189).

### 2.30.4 Hygienic Standards of Permissible Exposure

Standards have not been assigned.

## 2.31 Phenylenediamine Isomers

### 2.31.1 o-Phenylenediamine [CAS # 95-54-5]

The formula for o-phenylenediamine is $C_6H_4(NH_2)_2$; it is also known as 1,2-diaminobenzene, 1,2-phenylenediamine, 1,2-benzenediamine, 2-aminoaniline, Orthamine, and o-diaminobenzene.

### 2.31.1.1 Uses.
*o*-Phenylenediamine is used in the synthesis of dyes, dyestuffs, and fungicides and as an oxidative hair and fur dye.

### 2.31.1.2 Physical and Chemical Properties.
*o*-Phenylenediamine has the following properties:

| | |
|---|---|
| Physical state | Yellow crystals |
| Molecular weight | 108.14 |
| Melting point | 102 to 104°C |
| Boiling point | 256 to 258°C |
| Solubility | Soluble in benzene, alcohol, chloroform, ether; slightly soluble in water |

### 2.31.1.3 Toxicity.
The oral $LD_{50}$ in rats ranges from 660 to 1284 mg/kg, with females more sensitive than males (139).

*o*-Phenylenediamine (dihydrochloride) caused hepatocellular carcinomas to develop during an 18-month feeding study in male Charles River CD rats (2000 or 4000 ppm), but not in male or female CD-1 mice (6872 or 13,743 ppm) (190).

A commercial hair dye containing 1 percent *o*-phenylenediamine did not produce any evidence of embryo toxicity or teratogenicity when dermally applied to the shaved backs of CD female rats at a dose of 2 ml/kg during every third day from days 1 through 19, and mixed with an equal volume of 6 percent hydrogen peroxide prior to application (191).

Definitive dose-response carcinogenicity and teratogenicity tests have not been conducted.

### 2.31.1.4 Hygienic Standards of Permissible Exposure
A TLV–TWA of 0.1 mg/m$^3$ has been assigned and designated an A2—suspected human carcinogen (139).

## 2.31.2 m-Phenylenediamine [CAS # 108-45-2]

The formula is $C_6H_4(NH_2)_2$; the compound is also known as 1,3-diaminobenzene, 1,3-phenylenediamine, 1,3-benzenediamine, 3-aminoaniline, and *m*-diaminobenzene.

### 2.31.2.1 Uses.
It is used chiefly in the synthesis of dyes for leather, inks, and hair dye formulations and of other dyestuff intermediates, and as a rubber curing agent, epoxy resin curing agent, petroleum additive, corrosion inhibitor, photographic chemical, and analytical reagent.

### 2.31.2.2 Physical and Chemical Properties
*m*-Phenylenediamine has the following properties:

| | |
|---|---|
| Physical state | White rhombic crystals, becoming red on exposure to air |
| Molecular weight | 108.14 |
| Density | 1.1389 (5°C); 1.107 (58°C) |

Melting point  63 to 64°C
Boiling point  282 to 284°C
Refractive index  1.63390 (57.7°C)
Solubility  Soluble in water, methanol, chloroform, acetone, dimethylformamide, dioxane, methyl ethyl ketone; slightly soluble in ether, carbon tetrachloride; very slightly soluble in benzene, toluene, xylene
Flash point  187°C (closed cup)

*2.31.2.3 Toxicity.* The oral $LD_{50}$ in rats is reported to be 650 mg/kg (oil-in-water emulsion) (139). *m*-Phenylenediamine has also been reported to stain skin yellow. *m*-Phenylenediamine is a human sensitizer (192).

Methemoglobinemia has been demonstrated in dogs following dermal application of 1.5 g (in lauryl sulfate based gel) to dogs (193); and following 1 ml twice weekly for 13 weeks of a hair-dye preparation containing 1.5 percent *m*- and 1 percent *o*-phenylenediamine mixed 1–1 with hydrogen peroxide applied dermally to rabbits (191).

*m*-Phenylenediamine has not been reported to be a teratogen in animal studies. Using the same hair-dye formulation, the mixture was demonstrated to be nonteratogenic when applied dermally to shaved backs of Charles River CD rats at a dose of 2 ml/kg mixed 1–1 with 6 percent hydrogen peroxide just prior to application, on gestation days 1, 4, 7, 10, 13, 16, and 19 (191). *m*-Phenylenediamine administered by gavage to Sprague–Dawley rats at dose levels of 0, 10, 30, or 90 mg/kg on gestation days 6 to 15 was also reported to be nonteratogenic; however, the authors concluded that *m*-phenylenediamine is fetotoxic and could be a weak teratogen at the maternally toxic dose of 90 mg/kg (194). The absence of teratogenic effects was also confirmed in Sprague–Dawley rats administered *m*-phenylenediamine by gavage on gestation days 6 to 15 (45, 90, or 180 mg/kg) (195).

Definitive dose-response carcinogenic activity has not been reported. 1,3-Phenylenediamine dihydrochloride was not carcinogenic in a diet feeding study conducted at high dose levels by Weisburger et al. with male Charles River CD rats (1000 or 2000 ppm) or to male or female CD-1 mice (2000 or 4000 ppm) (190). In a follow-up drinking water study administered to C57BL/(6×C23H/He)$F_1$ mice at 0.02 or 0.04 percent for 78 weeks, *m*-phenylenediamine was also not carcinogenic (196). A skin painting study in which the base was dissolved in acetone and applied to male and female C3Hf/Bd and C57BL/6Bd mice (0.6 mg or 3.0 mg) also failed to produce carcinogenicity (197). An 18-month skin painting study using two groups of 50 male and 50 female Swiss-Webster mice, which received applications of 0.5 ml of a 1–1 mixture of a hair-dye formulation containing 0.17 *m*-phenylenediamine and 6 percent hydrogen peroxide, also did not demonstrate carcinogenicity; however, a nonsignificant increase in lung tumor incidence was reported (198). The IARC working group concluded that *m*-phenylenediamine was inadequately tested by this skin painting study, and no evaluation of the carcinogenicity could be made (192).

An IARC working group concluded that on the basis of lack of human data, and inadequate animal data, *m*-phenylenediamine was not classified as to its carcinogenicity to humans (192, 199).

*2.31.2.4 Hygienic Standards of Permissible Exposure.* A TLV–TWA of 0.1 mg/m$^3$ has been adopted (139).

### 2.31.3  p-Phenylenediamine [CAS # 106-50-3]

The formula is $C_6H_4(NH_2)_2$; the compound is also called *p*-aminoaniline, *p*-benzenediamine, 1,4-diaminobenzene, 1,4-phenylenediamine, 1,4-benzenediamine, *p*-diaminobenzene, 4-aminoaniline, and PPDA.

*2.31.3.1 Uses.* *p*-Phenylenediamine is used in the synthesis of dyestuffs and other intermediates, as a fur "dye," and as a component of hair-dye formulations producing a range of shades from golden-blond to black for disguising gray. PPDA stains skin brown or black. It is also used as a fine grain photographic developer (147).

*2.31.3.2 Physical and Chemical Properties (100).* The properties of *p*-phenylenediamine are as follows:

| | |
|---|---|
| Physical state | White to yellowish-white or slightly red, monoclinic crystals |
| Molecular weight | 108.14 |
| Melting point | 139.7°C |
| Boiling point | 267°C |
| Solubility | Slightly soluble in water; soluble in alcohol, chloroform, ether, acetone, benzene |
| Flash point | 155.5°C (closed cup) |

*2.31.3.3 Toxicity.* The oral $LD_{50}$ of *p*-phenylenediamine is 250 mg/kg in rabbits and 100 mg/kg in cats. An $LD_{50}$ of 80 mg/kg in rats was reported for an oil-in-water emulsion (192).

In humans, PPDA exposure reportedly causes vertigo, blindness, fatigue, weakness, and encephalitis in humans (129b). Other effects from accidental ingestion can include acute renal failure, methemoglobinemia, hemolysis, dyspnea, and swelling of the lips, tongue, and neck (200).

A case has been reported in which a woman apparently died following chronic application of a hair dye containing PPDA. The patient developed progressive neurological symptoms prior to her death including vertigo, gastritis, diplopia, and asthenia; in addition, exfoliative dermatitis developed. The liver and spleen also enlarged (201a). Gastrointestinal and nervous symptoms were reported in a patient who had used a PPDA hair-dye preparation (201b).

It was first documented by Mayer in 1928 that humans could be sensitized to PPDA (175a). PPDA is a potent sensitizer of the skin and respiratory tract and may cause asthma and cross-sensitization with hydroquinone, aniline (175a), the rubber antioxidant 4-isopropylaminodiphenylamine, and the antihistamine phenergan (202). Positive reactions to patch tests with PPDA may indicate previous contact with local anesthetics, sulfonamides, aniline dyes, or rubber antioxidants, as a result of cross-sensitization to "para-group" allergens (203).

At one time PPDA was used in the formulation of Lash Lure, resulting in sensitization of the lid skin and external ocular structures, and corneal ulceration with loss of vision; even a fatality was reported (114). "Nylon stocking" dermatitis occurred in individuals sensitive to azo dyes used in the manufacture of the stockings and to PPDA (204).

The mechanism of sensitization is based on the conversion of the prohapten (PPDA) into a hapten (benzoquinone) that can react directly with a protein. The mechanism of cross-allergy may be associated in some cases with a common hapten; that is, cross-sensitization between hydroquinone, aniline, and PPDA is due to the common biotransformation to benzoquinone (175a).

A safety assessment document of PPDA concluded that for those persons not sensitized, it is safe as a hair-dye ingredient at the current concentrations of use. In the final discussion, the authors stated that most researchers reported that PPDA was not teratogenic or carcinogenic. Application of hair dyes containing PPDA to the eyebrows and eyelashes can result in blindness or permanently damaged vision. PPDA is a sensitizer for guinea pigs and for human beings. Phototoxicity and photosensitization data are not available (205). PPDA exposure reportedly has no effect on sperm morphology (206).

Oral administration of PPDA in Sprague–Dawley rats administered 5, 10, 15, 20, and 30 mg/kg/day by gavage on days 6 through 15 of gestation produced no evidence of teratogenic or embryo toxic effects (207). An earlier dermal teratogenicity study of four commercially available hair-dye formulations containing 1, 2, 3, or 4 percent PPDA and several aromatic amine derivatives among their constituents, mixed 1–1 with 6 percent hydrogen peroxide just prior to use and applied topically to pregnant CD rats on gestation days 1, 4, 7, 10, 13, 16, and 19 also did not exhibit teratogenic effects. The authors noted that the one exception was a formulation containing 2 percent PPDA in which there were skeletal changes in 9/169 live fetuses (191).

A number of $N$-methylated $p$-phenylenediamines are known to cause necrosis of skeletal and cardiac muscle in rats, apparently via an in vivo reactive metabolic intermediate. Myotoxicity demonstrated by derivatives such as 2,3,5,6-tetramethyl- > 2,5-dimethyl- > 2,6-dimethyl- > 2-methyl-$p$-phenylenediamine appears to be a general property of certain substituted $p$-phenylenediamines (208).

PPDA was reportedly mutagenic on TA98 with S-9 metabolic activation in the Ames assay (209).

A number of bioassays have been reported based on dermal applications of PPDA in an organic solvent or in combination with hydrogen peroxide. For example: (1) an 85-week study in which female Swiss mice were treated with 5 or 10 percent PPDA in acetone, 0.02 ml/animal applied topically, did not cause any treatment-related effects, and there was no significant increase in tumor incidence in the treated mice (210). (2) A hair-dye formulation containing 1.5 percent PPDA mixed with equal volumes of 6 percent hydrogen peroxide was reported to be noncarcinogenic in a Swiss–Webster mice skin painting studies (198). The IARC working group concluded that an evaluation of the carcinogenicity of PPDA could not be made based upon this study (192). (3) Four commercially available hair-dye formulations containing 1, 2, 3, and 4 percent PPDA among other constituents were also tested with a 1–1 mixture with hydrogen peroxide in a Swiss–Webster skin painting carcinogenicity study with no effect (211). (4) The carcinogenicity of PPDA oxidized with 6 percent hydrogen peroxide was also evaluated in male and female Wistar rats. In the female rats, both topical application [0.5 ml of a 1–1 mixture of 5 percent *p*-phenylenediamine (in 2 percent ammonium hydroxide) and 6 percent hydrogen peroxide] and subcutaneous injection (0.1 ml of mixture) induced a statistically significant incidence of mammary gland and uterine tumors; no significant tumors were produced in male rats. The authors concluded that PPDA was oxidized to a carcinogenic derivative in the case of female rats (212).

The discrepancy between the above studies may be due either to species differences or to the inherent variables associated with solvent selected, stability of hydrogen peroxide, mixing procedures, and chemical interactions.

PPDA was not found to be carcinogenic when administered by diet to male and female F344 rats (625 or 1250 ppm) or male and female B6C3F1 mice (625 or 1250 ppm) (213). An IARC working group concluded that on the basis of lack of human data, and inadequate animal data, PPDA was not classifiable as to its carcinogenicity to humans (192, 199). However, the additional results of the above study had not been considered in the most recent IARC publication (199).

*2.31.3.4 Hygienic Standards of Permissible Exposure.* A TLV–TWA value of 0.1 mg/m$^3$ has been adopted. A caution against cutaneous or mucous membrane exposure is noted (139).

### 2.32 Phenylenediamine Derivatives

*2.32.1 2-Nitro-1,4-phenylenediamine [CAS # 5307-14-2]*

The formula of 2-nitro-1,4-phenylenediamine is $C_6H_3NO_2(NH_2)_2$; it is also known as 1-nitro-2,4-phenylenediamine and 2-nitro-*p*-phenylenediamine.

*2.32.1.1 Uses.* This compound is used in the formulation of semipermanent (lasting 5 to 6 shampoos) hair dyes requiring a red dye component, which does not require the use of hydrogen peroxide in the color development (214).

*2.32.1.2 Physical and Chemical Properties.* 2-Nitro-1,4-phenylenediamine has the following properties:

| | |
|---|---|
| Molecular weight | 153.14 |
| Melting point | 137 to 140°C |

*2.32.1.3 Toxicity.* 2-Nitro-1,4-phenylenediamine administered by diet to male (550 or 1100 ppm) and female (1100 or 2200 ppm) F344 rats or male (2200 or 4400 ppm) and female (2200 or 4400 ppm) B6C3F1 mice was reported to be carcinogenic for the liver of female mice (developing hepatocellular adenoma or hepatocellular carcinoma) (213).

*2.32.1.4 Hygienic Standards of Permissible Exposure.* Standards have not been assigned.

### 2.32.2  3-Nitro-1,2-phenylenediamine [CAS # 3694-52-8]

The formula for 3-nitro-1,2-phenylenediamine is $C_6H_3NO_2(NH_2)_2$; it is also known as 3-nitro-1,2-diaminobenzene and 3-nitro-*o*-phenylenediamine.

*2.32.2.1 Uses.* No specific applications were found in the literature.

*2.32.2.2 Physical and Chemical Properties.* No physical or chemical properties were located in the literature.

*2.32.2.3 Toxicity.* No toxicology information was found in the literature.

*2.32.2.4 Hygienic Standards of Permissible Exposure.* Standards have not been assigned.

### 2.32.3  4-Nitro-1,2-phenylenediamine [CAS # 99-56-9]

The formula for 4-nitro-1,2-phenylenediamine is $C_6H_3NO_2(NH_2)_2$; it is also known as 4-nitro-*o*-phenylenediamine, 4-nitro-1,2-diaminobenzene, 1,2-diamino-4-nitrobenzene, and NOP.

*2.32.3.1 Uses.* NOP is used in fur dyes, inks, and semipermanent yellow hair coloring formulations requiring a yellow color component that does not involve the use of hydrogen peroxide in the color development (214).

*2.32.3.2 Physical and Chemical Properties (100).* The properties of NOP are as follows:

| | |
|---|---|
| Molecular weight | 153.14 |
| Melting point | 199 to 201°C |

*2.32.3.3 Toxicity.* The oral $LD_{50}$ (rat) is reportedly 3720 mg/kg (100a). 4-Nitro-1,2-phenylenediamine has been reported to be devoid of carcinogenic activity in F344 rats (375 or 750 mg/kg diet) or B6C3F1 mice (3750 or 7500 mg/kg diet) (213), but has been reported to test positive consistently in the short-term in vitro genotoxicity assays (215).

## AROMATIC NITRO AND AMINO COMPOUNDS

*2.32.3.4 Hygienic Standards of Permissible Exposure.* Standards have not been assigned.

### 2.32.4  4-Chloro-1,2-phenylenediamine [CAS # 95-83-0]

The formula for 4-chloro-1,2-phenylenediamine is $C_6H_3Cl(NH_2)_2$; it is also known as 4-chloro-*o*-phenylenediamine.

*2.32.4.1 Uses.* This compound has been patented as a hair dye component, fur dyes, inks, and hair coloring formulations, and is believed to be used to produce 5-chlorobenzotriazole, an isomer of which is a photographic chemical (216).

*2.32.4.2 Physical and Chemical Properties (100).* 4-Chloro-1,2-phenylenediamine has the following properties:

| | |
|---|---|
| Molecular weight | 142.59 |
| Melting point | 70 to 73°C |
| Solubility | Slightly soluble in water; soluble in benzene; very soluble in ethanol |

*2.32.4.3 Toxicity.* In both male and female F344 rats, feeding 4-chloro-*o*-phenylenediamine at 0.5 or 1 percent in the diet led to a significant increase in rare bladder tumors. Similar levels of this compound (7000 or 14,000 ppm) in male and female B6C3F1 mice resulted in significant incidence of hepatocellular adenomas and carcinomas (217).

There is sufficient evidence for carcinogenicity in experimental animals. When administered in the diet, the technical-grade of this compound induced carcinomas of the urinary bladder in rats of both sexes and hepatocellular carcinomas of the liver in mice of both sexes (81). It has also been reported to be mutagenic (104).

*2.32.4.4 Hygienic Standards of Permissible Exposure.* EPA has proposed regulating 4-chloro-*o*-phenylenediamine as a hazardous constituent of waste under RCRA. OSHA regulates this compound as a chemical hazard in laboratories.

Workplace standards have not been assigned.

### 2.32.5  4-Chloro-1,3-phenylenediamine [CAS # 5131-60-2]

The formula for 4-chloro-1,3-phenylenediamine is $C_6H_3Cl(NH_2)_2$; it is also known as 4-chloro-*m*-phenylenediamine.

*2.32.5.1 Uses.* 4-Chloro-*m*-phenylenediamine is used as a dye intermediate and as a rubber processing agent (216).

*2.32.5.2 Physical and Chemical Properties (100).* The properties of 4-chloro-1,3-phenylenediamine are as follows:

Molecular weight 142.59
Melting point 91°C
Solubility Soluble in ethanol; slightly soluble in water

*2.32.5.3 Toxicity.* 4-Chloro-1,3-phenylenediamine, administered in the diet to male and female (2000 or 4000 ppm) F344 rats and male and female (7000 or 14,000 ppm) B6C3F1 mice was carcinogenic only for male F344 rats and female B6C3F1 mice (hepatocellular carcinomas), and adrenal tumors (pheochromocytomas) rather than bladder tumors were observed in male rats only (213). This compound has also been reported to be mutagenic (104).

*2.32.5.4 Hygienic Standards of Permissible Exposure.* Standards have not been assigned.

*2.32.6  2-Chloro-1,4-phenylenediamine [CAS # 615-66-7]*

The formula for 2-chloro-1,4-phenylenediamine is $C_6H_3Cl\,(NH_2)_2$; it is also known as 2-chloro-*p*-phenylenediamine.

*2.32.6.1 Uses.* 2-Chloro-1,4-phenylenediamine is used in hair-dye formulations.

*2.32.6.2 Physical and Chemical Properties (100).* 2-Chloro-1,4-phenylenediamine has a molecular weight of 142.59.

*2.32.6.3 Toxicity.* 2-Chloro-1,4-phenylenediamine sulfate, tested in a dietary feeding study of male and female F344 rats (1500 or 3000 ppm) and B6C3F1 mice (3000 or 6000 ppm) was completely inactive in either species tested (213).

*2.32.6.4 Hygienic Standards of Permissible Exposure.* Standards have not been assigned.

*2.32.7  2,6-Dichloro-1,4-phenylenediamine [CAS # 609-20-1]*

The formula for 2,6-dichloro-1,4-phenylenediamine is $C_6H_3(Cl)_2(NH_2)_2$; it is also known as 2,6-dichloro-*p*-phenylenediamine.

*2.32.7.1 Uses.* It is assumed that this compound has been used as a dye.

*2.32.7.2 Physical and Chemical Properties (100).* The molecular weight is 177.03.

*2.32.7.3 Toxicity.* An NCI dietary feeding study in male (1000 or 2000 ppm) and female (2000 or 6000 ppm) F344 rats and B6C3F1 male and female mice (1000 or 3000 ppm) demonstrated hepatocellular carcinomas in female and male mice only (213). However, a significant dose-related increase incidence of F344 rat pancreatic hepatocytes was observed in 2000-ppm males and 2000- and 6000-ppm females. The relationship is unclear at present (218).

# AROMATIC NITRO AND AMINO COMPOUNDS

Toxicity information was not located on the following homologues: 2,5-dichloro-1,4-phenylenediamine [CAS # 20103-09-7], also known as 2,5-dichloro-*p*-phenylenediamine, and 2,6-dichloro-1,2-phenylenediamine [CAS # 5348-42-5], also known as 2,6-dichloro-*o*-phenylenediamine.

*2.32.7.4 Hygienic Standards of Permissible Exposure.* Standards have not been assigned.

## 2.33 Diphenylamine [CAS # 122-39-4]

The formula for diphenylamine is $(C_6H_5)_2NH$; it is also called phenylaniline, anilinobenzene, *N*-Diphenylaniline, DPA, and *N*-Phenylbenzeneamine. Diphenylamine should not be confused with biphenylamine, which is the parent compound of some highly carcinogenic derivatives.

### 2.33.1 Uses

It is used in the synthesis of dyestuffs, other dyestuff intermediates, and pesticides, as a stabilizer of nitrocellulose explosives and celluloids, as an antioxidant to prevent scald on applies during storage, and as a fungicide, antihelmintic, and reagent in analytic chemistry.

### 2.33.2 Physical and Chemical Properties (100).

| | |
|---|---|
| Physical state | Colorless monoclinic leaflets |
| Molecular weight | 169.22 |
| Density | 1.159 (20/4°C) |
| Melting point | 53 to 54°C |
| Boiling point | 302°C |
| Solubility | Slightly soluble in water; soluble in alcohol, ether, methyl alcohol, acetone, benzene |

### 2.33.3 Toxicity

It is readily absorbed through the skin and respiratory tract, as well as orally. Human exposure reportedly results in bladder symptoms, tachycardia, hypertension, and eczema. Sensitization is unlikely and has been observed only as a consequence of cross-sensitization to *p*-phenylenediamine (175a). Exposure has also reportedly caused anorexia in humans (129b).

A subchronic dietary study in Sprague–Dawley rats caused cystic dilatation and interstitial nephritis at a threshold level of approximately 0.1 percent in the diet, but not at 0.01 percent (219). A 2-year chronic feeding study in dogs at dietary levels of 0.01, 0.1, or 1.0 percent was not carcinogenic; however, a slight increase in kidney and liver weight as well as mild homosiderosis of the spleen, kidneys, and bone marrow was reported (220). Nitrosation generates the weak bladder carcinogen nitrosodiphenylamine.

In a reproductive study, Sprague–Dawley rats were given a diet containing 1.5 or 2.5 percent of DPA during the last 7 days of gestation, and a second group received 20 mg of DPA aged 2 years, purified DPA or with 50 μg of one of three contaminants. Cystic lesions were confined to the proximal nephron in newborn rats, except in those rats gavaged with chromatographically pure diphenylamine. The impurity in commercial DPA that induces polycystic kidney disease in rats was identified as $N,N,N'$-triphenyl-$p$-phenylenediamine, which can be formed by heating or by aging DPA (221). This is interesting from the point of view that DPA has been used experimentally for years to study polycystic kidneys in animals, a human bilateral disorder, predominantly involving the giant nephrons, that affects approximately 200,000 to 400,000 persons in the United States; the most common form is inherited as an autosomal dominant trait, typically causing renal insufficiency by the fifth or sixth decade of life (222, 223). Researchers have also reported that the Syrian hamster appears to be more susceptible to the nephrotoxicity (renal papillary necrosis) of DPA than the Sprague–Dawley rat or the Mongolian gerbil (224).

*2.33.4 Hygienic Standards of Permissible Exposure*

The adopted TWA value is 10 mg/m$^3$ (138).

## 2.34  4-Aminophenyl Ether [CAS # 101-80-4]

The formula for 4-aminophenyl ether is $(H_2NC_6H_4)_2O$; it is also known as 4,4'-oxydianiline and 4,4'-diaminodiphenyl oxide.

*2.34.1  Uses*

4,4'-Oxydianiline is used primarily in the production of polyimide and poly(ester)imide resins. These resins are used in the manufacture of temperature-resistant products such as wire enamels, coatings, film, insulating varnishes, and flame-retardant fibers.

*2.34.2  Physical and Chemical Properties (100)*

| | |
|---|---|
| Molecular weight | 200.24 |
| Melting point | 190 to 192°C |
| Solubility | Insoluble in water, benzene, carbon tetrachloride; soluble in acetone |

*2.34.3  Toxicity*

4,4'-Oxydianiline administered in the diet at 0, 0.02, 0.04, and 0.05 percent increased adenomas and carcinomas in the thyroid and neoplastic nodules and carcinomas in the liver of male and female F344 rats; in addition to increasing thyroid adenomas in females and hepatocellular adenomas or carcinomas in male and female B6C3F1 mice, adenomas of the harderian gland in male and female mice were also significantly increased. Although the harderian gland is not an endocrine gland, development is affected by hormonal influences; a deficiency of thyroid hormone, for example, decreases its size in rats (170).

There is sufficient evidence for the carcinogenicity of 4,4'-oxydianiline in animals. When administered in the diet, the compound increased the incidence of adenomas of the harderian gland and hepatocellular adenomas or carcinomas in mice of both sexes, follicular cell adenomas in female mice, and hepatocellular carcinomas and follicular cell adenomas or carcinomas of the thyroid in F344 rats of both sexes. The population of thyrotrophs (thyrotropin-producing cells) was increased in the pituitary glands of the rats with neoplasms of the thyroid gland. These cells were distinct from prolactin cells and the increase in number suggested insufficient hormone production by the induced thyroid tumors. An IARC working group reported that there were no data available to evaluate the carcinogenicity in humans (81). (See Section 2.38.3 MDA for discussion of diphenyl methylene-bridged amines.)

### 2.34.4 Hygienic Standards of Permissible Exposure

Standards have not been assigned.

## 2.35 Triphenylamine [CAS # 603-34-9]

The formula for triphenylamine is $(C_6H_5)_3N$.

### 2.35.1 Uses

Triphenylamine is used as a primary photoconductor on polymer film; Eastman Kodak is apparently the only producer.

### 2.35.2 Physical and Chemical Properties (100)

| | |
|---|---|
| Molecular weight | 245.33 |
| Specific gravity | 0.774 at 0°C |
| Melting point | 127 to 129°C |
| Solubility | Soluble in benzene, ether |

### 2.35.3 Toxicity

The oral $LD_{50}$ (rat) is reportedly between 3200 and 6400 mg/kg; in mice, it is 1600 to 3200 mg/kg (139).

### 2.35.4 Hygienic Standards of Permissible Exposure

The TLV–TWA is 5 mg/m$^3$ (139).

## 2.36 Nitrobenzene [CAS # 98-95-3]

The formula for nitrobenzene is $C_6H_5NO_2$; it is also called nitrobenzol, oil of mirbane, and essence of mirbane. Nitrobenzene is manufactured by electroreduction.

### 2.36.1 Uses

Nitrobenzene is used as a chemical intermediate in organic synthesis, in the manufacture of explosives, aniline, benzidine, aniline dyes, and other chemicals, and as a solvent. It is also used in shoe and floor polishes, leather dressings, soaps, and paint solvents used to mask unpleasant odors, and in refining lubricating oils.

### 2.36.2 Physical and Chemical Properties (100)

| | |
|---|---|
| Physical state | Colorless to pale yellow liquid |
| Molecular weight | 123.11 |
| Density | 1.2037 (20/4°C) |
| Melting point | 5.7°C |
| Boiling point | 210.9°C |
| Density of vapor | 4.1 (air = 1) |
| Refractive index | 1.55291 (20°C) |
| Solubility | Slightly soluble in water; very soluble in alcohol, ether; soluble in benzene, oils |
| Flash point | 88°C (closed cup) |
| Odor | Oil of bitter almond |

### 2.36.3 Toxicity

Human exposure may cause eye and skin irritation. Clinical signs and symptoms of systemic toxicity are usually delayed for a period of about 1 to 4 hr. Nitrobenzene syndrome includes giddiness, fatigue, severe headache, tinnitus, vertigo, vomiting, general weakness, tachycardia, limb numbness, and in some cases severe depression, unconsciousness, and coma. The potential danger of developing methemoglobinemia, hemolytic anemia, and cyanosis, which appears when methemoglobin levels are approximately 15%, is of particular concern, the risk being greater in cases of hereditary G6PD deficiency, deficient levels of glutathione reductase, glutathione, or glutathione peroxidase activity. Conjunctival discoloration may even occur because of systemic effects. Chronic exposure can also lead to spleen and liver damage, jaundice, liver impairments, and hemolytic icterus. Anemia and Heinz bodies in the red blood cells have also been observed. Alcohol ingestion may increase the toxic effects (109, 110, 147, 225). The oral $LD_{50}$ in rats is reportedly 640 mg/kg (100a).

In vivo and in vitro studies have demonstrated that nitrobenzene is metabolically reduced to *p*-aminophenol and conjugated with glucuronic acid and sulfuric acid; the formation of the intermediate phenylhydroxylamine was initially postulated by

Meyer in 1905 (113). It is clear that a major reaction in the metabolism of nitrobenzene is nitro reduction. Rat isolated hepatocytes reduced nitrobenzene to nitrosobenzene and/or phenylhydroxylamine (the analysis used was not capable of differentiating) under aerobic conditions, but the rate at which reduction occurs in rat cecal contents is about 330 times as fast as the maximal rate observed in isolated hepatocytes under aerobic conditions, and about 150 times greater than in microsomes incubated under nitrogen. In vivo studies have also demonstrated that the intestinal bacteria are the most quantitatively important site of nitrobenzene reduction (16c). Aniline, the three isomeric nitrophenols, and 2- and 3-aminophenol combined account for about 25 percent of an oral dose of 250 mg/kg of nitrobenzene; 4-aminophenol is the major urinary metabolite, accounting for about 31 percent of a dose (16c). Metabolism studies of nitrobenzene have demonstrated both species as well as strain differences (16c). For example, urinary metabolites from Fischer 344 rats following an oral 225 mg/kg dose of nitrobenzene included 4-hydroxyacetanilide sulfate, 4-nitrophenyl sulfate, and 3-nitrophenyl sulfate; similar metabolites were found in CD rat urine, but 4-hydroxyacetanilide, 4-nitrophenol, and 3-nitrophenol were also present in free form and as glucuronide conjugates. CD rats excreted half as much of the nitrobenzene dose as 4-hydroxyacetanilide and 4-nitrophenol as the Fischer 344 rats, but the CD rats excreted nearly three times as much of the dose as very polar unidentified metabolites. 4-Aminophenol appears to be a major metabolite only in the B6C3F1 mouse and in rabbits; both F344 and CD rats apparently acetylate 4-aminophenol prior to excretion. The early data in rabbits are, however, confounded by the use of acid hydrolysis prior to analysis (acid hydrolysis converts 4-hydroxyacetanilide to 4-aminophenol) (16c). Nitrobenzene also inhibits microsomal P450 by reacting as a field ligand with the heme iron atom (151).

Significant species, strain, and sex differences in the response of rodents to nitrobenzene vapors have been demonstrated in F344 rats, Sprague–Dawley rats, and B6C3F1 mice exposed to 10 to 125 ppm vapor 6 hr/day, 5 days/week for 2 weeks. The mice and Sprague–Dawley rats showed signs of toxicity after 2 to 4 days, as well as increased mortality. F344 rats showed no signs of toxicity even at 125 ppm. The 125 ppm dose caused perivascular hemorrhages of the cerebellar peduncle in Sprague–Dawley rats and mice; no brain lesions were found in F344 rats. Male mice exposed to 125 ppm showed severe liver necrosis and less necrosis in the Sprague–Dawley rat. Splenic lesions were observed in all animals and dose groups. The authors concluded that B6C3F1 mice and Sprague–Dawley rats are more sensitive to the effects of nitrobenzene than F344 rats (226).

Animal studies have demonstrated adverse testicular effects following oral or inhalation exposure to nitrobenzene. Male F344 rats administered single oral doses of nitrobenzene resulted in liver lesions at 110 mg/kg and above and testicular lesions at 300 mg/kg and above; one rat receiving 450 mg/kg developed cerebellar lesions (227). In a follow-up study, it was confirmed through direct monitoring of sperm production by anastomosis of the vas deferentia with the urinary bladder that a single oral dose of nitrobenzene (300 mg/kg) induced testicular degeneration followed by substantial repair by 3 weeks post-treatment; by 100 days after treatment, there was more than 90 percent regeneration of the seminiferous epithelium (228).

Male F344 and Sprague–Dawley rats exposed by inhalation to 5, 16 or 50 ppm 6 hr/day, 5 days/week for 90 days developed a dose-related increase in methemoglobin along with secondary splenic changes; at 50 ppm degeneration of the testicular seminiferous epithelium was observed (229).

In a dominant lethal study, severe impairment of fertility was observed in Sprague–Dawley male rats administered nitrobenzene; however, the effect was transient, and the authors concluded that spermatogonia stem cells were probably not damaged and that the nitrobenzene was without effect (230).

However, in a follow-up two-generation reproduction inhalation study, fertility impairment and testicular atrophy were observed in rats. Male and female Sprague–Dawley CD rats ($F_0$ generation) were exposed to 1, 10, or 40 ppm for 6 hr/day, 5 days/week for 10 weeks. Significant differences between $F_0$ and control animals were observed as follows: decrease in relative and absolute testes and epididymides weights and testes size and increase in atrophy of the testes including degenerate spermatocytes in high-dose males. Differences between $F_1$ and control animals also occurred by increase in atrophy of testes and degenerate spermatocytes. Significant differences also included decreased fertility index in the high-dose group only. To examine the reversibility of testicular effects, the $F_1$ males from the 40-ppm group were allowed a 9-week nonexposure period and mated to naive females. An almost fivefold increase in the fertility index was observed, indicating at least partial functional reversibility upon removal from nitrobenzene exposure. The NOEL was selected as 10 ppm (231).

Nitrobenzene has been reported to delay embryogenesis and cause disorders of organogenesis when given prenatally in rats (232). However, no teratogenicity was observed in Sprague–Dawley rats exposed to 1, 10, and 40 ppm by inhalation during gestation days 6 through 15. Maternal body weight declined with 40 ppm, followed by recovery by the time of sacrifice on gestation day 21; increased spleen weight was demonstrated with 10 and 40 ppm. Skeletal variations observed did not show a clear dose–response relationship. No previous teratogenicity studies were cited (233).

Neither embryo toxicity nor teratogenicity was observed in the pups of mated New Zealand White rabbits exposed by inhalation to nitrobenzene at 10, 40, or 100 ppm (234). Nitrobenzene is negative in the Ames mutagenicity assay (102).

Biomonitoring studies of nitrobenzene, using $^{14}C$-labeled derivative orally administered to female Wistar rats, indicated that hemoglobin binding may be a better index of body burden than methemoglobin levels, and the authors suggested this binding activity may be a useful method for biomonitoring human exposure (235). Determination of total *p*-nitrophenol in urine specimens collected at the end of the work week was also recommended for monitoring nitrobenzene exposure (138).

### 2.36.4 Hygienic Standards of Permissible Exposure

The adopted TLV–TWA is 1 ppm, 5 mg/m$^3$, based on the prevention of methemoglobinemia. A caution against cutaneous or mucous membrane exposure is noted (138).

## 2.37 Dinitrobenzene Isomers

### 2.37.1 o-Dinitrobenzene [CAS # 528-29-0]

The formula is $C_6H_4(NO_2)_2$; it is also called 1,2-dinitrobenzene and 1,2-DNB.

*2.37.1.1 Uses.* Dinitrobenzene is usually manufactured as a mixture of three isomers, *o*-, *m*-, and *p*-nitrobenzene. It is used in the synthesis of dyestuffs and dyestuff intermediates, in explosives, and as a camphor substitute in celluloid production.

*2.37.1.2 Physical and Chemical Properties (100).* The properties of *o*-dinitrobenzene are as follows:

| | |
|---|---|
| Physical state | White to yellow, monoclinic plates |
| Molecular weight | 168.11 |
| Density | 1.565 (17/4°C) |
| Melting point | 117 to 118.5°C |
| Boiling point | 319°C (773 mm Hg) |
| Vapor density | 5.79 (air = 1) |
| Solubility | Slightly soluble in water; 3.8 in alcohol at 25°C; 27.1 in chloroform at 18°C; 5.0 in benzene at 18°C; soluble in methyl alcohol |
| Flash point | 302°F (closed cup) |

*2.37.1.3 Toxicity.* All three isomers cause anoxia owing to methemoglobin formation. They also cause irritation of the mucous membranes of the eyes, nose, and respiratory tract as well as liver damage; aplastic anemia and cyanosis have been reported as consequences of exposure to DNB. DNB is absorbed readily through the skin.

1,2-DNB does not cause testicular damage in Alpk/AP (Wistar derived) rats when administered as a single oral dose of 50 or 200 mg/kg in a study of the three isomers. The authors furthermore reported that 1,2-DNB is the least potent isomer with respect to generating methemoglobinemia, and was without effect when given at a dose level equivalent to that of 1,3- and 1,4-DNB in producing methemoglobinemia (236). See 1,3-DNB for discussion on metabolic differences in metabolism.

*2.37.1.4 Hygienic Standards of Permissible Exposure.* The adopted TLV–TWA value is 0.15 ppm, 1 mg/m$^3$ (skin) (139).

### 2.37.2 m-Dinitrobenzene [CAS # 99-65-0]

The formula for *m*-dinitrobenzene is $C_6H_4(NO_2)_2$; it is also called 1,3-dinitrobenzene and 1,3-DNB.

*2.37.2.1 Uses.* 1,3-DNB is used in the synthesis of *m*-nitroaniline and 1,3-phenylenediamine, in the synthesis of dye intermediates and explosives, and to some extent in the plastics industry. It is the most economically important isomer of dinitrobenzene.

*2.37.2.2 Physical and Chemical Properties (100).* m-Dinitrobenzene has the following properties:

| | |
|---|---|
| Physical state | Colorless to yellow, rhombic needles or plates |
| Molecular weight | 168.11 |
| Density | 1.571 (0/4°C) |
| Melting point | 89 to 90°C |
| Boiling point | 302.8°C (770 mm Hg) |
| Solubility | 0.0469 in water at 15°C, 0.32 at 100°C; 2.6 in alcohol at 20°C; 6.7 in ether at 15°C; soluble in benzene, toluene, chloroform, ethyl acetate |
| Flash point | 302°F (closed cup) |

*2.37.2.3 Toxicity.* Human symptoms of 1,3-DNB exposure can include nausea, headache, and symptoms of nervous dysfunction, methemoglobinemia, anemia, hypoxia, and cyanosis (237). 1,3-DNB is extremely toxic to humans; the estimated lethal dosage range is 5 to 50 mg/kg (238). It is considered to be the most potent of the three isomers with respect to methemoglobinemia (236). In addition to the risks from inhalation, it is also readily absorbed through the intact skin (239). Exposure has also resulted in human impairment of visual acuity and contraction of visual field, particularly for red and green colors; yellow discoloration of the conjunctiva and the sclera can also occur with exposure (240). Central nervous system effects in humans may be associated with impaired glucose metabolism. When 10 mg/kg of 1,3-DNB was administered orally to rats twice daily, methemoglobinemia and frequently ataxia was observed after four or five doses. Acute thiamine deficiency-like lesions were observed in the brain stems of both ataxic and apparently normal rats. It was postulated that 1,3-DNB interferes with intracellular redox mechanisms, resulting in impaired glucose oxidation (241a).

Investigations concerning 1,3-dinitrobenzene neurotoxicity have shown that the early stages of poisoning by 1,3-DNB both in vivo and in vitro produce glial swelling and vacuolation in certain brain stem nuclei, which are associated with sleeve-like arteriolar hemorrhages and finally with secondary neuronal death (241a); therefore the primary morphological targets of 1,3-DNB are the constituents of the blood–brain barrier (241b). An in vitro study carried out in both rat astrocytes and brain capillary endothelial cells showed a significantly larger enhancement of both glucose consumption and oxygen free-radical production (nitroblue tetrazolium method) over the first day of incubation with 0.05 m$M$ 1,3-DNB (241b). Both cell types showed increased lactate release at 0.5 m$M$ 1,3-DNB and a threshold cytotoxic concentration of 1.0 m$M$. The authors concluded that the results suggest that increased blood flow and functional disruption of the blood–brain barrier are prominent and possibly the causal feature of 1,3-DNB neurotoxicity (241b).

Metabolism studies indicated that the isomers are primarily metabolized by (1) nitro group reduction to yield aniline or substituted anilines; and (2) replacement of a nitro group by conjugation with glutathione. 1,3-DNB, which is the only isomer demonstrating testicular toxicity, is apparently metabolized exclusively by reduction and glucuronide conjugation in F344 rats; o-DNB is mainly metabolized via GSH, and p-DNB has pathways similar to both meta and ortho isomers (242a). When male F344 rats were given an oral dose (0.15 mmol/kg) of $^{14}$C-labeled 1,2-, 1,3-, or 1,4-DNB, 85, 60, and 75 percent of the 1,2-, 1,3- and 1,4-DNB doses, respectively, were recovered in 24 hr. Urine was the primary route of excretion. The major urinary metabolites of 1,2-, 1,3-, and 1,4-DNB were S-(2-nitrophenyl)-N-acetylcysteine (42 percent of dose), 3-aminoacetanilide (22 percent), and 2-amino-5-nitrophenyl sulfate (35 percent) (242b).

Differences in methemoglobin potency among the three isomers have also been demonstrated in vitro. In contrast to o-DNB and p-DNB, m-DNB was not metabolized by erythrocytes from F344 rats, rhesus monkeys, and humans. p-DNB produced the most methemoglobin in erythrocytes from all three species. o-DNB caused more methemoglobin production in human and monkey erythrocytes than m-DNB, and m-DNB was more toxic in rat erythrocytes than o-DNB. Erythrocytes from all three species metabolized o-DNB and p-DNB to S-(nitrophenyl)glutathione conjugates. This study demonstrated that glutathione does not provide (in vitro) protection from methemoglobinemia produced by DNBs (243).

Adverse male reproductive effects have been established by several animal studies. Testicular atrophy, depressed spermatogenesis, extensive damage to reproductive tissue, and reproductive failure in male rats after subchronic exposure via drinking water were initially reported by Cody et al. in 1981 (244). It was suggested by Lock et al. in 1985 that testicular damage appeared to affect the Sertoli cells (245).

Linder et al. then reported that the effect may be partially reversible. When weanling male Sprague–Dawley rats were gavaged 5 days/week with 1,3-DNB at dosages of 0. 0.75, 1.5, 3.0, and 6.0 mg/kg/day it was observed that none of the males sired litters; incomplete spermatogenesis occurred as well as decreased testicular weight. However, a 5-month post-treatment recovery period following 6 mg/kg/day indicated that partial recovery occurred (246).

In a subsequent rat study, a single oral dose of 1,3-DNB (48 mg/kg) was demonstrated to damage rat testes severely within 24 hr, followed by slow and incomplete recovery; normal fertilizing ability was restored in most animals by week 13, but two of seven rats remained infertile (247). Further studies with in vitro culture experiments suggested that in situ metabolism to reactive intermediates may be involved, postulated to be m-nitrosonitrobenzene via m-nitroaniline metabolism (248a). In a follow-up study, in vitro and in vivo experiments were conducted in which Sertoli cell lactate and pyruvate production was demonstrated to be significantly enhanced with 1,3-DNB in a dose-related manner, but no such changes occurred with the ortho or para isomers (248b).

In vivo testicular metabolites from 1,2- and 1,4-DNB administered to F344 rats reportedly included only glutathione conjugates and unchanged parent; in contrast 1,3-DNB generated five times more $^{14}$C (testes receive a larger percentage of 1,3-

DNB) and included not only the glutathione conjugate, but also four unidentified metabolites (249).

Testicular damage has also been observed in male Alpk/AP (Wistar derived) rats administered 15, 25, or 50 mg/kg by gavage; at 12 hr after a single dose of 25 mg/kg, 1,3-DNB produced testicular lesions limited to Stages VIII to XI of the spermatogenic cycle. By 24 hr widespread Sertoli cell damage was evident and in some tubules was associated with degeneration of primary spermatocytes. Ultrastructural examination at this time confirmed that there were effects on Sertoli cells in the absence of germ cell damage. Similar effects were seen 48 hr after a single oral dose of 15 mg/kg of 1,3-DNB. Doses of 5 or 10 mg/kg were without effect on the testis. Ultrastructural analyses of the testicular lesions suggested that the Sertoli cell was the primary target for 1,3-DNB toxicity and that germ cell damage occurred subsequent to Sertoli cell toxicity. In this isomer study, 1,2- and 1,4-DNB were without effect (236).

Age-dependent and species-dependent effects of 1,3-DNB on testicular function have also been investigated. Effects in prepubertal and pubertal mice appear to be less pronounced than in adult mice. Furthermore, following exposure to the same dosage, the effect of 1,3-DNB is less severe in adult mice than in adult rats (250). In a comparative study with Sprague–Dawley rats (75 or 105 days of age given a single oral dose of 0, 8, 16, 24, 32, or 48 mg/kg of $m$-DNB), mortality and neurotoxicity were observed at 48 mg/kg, but only in older animals. Adverse effects on epididymides and sperm were observed at 16 and 24 mg/kg and higher in the older and younger groups, and testicular weights were affected at 24 mg/kg and higher in both age groups. The study indicated that the lowest dosage level to produce reproductive changes was 16 mg/kg. The effects were observed to be apparently reversible in most animals after about 5 months following a dosage level of 48 mg/kg, although some rats remained infertile. Epididymal measurements demonstrated a no-effect dosage at 8 mg/kg. Rats 75 days old were less sensitive than 105-day-old rats, attributed to differences in metabolism (251). A study to evaluate species differences documented that the Sprague–Dawley rat is markedly more sensitive than the Syrian golden hamster to 1,3-DNB induced testicular toxicity and methemoglobinemia. The hamster showed no testicular lesions at dose levels up to 50 mg/kg whereas, as previously reported by others, damage to rat testicular tubules was readily apparent at a 25 mg/kg dose level. At the 25 mg/kg dose, peak levels of methemoglobin in the hamster were 15 percent compared with 80 percent in the rat. Mortality in the rat also occurred at lower doses than in the hamster (50 versus 100 mg/kg, respectively). The authors recommended that until the mechanism(s) underlying these species differences is elucidated, the rat model should not be used to extrapolate potential male reproductive damage, especially if humans respond to 1,3-DNB in a similar manner to the hamster (252).

In a follow-up pharmacokinetic study to investigate the mechanisms associated with the greater sensitivity of the rat, administration of [$^{14}$C]1,3-DNB (50 mg/kg i.p.) to the hamster resulted in 1,3-DNB blood levels that were similar to those found in the rat at 25 mg/kg. It was observed that no obvious testicular toxicity was apparent in the hamster even at the higher dose level, and therefore a direct effect of 1,3-DNB on the testes seems unlikely, although it is possible that hamster cells inherently lack sensitivity to toxicity. Other major differences between the two species were a more rapid initial elimination rate and much higher blood levels of nitroaniline in the rat. The rat also excreted more unconjugated metabolites and fewer phenolic metabolites compared with the hamster. The authors suggested that the capacity to form reductive metabolites may play an important role in susceptibility to toxicity (253).

As discussed above, all three dinitrobenzene isomers cause methemoglobinemia, but only 1,3-DNB produces testicular toxicity in rats.

1,3-DNB has also demonstrated positive mutagenicity in the Ames assay (123).

*2.37.2.4 Hygienic Standards of Permissible Exposure.* The adopted TLV–TWA value is 0.15 ppm, 1 mg/m$^3$ (skin) (139).

### 2.37.3 p-Dinitrobenzene [CAS # 100-25-4]

The formula for *p*-dinitrobenzene is $C_6H_4(NO_2)_2$; it is also called 1,4-dinitrobenzene, 1,4-DNB, and *p*-DNB.

*2.37.3.1 Uses.* Its uses are reportedly identical with those of *o*-dinitrobenzene (109).

*2.37.3.2 Physical and Chemical Properties (100).* *p*-Dinitrobenzene has the following properties:

| | |
|---|---|
| Physical state | Colorless to yellow monoclinic needles |
| Molecular weight | 168.11 |
| Density | 1.625 (20/4°C) |
| Melting point | 173 to 174°C |
| Boiling point | 298 to 299°C (777 mm Hg)—sublimable |
| Solubility | 0.18 in water at 100°C; 0.4 in alcohol at 20°C; 1.82 in chloroform at 18°C; 2.3 in benzene at 18°C |

*2.37.3.3 Toxicity.* 1,4-DNB administered to Alpk/AP (Wistar derived) rats, 50 mg/kg by gavage, did not cause testicular lesions; however, 1,4-DNB potency was similar to that of 1,3-DNB in producing cyanosis and splenic enlargement. 1,2-DNB was less potent in this regard and also did not cause testicular damage (236).

Prolonged exposure will induce methemoglobinemia and anemia (236).

*2.37.3.4 Hygienic Standards of Permissible Exposure.* The adopted TLV–TWA value is 0.15 ppm, 1 mg/m$^3$ (skin) (139).

## 2.38 Trinitrobenzene [CAS # 99-35-4]

The formula for trinitrobenzene is $C_6H_4(NO_2)_3$; it is also known as 1,3,5-trinitrobenzene and *sym*-trinitrobenzene.

### 2.38.1 Use

Trinitrobenzene is used for research purposes only. It is less sensitive than TNT to shock, and is a more powerful explosive than TNT, but it is more difficult to prepare and cannot be made in satisfactory yields. It can be prepared on a small scale for reagent purposes usually by decarboxylating 2,4,6-trinitrobenzoic acid, which is obtained by the oxidation of TNT.

### 2.38.2 Physical and Chemical Properties (100)

| | |
|---|---|
| Physical State | Orthorhombic bipyramidal plates from glacial acetic acid |
| Molecular weight | 213.11 |
| Melting point | 121 to 122.5°C |
| Boiling point | 315°C |
| Solubility | Soluble in acetone, benzene |

### 2.38.3 Toxicity

No toxicology reports were located in the literature.

### 2.38.4 Hygienic Standards of Permissible Exposure

Standards have not been assigned.

## 2.39 Chloronitrobenzene Isomers

### 2.39.1 o-Chloronitrobenzene [CAS # 88-73-3]

The formula for o-chloronitrobenzene is $ClC_6H_4NO_2$; it is also known as 1-chloro-2-nitrobenzene or 2-chloronitrobenzene, 2-chloro-1-nitrobenzene, 1-nitro-2-chlorobenzene, 2-nitrochlorobenzene, o-CNB, and ONCB.

*2.39.1.1 Uses.* It is used in preparation of dyes and dyestuffs.

*2.39.1.2 Physical and Chemical Properties (100).* The properties of o-chloronitrobenzene are as follows:

| | |
|---|---|
| Physical state | Yellow crystals |
| Molecular weight | 157.56 |
| Melting point | 34 to 35°C |
| Boiling point | 245 to 246°C |
| Solubility | Soluble in alcohol, benzene, ether, acetone |

*2.39.1.3 Toxicity.* A commission of the European Community reported short-term exposure effects including skin, eye, and respiratory tract irritation, pulmonary edema, methemoglobinemia, and neurotoxic effects; long-term exposure included dermatitis, skin sensitization, and hepatic, pancreatic, and renal disorders (254). Severe headache and methemoglobinemia have been reported in the workplace (147).

ONCB has been reported to be metabolized by reduction to $o$-chloroaniline and conjugated with GSH to $S$-(2-nitrophenyl)glutathione in the presence of cytosolic GSH (255).

Reproductive assessment by continuous breeding in mice (40, 80, 160 mg/kg) is under evaluation by the National Toxicology Program (NTP) (150). Although no testicular effects were reportedly observed in F344 rats following a 150 mg/kg dose administered by gavage (256), adverse reproductive effects on sperm motility and/ vaginal cytology were reported by NTP in 1990 (150).

$o$-CNB was reported by NTP in 1990 to be mutagenic in the Ames test (150). A carcinogenicity bioassay study was initiated by NTP 1991 at Battelle-Northwest (108).

*2.39.1.4 Hygienic Standards of Permissible Exposure.* Standards have not been assigned.

### 2.39.2 m-Chloronitrobenzene [CAS # 121-73-3]

The formula for *m*-chloronitrobenzene is $ClC_6H_4NO_2$; it is also called 1-chloro-3-nitrobenzene, 3-chloronitrobenzene, or *m*-CNB.

*2.39.2.1 Uses.* *m*-Chloronitrobenzene is of slight importance commercially and has limited use in the manufacturing of dyes.

*2.39.2.2 Physical and Chemical Properties (100).* *m*-Chloronitrobenzene has the following properties:

| | |
|---|---|
| Physical state | Plate yellow orthorhombic prisms from alcohol |
| Molecular weight | 157.56 |
| Melting point | 46°C |
| Boiling point | 235 to 236°C |
| Solubility | Soluble in hot alcohol, chloroform, ether, carbon disulfide, benzene; insoluble in water |

*2.39.2.3 Toxicity.* In contrast to *p*-chloronitrobenzene, *m*-CNB is not conjugated with GSH, as discussed under *p*-CNB below. *m*-CNB is reportedly nonmutagenic in the Ames test (145a). In contrast to 2-chloro- and 4-chloronitrobenzene, 3-chloronitrobenzene was not identified as a candidate for toxicology investigations by NTP for 1990–1991 (108, 150).

*2.39.2.4 Hygienic Standards of Permissible Exposure.* Standards have not been assigned.

### 2.39.3 p-Chloronitrobenzene [CAS # 100-00-5]

The formula for *p*-chloronitrobenzene is $ClC_6H_4NO_2$; it is also known as 1-chloro-4-nitrobenzene, 4-chloronitrobenzene, *p*-nitrochlorobenzene, PNCB, and *p*-CNB.

*2.39.3.1 Uses.* *p*-Chloronitrobenzene is used as an intermediate in the manufacture of dyes, rubber, and agricultural chemicals.

*2.39.3.2 Physical and Chemical Properties (100).* The properties of *p*-chloronitrobenzene are as follows:

| | |
|---|---|
| Physical state | Yellow crystals |
| Molecular weight | 157.56 |
| Specific gravity | 1.520 |
| Melting point | 82 to 84°C |
| Boiling point | 242°C |
| Vapor pressure | 0.009 torr at 25°C |
| Flash point | 127°C (closed cup) |
| Solubility | Sparingly soluble in water; soluble in ether, carbon disulfide, alcohol, acetone |

*2.39.3.3 Toxicity.* Oral, skin, or lung exposure can result in methemoglobinemia. Other symptoms may include headache, weakness, cyanosis, and anemia (257). Skin penetration is rapid, and *p*-CNB is more potent than aniline in terms of potential to produce cyanosis and anemia (258).

In vitro studies have demonstrated that *p*-chloronitrobenzene can be reduced to *p*-chloroacetanilide as well as *p*-chloroaniline, and that cytosolic GSH transferase is involved in the conjugation with GSH to form *S*-(4-nitrophenyl)glutathione. In contrast, *m*-CNB is not conjugated with GSH (255). Urinary metabolites of male Sprague–Dawley rats following a single intraperitoneal dose of 100 mg/kg of CNB diluted in olive oil included trace amounts of unchanged CNB, *p*-chloroaniline, 2,4-dichloroaniline, *p*-nitrothiophenol, 2-chloro-5-nitrophenol, 2-amino-5-chlorophenol, 4-chloro-2-hydroxyacetanilide, and a small amount of *p*-chloroacetanilide (259).

A slight decrease in pregnancy rate and male fertility in rats dosed at 5 mg/kg has been reported (260), and the NTP Annual Report for 1991 reveals positive results regarding evaluation of effects on sperm motility and/or vaginal cytology (150); however, no testicular effects were observed in F344 rats given 250 mg/kg by gavage (256).

A definitive teratogenicity study of PNCB was conducted in groups of 24 mated female Sprague–Dawley rats administered 5, 15, or 45 mg/kg/day by gavage on days 6 to 19 of gestation and in 18 New Zealand mated rabbits dosed with 5, 15, or 40 mg/kg/day on gestation days 7 to 19 via a stomach tube (148b). In the rat study, abnormal physical observations were limited to high-dosage level animals, and maternal toxicity in terms of decreased mean body weight gain was observed at the 45 mg/kg/day dosage level. The high-dosage level was embryo toxic in that a statistically significant increase in resorptions was observed. The incidence of skeletal anomalies in the high-dosage group fetuses was significantly higher than controls on both a per fetus and per litter basis and is thus considered related to treatment. In the rabbit study, death of 8 of the 18 rabbits dosed with 40 mg/kg/day and the loss of two litters because of abortion resulted in the termination of this dosage group. The small increase in the incidence of fetuses with skeletal malformations at the 5- and 15-mg/kg/day dosage level was not statistically significant. The authors concluded that PNCB should not be labeled as teratogenic (148b).

PNCB was reportedly not carcinogenic when administered to rats orally at 0.1, 0.7, or 5.0 mg/kg/day; the only predominant adverse effect was apparently significant methemoglobinemia observed at mid- and high-dose levels; $p$-CNB was also reportedly nonmutagenic in the Ames test (145a). However, NTP apparently initiated a carcinogenicity bioassay in 1991 at Battelle-Northwest (150).

*2.39.3.4 Hygienic Standards of Permissible Exposure.* In 1990 NIOSH classified $p$-CNB as a carcinogen (skin); the OSHA TLV–TWA is 1 mg/m$^3$ (skin) (40).

## 2.40 Chlorodinitrobenzene Isomers

Commercial chlorodinitrobenzene is usually a mixture of the six possible isomers, which are closely related chemically and physically to each other. General uses include the manufacture of dyestuffs, other dye intermediates, and certain explosives (109).

### 2.40.1  2-Chloro-1,3-Dinitrobenzene [CAS # 606-21-3]

The formula for 2-chloro-1,3-dinitrobenzene is $C_6H_3Cl(NO_2)_2$; it is also known as 1-chloro-2,6-dinitrobenzene.

*2.40.1.1 Uses.* No specific uses were reported in the literature.

*2.40.1.2 Physical and Chemical Properties (100).* The properties of 2-chloro-1,3-dinitrobenzene are as follows:

| | |
|---|---|
| Physical state | Yellow crystals |
| Molecular weight | 202.55 |
| Melting point | 86 to 87°C |
| Boiling point | 315°C |
| Solubility | Soluble in alcohol, ether, toluene |

*2.40.1.3 Toxicity.* No toxicology information was located in the literature.

*2.40.1.4 Hygienic Standards of Permissible Exposure.* Standards have not been assigned.

### 2.40.2  2-Chloro-1,4-dinitrobenzene [CAS # 619-16-9]

The formula for 2-chloro-1,4-dinitrobenzene is $C_6H_3Cl(NO_2)_2$; it is also known as 1-chloro-2,5-dinitrobenzene.

*2.40.2.1 Uses.* Its uses are reportedly identical with those of 4-chloro-1,2-dinitrobenzene (109).

*2.40.2.2 Physical and Chemical Properties (100).* The properties of 2-chloro-1,4-dinitrobenzene are as follows:

| | |
|---|---|
| Physical state | Light yellow crystals |
| Molecular weight | 202.55 |
| Melting point | 64°C |
| Solubility | Insoluble in water; soluble in alcohol, ether |

*2.40.2.3 Toxicity.* No toxicology information was located in the literature. Toxicity has been assumed to be identical with that of 4-chloro-1,2-dinitrobenzene (109).

*2.40.2.4 Hygienic Standards of Permissible Exposure.* Standards have not been assigned.

### 2.40.3  3-Chloro-1,2-dinitrobenzene [CAS # 602-02-8]

The formula is $C_6H_3Cl(NO_2)_2$; 3-chloro-1,2-dinitrobenzene is also known as 1-chloro-2,3-dinitrobenzene.

*2.40.3.1 Uses.* The uses of 3-chloro-1,2-dinitrobenzene are reportedly identical with those of 4-chloro-1,2-dinitrobenzene (109).

*2.40.3.2 Physical and Chemical Properties (100).* 3-Chloro-1,2-dinitrobenzene has the following properties:

| | |
|---|---|
| Physical states | Crystals from alcohol or ether |
| Molecular weight | 202.55 |
| Melting point | 78°C |
| Boiling point | 315°C |
| Refractive index | 1.6867 (16.5°C) |
| Solubility | Insoluble in water; soluble in alcohol, ether |

*2.40.3.3 Toxicity.* No toxicology information was located in the literature. Toxicity has been assumed to be identical with that of 4-chloro-1,2-dinitrobenzene (109).

*2.40.3.4 Hygienic Standards of Permissible Exposure.* Standards have not been assigned.

### 2.40.4   4-Chloro-1,2-dinitrobenzene [CAS # 610-40-2]

The formula for 4-chloro-1,2-dinitrobenzene is $C_6H_3Cl(NO_2)_2$; it is also known as 1-chloro-3,4-dinitrobenzene.

*2.40.4.1 Uses.* No specific uses were reported in the literature.

*2.40.4.2 Physical and Chemical Properties (100).* The properties of 4-chloro-1,2-dinitrobenzene are as follows:

| | |
|---|---|
| Physical state | Monoclinic prisms, needles |
| Molecular weight | 202.55 |
| Melting point | α, 36°C; β, 37°C; γ, 40 to 41°C |
| Boiling point | 315°C |
| Solubility | Insoluble in water; soluble in hot, difficultly soluble in cold alcohol; soluble in ether, benzene, carbon disulfide |

*2.40.4.3 Toxicity.* 4-Chloro-1,2-dinitrobenzene is a known sensitizer, causing contact dermatitis ranging from a few itching, vesicular papules to a generalized exfoliative dermatitis (109).

*2.40.4.4 Hygienic Standards of Permissible Exposure.* Standards have not been reported.

### 2.40.5   4-Chloro-1,3-dinitrobenzene [CAS # 97-00-7]

The formula for 4-chloro-1,3-dinitrobenzene is $C_6H_3Cl(NO_2)_2$; it is also known as 6-chloro-1,3-dinitrobenzene, 1,3-dinitro-4-chlorobenzene, 2,4-dinitrochlorobenzene, 1-chloro-2,4-dinitrobenzene, 2,4-dinitro-1-chlorobenzene, and DNCB.

*2.40.5.1 Uses.* DNCB is used in the manufacture of dyes and explosives, as a color-producing reagent for the detection of pyridine compounds, ethylenediamine, diethylenetriamine, and triethylenetetramine, and as an algicide. Raaf et al. (1976) reported successful medical use in combination with 5-fluorouracil for the topical treatment of squamous cell carcinoma, or "Bowen's disease," based upon developing an immune response to DNCB (109).

*2.40.5.2 Physical and Chemical Properties (100).* The properties of DNCB are as follows:

| | |
|---|---|
| Physical state | Yellow crystals |
| Molecular weight | 202.55 |

| | |
|---|---|
| Melting point | α, 53°C; β, 43°C |
| Boiling point | 315°C (762 mm Hg) |
| Solubility | Insoluble in water; readily soluble in hot alcohol; soluble in ether, benzene |

*2.40.5.3 Toxicity.* This chemical is a potent sensitizer and exposure can result in contact urticaria and yellow discoloration of the skin, as well as violent dermatitis (261, 202).

Adams, Jr., et al. (1971) reported a case in which 2,4-dinitro-1-chlorobenzene had been used as an algicide in the coolant water of air conditioning systems. Four repairmen working on these systems suffered severe contact dermatitis that was very difficult to treat. The conclusion was that because 1-chloro-2,4-dinitrobenzene is extremely allergenic, it should be used only in closed systems that afford no human contact (109). Oral $LD_{50}$ in rats is reportedly 1.07 g/kg (100).

DNCB has been used for many years to induce contact sensitivity experimentally and in allergenic cross-sensitization screening programs. One of the earliest reports regarding the use of DNCB in the study of allergic contact dermatitis is that by Wedroff, which appeared in 1927 (175a).

Metabolism studies have demonstrated that DNCB depletes GSH liver levels in the biotransformation displacement of chlorine to yield 1-SG-2,4-dinitrobenzene (262). A reactive intermediate may be involved for further studies have also demonstrated that DNCB was less mutagenic in a GSH-deficient derivative of *Salmonella typhimurium* TA100 [TA100/GSH-] than in TA100 itself, suggesting that the mutagenicity depends on GSH. Further investigations indicated that halogenated aromatics may react with bacterial DNA and produce premutagenic alterations according to two mechanisms: direct attack on the DNA through nucleophilic substitution ($SN_2$) of the halogen atoms, or activation through GSH conjugation and subsequent nitroreduction of the conjugate or its metabolic products to more reactive intermediates (263).

*2.40.5.4 Hygienic Standards of Permissible Exposure.* Standards have not been assigned.

*2.40.6  5-Chloro-1,3-dinitrobenzene [CAS # 618-86-0]*

The formula for 5-chloro-1,3-dinitrobenzene is $C_6H_3Cl(NO_2)_2$; it is also known as 1-chloro-3,5-dinitrobenzene.

*2.40.6.1 Uses.* Its uses are reportedly identical with those of 4-chloro-1,2-dinitrobenzene (109).

*2.40.6.2 Physical and Chemical Properties (100).* 5-Chloro-1,3-dinitrobenzene has the following properties:

| | |
|---|---|
| Physical state | Colorless needles |
| Molecular weight | 202.55 |

Melting point 59°C
Solubility Insoluble in water; Soluble in alcohol, ether

*2.40.6.3 Toxicity.* No toxicology information was located in the literature. Its toxicity has been assumed to be identical with that of 4-chloro-1,2-dinitrobenzene (109).

*2.40.6.4 Hygienic Standards of Permissible Exposure.* Standards have not been assigned.

### 2.41 Pentachloronitrobenzene [CAS # 82-68-8]

The formula for pentachloronitrobenzene is $C_6Cl_5NO_2$; it is also known as PCNB, Quintozene, Botrilex, Folosan, Terraclor, Tritisan, Tilicarex, and Brassico.

#### 2.41.1 Uses

PCNB has been used as a soil or seed fungicide for the control of *Botrytis* diseases, club root of crucifers, scab of potato, and *Rhizoctonia* damping-off disease of seedlings (264). It is also used as a turf fungicide to prevent root rotting.

#### 2.41.2 Physical and Chemical Properties (100).

| | |
|---|---|
| Physical state | Colorless solid |
| Molecular weight | 295.36 |
| Specific gravity | 1.718 at 25°C |
| Melting point | 142 to 145°C technical grade |
| Boiling point | 328°C at 760 mm Hg torr |
| Solubility | Soluble in carbon disulfide, benzene, chloroform |
| Other components | Hexachlorobenzene, tetrachlorobenzene, pentachlorobenzene |

#### 2.41.3 Toxicity

Exposure to PCNB can induce contact sensitization (202). Methemoglobinemia has been demonstrated in cats, which have an unusually high sensitivity due to the low rate of methemoglobin reductase activity, following a single high oral dose of 1600 mg/kg (265). The reported oral $LD_{50}$ for male rats (in corn oil) is 1740 mg/kg, and the $LD_{50}$ (dermal) for rabbits was found to be >4 g/kg (266). The reported $LC_{50}$ values are 1400 mg/m³ for rats and 2000 mg/m³ for mice (139).

PNCB has been demonstrated to be metabolized to pentachloroaniline and methyl pentachlorophenylsulfide by New Zealand rabbits, CD rats, and beagle dogs (266).

No adverse effects were reported in a three-generation reproductive study in which CD rats were administered a diet of PCNB at concentrations of 0, 5, 50, or 500 ppm [four groups, 25 males and females per group]. Additionally, no structural developmental defects were noted in any pup (266). A follow-up teratogenic study in Charles River strain albino rats, administered PCNB at dosages of 100 to 1563 ppm in corn oil, did not demonstrate any treatment-related developmental effects. Dilated renal pelvis, hydronephrosis, and hydroureter were found in both control and treated groups and appeared unrelated to treatment (267). Nonteratogenicity was also confirmed in Wistar rats following oral administration (268). In a study using contaminated PCNB (11 percent hexachlorobenzene) and purified PCNB (<20 ppm hexachlorobenzene), contaminated PCNB produced renal agenesis and cleft palates in C57B1/6 mice and cleft palates in CD-1 mice. Purified PCNB produced fewer cleft palates and no kidney malformations. Neither sample produced teratogenesis in CD rats. Pentachloroaniline (a metabolite of PCNB), pentachlorophenol, and tetrachloronitrobenzene were not teratogenic in rats. The authors concluded that the observed teratogenicity in mice was due to the HCB contamination (269). It should be noted, however, that tetrachloronitrobenzene isomers differ in their pharmacokinetics. For example, oral administration of 2,3,5,6-tetrachloronitrobenzene has been demonstrated to be only about 33 percent absorbed, whereas 2,3,4,5-tetrachloronitrobenzene is about 66 percent absorbed in rabbits. A major route of metabolism for 2,3,5,6-tetrachloronitrobenzene was presumably conjugation with glutathione, because 11 percent of the dose was found in the urine as $S$-(2,3,5,6-tetrachlorophenyl)-$N$-acetylcysteine. 2,3,5,6-Tetrachloroaniline accounted for 9 percent of the dose; 15 percent of the dose was converted to 2,3,5,6-tetrachloro-4-aminophenol and excreted as glucuronide (12 percent of dose) and sulfate (1 percent of the dose) conjugates or as free phenol (2 percent of dose). No acetylcysteine conjugate was found when rabbits were given 2,3,4,5-tetrachloronitrobenzene. All of the metabolites found in the urine had undergone nitro reduction. 2,3,4,5-Tetrachloroaniline accounted for 11 percent of the dose, and 2,3,4,5-tetrachloro-6-aminophenol excreted as the glucuronide (41 percent of the dose) or the sulfate (6 percent of the dose) accounted for the rest. Pentachloronitrobenzene and 2,3,4,6-tetrachloronitrobenzene were converted by rabbits to the mercapturic acids, $S$-(pentachlorophenyl)-$N$-acetylcysteine (4 to 14 percent of the dose) and $S$-(2,3,4,6-tetrachlorophenyl)-$N$-acetylcysteine (32 to 37 percent of the dose). Twelve to 14 percent of the dose of pentachloronitrobenzene and 24 to 31 percent of the dose of 2,3,4,6-tetrachloronitrobenzene were eliminated as the polychloroanilines (16c). Relatively few studies regarding the metabolism of tetrachloronitrobenzenes have been reported, compared with the more economically important PCNB.

A two-year dietary feeding study in four beagle dogs of each sex, at dietary levels of 0, 5, 30, 180, or 1080 ppm, demonstrated no treatment-related effect; 30 ppm was identified as the NOEL, and cholestatic hepatosis with secondary bile nephrosis was observed at 180 and 1080 ppm (266).

In a number of carcinogenic assays, PCNB has not demonstrated carcinogenicity at 25, 100, 300, 1000, or 2500 ppm (diet of a commercial mixture of 20 percent PCNB) (270). PCNB generated liver tumors in male B6AKF1 mice and female B6C3F1 mice; the purity of PCNB was not specified (271). A skin painting tumor initiation-promotion (croton oil) study in mice (purity not given) generated squamous cell carcinomas (272). Technical-grade PCNB, 97 percent + 1 percent hexachlorobenzene administered in the diet to Osborne–Mendel rats (5417 or 10,064 ppm for males, 7875 or 14,635 ppm for females) and B6C3F1 mice (2606 or 5213 ppm for males, 4093 or 8187 for females), was not carcinogenic (273). In a second NCI study in which B6C3F1 male and female mice were fed diets containing 2500 and 5000 ppm, PCNB was also noncarcinogenic (274). PCNB is not considered to be carcinogenic.

### 2.41.4 Hygienic Standards of Permissible Exposure

A TLV–TWA of 0.5 mg/m$^3$ has been proposed by ACGIH (139).

### 2.42 Benzidine [CAS # 92-87-5]

The formula is $NH_2C_6H_4C_6H_4NH_2$; benzidine is also known as 4,4'-biphenyldiamine, 4,4'-diaminobiphenyl, 4,4'-diphenylenediamine, p-diaminodiphenyl, and 4,4'-diaminodiphenyl.

### 2.42.1 Uses

Benzidine is used in the synthesis of dyes and dye intermediates, as a hardener for rubber, and as a laboratory reagent. The first successful synthetic direct dye was Congo Red, a diazo derivative prepared from benzidine by Boettiger in 1884. Nearly all direct dyes are azo products (275). Congo Red (CAS # 573-58-0, USP XV, sodium diphenyldiazo-bis-α-naphthylaminesulfonate) is used in humans intravenously for the medical diagnosis of amyloidosis. The basis for its use is an unexplained affinity for amyloid, which rapidly removes the dye from the blood. It is also used medically for the management of profuse capillary hemorrhage such as occurs in septicemias and in the terminal phases of leukemia.

### 2.42.2 Physical and Chemical Properties (100)

| | |
|---|---|
| Physical state | White or slightly reddish combustible powder or crystals |
| Molecular weight | 184.23 |
| Density | 1.250 (20/4°C) |
| Melting point | 115 to 120°C |
| Boiling point | 402°C |
| Solubility | Slightly soluble in hot water, boiling ethanol; soluble in ether, diethyl ether |

### 2.42.3 Toxicity

IARC working groups have concluded that there is sufficient evidence for carcinogenicity of benzidine in animals and in humans, it has been classified as a Group 1 carcinogen (32a). Case reports and follow-up studies of workers provide sufficient evidence that occupational exposure to benzidine is strongly associated with an increased risk of bladder cancer (81).

## 2.42.4 Hygienic Standards of Permissible Exposure

Standards have not been assigned.

## 2.43 3,3'-Dichlorobenzidine [CAS # 91-94-1]

The formula for 3,3'-dichlorobenzidine is $C_6H_3ClNH_2C_6H_3ClNH_2$; it is also called 4,4'-diamino-3,3'-dichlorobiphenyl and 3,3'-dichlorobiphenyl-4,4'-diamine.

### 2.43.1 Uses

3,3'-Dichlorobenzidine is used as a curing agent for isocyanate-containing resins and urethane plastics and also as a chemical intermediate for the preparation of dyes and yellow pigments.

### 2.43.2 Physical and Chemical Properties (100)

| | |
|---|---|
| Physical state | Grayish to purple crystals |
| Molecular weight | 253.11 |
| Melting point | 133°C |
| Boiling point | 368°C |
| Solubility | Insoluble in water; soluble in alcohol, ether, glacial acetic acid |

### 2.43.3 Toxicity

It has not always been clear in the literature whether the salt (dihydrochloride) or the free base was the compound under investigation, although only the salt is believed to be available commercially (81).

3,3'-Dichlorobenzidine is carcinogenic in animals (hepatomas in male mice, leukemia and Zymbal gland carcinomas in male rats, mammary adenocarcinomas in rats of both sexes, bladder carcinomas in hamsters, bladder carcinomas and hepatocellular carcinomas in female dogs); it is listed as a carcinogen by OSHA owing to its potent carcinogenicity in animals. An IARC working group reported that there were not adequate data to evaluate the carcinogenicity in humans. In three retrospective epidemiologic studies, no urinary bladder tumors were reported in men occupationally exposed to 3,3'-dichlorobenzidine, but the studies were judged to be statistically inadequate (32a, 81).

### 2.43.4 Hygienic Standards of Permissible Exposures

ACGIH, NIOSH, and OSHA have classified this compound as a suspected occupational human carcinogen (40).

Standards have not been assigned.

## 2.44 Nitro and Amino Biphenyls

### 2.44.1 2-Nitrobiphenyl [CAS # 86-00-0]

The formula is $C_6H_5C_6H_4NO_2$, and is also known as *o*-nitrobiphenyl or *o*-nitrodiphenyl.

*2.44.1.1 Uses.* 2-Nitrobiphenyl is used as a plasticizer for resins, cellulose acetate and nitrate, and polystyrene, as a fungicide for textiles, as a wood preservative, and as a dye intermediate.

*2.44.1.2 Physical and Chemical Properties (100).* 2-Nitrobiphenyl has the following properties:

| | |
|---|---|
| Physical state | Colorless crystals |
| Molecular weight | 199.21 |
| Boiling point | 320°C |
| Melting point | 36.7 to 37.2°C |
| Solubility | Soluble in methanol, ethanol, acetone, carbon tetrachloride, perchloroethylene |
| | Insoluble in water |

*2.44.1.3 Toxicity.* Toxicology information was not located in the literature.

*2.44.1.4 Hygienic Standards of Permissible Exposure.* Standards have not been assigned.

### 2.44.2 4-Nitrobiphenyl [CAS # 92-93-3]

The formula for 4-nitrobiphenyl is $C_6H_5C_6H_4NO_2$; it is also called 4-nitrodiphenyl, *p*-nitrobiphenyl, *p*-nitrodiphenyl, *p*-phenylnitrobenzene, 4-phenylnitrobenzene, and PNB.

*2.44.2.1 Uses.* 4-Nitrobiphenyl was formerly used in the synthesis of 4-aminodiphenyl. It is presently used as a plasticizer for resins, polystyrenes, and cellulose acetate and nitrate, as a fungicide for textiles, and as a preservative for wood.

*2.44.2.2 Physical and Chemical Properties (100).* The properties of 4-nitrobiphenyl are as follows:

| | |
|---|---|
| Physical state | Yellow crystals |
| Molecular weight | 199.21 |
| Melting point | 114 to 114.5°C |
| Boiling point | 340°C |
| Solubility | Soluble in hot water, alcohol, ether; insoluble in acetone, benzene |

*2.44.2.3 Toxicity.* 4-Nitrobiphenyl has produced bladder tumors in dogs and is a recognized carcinogen (32a, 40). The oral $LD_{50}$ for rabbits was reported to be 1.97 g/kg and for rats, 2.33 g/kg (139).

The IARC Working Group on the Evaluation of the Carcinogenic Risk of Chemicals to Man found no reports on carcinogenicity of PNB to humans and the animal data to be inadequate to classify the compound as a human carcinogen; consequently IARC assigned 4-nitrobiphenyl to Group 3, as not classifiable as to its carcinogenicity to humans (32a). However, it is considered an occupational carcinogen by NIOSH and OSHA (40), and as a "recognized carcinogen" (or Alb) by ACGIH (139).

*2.44.2.4 Hygienic Standards of Permissible Exposure.* Standards have not been assigned.

*2.44.3  4-Aminobiphenyl [CAS # 92-67-1]*

The formula is $C_6H_5C_6H_4NH_2$; other names include biphenyline, *p*-phenylaniline, *p*-xenylamine, 4-aminodiphenyl, 4-biphenylamine, *p*-aminodiphenyl, *p*-biphenylamine, or xenylamine.

*2.44.3.1 Uses.* 4-Aminobiphenyl is used for research purposes only; it was formerly used as rubber antioxidant and dye intermediate.

*2.44.3.2 Physical and Chemical Properties (100).* 4-Aminobiphenyl has the following properties:

| | |
|---|---|
| Molecular weight | 169.23 |
| Melting point | 53 to 54°C |
| Boiling point | 302°C |
| | 191°C (15 mm Hg) |
| Solubility | Slightly soluble in water; soluble in alcohol, ether, chloroform |

*2.44.3.3 Toxicity.* 4-Aminobiphenyl produces bladder cancers in humans and animals. Melick et al. in 1955 reported an incidence of 19 bladder tumors in 171 employees manufacturing 4-aminobiphenyl between 1935 and 1955 (276). In 1955, a surveillance program was initiated on workers reported to have been exposed to the chemical: during the following 14 years, 541 men were kept under surveillance by clinical and laboratory examinations; 86 had positive or suspicious cytology of the urinary sediment some time during the observation period, and 43 developed carcinoma of the bladder (32a). In a more recent survey of cancer mortality among workers at the Monsanto Company plant in Nitro, West Virginia (producing a variety of chemicals) a tenfold increase in mortality from bladder cancer was reported. All of the nine cases on which the excess was based had started work in the plant before 1949, and 4-aminobiphenyl was known to have been used from 1941 until 1952 (277).

4-Aminobiphenyl is metabolized via microsomal P450 enzyme system to the corresponding methemoglobin forming N-hydroxylamine (278, 172). It has been proposed that bladder cancer produced by 4-aminophenol (and 2-naphthylamine) occurs via hepatic N-hydroxylation followed by formation of the N-glucuronide conjugate, which is transported to the bladder where it is hydrolyzed in acidic urine, resulting in the release of the unstable carcinogenic N-hydroxylamine (107). The N-hydroxylamine is then believed to degrade to an electrophilic arylnitrenium ion that can covalently modify DNA and bladder epithelium. 4-Aminobiphenyl has been demonstrated to form DNA adducts in the bladder epithelium of dogs and protein adducts in serum albumin of rats treated in vivo (32a). Urine pH values are normally affected by diet; for example, a high protein diet usually produces acidic urine, whereas cereal or vegetable diets generally produce alkaline urine. In patients with proximal renal tubular acidosis, or bacterial bladder infections, urine pH can be markedly acidic with values as low as pH 4.5 to 6. The normal pH value for human urine is pH 4.5 to 8.5.

When administered by gavage, 4-aminobiphenyl induced carcinoma of the urinary bladder in mice and rabbits. When administered in the diet, it induced neoplasms at various sites in mice, including dose-related increases in the incidence of angiosarcomas. The compound also induced bladder carcinoma in dogs (81). An IARC working group reported that there is sufficient evidence for the carcinogenicity of 4-aminobiphenyl in humans and has classified the compound as a Group 1 carcinogen (32a).

4-Aminobiphenyl is considered an occupational carcinogen by NIOSH and OSHA (40); ACGIH considers the compound a recognized human carcinogen and assigned it an A1b classification (139).

*2.44.3.4 Hygienic Standards of Permissible Exposure.* Standards have not been assigned.

### 2.45 p-Aminodiphenylamine [CAS # 101-54-2]

The formula is $C_6H_5NHC_6H_4NH_2$, and it is also known as 4-aminodiphenylamine or N-phenyl-1,4-phenylenediamine.

### 2.45.1 Uses

The compound is used as an oxidation dye color in hair dyes, which as is the case with p-phenylenediamine, p-toluenediamine, and p-aminophenol, form a dye upon oxidation (hydrogen peroxide treatment) (214).

### 2.45.2 Physical and Chemical Properties (100)

Molecular weight    184.24
Melting point    73–75°C

### 2.45.3 Toxicity

*p*-Aminodiphenylamine has been reported to be nonteratogenic or embryo toxic when administered by gavage to pregnant Sprague–Dawley rats at dose levels of 50, 100, and 200 mg/kg (279).

### 2.45.4 Hygienic Standards of Permissible Exposure

Standards have not been assigned.

## 2.46 o-Aminoazotoluene [CAS # 97-56-3]

The formula for *o*-aminoazotoluene is $C_{14}H_{15}N_3$; it is also known as 4'-amino-2,3'-dimethylazobenzene, 4-amino-3:2'-azotoluene, and Solvent Yellow.

### 2.46.1 Uses

*o*-Aminoazotoluene is used in the manufacture of pigments and for coloring oils, fats, and waxes such as shoe and other wax polishes. It is not used in foods, drugs, or cosmetics.

### 2.46.2 Physical and Chemical Properties (100)

| | |
|---|---|
| Physical state | Golden crystals |
| Molecular weight | 225.28 |
| Melting point | 101 to 102°C |
| Solubility | Soluble in alcohol, ether, chloroform; practically insoluble in water |

### 2.46.3 Toxicity

There is sufficient evidence for the carcinogenicity in experimental animals, but no adequate data are available with which to evaluate the carcinogenicity in humans (81).

### 2.46.4 Hygienic Standards of Permissible Exposure

Standards have not been assigned.

## 2.47 4-Dimethylaminoazobenzene [CAS # 60-11-7]

The formula for 4-dimethylaminoazobenzene is $C_6H_5NNC_6H_4N(CH_3)_2$; it is also called *N,N*-dimethyl-4-(phenylazo)benzenamine, benzeneazodimethylaniline, Methyl

Yellow, Butter Yellow, N,N-dimethyl-4-phenylazoaniline, and N,N-dimethyl-4-aminoazobenzene.

### 2.47.1 Use

The compound is used as a chemical indicator for free hydrogen chloride and as a pH indicator; it was formerly used to color polishes and other wax products, polystyrene, and soap. It was used in the 1930s and early 1940s for coloring margarine and butter. In the 1940s the consumer was given the option of coloring white margarine at home by mixing a supplemental packet containing Butter Yellow.

### 2.47.2 Physical and Chemical Properties (100)

| | |
|---|---|
| Physical state | Yellow crystalline leaflets |
| Molecular weight | 225.28 |
| Melting point | 114 to 117°C |
| Solubility | Insoluble in water |
| | Soluble in alcohol, chloroform, ether, strong mineral acids, oils |

### 2.47.3 Toxicity

There is sufficient evidence for carcinogenicity in animals—lung tumors and hepatomas in mice, liver tumors in rats, bladder papillomas in dogs, and skin cancer induced by skin painting in mice. An IARC working group reported that there were no data available with which to evaluate the carcinogenicity in humans (81).

### 2.47.4 Hygienic Standards of Permissible Exposure

Standards have not been assigned.

## 2.48 Naphthylamines

### 2.48.1 α-Naphthylamine [CAS # 134-32-7]

The formula for α-naphthylamine is $C_{10}H_7NH_2$; it is also called 1-aminonaphthalene, 1-naphthylamine, 1-naphthalamine, and naphthalidine.

**2.48.1.1 Uses.** α-Naphthylamine is used as a chemical intermediate in manufacture of antioxidants, dyes, and herbicides. It is also used in the synthesis of many chemicals.

**2.48.1.2 Physical and Chemical Properties (100).** The properties of α-naphthylamine are as follows:

| | |
|---|---|
| Physical state | White to reddish needles, darken on oxidation |
| Molecular weight | 143.19 |

Density                 1.123 (25/25°C)
Melting point           50°C
Boiling point           300.8°C
Vapor density           4.93 (air = 1)
Refractive index        1.6703 (51.2°C)
Solubility              Slightly soluble in water; very soluble in alcohol, ether
Flash point             157°F (closed cup)

*2.48.1.3 Toxicity.* 1-Naphthylamine has been thought to cause human cancer, but animal experimentation has not demonstrated carcinogenicity (32a). Radomski (1979) also reviewed the evidence and concluded there was no evidence that 1-naphthylamine causes bladder tumors (109).

A key step in the bioactivation of species-specific carcinogenic arylamines, such as 2-naphthylamine, 4-aminobiphenyl, and 4,4'-diaminobiphenyl, is $N$-oxidation, and it has been demonstrated that 1-naphthylamine undergoes biotransformation to the $N$-oxide to only a limited extent (37). 1-Naphthylamine also does not form methemoglobin nor demonstrates liver toxicity in dogs (172). Documentation of methemoglobin-forming potential was not located in the literature. The compound reportedly did not demonstrate adverse effects on sperm morphology in rodents (206).

Manufacturing processes for the production of 1-naphthylamine can generate 2-naphthylamine as well as other carcinogenic aromatic amines, and the suspected carcinogenicity of 1-naphthylamine in humans may be due not to the pure chemical, but to production impurities (37).

Case et al. (1954) reported an excess of bladder cancer in workers who had been exposed to commercial 1-naphthylamine for 5 or more years, and who had not also been engaged in the production of 2-naphthylamine or benzidine (31). However, commercial 1-naphthylamine made at that time may have contained 4 to 10 percent 2-naphthylamine (32a). Among a cohort of 906 men employed for at least 1 year between 1922 and 1970 in a dyestuffs plant in Italy, Decarli and Peto (1985) reported that a considerable excess of bladder cancer deaths occurred (27 observed, 0.19 expected) among 151 workers involved in the manufacture of 1- and 2-naphthylamine and benzidine (32a). A case-control study of bladder cancer in the United Kingdom (R. W. Boyko et al., 1985) showed a significant exposure-related increased risk for dyestuff workers; even though 1-naphthylamine was of concern, it was not possible to single out any one compound from the combined exposure to arylamines (32a).

On the basis of inadequate animal and human data, the IARC working group has classified 1-naphthylamine as a Group 3 compound, not classifiable as to its carcinogenicity to humans (32a).

ACGIH has not provided an evaluation of this compound (138, 139). NIOSH and OSHA consider 1-naphthylamine to be an occupational carcinogen (40).

## AROMATIC NITRO AND AMINO COMPOUNDS

**2.48.1.4 Hygienic Standards of Permissible Exposure.** Standards have not been assigned.

### 2.48.2 β-Naphthylamine [CAS # 91-59-8]

The formula for β-naphthylamine is $C_{10}H_7NH_2$; it is also known as 2-naphthylamine and 2-aminonaphthalene.

**2.48.2.1 Uses.** β-Naphthylamine is used for research, and only rarely. It was formerly used in the manufacture of dyestuffs and as an antioxidant in the rubber industry.

**2.48.2.2 Physical and Chemical Properties (100).** β-Naphthylamine has the following properties:

| | |
|---|---|
| Physical state | Colorless leaflets that darken on oxidation |
| Molecular weight | 143.19 |
| Density | 1.061 (98/4°C) |
| Melting point | 113°C |
| Boiling point | 306.1°C |
| Refractive index | 1.64927 (98.4°C) |
| Solubility | Soluble in hot water, alcohol, ether, benzene |

**2.48.2.3 Toxicity.** 2-Naphthylamine has been tested for carcinogenicity by oral administration in animals; it induced bladder neoplasms in hamsters, dogs, and nonhuman primates and liver tumors in mice. A low incidence of bladder carcinomas was observed in rats after its oral administration (32a). High dosage levels induced cancer in the urinary bladder of hamsters in about a year, but 2 or more years are required in the dog or the monkey (37).

N-Hydroxylation occurs in monkeys given 2-naphthylamine and results in the excretion of N-hydroxy-2-naphthylamine in the urine as it does in the dog (109). The bioactivation step to N-hydroxy-2-naphthylamine occurs in the liver via the cytochrome P450 system, followed by conjugation with glucuronic acid which forms the proximate carcinogen, or aryl-N-O-glucuronide. The aryl-N-O-glucuronide is then transported by the circulation to its site of action in the bladder, where hydrolysis occurs under acidic conditions (pH 5 to 6), regenerating the corresponding unstable carcinogenic N-hydroxy-2-naphthylamine (107), which degrades to the ultimate carcinogenic species believed to be an electrophilic arylnitrenium ion. Acute poisoning reportedly can result in methemoglobinemia and/or acute hemorrhagic cystitis (139).

Two U.S. epidemiology studies that examined cancer incidence and mortality in a group of chemical workers exposed mainly to 2-naphthylamine demonstrated a significant increase incidence of bladder cancer not explained by smoking habits (280). Two reports on one occupational population at a dyestuffs plant in Italy documented a very high bladder cancer risk linked specifically to 2-naphthylamine production (six observed, 0.04 expected) (281a,b). A recent follow-up analysis of the bladder cancer mortality of the cohort of dyestuff workers exposed to aromatic amines (Turin, Northern Italy) (281b) has been published, and in relation to job category, manufacture of α- or β-naphthylamine or benzidine was associated with the highest risk, followed by fuchsin or safranine T manufacture and by use of or intermittent exposure to naphthylamine or benzidine (281c). Other case reports and studies have appeared that document the relationship between exposure to 2-naphthylamine and increased bladder cancer risk (32a).

According to The IARC working group evaluation, there is sufficient animal and human evidence to classify 2-naphthylamine as a Group 1 agent—carcinogenic to humans (32a, 81). ACGIH has classified this agent as a recognized carcinogen (Alb) (139); NIOSH and OSHA consider the compound to be an occupational carcinogen (40).

*2.48.2.4 Hygienic Standards of Permissible Exposure.* No standards have been assigned.

## 2.49  1,5-Diaminonaphthalene [CAS # 2243-62-1]

The formula for 1,5-diaminonaphthalene is $C_{10}H_8(NH_2)_2$; it is also known as 1,5-naphthalenediamine and 1,5-naphthylenediamine.

### 2.49.1  Uses

1,5-Naphthalenediamine is believed to be used almost exclusively as an intermediate for the manufacture of 1,5-naphthalene diisocyanate, in the production of polyurethane elastomers, and in organic dyes (216).

### 2.49.2  Physical and Chemical Properties (100)

| | |
|---|---|
| Molecular weight | 158.20 |
| Melting point | 190°C |
| Boiling point | Sublimes |
| Solubility | Soluble in alcohol, ether, chloroform |

### 2.49.3  Toxicity

An NCI bioassay (1978) in B6C3F1 mice of both sexes, fed diets containing 1000 or 2000 mg/kg for 103 weeks, demonstrated statistically significant increases in C-cell carcinomas of the thyroid gland in females, neoplasms of the thyroid gland (follicular cell adenomas and papillary adenomas) in males and females, hepatocellular carcinomas in females, and alveolar/bronchiolar adenomas and carcinomas in females. In F344 rats of both sexes fed diets containing 500 or 1000 mg/kg for 103 weeks, a statistically significant increase in the incidence of clitoral gland carcinomas was observed (282). 1,5-Naphthalenediamine is also mutagenic to *Salmonella typhimurium* strain TA100 without metabolic activation (216).

# AROMATIC NITRO AND AMINO COMPOUNDS

Based upon the lack of human data and limited evidence of carcinogenicity in animals, the IARC working group classified 1,5-naphthalenediamine as a Group 3 compound—not classifiable as to its carcinogenicity to humans (32a).

### 2.49.4 Hygienic Standards of Permissible Exposure

Standards have not been assigned.

## 2.50 Toluidines

### 2.50.1 o-Toluidine [CAS # 95-53-4]

The formula for o-toluidine is $CH_3C_6H_4NH_2$; the compound is also known as o-methylaniline, o-aminotoluene, 2-aminotoluene, 2-methylaniline, 1-methyl-2-aminobenzene, o-toluidine, 2-methylbenzenamine, and 1-amino-2-methylbenzene.

*2.50.1.1 Uses.* o-Toluidine and its hydrochloride salt are used primarily as an intermediate in the manufacture of dyes, including azo pigment dyes, triarylmethane dyes, sulfur dyes, and indigo compounds; it is also used as an intermediate for rubber vulcanizing chemicals, pharmaceuticals, and pesticides. The increasing use as a substitute for benzidine dyes may be questionable because it has also been found to be carcinogenic.

*2.50.1.2 Physical and Chemical Properties (100).* o-Toluidine has the following properties:

| | |
|---|---|
| Physical state | Light yellow to reddish brown liquid, darkens on exposure to air and light |
| Molecular weight | 107.16 |
| Density | 0.9984 (20/4°C) |
| Melting point | −14.7°C |
| Boiling point | 200.2°C |
| Refractive index | 1.57276 (20°C) |
| Solubility | Slightly soluble in water; soluble in alcohol, ether |
| Flash point | 85°C (closed cup) |

*2.50.1.3 Toxicity.* The oral $LD_{50}$ in rats is reportedly 900 to 940 mg/kg; that of the hydrochloride salt, diluted in water, in rats is 2951 mg/kg (139). Oral, dermal, or respiratory tract absorption of o-toluidine can result in methemoglobinemia, severe headache, skin and eye irritation, hematuria, and irritation of the kidneys and bladder (147, 139). There are no reported epidemiologic studies on workers who had been exposed only to o-toluidine (139).

There is sufficient evidence for the carcinogenicity of o-toluidine (as the hydrochloride salt) in experimental animals. When administered in the diet, o-toluidine organ-specific carcinogenesis included hepatocellular carcinomas or adenomas in female mice and hemangiosarcomas at multiple sites in male mice of one strain, hemangiosarcomas and hemangiomas of the abdominal viscera in both sexes of another strain, increased incidences of sarcomas of multiple organs in rats of both sexes and mesotheliomas in male rats, and carcinomas of the urinary bladder in female rats. An IARC working group reported that there were not adequate data to evaluate the carcinogenicity of *o*-toluidine hydrochloride in humans. Although an excess of bladder tumors has often been found in workers exposed to varying combinations of dyestuffs, no population of workers exposed to *o*-toluidine alone has been described, and either the data were insufficient, or insufficient follow-up time has prevented a clear association being made with the exposure (81).

An excess number of bladder cancers has recently been reported in workers exposed to *o*-toluidine and aniline; the authors concluded that it is more likely that *o*-toluidine is responsible for the observed excess number of cases of bladder cancer, although aniline may have played a role (283a). Investigations regarding possible biomonitoring methods have recently demonstrated that *o*-toluidine binds to both albumin and hemoglobin and that a linear dose relationship exists for hemoglobin (283b). Additionally, the biologic half-lives for the protein adducts are several times that reported for elimination of *o*-toluidine or its metabolites via the urine, thus providing evidence that these proteins may be valuable biomarkers of exposure to *o*-toluidine in the occupational setting (283b).

*o*-Toluidine is a metabolite of the local anesthetic prilocaine (284). A caution statement appears in the literature concerning prilocaine (Citanest, Astra); it warns only that a metabolite of *o*-toluidine, probably an *N*-hydroxy derivative, may produce methemoglobin particularly when 600 mg or more of prilocaine hydrochloride is administered in adults. The carcinogenicity of *o*-toluidine in animals is not currently mentioned; a correction is to be made in the 1993 edition (285a).

An IARC working group has classified this compound as a Group 2B agent—possibly carcinogenic to humans (32a); NIOSH considers this compound to be an occupational carcinogen (40), and ACGIH has identified the agent as a suspected human carcinogen—A2 classification (139).

*2.50.1.4 Hygienic Standards of Permissible Exposure.* The TLV–TWA value adopted by NIOSH is 2 ppm, or approximately 9 mg/m$^3$ (skin), and by OSHA, 5 ppm or 22 mg/m$^3$ (skin) by (40, 139).

## 2.50.2  4-Chloro-o-toluidine [CAS # 95-69-2]

The formula for 4-chloro-*o*-toluidine is $ClC_6H_3(CH_3)NH_2$; it is also known as *p*-chloro-*o*-toluidine and 4-chloro-2-methylaniline.

*2.50.2.1 Uses.* 4-Chloro-*o*-toluidine is used as an azo coupler in the synthesis of azo dyes used in the textile industry (286) and for the manufacture of the insecticide chlordimeform (285b).

*2.50.2.2 Physical and Chemical Properties (100).* The properties of 4-chloro-*o*-toluidine are as follows:

# AROMATIC NITRO AND AMINO COMPOUNDS

> Molecular weight    141.60
> Melting point    29 to 30°C
> Boiling point    241°C

*2.50.2.3 Toxicity.* No human data regarding carcinogenic risk are available; however, according to the IARC working group, sufficient animal data are available to classify this agent a Group 2B compound and therefore possibly carcinogenic to humans (32a).

For additional background, the reader is referred to a review with 50 references regarding the general toxicity in humans and laboratory animals; it discusses epidemiology investigations concerning a possible link between the incidence of bladder carcinomas and occupational exposure to 4-chloro-*o*-toluidine (285b).

*2.50.2.4 Hygienic Standards of Permissible Exposure.* Standards have not been assigned.

### 2.50.3 5-Chloro-o-toluidine [CAS # 95-79-4]

The formula for 5-chloro-*o*-toluidine is $ClC_6H_3(CH_3)NH_2$; it is also known as 5-chloro-2-methylaniline.

*2.50.3.1 Uses.* 5-Chloro-*o*-toluidine is used as an azo coupler in the synthesis of azo dyes used in the textile industry (286).

*2.50.3.2 Physical and Chemical Properties (100).* The properties of 5-chloro-*o*-toluidine are as follows:

> Molecular weight    141.60
> Melting point    20 to 22°C
> Boiling point    237°C

*2.50.3.3 Toxicity.* 5-Chloro-*o*-toluidine has reportedly been reviewed by an IARC working group; however, the monograph (Volume 48, 1990 per abstract) was not available for this review.

*2.50.3.4 Hygienic Standards of Permissible Exposure.* Standards have not been assigned.

### 2.50.4 6-Chloro-o-toluidine [CAS # 87-63-8]

Information on 6-chloro-*o*-toluidine was not located.

### 2.50.5 5-Nitro-o-toluidine [CAS # 99-55-8]

The formula for 5-nitro-*o*-toluidine is $CH_3C_6H_3(NO_2)NH_2$; it is also known as 2-methyl-5-nitroaniline and 5-nitro-2-methylaniline.

*2.50.5.1 Uses.* 5-Nitro-*o*-toluidine is used as a coupling component in the synthesis of organic textile dyes such as Naphthol Red M. The nitro moiety serves as a chromophore, (in common with other groups such as nitroso, carbonyl, thiocarbonyl, azo, azoxy, azomethine, and ethenyl, in which the double bonds contribute to the absorption of visible light); the amino group serves as an auxochrome (in common with other groups such as alkylamino, dialkylamine, methoxy, or hydroxy), which functions by intensifying or modifying the color (287).

*2.50.5.2 Physical and Chemical Properties (100).* 5-Nitro-*o*-toluidine has the following properties:

| | |
|---|---|
| Molecular weight | 152.15 |
| Melting point | 104 to 107°C |

*2.50.5.3 Toxicity.* When 5-nitro-*o*-toluidine was administered as a dietary feeding study to male and female F344 rats (50 or 100 ppm) and B6C3F1 mice (1200 or 2300 ppm), hepatocellular carcinomas were produced in male and female mice only (288a). According to the IARC working committee, *p*-chloro-*o*-toluidine is a probable carcinogen (288b).

*2.50.5.4 Hygienic Standards of Permissible Exposure.* Standards have not been assigned.

### 2.50.6  m-Toluidine [CAS # 108-44-1]

The formula for *m*-toluidine is $CH_3C_6H_4NH_2$; it is also known as 3-methylbenzenamine, 1-amino-3-methylbenzene, 3-methylaniline, and 3-aminotoluene.

*2.50.6.1 Uses.* The major uses of *m*-toluidine and its hydrochloride are as intermediates in the manufacture of dyes and other chemicals. Production has been limited because its nonplanar configuration, due to steric hindrance, limits its use in direct dyes. It is used in only 12 dyes; none is of major importance (289).

*2.50.6.2 Physical and Chemical Properties (100).* *m*-Toluidine has the following properties:

| | |
|---|---|
| Molecular weight | 107.16 |
| Melting point | 30.4°C |
| Boiling point | 203.3°C |
| Vapor pressure | 1 torr at 41°C |
| Solubility | Soluble in alcohol, ether, acetone, benzene |

*2.50.6.3 Toxicity.* Clinical signs of intoxication in humans include methemoglobinemia and hematuria; it is absorbed orally, dermally, and via the respiratory tract. There are no epidemiologic studies on workers who have been exposed only

# AROMATIC NITRO AND AMINO COMPOUNDS

to *m*-toluidine. The oral $LD_{50}$ of *m*-toluidine (rats) is reportedly 974 mg/kg (139). 200 mg/kg given orally to mice did not inhibit testicular DNA synthesis (290).

In an 18-month carcinogenicity diet evaluation in male CD rats (8000 ppm for 3 months, then 4000 ppm for an additional 15 months; or 16,000 ppm for 3 months, then 8000 ppm for an additional 15 months), and male and female CD-1 mice (16,000 ppm for 5 months, then 4000 ppm in males and 8000 ppm in females for an additional 13 months; or 32,000 ppm in both sexes for 5 months and then 8000 ppm in males and 16,000 ppm in females for additional 13 months), there was no evidence of a significant increase of incidence any kind of tumor in the rats, and only a significant increase in liver tumors in male mice (291). These studies did not demonstrate a dose/response effect. The IARC working groups have not evaluated *m*-toluidine.

*2.50.6.4 Hygienic Standards of Permissible Exposure.* A TLV–TWA value of 2 ppm has been adopted, with a skin notation (139).

## 2.50.7 p-Toluidine [CAS # 106-49-0]

The formula for *p*-toluidine is $CH_3C_6H_4NH_2$; the compound is also called *p*-methylaniline, *p*-aminotoluene, 4-aminotoluene, 4-methylbenzenamine, and 1-amino-4-methylbenzene.

*2.50.7.1 Uses.* *p*-Toluidine and its hydrochloride are used primarily in the synthesis of dyes and in the preparation of ion exchange resins.

*2.50.7.2 Physical and Chemical Properties (100).* *p*-Toluidine has the following properties:

| | |
|---|---|
| Physical state | Leaflets |
| Molecular weight | 107.16 |
| Melting point | 44 to 45°C |
| Boiling point | 200.5°C |
| Refractive index | 1.55324 (59.1°C) |
| Density | 0.9619 (20/4°C), 0.973 (50/50°C) |
| Solubility | 0.74 in water at 21°C; 156 in alcohol at 30°C; soluble in ether, methanol, carbon disulfide |
| Odor and warning properties | Aromatic, winelike odor; burning taste |
| Flash point | 86°C (closed cup) |
| Specific gravity | 1.046 (20°C) |
| Vapor pressure | 1 torr at 42°C |

*2.50.7.3 Toxicity.* Clinical signs of toxicity in humans include anoxic methemoglobinemia and also hematuria. It is absorbed orally, dermally, and via the respiratory tract. There are no epidemiologic studies reported on workers who have been exposed only to *p*-toluidine. The oral $LD_{50}$ of *p*-toluidine is 656 mg/kg in rats and 794 mg/kg in mice; its hydrochloride salt in water was 1285 mg/kg in rats; the $LD_{50}$ (rabbit dermal) is 890 mg/kg (139). *p*-Toluidine has also been reported to be a sensitizer in guinea pigs (292).

In an 18-month *p*-toluidine hydrochloride diet carcinogenicity study, male CD rats (1000 ppm and 2000 ppm for 18 months) did not develop statistically significant increases of tumors; however, CD-1 male and female mice (1000 ppm for 6 months and then 500 ppm for an additional 12 months; or 2000 ppm for 6 months and then 1000 ppm for an additional 12 months) showed significant increases in liver carcinomas, in males in both dose levels and in females in the high dose level (293).

*2.50.7.4 Hygienic Standards of Permissible Exposure.* A TLV–TWA of 2 ppm (approximately 9 mg/m$^3$) has been adopted, and *p*-toluidine is classified as a suspected human carcinogen (139).

## 2.51 Toluenediamines

The six possible toluenediamine isomers, or diaminotoluenes, are all components of the commercial synthetic mixture. Dinitration of toluene produces a mixture of approximately 78 percent 2,4-DNT, 18 percent 2,6-DNT, 3.5 percent 2,3-DNT and 3,4-DNT, 0.5 percent 2,5-DNT, and traces of 3,5-DNT. The most important reaction of toluenediamine is with phosgene to give toluene diisocyanate, or TDI (294). (No information was located on 3,5-DNT.)

### 2.51.1 2,3-Toluenediamine [CAS # 2687-25-4]

The molecular formula for 2,3-toluenediamine is $CH_3C_6H_3(NH_2)_2$; it is also known as 3-methyl-1,2-benzenediamine and 2,3-diaminotoluene.

*2.51.1.1 Uses.* No specific applications were located in the literature.

*2.51.1.2 Physical and Chemical Properties (100).* 2,3-Toluenediamine has the following properties:

| | |
|---|---|
| Molecular weight | 122.17 |
| Boiling point | 255°C |
| Melting point | 63 to 64°C |
| Solubility | Soluble in water, alcohol, ether |

*2.51.1.3 Toxicity.* No toxicology information was located in the literature.

*2.51.1.4 Hygienic Standards of Permissible Exposure.* Standards have not been assigned.

### 2.51.2 2,4-Toluenediamine [CAS # 95-80-7]

The formula is $CH_3C_6H_3(NH_2)_2$; 2,4-toluenediamine is also known as 2,4-diaminotoluene, 2,4-DAT, toluenediamine, TDA, and *m*-toluenediamine.

***2.51.2.1 Uses.*** The major use for TDA is in the manufacture of toluene diisocyanate (TDI), the predominant isocyanate in the flexible polyurethane foams and elastomers industry. It is also used in hair dye formulations.

***2.51.2.2 Physical and Chemical Properties (100).*** The properties of 2,4-toluenediamine are as follows:

| | |
|---|---|
| Molecular weight | 122.17 |
| Melting point | 99°C |
| Boiling point | 148 to 150°C |
| Solubility | Soluble in water, alcohol, ether |

***2.51.2.3 Toxicity.*** Clinical effects in humans include methemoglobinemia, especially when red blood cell-reducing mechanisms are impaired, such as in G6PD deficiency which occurs in humans in the absence of glutathione reductase, glutathione, or glutathione peroxidase. It is an eye irritant that may cause corneal damage (295), and delayed skin irritation reportedly occurs, which can result in blistering.

Reproductive toxicity in the rat has been demonstrated. Reduced fertility, arrested spermatogenesis, and diminished circulating testosterone levels have resulted in rats fed 0.03 percent 2,4-toluenediamine; electron microscopy revealed degenerative changes in Sertoli cells and a decrease in epididymal sperm reserves; after 3 weeks of 0.06 percent TDA feeding, sperm counts were further reduced and accompanied by a dramatic increase in testes weight, intense fluid accumulation, and ultrastructural changes occurred in Sertoli cells (296). In previous studies testicular atrophy, hormonal effects, and aspermatogenesis were also observed in Sprague–Dawley rats given a 0.1 percent diet for 9 weeks (297). However, epidemiology studies of workers exposed to commercial mixtures of dinitrotoluene and/or toluenediamine at three chemical plants indicated that the fertility of men had not been reduced significantly and reported no observable effects on the fertility of workers (298). However, other reports have suggested that human exposure may disrupt spermatogenesis and cause an excess of miscarriages (297).

2,4-Toluenediamine (and 2,6-toluenediamine) is mutagenic in the Ames assay requiring metabolic activation in the presence of S-9; it gave weakly positive results in the micronucleus test; however, this weak effect was detectable only at very toxic doses, and therefore the biologic relevance is questionable. It is interesting that the micronucleus test did not discriminate correctly between the carcinogenic 2,4- and the noncarcinogenic 2,6-toluenediamine (299, 300).

When 2,4-toluenediamine was administered in the diet to male and female F344 rats (79 ppm or 170 ppm) or B6C3F1 mice (100 ppm or 200 ppm), hepatocellular carcinomas were produced in female mice, hepatocellular carcinomas in male rats, and mammary adenomas or carcinomas in female rats, but no carcinomas in the male mice (301). Also male Wistar rats have reportedly been shown to develop liver hepatocarcinomas following treatment with 2,4-toluenediamine (302). A skin painting study in Swiss–Webster mice was reportedly noncarcinogenic (303). However, it has also been observed that mice appear to be less sensitive than rats; this difference may be based upon differences in metabolism (304). Biliary tract cancer has been reported in industrial workers (305).

The IARC working group considers that although there are no human data for evaluation, there are sufficient animal data to classify 2,4-toluenediamine a Group 2B compound—an agent possibly carcinogenic to humans (32a).

*2.51.2.4 Hygienic Standards of Permissible Exposure.* Standards have not been assigned.

### 2.51.3  2,5-Toluenediamine [CAS # 95-70-5]

The formula for 2,5-toluenediamine is $CH_3C_6H_3(NH_2)_2$; it is also known as 2-methyl-1,4-benzenediamine, *p*-toluenediamine, 2,5-diaminotoluene, and 2-methyl-1,4-phenylenediamine.

*2.51.3.1 Uses.* 2,5-Toluenediamine is used primarily in hair-dye formulations.

*2.51.3.2 Physical and Chemical Properties (100).* 2,5-Toluenediamine has the following properties:

| | |
|---|---|
| Molecular weight | 122.17 |
| Boiling point | 273 to 274°C |
| Melting point | 64°C |
| Solubility | Soluble in water, alcohol, ether |

*2.51.3.3 Toxicity.* 2,5-Toluenediamine sulfate (CAS # 6369-59-1) when administered in a diet feeding study to male and female F344 rats (600 or 2000 ppm) and B6C3F1 mice (600 or 1000 ppm) was noncarcinogenic (306). It has also been reported to be "weakly mutagenic" in the Ames test (104).

*2.51.3.4 Hygienic Standards of Permissible Exposure.* Standards have not been assigned.

### 2.51.4  2,6-Toluenediamine [CAS # 823-40-5]

The formula for 2,6-toluenediamine is $CH_3C_6H_3(NH_2)_2$; it is also known as 2-methyl-1,3-benzenediamine and 2,6-diaminotoluene.

*2.51.4.1 Uses.* 2,6-Toluenediamine is used primarily in the synthesis of toluene diisocyanate.

*2.51.4.2 Physical and Chemical Properties (100).* 2,6-Toluenediamine has the following properties:

Molecular weight 122.17
Melting point 106°C
Solubility Soluble in water, alcohol

*2.51.4.3 Toxicity.* Positive in the Ames assay (307), but noncarcinogenic (dihydrochloride) in an NCI feeding study using male and female F344 rats (250 or 500 ppm) and B6C3F1 mice (50 or 100 ppm) (308, 309), 2,6-diaminotoluene is another example of a "false positive" chemical that produces a dose-response mutagenicity and yet is noncarcinogenic (310).

*2.51.4.4 Hygienic Standards of Permissible Exposure.* Standards have not been assigned.

### 2.51.5  3,4-Toluenediamine [CAS # 496-72-0]

The formula for 3,4-toluenediamine is $CH_3C_6H_3(NH_2)_2$; it is also known as 4-methyl-1,2-benzenediamine, 3,4-diaminotoluene, and *o*-toluenediamine.

*2.51.5.1 Uses.* No specific applications were found in the literature.

*2.51.5.2 Physical and Chemical Properties (100).* 3,4-Toluenediamine has the following properties:

Molecular weight 122.17
Melting point 89 to 90°C
Boiling point 265°C (sublimes)
Solubility Soluble in water

*2.51.5.3 Toxicity.* 3,4-Toluenediamine is reportedly nonmutagenic (104). No additional toxicology information was located.

### 2.51.6  3,5-Toluenediamine [CAS # 108-71-4]

No information was located in the literature on this isomer.

## 2.52  Nitrotoluenes

Technical-grade nitrotoluene is a mixture of the three isomers *o*-, *m*-, and *p*-nitrotoluene.

### 2.52.1  o-Nitrotoluene [CAS # 88-72-2]

The formula for *o*-nitrotoluene is $CH_3C_6H_4NO_2$; it is also known as 2-nitrotoluene, 2-NT, methylnitrobenzene, nitrotoluol, and nitrophenylmethane.

*2.52.1.1 Uses.* *o*-Nitrotoluene is used in the synthesis of sulfur and azo dyes, rubber chemicals, agricultural chemicals, and explosives, and as a chemical intermediate.

*2.52.1.2 Physical and Chemical Properties (100).* o-Nitrotoluene has the following properties:

| | |
|---|---|
| Physical state | Yellow liquid |
| Molecular weight | 137.14 |
| Density | 1.163 (20/4°C) |
| Melting point | −9.5°C |
| Boiling point | 221.7°C |
| Vapor pressure | 1.6 mm Hg (60°C) |
| Refractive index | 1.54739 (20.4°C) |
| Solubility | 0.0652 in water at 30°C; soluble in alcohol, ether, benzene, chloroform |

*2.52.1.3 Toxicity.* o-Nitrotoluene produces methemoglobin causing hypoxia, but of a low potency. It is also suspected of causing anemia in chronic exposures (109).

The metabolism of the mononitrotoluenes had not been studied until the 1980s, when interest in the carcinogenicity of the dinitrotoluenes became of concern. Isolated rat hepatocytes converted all three isomers to nitrobenzyl alcohols by cytochrome P450. There was no evidence for the formation of reduced metabolites. 2-Nitrobenzyl alcohol was primarily conjugated with glucuronic acid and 3 percent of the nitrotoluene was oxidized to 2-nitrobenzoic acid (16c). 2-NT has been demonstrated to be biotransformed to o-nitrobenzyl alcohol in vivo.

In an immune suppression study [200, 400, and 600 mg/kg/day administered by gavage to B6C3F1 female mice], 2-NT did not depress IgM antibody-forming cell response to T-dependent sheep erythrocyte antigen, whereas 3-NT and 4-NT caused depression (311).

o-Nitrotoluene has been shown to be the only isomer of mononitrotoluene that binds to hepatic DNA (312). 2-NT is reportedly nonmutagenic in the Ames bacterial assay (102).

Comparative toxicity studies, including genotoxicity, reproductive assessment by continuous breeding, and carcinogenicity studies, are presently being conducted on the ortho, meta, and para isomers under the auspices of NTP (150).

*2.52.1.4 Hygienic Standards of Permissible Exposure.* The adopted TWA values are 2 ppm (11 mg/m$^3$) for the technical-grade isomeric mixture. A caution against cutaneous or mucous membrane exposure is noted (40).

## 2.52.2 m-Nitrotoluene [CAS # 99-08-1]

The formula for *m*-nitrotoluene is $CH_3C_6H_4NO_2$; it is also called 3-nitrotoluene and 3-NT.

*2.52.2.1 Uses.* m-Nitrotoluene is used in the synthesis of sulfur and azo dyes, rubber chemicals, agricultural chemicals, and explosives, and as a chemical intermediate.

*2.52.2.2 Physical and Chemical Properties (100).* m-Nitrotoluene has the following properties:

| | |
|---|---|
| Physical state | Pale yellow liquid |
| Molecular weight | 137.14 |
| Density | 1.157 (20/4°C) |
| Melting point | 15.5°C |
| Boiling point | 232.6°C |
| Vapor pressure | 1.0 mm Hg (60°C) |
| Refractive index | 1.5475 (20°C) |
| Solubility | 0.0498 in water at 30°C; soluble in alcohol, ether, benzene |

*2.52.2.3 Toxicity.* 3-Nitrobenzoic acid has been identified as the major metabolite of 3-nitrotoluene (56 percent of the metabolized substrate), but only 13 percent of the metabolized 3-nitrobenzoic acid was converted to 3-nitrobenzyl glucuronide (16c).

In an immune suppression study [200, 400, and 600 mg/kg/day administered by gavage to B6C3F1 female mice], 3-NT depressed IgM antibody-forming cell response to T-dependent sheep erythrocyte antigen, but demonstrated less immune suppression potential than 4-NT (311).

3-NT is nonmutagenic in the Ames bacterial mutagenicity test (313). 3-NT has been demonstrated to bind to hepatic DNA (312).

Comparative toxicity studies, including genotoxicity, reproductive assessment by continuous breeding, and carcinogenicity studies, are presently being conducted on the ortho, meta, and para isomers under the auspices of NTP (150).

*2.52.2.4 Hygienic Standards of Permissible Exposure.* The adopted TWA value is 2 ppm (11 mg/m$^3$), for the technical-grade isomeric mixture. A caution against cutaneous or mucous membrane exposure is noted (40).

### 2.52.3 p-Nitrotoluene [CAS # 99-99-0]

The formula for *p*-nitrotoluene is $CH_3C_6H_4NO_2$; it is also called 4-nitrotoluene and 4-NT.

*2.52.3.1 Uses.* p-Nitrotoluene is used in the synthesis of sulfur and azo dyes, rubber chemicals, agricultural chemicals, and explosives, and as a chemical intermediate.

*2.52.3.2 Physical and Chemical Properties (100).* p-Nitrotoluene has the following properties:

| | |
|---|---|
| Physical state | Yellowish rhombic needles |
| Molecular weight | 137.14 |
| Density | 1.286 (20°C) |
| Melting point | 54.5°C |

Boiling point 238.3°C
Density of vapor 4.72 (air = 1)
Vapor pressure 1.3 mm Hg (65°C)
Refractive index 1.5346 (62.5°C)
Solubility 0.0442 in water at 30°C; soluble in alcohol, benzene, acetone, chloroform; very soluble in ether
Flash point 106°C (closed cup)

*2.52.3.3 Toxicity.* The primary metabolite formed from 4-nitrotoluene in isolated rat hepatocytes was S-(4-nitrobenzyl)glutathione (68 percent of the metabolized nitrotoluene); smaller percentages of the substrate were metabolized to 4-nitrobenzoic acid (2 percent), 4-nitrobenzyl glucuronide (2 percent), and 4-nitrobenzyl sulfate (4 percent). The formation of S-(4-nitrobenzyl)glutathione was shown to proceed from 4-nitrobenzyl alcohol through the intermediate 4-nitrobenzyl sulfate, a substrate for glutathione S-transferase(s) (16c). Because no nitro reduction could be demonstrated in isolated hepatocytes, it was postulated that nitro group reduction occurred in the intestinal contents after biliary excretion (16c).

In an immune suppression study, 200, 400, and 600 mg/kg/day administered by gavage to B6C3F1 female mice, 4-NT showed greater depression of IgM-antibody forming cell response to T-dependent sheep erythrocyte antigen than 3-NT (311).

4-NT is reportedly negative in the Ames bacterial mutagenicity assay (102). *p*-Nitrotoluene binds to hepatic DNA (312).

Comparative toxicity studies, including genotoxicity, reproductive assessment by continuous breeding, and carcinogenicity studies, are presently being conducted on the ortho, meta, and para isomers under the auspices of NTP (150).

*2.52.3.4 Hygienic Standards of Permissible Exposure.* The adopted TWA value is 2 ppm (11 mg/m$^3$) for the technical-grade isomeric mixture. A caution against cutaneous or mucous membrane exposure is noted (40).

## 2.53 Polynitrotoluenes

The formula for each of the six isomers is $C_6H_3CH_3(NO_2)$; the isomers are 2,4-, 2,6-, 2,3-, 2,5-, 3,4-, and 3,5-dinitrotoluene.

### 2.53.1 Dinitrotoluene Technical Grade [CAS # 25321-14-6]

The commercial or technical grade is a mixture of approximately 76 percent 2,4-DNT, 19 percent 2,6-DNT, and 5 percent of the remaining isomers, which are 2,3-, 2,5-, 3,4-, and 3,5-dinitrotoluene; it is also known as DNT. The isomeric mixture is a combustible, oily liquid. Sufficient DNT may be absorbed through the skin to result in toxic effects (139).

*2.53.1.1 Uses.* The major use of DNT is in the production of toluene diisocyanate, which is primarily used in the production of polyurethane foams and polymers, in the manufacture of explosives, and as a modifier of smokeless powders.

*2.53.1.2 Physical and Chemical Properties (100).* The properties of technical-grade dinitrotoluene are as follows:

| | |
|---|---|
| Physical state | Oily liquid, combustible |
| Specific gravity | 1.3208 at 71°C |
| Molecular weight | 182.14 (isomers) |
| Boiling point | 250°C |
| Vapor pressure | 1 torr at 20°C |
| Flash point | 207°C |

*2.53.1.3 Toxicity.* Human exposure by oral, dermal, or inhalation routes results in hypoxia, dyspnea, and cyanosis produced by formation of methemoglobin. Jaundice and anemia have been reported from chronic exposure. In addition to methemoglobinemia, severe headache, optic neuritis, CNS/respiratory depression, fatigue, vertigo, dyspnea, dizziness, sleepiness, and joint pain have also been reported as a consequence of exposure (147, 139, 314). The rat oral $LD_{50}$ is for males, 568 mg/kg, and for females, 650 mg/kg, indicating a possible male sensitivity. The $LC_{50}$ (rat inhalation) is < 2 mg/l in 1 hr. There is also a species-dependent sensitivity in the order cats > rodents > dogs, and liver toxicity in dogs is greater in males than females (314).

A three-generation reproductive study (Ellis, 1979) in CD rats (34 mg/kg/day in males; 45 mg/kg/day in females) did not demonstrate any treatment-related effects from technical-grade DNT (315). In addition, the technical grade was reported to be nonteratogenic in F344 rats (gavage in corn oil, six dose levels from 14 to 150 mg/kg/day (316).

However, testicular atrophy in rats and aspermatogenesis in dogs has been reported by Lee in 1976 and subsequently also observed in CD rats administered (technical grade) 35 mg/kg/day, 5 days/week for 12 months (317). Feeding or gavage studies of technical-grade DNT, 2,4-DNT, or 2,6-DNT have all induced testicular atrophy, decreased spermatogenesis, or aspermatogenesis in treated rats, mice, and dogs. Nonfunctioning ovaries were also found in mice chronically fed technical-grade DNT (318). It has been suggested that the locus of dinitrotoluene isomer testicular effect is the Sertoli cells, disrupting spermatogenesis. The in vitro toxicity of DNT isomers for Sertoli cells in culture was ranked as follows: 3,4-DNT > 2,3-DNT > 2,4-DNT ≥ 2,6-DNT (319).

A NIOSH study to evaluate effects upon sperm counts in workers potentially exposed to technical-grade DNT or toluenediamine indicated a reduction in count among nine workers; a follow-up study of 84 workers did not reveal any effects upon sperm counts, and no significant differences in infertility rates were observed (320). An additional study of fertility rates of 579 workers exposed to DNT and toluenediamine also did not demonstrate any effects on fertility rates (298c, 321). Technical-grade DNT is considered a human reproductive health hazard in the workplace by NIOSH (322).

Four carcinogenicity studies have been conducted on dinitrotoluenes in rats, the first two (NCI, 2,4-DNT, 1978; U.S. Army, 2,4-DNT, 1979) and the fourth (NCI,

2,6-DNT versus DNT technical mixture, 1987) to be discussed under 2,4- and 2,6-dinitrotoluene, respectively.

The third study by CIIT (1982) involved feeding a representative technical-grade DNT to male and female F344 rats. The representative mixture contained 76.5 percent 2,4-DNT, 18.8 percent 2,6-DNT, 2.43 percent 3,4-DNT, 1.54 percent 2,3-DNT, 0.69 percent 2,5-DNT, and 0.04 percent 3,5-DNT. Male and female rats were fed approximately 3.5, 14, or 34 mg/(kg)/(day). The rats in this study received 5 to 10 times the dose of 2,6-DNT than the rats in the fourth study, although the dose of 2,4-DNT was approximately the same (323a, b). The incidence of hepatocellular carcinomas was 100 percent in males and 55 percent in females in the high dose group; only 11 percent of the male rats fed 3.5 mg/kg/day for 2 years had hepatocellular carcinomas (323c).

The results of this study demonstrated that 2,6-DNT is a complete hepatocarcinogen. In contrast, 2,4-DNT was not hepatocarcinogenic when fed at twice the high dose of 2,6-DNT over the same time period. 2,4-DNT may act as a promoter, but is not capable of initiating carcinogenesis. The hepatocarcinogenicity of technical-grade DNT is mainly due to 2,6-DNT.

Epidemiologic studies of explosives workers expected to have substantial exposure to technical-grade DNT indicated no correlation of human exposure with hepatic cancer; however, an unsuspected excess of mortality from ischemic heart disease was detected in the workers, correlated with exposure to dinitrotoluenes (325).

*2.53.1.4 Hygienic Standards of Permissible Exposure.* The TLV–TWA value is 1.5 mg/m$^3$ for the commercial grade of dinitrotoluene (139).

### 2.53.2  2,4-Dinitrotoluene [CAS # 121-14-2]

The formula of 2,4-dinitrotoluene is $CH_3C_6H_3(NO_2)_2$; it is also known as 1-methyl-2,4-dinitrobenzene, 2,4-dinitro-1-toluene, and dinitrotoluol.

*2.53.2.1 Uses.* 2,4-Dinitrotoluene is used primarily with 2,6-dinitrotoluene in the manufacture of toluene diisocyanate. The reaction sequence involves the nitration of toluene, which is hydrogenated to yield 2,4-diaminotoluene, which is then treated with phosgene to yield toluene 2,4-diisocyanate. The nitration step produces two isomers, 2,4-dinitrotoluene and 2,6-dinitrotoluene, the former predominating. Mixtures of the two isomers are also frequently used. Polyurethane foams are then formed by reacting toluene diisocyanate with glycerol (326).

It is also used for the manufacture of highly explosive plasticizers (which can be formed readily by hand into any desired shape for controlled detonation) (327), as well as dyes (109).

*2.53.2.2 Physical and Chemical Properties (100).* 2,4-Dinitrotoluene has the following properties:

| | |
|---|---|
| Physical state | Yellow or orange crystals |
| Molecular weight | 182.14 |
| Boiling point | 300°C |
| Melting point | 71°C |
| Density | 1.3208 |
| Solubility | Soluble in alcohol, ether, acetone, benzene |

*2.53.2.3 Toxicity.* The various isomers of DNT are considerably less toxic to the mouse, with a possible exception of 2,3-dinitrotoluene, indicating widely differing capacities for metabolism. The oral $LD_{50}$s for 2,4-dinitrotoluene are 268 (rat) versus 1625 (mouse) (139). 2,4-DNT is reportedly a nonsensitizer (315).

2,4-DNT is rapidly absorbed via skin exposure, and repeated dermal applications to rabbits have produced cyanosis, lipemic plasma, depressed hemoglobin and red blood cells, liver hyperplasia and focal necrosis, bone marrow damage, congested spleen, distended bladder, and brain edema; testicular atrophy and aspermatogenesis were reported in beagles (314). Adverse neuromuscular effects, tremors, and brain lesions in dogs following oral administration have also been reported (315).

2,4-Dinitrotoluene is also rapidly absorbed after oral administration. Little reduction of dinitrotoluene occurs in isolated perfused liver preparation, in isolated hepatocytes, or in microsomal preparations incubated under air (16c). The first step in metabolism in male or female Fischer 344 rats is oxidation at the methyl group to yield dinitrobenzyl alcohols, followed by conjugation with glucuronic acid, preparing the alcohol for bile excretion, which occurs to a much greater degree in male than in female rats (16c). Intestinal microflora hydrolyze the glucuronides and reduce one of the nitro groups (to an amino group via nitroso intermediates) (328); the reduced metabolites, aminonitrobenzyl alcohols, are then reabsorbed, whereby, it is postulated, the 2,6 isomer is activated (16c). Mori (1984) found that all six dinitrotoluene isomers were metabolized to aminonitrotoluenes by an *Escherichia coli* isolated from human intestinal contents (329), as well as by the intestinal microflora of rats and mice (328). The human urinary metabolites of 2,4-dinitrotoluene in volunteers exposed to dinitrotoluene are 2,4-dinitrobenzoic acid, 2-amino-4-nitrobenzoic acid, 2,4-dinitrobenzyl glucuronide, and 2-(*N*-acetyl)amino-4-nitrobenzoic acid. The first three of these are those found in rat urine; the last differs in the position of reduction and acetylation. The most abundant metabolites of the dinitrotoluenes in human urine were the dinitrobenzoic acids; in rats the most abundant metabolites were the dinitrobenzyl glucuronides. No metabolites were found in the urine of human volunteers who had been exposed to dinitrotoluene that had undergone reduction in both nitro groups. The appearance of a reduced metabolite of 2,4-dinitrotoluene suggests either that human hepatic enzymes are capable of nitro group reduction of dinitrotoluene or that 2,4-dinitrotoluene (or one of its metabolites) gains access to the intestinal microflora (16c). It was suggested that the biliary excretion/nitro reduction pathway for bioactivation of 2,6-DNT may occur to a lesser extent in humans than in rats, making an important difference in risk assessment (330).

In a dominant lethal mutation study of 2,4-DNT [60, 180, or 240 mg/kg/day for 5 days, Sprague-Dawley male rats by gavage], lethal mutations were not detected, and no changes were observed in the number of preimplantation losses or implantation sites; however, reproductive performance was adversely affected at the 240-mg dose level (331).

Reproductive toxicity evaluation of 2,4-dinitrotoluene in adult male rats fed 0.1 or 0.2 percent DNT for 3 weeks demonstrated a marked change in Sertoli cell morphology following 0.2 percent DNT exposure. Circulating levels of follicle-stimulating hormone (FSH) and luteinizing hormone (LH) were increased in 0.2 percent DNT-treated animals. Raised serum levels of FSH have been reported to be frequently, if not always, associated with Sertoli cell malfunction. Reduced weights of the epididymides and decreased epididymal sperm reserves were also observed. The authors concluded that DNT is capable of inducing testicular injury, of directly or indirectly disturbing pituitary function, and of exerting a toxic effect at the late stages of spermatogenesis. A direct effect on the hypothalamic–hypophyseal axis was not precluded according to the authors, because testosterone concentrations remained within the normal range, whereas LH levels were elevated (332).

Exposure to 2,4-DNT reportedly (Ahrenholz, 1980) decreased human sperm count and increased spontaneous abortions in the workers' wives (technical-grade dinitrotoluene is assumed) (315); however, in a follow-up epidemiology study by Hamill et al. in 1982, no effect on sperm levels was found in workers (320).

The initial carcinogenicity study in 1978 was conducted by NCI and involved a dietary feeding study of 2,4-DNT continuously for 18 months to male and female F344 rats (and B6C3F1 mice) at 0.008 or 0.02 percent (and 0.008 or 0.04 percent, respectively), followed by a 6-month observation period. Because the incidence of hepatic neoplasms in treated animals and control animals was not significantly different, the NCI bioassay was considered negative for hepatocarcinogenesis in both rats and mice. In this study, 2,4-DNT was the primary component (95 percent 2,4-DNT, <5 percent 2,6-DNT) (333). The study, however, was positive for other tumor types, with an increased incidence of fibroma of the skin and subcutaneous tissue in the high- and low-dose male rats and an increased incidence of mammary gland fibroadenoma in high-dose female rats (323c).

The second study (1979), sponsored by the U.S. Army, evaluated 2,4-DNT (containing approximately 2 percent 2,6-DNT) in Sprague–Dawley (CD) rats and mice (CD-1) for 2 years (324). The high doses were toxic to both mice and rats, with the life-span of mice shortened by 50 percent. About half of the high dose and approximately 25 percent of the control male rats died by the end of the 20th month. 2,4-DNT was hepatocarcinogenic, resulting in a 21 percent incidence of hepatocellular carcinomas in male rats of the high-dose (34 mg/kg/day) group dying or killed after 1 year of age. By comparison, high-dose female rats had a 53 percent incidence of hepatocellular carcinomas (324). The reason for the higher incidence of hepatocellular carcinoma in female rats compared with male rats in the Army study, the inverse of that observed in the CIIT study, is not clear; differences in strain of rat, the isomeric composition of the DNT, or other unspecified differences in protocols may be the basis for the differences in sex response between these two studies (323c).

The results of the third study by CIIT (discussed above under technical-grade DNT) demonstrated that 2,6-DNT is a complete hepatocarcinogen; in contrast, 2,4-DNT was not hepatocarcinogenic when fed at twice the high dose of 2,6-DNT over the same time period. The authors concluded that 2,4-DNT may act as a promoter, but is not capable of initiating carcinogenesis. The hepatocarcinogenicity of technical-grade DNT is mainly due to 2,6-DNT.

In an analytically controlled comparative bioassay conducted by NCI, discussed in detail below, under 2,6-dinitrotoluene, 2,4-dinitrotoluene was not found to be carcinogenic in male F344 rats (324); the lack of hepatocarcinogenicity of 2,4-DNT was consistent with the CIIT study conclusion that 2,4-DNT is not a hepatocarcinogen. The difference between these studies and the Army study could simply be due to strain differences, the greater amount of 2,6-DNT contaminating the 2,4-DNT used, the slightly higher dose of 2,4-DNT used (34 versus 27 mg), or the extended duration of feeding (2 years versus 1 year) (324). The previously reported positive response by 2,4-dinitrotoluene in the Ames assay is unclear (102), and may be the result of contamination by 2,6-DNT.

IARC, NIOSH, OSHA, and ACGIH have not classified this isomer as a potential carcinogenic risk.

*2.53.2.4 Hygienic Standards of Permissible Exposure.* The TWA–TLV for technical-grade dinitrotoluene has been established at 1.5 mg/m$^3$ (NIOSH, OSHA) with a skin warning; NIOSH considers this mixture to be an occupational carcinogen (40).

### *2.53.3 2,6-Dinitrotoluene [CAS # 606-20-2]*

The formula for 2,6-dinitrotoluene is $CH_3C_6H_3(NO_2)_2$; it is also known as 1-methyl-2,6-dinitrobenzene or dinitrotoluol.

*2.53.3.1 Uses.* 2,6-Dinitrotoluene is used primarily in the manufacture of toluene diisocyanate.

*2.53.3.2 Physical and Chemical Properties (100).* 2,6-Dinitrotoluene has the following properties:

| | |
|---|---|
| Physical state | Yellow rhombic crystals |
| Molecular weight | 182.14 |
| Melting point | 64 to 66°C |
| Density | 1.2833 |
| Solubility | Soluble in alcohol |

*2.53.3.3 Toxicity.* In vivo metabolic evaluation in F344 rats identified four metabolites: 2,6-dinitrobenzyl alcohol glucuronide (which is excreted in the bile and to the intestine, where it is hydrolyzed by bacteria to aminonitrobenzyl alcohol), 2,6-

dinitrobenzoic acid, and 2-amino-6-nitrobenzoic acid (315). 2,6-Dinitrobenzoic acid and 2,6-dinitrobenzyl glucuronide were the metabolites found in human urine from volunteers exposed to dinitrotoluene. These were also found in rat urine, but in addition rat urine contained 2-amino-6-nitrobenzoic acid (16c). Two metabolites of 2-amino-6-nitrobenzyl alcohol were also identified and quantified in microsomal incubations. These were 2-hydroxylamino-6-nitrobenzyl alcohol and 2-amino-5-hydroxy-6-nitrobenzyl alcohol (16c). Although a sulfate conjugate of 2-hydroxylamino-6-nitrobenzyl alcohol has not been isolated, it has been implicated in the final activation step of 2,6-dinitrotoluene through in vitro tests. This sulfate would be expected to be sufficiently unstable to decompose to an electrophilic nitrenium ion, which could then react with DNA (16c). It has also been demonstrated that 2,6-DNT is the only DNT isomer that binds to rat liver DNA in vitro (312).

Although metabolism and elimination by rats appear to be qualitatively similar to that in humans, epidemiologic studies of explosives workers have not shown a correlation of human exposure with hepatic cancer. However, a correlation of dinitrotoluene exposure with cardiac disease in workers has been documented (16c).

In order to evaluate the initiating versus promoting activity of DNT isomers and to test the hypothesis that 2,6-DNT is a complete hepatocarcinogen at the doses fed to rats in the CIIT study, a fourth study was conducted by NCI. A representative technical-grade DNT was prepared by mixing purified DNT isomers in a ratio that was representative of a standard technical grade, with a final composition of 76.5 percent 2,4-DNT, 18.8 percent 2,6-DNT, 2.43 percent 3,4-DNT, 1.54 percent 2,3-DNT, 0.69 percent 2,5-DNT, and 0.04 percent 3,5-DNT. Male (F344/CrlBR) rats were fed the following doses for 1 year: 2,6-DNT, 7 or 14 mg/kg/day; 2,4-DNT (containing 0.5 percent 2,6-DNT), 27 mg/kg/day; and technical-grade DNT, 35 mg/kg/day. 2.6-DNT induced hepatocellular carcinomas in twice as many animals as did technical-grade DNT. No liver carcinomas were found in F344 rats administered 2,4-DNT (324).

It is believed that 2,6-DNT had acted as an initiator, and 2,4-DNT as a promoter; however, neither 2,6- nor 2,4-dinitrotoluene is positive in the strain A mouse lung tumor assay, a bioassay for carcinogenicity that has proven useful for other compounds. Reasons for this are not clear, for the 2,4- and 2,6-dinitrotoluene metabolites found in the urine and formed in isolated hepatocyte or hepatic microsomes are similar between the two species (16c).

Adverse effects of 2,6-DNT upon male reproductive Sertoli cells in rodents has been demonstrated as discussed under technical-grade DNT. However, an IARC working group, OSHA, or ACGIH has yet to classify this isomer as to its potential carcinogenic risk.

*2.53.3.4 Hygienic Standards of Permissible Exposure.* The TWA–TLV for technical-grade dinitrotoluene has been established at 1.5 mg/m$^3$ (NIOSH, OSHA) with a caution against cutaneous or mucous membrane exposure; NIOSH considers this mixture to be an occupational carcinogen (40).

## 2.53.4  2,3-Dinitrotoluene [CAS # 602-01-7]

No information was located on this isomer.

## 2.53.5  2,5-Dinitrotoluene [CAS # 619-15-8]

No information was located on this isomer.

## 2.53.6  3,4-Dinitrotoluene [CAS # 610-39-9]

No information was located on this isomer.

## 2.53.7  3,5-Dinitrotoluene [CAS # 618-85-9]

No information was located on this isomer.

## 2.53.8  2,4,6-Trinitrotoluene [CAS # 118-96-7]

The formula for 2,4,6-trinitrotoluene is $C_6H_2CH_3(NO_2)_3$; it is commonly known as TNT, and also called trinitrotoluol, α-2,4,6-trinitrotoluene, 1-methyl-2,4,6-trinitrobenzene, methyltrinitrobenzene, and *sym*-trinitrotoluene. Five unsymmetrical isomers exist; these include 2,3,4-, 2,4,5-, 2,3,5-, 2,3,6-, and 3,4,5-trinitrotoluene (122). Numerous other compounds are formed during the manufacture of TNT including unsymmetrical isomers of TNT, tetranitromethane, nitrobenzoic acid, nitrocresol, and partially nitrated toluenes. Unless these impurities are removed, the TNT may be unstable at elevated temperatures (334a).

*2.53.8.1 Uses.* Trinitrotoluene (TNT) has been used extensively in explosives since 1902, and its use increased greatly during World War I, when severe toxicity from TNT was first noted (334b). It is relatively insensitive to shock and must be exploded by a detonator; it can therefore be poured into shells and allowed to solidify. TNT is a fire hazard, but detonates only when shocked or heated to its explosive temperature.

*2.53.8.2 Physical and Chemical Properties (100).* The properties of TNT are as follows:

| | |
|---|---|
| Physical state | Monoclinic prisms, crystals colorless to pale yellow |
| Molecular weight | 227.13 |
| Specific gravity | 1.654 (20°C) |
| Melting point | 80.75°C |
| Boiling point | 240°C |
| Vapor pressure | 0.046 mm Hg (82°C) |
| Solubility | 0.02 in water at 15°C, 0.15 in hot water; soluble in ether, benzene chloroform, carbon tetrachloride, toluene, acetone |
| Explosive temperature | 475°C |

*2.53.8.3 Toxicity.* Trinitrotoluene, or TNT, has been the most commonly used explosive in the munitions industry since World War I. During World War I and World War II, many deaths were reported in TNT workers from aplastic anemia and toxic hepatitis. By the end of 1915 more than 50 cases of fatal toxic hepatitis had been seen in England (334b). During World War II numerous cases of severe hepatitis and anemia were seen again with several fatalities (334b). The nature of toxic effects of TNT are well described in many articles published between the early 1920s and the early 1950s (109). Since World War II there have been only occasional reports of deaths due to TNT poisoning and there have been very few reports in the English literature of any problems related to TNT use in the past 20 years. In 1972 Goodwin reported only reversible liver damage occurring in a few percentage of workers in a shell loading plant in a period of over 20 years (109). Following World War II, in addition to liver and hematologic effects reported earlier, reports from the Soviet Union and other eastern European countries described cataracts and gastric irritation, as well as function-type symptoms related to the nervous system (334b). Lenticular opacities may develop slowly over several years when exposures regularly exceed 1.0 mg/m$^3$. Although they may occur in a few individuals as the first indication of TNT intoxication, most individuals who develop opacities have other symptoms of TNT poisoning (334b).

Statistically significant rises in serum glutamic oxalacetic transaminase and lactic dehydrogenase determinations occurred at exposures to TNT of 0.8 mg/m$^3$ and persisted at exposures of 0.6 mg/m$^3$ (109). Based on these findings, the adequacy of the previous threshold limit value for TNT (1.5 mg/m$^3$) was questioned and has subsequently been lowered. Subclinical effects of TNT exposure appear to be dose related, with red blood cell destruction first apparent at exposures exceeding 0.5 mg/m$^3$. At exposures exceeding 1.0 mg/m$^3$ substantial shortening of red blood cell survival occurs, although significant anemia is unlikely because of an effective response by the bone marrow manifested by markedly increased reticulocytosis. At levels of TNT exposure between 0.5 and 1.0 mg/m$^3$ elevations of liver function enzymes may be seen particularly in new employees or those recently exposed to higher levels of TNT (334b).

Exposure may cause irritation of the eyes, nose, throat, and skin. It may stain the skin, hair, and nails a yellowish color. Numerous fatalities have occurred in workers exposed to TNT. Clinical symptoms range from headaches, gastritis, and toxic hepatitis to bone marrow depression. TNT syndrome can include methemoglobinemia with cyanosis, weakness, drowsiness, dyspnea and unconsciousness, muscular pains, heart irregularities, renal irritation, cataracts, menstrual irregularities, and peripheral neuritis (122). Individuals with G6PD deficiency (see Section 2.55 below) can be at greater risk for hemolytic episodes. A report by Russian investigators in 1972 stated "One half of 360 people exposed occupationally to trinitrotoluene for several years developed some degree of cataracts" (109). In addition to aplastic anemia, pancytopenia can also occur from exposure where all three cell lines are affected, resulting in bone marrow depression [red cell, white cells (agranulocytosis), and platelets (thrombocytopenia)] (153).

A recent study reported that 200 mg/kg/day of TNT administered by gavage to Wistar rats for 6 weeks resulted in increased serum zinc levels owing to a sustained decrease in testicular zinc concentration. It was further noted that zinc concentration in hair samples from TNT male workers was higher than in control workers, and that premature ejaculation and impotence were greater in TNT-exposed male workers (335).

There are surprisingly few data on the metabolism of trinitrotoluenes by mammalian species. Rabbits given oral doses (about 150 mg/kg) of 2,4,6-trinitrotoluene excreted 2,6-dinitro-4-hydroxylaminotoluene, 2,6-dinitro-4-aminotoluene, and 2,4-dinitro-6-aminotoluene in the urine. The urine of rats fed 2,4,6-trinitrotoluene in the diet contained 2,4-diamino-6-nitrotoluene and 3-amino-5-nitroaniline as the major metabolites according to the workers in the 1940s, but more recent work has suggested that the major metabolites are 2,4-dinitro-6-aminotoluene and 2,6-dinitro-4-aminotoluene. Small amounts of 2,4-diamino-6-nitrotoluene were also found (16c).

*2.53.8.4 Hygienic Standards of Permissible Exposure.* The adopted TWA value is 0.5 mg/m$^3$, including a warning concerning skin exposure (139).

## 2.54 Ethylene Glycol Dinitrate [CAS # 628-96-6]

The formula is $O_2NOCH_2CH_2ONO_2$; the compound is also known as 1,2-ethanediol dinitrate, glycol dinitrate, ethylene dinitrate, 1,2-dinitroethane, nitroglycol, and EGDN.

### 2.54.1  Uses

EGDN is an explosive; however, it is primarily used to lower the melting point of nitroglycerin and reduces the hazard associated with the use of frozen dynamite; together these compounds are the major constituents of commercial dynamite, cordite, and blasting gelatin. Occupational exposures usually involves the mixture of the two, and they are considered together. EGDN is approximately 160 times more volatile than nitroglycerin; the usual mixtures in dynamite are 60 to 80 percent EGDN and 20 to 40 percent nitroglycerin (334a).

### 2.54.2  Physical and Chemical Properties (100)

| | |
|---|---|
| Physical state | Yellow liquid |
| Molecular weight | 152.06 |
| Melting point | $-22.3°C$ |
| Boiling point | 197 to 200°C |
| Density | 1.4918 |
| Solubility | Insoluble in water; soluble in alcohol, ether |
| Explosive temperature | 114°C |
| Vapor pressure | 0.05 torr at 20°C |

### 2.54.3  Toxicity

Absorption may occur readily through the lungs, skin, and digestive tract. Because higher absorption may result from skin exposure than from inhalation, measuring the concentration in the air of workplaces may not reflect the actual exposure level. It may cause severe, throbbing headache with small exposures, and nausea, palpitation, vomiting, angina, cyanosis, coma, respiratory arrest, and death with heavier exposures. Vertigo, paresthesias, and EEG changes have also been reported (129b). A temporary tolerance to the headache may develop, but this is lost after a few days without exposure. It is also slightly irritating to the eyes.

Vasodilatation occurs as a consequence of metabolic intermediates, causing a fall in blood pressure and acceleration of heart rate (129b). Cardiac arrhythmias were recently recognized as a potential occupational disease due to exposure (336). On some occasions a worker may have anginal pains a few days after discontinuing repeated daily exposure (129b).

Human exposure threshold levels for lowered blood pressure and headache among volunteers exposed to a mixture of nitroglycerin and EGDN showed a fall in blood pressure and marked headache at 2 $mg/m^3$ EGDN. A mean concentration of 0.7 $mg/m^3$ for 25 min will produce lowered blood pressure and slight headache (337). Other estimates of EGDN responsible for headaches and hypotension in workers reportedly range from 0.1 to 14 $mg/m^3$.

A single case of death of a dynamite worker who was exposed to concentrations of EGDN:nitroglycerin of between 0.3 and 1.4 $mg/m^3$ has been reported (338).

Ethylene glycol dinitrate is biotransformed into inorganic nitrate and ethylene glycol mononitrate (339). It has also been demonstrated that inorganic nitrite is initially produced and metabolized to nitrate in rat blood at 37°C (340).

Protective clothing should be worn to prevent skin exposure, and adequate ventilation provided.

### 2.54.4 Hygienic Standards of Permissible Exposure

The adopted TLV–TWA standard is 0.05 ppm (0.31 $mg/m^3$); caution against cutaneous or mucous membrane exposure is noted (138).

## 2.55 Nitroglycerin [CAS # 55-63-0]

The formula is $C_3H_5(ONO_2)_3$; nitroglycerin is also known as glyceryl trinitrate, nitroglycerol, trinitroglycerol, glonoin, trinitrin, blasting gelatin, blasting oil, trinitroglycerin, 1,2,3-propanetriol trinitrate, NG, and GTN.

### 2.55.1 Uses

Nitroglycerin and ethylene glycol dinitrate are the major constituents of dynamite; GTN is also used to make guncotton, cordite or other smokeless powders, and blasting gelatin. It is also employed in rocket propellants. It is a pharmaceutical agent used in treatment of angina. Occupational exposure usually involves the mixture of the two compounds and they are considered together.

## 2.55.2 Physical and Chemical Properties (100)

| | |
|---|---|
| Physical state | Pale yellow triclinic or rhombic crystals or pale yellow oily liquid |
| Molecular weight | 227.09 |
| Melting point | 13°C (stable form) |
| Boiling point | 256°C (explodes) |
| Density | 1.5931 |
| Solubility | Slightly soluble in water; soluble in alcohol, benzene; very soluble in ether, acetone |

## 2.55.3 Toxicity

Nitroglycerin is a potent vasodilator, and the route of absorption in the workplace is both respiratory and percutaneous. Nitroglycerin may cause severe throbbing headaches in workers with small exposures. With larger exposures there may be nausea, vomiting, cyanosis, coma, and death. The face and other extremities of dynamite workers can turn blue in color as a consequence of excessive exposure. A temporary tolerance to the headache may develop, but this is lost after a few days without exposure. On some occasions a worker may have anginal pains a few days after discontinuing repeated daily exposure. The mechanism associated with anginal pains is apparently due to decrease in myocardial oxygen consumption. Although GTN reflexly increases heart rate and myocardial contractility, which in turn increase myocardial oxygen consumption, the reduction in ventricular wall tension results in a net decrease in myocardial oxygen consumption. By decreasing myocardial oxygen consumption, GTN alters the imbalance of myocardial oxygen supply and consumption, which is thought to cause angina pectoris.

In addition to vascular smooth muscle, GTN relaxes bronchial, biliary, gastrointestinal, ureteral, and uterine smooth muscles, and is a functional antagonist of norepinephrine, acetylcholine, and histamine.

It has been proposed that organic nitrates enter the smooth muscle cells where they undergo metabolic activation to nitric oxide via catalytic facilitation by sulfhydryl compounds. The nitric oxide so formed can combine with sulfhydryl groups to form an active intermediate called S-nitrosothiol, which can then react with the enzyme guanylate cyclase to produce cyclic guanosine monophosphate and vasodilatation. It has also been suggested that nitric oxide itself can react directly with guanylate cyclase. Tolerance to nitrates is thought to arise from intracellular depletion of the sulfhydryl donors that are necessary for the metabolic activation step (341). It has been documented that nitric oxide free radical, as well as other free radicals and reactive species of oxygen (probably hydroxyl free radical) can activate guanylate cyclase. Guanylate cyclase is the only enzyme known to date whose activity is increased with free radicals (342).

Nitrovasodilators are classified as endothelium independent because vasodilatation occurs with and without the presence of endothelium. Acetylcholine, histamine, some hormones, ADP, ATP, hydralazine, serotonin, norepinephrine, epinephrine, xanthine oxidase, and guanosine triphosphate are classified as endothelium-dependent vascular smooth muscle relaxants. Both classes appear to stimulate vascular relaxation through activation of guanylate cyclase and accumulation of cyclic guanosine monophosphate (343a).

Endothelium-dependent vasodilators are thought to activate phospholipase A2 and cause the release of an unsaturated fatty acid. The released unsaturated fatty acid, or a metabolite, is thought to be the endothelial relaxant factor that interacts with the smooth muscle component to cause vascular relaxation. An oxidized fatty acid or a free radical (such as nitric oxide) has also been shown to activate smooth muscle guanylate cyclase and increase cGMP levels. Unsaturated fatty acids, lipid peroxides, hydroperoxides, and the nitric oxide free radical of nitrovasodilators can activate guanylate cyclase. The resulting increase in cGMP activates cGMP-dependent protein kinase (342).

Organic nitrate exposure headache is apparently the consequence of nitric oxide free-radical metabolism, which triggers a vasodilation response. Nitrovasodilator compounds are thought to mediate smooth muscle relaxation by activating guanylate cyclase and increasing intracellular concentrations of cGMP, which is associated with smooth muscle relaxation.

Withdrawal of nitroglycerin has caused sudden death in industrial workers who were exposed all week on the job and then from the chemical on the weekends. They were adapted and thus depended on NG to maintain a minimum level of coronary flow (343b). A recent case involved a 34-year-old munitions worker who suffered a cardiac arrest on a Sunday morning, which was ruled a withdrawal syndrome from occupational exposure to nitroglycerin (344); a previously reported death of a dynamite worker reportedly exposed to concentrations of nitroglycerin:EGDN between 0.3 and 1.4 mg/m$^3$ occurred in 1963 (338).

The pharmacokinetics of glyceryl trinitrate are still not completely understood (345). The major route of nitroglycerin degradation by liver enzyme was reported to be denitration, with formation of glyceryl dinitrate and inorganic nitrite in the presence of reduced GSH (346). It has also been demonstrated that nitroglycerin decreases ATP as well as GSH. The decrease in ATP is believed to be the result of the inhibitory effect of nitroglycerin on mitochondrial phosphorylation (347). Metabolism involves the utilization of GSH, and hence chronic exposure potentially lowers GSH levels (348). Instead of resulting in mercapturic acids, one molecule of nitroglycerin reacts with two GSH molecules, GSH is oxidized to GSSG, and nitroglycerin is reduced to the corresponding 1,3-glyceryl dinitrate, 1,2-glyceryl dinitrate, glyceryl mononitrate, glycerol, and inorganic nitrite, which is rapidly metabolized to nitrate ion; complete degradation of nitroglycerin to glycerol has not been demonstrated (349). In vitro rabbit liver homogenates have also demonstrated that nitroglycerin undergoes a spontaneous or nonenzymatic reaction with reduced glutathione in forming inorganic nitrite and oxidized glutathione (350). The antianginal mechanism(s) of action has been demonstrated to depend on the intact molecule of nitroglycerin, not on the formation of nitrate ion (351).

### 2.55.4 Hygienic Standards of Permissible Exposure

The adopted TLV–TWA value is 0.05 ppm (approximately 0.5 mg/m$^3$) for nitroglycerin, ethylene glycol dinitrate, or any mixtures of the two. A caution against cutaneous or mucous membrane exposure is noted (139).

## REFERENCES

1. (a) R. R. Bronaugh and H. I. Maibach, "Percutaneous Penetration of Nitroaromatic Compounds: Studies In Vivo and In Vitro in Animals and Humans, in *Toxicity of Nitroaromatic Compounds*, D. E. Rickert, Ed., Hemisphere, New York, 1985, pp. 141–148. (b) D. R. Hartter, "The Use and Importance of Nitroaromatic Chemicals in the Chemical Industry," in D. E. Rickert, *op. cit.*, pp. 1–13.
2. C. D. Klassen, "Absorption, Distribution and Excretion of Toxicants," in *Toxicology, The Basic Science of Poisons*, 2nd ed., J. Doull, C. D. Klassen, and M. O. Amdur, Eds., Macmillan, New York, 1980, pp. 28–56.
3. (a) W. W. Cox and W. B. Wendel, *J. Biol. Chem.*, **143**, 331 (1942), through R. R. Beard et al., "Aromatic Nitro and Amino Compounds," Chapter 34 in *Patty's Industrial Hygiene and Toxicology*, G. D. Clayton and F. E. Clayton, Eds., 3rd ed., Wiley, New York, 1981, pp. 2413–2489. (b) O. Bodansky, *Pharmacol. Rev.*, **3**, 144 (1951), through R. R. Beard et al., *op. cit.* (c) R. R. Beard et al., *op. cit.*
4. B. G. Reddy, I. R. Pohl, and G. Krishna, "The Requirement of the Gut Flora in Nitrobenzene-induced Methemoglobinemia in Rats," *Biochem. Pharmacol.*, **25**, 1119–1122 (1976).
5. M. Kiese, *Methemoglobinemia; A Comprehensive Treatise*, Chemical Rubber Co. Press, Cleveland, OH, 1974.
6. S. Charache, "Methemoglobinemia—Sleuthing for a New Cause," *New Engl. J. Med.*, **314**, 776–778 (1986).
7. E. M. Scott, I. W. Duncan et al., "The Reduced Pyridine Nucleotide Dehydrogenases of Human Erythrocytes," *J. Biol. Chem.*, **240**, 481–485 (1965), through R. P. Smith, "Toxic Responses of the Blood," Chapter 8 in *Casarett and Doull's Toxicology*, 3rd ed., C. D. Klaassen, M. O. Amdur, and J. Doull, Eds., Macmillan, New York, 1986, pp. 223–244.
8. P. W. Straub, *Proc. Eur. Soc. Study Drug Toxicol.*, **1973**, 116–124.
9. R. P. Smith, "Toxic Responses of the Blood," Chapter 8 in *Casarett and Doull's Toxicology*, C. D. Klaassen, M. O. Amdur, and J. Doull, Eds., 3rd ed., Macmillan, New York, 1986, pp. 223–244.
10. M. D. Sass, C. J. Caruso, and M. Farhangi, *Lab. Clin. Med.*, **70**, 760–767 (1976), through R. P. Smith, "Toxic Responses of the Blood," Chapter 8 in *Casarett and Doull's Toxicology*, 3rd ed., C. D. Klaassen, M. O. Amdur, and J. Doull, Eds., Macmillan, New York, 1986, pp. 223–244.
11. R. E. Gosselin, R. P. Smith, and H. C. Hodge, "Aniline," *Clinical Toxicology of Commercial Products*, Williams and Wilkins, Baltimore, 1984, pp. III-31 to III-36.
12. Center for Disease Control, "Morbidity and Mortality Reports, Methemoglobinemia

Due To Occupational Exposure to Dinitrobenzene—Ohio," *Arch. Dermatol.*, **124**, 1171–1172 (1988).
13. D. V. Parke, "Studies in Detoxication. 68. The Metabolism of (14C)Nitrobenzene in the Rabbit and Guinea Pig," *Biochem. J.*, **62**(2), 339–346 (1956).
14. D. E. Rickert, J. A. Bond, R. M. Long, and J. P. Chism, "Metabolism and Excretion of Nitrobenzene by Rats and Mice," *Toxicol. Appl. Pharmacol.*, **67**(2), 206–214 (1983).
15. L. M. Tremaine, G. L. Diamond, and A. J. Quebbemann, "In vivo Quantification of Renal Glucuronide and Sulfate Conjugation of 1-Naphthol and *p*-Nitrophenol in the Rat," *Biochem. Pharmacol.*, **33**, 419–427 (1984).
16. (*a*) D. D. Nystrom and D. E. Rickert, "Metabolism and Excretion of Dinitrobenzene by Male Fischer-344 Rats," *Drug Metab. Dispos.*, **15**, 821–825 (1987). (*b*) D. S. Hewick, "Reductive Metabolism of Nitrogen-Containing Functional Groups," W. B. Jakoby, J. R. Bend, and J. Caldwell, Eds., *Metabolic Basis of Detoxication, Metabolism of Functional Groups*, Academic Press, New York, 1982, pp. 151–170. (*c*) D. E. Rickert, "Metabolism of Nitroaromatic Compounds," *Drug Metab. Rev.*, **18**, 23–53 (1987).
17. M. J. Turner, Jr., R. J. Levine, D. D. Nystrom, Y. S. Crume, and D. E. Rickert, "Identification and Quantification of Urinary Metabolites of Dinitrotoluenes in Occupationally Exposed Humans," *Toxicol. Appl. Pharmacol.*, **80**, 166–174 (1985).
18. D. Guest, S. R. Schnell, D. E. Rickert, and J. G. Dent, "Metabolism of 2,4-Dinitrotoluene by Intestinal Microorganisms from Rat, Mouse, and Man," *Toxicol. Appl. Pharmacol.*, **64**, 160–168 (1982).
19. J. Kao, J. Faulkner, and J. W. Bridges, "Metabolism of Aniline in Rats, Pigs, and Sheep," *Drug Metab. Dispos.*, **6**, 549–555 (1978).
20. S. McLean et al., "Methaemoglobin Formation by Aromatic Amines," *Pharm. Pharmacol.*, **21**, 441–450 (1969).
21. C. Mate, A. J. Ryan, and S. E. Wright, "Metabolism of Some 4-Nitro-aniline Derivatives in the Rat," *Food Cosmet. Toxicol.*, **5**, 657–663 (1967).
22. K. C. Morton, K. M. King, and K. P. Baetcke, "Metabolism of Benzidine and Subsequent Nucleic Acid Binding and Mutagenicity," *Cancer Res.*, **39**, 3107–3113 (1979).
23. K. C. Morton, Y. W. Ching, and C. D. Garner, "Carcinogenicity of Benzidine, *N,N'*-Diacetylbenzidine and *N*-Hydroxy-*N,N'*-diacetylbenzidine for Female CD Rats," *Carcinogenesis*, **2**, 747–752 (1981).
24. R. P. Bos, R. M. E. Brouns, R. VanDoorn, J. L. G. Theuws, and P. T. Henderson, "The Appearance of Mutagens in Urine of Rats After Administration of Benzidine and Some Other Aromatic Amines," *Toxicology*, **21**, 223–233 (1981).
25. R. K. Lynn, C. T. Garvie-Gould, D. F. Milam, K. F. Scott, C. F. Eastman, A. M. Ilias, and R. M. Rodgers, "Disposition of the Aromatic Amine, Benzidine, in the Rat: Characterization of Mutagenic Urinary and Biliary Metabolites," *Toxicol. Appl. Pharmacol.*, **72**, 1–14 (1984).
26. R. K. Lynn, D. W. Donielson, A. M. Ilias, J. M. Kennish, K. Wong, and H. B. Matthews, "Metabolism of bis-Azobiphenyl Dyes Derived from Benzidine, 3,3'-Dimethylbenzidine or 3,3'-Dimethoxybenzidine to Carcinogenic Aromatic Amines in the Dog and Rat," *Toxicol. Appl. Pharmacol.*, **56**, 248–258 (1980).
27. E. Boyland, C. H. Kinder, and D. Manson, "Synthesis and Detection of Di(2-amino-10-naphthyl) Hydrogen Phosphate, a Metabolite of 2-Naphthylamine in the Dog," *Biochem. J.*, **78**, 175–188 (1961).

28. E. Boyland and D. Manson, "The Biochemistry of Aromatic Amines. 2-Formamido-1-naphthyl Hydrogen Sulfate, a Metabolite of 2-Naphthylamine," *Biochem. J.*, **99**, 189–199 (1966).

29. E. Brill, "The Role of Dog Bladder Mucosa in the N-oxidation of Arylamines," *Res. Commun. Chem. Pathol. Pharmacol.*, **16**, 73–84 (1977).

30. J. L. Radomski, G. M. Conzelman, Jr., A. A. Rey, and E. Brill, "*N*-Oxidation of Certain Aromatic Amines, Acetamides and Nitro Compounds by Monkeys and Dogs," *J. Natl. Cancer Inst.*, **50**, 989–995 (1973).

31. International Agency for Research on Cancer, "IARC Monographs on the Evaluation of Carcinogenic Risk of Chemicals to Man: Some Aromatic Amines, Hydrazine and Related Substances, *N*-Nitroso Compounds and Miscellaneous Alkylating Agents," Vol. 4, 1974, pp. 97–111.

32. (*a*) International Agency for Research on Cancer, Monographs on the Evaluation of Carcinogenic Risks to Humans, "*Overall Evaluations of Carcinogenicity: An Updating of IARC Monographs Volumes 1 to 42*," Suppl. 7, 1987. (*b*) U.S. Dept. of Health and Human Services, "Sixth Annual Report on Carcinogens: Summary," U.S. Public Health Service, 1991.

33. (*a*) D. C. McMillan et al., "Evaluation of Propanil and its *N*-Oxidized Derivatives for Genotoxicity in the *Salmonella typhimurium* Reversion, Chinese Hamster Ovary/hypoxanthine Guanine Phosphoribosyl Transferase, and Rat Hepatocyte/DNA Repair Assays," *Fundam. Appl. Toxicol.*, **11**, 429–439 (1988). (*b*) J. W. Cramer et al., *J. Biol. Chem.*, **235**, 885–888 (1960). (*c*) E. C. Miller, *Cancer Res.*, **38**, 1479–1496 (1978).

34. J. L. Radomski, "The Primary Aromatic Amines: Their Biological Properties and Structure–Activity Relationships," *Ann. Rev. Pharmacol. Toxicol.*, **19**, 129–157 (1979).

35. P. D. Lotlikar et al., "Role of Glutathione (GSH) and GSH S-Transferases in Conjugation of Reactive Metabolites of Chemical Carcinogens," *Indian J. Biochem. Biophys.*, **24**, 36–43 (1987).

36. L. A. Poirier and J. H. Weisburger, "Enzymic Reduction of Carcinogenic Aromatic Nitro Compounds by Rat and Mouse Liver Fraction," *Biochem. Pharmacol.*, **23**(3), 661–669 (1974), through R. R. Beard et al., "Aromatic Nitro and Amino Compounds," Chapter 34 in *Patty's Industrial Hygiene and Toxicology*, 3rd ed. G. D. Clayton and F. E. Clayton, Eds., Wiley, New York, 1981, pp. 2413–2489.

37. In G. M. Williams and J. H. Weisburger, "Chemical Carcinogens," Chapter 5 in *Casarett & Doull's Toxicology*, 3rd ed., C. D. Klaassen, M. O. Amdur, and J. Doull, Eds., Macmillan, New York, 1986.

38. E. F. Stula, H. Sherman, J. A. Zapp, Jr., and J. W. Clayton, Jr., "Experimental Neoplasia in Rats from Oral Administration of 3,3'-Dichlorobenzidine, 4,4'-Methylene-bis(2-chloroaniline), and 4,4'-Methylene-bis(2-methylaniline)," *Toxicol. Appl. Pharmacol.*, **1**, 159–176 (1975).

39. American Conference of Governmental Industrial Hygienists, "Threshold Limit Values for Chemical Substances and Physical Agents in the Work Environment and Biological Exposure Indices with Intended Changes for 1990–1991," Cincinnati, OH.

40. *NIOSH Pocket Guide to Chemical Hazards*, U.S. Department of Health and Human Services (Public Health Service, Centers for Disease Control, National Institute for Occupational Safety and Health), DHHS (NIOSH) Publication No. 90-117, June, 1990.

41. NIOSH, "Metabolic Precursors of a Known Carcinogen, beta-Naphthylamine," NIOSH

Current Intelligence Bulletin 16, U.S. Department of Health, Education and Welfare, 1976.
42. NIOSH, "Dinitrotoluenes," NIOSH Current Intelligence Bulletin 44, U.S. Department of Health and Human Services, 1985.
43. U.S. Environmental Protection Agency, "Health Advisory for 2,4- and 2,6-Dinitrotoluene (DNT)," U.S. Department of Health and Human Services, 1992.
44. (*a*) H. R. Imbus, "A Review of Regulatory Risk Assessment with Formaldehyde as an Example," *Reg. Toxicol. Pharmacol.*, **8**, 356–366 (1988). (*b*) D. J. Paustenbach, "Important Recent Advances in the Practice of Health Risk Assessment: Implications for the 1990s," *Reg. Toxicol. Pharmacol.*, **10**, 204–243 (1989). (*c*) C. R. Cothern, "Some Scientific Judgements in the Assessment of the Risk of Environmental Contaminants," *Toxicol. Ind. Health*, **5**, 479–491 (1989).
45. R. A. M. Case, M. E. Hosker, D. B. McDonald, and J. T. Pearson, "Tumors of the Urinary Bladder in Workmen Engaged in the Manufacture and Use of Certain Dyestuff Intermediates in the British Chemical Industry. I. The Role of Aniline, Benzidine, alpha-Naphthylamine and beta-Naphthylamine," *Brit. J. Ind. Med.*, **11**, 75 (1954).
46. L. J. Goldwater, A. J. Rosso, and M. Kleinfeld, "Bladder Tumors in a Coal-tar Dye Plant," *Arch. Environ. Health*, **11**, 814 (1965).
47. M. R. Zavon, M. B. Mattammal, H. J. Armbrecht, and B. B. Davis, "Benzidine Exposure as a Cause of Bladder Tumors," *Arch. Environ. Health*, **27**, 1–7 (1973).
48. T. F. Mancuso and A. A. El-Attar, "Cohort Studies of Workers Exposed to beta-Naphthylamine and Benzidine," *J. Occup. Med.*, **9**, 277–285 (1966).
49. E. Ward, W. Halperin, M. Thun, H. B. Grossman, B. Fink, L. Koss, A. M. Osorio, and P. Schulte, "Bladder Tumors in Two Young Males Occupationally Exposed to MBOCA," *Am. J. Ind. Med.*, **14**, 267–272 (1988).
50. J. A. Bond, J. P. Chism, D. E. Rickert, and J. A. Popp, "Induction of Hepatic and Testicular Lesions in Fisher-344 Rats by Single Oral Doses of Nitrobenzene," *Fundam. Appl. Toxicol.*, **1**, 389–394 (1981).
51. D. E. Dodd, E. H. Fowler, W. M. Snellings, I. M. Pritts, R. W. Tyl, J. P. Lyon, F. O. O'Neal, and G. Kimmerle, "Reproduction and Fertility Evaluations in CD Rats Following Nitrobenzene Inhalation," *Fundam. Appl. Toxicol.*, **8**, 493–505 (1987).
52. A. A. Levin, T. Bosakowski, L. L. Earle, and B. E. Butterworth, "The Reversibility of Nitrobenzene-Induced Testicular Toxicity: Continuous Monitoring of Sperm Output from Vasocystotomized Rats," *Toxicology*, **30**, 219–230 (1988).
53. T. E. Cody, S. Witherup, L. Hastings, K. Stemmer, and R. T. Christian, "1,3-Dinitrobenzene: Toxic Effect in Vivo and in Vitro," *Toxicol. Environ. Health*, **7**, 829–847 (1981).
54. R. E. Linder, R. A. Hess, and L. F. Strader, "Testicular Toxicity and Infertility in Male Rats Treated with 1,3-Dinitrobenzene," *J. Toxicol. Environ. Health*, **19**, 477–489 (1986).
55. R. A. Hess, R. E. Linder, L. F. Strader, and S. D. Perreault, "Acute Effects and Long-term Sequelae of 1,3-Dinitrobenzene (48 mg/kg) on the Rat. II. Quantitative and Qualitative Histopathology of the Testis," *J. Androl.*, **9**, 327–342 (1988).
56. D. P. Evenson, F. C. Janca, R. K. Baer, L. K. Jost, and D. S. Karabinus, "Effect of 1,3-Dinitrobenzene on Prepubertal, Pubertal and Adult Mouse Spermatogenesis," *J. Toxicol. Environ. Health*, **28**, 67–89 (1989).

57. D. M. Blackburn, "A Comparison of the Effects of the Three Isomers of Dinitrobenzene on the Testis in the Rat," *Toxicol. Appl. Pharmacol.*, **92**, 54–64 (1988).
58. M. F. Obasaju, D. F. Katz, and M. G. Miller, "Species Differences in Susceptibility to 1,3-Dinitrobenzene-Induced Testicular Toxicity and Methemoglobinemia," *Fundam. Appl. Toxicol.*, **16**, 257–266 (1991).
59. S. F. McEuen and M. G. Miller, "Metabolism and Pharmacokinetics of 1,3-Dinitrobenzene in the Rat and Hamster," *Drug. Metab. Dispos.*, **3**, 661–666 (1991).
60. D. A. Cave and P. M. D. Foster, "Modulation of *m*-Dinitrobenzene and *m*-Dinitrosobenzene Toxicity in Rat Sertoli-germ Cell Cocultures," *Fundam. Appl. Toxicol.*, **14**, 199–207 (1990).
61. J. C. Topham, "The Detection of Carcinogen-Induced Sperm Head Abnormalities in Mice," *Mutat. Res.*, **69**, 149–155 (1980).
62. C. C. Lee, H. V. Ellis, J. C. Dacre, and J. P. Glennon, "Subchronic and Chronic Toxicity Studies of 2,4-Dinitrotoluene. Part II. CD Rats," *J. Am. Coll. Toxicol.*, **4**, 243–256 (1985).
63. H. V. Ellis III, C. B. Hong, C. C. Lee, J. C. Dacre, and J. P. Glennon, "Subchronic and Chronic Toxicity Studies of 2,4-Dinitrotoluene," *J. Am. Coll. Toxicol.*, **4**, 233–242 (1985).
64. S. C. J. Reader and P. M. D. Fosterm, "The in Vitro Effects of Four Isomers of Dinitrotoluene on Rat Sertoli and Sertoli-germ Cell Cocultures: Germ Cell Detachment and Lactate and Pyruvate Production," *Toxicol. Appl. Pharmacol.*, **106**, 287–294 (1990).
65. E. Bloch, B. Gondos, M. Gatz, S. K. Varma, and B. Thysen, "Reproductive Toxicity of 2,4-Dinitrotoluene in the Rat," *Toxicol. Appl. Pharmacol.*, **94**, 466–472 (1988).
66. C. J. Price, R. W. Tyl, T. A. Marks, L. L. Paschke, T. A. Redoux, and J. R. Reel, "Teratologic and Postnatal Evaluation of Aniline Hydrochloride in the Fischer 344 Rat," *Toxicol. Appl. Pharmacol.*, **77**, 465–478 (1985).
67. J. V. Rutkowski and V. H. Ferm, "Comparison of the Teratogenic Effects of the Isomeric Forms of Aminophenol in the Syrian Golden Hamster," *Toxicol. Appl. Pharmacol.*, **63**, 264–269 (1982).
68. S. C. Reader, C. Shingles, and M. D. Stonard, "Acute Testicular Toxicity of 1,3-Dinitrobenzene and Ethylene Glycol Monomethyl Ether in the Rat: Evaluation of Biochemical Effect Markers and Hormonal Responses," *Fundam. Appl. Toxicol.*, **16**, 61–70 (1991).
69. M. Alexander and B. K. Lustigman, "Effect of Chemical Structure on Microbial Degradation of Substituted Benzenes," *J. Agric. Food Chem.*, **14**, 410–413 (1966).
70. L. E. Hallas and M. Alexander, "Microbial Transformation of Nitroaromatic Compounds in Sewage Effluent," *Appl. Environ. Microbiol.*, **45**, 1234–1241 (1983).
71. H. H. Tabak, W. C. Chambers, and P. W. Kabler, "Microbial Metabolism of Aromatic Compounds. I. Decomposition of Phenolic Compounds and Aromatic Hydrocarbons by Phenol Adapted Bacteria," *J. Bacteriol.*, **87**, 910–919 (1964).
72. N. G. McCormick, F. E. Feeherry, and H. S. Levinson, "Microbial Transformation of 2,4,6-Trinitrotoluene and Other Nitroaromatic Compounds," *Appl. Environ. Microbiol.*, **31**, 949–958 (1976).
73. H. C. Bailey and R. J. Spanggord, "The Relationship Between the Toxicity and Structure of Nitroaromatic Chemicals," *Proc. Aquatic Toxicology and Hazard As-*

*sessment, 6th Symposium*, W. E. Bishop, R. D. Cardwell, and B. B. Heidolph, Eds., American Society for Testing and Materials (ASTM) STP, Philadelphia, 1983, pp. 98–107.

74. C. Hermens and A. J. Leeuwangh, "Joint Toxicity of Mixtures of 8 and 24 Chemicals to the Guppy (Poecilla reticulata)," *Ecotoxicol. Environ. Safety*, **6**, 302–310 (1981).

75. G. P. Fitzgerald, G. C. Gerloff, and F. Skoog, "Studies on Chemicals with Selective Toxicity to Blue-green Algae," *Sewage Ind. Wastes*, **24**, 888–896 (1952).

76. J. W. Deneer, T. L. Sinnige, W. Seinen, and J. L. M. Hermens, "Quantitative Structure–Activity Relationships for the Toxicity and Bioconcentration Factor of Nitrobenzene Derivatives Towards the Guppy (Poecilia reticulata)," *Aquat. Toxicol.*, **10**, 115–129 (1987).

77. R. J. Spanggord, B. W. Gibson, R. G. Keck, D. W. Thomas, and J. J. Barkley, Jr., "Effluent Analysis of Waste Water Generated in the Manufacture of 2,4,6-Trinitrotoluene. I. Characterization Study," *Environ. Sci. Technol.*, **16**, 229–232 (1982).

78. D. H. W. Liu, H. C. Bailey, and J. G. Pearson, "Toxicity of a Complex Munitions Wastewater to Aquatic Organisms," *Proc. Aquatic Toxicology and Hazard Assessment, 6th Symposium*, W. E. Bishop, R. D. Cardwell, and B. B. Heidolph, Eds., American Society for Testing and Materials (ASTM) STP, Philadelphia, 1983, pp. 135–150.

79. W. H. van der Schalie, T. R. Shedd, and M. G. Zeeman, "Ventilatory and Movement Responses of Bluegills Exposed to 1,3,5-Trinitrobenzene," *Aquatic Toxicology and Hazard Assessment*, W. J. Adams, G. A. Chapman, and W. G. Landis, Eds., American Society for Testing and Materials (ASTM) STP, Philadelphia, 1988, pp. 307–315.

80. R. R. Beard and J. T. Noe, "Aromatic Nitro and Amino Compounds," Chapter 34 in *Patty's Industrial Hygiene and Toxicology*, 3rd ed., Vol. 2A, G. D. Clayton and F. E. Clayton, Eds., Wiley, New York, 1981, pp. 2413–2489.

81. *Sixth Annual Report on Carcinogens, Summary 1991*, U.S. Department of Health and Human Services, Public Health Service, pp. 79–80.

82. R. Oyasu et al., *J. Natl. Cancer Inst.*, **50**, 503–506 (1973).

83. B. Lonbardi et al., *Int. J. Cancer*, **23**, 565–570 (1979).

84. H. A. Dunsford et al., *Am. J. Pathol.*, **118**, 218–224 (1985).

85. E. J. Calabrese, Ed., *Principles of Animal Extrapolation*, Wiley, New York, 1983.

86. M. R. Juchau et al., "Redox Cycling As a Mechanism For Chemical Teratogenesis," *Environ. Health Perspect.*, **70**, 131–136 (1986).

87. J. V. Rutkowski and V. H. Ferm, "Comparison of the Teratogenic Effects of the Isomeric Forms of Aminophenol in the Syrian Golden Hamster," *Toxicol. Appl. Pharmacol.*, **63**, 264–269 (1982).

88. R. Feinland et al., "Hair Preparations," in H. F. Mark, D. F. Othmer, C. G. Overberger, and G. T. Seaborg, Eds., *Kirk–Othmer Encyclopedia of Chemical Technology*, 3rd ed., Vol. 12, Wiley, New York, 1980, pp. 80–117.

89. (*a*) T. A. Re et al., "Results of Teratogenicity Testing of *m*-Aminophenol in Sprague–Dawley Rats," *Fundam. Appl. Toxicol.*, **4**, 98–104 (1984). (*b*) In T. C. Daniels and W. D. Kumler, "Metabolic Changes of Drugs and Related Organic Compounds in the Body (Detoxication)," Chapter 3 in *Organic Chemistry in Pharmacy*, C. O. Wilson and O. Gisvold, Eds., Lippincott, Philadelphia, 1949, p. 44.

90. R. J. Flower, S. Moncada, and J. R. Vane, "Analgesic-Antipyretics and Anti-Inflammatory Agents; Drugs Employed in the Treatment of Gout," Chapter 29 in

Goodman and Gilman's *The Pharmacological Basis of Therapeutics*, 7th ed., A. Goodman Gilman, L. S. Goodman, T. W. Rall, and F. Murad, Eds., Macmillan, New York, 1985, pp. 692–694.

91. Joseph P. Remington, *The Practice of Pharmacy, A Treatise*, 5th ed., Lippincott, Philadelphia, 1907, p. 737.

92. *The National Formulary X*, American Pharmaceutical Association, Lippincott, Philadelphia, 1955, p. 13.

93. (*a*) Robert E. Willett, "Analgesic Agents," Chapter 17 in *Wilson and Gisvold's Textbook of Organic Medicinal and Pharmaceutical Chemistry*, 8th ed., Robert F. Doerge, Ed., Lippincott, 1982, pp. 645–648. (*b*) S. M. Sicardi et al., "Mutagenic and Analgesic Activities of Aniline Derivatives," *J. Pharm. Sci.*, **80**, 761–764 (1991).

94. F. Bernheim et al., "The Action of *p*-aminophenol on Certain Tissue Oxidations," *J. Pharmacol. Exp. Ther.*, **61**, 311–320 (1937).

95. C. M. Burnett, T. A. Re, S. Rodriguez, R. F. Leoehr, and W. E. Dressler, "The Toxicity of *p*-aminophenol in the Sprague–Dawley Rat: Effects on Growth, Reproduction and Fetal Development," *Food Chem. Toxicol.*, **27**, 691–698 (1989).

96. Z. L. Hegedus and U. Nayak, "*para*-Aminophenol and Structurally Related Compounds as Intermediates in Lipofuscin Formation and in Renal and Other Tissue Toxicities," *Arch. Int. Physiol. Biochim. Biophys.*, **99**, 99–105 (1991).

97. K. P. Gartland, F. W. Bonner, J. A. Timbrell, and J. K. Nicholson, "Biochemical Characterization of *para*-Aminophenol-induced Nephrotoxic Lesions in the F344 Rat," *Arch. Toxicol.*, **63**, 97–106 (1989).

98. J. F. Newton, M. Yoshimoto, J. Bernstein, G. F. Rush, and J. B. Hook, "Acetaminophen Nephrotoxicity in the Rat. II. Strain Differences in Nephrotoxicity and Metabolism of *p*-Aminophenol, a Metabolite of Acetaminophen," *Toxicol. Appl. Pharmacol.*, **69**, 307–318 (1983).

99. P. C. Jurs et al., *J. Med. Chem.*, **22**, 476 483 (1979).

100. (*a*) M. Windholz, Ed., *The Merck Index*, 10th ed., Merck, Rahway, NJ, 1983. (*b*) Robert C. Weast, Ed., *CRC Handbook of Chemistry and Physics*, 69th ed., CRC Press, Boca Raton, FL, 1988–1989. (*c*) Through Reference 109. (*d*) Reference 189. (*e*) Reference 138. (*f*) Reference 139.

101. R. P. Smith et al., "Chemically Induced Methemoglobinemias in the Mouse," *Biochem. Pharmacol.*, **16**, 317–328 (1967).

102. C. W. Chiu et al., *Mutat. Res.*, **58**, 11–22 (1978).

103. B. Testa and P. Jenner, Eds., *Drug Metabolism: Chemical and Biochemical Aspects*, Marcel Dekker, New York, 1976, p. 125.

104. M. M. Shahin, *Mutat. Res.*, **22**, 165–180 (1989).

105. L. A. Reinke and M. J. Moyer, "A Microsomal Oxidation which is Highly Inducible by Ethanol," *Drug Metab. Dispos.*, **13**, 548–552 (1984).

106. (*a*) J. P. Gorski et al., *J. Biol. Chem.*, **252**, 1336–1343 (1977). (*b*) K. W. Bock et al., *Drug Metab. Dispos.*, **12**, 93–97 (1984).

107. I. G. Sipes and A. J. Gandolfi, "Biotransformation of Toxicants," Chapter 4 in *Casarett and Doull's Toxicology, The Basic Science of Poisons*, 3rd ed., C. D. Klaassen, M. O. Amdur, and J. Doull, Eds., Macmillan, New York, 1986, pp. 64–98.

108. National Toxicology Program, "Review of Current DHHS, DOE, and EPA Research

Related to Toxicology, Fiscal year 1991," U.S. Department of Health and Human Services, June, 1991.
109. Cited in R. R. Beard and J. T. Noe, "Aromatic Nitro and Amino Compounds," Chapter 34 in *Patty's Industrial Hygiene and Toxicology*, 3rd ed., Vol. 2A, G. D. Clayton and F. E. Clayton, Eds., Wiley, New York, 1981, pp. 2413–2489.
110. R. D. Kimbrough, K. R. Mahaffey, P. Grandjean, S.-H. Sandoe, and D. D. Rutstein, Eds., *Clinical Effects of Environmental Chemicals*, Hemisphere, New York, 1989.
111. T. L. Kurt et al., *Vet. Hum. Toxicol.*, **28**, 574–575 (1986).
112. W. D. Horner, *Arch. Ophthalmol.*, **27**, 1097–1121 (1942), through A. M. Potts, "Toxic Responses of the Eye," Chapter 17 in *Casarett and Doull's Toxicology, The Basic Science of Poisons*, 3rd ed., C. D. Klaassen, M. O. Amdur, and J. Doull, Eds., Macmillan, New York, 1986, pp. 478–515.
113. R. Tecwyn Williams, Ed., *Detoxication Mechanisms*, Wiley, New York, 1947, pp. 131–136.
114. A. M. Potts, "Toxic Responses of the Eye," Chapter 17 in *Casarett and Doull's Toxicology, The Basic Science of Poisons*, 3rd ed., C. D. Klaassen, M. O. Amdur, and J. Doull, Eds., Macmillan, New York, 1986, pp. 478–515.
115. B. Issekutz, Jr., *Arch. Int. Pharmacodyn. Ther.*, **272**, 310–319 (1984).
116. *Merck's Index, An Encyclopedia For the Chemist, Pharmacist and Physician*, 4th ed., Merck, Rahway, NJ, 1930.
117. O. Gisvold, "Phenols and Their Derivatives," Chapter 13 in *Organic Chemistry in Pharmacy*, C. O. Wilson and O. Gisvold, Eds., Lippincott, Philadelphia, 1949, pp. 147–172.
118. E. F. Cook and E. W. Martin, Eds., *Remington's Practice of Pharmacy*, 10th ed., Mack Publishing, Easton, PA, 1951, p. 724.
119. L. Schwartz, *J. Am. Med. Assoc.*, **125**, 186–191 (1944).
120. H. Nishimoto et al., *Int. Arch. Allergy Appl. Immunol.*, **95**, 221–230 (1991).
121. W. M. Grant, Ed., *Toxicology of the Eye*, 2nd ed., Charles C Thomas, Springfield, IL, 1974, pp. 832–833.
122. "Occupational Disease, A Guide to their Recognition," NIOSH/CDC DHEW/PUB/NIOSH77-181, 1977.
123. C. W. Chiu et al., *Mutat. Res.*, **58**, 11–22 (1978).
124. H. C. Spencer et al., *J. Ind. Hyg. Toxicol.*, **30**, 10–28 (1948).
125. A. Froslie, *Acta Vet. Scand. Suppl.*, **49**, 1–61 (1974).
126. R. L. Baran, *Arch. Dermatol.*, **110**, 467 (1974).
127. T. L. Kurt et al., "Dinitrophenol in Weight Loss: the Poison Center and Public Health Safety," *Vet. Hum. Toxicol.*, **28**, 574–575 (1986).
128. (*a*) D. G. Harvey et al., *Brit. J. Med.*, **2**, 13–15 (1951). (*b*) R. D. Kimbrough et al., Eds. *Clinical Effects of Environmental Chemicals*, Hemisphere, New York, 1989.
129. (*a*) J. B. Cavanagh, *Arch. Pathol. Lab. Med.*, **103**, 659–664 (1979). (*b*) W. K. Anger et al., "Chemicals Affecting Behavior," in *Neurotoxicity of Industrial and Commercial Chemicals*, Vol. I, J. O'Donoghue, Ed., CRC Press, Cleveland, OH, 1985.
130. M. J. Kland, "Teratogenicity of Pesticides and Other Environmental Pollutants," in *Teratogens*, Vera Kolb Meyers, Ed., Elsevier, New York, 1988, p. 402.
131. S. D. Murphy, "Toxic Effects of Pesticides," Chapter 18 in *Casarett and Doull's*

*Toxicology, The Basic Science of Poisons*, 3rd ed., C. D. Klaassen, M. O. Amdur, and J. Doull, Eds., Macmillan, New York, 1986, pp. 555–556.

132. J. B. Cavanagh, *Arch. Path. Lab. Med.*, **103**, 659–664 (1979), abstract.
133. R. L. Sandridge and H. B. Staley, "Amines by Reduction," *Kirk-Othmer Encyclopedia of Chemical Technology*, 3rd ed., Vol. 2, M. F. Herman, D. F. Othmer, C. G. Overberger, and G. T. Seaborg, Eds., Wiley, New York, 1978, pp. 355–376.
134. K. H. Jacobson, "Acute Oral Toxicity of Mono- and Di-alkyl Ring-substituted Derivatives of Aniline," *Toxicol. Appl. Pharmacol.*, **22**, 153–154 (1972).
135. C. J. Price et al., "Teratologic and Postnatal Evaluation of Aniline Hydrochloride in the Fischer 344 Rat," *Toxicol. Appl. Pharmacol.*, **77**, 465–478 (1985).
136. D. G. Goodman et al., *J. Natl. Cancer Inst.*, **73**, 265–270 (1984).
137. E. Ward et al., "Excess Number of Bladder Cancers in Workers Exposed to *o*-Toluidine and Aniline," *J. Natl. Cancer Inst.*, **83**, 501–506 (1991).
138. ACGIH, *Documentation Of The Threshold Limit Values and Biological Exposure Indices*, 6th ed., American Conference of Governmental Industrial Hygienists, Inc., 1991.
139. ACGIH, *Documentation Of The Threshold Limit Values and Biological Exposure Indices*, 5th ed., American Conference of Governmental Industrial Hygienists, Inc., 1986.
140. H. F. Smyth, Jr., et al., "Range-finding Toxicity Data: List VI," *Am. Ind. Hyg. Assoc. J.*, **23**, 95–107 (1962).
141. (*a*) S. Hamill and D. Y. Cooper, "The Role of Cytochrome P-450 in The Dual Pathways Of *N*-Demethylation of *N,N'*-Dimethylaniline By Hepatic Microsomes," *Xenobiotica*, **14**, 139–149 (1984). (*b*) L. L. Poulsen and D. M. Ziegler, *J. Biol. Chem.*, **254**, 6449–6455 (1979).
142. B. D. Hardin et al., *Teratogen, Carcinogen, Mutagen*, **7**, 29–48 (1987).
143. K. M. Abdo, M. P. Jokinen, and R. Hiles, "Subchronic (13-Week) Toxicity Studies of *N,N'*-Dimethylaniline Administered to Fischer 344 Rats and B6C3F1 Mice," *J. Toxicol. Environ. Health*, **29**, 77–88 (1990).
144. NTP Technical Report 360, NIH Publication 89-2815, 1990, abstract.
145. (*a*) M. M. Shahin, *Mutat. Res.*, **221**, 165–180 (1989). (*b*) C. W. Chiu et al., *Mutat. Res.*, **58**, 11–22 (1978).
146. EPA/OTS Doc. 40-8576371, abstract.
147. N. H. Proctor, J. P. Hughes, M. L. Fischman, J. P., Eds., *Chemical Hazards of the Workplace*, 2nd ed., Lippincott, Philadelphia, 1988.
148. (*a*) R. S. Nair, C. S. Auletta, et al., "Chronic Toxicity, Oncogenic Potential, and Reproductive Toxicity of *p*-Nitroaniline in Rats," *Fundam. Appl. Toxicol.*, **15**, 607–621 (1990). (*b*) R. S. Nair, F. R. Johannsen, and R. E. Schroeder, "Evaluation of Teratogenic Potential of Para-Nitroaniline and Para-Nitrobenzene in Rats and Rabbits," Chapter 5 in *Toxicity of Nitroaromatic Compounds*, D. E. Rickert, Ed., Hemisphere, New York, 1985, pp. 61–85.
149. G. O. Rankin et al., "In vivo and in vitro Nephrotoxicity of Aniline and its Monochlorophenyl Derivatives in the Fischer 344 Rat," *Toxicology*, **38**, 269–284 (1986).
150. National Toxicology Program, Fiscal Year 1991 Annual Plan, Department of Health and Human Services, June, 1991.

151. B. Testa and P. Jenner, "Inhibitors of Cytochrome P-450s and Their Mechanism of Action," *Drug Metab. Rev.*, **12**(1), 1–117 (1981).
152. R. C. James et al., *Biochem. Pharmacol.*, **24**, 835–838 (1975).
153. R. P. Smith, "Toxic Responses of the Blood," Chapter 8 in *Casarett and Doull's Toxicology, The Basic Science of Poisons*, 3rd ed., C. D. Klaassen, M. O. Amdur, and J. Doull, Eds., Macmillan, New York, 1986, pp. 223–244.
154. J. R. Mitchell et al., *Gastroenterology*, **68**, 392–410 (1975).
155. W. J. Ehlhardt and J. J. Howbert, "Metabolism and Disposition of p-Chloroaniline in Rat, Mouse and Monkey," *Drug Metab. Dispos.*, **19**, 366–369 (1991).
156. J. Ashby and R. W. Tennant, "Chemical Structure, Salmonella Mutagenicity and Extent of Carcinogenicity as Indicators of Genotoxic Carcinogenesis Among 222 Chemicals Tested in Rodents by the U.S. NCI/NTP," *Mutat. Res.*, **204**, 17–115 (1988).
157. R. S. Chhabra et al., "Carcinogenicity of *p*-Chloroaniline in Rats and Mice," *Food Chem. Toxicol.*, **29**, 119–124 (1991).
158. S. D. Singleton and S. D. Murphy, "Propanil (3,4-dichloropropionanilide)-induced Methemoglobin Formation in Mice in Relation to Acylamidase Activity," *Toxicol. Appl. Pharmacol.*, **25**, 20–29 (1973).
159. S. D. Murphy, "Toxic Effects of Pesticides," in *Casarett and Doull's Toxicology, The Basic Science of Poisons*, 3rd ed., C. D. Klaassen, M. O. Amdur, and J. Doull, Eds., Macmillan, New York, 1986, pp. 519–581.
160. (*a*) EPA/OTS Doc. 878221308, abstract. (*b*) EPA/OTS Doc. 878221298, abstract.
161. J. P. Tindall, "Chloracne and Chloracnegens," *J. Am. Acad. Dermatol.*, **13**, 539–558 (1985).
162. S. J. Rinkus and M. S. Legator, "Chemical Characterization of 465 Known or Suspected Carcinogens and Their Correlation with Mutagenic Activity in the Salmonella typhimurium System," *Cancer Res.*, **39**, 3289–3318 (1979).
163. W. L. Hardy et al., *Am. Med. Assoc. Arch. Ind. Hyg. Occup. Med.*, **1**, 545–549 (1950).
164. H. B. Troup, *Brit. J. Ind. Med.*, **3**, 20–23 (1946).
165. E. Ward et al., "Bladder Tumors in Two Young Males Occupationally Exposed to MBOCA," *Am. J. Ind. Med.*, **14**, 267–272 (1988).
166. C. A. McQueen et al., *Mutat. Res.*, **239**, 133–142 (1990).
167. K. L. Cheever et al., "4,4'-Methylenebis(2-chloroaniline) (MOCA): The Effect of Multiple Oral Administration, Route, and Phenobarbital Induction on Macromolecular Adduct Formation in the Rat," *Fundam. Appl. Toxicol.*, **16**, 71–80 (1991).
168. J. Cocker et al., *Brit. J. Ind. Med.*, **47**, 154–161 (1990).
169. A. M. Osorio et al., *Am. J. Ind. Med.*, **18**, 577–590 (1990).
170. E. K. Weisburger et al., "Neoplastic Response of F344 Rats and B6C3F1 Mice to the Polymer and Dyestuff Intermediates 4,4'-Methylenebis(*N*,*N*'-dimethyl)-benzenamine, 4,4'-Oxydianiline and 4,4'-Methylenedianiline," *J. Natl. Cancer Inst.*, **72**, 1457–1463 (1984).
171. M. Stoltz et al., *Toxicologist*, **6**, 255 (1986).
172. J. L. Radomski, "The Primary Aromatic Amines: Their Biological Properties and Structure-Activity Relationships," *Ann. Rev. Pharmacol. Toxicol.*, **19**, 129–157 (1979).
173. B. K. J. Leong et al., "Retinopathy from Inhaling 4,4'-methylenedianiline Aerosols," *Fundam. Appl. Toxicol.*, **9**, 645–658 (1987).

174. (a) E. A. Emmett, "Allergic Contact Dermatitis in Polyurethane Plastic Moulders," *J. Occup. Med.*, **18**, 802–804 (1976). (b) TSCA 8[d] Report by Dow Chemical, through *Fed Reg.*, **48**(133), 31008 (1983).
175. (a) G. Dupuis and C. Benezra, Eds., *Allergic Contact Dermatitis to Simple Chemicals*, Marcel Dekker, New York, 1983, p. 93. (b) T. Van Joost et al., "Sensitization to Methylenedianiline and Para-Structures," *Contact Dermatitis*, **16**, 246–248 (1987).
176. M. J. LeVine, *Contact Dermatitis*, **9**, 488–490 (1983).
177. S. R. Cohen, *Arch. Dermatol.*, **121**, 1022–1027 (1985).
178. *IARC Monographs on the Evaluation of Carcinogenic Risks to Humans*, Vol. 4, 1974.
179. J. McLaughlin, Jr., et al., "The Injection of Chemicals into the Yolk Sac of Fertile Eggs Prior to Incubation as a Toxicity Test," *Toxicol. Appl. Pharmacol.*, **5**, 760–771 (1963).
180. (a) H. Kopelman et al., *Brit. Med. J.*, **1**, 514–516 (1966), through Reference 110. (b) D. B. McGill et al., *N. Engl. J. Med.*, **291**, 278–282 (1974). (c) Through Reference 173.
181. L. J. Brooks and J. M. Neale, "Acute Myocardiopathy Following Tripathway Exposure to Methylenedianiline," *J. Am. Med. Assoc.*, **242**, 1527–1528 (1979).
182. C. W. Roy et al., *Hum. Toxicol.*, **4**, 61–66 (1985).
183. W. B. Deichmann et al., *Toxicology*, **11**, 185–188 (1978), abstract.
184. Dow Chemical TSCA 8[d] Report, in *Fed. Reg.*, **48**(133), 31008 (1983).
185. A. Rannug et al., "Genotoxic Effects of Additives in Synthetic Elastomers with Special Consideration to the Mechanism of Action of Thiurames and Dithiocarbamates," *Prog. Clin. Biol. Res.*, **141**, 407–419 (1984).
186. NIOSH Current Intelligence Bulletin 47, July 25, 1986.
187. (a) D. B. McGill et al., *N. Engl. J. Med.*, **291**, 278–282 (1974). (b) G. W. Dunn et al., *Arch. Hig. Rada. Toksikol.* **30**(Suppl.), 639–645 (1979), in Reference 173.
188. R. D. Irwin, NTP Technical Report 334, # 19881988, abstract.
189. Aldrich Chemical Company Catalog, St. Louis, MO, 1992–1993.
190. E. K. Weisburger et al., "Testing of Twenty-one Environmental Aromatic Amines or Derivatives for Long-term Toxicity or Carcinogenicity," *J. Environ. Pathol. Toxicol.*, **2**, 325–356 (1978).
191. C. Burnett et al., *J. Toxicol. Environ. Health*, **1**, 1027–1040 (1976), abstract.
192. *IARC Monographs on the Evaluation of Carcinogenic Risks to Humans*, Volume 16, 1978.
193. M. Kiese et al., "The Absorption of Some Phenylenediamines Through the Skin of Dogs," *Toxicol. Appl. Pharmacol.*, **12**, 495–507 (1968).
194. EPA/OTS Doc. 40-8336154, 1981, abstract.
195. J. C. Picciano et al., "The Absence of Teratogenic Effects of Several Oxidative Dyes in Sprague-Dawley Rats," *J. Am. Coll. Toxicol.*, **1**, 125 (1983).
196. H. Amo et al., "Carcinogenicity and Toxicity Study of m-Phenylenediamine Administered in the Drinking Water to (C57BL/6 × C3H/He)F1 Mice," *Food Chem. Toxicol.*, **26**, 893–897 (1988).
197. J. M. Holland et al., *Cancer Res.*, **39**, 1718–1723 (1979).
198. C. Burnett et al., "Long-term Toxicity Studies on Oxidation Hair Dyes," *Food Cosmet. Toxicol.*, **13**, 353–357 (1975).

199. *IARC Monographs on the Evaluation of Carcinogenic Risks to Humans, Overall Evaluations of Carcinogenicity: An Updating of IARC Monographs Volumes 1 to 42*, World Health Organization, Suppl. 7, 1987.
200. (*a*) S. M. Suliman et al., *Hum. Toxicol.*, **2**, 633 (1983), abstract. (*b*) K. S. Chugh et al., *J. Med.*, **13**, 131 (1982), abstract. (*c*) F. Baud et al., *Lancet*, **1983**, 514 (August), Abstract.
201. (*a*) C. Davison, *Arch. Neurol. Psych.*, **49**, 254–265 (1943), through Reference 192. (*b*) W. J. Close, *Med. J. Austr.*, **1**, 53–54 (1932), in Reference 192.
202. E. Cronin, Ed., *Contact Dermatitis*, Churchill Livingstone, Edinburgh, 1983.
203. K. E. Malten et al., *Dermatologica*, **147**, 241–254 (1973).
204. S. Dobkevitch et al., *J. Invest. Dermatol.*, **9**, 203–211 (1947), in Reference 192.
205. R. L. Elder, Ed., "Safety Assessment of Cosmetic Ingredients," *J. Am. Coll. Toxicol.*, **4**, 203–266 (1985).
206. J. C. Topham, *Mutat. Res.*, **74**, 379–387 (1980), abstract.
207. T. A. Re et al., "The Absence of Teratogenic Hazard Potential of *p*-Phenylenediamine in Sprague–Dawley Rats," *Fundam. Appl. Toxicol.*, **1**, 421–425 (1981).
208. R. Munday et al., "Structure–Activity Relationships in the Myotoxicity of Ring-Methylated *p*-Phenylenediamines in Rats and Correlation with Autoxidation Rates in vitro," *Chem. Biol. Interact.*, **76**, 31–45 (1990).
209. M. Degawa et al., "Mutagenicity of Metabolites of Carcinogenic Aminoazo Dyes," *Cancer Lett.*, **8**, 71–76 (1979).
210. F. G. Stenback et al., "Non-Carcinogenicity of Hair Dyes: Lifetime Percutaneous Applications in Mice and Rabbits," *Food Cosmet. Toxicol.*, **15**, 601–606 (1977).
211. C. Burnett et al., *J. Toxicol. Environ. Health*, **6**, 247 (1980), abstract.
212. W. Rojanapo et al., *Carcinogenesis*, **7**, 1997 (1986), abstract.
213. H. A. Milman and C. Peterson, "Apparent Correlation Between Structure and Carcinogenicity of Phenylenediamines and Related Compounds," *Environ. Health Perspect.*, **56**, 261–273 (1984).
214. R. Feinland et al., "Hair Preparations," in *Kirk-Othmer Encyclopedia of Chemical Technology*, 3rd ed., Vol. 12, H. F. Mark, D. F. Othmer, C. G. Overberger, and G. T. Seaborg, Eds., Wiley, New York, 1980, pp. 80–117.
215. L. Soler-Niedziela et al., "Studies on Three Structurally Related Phenylenediamines with the Mouse Micronucleus Assay System," *Mutat. Res.*, **259**, 43–48 (1991).
216. IARC Monographs on the Evaluation of Carcinogenic Risks to Humans, "Some Aromatic Amines, Anthraquinones and Nitroso Compounds, and Inorganic Fluorides Used in Drinking-water and Dental Preparations," Vol. 27, 1982.
217. E. K. Weisburger et al., "Carcinogenicity of 4-Chloro-*o*-phenylenediamine, 4-Chloro-*m*-phenylenediamine, and 2-Chloro-*p*-phenylenediamine in Fischer 344 Rats and B6C3F1 Mice," *Carcinogenesis*, **1**, 495–499 (1980).
218. M. M. McDonald and G. A. Boorman, "Pancreatic Hepatocytes Associated with Chronic 2,6-Dichloro-*p*-phenylenediamine Administration in Fischer 344 Rats," *Toxicol. Pathol.*, **17**, 1–6 (1989).
219. J. O. Thomas et al., "Chronic Toxicity of Diphenylamine to Albino Rats," *Toxicol. Appl. Pharmacol.*, **10**, 362–374 (1967).
220. J. O. Thomas et al., "The Chronic Toxicity of Diphenylamine for Dogs," *Toxicol. Appl. Pharmacol.*, **11**, 184–194 (1967).

221. S. Clegg et al., *J. Environ. Sci. Health*, **16**, 125–130 (1981), abstract.
222. J. F. Crocker, S. R. Blecher, and S. H. Safe, "Chemically Induced Polycystic Kidney Disease," *Prog. Clin. Biol. Res.*, **140**, 281–296 (1983).
223. J. J. Grantham, "Polycystic Kidney Disease: A Predominance of Giant Nephrons," *Am. J. Physiol.*, **244**, 3–10 (1983).
224. S. D. Lenz and W. W. Carlton, "Diphenylamine-induced Renal Papillary Necrosis and Necrosis of the Pars Recta in Laboratory Rodents," *Vet. Pathol.*, **27**, 171–178 (1990).
225. (*a*) P. W. Straub, in *Proc. Eur. Soc. Study Drug Tox.* 1973, pp. 116–124. (*b*) B. A. Fay et al., in *Index of Signs and Symptoms of Industrial Diseases*, CDC/NIOSH, 1980. (*c*) P. S. Spencer and H. H. Schaumburg, "Organic Solvent Neurotoxicity," **11**(Suppl. 1), 53–60 (1985).
226. M. A. Medisky et al., in *Toxicity of Nitroaromatic Compounds*, D. E. Rickett, Ed., Hemisphere, New York, 1985.
227. J. A. Bond et al., "Induction of Hepatic and Testicular Lesions in Fischer-344 Rats by Single Oral Doses of Nitrobenzene," *Fundam. Appl. Toxicol.*, **1**, 389–394 (1981).
228. A. A. Levin et al., "The Reversibility of Nitrobenzene-induced Testicular Toxicity: Continuous Monitoring of Sperm Output from Vasocystotomized Rats," *Toxicology*, **30**, 219–230 (1988).
229. T. E. Hamm, Jr., et al., *Toxicologist*, **4**, 181–192 (1984), abstract.
230. C. J. Rushbrook et al., Society of Toxicology Annual Meeting Abstract 483, 1985.
231. D. E. Dodd et al., "Reproduction and Fertility Evaluations in CD Rats Following Nitrobenzene Inhalation," *Fundam. Appl. Toxicol.*, **8**, 493–505 (1987).
232. S. S. Kazanina, *Nauch. Tr. Novosib. Med. Inst.*, **48**, 42–44 (1968), through J. L. Schardein, Ed., *Chemically Induced Birth Defects*, Marcel Dekker, New York, 1985, p. 646.
233. R. W. Tyl et al., "Development Toxicity Evaluation of Inhaled Nitrobenzene in CD Rats," *Fundam. Appl. Toxicol.*, **8**, 482–492 (1987).
234. "Biodynamics," EPA/OTS Doc. 40-8424492, abstract.
235. W. Albrecht and H. G. Neumann, "Biomonitoring of Aniline and Nitrobenzene. Hemoglobin Binding in Rats and Analysis of Adducts," *Arch. Toxicol.*, **57**, 1–5 (1985).
236. D. M. Blackburn et al., "A Comparison of the Effects of the Three Isomers of Dinitrobenzene on the Testis in the Rat," *Toxicol. Appl. Pharmacol.*, **92**, 54–64 (1988).
237. (*a*) R. R. Beard et al., *Ind. Hyg. Toxicol.*, **2A**, 2413–2489 (1981). (*b*) T. Okubo et al., *Ind. Health*, **20**, 297–304 (1982).
238. M. N. Gleason et al., *Clinical Toxicology of Commercial Products*, 3rd ed., 1969.
239. N. Ishihara et al., *Int. Arch. Occup. Env. Health*, **36**, 161–168 (1976).
240. (*a*) H. Teir, *Acta Ophthalmol.*, Suppl. 161, 60–65 (1984). (*b*) W. E. von Oettingen, U.S. Public Health Service Bull. No. 271, 1941, pp. 99–103. (*c*) Through Reference 147.
241. (*a*) M. A. Philbert et al., "1,3-Dinitrobenzene-induced Encephalopathy in Rats," *Neuropathol. Appl. Neurobiol.*, **13**, 371–389 (1987). (*b*) I. A. Romero et al., "1,3-Dinitrobenzene, a Neurotoxin Acting at the Rat Blood-brain Barrier," *J. Physiol.*, **446**, 498P (1992).
242. (*a*) D. D. Nystrom et al., Society of Toxicology Annual Meeting Abstract 476, 1987.

(b) D. D. Nystrom and D. E. Rickert, "Metabolism and Excretion of Dinitrobenzenes by Male Fisher-344 Rats," *Drug Metab. Dispos.*, **15**, 821–825 (1987).

243. P. A. Cossum and D. E. Rickert, "Metabolism and Toxicity of Dinitrobenzene Isomers in Erythrocytes from F344 Rats, Rhesus Monkeys and Humans," *Toxicol. Lett.*, **37**, 157–163 (1987).

244. T. E. Cody et al., *J. Toxicol. Environ. Health*, **7**, 829–847 (1981).

245. E. A. Lock et al., Society of Toxicology Annual Meeting Abstract 482, 1985.

246. R. E. Linder et al., "Testicular Toxicity and Infertility in Male Rats Treated with 1,3-DNB," *J. Toxicol. Environ. Health*, **19**, 477–489 (1986).

247. (a) R. E. Linder et al., "Acute Effects and Long-term Sequelae of 1,3-Dinitrobenzene on Male Reproduction in the Rat. I. Sperm Quality, Quantity, and Fertilizing Ability," *J. Androl.*, **9**, 317–326 (1988). (b) R. A. Hess et al., "Acute Effects and Long-term Sequelae of 1,3-Dinitrobenzene on Male Reproduction in the Rat. II," *J. Androl.*, **9**, 327–342 (1988).

248. (a) P. M. D. Foster et al., Society of Toxicology Annual Meeting Abstract 574, 1987. (b) P. M. D. Foster, *Arch. Toxicol.*, Suppl. 13, 3–17 (1989), abstract.

249. D. D. Nystrom et al., Society of Toxicology Annual Meeting Abstract 726, 1988.

250. D. P. Evenson et al., "Effect of 1,3-DNB on Prepubertal, Pubertal and Adult Mouse Spermatogenesis," *J. Toxicol. Environ. Health*, **28**, 67–80 (1989).

251. R. E. Linder et al., "Reproductive Toxicity of a Single Dose of 1,3-Dinitrobenzene in Two Ages of Young Adult Male Rats," *Fundam. Appl. Toxicol.*, **14**, 284–298 (1990).

252. M. F. Obasaju et al., "Species Differences in Susceptibility to 1,3-Dinitrobenzene-induced Testicular Toxicity and Methemoglobinemia," *Fundam. Appl. Toxicol.*, **16**, 257–266 (1991).

253. S. F. McEuen and M. G. Miller, "Metabolism and Pharmacokinetics of 1,3-Dinitrobenzene in the Rat and the Hamster," *Drug Metab. Dispos.*, **19**, 661–666 (1991).

254. United Nations Report 1578, World Health Organization, 1990, abstract.

255. S. D. Held et al., Abstract 47, CIIT 11th Scientific Evening Report, September 8, 1987.

256. K. L. Mohr et al., Society of Toxicology Annual Meeting Abstract 58, 1988.

257. (a) G. Saita et al., *Med. Lav.*, **49**, 494, 1958, through *Bull. Hyg.*, **34**, 139, 1959, abstract. (b) R. S. Nair et al., "Subchronic Inhalation Toxicity of *p*-Nitroaniline and *p*-Nitrochlorobenzene in Rats," *Fundam. Appl. Toxicol.*, **6**, 618–627 (1986).

258. R. S. Nair et al., 5th CIIT Conference on Toxicology Report, 1982.

259. T. Yoshida et al., "Identification of Urinary Metabolites in Rats Treated with *p*-Chloronitrobenzene," *Arch. Toxicol.*, **65**, 52–58 (1991).

260. R. S. Nair et al., Society of Toxicology Annual Meeting Abstract 846, 1989.

261. E. Van Hecke et al., *Contact Dermatitis*, **12**, 282 (1985), abstract.

262. (a) W. H. Habig et al., *J. Biol. Chem.*, **249**, 7130–7139 (1974), abstract. (b) N. Ballatori and T. W. Clarkson, "Biliary Secretion of Glutathione and of Glutathione–Metal Complexes," *Fundam. Appl. Toxicol.*, **5**, 816–831 (1985).

263. P. R. Kerklaan et al., "Mutagenicity of Halogenated and Other Substituted Dinitrobenzenes in Salmonella typhimurium TA100 and Derivatives Deficient in Glutathione," *Mutat. Res.*, **176**, 171–178 (1987).

264. E. J. Butterfield and D. C. Torgeson, "Agricultural Fungicides," in *Kirk-Othmer*

*Encyclopedia of Chemical Technology*, Vol. 11, H. F. Mark, D. F. Othmer, C. G. Overberger, and G. T. Seaborg, Eds., Wiley, New York, 1980, pp. 490–498.

265. A. M. Schuhmann and J. F. Borzelleca, "An Assessment of the Methemoglobin and Heinz-body-inducing Capacity of Pentachloronitrobenzene in the Cat," *Toxicol. Appl. Pharmacol.*, **44**, 523–529 (1978).

266. J. F. Borzelleca et al., "Toxicologic and Metabolic Studies on Pentachloronitrobenzene," *Toxicol. Appl. Pharmacol.*, **18**, 522–534 (1971).

267. R. L. Jordan and J. F. Borzelleca, "Teratogenic Studies with Pentachloronitrobenzene in Rats," Society of Toxicology Annual Meeting Abstract 40, 1973.

268. K. S. Khera and D. C. Villeneuve, *Toxicology*, **5**, 117 (1975), abstract.

269. K. D. Courtney et al., "The Effects of Pentachloronitrobenzene, Hexachlorobenzene, and Related Compounds on Fetal Development," *Toxicol. Appl. Pharmacol.*, **35**, 239–256 (1976).

270. J. Finnegan et al., "Acute and Chronic Toxicity Studies on Pentachlorobenzene," *Arch. Int. Pharmacodyn.*, **114**, 38–52 (1958).

271. J. R. M. Innes et al., *J. Natl. Cancer Inst.*, **42**, 1011–1015 (1969), abstract.

272. C. Searle et al., *Cancer Res.*, **26**, 12–16 (1966), abstract.

273. NCI Technical Report No. 61, U.S. Dept. Health, Education and Welfare, Public Health Service, 1978, abstract.

274. NTP Technical Report No. 325, U.S. Dept. Health, Education and Welfare, Public Health Service, 1987, abstract.

275. R. G. Kuehni et al., "Dyes, Application and Evaluation," in *Kirk-Othmer Encyclopedia of Chemical Technology*, Vol. 8, H. F. Mark, D. F. Othmer, C. G. Overberger, and G. T. Seaborg, Eds., 1979, pp. 280–350.

276. (*a*) W. F. Melick et al., *J. Urol.*, **74**, 760–766 (1955), in References 81, 139. (*b*) *IARC Monographs on the Evaluation of Carcinogenic Risk of Chemicals to Man*, Vol. 1, 1972.

277. J. A. Zack and W. R. Gaffey, "A Mortality Study of Workers at the Monsanto Company Plant in Nitro, West Virginia," *Environ. Sci. Res.*, **26**, 575–591 (1983), through Reference 32a.

278. R. Von Jagow et al., *Biochem. Pharmacol.*, **15**, 899–910 (1966).

279. J. C. Picciano et al., "Evaluation of the Teratogenic Potential of the Oxidative Dye N-phenyl-*para*-phenylenediamine," *Drug Chem. Toxicol.*, **7**, 167–176 (1984).

280. P. A. Schulte et al., "Risk Assessment of a Cohort Exposed to Aromatic Amines," *J. Occup. Med.*, **27**, 115–121 (1985).

281. (*a*) G. F. Rubino et al., "The Carcinogenic Effect of Aromatic Amines: an Epidemiological Study on the Role of *o*-Toluidine and 4,4'-Methylenebis(2-methylaniline) in Inducing Bladder Cancer in Man," *Environ. Res.*, **27**, 241–254 (1982). (*b*) A. Decarli et al., "Bladder Cancer Mortality of Workers Exposed to Aromatic Amines: Analysis of Models of Carcinogenesis," *Brit. J. Cancer*, **51**, 707–712 (1985). (*c*) G. Piolatto et al., "Bladder Cancer Mortality of Workers Exposed to Aromatic Amines: an Updated Analysis," *Brit. J. Cancer*, **63**, 457–459 (1991).

282. NCI Report No. 143, DHEW Publ. No. (NIH) 78-1398, 1978, in Reference 216.

283. (*a*) E. Ward et al., "Excess Number of Bladder Cancers in Workers Exposed to *o*-Toluidine and Aniline," *J. Natl. Cancer Inst.*, **83**, 501–506 (1991). (*b*) D. G. DeBord,

et al., "Binding Characteristics of *ortho*-toluidine to Rat Hemoglobin and Albumin," *Arch. Toxicol.*, **66**, 231–236 (1992).

284. S. D. Nelson et al., *J. Med. Chem.*, **21**, 721–725 (1978), in Reference 37.
285. (*a*) *American Hospital Formulary Service*, American Society of Hospital Pharmacists, 1992, p. 1956 (carcinogenicity information recommended by personal communication to K. Litvak, ASHP, September 4, 1992). (*b*) M. J. Stasik, "Urinary Bladder Cancer from 4-chloro-*o*-toluidine," *Dtsch. Med. Wochenschr.*, **116**, 1444–1447 (1991) (in German); English abstract, *CA Selects*, Issue 4, 1992, p. 9.
286. S. C. Catino and R. E. Farris, "Azo Dyes," in *Kirk-Othmer Encyclopedia of Chemical Technology*, 3rd ed., Vol. 3, H. F. Mark, D. F. Othmer, C. G. Overberger, and G. T. Seaborg, Eds., Wiley, New York, 1978, pp. 387–433.
287. M. Fytelson, "Organic Pigments," in *Kirk-Othmer Encyclopedia of Chemical Technology*, 3rd ed., Vol. 17, H. F. Mark, D. F. Othmer, C. G. Overberger, and G. T. Seaborg, Eds., Wiley, New York, 1982, pp. 838–871.
288. (*a*) NCI NIH 78-1357, 1978, through Reference 213. (*b*) *IARC Monographs On the Evaluation of Carcinogenic Risks to Humans: Some Flame Retardants and Textile Chemicals and Exposures in the Textile Manufacturing Industry*, Vol. 48, 1990.
289. K. H. Ferber, "Benzidine and Related Biphenyldiamines," in *Kirk-Othmer Encyclopedia of Chemical Technology*, 3rd ed., Vol. 3, H. F. Mark, D. F. Othmer, C. G. Overberger, and G. T. Seaborg, Eds., Wiley, New York, 1978, pp. 772–777.
290. J. P. Seller, *Mutat. Res.*, **46**, 305–310 (1977), abstract.
291. E. K. Weisburger et al., *J. Environ. Pathol. Toxicol.*, **2**, 325–356 (1978), abstract.
292. D. Kleniewska and H. Maibach, *Dermatosen, Beruf, Umwwwelt*, **28**, 11–13 (1980), abstract.
293. E. K. Weisburger et al., *J. Environ. Pathol. Toxicol.*, **2**, 325–356 (1978), abstract.
294. B. Milligan and K. E. Gilbert, "Diaminotoluenes," in *Kirk-Othmer Encyclopedia of Chemical Technology*, 3rd ed., Vol. 2, H. F. Mark, D. F. Othmer, C. G. Overberger, and G. T. Seaborg, Eds., Wiley, New York, 1978, pp. 321–329.
295. Mobay Chemical Co. MSDS, 1981.
296. S. K. Varma et al., "Reproductive Toxicity of 2,4-Toluenediamine in the Rat," 3. Effects on Androgen-Binding Seminiferous Tubule Characteristics and Spermatogenesis," *J. Toxicol. Environ. Health*, **25**, 435–452 (1988).
297. (*a*) B. Thysen et al., "Reproductive Toxicity of 2,4-Toluenediamine in the Rat. 1. Effect on Male Fertility," *J. Toxicol. Environ. Health*, **16**, 753–761 (1985). (*b*) B. Thysen et al., "Reproductive Toxicity of 2,4-Toluenediamine in the Rat. 2. Spermatogenic and Hormonal Effects," *J. Toxicol. Environ. Health*, **16**, 763–769 (1985).
298. (*a*) R. J. Levine et al., CIIT Report, December 1982. (*b*) P. V. V. Hamill et al., "The Epidemilogic Assessment of Male Reproductive Hazard from Occupational Exposure to Toluenediamine and Dinitrotoluene," *J. Occup. Med.*, **24**, 985–993 (1982). (*c*) R. J. Levine, R. D. Dal Corso, and P. B. Blunden, "Fertility of Workers Exposed to Dinitrotoluene and Toluenediamine at Three Chemical Plants," Chapter 16 in *Toxicity of Nitroaromatic Compounds*, D. E. Rickert, Ed., Hemisphere, New York, 1985, pp. 243–254.
299. E. George and C. Westmoreland, "Evaluation of the in vivo Genotoxicity of the Structural Analogues 2,6-Diaminotoluene and 2,4-Diaminotoluene Using the Rat Micronucleus Test and Rat Liver UDS Assay," *Carcinogenesis*, **12**, 2233–2237 (1991).

300. M. L. Cunningham et al., "Evidence for an Acetoxyarylamine as the Ultimate Mutagenic Reactive Intermediate of the Carcinogenic Aromatic Amine 2,4-Diaminotoluene," *Mutat. Res.*, **242**, 101–110 (1990).
301. NCI Report NIH 79-1718, in Reference 213.
302. N. Ito et al., *Cancer Res.*, **29**, 1137–1139 (1969), abstract.
303. A. L. Giles et al., *J. Toxicol. Environ. Health*, **1**, 433–440 (1976), abstract.
304. T. Glinsukon et al., *Xenobiotica*, **5**, 475–484 (1975), abstract.
305. P. W. Brandt-Rauf and J. A. Hathaway, "Biliary Tract Cancer in the Chemical Industry A Proportional Mortality Study," *Brit. J. Ind. Med.*, **43**, 716–717 (1986).
306. NCI Report No. NIH 78-1381, 1978, in Reference 213.
307. J. Ashby and R. W. Tennant, "Chemical Structure, Salmonella Mutagenicity and Extent of Carcinogenicity as Indicators of Genotoxic Carcinogenesis Among 222 Chemicals Tested in Rodents by the U.S. NCI/NTP," *Mutat. Res.*, **204**, 17–115 (1988).
308. M. L. Cunningham et al., "Identification and Mutagenicity of the Urinary Metabolites of the Mutagenic Noncarcinogen 2,6-Diaminotoluene," *J. Liq. Chromatogr.*, **12**, 1407–1416 (1989).
309. NIH Report No. 80-1756 (1980), through Reference 213.
310. M. L. Cunningham et al., "Correlation of Hepatocellular Proliferation with Hepatocarcinogenicity Induced by the Mutagenic Noncarcinogen:Carcinogen Pair—2,6- and 2,4-Diaminotoluene," *Toxicol. Appl. Pharmacol.*, **107**, 562–567 (1991).
311. H. H. Lysy et al., Society of Toxicology Annual Meeting Abstract 322, 1988.
312. J. K. P. Chism et al., Abstract 45, 11th CIIT Scientific Evening, September 8, 1987.
313. J. Suzuki et al., "Mutagenicities of Mono-nitrobenzene Derivatives in the Presence of Norharman," *Mutat. Res.*, **120**, 105–110 (1983).
314. CIIT Technical Dinitrotoluene Report, May 1977.
315. D. E. Rickert, B. E. Butterworth, and J. A. Popp, "Dinitrotoluene Acute Toxicity Oncogenicity Genotoxicity and Metabolism," *CRC Crit. Rev. Toxicol.*, **13**, 217–234 (1984).
316. C. J. Price et al., "Teratologic Evaluation of Dinitrotoluene in the Fischer 344 Rat," *Fundam. Appl. Toxicol.*, **5**, 948–961 (1985).
317. H. V. Ellis, III et al., "Chronic Toxicity of 2,4-Dinitrotoluene in the Rat," *Toxicol. Appl. Toxicol.*, **45**, 245–246 (1978).
318. NIOSH Current Intelligence Bulletin 44, Publ. No. 85-109, NTIS Publ. No. PB-86-105-913, 1985.
319. S. C. J. Reader and R. M. D. Foster, "The in vitro Effects of Four Isomers of Dinitrotoluene on Rat Sertoli and Sertoli-germ Cell Cocultures: Germ Cell Detachment and Lactate and Pyruvate Production," *Toxicol. Appl. Pharmacol.*, **106**, 287–294 (1990).
320. P. V. V. Hamill et al., "The Epidemiologic Assessment of Male Reproductive Hazard from Occupational Exposure to Toluenediamine and Dinitrotoluene," *J. Occup. Med.*, **24**, 985–991 (1982).
321. D. E. Rickert, Ed., *Toxicity of Nitroaromatic Compounds*, Hemisphere Publishing Corp., Washington, DC, 1985.
322. NIOSH OTA-BA-266, December 1985, abstract.
323. (*a*) M. E. Graichen et al., *Toxicologist*, **3**, 92–96 (1983), abstract. (*b*) T. B. Leonard,

M. E. Graichen, and J. A. Popp, "Dinitrotoluene Isomer-Specific Hepatocarcinogenesis in F344 Rats," *J. Natl. Cancer Inst.*, **79**, 1313–1319 (1987). CIIT Project No. 2010-101, Hazelton Laboratories, 1982, through T. B. Leonard et al., op. cit. (*c*) J. A. Popp and T. B. Leonard, "The Hepatocarcinogenicity of Dinitrotoluenes," Chapter 4 in *Toxicity of Nitroaromatic Compounds*, D. E. Rickert, Ed., Hemisphere, New York, 1985, pp. 53–60.

324. T. B. Leonard, M. E. Graichen, and J. A. Popp, "Dinitrotoluene Isomer-Specific Hepatocarcinogenesis in F344 Rats," *J. Natl. Cancer Inst.*, **79**, 1313–1319 (1987).

325. R. J. Levine et al., "Heart Disease in Workers Exposed to Dinitrotoluene," *J. Occup. Med.*, **28**, 811–816 (1986).

326. M. C. Hoff, "Toluene," in *Kirk-Othmer Encyclopedia of Chemical Technology*, 3rd ed., Vol. 23, H. F. Mark, D. F. Othmer, C. G. Overberger, and G. T. Seaborg, Eds., Wiley, New York, 1983, pp. 246–273.

327. J. K. Sears and N. W. Touchette, "Plasticizers," in *Kirk-Othmer Encyclopedia of Chemical Technology*, 3rd ed., Vol. 18, H. F. Mark, D. F. Othmer, C. G. Overberger, and G. T. Seaborg, Eds., Wiley, New York, 1982, pp. 111–183.

328. D. Guest et al., "Metabolism of 2,4-Dinitrotoluene by Intestinal Microorganisms from Rat, Mouse and Man," *Toxicol. Appl. Pharmacol.*, **64**, 160–168 (1982).

329. M.-A. Mori et al., "Metabolism of Dinitrotoluene Isomers by Escherichia Coli Isolated from Human Intestin," *Chem. Pharm. Bull. (Tokyo)*, **32**, 4070–4075 (1984).

330. D. E. Rickert, "Identification and Quantification of Urinary Metabolites of Dinitrotoluenes in Occupationally Exposed Humans," *Toxicol. Appl. Pharmacol.*, **80**, 166–174 (1985).

331. R. W. Lane et al., "Reproductive Toxicity and Lack of Dominant Lethal Effects of 2,4-Dinitrotoluene in the Male Rat," *Drug Chem. Toxicol.*, **8**, 265–280 (1985).

332. E. Bloch et al., "Reproductive Toxicity of 2,4-Dinitrotoluene in the Rat," *Toxicol. Appl. Pharmacol.*, **94**, 466–472 (1988).

333. National Cancer Institute, DHEW Publ. No. NIH 78-1360, NCI NIH 78-1360, 1978, in Reference 213.

334. (*a*) V. Lindner, "Explosives and Propellants," in *Kirk-Othmer Encyclopedia of Chemical Technology*, Vol. 9, H. F. Mark, D. F. Othmer, C. G. Overberger, and G. T. Seaborg, Eds., Wiley, New York, 1980, pp. 561–671. (*b*) J. A. Hathaway, "Subclinical Effects of Trinitrotoluene: A Review of Epidemiology Studies," Chapter 17 in *Toxicity of Nitroaromatic Compounds*, D. E. Rickert, Ed., Hemisphere, New York, 1985, pp. 255–274.

335. J. Quan-Guan et al., "The Effects of Trinitrotoluene Toxicity on Zinc and Copper Metabolism," *Toxicol. Lett.*, **55**, 343–349 (1991).

336. K. D. Bergert et al., "Cardiovascular Diseases in Exposure to the Saltpeter Acid Ester Ethylene Glycol Dinitrate," *Z. Ges. Inn. Med.*, **42**, 581–583 (1987), abstract.

337. D. C. Trainor et al., *Arch. Env. Health*, **12**, 231–234 (1966), abstract.

338. S. Bille et al., *Nord. Med.*, **70**, 842–843 (1963).

339. H. Tsuruta et al., *Ind. Health*, **8**, 99–118 (1970).

340. D. G. Clark and M. H. Litchfield, "The Toxicity, Metabolism, and Pharmacologic Properties of Propylene Glycol 1,2-Dinitrate," *Toxicol. Appl. Pharmacol.*, **15**, 175–184 (1969).

341. L. J. Ignarro et al., *J. Pharmacol. Exp. Ther.*, **218**, 739–749 (1981), abstract.

342. R. M. Rapoport et al., *J. Cyclic Nucleotide Protein Phosphorylation Res.*, **9**, 281–296 (1983).
343. (*a*) S. A. Waldman and F. Murad, "Cyclic GMP Synthesis and Function," *Pharmacol. Rev.*, **39**, 163–195 (1987). (*b*) T. Balazs, J. P. Hanig, and E. H. Herman, "Toxic Responses of the Cardiovascular System," Chapter 14 in *Casarett and Doull's Toxicology, The Basic Science of Poisons*, 3rd ed., C. D. Klaassen, M. O. Amdur, and J. Doull, Eds., Macmillan, New York, 1986, pp. 387–411.
344. A. Ben-David, "Cardiac Arrest in an Explosives Factory Worker Due to Withdrawal from Nitroglycerin Exposure," *Am. J. Ind. Med.*, **15**, 719–722 (1989).
345. M. G. Bogaert, "Pharmacokinetics of Organic Nitrates in Man: an Overview," *European Heart J.*, **9**, 33–37 (1988).
346. P. Needleman et al., *Molec. Pharmacol.*, **1**, 77–86 (1965).
347. P. Needleman et al., *Biochem. Pharmacol.*, **20**, 1865–1876 (1971).
348. (*a*) D. J. Kornbrust and J. S. Bus, "Glutathione Depletion by Methyl Chloride and Association with Lipid Peroxidation in Mice and Rats," *Toxicol. Appl. Pharmacol.*, **72**, 388–399 (1984). (*b*) C. D. Klaassen et al., "Role of Sulfhydryls in the Hepatotoxicity of Organic and Metallic Compounds," *Fundam. Appl. Toxicol.*, **5**, 806–815 (1985). (*c*) W. B. Jakoby, "The Glutathione *S*-transferases: a Group of Multifunctional Detoxification Proteins," *Adv. Enzymol.*, **46**, 383–414 (1978). (*d*) W. H. Habig et al., *J. Biol. Chem.*, **249**, 7130–7139 (1974), abstract.
349. (*a*) Through Reference 346. (*b*) L. A. Heppel et al., *J. Biol. Chem.*, **251**, 6183 (1976), Abstract. (*c*) S. Someroja and H. Savolainen, "Neurochemical Effects of Ethylhexyl Nitrate in Rats," *Toxicol. Lett.*, **19**, 189–193 (1983). (*d*) L. A. Heppel et al., *J. Biol. Chem.*, **183**, 129–138 (1965), abstract.
350. In Reference 349d.
351. G. H. Cocolas, "Cardiovascular Agents," Chapter 14 in *Textbook of Organic Medicinal and Pharmaceutical Chemistry*, 8th ed., R. F. Doerge, Ed., Lippincott, Philadelphia, 1982, pp. 521–562.

CHAPTER SEVENTEEN

# Aliphatic and Alicyclic Amines

Theodore J. Benya, Ph.D., D.A.B.T., and Raymond D. Harbison, Ph.D., D.A.B.T.

## 1 GENERAL CONSIDERATIONS

This chapter presents information on nonaromatic amines represented by aliphatic amines and cyclic compounds having aliphatic amine properties.

The aliphatic amines are highly alkaline and tend to be fat soluble, characteristics that in combination result in considerable irritation to skin and mucous membranes. Corrosive burns as well as marked allergic sensitization may occur (1). The most volatile amines, characterized by boiling points lower than 100°C, are highly irritating and include methylamine, dimethylamine, trimethylamine, ethylamine, diethylamine, triethylamine, *n*-propylamine, isopropylamine, diisopropylamine, allylamine, *n*-butylamine, isobutylamine, *sec*-butylamine, *tert*-butylamine, and dimethylbutylamine. These properties account in large measure for the hazards they present in the workplace. Toxicity information in humans continues to be limited. During the past decade, important advances have been made in understanding mechanisms and determinants of carcinogenesis, yet answers to some of the more critical questions remain elusive; and controversies regarding potential aliphatic amine carcinogenicity are far from being resolved. In particular, further investigations are needed to determine relative risks for nitrosamine formation, which is both compound specific and pH dependent.

### 1.1 Physical and Chemical Properties

The lower aliphatic amines are highly alkaline derivatives of ammonia where one, two, or three of the hydrogen atoms are replaced by alkyl or alkanol radicals of

---

*Patty's Industrial Hygiene and Toxicology, Fourth Edition, Volume 2, Part B*, Edited by George D. Clayton and Florence E. Clayton.
ISBN 0-471-54725-5 © 1994 John Wiley & Sons, Inc.

six carbons or fewer. However, primary and secondary amines can also act as very weak acids ($K_a$ approximately $10^{-33}$) (2). Many of these lower aliphatic amines have low flash points and are flammable liquids or gases. Branching of the alkyl chain tends to enhance volatility, whereas hydroxy substitution as in the alkanolamines decreases volatility (1). Methylamine solutions are good solvents for many inorganic and organic compounds.

Most aliphatic amines have a distinctly unpleasant odor. The fishy or fishlike odor of methylamines increases from mono- to trimethylamine, and in high concentrations they all have the odor of ammonia (2). In the case of methylamine, the odor is faint but readily detectable at less than 10 ppm, becomes strong at 20 to 100 ppm, and is intolerably ammoniacal at 100 to 500 ppm.

Olfactory fatigue occurs readily. No symptoms of irritation are produced from chronic exposures to less than 10 ppm (1). Deodorization of ammonia and amines can be achieved (K. Hasagawa, 1976) with the use of dihydroxyacetone, which is reportedly nontoxic to humans and domestic animals and reacts rapidly with ammonia or amines (1).

J. Amoore and L. Furrester (1976) reported that about 7 percent of humans are unable to smell (anosmic) trimethylamine. Odor threshold measurements on 16 aliphatic amines were made with panels of specific anosmics and normal observers. The anosmia was most pronounced with low-molecular-weight tertiary amines, but was also observed to a lesser degree with primary and secondary amines. This specific anosmia apparently corresponds with the absence of a new olfactory primary sensation, the fishlike odor (1).

The aliphatic amines are conveniently classified as primary, secondary, and tertiary amines according to the number of substitutions on the nitrogen atom. If only one radical is substituted, the amine is a primary amine, even though the alkyl substituent may have a secondary or tertiary structure (1).

The alkylamines and alkanolamines behave as bases in organic solvents and aqueous solutions. Primary and secondary amines can also act as very weak acids. One way to express the basicity of an amine focuses on its reaction with water. The equilibrium constant for the acid–base reaction of an amine with water is called $K_b$:

$$RNH_2 + H_2O \rightleftharpoons RNH_3^+ + OH^-$$

$$K_b = \frac{[RNH_3^+][OH^-]}{[RNH_2]}$$

and this is conveniently converted to $pK_b$, where

$$pK_b = -\log_{10} K_b$$

The more basic the compound, the lower the $pK_b$. Measured in this way, the alkylamines have $pK_b$ values of 3 to 5; arylamines have $pK_b$ values of 9 to 10; and

**Table 17.1.** Basicity of Ammonia and Some Amines[a]

| Amine | Formula | $pK_b$ | $pK_a$ |
|---|---|---|---|
| Ammonia | $NH_3$ | 4.76 | 9.24 |
| Methylamine | $CH_3NH_2$ | 3.35 | 10.65 |
| Dimethylamine | $(CH_3)_2NH$ | 3.32 | 10.68 |
| Trimethylamine | $(CH_3)_3N$ | 4.22 | 9.78 |
| Ethylamine | $CH_3CH_2NH_2$ | 3.29 | 10.71 |
| Diethylamine | $(CH_3CH_2)_2NH$ | 3 | 11 |
| Triethylamine | $(CH_3CH_2)_3N$ | 3.25 | 10.75 |
| n-Propylamine | $CH_3CH_2CH_2NH_2$ | 3.39 | 10.61 |
| Di-n-propylamine | $(C_3H_7)_2NH$ | 3 | 11 |
| Isopropylamine | $(CH_3)_2CHNH_2$ | 3.4 | 10.6 |
| n-Butylamine | $C_4H_9NH_2$ | 3.32 | 10.68 |
| Cyclohexylamine | $C_6H_{11}NH_2$ | 3.3 | 10.7 |
| Hexamethylenediamine | $NH_2(CH_2)_6NH_2$ | 3.3 | 10.7 |

[a]Modified from H. K. Hall, Jr., *J. Phys. Chem.*, **60**, 63 (1956), and A. Streitwieser, Jr., and C. H. Heathcock, *Introduction to Organic Chemistry*, Macmillan, New York, 1976.

pyrrole has a $pK_b$ of 13.6. The $pK_b$ of ammonia is 4.76 (1). $pK_b$ values for some of the more common amines are given in Table 17.1.

In general, primary amines are stronger bases than ammonia, and secondary amines are stronger bases than tertiary amines. As the length of the chain increases up to four or five carbon atoms, the base strength tends to decrease. Aqueous solutions of alkylamines attack glass as a consequence of the basic properties of the amines in water (2).

The polyamines are usually slightly viscous liquids with a strong ammoniacal odor; most are completely miscible in water (except pentaethylenehexamine, which forms a gel), alcohol, acetone, benzene, and ethyl ether (3).

Diamines such as ethylenediamine also behave a strong bases. The alkanolamines are weaker bases than the corresponding substituted amines (1).

Neurath et al. (1977) reported the presence of a great variety of amines in the human environment. Forty primary and secondary amines with different gas chromatographic properties were detected in samples of fresh vegetables, preserves, mixed pickles, fish and fish products, bread, cheese, stimulants, animal feedstuffs, and surface waters, and 21 of these were identified by mass spectrometry. Secondary amines were generally found in concentrations below 10 ppm, although higher concentrations occurred in herring preparations, some cheeses, and samples of large radish and red radish. The highest content of secondary amines found so far was in red radishes (38 ppm pyrrolidine, 20 ppm pyrroline, 5.4 ppm N-methylphenethylamine, and 1.1 ppm dimethylamine) (1).

The physical properties of some of the aliphatic and alicyclic amines are given in Tables 17.2 and 17.3 (1,5–8).

The higher-molecular-weight amines prepared from longer-chain fatty acids, referred to as simply fatty amines, are either liquids or solids and essentially insoluble in water (4). Commercially offered fatty amines in general contain mixed

Table 17.2. Physical and Chemical Properties of Aliphatic and Alicyclic Monoamines

| Name | Formula | Mol Wt. | M.P. (°C) | B.P. (°C) | Density (g/ml) | Solubility in Water (g/100 ml) | Vapor Pressure (torr) (°C) | Vapor Density (Air = 1) | Flash Point[a] (°F) | Conversion Units | |
|---|---|---|---|---|---|---|---|---|---|---|---|
| | | | | | | | | | | 1 mg/l (ppm) | 1 ppm (mg/m³) |
| Methylamine | CH₃NH₂ | 31.06 | −93.5 | −6.3 | 0.7691 (−70/4) | Ver sol. | 2 atm (25) | 1.07 | 34 (30% soln.) | 783 | 1.27 |
| Dimethylamine | (CH₃)₂NH | 45.08 | −93 | 7.4 | 0.6804 (0/4) | Very sol. | 2 atm (10) | 1.55 | 54 (25% soln.) | 542 | 1.84 |
| Trimethylamine | (CH₃)₃N | 59.11 | −117.2 | 2.87 | 0.6709 (0/4) | Very sol. | 760 (2.9) | 2.04 | 38 (25% soln.) | 414 | 2.42 |
| Ethylamine | CH₃CH₂NH₂ | 45.08 | −81 | 16.6 | 0.6836 (20/20) | Complete | 400 (2.9) | 1.55 | <0.0 | 542 | 1.84 |
| Diethylamine | (CH₃CH₂)₂NH | 73.14 | −48 | 56.3 | 0.7400 (20/4) | Complete | 195 (20) | 2.52 | −9 | 334 | 2.99 |
| Triethylamine | (CH₃CH₂)₃N | 101.19 | −115.3 | 89.5 | 0.7275 (20/4) | Sol. | 53.5 (20) | 3.49 | 20 | 242 | 4.14 |
| Propylamine | CH₃CH₂CH₂NH₂ | 59.11 | −83 | 48.7 | 0.719 (20/20) | Sol. | 400 (31/5) | 2.04 | −35 | 414 | 2.42 |
| Di-n-propylamine | (CH₃CH₂CH₂)₂NH | 101.19 | 63 | 110.7 | 0.7400 (20/4) | Sol. | 30 (25) | 3.49 | 63 | 242 | 4.14 |
| Isopropylamine | (CH₃)₂CHNH₂ | 59.08 | −101.2 | 34 | 0.694 (15/4) | Complete | 460 (20) | 2.04 | −35 | 414 | 2.42 |
| Diisopropylamine | [(CH₃)₂CH]₂NH | 101.19 | −61 | 84 | 0.720 (20/20) | Sl sol. | 70 (20) | 3.49 | 30 | 242 | 4.14 |
| n-Butylamine | CH₃(CH₂)₃NH₂ | 73.14 | −50.5 | 77.8 | 0.740 (20/4) | Complete | 72 (20) | 2.52 | 10 | 334 | 2.99 |
| Di-n-butylamine | [CH₃(CH₂)₃]₂NH | 129.24 | −60 | 159.6 | 0.7613 (20/20) | Sol. | 1.9 (20) | 4.46 | 117 | 189 | 5.29 |
| Tri-n-butylamine | [CH₃(CH₂)₃]₃N | 185.34 | <−70 | 214 | 0.775 (20/20) | Insol. | 20 (100) | 6.39 | 187 | 132 | 7.58 |
| Isobutylamine | (CH₃)₂CHCH₂NH₂ | 73.14 | −85.5 | 68 | 0.724 (25/4) | Complete | 100 (18.8) | 2.52 | <15 | 334 | 2.99 |
| n-Amylamine | CH₃(CH₂)₄NH₂ | 87.17 | −55 | 104 | 0.7782 (20/20) | Sol. | | 3.01 | 30 | 281 | 3.56 |
| Isoamylamine | (CH₃)₂CHCH₂CH₂NH₂ | 87.17 | | 95 | 0.7505 (20/4) | Sol. | | 3.01 | | 281 | 3.56 |
| n-Hexylamine | CH₃(CH₂)₅NH₂ | 101.19 | −19 | 132.7 | 0.767 | 1.2 | 6.5 (20) | 3.49 | 85 | 242 | 4.14 |
| 2-Ethylbutylamine | (CH₃CH₂)₂CHCH₂NH₂ | 101.19 | | 125 | 0.776 (20/20) | | | 3.49 | 70 | 242 | 4.14 |
| n-Heptylamine | CH₃(CH₂)₆NH₂ | 115.22 | −18 | 156.9 | 0.7754 (20/4) | Sl sol. | | 3.97 | 130 | 212 | 4.71 |
| Di-n-heptylamine | [CH₃(CH₂)₆]₂NH | 213.4 | 30 | 271 | | Sl sol. | | 7.35 | | 115 | 8.73 |
| 2-Ethylhexylamine | CH₃(CH₂)₃(CH₃CH₂)CHCH₂NH₂ | 130.23 | | 142.3 | 0.7894 (20/20) | 0.25 | | 4.46 | 140 | 189 | 5.29 |
| Di(2-ethylhexyl)amine | [CH₃-(CH₂)₃(CH₃CH₂)—CHCH₂]₂NH | 214.45 | | 280.7 | 0.8062 (20/20) | Insol. | <0.01 (20) | 8.33 | 270 | 101 | 9.88 |
| Octadecylamine | CH₃(CH₂)₁₇NH₂ | 269.5 | | 232 | | Insol. | | 9.29 | | 91 | 11.02 |
| Allylamine | CH₂CHCH₂NH₂ | 57.09 | −88.4 | 58 | 0.7621 (20/4) | Complete | | 1.97 | −20 | 428 | 2.23 |
| Diallylamine | (CH₂CHCH₂)₂NH | 97.16 | <−70 | 111 | 0.7627 (10/4) | 8.6 | | 3.35 | 60 | 252 | 3.97 |
| Triallylamine | (CH₂CHCH₂)₃N | 137.22 | | 155.6 | 0.809 (20/4) | 0.25 | | 4.73 | 103 | 178 | 5.61 |
| Cyclohexylamine | C₆H₁₁NH₂ | 99.17 | | 134 | 0.8191 (20.4) | Sol. | | 3.42 | 88 | 247 | 4.06 |
| Dicyclohexylamine | (C₆H₁₁)₂NH | 181.31 | 20 | 254 | 0.9104 (25/25) | Sl sol. | | 6.25 | >210 | 135 | 7.42 |
| N,N-Dimethylcyclohexylamine | C₆H₁₁N(CH₃)₂ | 127.22 | <−77 | 159 | 0.849 (20/20) | 1.1 | 3 (25) | 4.39 | 110 | 192 | 5.20 |

**Table 17.3.** Physical and Chemical Properties of Aliphatic and Alicyclic Polyamines

| Name | Formula | Mol Wt. | M.P. (°C) | B.P. (°C) | Density (g/ml) | Solubility in Water (g/100 ml) | Vapor Pressure (torr) (°C) | Vapor Density (Air=1) | Flash Point[a] (°F) | Conversion Units 1 mg/l (ppm) | 1 ppm (mg/m³) |
|---|---|---|---|---|---|---|---|---|---|---|---|
| Ethylenediamine | $NH_2CH_2CH_2NH_2$ | 60.10 | 8.5 | 116.1 | 0.898 (25/4) | Sol. | 10 (21.5) | 2.07 | 93 | 407 | 2.46 |
| N,N-Diethylethylenediamine | $(CH_3CH_2)_2NCH_2CH_2NH_2$ | 116.20 | | 145.2 | 0.8211 | Very sol. | 4.1 (20) | 4.01 | 115 | 210 | 4.75 |
| Trimethylenediamine | $NH_2(CH_2)_3NH_2$ | 74.13 | −23.5 | 135.5 | 0.884 (25/4) | Sol. | | 2.56 | 75 | 330 | 3.03 |
| 1,2-Propanediamine | $CH_3CH(NH_2)CH_2NH_2$ | 74.13 | −37.2 | 120.9 | 0.864 (20/20) | Complete | 8.0 (20) | 2.56 | 92 | 330 | 3.03 |
| Tetramethylenediamine | $NH_2(CH_2)_4NH_2$ | 88.15 | 27 | 158 | | Very sol. | | 3.04 | | 277 | 3.60 |
| 1,3-Butanediamine | $CH_3CH(NH_2)CH_2CH_2NH_2$ | 88.15 | | 142–150 | 0.85 | | | 3.04 | 125 | 277 | 3.60 |
| Pentamethylenediamine | $NH_2(CH_2)_5NH_2$ | 102.18 | 9 | 178–180 | 0.9174 (0/4) | Sol. | | 3.52 | | 239 | 4.18 |
| Hexamethylenediamine | $NH_2(CH_2)_6NH_2$ | 116.21 | 41–42 | 204–205 | | Sol. | | 4.01 | | 210 | 4.75 |
| Diethylenetriamine | $(NH_2CH_2CH_2)_2NH$ | 103.17 | −3 | 207.1 | 0.9586 (20/20) | Complete | 0.2 (20) | 3.56 | 215 | 237 | 4.22 |
| Triethylenetetramine | $NH_2(CH_2CH_2NH)_2CH_2CH_2NH_2$ | 146.24 | 12 | 266–267 | 0.9818 (20/20) | Complete | <0.01 (20) | 5.04 | 290 | 167 | 5.98 |
| Tetraethylenepentamine | $NH_2(CH_2CH_2NH)_3CH_2CH_2NH_2$ | 189.31 | | 340.3 | 0.9980 (20/20) | Complete | <0.01 (20) | 6.53 | 325 | 129 | 7.74 |

[a]Open cup.

alkyl chain lengths based on the fatty acids occurring in nature. This has led to some difficulty in naming the products. Although systematic names have been used, such as 1-dodecanamine, 1-octadecanamine, and 9-octadecenamine for fatty acids corresponding to lauric, stearic, and oleic acids, respectively, trade names are more commonly employed (4).

## 1.2 Manufacture

The lower aliphatic amines are made from a variety of starting materials, mainly ammonia with alcohols, aldehydes, ketones, or alkyl halides, or hydrogen cyanide with an alkene (olefin). The most used methods are as follows: (*a*) Ammonia and an alcohol are passed continuously over a catalyst at a temperature of 300 to 500°C and a pressure of 790 to 3550 k$P_a$ (100 to 500 psig), producing a good yield of a mixture of primary, secondary, and tertiary amines with negligible by-products. (*b*) Ammonia, hydrogen, and alcohol are passed continuously over a different catalyst at 130 to 250°C and high pressure as before, giving high conversion, again with a mixed amine product that must be separated. (*c*) Ammonia and aldehyde or ketone and hydrogen are passed over a hydrogenation catalyst under conditions similar to those of method 2. (*d*) Hydrogen cyanide is reacted with an alkene $R_2C = CH_2$ to yield $R_3CNH_2$, at a temperature of 30 to 60°C, followed by further heating to hydrolyze the intermediate amide, $R_3CNHCNO$. Many other methods have been used (1). Protective butyl rubber gloves, aprons, chemical face shields, and self-contained breathing apparatus should be used by all personnel handling alkylamines (2).

## 1.3 Uses

In 1976 the commercial production of alkylamines in the United States was in excess of 232 thousand tons. Unspecified amounts of fatty amines and cyclic amines are produced in the United States (1).

In 1989, production of alkylamines in the United States was 877,229 kg. This included more than 13,000 kg of butylamines, more than 300,000 kg of ethanolamines, more than 34,000 kg of ethylenediamine, and more than 45,000 kg of dimethylamine (6).

Alkylamines are used extensively as beginning materials for chemical syntheses, as intermediates, and also as solvents. Methylamine and dimethylamine alone are the bases for more than 20 products, including antispasmodic, analgesic, anesthetic, and antihistaminic pharmaceuticals, insecticides, a soil sterilizer, surfactants, a photographic developer, several solvents, an explosive, a fungicide, a rubber accelerator, a rocket propellant, an ion-exchange resin, a plastic monomer, and a catalyst. Other amines are used as catalysts for polymerization reactions, a poultry feed, a bactericide, corrosion inhibitors, drugs, and herbicides (1).

Butylamines and cyclohexylamines are readily soluble in water and can generally be steam distilled; these properties lead to uses in soaps for water and oil emulsions, and as corrosion inhibitors in steam boiler applications (2). In boiler treatment, so-called filming amines, such as octadecylamine and hexadecylamine, are effective in retarding corrosion by carbonic acid (5).

The aliphatic polyamines are used in a wide variety of manufacturing processes, including the synthesis of fungicides, chelating agents, thermosetting resins, epoxy curing agents, polyamide resins, surfactants, softeners, corrosion inhibitors, ashless dispersant-detergent additives for motor oils, asphalt emulsifiers, and products useful in the preparation of durable press and crease-resistant textiles (3).

The main uses of fatty amines and their derivatives depend on their cationic functionality, which encourages attachment to negatively charged surfaces. They are employed in mining operations involving flotation processes and as fabric softeners, anticorrosive agents, biocides, defoamers, foamers, de-emulsifiers, antistatics, detergents, wetting agents, waterproofing agents, and emulsifiers (ethoxylated amines) for herbicidal formulations (4).

## 2 ABSORPTION

A number of aliphatic amines have been identified as normal constituents of mammalian and human urine. These include methylamine, dimethylamine, trimethylamine, ethanolamine, ethylamine, and isoamylamine, as well as the catecholamines (hydroxytryramine and norepinephrine), histamine, and piperidine (1). Rechenberger (1940) reported that humans excrete approximately 10 mg of volatile alkyl amine nitrogen per day (1). Davies (1954) found as many as eight aliphatic and ring-substituted primary amines in urine, which were excreted in amounts up to 100 μg/day (1). The origin of these amines are partly from endogenous sources as well as environmental.

For example, methylamines are biosynthesized endogenously by degradation of muscle sarcosine and creatine, occur in dietary choline (pat of lecithin), and are generated by intestinal bacteria from trimethylamine oxide present in fish. It has also been suggested that they may arise from the absorption of primary amines formed by decarboxylation of amino acids by intestinal bacteria. Simenhoff (1952) has shown that amine excretion is remarkably reduced in germfree animals or in those whose intestinal flora have been destroyed (1). Methylamine can also be produced environmentally, for example, in the case of spontaneous degradation of methyl isocyanate (7).

The amines are well absorbed from the gut and the respiratory tract, and the simple aliphatic amines can produce lethal effects by percutaneous absorption; the $LD_{50}$ by this route is often the same as that for oral administration (1).

## 3 METABOLISM, DISTRIBUTION, AND ELIMINATION

### 3.1 Background

In 1928 Hare reported the discovery of a new enzyme system in mammalian livers, namely, tyramine oxidase responsible for catalyzing the transformation of two amines by rabbit liver, tyramine, and phenylethylamine, as observed by Ewins and Laidlaw (1910–1911) and Guggenheim and Loffler (1916) (8a). Later, two other enzymes called amine oxidase (Pugh, 1937) and adrenaline oxidase (Blaschko, 1937) were found to catalyze the oxidative deamination of aliphatic amines and epinephrine, respectively. These authors eventually concluded that the enzymes tyramine oxidase, amine oxidase, and adrenaline oxidase were identical (8a).

To differentiate these enzymes from diamine oxidases, which oxidize diamines to corresponding aminoaldehydes (8b), Zeller et al. (1939) proposed the name monoamine oxidase (MAO) (8a).

The term MAO covers more than one enzyme entity, and exists as a great variety of similar mitochondrial flavoproteins with overlapping substrate specificities, located in various tissues such as liver, kidney, and brain (8a).

In animals, the lower aliphatic amines are mainly metabolized to the corresponding carboxylic acid and urea; the intermediate compounds have been shown by in vitro experiments to be the corresponding aldehyde and ammonia. MAOs oxidatively deaminate primary, secondary, and tertiary amines. However, compounds that have a substituted methyl group on the alpha carbon are not metabolized by MAOs (9). To serve as a substrate, the amine group must be attached to an unsubstituted methylene group ($\alpha$-carbon hydrogen prerequisite). For example, MAO cannot catalyze the oxidation of aniline, amphetamine, or methylamine. Ethylamine and propylamine have also been demonstrated to be poor substrates for MAO. The rate of enzyme oxidation activity increases with increase in aliphatic chain length to an optimum of $C_5$ or $C_6$, followed by a decline in activity to a maximum at 13 methylene groups (1).

For example, butyl-, amyl-, isoamyl-, and heptylamines are readily metabolized to aldehyde and ammonia (10); but compounds of longer chain lengths, for example, octadecylamine, may inhibit the enzyme. Deamination rate also diminishes with secondary and tertiary amines (1).

By the 1950s it became rather well accepted that the lower-molecular-weight secondary amines, which were more resistant to metabolism, tended to be excreted unchanged, and the first step of metabolism when it occurred was believed to be dealkylation to a primary amine, according to the following reaction scheme (10a):

$$RCH_2NHCH_2R' \rightarrow R'CHO + RCH_2NH_2 \rightarrow RCHO + NH_3$$

In 1941 Green argued that amine oxidase catalytically oxidizes aliphatic secondary amines to an aldehyde and a lower amine according to the following reaction (again indicating the presence of an available $\alpha$-carbon hydrogen) (10a):

$$RCH_2NH_2R' + H_2O + O_2 \rightarrow RCHO + NH_3R' + H_2O_2$$

The monamine oxidase deamination equation for primary, secondary, and tertiary amines is as follows (10a):

$$RCH_2NR'R'' + O_2 + H_2 \rightarrow RCHO + NHR'R'' + H_2O_2$$

The corresponding equation for primary aliphatic amine oxidative deamination by MAO is as follows (10b):

$$RCH_2NH_2 + O_2 + H_2O \rightarrow RCHO + NH_3 + H_2O_2$$

The aldehyde is eventually oxidized to the corresponding carboxylic acid and excreted; ammonia when formed would be eliminated as urea.

Deamination by monamine oxidase is in effect equivalent to a dealkylation reaction; the intermediate is assumed to be a carbinolamine, and upon loss of ammonia, yields a carbonyl derivative.

MAO also functions endogenously in the metabolic modulation of neurotransmitter amines such as noradrenaline, dopamine, and serotonin, metabolizes catecholamines in conjunction with COMT, and deaminates long-chain diamines (10b).

MAO stayed in relative obscurity until 1952 when the first in vivo inhibitor of MAO appeared, namely, iproniazid, and then interest in this group of enzymes expanded rapidly, as well as in aromatic amine research. By the late 1950s it was becoming clear that secondary and tertiary amines, which are xenobiotics, and only slowly oxidized by MAO, could also be metabolized by a microsomal enzyme system that was not identical with MAO. This system was later identified as P450 monooxygenase, which was initially demonstrated to catalyze the $N$-oxidation of amines. The microsomal enzyme system deaminates MAO inhibitors, aromatic amines, but does not act on MAO substrates such as benzylamine and isoamylamine (11).

Many of the biologically and pharmacologically important secondary and tertiary substituted amines are metabolized by dealkylation, which may be carried out through the function of an enzyme system that is different from monoamine oxidase, and is located in the microsomes of liver cells (1).

In 1962 Ziegler demonstrated the existence of still another amine oxidation enzyme system, a microsomal flavoprotein that did not depend upon cytochrome P450. This enzyme not only $N$-oxidized aromatic amines, such as dimethylaniline, but also trimethylamine (9, 12).

MAO is involved only to a limited extent in the metabolism of aliphatic amines (10b), whereas microsomal enzymes can be responsible for multiple amine biotransformations. Microsomal versus mitochrondrial substrate affinity depends not only upon structural characteristics, but also upon the degree of lipophilicity. Microsomal enzyme-catalyzed aliphatic and alicyclic amine biotransformation reactions can include deamination, methylation, $N$-dealkylation, $N$-oxidation, $N$-acetylation, cyclization, $N$-hydroxylation, and nitrosation (13–15).

For example, microsomal $N$-oxidation of secondary aliphatic amines generates the corresponding, fairly reactive hydroxylamines. These in turn are susceptible to further oxidation to nitrones. Primary aliphatic amines, with an available α-carbon hydrogen, are $N$-oxidized to a nitroso compound (via possible nitroxide intermediate), followed by an oxime. Nitroxide intermediates can form stable complexes with P450 (16). Both types of "immonium $N$-oxides," nitrones and oximes, are particularly susceptible to nucleophilic attack and are readily hydrolyzed in aqueous media to yield the corresponding carbonyl compound, primary hydroxylamine, or ammonia, or possibly reduced to a primary amine (17–20). In contrast, N-oxidation of tertiary amines yields a stable tertiary amine oxide.

In summary, the principal routes of biotransformation of aliphatic and alicyclic amines are N-dealkylation or oxidative deamination and N-oxidation. (*a*) *Primary* aliphatic amines are susceptible to N-oxidation only if the α-carbon is substituted. This prevents the formation of a carbinolamine intermediate, and N-oxidation occurs, resulting in formation of primary hydroxylamines, which can then undergo further oxidation to nitroso and oxime metabolites. (*b*) *Secondary* acyclic and cyclic aliphatic amines are susceptible to microsomal oxidative N-dealkylation, oxidative deamination, and N-oxidation to hydroxylamines and nitrone; this is a recognized route for metabolism of many medicinal secondary amines. (*c*) *Tertiary* aliphatic amines are primarily N-oxidized to stable amine N-oxides (21, 22). Additional metabolic pathways are discussed below in the specific chemical section, including methylation, C-dealkylation, cyclization, N-hydroxylation, and nitrosation.

In contrast to secondary amines, and tertiary or quaternary alkyl amines following N-dealkylation, nitrosation by nitrite ions of primary amines yields essentially nonreactive alcohols and olefins. However, secondary amines are prone to react with nitrite, depending on the pH of the media, to form nitrosamines, some of which are potent animal carcinogens, as discussed in Chapter 2, Part A. Carcinogenicity, however, reportedly depends on the availability of an α-carbon hydrogen, which is essential for generating a reactive, alkylating species (23, 24).

Some studies have suggested the possibility of in vivo biosynthesis of carcinogenic nitrosamines within the acidic environment of the stomach following ingestion of nitrites and secondary amines present in food (1). Amines absorbed by pathways other than food may also reach the gastrointestinal tract via absorption from the systemic circulation. Dimethylamine appeared in gastric fluid within minutes of intravenous injection into dogs and ferrets. Concentrations of dimethylamine continued to rise in gastric fluid for 3 hr after injection, and remained elevated for more than 5 hr (25). Friedman et al. (1975) found that oral administration of sodium nitrite within 1 hr of oral dimethylamine to mice yielded a marked dose-response inhibition of liver nuclear RNA synthesis (1), a known effect of nitrosamines (6). No inhibition of nuclear RNA synthesis was observed when sodium nitrite was given 30 min prior to the dimethylamine (1).

It is believed that the principal source of exposure to humans to N-nitroso compounds seems to be their formation in the gastrointestinal tract, most notably the stomach. However, as indicated above, direct nitrosation of a secondary amine by a nitrite ion is a pH-dependent reaction, and depending on the basicity of the amine, may not occur very easily except in the presence of very high concentrations of nitrite ion. Moreover, the nitrite ion is unstable at low pH and the highest yields of nitrosamine formation are usually at pH of 3 or higher (20). The normal human pH of gastric fluid is <2.0 (26), and normally ranges from 1 to 2.5 (27).

There is also a diurnal cycle of gastric acidity in humans; during the night the stomach contents are usually highly acidic (pH about 1.3) (27). Individuals who routinely present with a gastric secretory pH equal to or greater than 3.5 are considered to be clinically achlorhydric or anacidic (26). During digestion, by counteracting the alkalinizing effect of pancreatic juice and bile, gastric hydrochloric acid also normally maintains the duodenal contents at about pH 6 (28), and ranges from a pH of 5 to 7 (27). Recovery of stomach acidity after a meal usually occurs quite rapidly (27).

As pointed out above, high nitrite concentration is essential for the formation of nitrosamines, and the low concentration of nitrite in the saliva of a nonsmoking individual, formed by bacterial reduction of nitrate, is considered to be quantitatively less important than the nitrite present in cured meats and fish. Nitrosation reaction rates of secondary amines have been shown to be proportional to the square of the nitrite concentration (29).

It has also been demonstrated in animals that nitrosation of diethylamine and dimethylamine in vivo is a very slow process because of the strong basicity of these amines; consequently, as has been suggested, insufficient nitrosodiethylamine was probably formed during an experiment in which diethylamine fed to rats as the hydrochloride together with sodium nitrite for 2 years did not lead to induction of tumors typical of treatment of rats with nitrosodiethylamine (30). See Chapter 11, Part A, entitled $N$-Nitrosamines for a further detailed discussion.

There are five types of amino groups that can be conjugated by acetylation in humans and other primate species; these include the aromatic and aliphatic amino group (10a), as well as the sulfonamide group, the amino group of amino acids, and the hydrazine group (31). These reactions are characterized by the transfer of an acetyl moiety, the donor generally being acetyl coenzyme A, whereas the accepting chemical group has a primary amino group. Although in comparison to oxidative deamination, $N$-acetylation conjugation represents only a very minor pathway in the metabolism of aliphatic amines, acetylator genotype may play a role in individual susceptibility to toxicity. For example, in considering relative susceptibility to isoniazid polyneuritis as a consequence of slow acetylation, about 50 percent of Caucasians and Negroes inactivate INH slowly and 50 percent rapidly, whereas about 90 percent of Japanese and Eskimos are rapid acetylators. The acetylation is controlled by an autosomal pair of genes with a dominant effect. Racial variations in the prevalence of slow acetylators are as follows: Hindu Indians, 60 percent; Jews, 55 to 75 percent; American Caucasians, 45 to 60 percent; American Negroes, 45 percent; Thais, 28 percent; Chinese 15 percent; Japanese, 10 percent; and Eskimos, 5 percent (32).

It was recognized as early as 1953 that humans can be divided into groups according to their ability to acetylate isoniazid (33). The acetylation is controlled by an autosomal pair of genes with a dominant effect, and it has recently been demonstrated that the capacity for $N$-acetylation is genetically regulated in humans at the NAT2 gene locus (34). Human epidemiologic evidence suggests a correlation between acetylator phenotype and the incidence and severity of tumors associated with arylamine exposures (35). See Chapter 2, Part A, entitled Occupational Carcinogens for further background.

In addition to potential adverse effects due to slow $N$-acetylation of amines, there are also persons who possess a relatively defective $N$-oxidation capacity of trimethylamine, and consequently excrete large quantities of trimethylamine in their urine after eating fish or choline rich diets. These patients generate what is known as the "fish odor syndrome" (36). This a not a major problem, however, because impairment of N-oxidation of trimethylamine is thought to involve a rare allele defect, and only a few affected individuals and their families have been identified (37).

### 3.2 Methylamine

When administered to humans, dogs, or rabbits as the hydrochloride, methylamine is rapidly absorbed and only traces of the unchanged compound was observed to be excreted in the urine by early investigators (Salkowski, 1877; Schiffer, 1880) (10a). Salkowski claimed that very small amounts of methylurea were excreted by rabbits, but this was not observed in the case of dogs by Schmiedeberg (1878) (10a). In 1893, Pohl claimed that small amounts of formate were excreted by dogs receiving methylamine hydrochloride (10a), and urea formation was demonstrated in a perfused dog liver experiment by Loffler in 1918 (10a). Kapellar-Adler and Krael (1931) were the first to postulate that the amino group was probably split off as ammonia (38).

In 1937 Richter put the two concepts together and proposed that the reaction may be split into two stages in which the amine is first dehydrogenated to the corresponding intermediate imine which reacts spontaneously with water forming an aldehyde and ammonia. Richter believed that most amines were oxidized by amine oxidase, according to the following reaction in the case of primary amines (38):

$$RCH_2NH_2 + O_2 \rightarrow RCH=NH + H_2O_2$$

$$RCH=NH + H_2 \rightarrow RCHO + NH_3$$

Methylamine and trimethylamine are also reportedly metabolized to a small extent to dimethylamine in the body (1, 7), and Alles and Heegaard (1943) demonstrated that methylamine could be oxidized, although with difficulty, in vitro by MAO (8).

### 3.3 Dimethylamine

Rechenberger (1940) demonstrated that dimethylamine (DMA) is almost quantitatively excreted unchanged by humans (10a) and is also a normal constituent of human urine (39). Although oxidative detoxification is the major route of elimination, as discussed above, the presence of unchanged DMA in cases of chronic urinary tract infections, which may also lower the normal pH, has been proposed as a risk factor for bladder cancer. The concept involves the presence of bacterial flora, which may contribute to the reduction of urinary nitrate to nitrite in an acidic environment resulting in the formation of carcinogenic nitrosamines (39).

## 3.4 Trimethylamine

TMA can be detected in mammalian urine and may arise from choline. When administered to humans, dogs, or rabbits (Linzel, 1934; Hoppe-Seyler, 1934), it is partly degraded to ammonia and subsequently to urea, and oxidized to trimethylamine oxide (10a); according to Muller (1940) about 50% of the dose of trimethylamine hydrochloride (administered orally to dogs) was eliminated unchanged together with traces of dimethylamine, suggesting that trimethylamine was $N$-dealkylated (38). In humans, $N$-oxidation to trimethylamine oxide is apparently the major route of metabolism; although $N$-demethylation to dimethylamine can also occur, it is a virtually negligible minor route (7, 40).

In 1941, Green reported that tertiary amines are also oxidized by amine oxidase to the corresponding formaldehyde and secondary amine and proposed the following reaction (38):

$$RCH_2NR'R'' \rightarrow RCHO + R'R''NH$$

It was also proposed that the secondary amine can then form a primary amine and this eventually would yield ammonia. In vivo, therefore, the nitrogen of a tertiary amine could be expected to appear in part as urea.

The second reaction involved the oxidation of the amine to the amine oxide as follows (10a):

$$RR'R''N \rightarrow RR'R''NO$$

$N$-dealkylation can be expected to occur with most amines, with an available $\alpha$-carbon hydrogen. The reaction is essentially the same as oxidative deamination, that is, cleavage of the carbon–nitrogen bond with transfer of two electrons with the formation of a carbonyl compound and a dealkylated amine. It is therefore unlikely that tertiary aliphatic amines will be easily $N$-dealkylated to secondary amines. However, it may depend on the size of the alkyl groups.

For example, trimethylamine has been reported to be $N$-dealkylated to dimethylamine, and studies have also shown that a $t$-butyl group can be removed through initial hydroxylation of one of the $t$-methyl groups, resulting in $C$-dealkylation to the corresponding carbinol or alcohol product (41).

In cases, therefore, where tertiary amines have a small, easily removed group (methyl, ethyl, isopropyl), then oxidative $N$-dealkylation can probably proceed with preferential removal of the smaller substituents, dissociating into a secondary amine and aldehyde via microsomal P450 (42).

Trimethylamine oxide occurs in large amounts in the tissues of fish, and its transformation into trimethylamine by bacteria containing trimethylamine oxide reductase is largely responsible for the spoiling of fish (10a).

## 3.5 Ethylamine

Monoethylamine (MEA) is less readily metabolized than methylamine, and although a large proportion may be destroyed, its nitrogen being converted into urea

(Schmiedeberg, 1878; Loffler, 1918), nearly a third of the dose may be excreted unchanged by humans when administered as the hydrochloride. Schmiedeberg suggested that dogs might excrete traces of ethylurea after ethylamine (10a).

### 3.6 Diethylamine

Diethylamine (DEA) is mainly excreted unchanged by humans when it is administered as the hydrochloride (10a).

### 3.7 Triethylamine

Biotransformation metabolites eliminated in human urine following inhalation at 10, 20, 35, and 50 mg/m$^3$ for 4 or 8 hr included unchanged TEA and triethylamine N-oxide (43).

Pharmacokinetics in four male volunteers has been studied. The doses of TEA and TEAO was 25 mg and 15 mg orally and 15 mg/hr × 1 hr intravenously, respectively. TEA was efficiently absorbed from the gastrointestinal (GI) tract (90 to 97 percent), no significant first-pass metabolism. The apparent volumes of distribution during the elimination phase were 192 liters for TEA and 103 liters for TEAO. TEA was metabolized into triethylamine N-oxide (TEAO) by addition of oxygen to the nucleophilic nitrogen. TEAO was also absorbed from the GI tract. Both TEA and diethylamine (DEA) appeared in the urine after ingestion of TEAO, but not after an intravenous route of administration. This indicates that TEAO is reduced into TEA or dealkylated into DEA only within the GI tract. The TEA and TEAO had plasma half-lives of about 3 and 4 hr, respectively. Exhalation of TEA was minimal; more than 90 percent of the dose was recovered in the urine as TEA and TEAO. The authors recommend that the sum of TEA and TEAO be used for biologic monitoring (44). Previously reported symptoms of visual disturbances in humans (foggy vision, blue haze) (45) were not observed following systemic administration (44).

### 3.8 n-Propylamine

n-Propylamine appears to be more readily metabolized than ethylamine in humans, and in dogs Bernhard (1938) found none of the administered amine to be excreted unchanged (10a).

### 3.9 n-Butylamine

n-Butylamine is as readily metabolized in humans as methylamine, and Pugh (1937) demonstrated in guinea pig liver slices that one of its metabolites is acetoacetic acid (10a).

## 3.10  n-Amylamine

n-Amylamine is converted to acetone, valeric acid, and urea (Sachs, 1910; Loffler, 1918); and Richter (1938) isolated in guinea pig liver slices the corresponding aldehydes (10a).

## 3.11  Isoamylamine

Isoamylamine is converted to isovaleric acid and urea (Sach, 1910; Loffler, 1918); and Richter (1938) isolated in guinea pig liver slices the corresponding aldehyde and isoamyl alcohol (10a). When 100 mg was administered orally to humans (Richter, 1938), only metabolites of the parent compound were excreted (10a).

Isoamylamine has been shown to be deaminated by MAO but not only by microsomal deaminase (11). This may be of clinical significance in the event concurrent therapy is undertaken with MAO inhibitors, resulting in potential chemical/drug interaction(s).

## 3.12  Long- and Branched-Chain Aliphatic Amines

Structure–activity relationships, corresponding potency of nitroso derivatives (46, 47), and P450 inhibition by the longer-chain aliphatic amines have been extensively studied (48–51). It has been observed that as the length of the carbon chain increases with aliphatic amines, the carcinogenic potency of the corresponding nitrosoamine decreases, and branching of an alkyl chain at the alpha or beta carbon (diisopropylamine, diisobutylamine, or di-sec-butylamine) results in either decrease or loss of carcinogenic activity (52). However, increased carbon chain length, for example, does not necessarily destroy carcinogenicity potential of the corresponding nitrosoamine. For example, nitrosoethyl-N-hexylamine is a potent esophageal carcinogen in rats (46), nitrosomethyl-N-octylamine is a liver and urinary bladder carcinogen in experimental animals (47), dodecyldimethylamine (lauryldimethylamine) when fed to rats with nitrite reportedly produced bladder and forestomach tumors (53), and nitrosomethyltetradecylamine caused bladder cancer in rats (46). However, nitroso-di-n-octylamine is reportedly noncarcinogenic in animals (52).

## 3.13  Cycloamines

The only example of a metabolism study involving a cycloamine reported in the literature involved the highly polar cyclohexylamine, which was essentially unmetabolized and eliminated unchanged (54).

## 3.14  Aliphatic Polyamines

In 1904 Mayer reported that both amino groups were removed from diaminopropionic acid, yielding $\alpha,\beta$-dihydroxypropionic acid, and hence deamination was shown to occur with diamines and was not confined to monoamines (38). However, diamine oxidase, or histaminase, acts only on the short-chain polymethylene diamines in vitro. The four-carbon diamine, putrescine, is most rapidly oxidized. The rate drops off with increasing chain lengths to a minimum around $C_{10}$ (1). Diamines with longer chains are oxidized by MAO with a maximum rate occurring at $C_{13}$ (tridecamethylenediamine) (38).

The enzyme oxidizes diamines with molecular oxygen producing an aminoaldehyde, ammonia, and hydrogen peroxide according to the following scheme (55):

$$H_2N[CH_2]_nNH_2 + O_2 + H_2O \rightarrow H_2N[CH_2]_{n-1}CHO + H_2O_2 + NH_3$$

In these reactions the ammonia that eventually is formed is converted to urea. The hydrogen peroxide ($H_2O_2$) is acted on by catalase (generating $CO_2$ and water), and the aldehyde formed is probably converted to the corresponding carboxylic acid by the action of aldehyde oxidase.

Diamine oxidase (or histaminase) occurs widely in animal tissues, particularly in the intestine, where it can destroy such diamines as putrescine, cadaverine, and agmatine, which may be formed in the gut by bacterial action upon basic amino acids. The diamines ethylenediamine (EDA), trimethylenediamine, putrescine, and cadaverine are deaminated by this enzyme, yielding ammonia and an aldehyde (10a, 56).

In an experiment to investigate pharmacokinetic parameters in male Hilltop Wistar rats administered doses of 5, 50, or 500 mg/kg of [$^{14}$C]ethylenediamine · 2HCl by the oral, endotracheal, or intravenous route dose-dependent pharmacokinetics were observed. As the dosage increased from 5 to 500 mg/kg, the proportion of urinary excretion appeared to decrease, whereas the fecal excretion increased. Following endotracheal dosing, at 5 mg/kg, the fecal excretion rate was higher than that of the orally dosed animals. At a 500-mg/kg intravenous dose, the amounts of radiochemical(s) excreted via both urinary and fecal routes were relatively high during the 24 to 48 hr period, suggesting the possibility that the large dose, particularly as an intravenous bolus, overwhelmed the excretory capacity of the animal. Distribution among tissues analyzed was relatively proportional to the dose; however, the thyroid and bone marrow contained relatively high levels of radioactivity, and in some instances their levels exceeded the levels of those organs that are responsible for metabolic and excretory functions of xenobiotics. $N$-Acetylation is a major metabolic pathway for EDA in the rat, and $N$-acetylethylenediamine was identified as the major urinary metabolite; the urinary profiles from the rats dosed endotracheally or intravenously were in general comparable to those from orally treated rats, and depending on the dosage level, 2 to 49 percent of the radioactivity was unchanged parent compound. At 5 mg/kg the bioavailability of EDA was considerably lower (approximately 60 percent) for the oral and endotracheal dosing, which was explained in terms of the influence of a "first-pass effect" of the liver or the lung. It is conceivable that liver and lung possess enzyme systems that metabolize EDA, and at a nonoverwhelming low dose significantly reduced the bioavailability but had a lesser effect at higher dose levels (near 100 percent bioavailability at 50 mg/kg). Terminal half-lives were as follows depending upon dose levels: 7.5, 5.3, and 4.1 hr for 5, 50, and 500 mg/kg orally, respectively; 7.5, 6.9, and 5.1 hr for 5, 50, and 500 mg/kg endotracheally, respectively; and 8, 8, and 6.9 hr for 5, 50, and 500 mg/kg intravenously, respectively. Oral bioavailability dose-dependent kinetics were as follows (57): 60, 95, and 82 percent at 5, 50, and 500 mg/kg, respectively.

### 3.15 Ethylenediaminetetraacetic Acid

Ethylenediaminetetraacetic acid (EDTA) is reportedly eliminated essentially unchanged (58).

### 3.16 Tetramethylenediamine

Bacterial amino acid decarboxylases decompose the amino acids ornithine or arginine to produce tetramethylenediamine (putrescine). Putrescine is classified as a ptomaine (alkaloidal or basic product of the putrefaction of animal or vegetable matter); however, putrescine per se, like cadaverine, does not cause food poisoning.

### 3.17 Triethylenetetramine

This compound is reported to cause increased enzyme activity in the kidneys of pregnant guinea pigs and in the livers of nonpregnant guinea pigs (1).

### 3.18 Ethanolamine

This compound is an endogenous alkanolamine biosynthesized from serine by decarboxylation. It is a principal precursor of phosphoglycerides, which are important elements in the structure of biologic membranes, and choline (59).

Forty percent of $^{15}$N-labeled ethanolamine appears as urea within 24 hr when it is given to rabbits, suggesting that it is deaminized. In rat liver homogenates, ethanolamine undergoes demethylation yielding formaldehyde (60).

Ethanolamine is also a normal constituent of human urine. The excretion rate of ethanolamine in men varies between 4.8 and 22.9 mg/day, with a mean of 0.162 mg/kg. Eleven women were observed to excrete larger amounts, varying between 7.7 and 34.9 mg/day, with a mean excretion rate of 0.492 mg/kg. The excretion rates in animals were, approximately, for cats, 0.47 mg/kg/day; for rats, 1.46 mg/kg/day; and for rabbits, 1.0 mg/kg/day (1).

### 3.19 Dimethylethanolamine

This tertiary amine can be methylated in the body to the quarternary base choline (61).

### 3.20 Diethylethanolamine

Some 33 percent of diethylaminoethanol injected into humans in 1-g doses is excreted unchanged. The transformation of the remaining portion is unknown. It could be deethylated to ethanolamine and thus enter the normal metabolic pathways (1).

### 3.21 Triethanolamine

Hoshino and Tanooka (1978) found that triethanolamine reacted with sodium nitrite to produce $N$-nitrosodiethanolamine and that the product caused mutagenesis in bacteria (1).

## 4 PHYSIOLOGICAL AND PATHOLOGICAL EFFECTS IN ANIMALS

In 1944 Smyth and Carpenter at the Mellon Institute initiated a series of acute toxicity "range finding tests" on previously untested compounds; these established the initial data base, which included a considerable number of amines (1). Their reports are the main sources of the information summarized in Table 17.4. It should be understood that the figures cited and the indexes quoted are not the carefully refined products of detailed study, but are approximations. In their fifth publication (1954), Smyth et al. described the procedures as they had become established over a span of several years (1):

> Single dose oral toxicity for rats is estimated by intubation of dosages in a logarithmic series to groups of five male rats or at times to female rats. The animals are Carworth-Wistar rats, raised in our own colony, fed Rockland rat diet complete, weighing 90 to 120 gm., and not fasted before dosing.
>
> Chemicals are diluted with water, corn oil, or a 1% solution of sodium 3,9-diethyl-6-tridecanol sulfate (Tergitol Penetrant 7) when necessary to bring the volume given one rat to between 1 and 10 ml. Fourteen days after dosing, morbidity is considered complete. The most probable $LD_{50}$ value and its fiducial range are estimated by the method of Thompson using the tables of Weil. When fractional mortality is not observed, the $LD_{50}$ value is recorded without a fiducial range.
>
> Penetration of rabbit skin is estimated by a technique essentially the one-day cuff method of Draize and associates, using groups of four male New Zealand giant albino rabbits weighing 2.5 to 3.5 kg. The fur is closely clipped over the entire trunk, and the dose, retained beneath an impervious plastic film, contacts about 1/10 of the body surface. Dosages greater than 20 ml per kilogram of body weight cannot be retained in contact with the skin. After 24 hours' contact the film is removed, and mortality is considered complete after 14 additional days . . . .
>
> Earlier papers . . . have spoken of saturated vapor inhalation . . . . we now speak of concentrated vapor inhalation. This consists in exposing six male albino rats to a flowing stream of air approaching saturation with vapors. The stream is prepared by passing dried air through a fritted disc gas washing bottle initially at room temperature. When carbon dioxide will react with a sample, the air stream is freed of that gas.
>
> Inhalations are continued for periods of time in an essentially logarithmic series with a ratio of two, extending to eight hours, until the period killing half the rats within

**Table 17.4.** Acute Toxicity of Selected Amines (1)

| Name | CASRN | Acute Oral Toxicity LD$_{50}$ Rat[a] (g/kg) | Acute Skin Toxicity LD$_{50}$ Rabbit[a] (ml/kg) | Inhalation Toxicity (Rats except as Noted) | | | | Rabbit Skin Irritation | Rabbit Eye Irritation | Other Manifestations | e |
|---|---|---|---|---|---|---|---|---|---|---|---|
| | | | | ppm | Time | Mortality | "Saturated" Vapor Time for 0 Deaths | | | | |
| Methylamine | 74-89-5 | 0.1–0.20 (10% soln.) | 0.1 ml survived 40 % 1 died gp | | | | | 40% soln. 0.1 ml, necrosis | 40% soln. corneal damage After 50 mg, 5 min | | * |
| Dimethylamine | 124-40-3 | 0.698 0.240 r 0.240 gp | | | | | | | | | |
| Ethylamine | 75-04-7 | 0.4 0.4–0.8 (70% soln.) | 0.39 | 8000 | 4 | 2/6 | 2 min; all died | Grade 1 Necrosis from 70% gp | Grade 9 | | * |
| Diethylamine | 119-89-7 | 0.54 0.649 m | 0.82 | 4000 | 4 | 3/6 | 5 min; all died | Grade 4 | Grade 10 | Lung irritation severe at 50 ppm | * |
| Triethylamine | 121-44-8 | 0.46 | 0.57 | 1000 1000 | 4 4 | 1/6 2/6 gp | | Grade 2 | Grade 9 Severe at 50 ppm | Lung irritation severe at 50 ppm | * |
| Propylamine | 107-10-8 | 0.57 | 0.56 | 8000 | 4 | 5/6 | 2 min | Grade 6 | Grade 9 | | * |
| Dipropylamine | 142-84-7 | 0.93 | 125 r | 1000 | 4 | 2/6 | 5 min | Grade 6 | Grade 9 | | * |
| Tripropylamine | 102-69-2 | 0.096 r | 0.57 | 250 | 4 | 3/6 | 8 hr | Grade 4 | Grade 1 | | * |
| Isopropylamine | 75-31-0 | 0.82 r | 0.55 | 4000 8000 | 4 4 | 0/6 6/6 | 2 min | Grade 6 | Grade 10 | | * |
| Diisopropylamine | 108-18-9 | 0.77 | | 8000 261 597 261 | 4 [b] [c] [d] | 2/6 4/5 r 1/2 1/2 | 5 min | Grade 1 | Grade 8 | | * |
| Butylamine | 109-73-9 | 0.5 | 0.5 gp | 4000 | 4 | 2-4/6 | 2 min | | | | * |
| Dibutylamine | 11-92-2 | 0.55 | 1.01 | 250 | 4 | 0.6 | | Grade 5 | Grade 9 | | * |
| Tributylamine | 102-89-9 | 0.54 | 0.25 | 75 | 4 | 4/6 | 50 min | Grade 4 | Grade 1 | | * |
| Diisobutylamine | 110-96-3 | 0.258 0.629 m 0.620 gp | | | | | | | | | * |

Table 17.4. (Continued)

| Name | CASRN | Acute Oral Toxicity LD$_{50}$ Rat[a] (g/kg) | Acute Skin Toxicity LD$_{50}$ Rabbit[a] (ml/kg) | Inhalation Toxicity (Rats except as Noted) | | | | "Saturated" Vapor Time for 0 Deaths | Rabbit Skin Irritation | Rabbit Eye Irritation | Other Manifestations | e |
|---|---|---|---|---|---|---|---|---|---|---|---|---|
| | | | | ppm | Time | Mortality | | | | | | |
| Amylamine, mixed isomers | 110-58-71 | 0.47 | 0.65 | 2000 | 4 | 4/6 | | 30 min | Grade 6 | Grade 9 | | * |
| Pentyl-1-pentamine, (dipentylamine) | 2050-92-2 | 0.27 | 0.35 | 63 | 4 | 4/6 | | 30 min | Grade 6 | Grade 5 | | * |
| 2,2'-Diethyldihexylamine | 106-20-7 | 1.64 | | | | | | 8 hr | Grade 5 Extreme irritation | Grade 8 Extreme irritation | | * |
| Allylamine (propenylamine) | 107-11-9 | 0.106 | 0.035 | LC$_{50}$ | | 286 ppm | | 4 hr | | | Chronic inhalation effect on liver at 5 ppm | * |
| Diallylamine | 24-02-7 | 0.578 | 0.356 | LC$_{50}$ | 2755 | 4 hr | | Severe irritation | Severe irritation | Chronic inhalation effects on several organs at 50 ppm, deaths at 200 ppm | | |
| Triallylamine | 1102-75-5 | 1.31 | 2.25 | LC$_{50}$ | 828 ppm | 4 hr | | Mild irritation | Mild irritation | Chronic inhalation effects on several organs at 50 ppm, deaths at 200 ppm | | |
| 2-Ethylbutylamine | 617-79-8 | 0.39 | 2.00 | 2200 | 4 | 0/6 | | 422 min | Grade 6 | Grade 9 | | * |
| Cyclohexylamine | 108-91-8 | | | | | | | | | | Human skin severely irritated by 125 mg 48 hr. | |
| Dicyclohexylamine | 101-83-7 | 0.71 | 0.32 | 4000 | 4 | 0/6 | | | Grade 7 Severe irritation | Grade 10 | | * |
| 2-Aminoethanol (ethanolamine) | 141-43-5 | 2.1 | | | | | | | | | | |
| 2,2'-Iminodiethanol (diethanolamine) | 111-42-2 | 10.20 12.76 | | | | | | Mild irritation | Severe irritation | | | * |

| Compound | CAS No. | | | | | | | | |
|---|---|---|---|---|---|---|---|---|---|
| 2,2′,2″-Nitrilotriethanol (triethanolamine) | 102-71-6 | 8.0 | | | | | Mild irritation | Grade 5 | * |
| Isopropanolamine (1-amino-2-propanol) | 78-96-6 | 4.26<br>5.24<br>1.52 gp | 1.61 | | | 8 hr | Grade 3 | Grade 9 | * |
| Triisopropanolamine | 122-20-3 | 6.5 | | | | | | Grade 6 | * |
| 2-Methylaminoethanol | 109-83-1 | 2.34 | 1.37 | | | 8 hr | Grade 3 | Grade 9 | * |
| 2-Dimethylaminoethanol | 109-01-0 | 2.31 | 0.36 | | | 8 hr | Grade 1 | Grade 8 | * |
| 2-Ethylaminoethanol | 110-73-6 | 1.48<br>1.0 | | | | 8 hr | Grade 2 | Grade 9 | * |
| 2-Diethylaminoethanol | 100-37-8 | 1.3 | 1.08 gp | | | 4 hr | | | * |
| 2-Dibutylaminoethanol | 102-81-8 | 1.07 | 1.68 | | | 8 hr | Grade 6 | Grade 5 | * |
| 2,2′-(Methylimino)diethanol | 105-59-9 | 4.57 | | | | 8 hr | Grade 2 | Grade 8 | * |
| 2,2′-(Ethylimino)diethanol | 139-87-7 | 4.25 | | | | 8 hr | Grade 2 | Grade 8 | * |
| 3-Amino-1-propanol | 156-87-6 | 2.83 | 1.25 | | | 8 hr | Grade 5 | Grade 9 | * |
| Ethylenediamine | 107-15-3 | 1.46<br>0.47 gp | 0.73 | 2000<br>4000<br>255 | 8<br>8<br>7 × 30 | 0/6<br>6/6<br>11/15 | Grade 6 | Grade 8 | 200 ppm, 10 sec mild eye irritation in humans; 400 ppm intolerable |
| 1,3-Propanediamine | 109-76-2 | 0.35 | 0.20 | | | 8 hr | Grade 7 | Grade 8 | * |
| 1,2-Propanediamine (propylenediamine) | 78-90-0 | 2.23 | 0.50 | | | 8 hr | Grade 6 | Grade 9 | * |
| 1,3-Butanediamine | 590-88-5 | 1.35 | 0.43 | | | 8 hr | Grade 6 | Grade 9 | * |
| Diethylenetriamine | 111-40-0 | 2.33<br>1.08 | 1.09 | | | 8 hr | Grade 6 | Grade 8 | * |
| Diethylenediamine piperazine | 110-85-0 | | | | | | | Severe irritation | |
| Triethylenetetramine | 112-24-3 | 4.34 | 0.82 | | | 4 hr | | Grade 6 | Grade 5 | * |

[a] = Except as noted. r = rabbit; gp = guinea pig; m = mouse.
[b] = 7 hr/day, 2 to 20 days.
[c] = 7 hr/day, 5 days.
[d] = 7 hr/day, 19 days.

[e] Asterisks identify those entries derived from reports that use the methods and codes described by Smyth et al. and quoted here in the text, except that the fiducial ranges for oral toxicity have not been included.

14 days of inhalation is defined. The Table records the longest period which allowed all rats to survive 14 days or, in a few instances marked by a symbol, the shortest period tried when this killed all rats.

Inhalation of known vapor concentrations by rats is conducted with flowing streams of vapors prepared by various styles of proportioning pumps. Nominal concentrations are recorded, not confirmed by analytical methods. Exposures are four hours long, although rarely an eight-hour period is used.

Concentrations are in an essentially logarithmic series with a factor of two, and the Table records the concentration yielding fractional mortality among six rats within 14 days. Where no fractional mortality was observed, both the concentration yielding no mortality and that yielding complete mortality are indicated.

Primary skin irritation on rabbits records in a 10-grade ordinal series the severest reaction on the clipped skin of any of five albino rabbits within 24 hours of application of 0.01 ml. of undiluted sample or of dilutions in water, propylene glycol, or kerosene. Grade 1 in the Table indicates the least visible capillary injection from undiluted chemical. Grade 6 indicates necrosis when undiluted, and Grade 10 indicates necrosis from a 0.01% solution.

Eye injury in rabbits records the degree of corneal necrosis from various volumes and concentrations of chemical as detailed by Carpenter and Smyth. Grade 1 in the Table indicates at most a very small area of necrosis resulting from 0.5 ml of undiluted chemical in the eye; Grade 5 indicates a so-called severe burn from 0.05 ml, and Grade 10 indicates a severe burn from 0.5 ml of a 1% solution in water or propylene glycol.

The warning of the authors should not be overlooked or forgotten: "It should be emphasized that the range-finding test is relied upon only to allow predictions of the comparative hazards of handling new chemicals. Acute toxicity studies, no matter how carefully planned, yield no more than indications of the degree of care necessary to protect exposed workmen and indications that certain technically feasible applications of a chemical may or may not eventually be proved safe."

From an industrial hygiene point of view, the most important action of the amines is the strong local irritation they produce. Animals exposed to concentrated vapors exhibit signs and symptoms of mucous membrane and respiratory tract irritations. Single exposures to near-lethal concentrations and repeated exposures to sublethal concentrations result in tracheitis, bronchitis, pneumonitis, and pulmonary edema. For most of the amines listed in Table 17.5, a single skin application will cause deep necrosis, and a drop in a rabbit's eye results in severe corneal damage or complete eye destruction. The acute oral toxicities range from moderately high to slight. Some of the effects observed result from the local corrosive action of the bases in the gastrointestinal tract. The salts are less irritating and therefore less toxic by mouth. They also show less skin and eye irritation when applied as solutions (1).

Interest in the pharmacology of the simple aliphatic amines was initially stimulated by their structural relationship with epinephrine (adrenalin) (1). Barger and Dale (1910–1911) introduced the term sympathomimetic to describe effects similar

Table 17.5. Relative Toxicity of Amine Bases and Their Hydrochloride Salts in Aqueous Solutions (37)

| Amine | Base[a] | | Hydrochloride Salt[b] | |
|---|---|---|---|---|
| | Approximate Oral $LD_{50}$, Rats (g/kg) | Eye irritation, Rabbit, 1 Drop | Approximate Oral $LD_{50}$, Rats (g/kg) | Eye Irritation, Rabbit, 1 Drop |
| Methylamine | 0.1–2 (40%) | Immediate, severe (40%) | 1.6–3.2 (40%) | Mild, normal in 24 hr (40%) |
| Ethylamine | 0.4–0.8 (70%) | Immediate, severe (70%) | >3.2 (10%) | Moderate, normal at 14 days (70%) |
| n-Propylamine | 0.2–0.4 (10%) | Immediate, severe (undiluted) | 3.2–6.4 (25%) | Mild, normal in 24 hr (crystals) |
| n-Butylamine | 0.2–0.4 (10%) | Immediate, severe (undiluted) | 1.6–3.2 (10%) | Moderate, normal at 14 days (crystals) |
| Ethylenediamine | 0.7–1.4 (85%) | Severe (undiluted) | 1.6–3.2 (10%) | Mild, normal at 14 days (crystals) |

[a]Bases = vol/vol.
[b]Salts = wt/vol.

to those of epinephrine, which serve to stimulate the sympathetic branch of the autonomic nervous system, leading to elevation of blood pressure, contraction of smooth muscle, salivation, and dilatation of the pupil of the eye. They found that aliphatic amine hydrochlorides, given intravenously, caused increasing blood pressure responses with increasing carbon chain length up to $C_6$. Branched-chain members were less active. There was decreasing sympathomimetic activity above $C_7$, with increasing cardiac depression. Studies by Swanson et al. (1946) and Ahlquist (1945) on the influence of structure of epinephrine-like activity indicate that primary amines have somewhat more pressor activity than secondary and tertiary; straight chains are more active than branched; a second amine group in the chain increases this activity; and an amine group on the second carbon gives maximum activity (1). It has been generally observed that when amines are administered repeatedly, cardiac stimulation is replaced by vasodilatation and cardiac depression (1). Convulsions frequently occur after near-fatal doses.

Simple aliphatic amines can be both inhibitors and substrates of amine oxidases. Aliphatic amines can cause the release of histamine and can potentiate its action. The monoamines produce a typical histaminelike "triple response" (white vasoconstriction, red flare, wheal) in human skin at concentrations sufficient to cause release of histamine from guinea pig lung. Maximum histamine release from the series $C_nH_{2n+1}NH_2 \cdot HCl$ is at $C_{10}$. Straight-chain diamines show increasing histamine release from $C_6$ to about $C_{14}$. Potent histamine releasors, such as Compound 48/80 and octylamine, cause a decrease in blood pressure, tachycardia, headache, itching, erythema, urticaria, and facial edema when administered intravenously in humans, as does histamine. It is possible that histamine-releasing agents will produce bronchoconstriction and wheezing by inhalation, for histamine aerosol has this effect (1).

Pathological changes in the lungs, liver, kidneys, and heart have been observed following administration of aliphatic amines. Brieger et al. (1951) reported that pulmonary edema has been produced, with hemorrhage and bronchopneumonia, nephritis, liver degeneration, and degeneration of heart muscle in rabbits repeatedly exposed to ethylamines (1). With the exception of myocardial degeneration, Pozzani et al. (1954) reported similar effects occur in animals exposed to ethylenediamine (1). Hine et al. described myocardial damage after vapor exposures to allylamines (1). Tabor and Resenthal (1956) have shown that spermine (diaminopropyltetramethylenediamine) has a high degree of nephrotoxicity. Monoamines ($C_1$ to $C_{10}$), diamines ($C_4$ to $C_{10}$), and diethylenetriamine, triethylenetriamine, and tetraethylenepentamine were inactive. Ethylenediamine, ethylenimine, 1,3,-diaminopropane, and 1,2-diaminopropane produced proteinuria and tubular damage of a lesser degree than spermine (1).

## 5 PHYSIOLOGICAL AND PATHOLOGICAL EFFECTS IN HUMANS

Most studies of the effects of aliphatic amines have described their local action as primarily irritative and sensitizing. Vapors of the volatile amines cause eye irritation with lacrimation, conjunctivitis, and corneal edema, which result in "halos" around light (1). Inhalation causes irritation of the mucous membranes of the nose and throat and lung irritation with respiratory distress and cough. The vapors may also produce primary skin irritation and dermatitis (1). Direct local contact with the liquids is known to produce severe and sometimes permanent eye damage, as well as skin burns. Cutaneous sensitization has been recorded, chiefly due to the ethylenamines. Systemic symptoms from inhalation are headache, nausea, faintness, and anxiety. These systemic symptoms are usually transient and are probably related to the pharmacodynamic action of the amines.

Eckardt (1976) reported that a "variety of dermatitises and acute and chronic pulmonary problems have been associated with the use of two plastics, namely epoxides and polyurethanes, especially polyurethane foams. In the case of epoxy resins, these effects have been associated with the curing agents namely ethylenediamine, diethylenetriamine and triethylenetriamine. These are highly alkaline compounds capable of causing extensive corrosive skin reactions. They are also known as skin sensitizers, and hence may cause allergic skin reactions. Because the reaction is exothermal, fumes may be produced that in sensitized individuals can lead to bronchial asthma. Because the curing agents may be used in excess, in order to drive the polymerization to completion, subsequent grinding, sanding, or

Table 17.6. Toxicity of Allylamines for Rats (27)

|  | Monoallylamine | Diallylamine | Triallylamine |
|---|---|---|---|
| Oral $LD_{50}$, mg/kg | 106 | 578 | 1310 |
| Percutaneous $LD_{50}$ (rabbit, mg/kg) | 35 | 356 | 2250 |
| Inhalation $LC_{50}$, ppm[a] | | | |
|   4 hr | 286 | 2755 | 828 |
|   8 hr | 177 | 795 | 554 |
| Repeated inhalation, 7 hr × 50, ppm[b] | | | |
|   Change in liver or kidney weights | 5 | 200 | 100 |
|   Reduced growth | 10 | 200 | 200 |
|   Deaths | 40 | 200 | 200[c] |

[a]Calculated.
[b]Measured.
[c]1/15 occurred at 100 ppm.

polishing of epoxy resins may produce dusts and fumes, with all of the above-described reactions occurring" (1).

Brubaker et al. (1979) described a dose-related reduction of several pulmonary function test results and symptoms of cough, phlegm, wheezing, and chest tightness associated with exposure to 3-(dimethylamino)propylamine at 0.9 ppm or lower average concentration. Pulmonary functions declined during the course of a 10-hr work shift. Improved ventilation reduced the exposure level to 0.13 ppm. After that, the workers showed improvement of pulmonary functions during a work shift and lower respiratory symptoms disappeared (1).

## 6 SPECIFIC COMPOUNDS

This section is intended to summarize the animal and human toxicology information from the literature, including physical and chemical properties, synonyms, uses, and hygienic standards. In addition, compounds that were not cited in the literature on toxicology studies or reports from 1980 to 1992, based on a computer literature search, are also indicated (62). See Table 17.5 regarding relative toxicity of amine bases and corresponding hydrochloride salts, Table 17.6 regarding toxicity of allylamines, and Table 17.7 for physical characteristics and acute toxicity values for some alkanolamines.

### 6.1 Methylamines

#### 6.1.1 Methylamine [CAS # 74-89-5]

The molecular formula for methylamine is $CH_3NH_2$; it is also known as monomethylamine, anhydrous monomethylamine, aminomethane, methanamine, and MMA. Methylamine is a colorless, flammable gas at room temperature and at-

**Table 17.7. Physical Characteristics and Toxicologic Levels of Some Alkanolamines (1)**

| Name | Formula | Mol. Wt. | B.P. (°C) | Specific Gravity (20/4) | Flash Point[a] (°F) | Solubility in Water | Approximate Oral $LD_{50}$, Rats (g/kg) | Approximate Percutaneous $LD_{50}$ (ml/kg) |
|---|---|---|---|---|---|---|---|---|
| Ethanolamine | $HOCH_2CH_2NH_2$ | 61.08 | 170.5 | 1.018 | 185 | Complete | 2.1 | |
| Diethanolamine | $(HOCH_2CH_2)_2NH$ | 105.14 | 268 | 1.09664 | 280 | 96.4% w/w | 1.82 | |
| Triethanolamine | $(HOCH_2CH_2)_3N$ | 149.19 | 335 | 1.124 | 355 | Complete | 9.11 | |
| 3-Amino-1-propanol | $HOCH_2CH_2CH_2NH_2$ | 75.11 | 187–188 | 0.9824 | 175 | Miscible | 2.83 | 1.25 |
| Isopropanolamine (1-amino-2-propanol) | $CH_3CH(OH)CH_2NH_2$ | 75.11 | 160 | 0.9611 | 171 | Complete | 4.26 | 1.64 |
| Triisopropanolamine | $[CH_3CH(OH)CH_2]_3N$ | 191.27 | 305 | 1.0 | 320 | Ver sol. | 6.50 | |
| 2-Methylaminoethanol (methylethanolamine) | $CH_3NHCH_2CH_2OH$ | 75.11 | 158 | 0.937 | 165 | Complete | 2.34 | Absorbed |
| 2-Dimethylaminoethanol (dimethylethanolamine) | $(CH_3)_2NCH_2CH_2OH$ | 89.14 | 134 | 0.8864 | 105 | Complete | 2.34 | 1.37 |
| 2-Ethylaminoethanol (ethylethanolamine) | $CH_3CH_2NHCH_2CH_2OH$ | 89.14 | 169–170 | 0.914 | 160 | Complete | 1.48<br>1.00 | 0.36 |
| 2-Diethylaminoethanol (diethylethanolamine) | $(CH_3CH_2)_2NCH_2CH_2OH$ | 117.19 | 163 | 0.8921 | 140 | Complete | 1.3 | 1.0 |
| 2-Dibutylaminoethanol (dibutylethanolamine) | $(CH_3(CH_2)_3)_2NCH_2CH_2OH$ | 173.29 | 229.7 | 0.859 | 200 | 0.4% w/w | 1.07 | 1.68 |
| N-Methyl-2,2'-iminodiethanol (methyldiethanolamine) | $CH_3N(CH_2CH_2OH)_2$ | 119.16 | 247.2 | 1.0418 | 260 | Complete | 4.78 | 5.99 |
| N-Ethyl-2,2'-iminodiethanol (ethyldiethanolamine) | $CH_3CH_2N(CH_2CH_2OH)_2$ | 133.19 | 146–252 | 1.02 | 280 | Complete | 4.57 | |
| 1-Dimethylamino-2-propanol (dimethylpropanolamine) | $(CH_3)_2NCH_2CH(OH)CH_3$ | 103.16 | 121–127 | 0.85 | | Complete | 1.89 | |

[a] Open cup.

mospheric pressure. It has a characteristic fishlike odor in lower concentrations, readily detectable at ≃10 ppm, becomes strong at 20 to 100 ppm, and becomes intolerably ammoniacal at 100 to 500 ppm (1). The odor threshold reportedly ranges from 0.0009 to 4.68 ppm, and becomes irritating at 24 mg/m$^3$ (63d, e). It is readily liquefied and is shipped as a liquefied gas under its own vapor pressure.

*6.1.1.1 Uses.* Methylamine is used in tanning, fuel additives, photographic developers, and rocket propellants (63a), in the manufacture of dyestuffs, in the treatment of cellulose acetate rayon, and in organic synthesis.

*6.1.1.2 Toxicity.* Brief exposures to 20 to 100 ppm produce transient eye, nose, and throat irritation; no symptoms of irritation are produced from longer exposures at less than 10 ppm; olfactory fatigue occurs readily (1).

In one plant a case of allergic or chemical bronchitis occurred in a worker exposed to 2 to 60 ppm, although masks or respirators were reportedly worn during "greatest exposures" (63a).

Methylamine has been reported to cause liver toxicity in laboratory animals (64). A drop of 5 percent solution in water applied to animal eyes was reported to cause hemorrhages in the conjunctiva, superficial corneal opacities, and edema (65a). Treatment of CD-1 mice by daily intraperitoneal injection of 0.25, 1, 2.5, or 5 mmol/kg from day 1 to day 17 of gestation demonstrated no adverse reproductive effects. In the same experiment, dimethylamine was also without effect; trimethylamine decreased fetal weight without affecting maternal body weight gain. None of the amines caused a significant increase in external or internal organ or skeletal abnormalities (7).

*6.1.1.3 Hygienic Standards of Permissible Exposure.* A threshold limit value (TLV)–time-weighted average (TWA) value of 5 ppm (6.4 mg/m$^3$) and a short-term exposure limit (STEL) of 15 ppm (19 mg/m$^3$) have been adopted (63c).

*6.1.2 Dimethylamine [CAS # 124-40-7]*

The molecular formula for dimethylamine is $(CH_3)_2NH$, with a molecular weight of 45.08. Dimethylamine is also known as DMA and *N*-methylmethanamine. DMA is a flammable, alkaline, colorless gas at room temperature and atmospheric pressure (66). It has a characteristic rotten fishlike odor in lower concentrations (63d, e). The odor threshold ranges from 0.00076 to 1.6 ppm, and becomes irritating at 175 mg/m$^3$ (63d, e). In higher concentrations (100–500 ppm) the fishlike odor is no longer detectable and the odor is more like that of ammonia. It is readily liquefied and is shipped in steel cylinders as a liquefied gas under its own vapor pressure (67).

*6.1.2.1 Uses.* DMA has been used as an accelerator in vulcanizing rubber, in the manufacture of detergent soaps, and for attracting boll weevils for extermination (66). It has also been used as a depilating agent in tanning, as an acid gas absorbent, in dyes, as a flotation agent, as a gasoline stabilizer, in pharmaceuticals, in soaps and cleaning compounds, in the treatment of cellulose acetate rayon, in organic syntheses, and as an agricultural fungicide (67).

***6.1.2.2 Toxicity.*** Dimethylamine is irritating and corrosive to both the eyes and skin of test animals (63b) and is considered to be a severe lung and skin irritant in humans (68).

Occupational exposure to vapors of DMA at concentrations too low to cause discomfort or disability during several hours of exposure have been associated with misty vision and halos that appeared several hours after the exposure occurred. Edema of the corneal epithelium, the effect principally responsible for disturbance of vision, usually clears without treatment within 24 hr. However, after exceptionally intense exposures, the edema and blurring have taken several days to clear and have been accompanied by photophobia and discomfort from roughness of the corneal surface (69).

Several DMA inhalation studies have been reported. Exposure of rats, mice, guinea pigs, and rabbits to 97 or 185 ppm of DMA 7 hr/day, 5 days/week for 18 to 20 weeks revealed corneal injury in eyes of guinea pigs and rabbits as well as central lobular fatty degeneration and necrosis of the liver in all species (70). However, histopathological examination of rats, guinea pigs, rabbits, monkeys, and dogs exposed to 9 $mg/m^3$ of DMA continuously for 90 days produced only mild inflammatory changes in the lungs of all species; dilated bronchi in rabbits and monkeys were noted (71).

Steinhagen et al. (72) reported an acute 6-hr whole-body exposure of male F344 rats to DMA concentrations ranging from 600 to 6000 ppm for 6 hr, which produced a spectrum of pathological changes in the nasal passages, including severe congestion, ulcerative rhinitis, and necrosis of the nasal turbinates. Lesions outside the respiratory tract were evident in livers of rats exposed to 2500 to 6000 ppm, and corneal edema was observed in the eyes of rats at 1000 ppm; corneal ulceration, keratitis, edema, and loss of Descement's membrane was present at 2500 to 6000 ppm; in addition, many rats exposed to 4000 or 6000 ppm had necrosis of the iris and severe degeneration of the lens, suggestive of acute cataract formation. From a 10-min head-only exposure of male F344 rats to DMA concentrations ranging from 49 to 1576 ppm, a concentration–response curve indicated that a 10-min exposure of 600 ppm would be expected to result in a 51 percent decrease in respiratory rate; the corresponding calculated response dose, $RD_{50}$, was 573 ppm.

In an experiment to determine whether pathological changes occurred in the respiratory tract of mice after inhalation exposure to various sensory irritants at their respective $RD_{50}$ concentrations, 16 to 24 male Swiss–Webster mice were exposed to 510 ppm DMA, 6 hr/day for 5 days; lesions induced in the respiratory epithelium ranged from epithelial hypertrophy or hyperplasia to erosion, ulceration, inflammation, and squamous metaplasia (73).

In a 1-year inhalation study, male and female F344 rats and B6C3F1 mice were exposed to 0, 10, 50, or 175 ppm DMA for 6 hr/day, 5 days/week for 12 months. The mean body weight gain of rats and mice exposed to 175 ppm was depressed to 90 percent of control after 3 weeks of exposure. The only other treatment-related effects were the dose-related lesions confined to the nasal passages, which were very similar in rats and mice; however, after 12 months of exposure, rats had more extensive olfactory lesions (74).

In a subsequent 2-year inhalation study in male F344 rats exposed to 175 ppm, no evidence of carcinogenicity was observed, and in addition, despite severe tissue destruction in the anterior nose following a single 6-hr exposure, the nasal lesions exhibited very little evidence of progression, even at 2 years of exposure; it was concluded that this indicated possible regional susceptibility to DMA toxicity or a degree of adaptation by the rat to continued DMA exposure. A detailed evaluation of mucociliary apparatus function and response to alterations of nasal structure was presented by the authors (75).

As reported above under methylamine, DMA was not shown to have reproductive effects in mice by intraperitoneal injection (7).

*6.1.2.3 Hygienic Standards of Permissible Exposure.* A TLV–TWA value of 5 ppm (9.2 mg/m$^3$) and a STEL of 15 ppm (27.6 mg/m$^3$) have been adopted (63c).

*6.1.3. Trimethylamine [CAS # 75-50-3]*

The formula for trimethylamine (TMA) is $(CH_3)_3N$, with a molecular weight of 59.11. It is readily soluble in water and is also soluble in ether, benzene, toluene, xylene, ethylbenzene, and chloroform. TMA is a flammable, alkaline, colorless gas at ambient temperature and atmospheric pressure. The odor threshold reportedly ranges from 0.00011 to 0.87 ppm (63e) and is a characteristic pungent, fishlike odor in low concentrations, but in higher concentrations (100 to 500 ppm) the fishy odor is no longer detectable and the odor is more like that of ammonia (67). In conjugated form, it is widely distributed in animal tissue and especially in fish (66). It is also found in nature as a degradation product of nitrogenous plant and animal residues and is responsible for the odor of rotting cartilaginous marine fish (63a). Trimethylamine is shipped as a liquefied compressed gas in cylinders, tank cars, and cargo tank trucks.

*6.1.3.1 Uses.* It is used primarily in the manufacture of quaternary ammonium compounds, in the manufacture of disinfectants, as a corrosion inhibitor, in the preparation of choline chloride, in various organic syntheses (67), and as a warning agent for natural gas (63a).

*6.1.3.2 Toxicity.* A 1 percent solution applied to animal eyes resulted in severe irritation, 5 percent causes hemorrhagic conjunctivitis, and 16.5 percent causes a severe reaction with conjunctival hemorrhages, corneal edema, and opacities, followed by some clearing but much vascularization (65b).

As reported above, TMA, in contrast to MMA and DMA has been demonstrated to cause embryo toxicity in mice (decreased fetal weight of CD-1 pups) (7).

*6.1.3.3 Hygienic Standards of Permissible Exposure.* A TLV–TWA value of 5 ppm (12 mg/m$^3$) and a STEL of 15 ppm (36 mg/m$^3$) have been adopted (63c).

## 6.2  Ethylamines

### 6.2.1  Ethylamine [CAS # 75-04-7]

The formula for ethylamine is $CH_3CH_2NH_2$. It is also known as MEA, monoethylamine, or aminoethane. Its molecular weight is 45.08, and it is a flammable, colorless gas. The odor threshold reportedly ranges from 0.027 to 3.5 ppm, having a sharp fishy, ammoniacal odor, which becomes irritating at 180 mg/m$^3$ (63d, e). It is miscible with water, alcohol, and ether (66, 69).

*6.2.1.1 Uses.* Ethylamine is used in resin chemistry, as a stabilizer for rubber latex, as an intermediate for dyestuffs, and in pharmaceuticals, and in oil refining and organic syntheses (66).

*6.2.1.2 Toxicity.* Ethylamine is a primary skin irritant (68), as well as an eye and mucous membrane irritant (1), and when tested on rabbit eyes, was severely damaging (69). A 70 percent solution applied to the skin of guinea pigs resulted in prompt skin burns leading to necrosis; when it was held in contact with guinea pig skin for 2 hr, there was severe skin irritation with extensive necrosis and deep scarring (1).

Ethylamine vapor has also reportedly produced vision disturbances and decreased olfaction sensitivity in humans. (76); therefore, the warning properties associated with odor should not be relied upon.

In a comparative inhalation study, Brieger and Hodes (79c) exposed rabbits repeatedly to measured concentrations of ethylamine, diethylamine, and triethylamine. The three amines produced lung, liver, and kidney damage at 100 ppm. The triethylamine produced definite degenerative changes in the heart at 100 ppm, whereas this was an inconstant finding with ethylamine and diethylamine; 50 ppm of these amines was sufficient to produce lung irritation and corneal injury (delayed until 2 weeks with ethylamine).

Ethylamine has been reported to cause adrenal cortical gland necrosis, which, according to Ribelin (77), can also be expected following exposure to short three- or four-carbon chain alkyl compounds bearing terminal electronegative groups. Of the three adrenal gland areas, medulla, zona glomerulosa, and zona fasciculata/reticularis, the last is most sensitive to toxic injury, which is the case with ethylamine. Sensitivity of endocrine glands to toxic insult occurs in the following decreasing order: adrenal, testis, thyroid, ovary, pancreas, pituitary, and parathyroid (77).

*6.2.1.3 Hygienic Standards of Permissible Exposure.* A TLV–TWA value of 10 ppm (18 mg/m$^3$) has been adopted; no STEL has been recommended until additional toxicologic data and industrial hygiene experience become available (63b, c).

### 6.2.2 Diethylamine [CAS # 109-89-7]

The molecular formula for diethylamine is $(C_2H_5)_2NH$, with a molecular weight of 73.14. It is also known as DEA, $N,N$-diethylamine, diethamine, and $N$-ethylethanamine. It is a colorless, flammable, strongly alkaline liquid, miscible with water and alcohol (66). The odor threshold ranges from 0.02 to 14 ppm, having a fishlike odor that becomes ammoniacal and irritating at 150 mg/m$^3$ (63d, e).

*6.2.2.1 Uses.* DEA is used as a solvent, as a rubber accelerator, in the organic synthesis of resins, dyes, pesticides, and pharmaceuticals, in electroplating, and as a polymerization inhibitor (66). Other applications include uses as a corrosion inhibitor. It was reported noneffective as a skin depigmentator (78).

*6.2.2.2 Toxicity.* DEA is an eye and mucous membrane irritant and primary skin irritant (68). Smyth et al. (79a) reported an acute 4-hr $LC_{50}$ value of 4000 ppm, the oral $LD_{50}$ for rats was reported as 540 mg/kg, and the dermal $LD_{50}$ in rabbits was reported to be 580 mg/kg; a 10 percent dilution caused severe eye burns. Adverse ocular effects have also been reported following human exposure (76).

Drotman and Lawhorn (79b) reported histological evidence of liver damage and elevations of serum enzymes following intraperitoneal administration to rats.

In addition to these acute studies, Brieger and Hodes (79c) exposed rabbits at 50 and 100 ppm DEA vapor for 7 hr/day, 5 days/week for 6 weeks and all animals survived. Lungs exposed to 100 ppm exhibited cellular infiltration and bronchopneumonia, livers showed marked parenchymatous degeneration, and kidneys showed nephritis. Similar changes to a lesser degree were observed in animals exposed to 50 ppm. There was slight cardiac muscle degeneration at 50 ppm.

Lynch et al. (80), in a follow-up study to investigate heart muscle degeneration reported by Brieger and Hodes (79c), were unable to find any evidence of cardiac muscle degeneration or any changes in electrocardiograms in F344 rats of both sexes exposed to DEA vapor concentrations of 25 or 250 ppm for 6.5 hr/day, 5 days/week for 30, 60, or 120 exposure days. Rats exposed at 25 ppm showed no effect in any measured parameter. In contrast to DMA observations previously discussed (72), the incidence of histopathological effects (bronchiolar lymphoid hyperplasia) was noted in both controls and the 25-ppm dosage groups; however, the incidence was not dose-related and did not increase with increasing DEA exposure. The authors therefore concluded that the lesions of the respiratory epithelium (consisting of squamous metaplasia, suppurative rhinitis, and lymphoid hyperplasia) were not considered to be of toxicologic significance but were considered a result of irritation. No evidence of cardiac muscle degeneration or of changes in electrocardiograms were seen for up to 24 weeks. It was noted that these negative findings may reflect a species difference in susceptibility to DEA cardiotoxicity previously reported by Brieger (80).

In a follow-up parallel inhalation study with DEA and triethylamine, no respiratory lesions were observed with triethylamine in F344 rats exposed to 25 or 250 ppm (81).

***6.2.2.3 Hygienic Standards of Permissible Exposure.*** A TLV–TWA value of 10 ppm (30 mg/m$^3$) and a STEL of 25 ppm (75 mg/m$^3$) have been adopted (63c).

### 6.2.3 Triethylamine [CAS # 121-44-8]

The molecular formula for triethylamine is $(C_3H_5)N$, with a molecular weight of 101.19; it is also known as *N,N*-diethylethanamine and TEA. It is a colorless, flammable, volatile liquid. The odor threshold reportedly ranges from 0.10 to 29 ppm, having a fishlike, pungent odor, which becomes ammoniacal and irritating at 200 mg/m$^3$ (63d, e). It is slightly soluble in water and miscible with alcohol and ether (66).

***6.2.3.1 Uses.*** TEA is used as a solvent and in the synthesis of quaternary ammonia wetting agents (66); it is also reportedly used in penetrating and waterproofing agents, as a corrosion inhibitor, and in the synthesis of other organic compounds (66).

***6.2.3.2 Toxicity.*** TEA is a skin and eye irritant (68). The acute oral LD$_{50}$ in rats has been reported to be 0.46 g/kg; the rabbit dermal LD$_{50}$, 0.57 ml/kg; and the acute 4-hour LC$_{50}$, apparently 500 ppm (63a). Chronic exposure of rabbits to TEA vapors at concentrations as low as 50 ppm causes multiple erosions of the cornea and conjunctiva, as well as injuries of the lungs in the course of 6 weeks (82).

Male and female F344 rats exposed to vapor concentrations of 0, 25, or 250 ppm up to 28 weeks showed no physiological or pathological evidence of cardiotoxicity; no treatment-related effects were observed (83). However, as in the case with diethylamine, this may be species dependent, and it has also been reported that TEA inhalation in humans has resulted in EEG changes (76).

***6.2.3.3 Hygienic Standards of Permissible Exposure.*** A TLV–TWA value of 10 ppm (41 mg/m$^3$) and a STEL of 15 ppm (62 mg/m$^3$) have been adopted (63c).

## 6.3 Propylamines

### 6.3.1 Isopropylamine [CAS # 75-31-0]

The molecular formula for isopropylamine is $(CH_3)_2CHNH_2$, with a molecular weight of 59.08; it is also known as MIPA, 2-aminopropane, 2-propanamine, monisopropylamine, 2-propylamine, and *sec*-propylamine. It is a colorless, flammable liquid. The odor threshold reportedly ranges from 0.21 to 0.70 ppm; the pungent, ammoniacal odor becomes irritating at 24 mg/m$^3$ (63,e). Isopropylamine is miscible with water, alcohol, and ether (66).

## ALIPHATIC AND ALICYCLIC AMINES

**6.3.1.1 Uses.** Isopropylamine is used as a solvent, as a depilating agent (63a), in the chemical synthesis of dyes and pharmaceuticals, and in other organic syntheses (1).

**6.3.1.2 Toxicity.** Humans exposed briefly to isopropylamine at 10 to 20 ppm experienced irritation of the nose and throat. Workers complained of transient visual disturbances (halos around lights) after exposure to the vapor for 8 hr, probably owing to mild corneal edema, which usually cleared within 3 to 4 hr. The liquid can cause severe eye burns and permanent visual impairment. Isopropylamine in either liquid or vapor form is irritating to the skin and may cause burns; repeated lesser exposures may result in dermatitis (1, 68). The acute oral $LD_{50}$ in rats is reportedly 820 mg/kg (84).

Dudek (85) reported a no observable effect level (NOEL) in Sprague–Dawley rats to be 0.1 mg/l in air following a 30-day inhalation study.

**6.3.1.3 Hygienic Standards of Permissible Exposure.** A TLV–TWA value of 5 ppm (12 mg/m$^3$) and a STEL of 10 ppm (24 mg/m$^3$) have been adopted (63c).

### 6.3.2 n-Propylamine [CAS # 107-10-8]

The molecular formula for *n*-propylamine is $CH_3CH_2CH_2NH_2$, with a molecular weight of 59.11; it is also known as 1-aminopropane and MNPA. It is a colorless, flammable, alkaline liquid with a strong ammoniacal odor, miscible with water, alcohol, and ether (66).

**6.3.2.1 Uses.** *n*-Propylamine is used as a solvent and in organic syntheses.

**6.3.2.2 Toxicity.** During lethal exposure in rats, *n*-propylamine caused clouding of the cornea at 800 ppm (86). Application to rabbit eyes caused severe injury after 24 hr (87). The acute oral $LD_{50}$ in rats is reportedly 0.57 g/kg (88). It is a possible skin sensitizer (66). Additional toxicology information was not located in the literature (62).

**6.3.2.3 Hygienic Standards of Permissible Exposure.** Standards have not been adopted (63c).

### 6.3.3 n-Dipropylamine [CAS # 142-84-7]

The molecular formula for *n*-dipropylamine is $(CH_3CH_2CH_3)_2NH$, with a molecular weight of 101.19; it is also known as di-*n*-propylamine and DNPA. It is a colorless, flammable liquid. The odor threshold reportedly ranges from 0.08 to 227 mg/m$^3$; the odor is ammoniacal (63d). *n*-Dipropylamine is freely soluble in water and alcohol (66).

**6.3.3.1 Uses.** Specific uses were not identified in the literature (62).

***6.3.3.2 Toxicity.*** Dipropylamine is a severe eye irritant (69), and the acute oral $LD_{50}$ in rats is reportedly 0.93 g/kg (88); it may also be a possible skin sensitizer (66). Additional toxicology information was not located in the literature (62).

***6.3.3.3 Hygienic Standards of Permissible Exposure.*** Standards have not been adopted (63c).

### 6.3.4  Diisopropylamine [CAS # 108-18-9]

The molecular formula for diisopropylamine is $(CH_3)_2CHNHCH(CH_3)_2$, with a molecular weight of 101.19; it is also known as *N,N*-diisopropylamine and DIPA. It is a flammable, strongly alkaline liquid. The odor threshold ranges from 0.017 to 4.2 ppm; the fishlike odor becomes irritating at 100 mg/m$^3$ (63d, e, 69). Diisopropylamine is soluble in water and alcohol (66).

***6.3.4.1 Uses.*** Diisopropylamine is used as a solvent and in the chemical synthesis of dyes, pharmaceuticals, and other organic syntheses (1, 66).

***6.3.4.2 Toxicity.*** The acute oral rat $LD_{50}$ is reportedly 0.77 g/kg (89); the compound is irritating to the eyes, skin, and mucous membranes (66, 68).

Temporary impairment of vision has occurred in humans after exposure to vapor concentrations possibly as low as 25 to 50 ppm during distillation of diisopropylamine (90, 91).

Experimental exposures of rabbits, guinea pigs, rats, and cats have established that the vapor is injurious to the corneal epithelium; this presumably is the cause of the visual disturbances observed in humans. Exposure of experimental animals to vapor concentrations from 260 to 2200 ppm for several hours causes clouding of the cornea from epithelial injury and stromal swelling. The highest concentrations were lethal in many instances, owing to severe pulmonary damage, but the corneas of surviving animals ultimately returned to normal (90).

Diisopropylamine is considered to be a severe pulmonary irritant in animals and humans (68). Workers exposed to concentrations between 25 and 50 ppm complained of disturbances of vision described as "haziness," or "halos" around lights (68). There have also been reports of nausea and headache (1). Prolonged skin contact is likely to cause dermatitis (1). Exposure of several species of animals to 2207 ppm for 3 hr was fatal; effects were lacrimation, corneal clouding, and severe irritation of the respiratory tract. At autopsy, findings were pulmonary edema and hemorrhage (1).

***6.3.4.3 Hygienic Standards of Permissible Exposure.*** A TLV–TWA value of 5 ppm (21 mg/m$^3$) has been adopted (63c).

### 6.3.5  Tripropylamine [CAS # 102-69-2]

The molecular formula for tripropylamine is $(CH_3CH_2CH_2)_3N$, with a molecular weight of 143.27. It is a flammable liquid, and is also known as TNPA.

## ALIPHATIC AND ALICYCLIC AMINES

**6.3.5.1 Uses.** No specific uses were located in the literature (62).

**6.3.5.2 Toxicity.** No toxicology information was located in the literature (62).

**6.3.5.3 Hygienic Standards of Permissible Exposure.** Standards have not been adopted (63c).

### 6.4 Butylamines

Butylamines are used in pharmaceuticals, dyestuffs, emulsifying agents, photography, desizing agents for textiles, synthetic tanning agents, and pesticides (1). In particular, the four isomeric monobutylamines are finding increasing use in the manufacture of textiles, plastics, dyestuffs, corrosion inhibitors, lubricating oil additives, antioxidants, fungicides, herbicides, rubber chemicals, and emulsifying agents (92).

#### 6.4.1 n-Butylamine [CAS # 109-73-9]

The molecular formula for *n*-butylamine is $CH_3CH_2CH_2CH_2NH_2$, with a molecular weight of 73.14. It is also known as MNBA, 1-butanamine, and 1-aminobutane. *n*-Butylamine is a flammable volatile liquid. The odor threshold ranges from 0.24 to 13.9 ppm; the sour, fishlike, ammoniacal odor becomes irritating at 30 mg/m$^3$ (63d, e). It is miscible with water, alcohol, and ether (66).

**6.4.1.1 Uses.** *n*-Butylamine is used as a solvent and as an intermediate for pharmaceuticals, dyestuffs, rubber chemicals, emulsifying agents, insecticides, and synthetic tanning agents (66). It also has been reported to be effective (0.1 *M*) in inhibiting the corrosion of iron in concentrated perchloric acid (5).

**6.4.1.2 Toxicity.** *n*-Butylamine vapor is only mildly irritating to the eyes, but the liquid tested on animals' eyes is as severely injurious as ammonium hydroxide; the injurious effect seems to be attributable to the alkalinity of butylamine because, as with other amines, its damaging effects on the cornea is prevented if it is neutralized with acid before testing (93). Other references indicate that *n*-butylamine is a potent skin, eye, and mucous membrane irritant, and direct skin contact causes severe primary irritation and blistering (66). The acute oral LD$_{50}$ for Sprague-Dawley rats (male + female) has been reevaluated and reported as 371 mg/kg (92).

*n*-Butylamine at measured concentrations of 3000 to 5000 ppm produces an immediate irritant response, labored breathing, and pulmonary edema, with death of all rats in minutes to hours. Ten and 50 percent vol./vol. aqueous solutions and the undiluted base produce severe skin and eye burns in animals. The immediate skin and eye reactions are not appreciably altered by prolonged washing or attempts at neutralization when these are commenced within 15 sec after application. Direct skin contact with the liquid causes severe primary irritation and deep second-degree burns (blistering) in humans. The odor of butylamine is slight at less than 1 ppm, noticeable at 2 ppm, moderately strong at 2 to 5 ppm, strong at 5 to 10 ppm, and strong and irritating at concentrations exceeding 10 ppm. Workers with daily exposures of from 5 to 10 ppm complain of nose, throat, and eye irritation, and headaches. Concentrations of 10 to 25 ppm are unpleasant to intolerable for more than a few minutes. Daily exposures to less than 5 ppm (most often between 1 and 2 ppm) produce no complaints or symptoms (1). No additional toxicology information was located in the literature (62).

### 6.4.2  2-Butylamine [CAS # 13952-84-6]

The molecular formula for 2-butylamine is $CH_3CH_2CH(NH_2)CH_3$, with a molecular weight of 73.14. It is a flammable liquid, also known as 2-aminobutane, *sec*-butylamine, 2-butanamine, 1-methylpropylamine, or Butafume. The compound is available as (*R*) B.P. 63°C, (+/−) B.P. 63°C, and (*S*) B.P. 62.5°C; the corresponding CASRNs are 13250-12-9, 33966-50-6, and 513-49-5 (66).

*6.4.2.1  Uses.*  2-Butylamine has been used as a fungistat (66).

*6.4.2.2  Toxicity.*  The acute oral $LD_{50}$ for Sprague–Dawley rats (males + females) has been reported to be 152 mg/kg (92). An earlier acute study reported an $LD_{50}$ value of 380 mg/kg (female, Harlan Wistar rats) (94).

2-Butylamine is irritating to the skin and mucous membranes (66). Seven male rats exposed to saturated vapor (280 mg/l) for 5 hr exhibited intense eye and nose irritation, respiratory difficulty, and convulsions; all died. Cornea lesions were opaque and white, and all other organs appeared normal; all amines examined showed irritant and central nervous system stimulant effects; these increased with the degree of substitution, but the higher members had too low a volatility to present a significant vapor hazard. Other amino compounds studied included di-*sec*-butylamine, tributylamine, nonylamine, dinonylamine, trinonylamine, hexamethylenediamine, diethylenetriamine, 2-aminobutan-1-ol and 1-diethylaminopentan-2-one (95). No additional toxicology information was located in the literature (62).

*6.4.2.3  Hygienic Standards of Permissible Exposure.*  Standards have not been adopted (63c).

### 6.4.3  Isobutylamine [CAS # 78-81-9]

The molecular formula for isobutylamine is $(CH_3)_2CHCH_2NH_2$, with a molecular weight of 73.14. It is a flammable liquid, miscible with water, alcohol, and ether (66), and is also known as IBA, 1-amino-2-methylpropane, and 2-methylpropylamine (66).

*6.4.3.1  Uses.*  Specific uses were not located in the literature (62).

**6.4.3.2 Toxicity.** The acute oral $LD_{50}$ for Sprague–Dawley rats (males + females) has been reported to be 228 mg/kg (92). Skin contact can result in erythema and blistering; inhalation causes headache and dryness of the nose and throat (66).

**6.4.3.3 Hygienic Standards of Permissible Exposure.** Standards have not been adopted (63c).

### 6.4.4 Diisobutylamine [CAS # 110-96-3]

The molecular formula for diisobutylamine is $[(CH_3)_2CHCH_2]_2NH$, with a molecular weight of 129.24; it also known as DIBA. It is a colorless, flammable liquid, very slightly soluble in water and soluble in alcohol or ether (66).

**6.4.4.1 Uses.** Specific uses were not located in the literature (62).

**6.4.4.2 Toxicity.** Toxicology information was not located in the literature (62).

**6.4.4.3 Hygienic Standards of Permissible Exposure.** Standards have not been adopted (63c).

### 6.4.5 Di-n-butylamine [CAS # 111-92-2]

The molecular formula for di-$n$-butylamine is $(C_4H_9)_2NH$, with a molecular weight of 129.24; it is also known as dibutylamine, $n$-dibutylamine, and DNBA. It is a colorless liquid, soluble in water, alcohol, and ether (66). The odor threshold reportedly ranges from 0.42 to 2.5 mg/m$^3$; the fishlike odor becomes ammoniacal at higher concentrations (63d).

**6.4.5.1 Uses.** Dibutylamine is used as a solvent and in organic syntheses.

**6.4.5.2 Toxicity.** Animal studies have demonstrated that dibutylamine is severely irritating to the eyes (69). An acute oral rat $LD_{50}$ value of 550 mg/kg has been reported (89).

**6.4.5.3 Hygienic Standards of Permissible Exposure.** Standards have not been adopted (63c).

### 6.4.6 Tri-n-butylamine [CAS # 102-89-9]

The molecular formula for tri-$n$-butylamine is $(CH_3CH_2CH_2CH_2)_3N$, with a molecular weight of 185.34; it is also known as tributylamine and TNBA. It is a hygroscopic liquid, sparingly soluble in water and very soluble in alcohol and ether (66).

**6.4.6.1 Uses.** Specific uses were not located in the literature (62).

*6.4.6.2 Toxicity.* Exposure of four male and four female rats to 120 ppm for 19 6-hr exposures caused nose irritation, restlessness, incoordination, and tremors; organs appeared normal at autopsy (95). An $LD_{50}$ (rat oral) of 540 mg/kg has been reported (96). Tributylamine has reportedly caused central nervous system (CNS) stimulation and skin irritation and sensitization (66).

*6.4.6.3 Hygienic Standards of Permissible Exposure.* Standards have not been adopted (63c).

### 6.4.7 tert-Butylamine [CAS # 75-64-9]

The molecular formula for *tert*-butylamine is $(CH_3)_3CNH_2$, with a molecular weight of 73.14; it is also known as trimethylcarbinylamine, *t*-butylamine, 2-aminoisobutane, 1,1-dimethylethylamine, and trimethylaminoethane. It is a colorless, flammable liquid, miscible with alcohol (66).

*6.4.7.1 Uses.* *tert*-Butylamine is used as a solvent and in organic syntheses.

*6.4.7.2 Toxicity.* The acute oral $LD_{50}$ for Sprague–Dawley rats (males + females) has been reported to be 80 mg/kg (92). Liver toxicity NOEL was reported to be 0.20 mg/l for female Sprague–Dawley rats (96).

### 6.4.8 Triisobutylamine [CAS # 1116-40-1]

The molecular formula for triisobutylamine is $[(CH_3)_2CHCH_2]_3N$; it is a liquid having a molecular weight of 185.36, and it is also known as TIBA (66).

*6.4.8.1 Uses.* Specific uses were not located in the literature (62).

*6.4.8.2 Toxicity.* No toxicology information was located in the literature (62).

*6.4.8.3 Hygienic Standards of Permissible Exposure.* Standards have not been adopted (63c).

### 6.4.9 Ethyl-n-butylamine [CAS # 13360-63-9]

The molecular formulas for ethyl-n-butylamine is $CH_3(CH_2)_3(C_2H_5)NH$, with a molecular weight of 101.19; it is a flammable liquid and is also known as *N*-ethylbutylamine and EBA.

*6.4.9.1 Uses.* Specific uses were not identified in the literature (62).

*6.4.9.2 Toxicity.* Toxicology information was not located in the literature (62).

# ALIPHATIC AND ALICYCLIC AMINES

***6.4.9.3 Hygienic Standards of Permissible Exposure.*** Standards have not been adopted (63c).

## 6.4.10 Dimethylbutylamine [CAS # 927-62-8]

The molecular formula of dimethylbutylamine is $CH_3(CH_2)_3(CH_3)_2N$, with a molecular weight of 101.19; it is a flammable liquid and is also known as *N,N*-dimethylbutylamine and DMBA.

***6.4.10.1 Uses.*** Specific uses were not identified in the literature (62).

***6.4.10.2 Toxicity.*** Toxicology information was not located in the literature (62).

***6.4.10.2 Hygienic Standards of Permissible Exposure.*** Standards have not been adopted (63c).

## 6.5 Amylamines

### 6.5.1 n-Amylamine [CAS # 110-58-7]

The molecular formula for *n*-amylamine is $CH_3(CH_2)_4NH_2$; it has a molecular weight of 87.16 and is also known as pentylamine and 1-aminopentane. It is a flammable liquid, very soluble in water, soluble in alcohol, and miscible with ether (66).

***6.5.1.1 Uses.*** *n*-Amylamine is used as a solvent and in organic syntheses.

***6.5.1.2 Toxicity.*** Direct contact with the liquid causes first and second degree burns of the skin. Inhalation results in irritation of the respiratory tract and mucous membranes (1). Additional toxicology information was not located in the literature (62).

***6.5.1.3 Hygienic Standards of Permissible Exposure.*** Standards have not been adopted (63c).

### 6.5.2 di-n-Amylamine [CAS # 2050-92-2]

The molecular formula for di-*n*-amylamine is $[CH_3(CH_2)_4]_2NH$, with a molecular weight of 157.30; it is a corrosive liquid, also known as dipentylamine (66).

***6.5.2.1 Uses.*** di-*n*-Amylamine is used as a solvent and in organic syntheses.

***6.5.2.2 Toxicity.*** Direct contact causes severe burns, as indicated above. Additional toxicology information was not located in the literature (62).

***6.5.2.3 Hygienic Standards of Permissible Exposure.*** Standards have not been adopted (63c).

### *6.5.3 tri-n-Amylamine [CAS # 621-77-2]*

The molecular formula for tri-n-amylamine is $[CH_3(CH_2)_4]_3N$; it is a liquid, a mixture of isomers, with a molecular weight of 227.44. It is soluble in alcohol and either, and is also known as tripentylamine (66).

***6.5.3.1 Uses.*** tri-*n*-Amylamine is used as a solvent and in organic syntheses.

***6.5.3.2 Toxicity.*** Direct contact will be irritating, and may result in severe burns. Additional toxicology information was not located in the literature (62).

***6.5.3.3 Hygienic Standards of Permissible Exposure.*** Standards have not been adopted (63c).

### *6.5.4 Isoamylamine [CAS # 107-85-7]*

The molecular formula for isoamylamine is $(CH_3)_2CHCH_2CH_2NH_2$, with a molecular weight of 87.16; it is also known as isobutylcarbylamine, 3-methylbutylamine, isopentylamine, and 1-amino-3-methylbutane. It is a colorless, flammable liquid, having a strong ammonia odor; it is miscible with water, alcohol, chloroform, and ether (66).

***6.5.4.1 Uses.*** Isoamylamine is used in organic syntheses.

***6.5.4.2 Toxicity.*** Isoamylamine is irritating to the skin and mucous membranes (66). Sympathomimetic amine pharmacological activity has been demonstrated in humans; when isoamylamine is injected intravenously, flushing and evidence of apprehension are also observed. The reaction is reportedly mild at 83 mg/kg (97a, b). It has also been reported to stimulate salivary and lacrimal secretions and smooth muscle (98). Hartung (1931) reported that 250 mg/kg is not toxic to rabbits, 1.5 g of the sulfate kills rats, and 1.8 g of the hydrochloride kills rabbits (1). Additional toxicology information was not located in the literature (62).

***6.5.4.3 Hygienic Standards of Permissible Exposure.*** Standards have not been adopted (63c).

## 6.6 Hexylamines

### *6.6.1 n-Hexylamine [CAS # 111-26-2]*

The molecular formula is $CH_3(CH_2)_5NH_2$ having a molecular weight of 101.19; also known as 1-aminohexane. It is a colorless, flammable liquid, slightly soluble in water and soluble in alcohol and ether (66).

## ALIPHATIC AND ALICYCLIC AMINES

*6.6.1.1 Uses.* n-Hexylamine is used in organic syntheses.

*6.6.1.2 Toxicity.* n-Hexylamine is an irritant for the skin, eyes, and mucous membranes. It is an active sympathomimetic agent (98). The vapor toxicity is somewhat higher than that of the butylamines and skin irritation is at least as great (1). Additional toxicology information was not located in the literature (62).

*6.6.1.3 Hygienic Standards of Permissible Exposure.* Standards have not been adopted (63c).

### 6.6.2 Isohexylamine [CAS # 617-79-8]

The molecular formula for isohexylamine is $CH_3CH_2CH(CH_2CH_3)CH_2NH_2$, and the molecular weight is 101.19; it is a liquid and is also known as 2-ethylbutylamine and 1-amino-2-ethylbutane.

*6.6.2.1 Uses.* Isohexylamine is used in organic syntheses.

*6.6.2.2 Toxicity.* See acute toxicity table. Additional toxicology information was not located in the literature (62).

*6.6.2.3 Hygienic Standards of Permissible Exposure.* Standards have not been adopted (63c).

### 6.6.3 Dihexylamine [CAS # 143-16-8]

The molecular formula for dihexylamine is $[CH_3(CH_2)_5]_2NH$, with a molecular weight of 185.36; a corrosive liquid, it is also known as di-n-hexylamine and N,N-dihexylamine (66).

*6.6.3.1 Uses.* Dihexylamine is used in organic syntheses.

*6.6.3.2 Toxicity.* It is reported to be a severe eye and skin irritant (99). Additional toxicology information was not located in the literature (62).

*6.6.3.3 Hygienic Standards of Permissible Exposure.* Standards have not been adopted (63c).

### 6.6.4 2-Ethylhexylamine [CAS # 104-75-6]

The molecular formula for 2-ethylhexylamine is $CH_3(CH_2)_3CH(C_2H_5)CH_2NH_2$, with a molecular weight of 129.25; it is a corrosive liquid (66).

*6.6.4.1 Uses.* 2-Ethylhexylamine is used in organic syntheses.

*6.6.4.2 Toxicity.* 2-Ethylhexylamine exhibits a high degree of acute vapor toxicity for rats and is a potent skin and eye irritant (1). Additional toxicology information was not located in the literature (62).

### 6.6.4.3 Hygienic Standards of Permissible Exposure. Standards have not been adopted (63c).

## 6.7 Heptylamines

### 6.7.1 1-Aminoheptane [CAS # 111-68-21]

The molecular formula for 1-aminoheptane is $CH_3(CH_2)_6NH_2$; it is a flammable liquid, having a molecular weight of 115.22, and is also known as $n$-heptylamine or heptylamine (66).

#### 6.7.1.1 Uses. Heptylamine is used in organic syntheses.

#### 6.7.1.2 Toxicity. The acute intraperitoneal $LD_{50}$ in mice is reportedly 60 to 100 mg/kg (1). Additional toxicology information was not located in the literature (62).

#### 6.7.1.3 Hygienic Standards of Permissible Exposure. Standards have not been adopted (63c).

### 6.7.2 2-Aminoheptane [CAS # 123-82-0]

The molecular formula for 2-aminoheptane is $CH_3(CH_2)_4CH(NH_2)CH_3$, with a molecular weight of 115.22; it is also known as 1-methylhexylamine, 2-heptanamine, and tuaminoheptane. It is a volatile liquid, slightly soluble in water and freely soluble in alcohol, ether, petroleum ether, chloroform, and benzene; the sulfate is readily soluble in water (66).

#### 6.7.2.1 Uses. 2-Aminoheptane is a potent adrenergic vasoconstrictor that has been used as a pharmaceutical preparation (Tuamine); it is also used in organic syntheses.

#### 6.7.2.2 Toxicity. The acute intraperitoneal $LD_{50}$ in mice is reportedly 60 to 100 mg/kg (1). In rats, the sulfate has an intraperitoneal $LD_{50}$ of 42 mg/kg. 2-Aminoheptane produces local vasoconstrictive action similar to that of other sympathomimetic drugs. Systemically, 2-aminoheptane produces a sustained elevation of blood pressure in dogs (1). In humans 2 mg/kg by mouth results in palpitation, dry mouth, and headache, with slight rise in blood pressure (100).

Exposure may be contraindicated in individuals who have cardiovascular disease (101). Additional toxicology information was not located in the literature (62).

#### 6.7.2.3 Hygienic Standards of Permissible Exposure. Standards have not been adopted (63c).

### 6.7.3 3-Aminoheptane [CAS # 2829-42-4]

The molecular formula for 3-aminoheptane is $CH_3(CH_2)_3CH(NH_2)C_2H_5$, with a molecular weight of 115.22; it is also known as 1-ethylpentylamine (66).

*6.7.3.1 Uses.* 3-Aminoheptane is used in organic synthesis.

*6.7.3.2 Toxicity.* The acute intraperitoneal $LD_{50}$ in mice is reportedly 60 to 100 mg/kg (1). Additional toxicology information was not located in the literature (62).

*6.7.3.3 Hygienic Standards of Permissible Exposure.* Standards have not been adopted (63c).

### 6.7.4  4-Aminoheptane [CAS # 16751-59-0]

The molecular formula of 4-aminoheptane is $CH_3(CH_2)_2CH(NH_2)C_3H_7$, with a molecular weight of 115.22; it is also known as 1-propylbutylamine and isoheptylamine (66).

*6.7.4.1 Uses.* 4-Aminoheptane is used in organic synthesis.

*6.7.4.2 Toxicity.* The acute intraperitoneal $LD_{50}$ in mice is reportedly 60 to 100 mg/kg (1). Additional toxicology information was not located in the literature (62).

*6.7.4.3 Hygienic Standards of Permissible Exposure.* Standards have not been adopted (63c).

### 6.7.5  Di-n-Heptylamine [CAS # 2470-68-0]

The molecular formula for di-*n*-heptylamine is $[CH_3(CH_2)_6]_2NH$, with a molecular weight of 213.24; it is also known as diheptylamine.

*6.7.5.1 Uses.* Diheptylamine is used in organic syntheses.

*6.7.5.2 Toxicity.* Di-*n*-heptylamine has an approximate oral $LD_{50}$ for rats (undiluted) of 0.2 to 0.4 g/kg; the approximate oral $LD_{50}$ for mice (5 percent in corn oil) is 0.2 to 0.4 g/kg; deaths occur within a few minutes with dyspnea and convulsions (1). One drop of the undiluted amine causes strong irritation of the eye and surrounding tissues and permanent corneal damage; it is also a strong primary skin irritant (1). Additional toxicology information was not located in the literature (62).

*6.7.5.3 Hygienic Standards of Permissible Exposure.* Standards have not been adopted (63c).

### 6.8  Higher Alkylamines ($C_8$–$C_{30}$)

The longer-chain fatty amines and their salts have found significant applications in ore flotation, corrosion inhibition as petroleum additives, fabric softeners, defoamers, foamers, de-emulsifiers, emulsifiers, and the synthesis of disinfectants and antiseptics. The commercially available fatty amines include octanamine, or

1-octylamine, and higher-carbon chains including branched, secondary, tertiary, and diamines. These fatty amines are essentially insoluble in water and are weakly basic. The main uses of fatty amines depend on their cationic functionality, which encourages attachment to negatively charged surfaces (4).

Little toxicity information is available on amines with alkyl chains containing eight or more carbons. It is assumed that these higher alkylamines would be strong local irritants for eyes, skin, and mucous membranes; for example, laurylamine (dodecylamine) is classified by Fleming et al. (1954) with compounds that produce severe burns and vesication of the skin (1). Their low vapor pressure, however, should decrease the hazard from vapor exposures (1).

These compounds, however, can exhibit biologic activity, depending on chain length. For example, in terms of maximum in vitro toxicity to bacteria, maximum activity is reached around $C_{12}$, or dodecylamine; a rapid falling off occurs with additional carbons to the long paraffinic side chain, apparently where micelle formation is beginning to increase rapidly with further increase in chain length (102).

### 6.8.1  1-Octylamine [CAS # 111-86-4]

The molecular formula for 1-octylamine is $CH_3(CH_2)_7NH_2$, with a molecular weight of 129.25; it is a corrosive liquid, also known as 1-aminooctane, octylamine, and octanamine (66).

*6.8.1.1 Uses.* Specific uses were not located in the literature (62). 1-Octylamine is manufactured under the trade name Armeen 8D (Armak) (4).

*6.8.1.2 Toxicity.* Octylamine has been demonstrated to exhibit a Type II P450 ultraviolet binding absorption spectrum, which results from a reversible nitrogen ligand complex with heme iron, producing a lambda peak at 425 to 430 nm (50). This characteristic has been used as a tool to assess the contribution of different forms of cytochrome P450 on regioselectivity in xenobiotic metabolism (51). Toxicology information was not located in the literature (62).

*6.8.1.3 Hygienic Standards of Permissible Exposure.* Standards have not been adopted (63c).

### 6.8.2  2-Octylamine [CAS # 693-16-3]

The molecular formula for 2-octylamine is $CH_3(CH_2)_5CHCH_3NH_2$, with a molecular weight of 129.25; it is a liquid and is also known as 1-methylheptylamine and 2-aminooctane (66).

*6.8.2.1 Uses.* Specific uses were not identified in the literature (62).

*6.8.2.2 Toxicity.* 2-Aminooctane produces elevation of blood pressure in dogs at 1 mg/kg. The minimum lethal dose by injection is 0.135 g/kg in mice. Lethal doses

# ALIPHATIC AND ALICYCLIC AMINES

result in dyspnea, excitation, convulsions, and death in respiratory paralysis (1). 2-Aminooctane is classified as an irritant (66).

### 6.8.2.3 Hygienic Standards of Permissible Exposure. Standards have not been adopted (63c).

### 6.8.3 Dioctylamine [CAS # 1120-48-5]

The molecular formula for dioctylamine is $[(CH_3(CH_2)_7]_2NH$, with a molecular weight of 241.46; it is a liquid and is also known as di-(2-ethylhexyl)amine and di-$n$-octylamine (66).

### 6.8.3.1 Uses. Specific uses were not identified in the literature (62).

### 6.8.3.2 Toxicity. Dioctylamine shows a low degree of acute oral toxicity for rats ($LD_{50}$ of 1.64 g/kg, unneutralized) and is not severely irritating to the skin and eye. A single saturated vapor exposure in rats produced no deaths in 8 hr (1).

### 6.8.3.3 Hygienic Standards of Permissible Exposure. Standards have not been adopted (63c).

### 6.8.4 11-Aminoundecanoic Acid [CAS # 2432-99-7]

The molecular formula of 11-aminoundecanoic acid is $H_2N(CH_2)_{10}COOH$, with a molecular weight of 201.31; it is a solid (66).

### 6.8.4.1 Uses. Specific uses were not identified in the literature (62).

### 6.8.4.2 Toxicity. Reportedly a cancer suspect agent (66b).

### 6.8.4.3 Hygienic Standards of Permissible Exposure. Standards have not been adopted (63c).

### 6.8.5 Dodecylamine [CAS # 124-22-1]

The molecular formula for dodecylamine is $CH_3(CH_2)_{11}NH_2$, with a molecular weight of 185.36. It is a solid and is also known as laurylamine or dodecanamine (66).

### 6.8.5.1 Uses. Specific uses were not identified in the literature (62). Dodecylamine is manufactured under the trade names of Adogen 163D (Ashland), Alamine (General Mills), Armeen 12D (Armak), and Jetamine 12 (Jetco) (4).

### 6.8.5.2 Toxicity. Laurylamine (dodecylamine) is classified by Fleming et al. (1954) with compounds that produce severe burns and vesication of the skin (1).

***6.8.5.3 Hygienic Standards of Permissible Exposure.*** Standards have not been adopted (63c).

### 6.8.6 Octadecylamine [CAS # 124-30-1]

The molecular formula for octadecylamine is $CH_3(CH_2)_{17}NH_2$, with a molecular weight of 269.52; it is a solid and is also known as octadecanamine (66).

***6.8.6.1 Uses.*** Specific uses were not identified in the literature (62); octadecylamine is manufactured under the trade names of Adogen 142D (Ashland), Alamine 7D (General Mills), Armeen 18D (Armak), Kemamine P-990D (Humko), and Jetamine 18D (Jetco) (4).

***6.8.6.2 Toxicity.*** Octadecylamine has been studied by Deichmann et al. (1958) in connection with its possible use as an anticorrosive agent in live steam that could be used to cook food. Rats fed levels of 0 to 500 ppm in the diet for 2 years showed no detectable effects on growth, food consumption, hematology, or microscopic pathology. At 3000 ppm there was anorexia, weight loss, and some histological changes in the gastrointestinal tract, mesenteric nodes, and liver (1). The acute oral $LD_{50}$ for mice and rats is approximately 1 g/kg. Octadecylamine is a primary skin irritant (1).

***6.8.6.3 Hygienic Standards of Permissible Exposure.*** Standards have not been adopted (63c).

### 6.8.7 Other Higher Alkylamines

Toxicology information was not located on the following fatty amines: 1-aminononoame [CAS # 112-20-9], decylamine [CAS # 2016-57-1], didecylamine [CAS # 1120-49-6], tridecylamine [CAS # 2869-34-3], undecylamine [CAS # 7307-55-3], decyldimethylamine [CAS # 1120-24-7], dodecyldimethylamine [CAS # 112-18-5], tetradecylamine [CAS # 2016-42-4], tetradecyldimethylamine [CAS # 112-75-4], hexadecylamine [CAS # 143-27-1], hexadecyldimethylamine [CAS # 112-69-6], and octadecyldimethylamine [CAS # 124-24-7] (62).

## 6.9 Allylamines

Hine (1960) observed decreasing acute oral, dermal, and inhalation toxicity from monoallylamine to triallylamine with the exception of triallylamine, which demonstrated an increased relative toxicity as compared to diallylamine on inhalation (See Table 17.6) (1). The pathological changes observed in rats following repeated inhalation included chemical pneumonias and some liver and kidney damage. The most prominent pathological effect described was myocarditis, which occurred after repeated inhalation of all three allylamines (1). Both mono- and diallylamines are severely irritating to skin, and triallylamine was reported to be mildly irritating (1).

### 6.9.1 Allylamine [CAS # 107-11-9]

The molecular formula for allylamine is $CH_2\!\!=\!\!CHCH_2NH_2$, with a molecular weight of 57.09. It is a flammable liquid, also known as 3-aminopropylene, 3-aminopentene, and monoallylamine (66). The odor threshold is reportedly 14.5 mg/m$^3$, and the odor becomes irritating at approximately 187 mg/m$^3$ (63d).

**6.9.1.1 Uses.** Allylamine is used as a solvent and in organic syntheses including the synthesis of rubber, mercurial diuretics, sedatives, and antiseptics (103).

**6.9.1.2 Toxicity.** Hart (1938) reported that inhalation of allylamine in mice at 1.27 m$M$ was lethal to almost all the animals within 10 min; the investigators experienced transient irritation of the mucous membranes of the nose, eyes, and mouth, with lacrimation, coryza, and sneezing after accidental exposure to an unspecified concentration of the vapor of allylamine (1). Based upon human exposure, allylamine can be detected at 2.5 ppm, and mucous membrane irritation and chest discomfort occurs in some persons at 2.5 ppm; allylamine is intolerable to most at 14 ppm. (1).

Calandra (1959) conducted a 1-year chronic vapor inhalation study with monoallylamine in which rats, rabbits, and dogs were exposed for 8 hr/day, five days/week, to 5 ppm or 20 ppm. No adverse effects on growth, behavioral reactions, or abnormal blood or urine changes were observed. Deaths from pneumonia occurred in three of six rabbits exposed to 20 ppm. Lung changes consistent with chronic irritation were found at both exposure levels. Periodic liver and kidney function tests, transaminase determinations, and electrocardiographic examinations of dogs did not reveal any abnormalities. Congestive changes in the liver and kidneys were noted in dogs at both exposure levels. Myocardial damage was reportedly not observed in rabbits or dogs, and only a few rats showed slight changes, which were not considered different from those expected in unexposed rats (1).

However, allylamines are known cardiotoxins. They can cause acute myocardial necrosis that is cumulative with repeated administrations, as well as vascular proliferative lesions after several weeks of treatment. Rats given 10.7 m$M$ allylamine in drinking water for 1 to 7 days developed alterations in the myocardium that could be detected by light microscopy within 2 days of treatment. Alterations were apparent in several organelles including the endoplasmic reticulum, mitochondria, Golgi complex, and nucleus. Interstitial and endothelial cells underwent frequent mitosis. Large areas of myocardial necrosis showed extravasated red blood cells and minimal intra- and extracellular deposition of fibrin (104).

Cardiotoxicity was described in 1935 by Mellon, who observed localized arteritis following intradermal injection of allylamine (103). Waters extensively studied canine arterial lesions that closely resemble atherosclerosis in humans, after inducing the initial arterial injury with allylamine, but the underlying mechanism by which allylamine injured the heart and blood vessels remains obscure (103). Allylamine specifically causes acute myocardial and vascular toxic effects that are cumulative with chronic treatment. It is believed that vascular smooth muscle metabolism of allylamine is related to localized toxicity (103). The cardiovascular toxicity of allylamine has been postulated to result from metabolic biotransformation to acrolein and subsequent formation of 3-hydroxypropylmercapturic acid, which may initially occur in the media of arteries, followed by glutathione (GSH) conjugation (105).

Acrolein has been detected in homogenates of aorta, lung, skeletal muscle, and heart incubated with allylamine. Hydrogen peroxide was generated during allylamine oxidation, suggesting that the reaction was an oxidative deamination. Acrolein was also produced by bovine plasma amine oxidase and porcine kidney diamine oxidase, but not by rat liver or brain homogenates, suggesting that the responsible oxidase was benzylamine oxidase rather than monoamine oxidase (106). Allylamine has been shown to be roughly 40 times more toxic to myocytes than to fibroblasts. Semicarbazide, a benzylamine oxidase inhibitor, protected myocytes from toxic effects of allylamine, including cell lysis, arrest of beating activity, and reduced ATP content. On the other hand, clorgyline, a monoamine oxidase inhibitor, was ineffective at altering the toxicity of allylamine (107).

In contrast to a number of allyl derivatives that have been shown to be suicide inhibitors of P450, via heme alkylation resulting in formation of "green pigment" porphyrins, including allyl alcohol, allyl bromide, allyl chloride, allyl cyanide, allyl mercaptan, allyl isocyanate, and 5-allyl-barbiturates, allylamine is inactive (108).

*6.9.1.3 Hygienic Standards of Permissible Exposure.* Standards have not been adopted (63c).

### 6.9.2  Diallylamine [CAS # 124-02-7]

The molecular formula for diallylamine is $(CH_2{=}CHCH_2)_2NH$, with a molecular weight of 97.16. It is a flammable liquid and is also known as di-2-propenylamine and di-2-pentenamine (66).

*6.9.2.1 Uses.* Diallylamine is used as a solvent and in organic syntheses.

*6.9.2.2 Toxicity.* Relative to monoallylamine, diallylamine appears to exhibit decreased acute toxicity (1). Based upon human exposure, diallylamine can be detected, but it is not unpleasant at 2 to 9 ppm; mucous membrane irritation and chest discomfort occur in a few subjects at 22 ppm, but it is not intolerable at 70 ppm (1). Additional toxicology information was not located in the literature (62).

*6.9.2.3 Hygienic Standards of Permissible Exposure.* Standards have not been adopted (63c).

### 6.9.3  Triallylamine [CAS # 102-70-5]

The molecular formula for triallylamine is $(CH_2{=}CHCH_2)_3N$. It is a flammable liquid, with a molecular weight of 137.25. Triallylamine is also known as tripentenamine (66).

***6.9.3.1 Uses.*** Triallylamine is used as a solvent and in organic syntheses.

***6.9.3.2 Toxicity.*** Based upon human exposure, triallylamine can be detected at 0.5 ppm; there are mucous membrane irritation or chest discomfort in some at 12.5 ppm, increasingly frequent symptoms to 50 ppm, and irritant symptoms more severe at 75 to 100 ppm, with unpleasant systemic symptoms including nausea, vertigo, and headache (1).

Human exposure to triallylamine has reportedly produced vertigo (76).

Additional toxicology was not located in the literature (62).

### 6.10 Cyclohexylamines

*6.10.1 Cyclohexylamine [CAS # 108-91-8]*

The molecular formula for cyclohexylamine is $C_6H_{11}NH_2$, with a molecular weight of 99.17. It is a flammable liquid, having a strong, fishlike odor; completely miscible with water, alcohols, ethers, ketones, esters, and aliphatic hydrocarbons; and also known as aminocyclohexane, cyclohexanamine, hexahydroaniline, and CHA (66). The odor threshold reportedly ranges from 106 to 448 $mg/m^3$ (63d).

***6.10.1.1 Uses.*** Cyclohexylamine is used as a solvent, as a corrosion inhibitor in boiler water (for binding of carbon dioxide), as a rubber accelerator, and as an intermediate in chemical synthesis of insecticides, emulsifying agents (1), dry-cleaning soaps, and acid gas absorbents, and as a chemical intermediate in plasticizers, dyestuffs, textile chemicals, and to a limited extent cyclamates (66, 109).

***6.10.1.2 Toxicity.*** Cyclohexylamine produced convulsant deaths in rabbits when injected in olive oil at doses of 0.5 g/kg (1). The $LD_{50}$ (intraperitoneal) in mice is 620 mg/kg, and in rats about 350 mg/kg; the $LD_{50}$ (intravenous) in dogs is about 200 mg/kg (110). When it was administered daily for 82 days in the drinking water at 100 mg/kg (Carswell et al. 1937), pathological findings or weight loss appeared in rabbits, guinea pigs, and rats (1). Cyclohexylamine can cause irritation and sensitization (66). Its acute animal toxicity is summarized in Table 17.4.

Watrous and Schulz (1950) exposed rabbits, guinea pigs, and rats to cyclohexylamine vapors 7 hr/day, 5 days/week, at average concentrations of 150, 800, or 1200 ppm. At 1200 ppm, all animals except one rat showed extreme irritation and died after a single exposure. Fractional mortality occurred after repeated exposure at 800 ppm. At 150 ppm, four of five rats and two guinea pigs survived 70 hr of exposure, but one rabbit died after 7 hr. The chief effects were irritation of the respiratory tract and eye irritation with the development of corneal opacities. No convulsions were observed (1).

Watrous and Schultz (1950) also reported three cases of transitory systemic toxic effects from acute accidental industrial exposures. The symptoms were lightheadedness, drowsiness, anxiety and apprehension, and nausea. Slurred speech, vomiting, and pupillary dilatation occurred in one case. Operators exposed to 4 to 10 ppm had no symptoms (1). Mallette et al. (1952) reported that in human patch tests a 25 percent solution produced severe skin irritation and possible skin sensitization (1). The same authors reported the intraperitoneal $LD_{50}$ in rats to be 200 mg/kg; they considered cyclohexylamine to be highly toxic and noted signs of severe shock in injected animals and observed degenerative changes in the brain, liver, and kidney (63b). Tests in guinea pigs did not give evidence of skin sensitization (1).

Reports that sodium cyclamate is metabolized to cyclohexylamine in humans and dog (111, 1), and that cyclohexylamine produces chromosome aberrations following intraperitoneal injection in rats (112), resulted in intensive investigations to reevaluate the human risk of cyclohexylamine exposure. Biotransformation of cyclamate to cyclohexylamine had also been demonstrated to be carried out by intestinal microorganisms which cleave the N–S bond via hydrolysis yielding cyclohexylamine (108). However, earlier claims that cyclamate was carcinogenic have yet to be substantiated (113).

In 1970 Khera and Stoltz reported that male rats given 220 mg/kg/day of cyclohexylamine, a dose that had no apparent effect on growth or behavior, fathered significantly fewer litters than control rats and the litters were smaller (1).

Becker and Gibson found cyclohexylamine to be embryo toxic following single intraperitoneal injections of 61, 77, or 122 mg/kg administered to Swiss–Webster mice on day 11 of pregnancy. Examination of 49 hand-sectioned fetuses revealed two malformed fetuses in the group given 61 mg/kg, one exencephalus and one hydronephrosis; and the authors recommended that a more extensive study incorporating days 8 through 13 may possibly show a more clear-cut effect on the kidney. The authors concluded, however, that no statistically significant anomalies were observed (114).

Price et al. (115) reported the development of bladder tumors in Charles River rats fed cyclohexylamine sulfate for 2 years at doses of 0, 0.15, 1.5, and 15 mg/kg/day, 25 male and 25 female per dosage group. During the first year, there was only a slight depression of weight gain in the males of the high dose group; no other signs of toxicity were observed. At the end of 2 years (104 weeks) 13 to 16 animals were still alive in the 0.15 and 1.5 mg/kg groups and eight males and nine females in the 15 mg/kg group. An invasive transitional-cell carcinoma of the bladder was observed in one of eight male survivors of the high-dose group. The author noted that spontaneous bladder tumors were very rare in the strain of rats used. No other relevant findings were noted. In another experiment, rats were fed a mixture of cyclamate and saccharin in their food. Many of the rats were found to convert cyclamate to cyclohexylamine. After 78 weeks, half of the animals were given supplementary feedings of cyclohexylamine in doses from 25 to 125 mg/kg/day. Among 240 rats receiving 2500 mg/kg/day of cyclamate–saccharin mixture, seven males and one female showed papillary tumors of the bladder; all but one of these rats had been shown to convert cyclamate to cyclohexylamine. Three of the animals had received additional cyclohexylamine (1, 115).

Khera et al. reported that when cyclohexylamine sulfate was administered orally to Wistar rats (approximately 22, 44, 89, or 178 mg/kg), the treatment had no adverse effect on female fertility, and was not associated with any adverse effect on postimplantation embryonal viability; however, the male fertility appeared to be impaired in the first of three breeding trials, but this impairment was not evident in the succeeding two of the three trials. No deleterious effect was noted on embryo viability, litter size, litter weight, postnatal viability, weight gain of the pups, or somatic cell chromosomes of pups and dams (116).

No teratogenic effects were observed in ICR mice treated orally with 20, 50, or 100 mg/kg of cyclohexylamine either from day 0 to day 5 or from day 6 to day 11 of pregnancy. There was pronounced fetal growth retardation with the highest dose, which was lethal to some dams. The number of resorptions was increased in some treated groups, but was statistically significant only with the highest dose (117).

In rhesus monkeys, no significant teratogenic or embryo toxic effects were seen from cyclohexylamine administered in doses of 25, 50, or 75 mg/kg body weight for different 4-day periods during the phase of organogenesis (20 to 45 days of pregnancy) (118).

Oser et al. (119) conducted an extensive study involving five groups of 30 male and 30 female FDRL rats given 0, 15, 50, 100 and 150 mg/kg cyclohexylamine hydrochloride in the diet for 2 years. Mucosal thickening of the bladder was seen in animals given the 50 and 150 mg levels; no neoplasms of the urinary bladder were observed. *Trichosomoides crassicauda* was present in all rats surviving more than 65 weeks. The authors concluded, "Except for some nonprogressive growth retardation in the higher dosage groups, due to lower food consumption, the physical and clinical observations in the test groups fell substantially within normal limits and were not significantly different from the untreated controls." However, a significant incidence of testicular atrophy and reduction of litter size were seen at the highest dosage level. No tumors were seen (1, 119).

Gaunt et al. (1974) fed rats diets containing 0, 600, 2000, or 6000 ppm of cyclohexylamine chloride for 13 weeks. At the highest dosages, there was a reduced rate of weight gain, not fully explained by diminished food intake. No specific organ changes were observed except that the testes showed reduced spermatogenesis. Reduction of testis weight was evident. However, the rats remained fertile and their offspring appeared normal (1).

Gaunt et al. (1976) used the same dosages for a 2-year diet feeding study in Wistar rats. They observed no evidence of carcinogenicity, slight anemia, failure to produce normally concentrated urine, and an increase in the number of animals with foamy macrophages in the pulmonary alveoli at the highest dosage level. Decreased food intake caused the lessened body weight gain and organ weights as compared to the controls. Animals that received 2000 or 6000 ppm showed testicular atrophy or tubules with few spermatids. The no-untoward-effect level in both of these studies was 600 ppm, equivalent to an intake of about 30 mg/kg/day (1, 109).

Hardy et al. (120), working in the same laboratory, gave diets to groups of 50 female and 48 male ASH-CS1 SPF mice containing 0, 300, 1000, or 3000 ppm of cyclohexylamine hydrochloride for 80 weeks. Except for some depression of weight gain in the males and minor hepatitis in the females at 3000 ppm, there was no evidence of carcinogenicity, although some mice developed tumors not seen in controls; these were considered to be sporadic findings and within the normal range of spontaneous tumors found in this strain of mice; there was no testicular atrophy or degeneration (1, 120).

Mason and Thompson (1977) fed cyclohexylamine to rats in concentrations of 600, 2000, and 6000 ppm for 90 days. Only at the highest dosage did they observe testicular atrophy and reduction of spermatogenesis, one strain of rats being affected more than another. Mice exposed to cyclohexylamine for 80 weeks and dogs exposed for 8½ years showed no testicular lesions. They suggested that "the seminiferous epithelium of the rat is unusually susceptible to chemically-induced damage" (1).

Kroes et al. (121) conducted a six-generation study in which Swiss SPF mice were exposed to 0.5 percent of cyclohexylamine in the diet for 84 weeks. The $F_0$, $F_{3b}$, and $F_{6a}$ generations, consisting of 50 males and 50 females, were used for carcinogenicity studies. There were no differences in tumor incidences between treated and control animals. One female control animal developed an anaplastic carcinoma of the urinary bladder at 82 weeks. Urinary bladder calculi were observed in all groups. It was reported that pregnancy rate, the number of live-borne fetuses, the number of postnatal survivors, and the body weight of the offspring were all affected unfavorably, and the proportion of male offspring was diminished; no evidence of carcinogenicity or teratogenicity was observed (1, 121).

Lorke and Machemer (1974) looked for dominant lethal mutations in mice after treatment with 102 mg/kg/day of cyclohexylamine sulfate for 5 days, and found no indication of mutagenic action (1).

Cattanach reported a very extensive study of cyclohexylamine, cyclamate, and saccharin (1976) that concluded that none of the compounds could be considered mutagenic (1).

There are several reports regarding the cardiac and blood pressure effects of cyclohexylamine. Eichelbaum et al. (1974), in a study of humans, found a close correlation between plasma levels of cyclohexylamine and arterial blood pressure. However, it was some orders of magnitude less potent than related sympathomimetic substances (1).

Gondry (1972) reported that cyclohexylamine has a diabetogenic effect when fed to rats at a level of 1 percent in the diet. It inhibits the growth of mice at a level of 0.1 percent, and shows marked growth inhibition at higher doses, up to 1 percent. Continued over six generations, growth inhibition persists, but a normal growth pattern is resumed when cyclohexylamine administration is discontinued (1).

According to the International Agency for Research on Cancer (IARC) working group, there is no evidence that cyclohexylamine is teratogenic or carcinogenic (109).

Cyclamate (cyclohexylsulfamic acid calcium salt) has been classified by IARC as a Group 3 carcinogen: "not classifiable as to its carcinogenicity to humans," because the evidence that a risk of cancer is increased among users of artificial sweeteners is inconsistent. IARC also states that the evidence for carcinogenicity to animals is limited (122).

*6.10.1.3 Hygienic Standards of Permissible Exposure.* A TLV–TWA of 10 ppm (41 mg/m$^3$) has been adopted (63c).

### 6.10.2 Dicyclohexylamine [CAS # 101-83-7]

The molecular formula of dicyclohexylamine is $(C_6H_{11})_2NH$, with a molecular weight of 181.32; it is a liquid with a faint fishlike odor, sparingly soluble in water and soluble in usual organic solvents. Dicyclohexylamine is also known as dodecahydrodiphenylamine (66).

*6.10.2.1 Uses.* Dicyclohexylamine is used as a solvent and in organic syntheses. It is reportedly used as a chemical intermediate for the synthesis of corrosion inhibitors, rubber vulcanization accelerators, textiles, and varnishes (109). Additional specific uses were not located in the literature (62).

*6.10.2.3 Toxicity.* Dicyclohexylamine appears to be somewhat more toxic than cyclohexylamine (1). Symptoms and death appear earlier in rabbits (Watrous et al., 1950) after injection of 0.5 g/kg (1). Doses of 0.25 g/kg are just sublethal, causing convulsions and temporary paralysis (1). The oral LD$_{50}$ of dicyclohexylamine in rats is reportedly 373 mg/kg (123). It is also a skin irritant (1). A number of animal studies have been reported; however, the IARC working group noted deficiencies among all of them (109). Additional toxicology information was not located in the literature (62).

*6.10.2.4 Hygienic Standards of Permissible Exposure.* Standards have not been adopted (63c).

### 6.10.3 Dimethylcyclohexylamine [CAS # 98-94-2]

The molecular formula is $C_6H_{11}N(CH_3)_2$, with a molecular weight of 127.23; dimethylcyclohexylamine, a liquid, is also known as N,N-dimethylcyclohexylamine (66).

*6.10.3.1 Uses.* N,N-Dimethylcyclohexylamine is finding use in polyurethane plastics and textiles and as a chemical intermediate (1).

*6.10.3.2 Toxicity.* Acute toxicity tests show that it is somewhat less irritating than cyclohexylamine and less toxic on oral and intraperitoneal administration in rats and mice. The symptoms produced by effective doses are similar: weakness, tremor, salivation, gasping, and convulsions. Inhalation of either compound causes respiratory irritation. Repeated skin applications of the diluted amine (1 percent) do not give evidence of sensitization in guinea pigs (1). Additional toxicology information was not located in the literature (62).

## 6.11 Aliphatic Polyamines

The diamines are strong bases and exhibit skin and eye irritant properties similar to the monoamines. In some cases (ethylenediamine, hexamethylenediamine) they exhibit skin sensitization properties not experienced with the corresponding monoamines. They are absorbed through the skin. The acute percutaneous toxicity is often approximately equivalent to that of the corresponding monoamine (1).

Systemic absorption of diamines can potentially result in either sympatholytic or sympathomimetic effects, depending on carbon chain length. For example, the shorter-chain diamines have a sympatholytic effect upon blood pressure rather than a sympathomimetic effect: ethylenediamine, tetramethylenediamine (1,4-butanediamine, putrescine), and pentamethylenediamine (1,5-pentanediamine, cadaverine) cause depression of the blood pressure in animals; the longer-chain diamines, such as hexadecylmethylenediamine (Blaschko, 1952) may exhibit sympathomimetic activity (1). The histamine-releasing activity of the diamines is slight at $C_4$ (tetramethylenediamine) and increases to a maximum at $C_{10}$ (1,10-decanediamine). Mongar and Schild (1953) demonstrated increasing toxicity to paramecia from $C_6$ to $C_{15}$ (1).

Diamines and polyamines have been shown to act as acyl acceptors for transglutaminase, a $Ca^{2+}$-dependent enzyme of the eye lens which catalyzes the formation of isopeptide cross-links with specific glutamine residues of beta crystalline. This reaction is believed to be associated with the development of senile cataract. Putrescine, spermine, and spermidine have been identified in bovine lens and aqueous humor; elevated levels of these compounds, both free and bound, were found in cataractous, relative to normal, human lenses (124).

One of the earlier theories proposed concerning the mechanism of thalidomide teratogenicity suggested that this compound may be a biologic acylating agent, with an affinity toward nucleophilic compounds, and form stable products with putrescine, cadaverine, or other polyamines, depriving the embryo of the biogenic amines resulting in embryo toxic effects (125).

A number of other theories have been proposed to explain the biochemical and cellular mechanisms of thalidomide teratogenicity, but none has been adequately substantiated or definitively disproved, and the mode of action remains unknown (126).

The triamines [diethylenetriamine (DETA), triethylenetetramine (TETA), tetraethylenepentamine (TEPA)] and polyamine HPA No. 2 have a wide range of industrial applications including hardeners and stabilizers for epoxy resins, as solvents for sulfur, acid gases, natural resins, and dyes, in organic synthesis as saponification agents for acidic materials, as intermediates in the synthesis of detergents, softeners, dyestuffs, emulsifiers, plastics, and pharmaceuticals, and in the vulcanization of rubber (127).

The physical properties of the aliphatic polyamines are given in Table 17.3. Acute animal toxicity data are in Table 17.4.

### 6.11.1 Ethylenediamine [CAS # 107-15-3]

The molecular formula for ethylenediamine is $H_2NCH_2CH_2NH_2$, with a molecular weight of 60.10; it is a flammable, corrosive, colorless, hygroscopic, fuming thick

# ALIPHATIC AND ALICYCLIC AMINES

liquid, having an ammonia odor, freely soluble in water and alcohol. It is also known as 1,2-ethanediamine, 1,2-diaminoethane, and dimethylenediamine (1, 66). The odor threshold range is reportedly 25 to 28 mg/m$^3$; the musty, ammoniacal odor becomes irritating at 250 to 500 mg/m$^3$ (63d).

**6.11.1.1 Uses.** Ethylenediamine is used as a solvent for shellac, sulfur, casein, and albumin; an emulsifier; a stabilizer in rubber latex; a corrosion inhibitor in antifreeze solutions, and a textile lubricant (66); and also in the manufacture of chelating agents, fungicides, synthetic waxes, polyamide resins, corrosion inhibitors, gasoline additives (57), dyes, and pharmaceuticals (1).

**6.11.1.2 Toxicity.** Animal studies have demonstrated that ethylenediamine vapors are irritating to the eyes, mucous membranes, and respiratory tract, and the liquid can cause severe skin corrosion and corneal injury (1). When neutralized, as ethylenediamine hydrochloride, it was found not to injure rabbit eyes that were irrigated with a 0.02 $M$ solution for 10 min after mechanical removal of the epithelium to facilitate penetration; the injuriousness of the free base seems to be attributable principally to its strong alkalinity (128). Ethylenediamine has also been reported to cause severe eye damage in humans (1); vapor exposure may also cause delayed corneal epithelial edema in humans resulting in visualization of halos around lights (68). A fatal poisoning is reported to have occurred when a worker was splashed by EDA and exposed both dermally and by inhalation; clinical hematology included hemolysis, which induced tubulonephritis with anuria and lethal hyperkalemia; death occurred from cardiac collapse 55 hr after the accident (129).

In contrast, the dihydrochloride salt is reportedly less irritating to the eye or skin, and less toxic dermally, presumably because of the neutralizing effect of the hydrochloride; however, the salt form was not found to be less acutely toxic following oral administration in rodents (130).

Repeated exposures of rats to measured concentrations of ethylenediamine vapors produced hair loss and lung, kidney, and liver damage at 484 ppm, with lesser degrees of injury at 225 and 132 ppm (1). No injury was observed at 125 ppm continued for 37-hr exposure (1). Renal tubular damage and proteinuria are produced in rats from intraperitoneal doses of 300 mg/kg (1). Voluntary vapor inhalation for 5 to 10 sec produced tingling of the face and irritation of the nasal mucosa at 200 ppm and severe nasal irritation at 400 ppm (1). Central nervous system (CNS) disorders have been reported in humans exposed to ethylenediamine (76); it has also been demonstrated that ethylenediamine has powerful depressant actions on central neuronal excitability. In rat studies, ethylenediamine exhibited γ-aminobutyric acid-like CNS depression (GABAmimetric activity) (131).

*Note*: γ-aminobutyric acid (GABA) was the first amino acid shown conclusively to function as a neurotransmitter in both vertebrate and invertebrate nervous systems. GABA most often exerts inhibitory or hyperpolarizing effects, but it has also appeared to have excitatory or depolarizing actions (132). Studies have also suggested that glutamate and GABA may be among the major neurotransmitters in the CNS (133).

In a comparative study of a homologous series of diamines (ethylenediamine, 1,2-diaminopropane, 1,3-diaminopropane, 1,4-diaminobutane, 1,5-diaminopentane, and 1,6-diaminohexane) tested by injection into the lateral ventricle of rats and responses documented as changes in behavior and electroencephalograms (EEG), three patterns were seen ranging from prostration and EEG depression, to EEG seizures and convulsions, to a mixture of the patterns. All compounds were acutely lethal after micromolar doses. The most toxic of the diamines tested was 1,3-diaminopropane; similar doses of 1,2-diaminopropane did not cause seizures. An intact blood–brain barrier appears necessary for protection from the neurotoxic effects. Toxicity decreased with increasing or decreasing methylene groups between the amine groups. Toxicity appeared to increase again when amine separation reached the one and six positions (134). Ethylenediamine has been shown by other investigators to be GABA-mimetic and blockable by bicuculline (135). The sympatholytic effect by EDA on blood pressure has been previously discussed.

Ethylenediamine can apparently mimic the effects of $\gamma$-aminobutyric acid (GABA), a neurotransmitter in the brain. Ethylenediamine enhanced binding of 3H-diazepam to rat forebrain membrane preparation in a bicuculline-sensitive manner, though it was much less potent in this effect than GABA (136). This GABA-mimetic effect may be mediated by stimulation of release of GABA from glial cells (137). Ethylenediamine inhibited the convulsive threshold to the intravenous infusion of three convulsants in the order of pentylenetetrazole > bicuculline > strychine (138).

However, in an acute and subchronic toxicity study, in which ethylenediamine dihydrochloride was administered in the diet for 7 days at up to 2.7 g/kg to F344 rats or B6C3F1 mice, as well as during a 3-month dietary feeding study at dose levels of 0.05, 0.25, and 1 g/kg, neither overt signs of neurotoxicity nor any microscopic lesion in the CNS was observed; the authors speculated that perhaps the blood–brain barrier in the animals prevented brain entry of EDA, or possibly biotransformation resulted in very little EDA reaching the site of action to exert its inhibitory power against neuronal firing (133).

Dernehl (1951) reported that dermatitis occurred in a high proportion of exposed operating personnel manufacturing mixed ethyleneamines, which included ethylenediamine, and concluded that both primary irritation and sensitization probably occurred; respiratory irritation and asthmatic symptoms may also follow exposures to low vapor concentrations (1). Lam et al. (1980) have also reported an incidence of asthmatic sensitization to EDA (1); such hypersensitive individuals may not be adequately protected at or below the recommended TLVs (63b). Hypersensitivity induction in guinea pigs has also been demonstrated (139). Skin sensitization has been observed in a number of instances because of the use of ethylenediamine as a stabilizer in pharmaceutical skin creams. Sensitization is less likely in industrial exposures because the contact is less intimate and because damaged skin is not usually involved (1). According to the Occupational Safety and Health Administration (OSHA) (1978), ethylenediamine can cause contact dermatitis, sensitization, rash, and asthma (68). This compound can therefore potentially aggravate asthma or fibrotic pulmonary disease.

Ethylenediamine reduced birth weights and weight gain in pups when administered to pregnant mice at 400 mg/(kg)(day) by gavage on days 6 to 13 of pregnancy.

No maternal toxicity was apparent at this dose (140), other than reduced maternal weight gain, fetal size, and missing or shortened length of the innominate artery (probably not a specific nor irreversible effect); the authors concluded that this anomaly was related to dietary deficiency of folic acid, or vitamin A. A shortened innominate artery is not considered a teratological effect because it would result in no functional deficit; the authors concluded that there was no evidence of teratogenicity in Fischer 344 rats following ingestion of ethylenediamine dihydrochloride, at 0, 0.05, 0.25 and 1.0 g/kg/day on gestation days 6 to 15 by pregnant dams during the period of organogenesis (141).

A two-generation reproduction study in Fischer 344 rats at dietary doses of 0, 0.05, 0.15, and 0.5 g/kg/day EDA dihydrochloride did not show any adverse reproductive effects; and no gross lesion of biologic significance was seen in any of the three groups ($F_1$ and $F_2$ weanlings, $F_1$ adults) (142).

In a 90-day gavage study of EDA dihydrochloride in mice, the no-effect level was reportedly 100 mg/kg/day (free base equivalent), with kidney injury the only consistent finding at higher doses; in a gavage study in rats (100, 600, and 800 mg/kg/day corresponding to free base equivalents for 90 days), kidney and uterine effects were observed at doses of 600 and 800 mg/kg/day, 65 percent mortality at the 800-mg dose level, and cataracts, conjunctivitis, cloudy cornea, and retinal atrophy at all dose levels in rats (143).

In contrast, no eye lesions were apparently observed in any F344 rats fed EDA dihydrochloride in the diet for 3 months at doses of 0.05, 0.25, and 1.0 g/kg/day; mild hepatocellular degeneration was observed at the high dose, marginal effects at the intermediate dose level, and no effects at the lowest dose (130). Subsequently, a pathology report confirmed (Garman, 1984) that no eye lesions were present, and the difference between the above two studies may have been a result of a "pulsed" blood level following a gavage dose (63b) (blood levels were not taken in either study).

Contrary to published reports of mutagenic effects of EDA in the Ames (*Salmonella*) mutation test, EDA was not genotoxic in the following battery of mammalian in vitro and in vivo mutation tests: Chinese hamster ovary (CHO) gene mutation assay, the sister-chromatid exchange (SCE) test with CHO cells, unscheduled DNA synthesis (UDS) assays with primary rat hepatocytes, and a dominant lethal study with Fischer 344 rats (144).

EDA was not carcinogenic when applied to the skin of C3H/HeJ mice at 1 percent three times weekly for 18 months (145).

A chronic dietary feeding study in rats at doses of EDA dihydrochloride up to 350 mg/kg/day (158 mg free base/kg/day) also failed to demonstrate carcinogenicity (146).

Available data indicate that EDA does not possess genotoxic potential.

*6.11.1.3 Hygienic Standards of Permissible Exposure.* A TLV–TWA of 10 ppm (25 mg/m$^3$) has been adopted, with a skin notation (63c).

### 6.11.2 Ethylenediaminetetraacetic Acid [CAS # 60-00-4]

The molecular formula for ethylenediaminetetraacetic acid is $(HO_2CCH_2)_2$ $NCH_2CH_2N(CH_2CO_2H)_2$, with a molecular weight of 292.24; a solid, it is also known as EDTA or (ethylenedinitrilo)tetraacetic acid.

***6.11.2.1 Uses.*** EDTA or ethylenediaminetetraacetic acid, a substituted ethylenediamine, has been used extensively in medicine as a chelating agent for the removal of toxic heavy metals, usually in the form of the calcium disodium salt (calcium disodium edetate).

***6.11.2.2 Toxicity.*** Large or repeated doses may cause kidney injury. Gastrointestinal upset, pain at the injection site, transient bone marrow depression, mucocutaneous lesions, fever, muscle cramps, and histamine-like reactions (sneezing, lacrimation, nasal congestion) have been reported (1).

The trisodium salt of EDTA has been demonstrated to be nonsensitizing to guinea pigs (147).

EDTA has been reported to be teratogenic in rats by subcutaneous injection; teratogenicity was prevented by zinc diet supplementation (148).

***6.11.2.3 Hygienic Standards of Permissible Exposure.*** Standards have not been adopted (63c).

### 6.11.3 1,2-Propanediamine [CAS # 78-90-0]

The molecular formula for 1,2-propanediamine is $CH_3CH(NH_2)CH_2NH_2$, with a molecular weight of 74.13; it is a flammable, corrosive liquid and is also known as 1,2-propanediamine, 1,2-diaminopropane, and propylenediamine (66).

***6.11.3.1 Uses.*** Specific uses were not identified in the literature (62).

***6.11.3.2 Toxicity.*** Renal tubular damage and proteinuria is produced by intraperitoneal injections in rats of 1,3-propanediamine and 1,2-propanediamine. Similar effects are not produced by tetramethylenediamine, pentamethylenediamine, hexamethylenediamine, or decamethylenediamine (1).

In rat studies, via lateral ventricle injection, 1,2-diaminopropane exhibited GABA-like neurotoxicity but less potent than 1,3-diaminopropane, which was the most powerful of the series tested including GABA. With equal doses, 1,3-diaminopropane also caused seizures, but similar doses of 1,2-diaminopropane did not (134).

***6.11.3.3 Hygienic Standards of Permissible Exposure.*** Standards have not been adopted (63c).

### 6.11.4 1,3-Propanediamine [CAS # 109-76-2]

The molecular formula for 1,3-propanediamine is $H_2N(CH_2)_3NH_2$, with a molecular weight of 74.13; a corrosive liquid, it is also known as 1,3-diaminopropane (66).

***6.11.4.1 Uses.*** Specific uses were not identified in the literature (62).

***6.11.4.2 Toxicity.*** As indicated above, 1,3-diaminopropane is a potent GABA-mimetric agent in rats (134).

## ALIPHATIC AND ALICYCLIC AMINES

*6.11.4.3 Hygienic Standards of Permissible Exposure.* Standards have not been adopted (63c).

### 6.11.5  Tetramethylenediamine [CAS # 110-60-1]

The molecular formula for tetramethylenediamine is $H_2N(CH_2)_4NH_2$, with a molecular weight of 88.15; it has a solid "stench" at room temperature and is corrosive. It is also known as 1,4-butanediamine, 1,4-diaminobutane, and putrescine (66). An odor threshold of 79 mg/m$^3$ has been reported (63d).

*6.11.5.1 Uses.* Putrescine is used as a tool in biochemical research (66).

*6.11.5.2 Toxicity.* Putrescine does not appear to be nephrotoxic (1). Bacterial amino acid decarboxylases decompose the amino acids ornithine or arginine to produce putrescine. Putrescine is classified as a ptomaine (alkaloidal or basic product of the putrefaction of animal or vegetable matter); however, like cadaverine, putrescine per se does not cause food poisoning.

As noted above, putrescine has been shown to have weak GABAnergic properties, as well as antihypertensive activity as previously discussed.

The diamines putrescine and the tetramine spermine are involved in cell growth: elevated levels of polyamines in cancer patients have been associated with increased cell proliferations in rapidly growing tissues; elevated levels of ornithine decarboxylase and/or putrescine, spermidine, and spermine have been correlated with the rate of proliferation of cancer cells; and it has been hypothesized that inhibitors of ornithine decarboxylase in turn may retard the growth of cancer cells (149).

*6.11.5.3 Hygienic Standards of Permissible Exposure.* Standards have not been adopted (63c).

### 6.11.6  Pentamethylenediamine [CAS # 462-94-2]

The molecular formula for pentamethylenediamine is $H_2N(CH_2)_5NH_2$, with a molecular weight of 102.18; it is a corrosive liquid having a pronounced stench (66). A so-called ptomaine (which includes putrescine as well as methylamine), it does not cause "food poisoning" per se, but its presence indicate possible bacterial toxins. It is also known as pentamethyldiamine, 1,5-diaminopentane, 1,5-pentanediamine, or cadaverine. Bacterial amino acid decarboxylases decompose the amino acid lysine to produce cadaverine.

*6.11.6.1 Uses.* Specific uses were not identified in the literature (62).

*6.11.6.2 Toxicity.* As noted above, cadaverine has been shown to have weak GABAnergic properties; antihypertensive activity has been previously discussed.

Pentamethylenediamine does not appear to be nephrotoxic (1).

### 6.11.7  Hexamethylenediamine [CAS # 124-09-4]

The molecular formula for hexamethylenediamine is $H_2N(CH_2)_6NH_2$, with a molecular weight of 116.21; it is a corrosive solid, also known as 1,6-hexanediamine or 1,6-diaminohexane (66).

*6.11.7.1 Uses.* Specific uses were not identified in the literature (62).

*6.11.7.2 Toxicity.* Hexamethylenediamine causes anemia, weight loss, and degenerative microscopic changes in the kidneys and liver and to a lesser degree in the myocardium of guinea pigs after repeated doses (1). Conjunctival and upper respiratory tract irritation have been observed in workers handling hexamethylenediamine. One worker, out of 20 studied, developed acute hepatitis followed by dermatitis that were attributed to hexamethylenediamine. No anemia was observed. Air concentrations varied from 2 and 5.5 mg/m$^3$ during normal operations to 32.7 and 131.5 mg/m$^3$ during autoclave operations in two plants (1). Hexamethylenediamine does not appear to be nephrotoxic (1).

*6.11.7.3 Hygienic Standards of Permissible Exposure.* Standards have not been adopted (63c).

### 6.11.8 Decamethylenediamine [CAS # 646-25-3]

The molecular formula for decamethylenediamine is $H_2N(CH_2)_{10}NH_2$, with a molecular weight of 172.32; a solid, it is also known as 1,10-decanediamine or decanediamine (66).

*6.11.8.1 Uses.* Specific uses were not identified in the literature (62).

*6.11.8.2 Toxicity.* Decanediamine has been classified as an irritant (66). The compound apparently does not have any nephrotoxicity (1).

*6.11.8.3 Hygienic Standards of Permissible Exposure.* Standards have not been adopted (63c).

### 6.11.9 Hexadecylmethylenediamine [CAS # 929-94-2]

Toxicology information was not located in the literature (62).

### 6.11.10 Diethylenetriamine [CAS # 111-40-0]

The molecular formula for diethylenetriamine is $H_2NCH_2CH_2NHCH_2CH_2NH_2$, with a molecular weight of 103.17; it is a corrosive liquid and a solvent, also known as DETA, 1,4,7-triazaheptane, 2,2'-diaminodiethylamine, and bis-2-aminoethylamine (66).

*6.11.10.1 Uses.* Diethylenetriamine is used as a solvent, in organic syntheses, and in a variety of industrial applications including use as a fuel component (63c).

*6.11.10.2 Toxicity.* Diethylenetriamine is a skin, eye, and respiratory irritant; severe corneal injury can occur (1). The acute toxic effects for animals are listed in Table 17.4. Hine and associates found the oral $LD_{50}$ for rats to be 1.08 g/kg (1).

Application to rabbit skin produced maximum irritation (1). Rats exposed to concentrated vapors and to 300 ppm showed no effects (1). Solutions of 15 percent or undiluted caused severe corneal injury, but a 5 percent solution caused only minor injury (1).

It is known to produce skin sensitization and probably pulmonary sensitization. Human skin sensitization has been observed repeatedly, particularly during the use of diethylenetriamine as a catalyst for epoxy resins (1). It has been stated that in view of the relatively high frequency of cutaneous and pulmonary sensitization, great care must be used in handling diethylenetriamine; if a definite odor can be detected, process control may be inadequate (1).

Diethylenetriamine was pharmacologically screened for possible GABAnergic activity, because it is a likely contaminant of ethylenediamine, but its potential CNS activity, as well as that of cadaverine and putrescine, to mimic GABA activity and depolarize sympathetic ganglia is only minimal (131).

Dermal application of DETA-HP (high purity) and DETA-C (commercial grade), 5 percent aqueous solutions in 25-μl aliquots to the skin of male C3H/HeJ mice three times weekly for life, did not result in carcinogenicity (127).

*6.11.10.3 Hygienic Standards of Permissible Exposure.* A TLV–TWA of 1 ppm (4.2 mg/m$^3$) with a skin notation has been adopted (63c).

### 6.11.11  N-(Hydroxyethyl)diethylenetriamine [CAS # 1965-29-3]

*6.11.11.1 Uses.* No specific uses were identified in the literature (62).

*6.11.11.2 Toxicity.* N-(Hydroxyethyl)diethylenetriamine is reportedly less toxic orally and intraperitoneally than diethylenetriamine and less irritating to the skin of rabbits on a single or repeated applications (1). No additional toxicology information was located in the literature (62).

*6.11.11.3 Hygienic Standards of Permissible Exposure.* Standards have not been adopted (62c).

### 6.11.12  N-(Cyanoethyl)diethylenetriamine

*6.11.12.1 Uses.* Specific uses were not located in the literature (62).

*6.11.12.2 Toxicity.* *N*-(Cyanoethyl)diethylenetriamine is reportedly less toxic orally and intraperitoneally than diethylenetriamine and less irritating to the skin of rabbits on a single or repeated applications (1). No additional toxicology information was located in the literature (62).

*6.11.12.3 Hygienic Standards of Permissible Exposure.* Standards have not been adopted (62c).

### 6.11.13  Diethylenetriaminepentaacetic Acid [CAS # 67-43-6]

The molecular formula for diethylenetriaminepentaacetic acid is $(HO_2CCH_2)_2$ $NCH_2CH_2N(CH_2CO_2H)CH_2CH_2N(CH_2CO_2H)_2$, with a molecular weight of 393.35; it is a solid, also known as DPTA.

***6.11.13.1 Uses.*** DPTA has been recommended for use as a chelating agent for the removal of radionuclides deposited in the respiratory tract. Dudley et al. (1980) found that this material can be administered effectively by inhalation in beagle dogs, and that the aerosol particles need not penetrate deeply into the lung to assure adequate absorption (1).

***6.11.13.2 Toxicity.*** DPTA has been noted to be a suspected cancer agent and mutagen (66b). No additional toxicology information was located in the literature (62).

***6.11.13.3 Hygienic Standards of Permissible Exposure.*** Standards have not been adopted (62c).

### 6.11.14  Triethylenetetramine [CAS # 112-24-3]

The molecular formula for triethylenetetramine is $H_2NCH_2CH_2NHCH_2CH_2NHCH_2CH_2NH_2$, with a molecular weight of 146.24; a corrosive liquid, it is also known as TETA (66).

***6.11.14.1 Uses.*** Triethylenetetramine is a copper chelator and, as in the case with *d*-penicillamine, causes teratogenicity through copper deficiency; copper supplementation prevents teratogenicity (150). TETA has a variety of industrial applications as discussed previously, as well as a plasticizer in plastics production (1).

***6.11.14.2 Toxicity.*** In common with the other ethylenamines, triethylenetetramine causes skin sensitization as well as primary irritation. Exposure to the hot vapor results in respiratory tract irritation and itching of the face with erythema and edema (1). Grandjean was unable to detect triethylenetetramine in the air of a workroom where dermatitis was occurring. He concluded that the control problem was primarily one of preventing direct skin contact. Successful control requires good personnel training and scrupulous handling technique (1).

Skin application daily for 10 days and every second day for 45 days caused cachexia, cutaneous alterations at the site of application, liver degeneration, and congestion of the kidneys and brain. It caused necrotic changes in the placenta and miscarriages or fetal death in pregnant animals (1).

***6.11.14.3 Hygienic Standards of Permissible Exposure.*** Standards have not been adopted (62c).

### 6.11.15  Tetraethylenepentamine [CAS # 112-57-2]

The molecular formula for tetraethylenepentamine is $HN(CH_2CH_2NHCH_2CH_2NH_2)_2$, having a molecular weight of 189.31; it is a liquid and is also known as TEPA (66).

***6.11.15.1 Uses.*** Specific uses were not located in the literature (62).

## ALIPHATIC AND ALICYCLIC AMINES

*6.11.15.2 Toxicity.* TEPA is a corrosive irritant. 25 percent aqueous solution applied three times weekly to the skin of male C3H/HeJ for life did not cause cancer (127). No additional toxicology information was located in the literature (62).

*6.11.15.3 Hygienic Standards of Permissible Exposure.* Standards have not been adopted (62c).

### 6.12 Alkanolamines and Alkylalkanolamines

The alkanolamines (amino alcohols) are substituted primary, secondary and tertiary amines, prepared from ammonia and ethylene oxide, propylene oxide, or butylene oxide leading to three major groups as follows:

| Alkanolamine | CAS # |
|---|---|
| *Ethanolamines* | |
| Monoethanolamine | [141-43-5] |
| Diethanolamine | [111-42-2] |
| Triethanolamine | [102-71-6] |
| *Isopropanolamines* | |
| Monoisopropanolamine | [78-96-6] |
| Diisopropanolamine | [110-97-4] |
| Triisopropanolamine | [122-20-3] |
| *Secondary Butanolamines* | |
| Mono-*sec*-butanolamine | [13552-21-1] |
| Di-*sec*-butanolamine | [21838-75-5] |
| Tri-*sec*-butanolamine | [2421-02-5] |

The alkylalkanolamines, which are prepared from amines and ethylene oxide or propylene oxide (151), include the following compounds (151):

| Alkylalkanolamines | CAS # |
|---|---|
| Monomethylethanolamine (2-methylaminoethanol) | [109-83-1] |
| Dimethylethanolamine (2-dimethylaminoethanol) | [108-01-0] |
| Monoethylethanolamine (2-ethylaminoethanol) | [110-73-6] |
| Diethylethanolamine (2-diethylaminoethanol) | [100-37-8] |
| Monoisopropylethanolamine (2-isopropylaminoethanol) | [109-56-8] |
| Diisopropylethanolamine (2-diisopropylaminoethanol) | [96-80-0] |
| Monobutylethanolamine | [111-75-1] |

(2-butylaminoethanol)
Dibutylethanolamine [102-81-8]
(2-dibutylaminoethanol)
Monomethyldiethanolamine [105-59-9]
[2,2'-(methylimino)diethanol]
Monoethyldiethanolamine [139-87-7]
[2,2'-(ethylimino)diethanol]
Dimethylisopropanolamine [108-16-7]
(1-dimethylamino-2-propanol)
N-acetylethanolamine [142-26-7]
(N-2-hydroxyethylacetamide)
Aminoethylethanolamine [111-41-1]
[2-(2-aminoethylamine)ethanol]

The alkanolamines are used in the chemical and pharmaceutical industries as intermediates for the production of emulsifiers, detergents, solubilizers, cosmetics, drugs, and textile-finishing agents (1). The alkyl-substituted alkanolamines have been used as absorbants, emulsifying agents, flotation agents, and resin-curing agents; as such, they find wide application in the manufacture of cosmetics, shampoos, waxes, polishes, lubricants, resins, and a variety of related compounds (152, 153a). The physical properties of some of the more common amino alcohols appear in Table 17.7.

The immediate adverse effects of these compounds following exposure in animals or humans are primarily related to dermal, ocular, and pulmonary irritancy effects, which can range from slightly to moderately severe and/or corrosive. Their salts show reduced local irritant activity. Primary irritation studies in animals have demonstrated that skin irritancy is enhanced by repeated application of the material under an occlusive dressing. These compounds can be absorbed through the skin, and when held in contact with the skin of small animals, may cause death in doses that are less than those that produce death when given by mouth (Table 17.5). The acute oral toxicity levels for laboratory animals are generally low. Concentrated unneutralized solutions of the more soluble alkanolamines cause intense gastrointestinal irritation with hemorrhage and congestion of the intestine. Adhesions of visceral organs are frequently found in the survivors. Neutralization (with hydrochloric acid) and increasing dilution reduce the oral toxicity. In the series reported by Smyth et al., single exposures of rats to "saturated" vapors seldom produced deaths in less than 8 hr. The generally low vapor pressures of these compounds reduce the inhalation hazard in industry (1).

Many of the physiological effects of the alkyl-substituted 2-aminoethanols appear to be related to the normal functions of choline and its metabolites (153a).

### 6.12.1 Monoethanolamine [CAS # 141-43-5]

The molecular formula for monoethanolamine is $H_2NCH_2CH_2OH$, with a molecular weight of 61.08; a corrosive liquid, it is also known as 2-aminoethanol, aminoethanol, monoethanolamine, 2-N-ethyaminoethanol, 2-hydroxyethylamine, and EA (66). A "sensation" threshold ranging from 2 to 3.3 ppm, as well as a threshold odor of 25 ppm, described as musty, ammoniacal, or foreign, have been reported (154).

*6.12.1.1 Uses.* Ethanolamine is used as a solvent, in the synthesis of surface-active agents, in emulsifiers, polishes, and waving solutions for hair; as an agricultural chemical dispersing agent (63b); to remove $CO_2$ and $H_2S$ from natural gas and other gases; and as a softening agent for hides (66). Ethanolamine and other amines appear to be potentially useful components of topical formulations used to decontaminate and protect the skin against chemical warfare agents (155). As a pharmaceutical adjuvant, monoethanolamine N.F. is used as a solvent for fats and oils, and in combination with fatty acids forms soaps in the formulations of various types of emulsions such as lotions and creams (156). No additional information was found in the literature (62).

The chemical reactivity of 2-aminoethanol lends itself to the formation of a wide variety of chemical compounds. Single or double substitution on the amine group results in the formation of compounds such as 2-ethylaminoethanol, 2-diethylaminoethanol, or the corresponding alkylalkanolamines, such as propyl, isopropyl, and butyl (152).

*6.12.1.2 Toxicity.* The acute oral and intraperitoneal $LD_{50}$ toxicity of neutralized ethanolamine in male Sprague–Dawley rats was reported to be 3.32 g/kg and 981 mg/kg, respectively (153a). Smyth demonstrated in a 90-day subacute oral toxicity study of unneutralized ethanolamine in rats that a maximum daily dose of 0.32 g/kg resulted in no effects; 0.64 mg/kg/day resulted in altered liver or kidney weight; and at 1.28 g/kg deaths occurred (1). Ethanolamine is considered to be liver toxin (157).

Treon (1958) studied the inhalation toxicity of monoethanolamine. The conditions were such that an unknown proportion of the ethanolamine was converted to the carbonate in the exposure chamber. Dogs and cats survived the exposures to concentrations of 2.47 mg/l for 7 hr on each of four successive days. Four of six guinea pigs died following exposure to a concentration of 0.58 mg/l for 1 hr. Rats, rabbits, and mice were less susceptible than guinea pigs, but more susceptible than cats or dogs. Sixty of 61 animals survived exposure to the inhalation of concentrations of 0.26 to 0.27 mg/l for 7 hr on each of five consecutive days, and 25 of 26 animals survived 25 7-hr exposures (over a period of 5 weeks) to concentrations of 0.26 mg/l (104 ppm). The observed effects were primarily those of respiratory tract irritation. Eye irritation was negligible, presumably due to the formation of carbonate under the conditions of the experiment. Pathological changes in those animals exposed to higher concentrations had normal autopsy findings (1). Ethanolamine is a recognized respiratory irritant (68) and a potential neurotoxin (76).

Weeks (1958) reported that dogs, rats, and guinea pigs survived inhalation of 12 to 25 ppm for 90 days, whereas fractional mortality occurred in 24 to 30 days at 100 ppm (dogs) and 66 to 75 ppm (rodents). Skin irritation and lethargy occurred at 5 and 12 ppm (1).

The acute dermal toxicity is reportedly more toxic than by oral administration with an $LD_{50}$ of 1 mg/kg, in which the undiluted ethanolamine caused redness and swelling when applied to the skin of the rabbit (158). Ethanolamine has also been reported (Hinglais, 1947) to cause irritation and necrosis when applied to the skin of the rabbit, comparable to a mild first degree burn (63b), and to cause severe injury when instilled in the rabbit eye (99, 68).

Ethanolamine has also been demonstrated not to injure the skin in low concentrations (155), and it is also a normal tissue metabolite as well as an essential component of tissue phospholipids (159). Browning (1953) also observed that when undiluted monoethanolamine is applied to human skin on gauze for 1½ hr, only marked redness and infiltration of the skin result (1).

Topical ethanolamine has been demonstrated to penetrate excised pig skin and is readily metabolized (via human skin grafted onto athymic nude mice and in nude mouse skin itself); absorbed radiolabeled ethanolamine was demonstrated to be oxidized to carbon dioxide and extensively metabolized as indicated by the appearance of $^{14}C$ radioactivity in skin and hepatic amino acids, in proteins, and by incorporation into phospholipids (155). In liver tissue, ethanolamine has also been demonstrated to be converted into amino acids, which in turn were incorporated into hepatic proteins. Hepatic ethanolamine was methylated to choline and converted to serine. All three compounds were incorporated into hepatic phospholipids (155). Phosphatidylethanolamine has been reported to be the most abundant phosphoglyceride in the skin (155).

Smyth's laboratory (1956) demonstrated that administration of monoethanolamine by the intravenous route in dogs produced increased blood pressure, diuresis, salivation, and pupillary dilatation, corresponding to symptoms produced by the pharmacologically active aliphatic amines; monoethanolamine was more active than triethanolamine (1).

An investigation to evaluate the potential of choline precursors (alkyl-2-aminoethanols) to inhibit cholinesterase activity relative to choline, in vitro using plasma cholinesterase, also demonstrated that ethanolamine exhibited weak inhibitory activity (153a).

The excretion rate in men was found to vary between 4.8 and 22.9 mg/day with a mean of 0.162 mg/kg. Eleven women were observed to excrete larger amounts, varying between 7.7 and 34.9 mg/day with a mean excretion rate of 0.492 mg/kg/day. The excretion rates in animals were approximately, for cats, 0.47 mg/kg/day; for rats, 1.46 mg/kg/day; and for rabbits, 1.0 mg/kg/day. From 6 to 47 percent of methanolamine administered to rats can be recovered in the urine (1).

Monoethanolamine has been demonstrated to be nonmutagenic in the Ames *Salmonella typhimurium* assay, with and without S9 microsomal metabolic activation, using strains TA1535, TA1537, TA1538, TA98, and TA100; and also negative in the *E. coli* assay, *Saccharomyces* gene conversion assay, and rat liver chromosome assay. The authors noted that the lack of mutagenic activity was in accord with the absence of electrophilic reactivity (160).

### 6.12.1.3 Hygienic Standards of Permissible Exposure.
A TLV–TWA standard of 3 ppm (7.5 mg/m$^3$) and a STEL of 6 ppm (15 mg/m$^3$) have been adopted (63c).

### 6.12.2 Diethanolamine [CAS # 111-42-2]

The molecular formula for diethanolamine is (HOCH$_2$CH$_2$)$_2$NH, with a molecular weight of 105.14; a corrosive, hygroscopic solid, it is also known as 2,2'-iminodiethanol, 2,2'-iminobisethanol, bis(hydroxyethyl)amine, 2,2'-dihydroxydiethylamine, and DEA (66).

#### 6.12.2.1 Uses.
Diethanolamine is used to scrub gases including cracking gases and coal or oil gases that contain carbonyl sulfide, as a rubber chemical intermediate, in the manufacture of surface-active agents used in textile specialities, herbicides, and petroleum demulsifiers, as an emulsifier and dispersing agent in various agricultural chemicals, cosmetics, and pharmaceuticals (66), as a detergent in paints, cutting oils, and shampoos, and in the manufacture of resins and plasticizers (63b).

#### 6.12.2.2 Toxicity.
The acute oral LD$_{50}$ in rats has been reported to be 0.71 ml/kg (161), 1.82 g/kg (1), and 2.83 g/kg (68).

The maximum daily dose having no effect in rats over a 90-day period is 0.02 g/kg; a daily dose of 0.17 g/kg over the same period produces microscopic pathology and deaths, and 0.09 g/kg causes changes in liver and kidney weights (1). Intraperitoneal administration of doses of 100 or 500 mg/kg in rats resulted in liver and kidney lesions (cytoplasmic vacuolization); the higher dose caused renal tubular degeneration (68).

The undiluted liquid and 40 percent solutions produce severe eye burns, whereas 15 percent produces only minor damage (1).

Ten percent solution applied to rabbit skin caused redness; higher concentrations caused increasing injury (1); under semiocclusion for 24 hr on 10 consecutive days, liquid diethanolamine (concentration not specified) caused only minor if any irritation (68). Clinical skin testing of cosmetic products containing DEA showed mild skin irritation in concentrations above 5% (68).

Smyth's laboratory (1956) demonstrated that administration of diethanolamine by the intravenous route in dogs produced increased blood pressure, diuresis, salivation, and pupillary dilatation, corresponding to symptoms produced by the pharmacologically active aliphatic amines (1).

Diethanolamine has been demonstrated to be nonmutagenic in the Ames *Salmonella typhimurium* assay, with and without S9 microsomal metabolic activation, using strains TA1535, TA1537, TA1538, TA98, and TA100; and also negative in the *E. coli* assay, Saccharomyces gene conversion assay and in the rat liver chromosome assay. The authors noted that the lack of mutagenic activity was in accord with the absence of electrophilic reactivity (160).

*6.12.2.3 Hygienic Standards of Permissible Exposure.* A TLV–TWA standard of 3 ppm (13 mg/m$^3$) has been adopted (63c).

### 6.12.3 Triethanolamine [CAS # 102-71-6]

The molecular formula for triethanolamine is $(HOCH_2CH_2)_3N$, with a molecular weight of 149.19; it is a pale yellow, viscous, hygroscopic, irritant liquid having a slight ammoniacal odor. It is also known as 2,2′,2″-nitrilotriethanol, trihydroxytriethylamine, tris(hydroxyethyl)amine, triethylolamine, and Trolamine, a mixture consisting largely of triethanolamine admixed with various proportions of diethanolamine and monoethanolamine (66, 162).

*6.12.3.1 Uses.* Triethanolamine is used in the manufacture of surface-active agents, textile specialties, waxes, polishes, herbicides, petroleum demulsifiers, toilet goods, and cement additives; in making emulsions with mineral and vegetable oils, paraffin, and waxes; as a solvent for casein, shellac, and dyes; in the manufacture of synthetic resins; for increasing the penetration of organic liquids into wood and paper; in the production of lubricants for the textile industry (66); in the formulations of various cosmetics (163); and in the preparation of flame-retardant fabrics (164). Triethanolamine salicylate has also been used as a nonsteroidal anti-inflammatory agent (165, 166). Triethanolamine has been identified in cutting fluids (167). Triethanolamine USP is also used as a pharmaceutical adjuvant or alkalizing agent, and in combination with a fatty acid (e.g., oleic acid, stearic acid) as an emulsifier (the triethanolamine soap formed lowers the surface tension of the aqueous phase) (156, 162).

*6.12.3.2 Toxicity.* Triethanolamine is generally considered to have a low acute and chronic toxicity. Lehman (1950) states that if deleterious effects were to occur in humans from triethanolamine, these would probably be acute in nature and would be due to its alkalinity rather than its inherent toxicity (1). Kindsvatter (1940) found the acute oral $LD_{50}$ in rats and guinea pigs to be 8 g/kg. The effects observed were confined to the gastrointestinal tract. He felt the toxic effects were probably from the alkaline irritation, because larger doses of the neutralized material produced no symptoms at levels where the free base would cause 100 percent mortality. Repeated feeding produced only slight, reversible pathology in the liver and kidneys. Applications on the skin gave evidence of skin absorption (1). Smyth and Carpenter (1944) also found low acute oral toxicity, with the $LD_{50}$ for rats being 9.11 g/kg. In a 90-day subacute feeding experiment with rats, the maximum dose producing no effect was 0.08 g/kg. Microscopic lesions and deaths occurred at 0.73 g/kg, and 0.17 g/kg produced alterations in liver and kidney weights (1).

Triethanolamine is considered an innocuous or slight, transient eye irritant following instillation (98 percent concentrated) in rabbit eyes; average Draize scores reported following 0.1 ml application were 4 on day 1, 2 on day 3, 2 on day 7, and 0 on day 14 (168). Other investigators found it to be a moderate eye irritant (99).

Applications of 5 or 10 percent solution to rabbit or rat skin did not produce irritation (1). Triethanolamine appears to be free of skin sensitization effects in its extensive use in cosmetics (1).

In view of its low vapor pressure (less than 0.01 torr), significant exposure by inhalation appears unlikely, and the chief risk in industry would be from direct local contact of the skin or eyes with the undiluted, unneutralized fluid (1).

Triethanolamine has been demonstrated to be nonmutagenic in the Ames *Salmonella typhimurium* assay, with and without S9 microsomal metabolic activation, using strains TA1535, TA1537, TA1538, TA98, and TA100; and also negative in the *E. coli* assay, *Saccharomyces* gene conversion assay, and the rat liver chromosome assay. The authors noted that the lack of mutagenic activity was in accord with the absence of electrophilic reactivity (160).

Triethanolamine did not produce any morphological transformation in Chinese hamster embryo cells at concentrations of 25 to 500 µg/ml and was also inactive in the Ames *S. typhimurium* and *E. coli* tests, as well as in the chromosomal aberration test in Chinese hamster cells; however, the assay was not carried out with microsomal metabolic activation (169).

Results of carcinogenicity studies have been controversial. Hoshino et al. (1978) reported that triethanolamine in the diet of mice at levels of 0.03 or 0.3 percent caused a significant increase in the occurrence of tumors, both benign and malignant. Females showed a 32 percent increase, mostly of thymic lymphomas. The increase of all other tumors, in both sexes, was 8.2 percent. They also found that triethanolamine reacted with sodium nitrite to produce *N*-nitrosodiethanolamine and that the product caused mutagenesis in bacteria (1).

Maekawa et al. (1986) reported that no carcinogenic activity was found when given orally to rats in drinking water at concentrations of 1 and 2 percent, for 2 years. However, the dosage to females was halved after week 69 of treatment owing to nephrotoxicity. Histological examination of renal damage in treated animals revealed acceleration of chronic nephropathy, mineralization of the renal papilla, nodular hyperplasia of the pelvic mucosa, and pyelonephritis with or without papillary necrosis. Nephrotoxicity seemed to affect life-span adversely, especially in females. Tumor incidence and histology were the same in the treated group as in controls (170).

A more recent study by Konishi et al. (1992) in which triethanolamine was administered to male and female B6C3F1 mice in drinking water at 0, 1, and 2 percent for 82 weeks resulted in a low incidence of neoplasms in all groups, including the control group, but no dose-related increase of any tumor was observed in treated groups of both sexes. These dosages were the same as those that produced nephrotoxicity in rats (Maekawa et al., 1986), but produced no chronic toxicity in mice. The neoplasms were noted to be the most common ones that occur spontaneously in the liver of male B6C3F1 mice. The authors noted that in the case of the study by Hoshino et al. (1978), the increased incidences of lymphomas occurred in ICR female mice only, and the investigators attributed the carcinogenicity to be due either to endogenous nitrosation of triethanolamine or to other endogenous reactions that may form some other potent carcinogen with triethanolamine. Konishi et al. concluded, however, that the formation of *N*-nitrosodiethanolamine is unlikely to account for lymphoma only in female mice, because this carcinogen induces primarily liver cancers. The discrepancy between these studies was explained on the basis of strain differences; it was concluded that the investigation documented a lack of carcinogenic activity of triethanolamine in B6C3F1 mice (171).

*6.12.3.3 Hygienic Standards of Permissible Exposure.* Standards have not been adopted (63c).

### 6.12.4  Monoisopropanolamine [CAS # 78-96-6]

The molecular formula for monoisopropanolamine is $CH_3CH(OH)CH_2NH_2$, with a molecular weight of 75.11; a corrosive liquid, it is also known as isopropanolamine or 1-amino-2-propanol (66).

*6.12.4.1 Uses.* Synthetic and semisynthetic machine lubricating/cutting oil formulations have included isopropanolamine, diisopropanolamine, and triethanolamine, together with nitrites (172); consequently the EPA has promulgated standards regarding the composition of cutting fluids. No additional uses were found in the literature (62).

*6.12.4.2 Toxicity.* No information was located in the literature (62).

*6.12.4.3 Hygienic Standards of Permissible Exposure.* Standards have not been adopted (63c).

### 6.12.5  Diisopropanolamine [CAS # 110-97-4]

The molecular formula for diisopropanolamine is $[CH_3CH(OH)CH_2]_2NH$, with a molecular weight of 133.19; it is a corrosive, hygroscopic solid (66).

*6.12.5.1 Uses.* As indicated above, diisopropylamine cutting oil formulations have included diisopropanolamine together with nitrites. No additional uses were found in the literature (62).

*6.12.5.2 Toxicity.* No information was located in the literature (62).

*6.12.5.3 Hygienic Standards of Permissible Exposure.* Standards have not been adopted (63c).

### 6.12.6  Triisopropanolamine [CAS # 122-20-3]

The molecular formula for triisopropanolamine is $[CH_3CH(OH)CH_2]_3N$, with a molecular weight of 191.27; it is a corrosive, hygroscopic solid, also known as TIPA (66).

*6.12.6.1 Uses.* No specific uses were located in the literature (62).

# ALIPHATIC AND ALICYCLIC AMINES

**6.12.6.2 Toxicity.** No toxicology information was located in the literature (62).

**6.12.6.3 Hygienic Standards of Permissible Exposure.** Standards have not been adopted (63c).

## 6.12.7 Mono-sec-Butanolamine [CAS # 13552-21-1]

The molecular formula for mono-*sec*-butanolamine is $NH_2CH_2CHOHC_2H_5$, having a molecular weight of 89.14; it is also known as 1-amino-2-butanol (66).

**6.12.7.1 Uses.** No specific uses were found in the literature (62).

**6.12.7.2 Toxicity.** No toxicology information was found in the literature (62).

**6.12.7.3 Hygienic Standards of Permissible Exposure.** Standards have not been adopted (63c).

## 6.12.8 Di-sec-Butanolamine [CAS # 21838-75-5]

The molecular formula for di-*sec*-butanolamine is $NH(CH_2CHOHC_2H_5)_2$, with a molecular weight of 161.24; it is also known as 1,1'-iminodi-2-butanol (66).

**6.12.8.1 Uses.** No specific uses were found in the literature (62).

**6.12.8.2 Toxicity.** No toxicology information was found in the literature (62).

**6.12.8.3 Hygienic Standards of Permissible Exposure.** Standards have not been adopted (63c).

## 6.12.9 Tri-sec-Butanolamine [CAS # 2421-02-5]

The molecular formula for tri-*sec*-butanolamine is $N(CH_2CHOHC_2H_5)_3$, having a molecular weight of 233.35; it is also known as 1,1',1''-nitrilotris-2-butanol (66).

**6.12.9.1 Uses.** No specific uses were found in the literature (62).

**6.12.9.2 Toxicity.** No toxicology information was found in the literature (62).

**6.12.9.3 Hygienic Standards of Permissible Exposure.** Standards have not been adopted (63c).

## 6.12.10 Monomethylethanolamine [CAS # 109-83-1]

The molecular formula for monomethylethanolamine is $CH_3NHCH_2CH_2OH$, with a molecular formula of 75.11; an irritant liquid, it is also known as 2-(methylamino)ethanol, methylethanolamine, and *N*-methylethanolamine (66).

*6.12.10.1 Uses.* No specific uses were found in the literature (62).

*6.12.10.2 Toxicity.* The acute oral and intraperitoneal $LD_{50}$ toxicity in male Sprague–Dawley rats was reported as 3.36 g/kg and 1.33 g/kg, respectively (153a), and is also irritating to the skin, eyes, and mucous membranes (66).

Monomethylethanolamine has been reported to function as a choline precursor (1, 173); in theory it should therefore, like choline, inhibit cholinesterase. Accordingly methylaminoethanol was demonstrated to inhibit plasma cholinesterase in vitro in a comparative study of mono- and dimethyl, ethyl, and butyl aminoethanols; methylaminoethanol demonstrated relatively weak activity; however, potency increased dramatically with increase in the size of the alkyl substituent; in addition, intraperitoneal administration in Sprague–Dawley rats of ethylaminoethanol, dimethylaminoethanol, and dibutylaminoethanol depressed brain cholinesterase following $LD_{50}$ doses (1.17 g/kg, 1.08 g/kg, and 0.144 g/kg, respectively) (153a). No additional toxicology information was found in the literature (62).

*6.12.10.3 Hygienic Standards of Permissible Exposure.* Standards have not been adopted (63c).

### 6.12.11  Dimethylethanolamine [CAS # 108-01-0]

The molecular formula for dimethylethanolamine is $HOCH_2CH_2N(CH_3)_2$, having a molecular weight of 89.14; a corrosive liquid and lacrimator, it is also known as *N,N*-dimethylethanolamine, 2-dimethylaminoethanol, (2-hydroxyethyl)dimethylamine, DMEA, and Deanol (66).

*6.12.11.1 Uses.* Deanol acetamidobenzoate [CAS # 3635-74-3], or Deaner (3M Rikert) (166), has been used medicinally as a psychostimulant in the treatment of behavioral problems and learning difficulties (174).

*6.12.11.2 Toxicity.* The acute oral and intraperitoneal $LD_{50}$ toxicity in male Sprague–Dawley rats was reported to be 6.0 g/kg and 1.08 g/kg, respectively, following neutralization with hydrochloric acid (153a).

Dimethylethanolamine is methylated in the body to the quarternary base, choline, or trimethylethanolamine (61), which is a precursor of acetylcholine.

The biologic properties of dimethylaminoethanol have been studied more extensively than those of the other amino alcohols because of interest in its potential usefulness as a central nervous system stimulant. Pfeiffer et al. (1957) found that large doses of the tartrate salt result in depression and pulmonary edema in rats. Intravenous injection in anesthetized dogs produces a transient fall in blood pressure with moderate doses whereas larger doses (greater than 30 mg/kg) cause a pressor effect. On chronic administration to rats in doses of 500 mg/kg/day, CNS stimulation appears, with a lowered threshold for audiogenic seizures. Occasional deaths from maximal convulsions occurred after 3 to 4 weeks. In humans, oral doses of 10 to 20 mg of the base, as tartrate salt, produce mild mental stimulation.

At 20 mg/day there is a gradual increase in muscle tone and an apparent increased frequency of convulsions in susceptible individuals. Large doses produce insomnia, muscle tenseness, and spontaneous muscle twitches (1).

The mechanism of Deaner's psychostimulant effects is thought to be associated with relief of a cholinergic deficit in the CNS; the compound penetrates the blood–brain barrier and acts as a precursor of acetylcholine (174). Although there was some initial controversy as to whether Deanol can raise levels of acetylcholine in the brain (175), 2-dimethylaminoethanol was used clinically in the treatment of anxiety, on the rationale that it would increase endogenous acetylcholine (176).

2-Dimethylaminoethanol has also been demonstrated to inhibit plasma cholinesterase in vitro (potency > choline) and brain cholinesterase in rat in vivo as discussed above (153a).

Administration of Deanol (dimethylaminoethanol) to mice has been demonstrated to increase both the concentration and the rate of turnover of free choline in blood. Deanol also increased levels of choline in kidneys and decreased rates of oxidation and phosphorylation of $3H$-methylcholine administered intravenously. Intravenous administration of Deanol inhibited the rate of phosphorylation of $3H$-methylcholine in liver but did not inhibit its rate of oxidation or increase its level of free choline. The increase in blood choline concentration may have resulted from inhibition of choline metabolism in tissues (177).

Dimethylaminoethanol is reportedly nonteratogenic in F344 rats by inhalation: 0, 10, 30, and 100 ppm (178). Dimethylaminoethanol has also been reported to be noncarcinogenic in two strains of mice and actually reduced lipofuscin pigment, suggesting an anti-aging effect, also noted with similar chemicals (179).

*6.12.11.3 Hygienic Standards of Permissible Exposure.* Standards have not been adopted (63c).

### 6.12.12   Monoethylethanolamine [CAS # 110-73-6]

The molecular formula for monoethylethanolamine is $C_2H_5NHCH_2CH_2OH$, with a molecular weight of 89.14; a corrosive liquid, it is also known as 2-(ethylamino)ethanol (66).

*6.12.12.1 Uses.* Specific uses were not found in the literature (62).

*6.12.12.2 Toxicity.* The acute oral and intraperitoneal $LD_{50}$ toxicity of neutralized ethylethanolamine in male Sprague–Dawley rats was reported as 1.45 g/kg and 1.17 g/kg, respectively (153a).

Ethylaminoethanol has been demonstrated to undergo methylation in vivo generating the corresponding analogue of choline (180); Hauschild (in 1943) reported reductions in blood pressure after monoethylaminoethanol and diethylaminoethanol were administered to cats (181); and ethylaminoethanol has also been demonstrated to inhibit plasma cholinesterase in vitro and rat (Sprague–Dawley) brain cholinesterase in vivo, as discussed above (153a). The biologic activity of $N$-ethylaminoethanols has been of considerable interest because of their possible action as antimetabolites owing to the structural relationship of these compounds to choline (*N*-trimethylaminoethanol). Choline has vitamin-like activity in animals and deficiencies may result in fatty liver, anemia, and hypoproteinemia. Moyer & du Vigneaud (1942) demonstrated that *N*,*N*-dimethylethylaminoethanol functioned as choline in promoting the growth of rats on a diet deficient in choline and supplemented with homocysteine; *N*,*N*-diethylmethylaminoethanol was found to be quite toxic (153b). It was hypothesized that because both of these compounds could be formed by the in vivo methylation of methylethanolamine (MEAE) and diethylethanolamine (DEAE), it was possible that the acute toxicities of MEAE and DEAE are related to their rapid conversion to the *N*,*N*-dimethylethyl- and *N*,*N*-diethylmethylamino ethanol, respectively (153b). N-triethylaminoethanol also apparently substitutes for choline, and hence *N*-ethyl analogues of choline may interfere with the normal metabolism of choline (153b). The acute oral toxicity of monoethylaminoethanol, but not that of 2-diethylaminoethanol (DEAE) was demonstrated to be largely reversed by simultaneous or subsequent ingestion of choline in Sprague–Dawley rats; however, no reversal of depressed growth rate was noted in choline-deficient animals whose diets were supplemented with either monoethylaminoethanol or diethylaminoethanol (153b).

It was also demonstrated that the increased gastric secretion in rats observed after a single lethal oral dose of monoethylaminoethanol was probably not due to osmotic effects, because equivalent amounts of 2-diethylaminoethanol did not produce the same result; and it appeared that ethylaminoethanol is able either to affect fluid transport in the gastrointestinal tract directly, or to influence the nervous or hormonal control of gastric secretion (153c). Because it has been shown (Bell et al., 1964) that ethylaminoethanol can be incorporated into phospholipids and methylated to the corresponding phosphatidylcholine analogue, and because phospholipids appear to be involved in the transport of substances across cell membranes, it was postulated by the authors that ethylaminoethanol may interfere with the normal functioning of phospholipids and affects the transport of fluids (153c). No additional toxicology information was located in the literature (62).

*6.12.12.3 Hygienic Standards of Permissible Exposure.* Standards have not been adopted (63c).

*6.12.13 Diethylethanolamine [CAS # 100-37-8]*

The molecular formula for diethylethanolamine is $HOCH_2CH_2N(C_2H_5)_2$, with a molecular weight of 117.19; a corrosive liquid, it is also known as 2-diethylaminoethanol, *N*,*N*-diethylethanolamine, *N*-diethylethanolamine, diethylaminoethanol, and DEAE (66). The sharp, ammoniacal odor has a threshold range from 0.01 to 0.25 ppm (63e).

*6.12.13.1 Uses.* Specific uses were not located in the literature (66).

*6.12.13.2 Toxicity.* The acute oral and intraperitoneal $LD_{50}$ toxicity for DEAE neutralized with hydrochloric acid in male Sprague–Dawley rats was reported as 5.65 g/kg and 1.22 g/kg, respectively (152, 153a). The oral $LD_{50}$ reported by Smyth et al. for nonneutralized DEAE was 1.48 g/kg; Smyth's report also indicated that the skin irritation was slight whereas eye irritation was severe (182). DEAE has also been reported to be a mild eye irritant and primary skin irritant (68). Interest in the biologic activity of 2-diethylaminoethanol (DEAE) was stimulated by the report that procaine hydrolyzes rather rapidly in the body to yield DEAE (183).

Subsequently, a number of physiological effects have been reported by C. Kraatz (1950), including stimulation of excised rabbit uterus or intestine, bronchoconstriction of isolated perfused guinea pig lung, and a slight reduction in arterial blood pressure in the anesthetized dog (184). DEAE has been demonstrated to reduce blood pressure in cats (181) and inhibit (potency of choline) plasma cholinesterase in vitro (as discussed above) (153a). Single doses of the hydrochloride ranging from 0.5 to 5 g have been utilized in studies of its effects on various cardiac arrhythmias in human subjects (185).

DEAE has been shown to enter into phospholipid synthesis as phosphatidyl DEAE (180). If biologically methylated, it would yield an analogue of choline that could competitively interfere with normal choline functions; one of the features of choline deficiency in animals is the development of fatty liver and other alterations of lipid metabolism (152).

A 6-month feeding study in Sprague–Dawley rats, carried out by adding neutralized DEAE to the drinking water at concentrations of 200 and 400 mg/100 ml (rats consumed 25 to 30 ml/day; thus the two dose levels were 50 or 100 mg/day per rat, which weighed 200 to 250 g) demonstrated no alteration of serum or liver lipids or of cholesterol distribution; ratios of kidney weight to body weight were slightly elevated at both dose levels (152).

Exposure of Sprague–Dawley rats to 500 ppm, 6 hr daily for 5 days, resulted in severe weight loss and high mortality. During exposure at 200 ppm for a period of 1 month, seven of 50 rats died from lung irritation culminating in bronchopneumonia. A single human exposure for a few seconds to a level considerably below 200 ppm resulted in nausea and vomiting (152).

Approximately 33 percent of diethylaminoethanol injected into humans in 1-g doses is excreted unchanged (1). The biotransformation of the remaining portion is unknown, and it has been suggested by Williams (1959) that it could be deethylated to ethanolamine and thus enter normal metabolic pathways (1). However, the major metabolic pathway of this compound remains obscure (152).

*6.12.13.3 Hygienic Standards of Permissible Exposure.* A TLV–TWA of 100 ppm (48 mg/m$^3$), with a skin notation has been adopted (63c).

*6.12.14  Monoisopropylethanolamine [CAS # 109-56-8]*

The molecular formula for monoisopropylethanolamine is $NH(C_3H_7)C_2H_4OH$, having a molecular weight of 103.16.

*6.12.14.1 Uses.* Specific uses were not identified in the literature (62).

*6.12.14.2 Toxicity.* Toxicology information was not located in the literature (62).

*6.12.14.3 Hygienic Standards of Permissible Exposure.* Standards have not been adopted (63c).

### 6.12.15 Diisopropylethanolamine [CAS # 96-80-0]

The molecular formula for diisopropylethanolamine is $[(CH_3)_2CH]_2NCH_2CH_2OH$, having a molecular weight of 145.25; it is a corrosive liquid and is also known as 2-(diisopropylamino)ethanol (66).

*6.12.15.1 Uses.* Specific uses were not identified in the literature (62).

*6.12.15.2 Toxicity.* Toxicology information was not located in the literature (62).

*6.12.15.3 Hygienic Standards of Permissible Exposure.* Standards have not been adopted (63c).

### 6.12.16 Monobutylethanolamine [CAS # 111-75-1]

The molecular formula for monobutylethanolamine is $NH(C_4H_9)C_2H_4OH$, having a molecular weight of 117.19; it is also known as 2-butylaminoethanol and butylaminoethanol.

*6.12.16.1 Uses.* Specific uses were not located in the literature (62).

*6.12.16.2 Toxicity.* The acute oral and intraperitoneal $LD_{50}$ toxicity in male Sprague–Dawley rats was reported as 7.27 and 0.84 g/kg, respectively, for neutralized butylaminoethanol (153a). Butylaminoethanol has also been demonstrated to inhibit plasma cholinesterase in vitro, but at approximately half the potency of choline, as discussed above (153a). No additional toxicology information was found in the literature (62).

*6.12.16.3 Hygienic Standards of Permissible Exposure.* Standards have not been adopted (63c).

### 6.12.17 Dibutylethanolamine [CAS # 102-81-8]

The molecular formula for dibutylethanolamine is $[CH_3(CH_2)_3]_2NCH_2CH_2OH$, having a molecular weight of 173.30; it is a corrosive liquid and is also known as 2-(dibutylamino)ethanol, 2-*N*-dibutylaminoethanol, and 2-*N*-di-*n*-butylaminoethanol (66).

*6.12.17.1 Uses.* Dibutylaminoethanol has found use as a conditioning agent for cellulose acetate filaments to facilitate textile manufacture, a catalyst in polyurethane foam manufacture, and an anticorrosion additive for lubricants and hydraulic fluids, in the manufacture of emulsifying and dispersing agents, and in the curing of silicone resins (153d).

**6.12.17.2 Toxicity.** The acute oral and intraperitoneal $LD_{50}$ toxicity in male Sprague–Dawley rats was reported as 1.78 and 0.144 g/kg, respectively, for neutralized dibutylethanolamine (153a). Cornish et al. (1969) found that neutralized dibutylaminoethanol in drinking water at 0.2 and 0.4 g/kg/day in male and female Sprague–Dawley rats resulted in reduced body weight and slight increases in kidney weight without accompanying pathology. Inhalation studies at 70 ppm for 6 hr daily for 5 days resulted in weight losses, chromodacryorrhea, tremors, and some convulsions. At that dose level there was 20 percent mortality (153e). The authors pointed out that with the high boiling point of this compound and its relatively low vapor pressure, it is doubtful that vapor concentrations such as those utilized in their study would be encountered at room temperature. At elevated temperatures, or where large surface areas of the material are exposed, such concentrations might be attained; however, the nauseating odor of dibutylaminoethanol makes it doubtful that individuals would stay in badly contaminated areas for any length of time (153e).

Dibutylethanolamine has been demonstrated to inhibit plasma cholinesterase in vitro, having a potency greater than choline or any of the lower mono- and dialkyl-substituted 2-aminoethanols, as discussed above (153a). Dibutylaminoethanol has also been demonstrated to induce strong clonic–tonic convulsions in Sprague–Dawley rats after oral or intraperitoneal administration, as well as inhalation. The immediate cause of death appears to be respiratory arrest, which follows immediately after the convulsions. An investigation into the possible mechanisms demonstrated that the respiratory arrest is apparently due to neuromuscular blockade as shown by in vivo phrenic nerve-diaphragm preparations. The same investigation demonstrated that the blockade appears to be a result of direct action on the central rather than peripheral nervous system, for peripherally acting drugs selected (atropine and edrophonium) for possible antagonistic action had no effect, whereas two centrally acting anticonvulsants (mephenesin and diphenylhydantoin) were effective in counteracting both the convulsive seizures and the respiratory arrest produced by dibutylaminoethanol. The authors postulated that dibutylaminoethanol appears to possess a duality in its mode of action. First, it produces a greatly increased CNS activity which results in strong tonic–clonic convulsions. Respiratory arrest due to the neuromuscular blockade develops during this period of elevated CNS activity. Because dibutylaminoethanol is also a moderate acetylcholinesterase inhibitor, this increased nerve activity, with its simultaneous increase in release of acetylcholine at the junction, may result in a neuromuscular blockade due to excess acetylcholine at that site (153d).

No additional toxicology information was found in the literature (62).

***6.12.17.3 Hygienic Standards of Permissible Exposure.*** A TLV–TWA of 2 ppm (14 mg/m$^3$) with a skin notation has been adopted (63c).

### 6.12.18 Monomethyldiethanolamine [CAS # 105-59-9]

The molecular formula for monomethyldiethanolamine is $(HOCH_2CH_2)_2NCH_3$, with a molecular weight of 119.16; an irritant liquid, it is also known as 2,2'-methyliminodiethanol and *N*-methyldiethanolamine (66).

***6.12.18.1 Uses.*** Specific uses were not found in the literature (62).

***6.12.18.2 Toxicity.*** No toxicology information was found in the literature (62).

***6.12.18.3 Hygienic Standards of Permissible Exposure.*** Standards have not been adopted (63c).

### 6.12.19 Monoethyldiethanolamine [CAS # 139-87-7]

The molecular formula for monoethyldiethanolamine is $(HOCH_2CH_2)_2NC_2H_5$, with a molecular weight of 133.19; an irritant liquid, it is also known as 2,2'-ethyliminodiethanol and *N*-ethyldiethanolamine (66).

***6.12.19.1 Uses.*** Specific uses were not found in the literature (62).

***6.12.19.2 Toxicity.*** No toxicology information was found in the literature (62).

***6.12.19.3 Hygienic Standards of Permissible Exposure.*** Standards have not been adopted (63c).

### 6.12.20 Dimethylisopropanolamine [CAS # 108-16-7]

The molecular formula for dimethylisopropanolamine is $(CH_3)_2NCH_2CH(OH)CH_3$, with a molecular weight of 103.17; it is a flammable, corrosive liquid and is also known as 1-dimethylamino-2-propanol (66).

***6.12.20.1 Uses.*** Specific uses were not found in the literature (62).

***6.12.20.2 Toxicity.*** No toxicology information was found in the literature (62).

***6.12.20.3 Hygienic Standards of Permissible Exposure.*** Standards have not been adopted (63c).

### 6.12.21 N-Acetylethanolamine [CAS # 142-26-7]

The molecular formula for *N*-acetylethanolamine is $CH_3CONHCH_2CH_2OH$, having a molecular weight of 103.12; it is also known as *N*-2-hydroxyethylacetamide (66).

# ALIPHATIC AND ALICYCLIC AMINES 1165

*6.12.21.1 Uses.* Specific uses were not found in the literature (62).

*6.12.21.2 Toxicology.* No toxicology information was found in the literature (62).

*6.12.21.3 Hygienic Standards of Permissible Exposure.* Standards have not been adopted (63c).

### 6.12.22 Aminoethylethanolamine [CAS # 111-41-1]

The molecular formula for aminoethylethanolamine is $NH_2C_2H_4NHC_2H_4OH$, with a molecular weight of 104.15; it is a corrosive liquid lacrimator and is also known as 2-(2-aminoethylamino)ethanol (66).

*6.12.22.1 Uses.* Specific uses were not found in the literature (62).

*6.12.22.2 Toxicity.* Inhalation of vapor and/or particles containing aminoethylethanolamine has reportedly been associated with so-called solderers lung disease (186). No additional information was found in the literature (62).

*6.12.22.3 Hygienic Standards of Permissible Exposure.* Standards have not been adopted (63c).

## REFERENCES

1. R. R. Beard and J. T. Noe, "Aliphatic and Alicyclic Amines," Chapter 44, in *Patty's Industrial Hygiene and Toxicology*, 3rd ed., Vol. 2B, G. D. Clayton and F. E. Clayton, Eds., Wiley, New York, 1981, pp. 3135–3173.
2. A. E. Schweizer et al., "Amines," in *Kirk-Othmer Encyclopedia of Chemical Technology*, 3rd ed., Vol. 2, H. F. Mark, D. F. Othmer, C. G. Overberger, and G. T. Seaborg, Eds., Wiley, New York, 1978, pp. 272–283.
3. R. D. Spitz, "Diamines and Higher Amines, Aliphatic," in *Kirk-Othmer Encyclopedia of Chemical Technology*, 3rd ed., Vol. 7, H. F. Mark, D. F. Othmer, C. G. Overberger, and G. T. Seaborg, Eds., Wiley, New York, 1979, pp. 580–602.
4. H. B. Bathina and R. A. Reck, "Fatty Amines," in *Kirk-Othmer Encyclopedia of Chemical Technology*, 3rd ed., Vol. 2, H. F. Mark, D. F. Othmer, C. G. Overberger, and G. T. Seaborg, Eds., Wiley, New York, 1978, pp. 283–295.
5. R. T. Foley and B. F. Brown, "Corrosion and Corrosion Inhibitors," in *Kirk-Othmer Encyclopedia of Chemical Technology*, 3rd ed., Vol. 7, H. F. Mark, D. F. Othmer, C. G. Overberger, and G. T. Seaborg, Eds., Wiley, New York, 1979, pp. 113–142.
6. Provided by R. D. Harbison, Director, Center for Environmental and Human Toxicology, University of Florida, Alachua, FL, August 11, 1992.
7. I. Guest and D. R. Varma, "Developmental Toxicity of Methylamines in Mice," *J. Toxicol. Env. Health*, **32**, 319–330 (1991).
8. (*a*) E. A. Zeller, "Amine Oxidases," Chapter 52 in *Concepts in Biochemical Pharmacology*, Part 2, B. B. Brodie and J. R. Gillette, Eds., Springer-Verlag, Berlin, 1971,

pp. 519–535. (b) A. M. Equi et al., "Chain-length Dependence of the Oxidation of Alpha,Omega-diamines by Diamine Oxidase," *J. Chem. Res.*, **1992**, 94–95.

9. E. Hodgson and W. C. Dauterman, "Metabolism of Toxicants, Phase I Reactions," Chapter 4 in *Introduction to Biochemical Toxicology*, E. Hodgson and F. E. Guthrie, Eds., Elsevier, Amsterdam, 1980, pp. 67–91.

10. (a) R. Tecwyn Williams, "The Metabolism of Aliphatic Amines and Amides and Various Compounds Derived from Them," Chapter 6 in *Detoxication Mechanisms*, R. T. Williams, Ed., Wiley, New York, 1959, pp. 127–187. (b) K. F. Tipton, "Monoamine Oxidase," Chapter 17 in *Enzymatic Basis of Detoxication*, Vol. I, W. B. Jakoby, Ed., Academic Press, New York, 1980, pp. 355–370.

11. B. B. Brodie, J. R. Gillette, and B. N. LaDu, "Enzymatic Metabolism of Drugs and Other Foreign Compounds," *Ann. Rev. Biochem.*, **27**, 427–454 (1958).

12. S. W. Cummings and R. A. Prough, "Metabolic Formation of Toxic Metabolites," Chapter 1 in *Biological Basis of Detoxication*, J. Caldwell and W. B. Jakoby, Eds., Academic Press, New York, 1983, pp. 1–30.

13. In *Biological Basis of Detoxication*, J. Caldwell and W. B. Jakoby, Eds., Academic Press, New York, 1983.

14. In *Enzymatic Basis of Detoxication*, Vol. I, W. B. Jakoby, Ed., Academic Press, New York, 1980.

15. In *Enzymatic Basis of Detoxication*, Vol. II, W. B. Jakoby, Ed., Academic Press, New York, 1980.

16. R. C. James et al., *Biochem. Pharmacol.*, **24**, 835–838 (1975), abstract.

17. D. M. Ziegler, "Microsomal Flavin-Containing Monooxygenase: Oxygenation of Nucleophilic Nitrogen and Sulfur Compounds," Chapter 9 in *Enzymatic Basis of Detoxication*, Vol. I, W. B. Jakoby, Ed., Academic Press, New York, 1980, pp. 201–227.

18. B. Lindeke and A. K. Cho, "N-Dealkylation and Deamination," Chapter 6 in *Metabolic Basis of Detoxication*, W. B. Jakoby, J. R. Bend, and J. Caldwell, Eds., Academic Press, New York, 1982, pp. 105–126.

19. B. Testa and P. Jenner, "Inhibitors of Cytochrome P-450s and Their Mechanism of Action," *Drug Metab. Rev.*, **12**, 1–117 (1981).

20. R. C. James et al., *Biochem. Pharmacol.*, **24**, 835–838 (1975), abstract.

21. M. H. Bickel, *Pharmacol. Rev.*, **21**, 325–355 (1969).

22. R. T. Coutts and A. H. Beckett, *Drug Metab. Rev.*, **6**, 51–104 (1977).

23. G. H. Loew et al., *Int. J. Quantum Chem.*, **10**, 201–213 (1983).

24. S. S. Mirvish, "Formation of N-Nitroso Compounds: Chemistry, Kinetics and in Vivo Occurrence," *Toxicol. Appl. Pharmacol.*, **31**, 325–351 (1975).

25. S. H. Zeisel, K-A. DaCosta, B. M. Edrise, and J. G. Fox, "Transport of Dimethylamine, a Precursor of Nitrosodimethylamine, into Stomach of Ferret and Dog," *Carcinogenesis*, **7**(5), 775–778 (1986).

26. In *Todd-Sanford Clinical Diagnosis By Laboratory Methods*, 15th ed., I. Davidsohn and J. B. Henry, Eds., Saunders, 1974.

27. J. G. Wagner, "Introduction to Biopharmaceutics," Chapter 1 in *Biopharmaceutics and Relevant Pharmacokinetics*, 1st ed., J. G. Wagner, Ed., Drug Intelligence Publications, 1971, pp. 1–5.

28. M. J. Lynch et al., Eds., *Medical Laboratory Technology and Clinical Pathology*, 2nd ed., Saunders, 1969.

29. J. H. Ridd, *Q. Rev.*, **15**, 418–441 (1961), W. Lijinsky, "Formation of *N*-Nitroso Compounds and Their Significance," Chapter 1 in *Genotoxicology of* N-*Nitroso Compounds*, T. K. Rao, W. Lijinsky, and J. L. Epler, Eds., Plenum Press, New York, 1984, pp. 1–11.
30. H. Druckrey et al., "Prufung von Nitrit auf toxische Wirkung an Ratten," *Arzneim-Forsch*, **13**, 320–323 (1963), in W. Lijinsky, "Formation of *N*-Nitroso Compounds and Their Significance," Chapter 1 in *Genotoxicology of* N-*Nitroso Compounds*, T. K. Rao, W. Lijinsky and J. L. Epler, Eds., Plenum Press, New York, 1984, pp. 1–11.
31. E. J. Calabrese, "Comparative Metabolism: The Principal Cause of Differential Susceptibility To Toxic and Carcinogenic Agents," Chapter 5 in *Principles of Animal Extrapolation*, E. J. Calabrese, Ed., Wiley, New York, 1983, pp. 203–281.
32. I. H. Porter, "The Genetics of Drug Susceptibility," *Dis. Nerv. Syst.*, **27**, 25–36 (1966).
33. (*a*) R. Bonicke and W. Reif, "Enzymatische inactivierung von Isonicotinsaurhydrazid in menschlichen und tierschen organismus," Nauyen-Schmiedeberg's *Arch. Pharmacol.*, **220**, 321–333 (1953). (*b*) H. B. Hughes et al., "Metabolism of Isoniazid in Man as Related to the Occurrence of Peripheral Neuritis," *Am. Rev. Tuberc. Pulm. Dis.*, **70**, 266–273 (1954).
34. M. Blum et al., "Human Arylamine *N*-Acetyltransferase Genes: Isolation, Chromosomal Location, and Functional Expression," *DNA Cell. Biol.*, **9**, 193–203 (1990).
35. W. W. Weber, Ed., *The Acetylator Genes and Drug Response*, Oxford University Press, New York, 1987.
36. E. Spellacy et al., *J. Inher. Metab. Dis.*, **2**, 85 (1979).
37. M. S. Lennard, G. T. Tucker, and H. F. Woods, "Inborn 'Errors' of Drug Metabolism, Pharmacokinetic and Clinical Implications," *Clin. Pharmacokinet.*, **19**, 257–263 (1990).
38. R. T. Williams, "The Metabolism of the Aromatic Nitro, Amino and Azo Compounds," Chapter 9 in *Detoxication Mechanisms, The Metabolism of Drugs and Allied Organic Compounds*, R. T. Williams, Ed., Wiley, New York, 1947, pp. 130–154.
39. T. Mizutani et al., *Sci. Total Environ.*, **80**, 161–165 (1989), abstract.
40. L. L. Poulsen and D. M. Ziegler, *J. Biol. Chem.*, **254**, 6449–6455 (1979).
41. J. J. Kamm et al., *J. Pharmacol. Exp. Ther.*, **184**, 729–733 (1973).
42. (*a*) P. G. Wislocki et al., "Reactions Catalyzed by the Cytochrome P-450 System," Chapter 7 in *Enzymatic Basis of Detoxication*, Vol. I, W. B. Jakoby, Ed., Academic Press, New York, 1980, pp. 135–182. (*b*) A. H. Beckett and S. Al-Sarraj, "The Mechanism of Oxidation of Amphetamine Enantiomorphs by Liver Microsomal Preparations from Different Species," *J. Pharm. Pharmacol.*, **24**, 174–176 (1972).
43. B. Akesson et al., "Experimental Study on the Metabolism of Triethylamine in Man," *Brit. J. Ind. Med.*, **45**, 262–268 (1988).
44. B. Akesson et al., "Pharmacokinetics of Triethylamine and Triethylamine-*N*-Oxide in Man," *Toxicol. Appl. Pharmacol.*, **100**, 529–538 (1989).
45. B. Akesson et al., "Visual Disturbances by Industrial Triethylamine Exposure," *Int. Arch. Occup. Environ. Health*, **57**, 297–302 (1986).
46. W. Lijinsky, "Structure–Activity Relations in Carcinogenesis by *N*-Nitroso Compounds," Chapter 10 in *Genotoxicity of N-Nitroso Compounds*, T. K. Rao, W. Lijinsky and J. L. Epler, Eds., Plenum Press, New York, 1984, pp. 189–233.
47. W. Lijinsky, J. E. Saavedra, and M. D. Reuber, "Induction of Carcinogenesis in Fischer Rats by Methylalkylnitrosamines," *Cancer Res.*, **41**, 1288–1292 (1981).

48. J. W. Gorrod and L. A. Damani, "The Effect of Various Potential Inhibitors, Activators and Inducers on the N-Oxidation of 3-Substituted Pyridines in vitro," *Xenobiotica*, **9**, 219–226 (1979).
49. M. Sugiura et al., *Mol. Pharmacol.*, **12**, 322–334 (1976).
50. K. Kumaki and D. W. Nebert, *Pharmacology*, **17**, 262 (1978), in G. J. Mannering, "Hepatic Cytochrome P-450-Linked Drug-Metabolizing Systems," in *Concepts in Drug Metabolism*, Part B, P. Jenner and B. Testa, Eds., Marcel Dekker, New York, 1981, p. 95.
51. C. R. Jefcoate, "Integration of Xenobiotic Metabolism in Carcinogen Activation and Detoxication," Chapter 2 in *Biological Basis of Detoxication*, J. Caldwell and W. B. Jakoby, Eds., Academic Press, New York, 1983, pp. 31–76.
52. T. R. Juneja et al., *J. Sci. Ind. Res.*, **48**, 100–110 (1989), abstract.
53. W. Lijinsky, "Formation of N-Nitroso Compounds and Their Significance," Chapter 1 in *Genotoxicity of* N-*Nitroso Compounds*, T. K. Rao, W. Lijinsky, and J. L. Epler, Eds., Plenum Press, New York, 1984, pp. 1–11.
54. A. G. Renwick and R. T. Williams, *Biochem. J.*, **129**, 857–867 (1972).
55. E. A. Zeller, in *The Enzymes*, Vol. II, Part I, J. B. Sumner and K. Myrback, Eds., 1951, in Reference 10a.
56. (*a*) N. R. Stephenson, *J. Biol. Chem.*, **149**, 169 (1943), in Reference 10a. (*b*) H. Tabor, *Fed. Proc.*, **9**, 319 (1950), in Reference 10a.
57. R. S. H. Yang and M. J. Tallant, "Metabolism and Pharmacokinetics of Ethylenediamine in the Rat Following Oral, Endotracheal or Intravenous Administration," *Fundam. Appl. Toxicol.*, **2**, 252–260 (1982).
58. H. Foreman et al., *J. Biol. Chem.*, **203**, 1045–1053 (1953).
59. L. O. Pilgeram et al., *J. Biol. Chem.*, **204**, 367 (1953), in Reference 10a.
60. J. M. Johnston et al., *J. Biol. Chem.*, **221**, 301 (1956), in Reference 10a.
61. R. T. Williams, "Pathways of Drug Metabolism," in *Concepts in Biochemical Pharmacology*, Part 2, B. B. Brodie and J. R. Gillette, Eds., Springer-Verlag, 1971, p. 235.
62. Center for Environmental and Human Toxicology, University of Florida, computer literature search information provided by Raymond D. Harbison and Christopher J. Borgert, August 11, 1992.
63. (*a*) *Documentation of the Threshold Limit Values and Biological Exposure Indices*, 5th ed., American Conference of Governmental Industrial Hygienists, Inc., 1986. (*b*) *Documentation of the Threshold Limit Values and Biological Exposure Indices*, 6th ed., American Conference of Governmental Industrial Hygienists, Inc., 1991. (*c*) *Threshold Limit Values and Biological Exposure Indices*, American Conference of Governmental Industrial Hygienists, 1992–1993. (*d*) Jon H. Ruth, "Odor Thresholds and Irritation Levels of Several Chemical Substances: A Review," *Am. Ind. Hyg. Assoc. J.*, **47**, A142–A151 (1986). (*e*) "Odor Thresholds for Chemicals with Established Occupational Health Standards," American Industrial Hygiene Association, 1989.
64. P. Kestell, A. Gescher, and J. A. Slack, "The Fate of N-Methylformamide in Mice, Routes of Elimination and Characterization of Metabolites," *Drug Metab. Dispos.*, **13**, 587–592 (1985).
65. (*a*) W. Friemann and W. Overhoff, "Keratitis al. Berufserkrankung in der Olheringsficherei," *Klin. Mbl. Augenheilk*, **128**, 425–438 (1956), through W. Morton Grant,

Ed., *Toxicology of the Eye*, 2nd ed., Charles C Thomas, 1974, p. 680. (*b*) *Ibid.*, p. 1060.

66. (*a*) M. Windholz, S. Budavari, R. Blumetti, and E. Otterbein, Eds., *The Merck Index*, 10th ed., Merck & Co., 1983. (*b*) *Aldrich Chemical Company Catalog Handbook of Fine Chemicals*, Aldrich Chemical Company, St. Louis, Mo., 1992–1993.

67. W. Braker and A. L. Mossman, Eds., *Matheson Gas Data Book*, 6th ed., Matheson Gas Products, Division Searle Medical Products, Inc., Lyndhurst, NJ, 1980.

68. N. H. Proctor, J. P. Hughes, and M. L. Fischman, Eds., *Chemical Hazards of the Workplace*, 2nd ed., Lippincott, Philadelphia, 1988.

69. W. M. Grant, Ed., *Toxicology of the Eye*, 3rd ed., Charles C Thomas, 1986, p. 76, HSDB abstract, Reference 62.

70. R. L. Hollingsworth and V. K. Rowe, Dow Chemical Company, 1964, unpublished, in Reference 71.

71. R. A. Coon et al., "Animal Inhalation Studies on Ammonia, Ethylene Glycol, Formaldehyde, Dimethylamine and Ethanol," *Toxicol. Appl. Pharmacol.*, **16**, 646–655 (1970).

72. W. H. Steinhagen et al., "Acute Inhalation Toxicity and Sensory Irritation of Dimethylamine," *Am. Ind. Hyg. Assoc. J.*, **43**, 411–417 (1982).

73. L. A. Buckley et al., "Respiratory Tract Lesions Induced by Sensory Irritants at the RD50 Concentration," *Toxicol. Appl. Pharmacol.*, **74**, 417–429 (1984).

74. L. S. Buckley et al., "The Toxicity of Dimethylamine in F-344 Rats and B6C3F1 Mice Following a 1-year Inhalation Exposure," *Fundam. Appl. Toxicol.*, **5**, 341–352 (1985).

75. E. A. Gross et al., "Effects of Acute and Chronic Dimethylamine Exposure on the Nasal Mucociliary Apparatus of F-344 Rats," *Toxicol. Appl. Pharmacol.*, **90**, 359–376 (1987).

76. W. K. Anger and B. L. Johnson, "Chemicals Affecting Behavior," Chapter 3 in *Neurotoxicity of Industrial and Commercial Chemicals*, J. O'Donoghue, Ed., CRC Press, Boca Raton, FL, 1985, pp. 51–148.

77. W. E. Ribelin, "The Effects of Drugs and Chemicals Upon the Structure of the Adrenal Gland," *Fundam. Appl. Toxicol.*, **4**, 105–119 (1984).

78. F. N. Marzulli and H. I. Maibach, Eds., *Dermatotoxicology*, 2nd ed., Hemisphere, New York, 1983.

79. (*a*) H. F. Smyth, Jr., C. P. Carpenter, and C. S. Weil, "Range-finding Toxicity Data, List IV," *Arch. Ind. Hyg. Occup. Med.*, **4**, 119–122 (1951), through D. W. Lynch et al., "Subchronic Inhalation of Diethylamine Vapor in Fischer-344 Rats: Organ System Toxicity," *Fundam. Appl. Toxicol.*, **6**, 559–565 (1986). (*b*) R. B. Drotman and G. T. Lawhorn, "Serum Enzymes as Indicators of Chemically Induced Liver Disease," *Drug Chem. Toxicol.*, **1**, 163–171 (1978), through D. W. Lynch, *op. cit.* (*c*) H. Brieger and W. A. Hodes, "Toxic Effects of Exposure to Vapors of Aliphatic Amines," *Arch. Ind. Hyg. Occup. Med.*, **3**, 287–291 (1951), in D. W. Lynch, *op. cit.*

80. D. W. Lynch et al., "Subchronic Inhalation of Diethylamine Vapor in Fischer-344 Rats: Organ System Toxicity," *Fundam. Appl. Toxicol.*, **6**, 559–565 (1986).

81. D. W. Lynch et al., Society of Toxicology Annual Meeting, 1987, Abstract 759.

82. H. Brieger and W. A. Hodes, "Toxic Effects of Exposure to Vapors of Aliphatic Amines," *Arch. Ind. Hyg.*, **3**, 287–291 (1951), in Reference 69.

83. D. W. Lynch et al., "Subchronic Inhalation of Triethylamine Vapor in Fischer-344 Rats: Organ System Toxicity," *Toxicol. Ind. Health*, **6**, 403–414 (1990).
84. H. F. Smyth et al., *Arch. Ind. Hyg. Occup. Med.*, **4**, 119 (1951), in Reference 66.
85. R. B. Dudek, Society of Toxicity Annual Meeting, 1991, Abstract 265.
86. D. H. Hine et al., "The Toxicity of Allylamines," *Arch. Environ. Health*, **1**, 343–352 (1960), in Reference 69.
87. H. F. Smyth, Jr., et al., "Range-finding Toxicity Data: List VI," *Am. Ind. Hyg. Assoc. J.*, **23**, 95–107 (1962), in Reference 69.
88. H. F. Smyth et al., *Am. Ind. Hyg. Assoc. J.*, **23**, 95 (1962), in Reference 66.
89. H. F. Smyth et al., *Arch. Ind. Hyg. Occup. Med.*, **10**, 61 (1954), in Reference 66.
90. J. F. Treon et al., "The Physiological Response of Animals to Respiratory Exposure to the Vapors of Diisopropylamine," *J. Ind. Hyg.*, **31**, 142–145 (1949), in Reference 69.
91. H. B. Elkins, *The Chemistry of Industrial Toxicology*, 2nd ed., Wiley, New York, 1958, in Reference 69.
92. K. L. Cheever et al., "The Acute Oral Toxicity of Isomeric Monobutylamines in the Adult Male and Female Rat," *Toxicol. Appl. Pharmacol.*, **63**, 150–152 (1982).
93. (a) P. J. Hanzlik, "Toxicity and Actions of the Normal Butylamines," *J. Pharmacol. Exp. Ther.*, **20**, 435 (1923), in Reference 69. (b) H. Herrmann and F. H. Hickman, "The Adhesion of Epithelium to Stroma in the Cornea," *Bull. Johns Hopkins Hosp.*, **82**, 182–207 (1948), through Reference 69. (c) H. F. Smyth, Jr. and C. P. Carpenter, "Further Experience with the Range-finding Test in the Industrial Toxicology Laboratory," *J. Ind. Hyg.*, **30**, 63–68 (1948), in Reference 69.
94. E. I. Goldenthal, "A Compilation of $LD_{50}$ Values in Newborn and Adult Animals," *Toxicol. Appl. Pharmacol.*, **18**, 185–207 (1971).
95. J. C. Gage, "The Subacute Inhalation Toxicity of 109 Industrial Chemicals," *Brit. J. Ind. Med.*, **27**, 1–18 (1970).
96. B. R. Dudek et al., Society of Toxicology Annual Meeting, 1987, Abstract 763.
97. (a) D. F. Davies, *J. Lab. Clin. Med.*, **43**, 620 (1954), in Reference 1. (b) D. Richter, *Biochem. J.*, **32**, 1763 (1938), in Reference 1.
98. G. Barger and H. H. Dale, *J. Physiol.*, **41**, 19 (1910–1911), in Reference 1.
99. C. P. Carpenter and H. F. Smyth, Jr., "Chemical Burns of the Rabbit Cornea," *Am. J. Ophthalmol.*, **29**, 1363–1372 (1946).
100. D. F. Marsh, *J. Pharmacol. Exp. Therap.*, **94**, 225 (1948), in Reference 1.
101. P. E. Hanna, "Adrenergic Agents," Chapter 11 in *Wilson and Gisvold's Textbook of Organic Medicinal and Pharmaceutical Chemistry*, 8th ed., R. F. Doerge, Ed., Lippincott, Philadelphia, 1982, pp. 401–432.
102. A. Fuller, *Biochem. J.*, **36**, 548–555 (1942).
103. D. Kumar et al., "Allylamine and beta-Aminopropionitrile-Induced Vascular Injury: An in Vivo and in Vitro Study," *Toxicol. Appl. Pharmacol.*, **103**, 288–302 (1990).
104. P. J. Boor and V. J. Ferrans, "Ultrastructural Alterations in Allylamine-induced Cardiomyopathy. Early Lesions," *Lab. Invest.*, **47**, 76 (1982), HSDB abstract, Reference 62.
105. P. J. Boor et al., Society of Toxicity Annual Meeting, 1987, Abstract 371.
106. T. J. Nelson and P. J. Boor, "Allylamine Cardiotoxicity IV. Metabolism to Acrolein

by Cardiovascular Tissues," *Biochem. Pharmacol.*, **31**, 509 (1982), HSDB abstract, Reference 62.

107. M. Toraason et al., *Toxicology*, **56**, 107–117 (1989), HSDB abstract, Reference 62.

108. B. Testa and P. Jenner, "Inhibitors of Cytochrome P-450s and Their Mechanism of Action," *Drug Metab. Rev.*, **12**, 1–117 (1981).

109. International Agency for Research on Cancer, *IARC Monographs on the Evaluation of the Carcinogenic Risk of Chemicals to Humans, "Some Non-Nutritive Sweetening Agents,"* Vol. 22, 1980, pp. 64–109.

110. T. Miyata et al., "Pharmacological Characteristics of Cyclohexylamine, One of the Metabolites of Cyclamate," *Life Sci.*, **8**, 843–853 (1969), through *IARC Monographs*, Vol. 22.

111. S. Kojima and H. Ichibagase, "Studies on Synthetic Sweetening Agents. VIII. Cyclohexylamine, a Metabolite of Sodium Cyclamate," *Chem. Pharm. Bull.*, **14**, 971–974 (1966), through K. S. Khera et al., "Reproduction Study in Rats Orally Treated with Cyclohexylamine Sulfate," *Toxicol. Appl. Pharmacol.*, **18**, 263–268 (1971).

112. M. S. Legator et al., "Cytogenetic Studies in Rats of Cyclohexylamine, a Metabolite of Cyclamate," *Sci.*, **165**, 1139–1140 (1969), in K. S. Khera et al., *Toxicol. Appl. Pharmacol.*, **18**, 263–268 (1971).

113. B. A. Bopp et al., *CRC Crit. Rev. Toxicol.*, **16**, 213–289 (1986).

114. B. A. Becker and J. E. Gibson, "Teratogenicity of Cyclohexylamine in Mammals," *Toxicol. Appl. Pharmacol.*, **17**, 551–552 (1970).

115. J. M. Price et al., "Bladder Tumors in Rats Fed Cyclohexylamine or High Doses of a Mixture of Cyclamate and Saccharin," *Science*, **167**, 1131–1132 (1970), in *IARC Monographs*, Vol. 22, 1980, pp. 80–81.

116. K. S. Khera et al., "Reproduction Study in Rats Orally Treated with Cyclohexylamine Sulfate," *Toxicol. Appl. Pharmacol.*, **18**, 263–268 (1971).

117. K. Takano and M. Suzuki, "Cyclohexylamine, a Chromosome-aberration Inducing Substance: No Teratogenicity in Mice (Jpn.)," *Congenital Anomalies*, **11**, 51–57 (1971), in *IARC Monographs*, Vol. 22.

118. J. G. Wilson, "Use of Primates in Teratological Investigations," in *Medical Primatology*, E. I. Goldsmith and J. Moor-Jankowski, Eds., Karger, Basel, 1972, pp. 286–295, in *IARC Monographs*, Vol. 22.

119. B. L. Oser et al., "Long-term and Multigeneration Toxicity Studies with Cyclohexylamine Hydrochloride," *Toxicology*, **6**, 47–65 (1976), in *IARC Monographs*, Vol. 22, 1980, pp. 80–81.

120. J. Hardy et al., "Long-term Toxicity of Cyclohexylamine Hydrochloride in Mice," *Food Cosmet. Toxicol.*, **14**, 269–276 (1976), through *IARC Monographs*, Vol. 22, 1980, pp. 80–81.

121. R. Kroes et al., "Long-term Toxicity and Reproduction Study (including a teratogenicity study) with Cyclamate, Saccharin and Cyclohexylamine," *Toxicology*, **8**, 285–300 (1977), in *IARC Monographs*, Vol. 22.

122. International Agency for Research on Cancer: *IARC Monographs on the Evaluation of Carcinogenic Risks to Humans. Overall Evaluations of Carcinogenicity: An Updating of the IARC Monographs*, Vols. 1–42, Suppl. 7, 1987, pp. 178–182.

123. J. Marhold et al., "On the Carcinogenicity of Dicyclohexylamine," *Neoplasma*, **14**, 177–180 (1967), in *IARC Monographs*, Vol. 22, 1980.

124. M. J. Crabbs, *Exp. Eye Res.*, **41**, 777–778 (1985), abstract.
125. S. Fabro, "Biochemical Basis of Thalidomide Teratogenicity," Chapter 5 in *The Biochemical Basis of Chemical Teratogenesis*, M. R. Juchau, Ed., Elsevier/North-Holland, New York, 1981, pp. 161–178.
126. J. M. Manson, "Teratogens," Chapter 7 in *Casarett and Doull's Toxicology, The Basic Science of Poisons*, 3rd ed., C. D. Klaassen, M. O. Amdur, and J. Doull, Eds., Macmillan, New York, 1986, p. 197.
127. L. R. DePass et al., "Dermal Oncogenicity Studies on Various Ethyleneamines in Male C3H Mice," *Fundam. Appl. Toxicol.*, **9**, 807–811 (1987).
128. In *Toxicology of the Eye*, 2nd ed., W. M. Grant, Ed., Charles C Thomas, 1974.
129. J. Niveau and J. Painchaux, "Fatal Poisoning by Ethylenediamine," *Serv. Med. Houilleres Bassin Lorraine, Freyming Merlebach.*, *Arch. Mal. Prof.*, **34**, 523 (1973), abstract in *Excerpta Medica* 4.9: Section 35, 548 (1973), in Reference 63b.
130. R. S. H. Yang et al., "Acute and Subchronic Toxicity of Ethylenediamine in Laboratory Animals," *Fundam. Appl. Toxicol.*, **3**, 512–520 (1983).
131. M. N. Perkins and T. W. Stone, "Comparison of the Effects of Ethylenediamine Analogues and Gamma-aminobutyric Acid on Cortical and Pallidal Neurols," *Brit. J. Pharmacol.*, **75**, 93–99 (1982).
132. E. Roberts et al., "Amino Acid Transmitters," in *Basic Neurochemistry*, 2nd ed., G. J. Siegel et al., Eds., Little, Brown, Boston, 1976, 218–246.
133. H. S. Maker et al., "Intermediary Metabolism of Carbohydrates and Amino Acids," in *Basic Neurochemistry*, 2nd ed., G. J. Siegel et al., Eds., Little, Brown, Boston, 1976, pp. 279–307.
134. G. M. Strain and W. Flory, "Simple Aliphatic Diamines: Acute Neurotoxicity," *Res. Commun. Chem. Pathol. Pharmacol.*, **64**, 489–492 (1989).
135. M. N. Perkins et al., *Neurosci. Lett.*, **23**, 325–327 (1981), in Reference 134.
136. L. P. Davies et al., *Neurosci. Lett.*, **29**(1): 57–61 (1982), HSDB abstract, Reference 62.
137. P. V. Sarthy, *J. Neurosci.*, **3**, 2494–2503 (1983), HSDB abstract, Reference 62.
138. P. F. Morgan and T. W. Stone, *Brit. J. Pharmacol.*, **77**, 525–529 (1982), HSDB abstract, Reference 62.
139. C. Babiuk et al., "Induction of Ethylenediamine Hypersensitivity in the Guinea Pig and the Development of ELISA and Lymphocyte Blastogenesis Techniques for its Characterization," *Fundam. Appl. Toxicol.*, **9**, 623–634 (1987).
140. B. D. Hardin et al., *Teratogenesis, Carcinogenesis and Mutagenesis*, **7**, 29–48 (1987), HSDB abstract, Reference 62.
141. L. R. DePass et al., "Evaluation of the Teratogenicity of Ethylenediamine Dihydrochloride in Fischer 344 Rats by Conventional and Pair-feeding Studies," *Fundam. Appl. Toxicol.*, **9**, 687–697 (1987).
142. R. S. H. Yang et al., "Two-generation Reproduction Study of Ethylenediamine in Fischer 344 Rats," *Fundam. Appl. Toxicol.*, **4**, 539–546 (1984).
143. Battelle NTP Report Contract (February 26, 1982), in Reference 63b.
144. R. S. Slesinski et al., "Assessment of Genotoxic Potential of Ethylenediamine: in Vitro and in Vivo Studies," *Mutat. Res.*, **124**, 299–314 (1983).
145. L. R. DePass et al., "Dermal Oncogenicity Studies on Ethylenediamine in Male C3H Mice," *Fundam. Appl. Toxicol.*, **4**, 641–645 (1984).

146. R. S. H. Yang et al., "Chronic Toxicity/Carcinogenicity Study of Ethylenediamine in Fischer 344 Rats," *Toxicologist*, **4**, 53 (1984), in Reference 63b.
147. K. S. Rao, J. E. Betso, and K. J. Olson, "A Collection of Guinea Pig Sensitization Test Results—Grouped By Chemical Class," *Drug Chem. Toxicol.*, **4**, 331–351 (1981).
148. R. M. Craig et al., *Poultr. Sci.*, **47**, 1164–1665 (1968).
149. G. Sosnovsky et al., *J. Med. Chem.*, **28**, 1350–1354 (1985), abstract.
150. C. L. Keen et al., *Teratology*, **29**, 8A (1984), abstract.
151. R. M. Mullins, "Alkanolamines," in *Kirk-Othmer Encyclopedia of Chemical Technology*, 3rd ed., Vol. 1, H. F. Mark, D. F. Othmer, C. G. Overberger, and G. T. Seaborg, Eds., Wiley, New York, 1978, pp. 944–960.
152. Herbert H. Cornish, "Oral and Inhalation Toxicity of 2-Diethylaminoethanol," *Am. Ind. Hyg. Assoc. J.*, **26**, 479–484 (1965).
153. (*a*) R. Hartung and H. H. Cornish, "Cholinesterase Inhibition in the Acute Toxicity of Alkyl-substituted 2-aminoethanols," *Toxicol. Appl. Pharmacol.*, **12**, 486–494 (1968). (*b*) H. H. Cornish and J. Adefuin, "Effect of 2-*N*-Mono- and 2-*N*-Diethylaminoethanol on Normal and Choline-deficient Rats," *Food Cosmet. Toxicol.*, **5**, 327–332 (1967). (*c*) R. Hartung and H. H. Cornish, "Acute and Short-term Oral Toxicity of 2-*N*-Ethylaminoethanol in Rats," *Food Cosmet. Toxicol.*, **7**, 595–602 (1969). (*d*) R. Hartung, L. B. Pittle, and H. H. Cornish, "Convulsions Induced by 2-*N*-Di-*n*-Butylaminoethanol," *Toxicol. Appl. Pharmacol.*, **17**, 337–343 (1970). (*e*) H. H. Cornish, T. Dambrauskas, and L. D. Beatty, "Oral and Inhalation Toxicity of 2-*N*-Dibutylaminoethanol," *Am. Ind. Hyg. Assoc. J.*, **30**, 46–51 (1969).
154. M. H. Weeks et al., "The Effects of Continuous Exposure of Animals to Ethanolamine Vapor," *Am. Ind. Hyg. Assoc. J.*, **21**, 374–381 (1960).
155. G. J. Klain et al., "Distribution and Metabolism of Topically Applied Ethanolamine," *Fundam. Appl. Toxicol.*, **5**, S127–S133 (1985).
156. E. W. Martin and E. F. Cook, Eds., *Remington's Practice of Pharmacy*, 11th ed., Mack Publishing Co., Easton, PA, 1956.
157. S. M. Pond, "Effects on the Liver of Chemicals Encountered in the Workplace," *Western J. Med.*, **137**, 506–514 (1982).
158. Union Carbide Monoethanolamine Toxicology Study Report, 1958, in Reference 63b.
159. R. M. C. Dawson, "The Animal Phospholipids: Their Structure, Metabolism and Biological Significance," *Biol. Rev.*, **32**, 188–229 (1957), in Reference 154.
160. B. J. Dean et al., "Genetic Toxicology Testing of 41 Industrial Chemicals," *Mutat. Res.*, **153**, 57–77 (1985).
161. H. F. Smyth, Jr. et al., "An Exploration of Joint Toxic Action," *Toxicol. Appl. Pharmacol.*, **17**, 498–503 (1970).
162. In *The United States Pharmacopeia, 19th Revision, The National Formulary XVII*, United States Pharmacopeial Convention, Inc., 1975.
163. H. Isacoff, "Cosmetics," in *Kirk-Othmer Encyclopedia of Chemical Technology*, 3rd ed., Vol. 7, H. F. Mark, D. F. Othmer, C. G. Overberger, and G. T. Seaborg, Eds., Wiley, New York, 1979, pp. 143–176.
164. G. L. Drake, Jr., "Flame Retardants for Textiles," in *Kirk-Othmer Encyclopedia of Chemical Technology*, 3rd ed., Vol. 10, H. F. Mark, D. F. Othmer, C. G. Overberger, and G. T. Seaborg, Eds., Wiley, New York, 1980, pp. 420–444.

165. In *American Hospital Formulary Service, Drug Information*, American Society of Hospital Pharmacists, 1992.
166. In *United States Adopted Names (USAN) and the USP Dictionary of Drug Names*, United States Pharmacopeial Convention, Inc., 1992.
167. R. N. Loeppky et al., "Reducing Nitrosamine Contamination in Cutting Fluids," *Food Chem. Toxicol.*, **21**, 607–613 (1983), through Y. Konishi et al., "Chronic Toxicity Carcinogenicity Studies of Triethanolamine in B6C3F1 Mice," *Fundam. Appl. Toxicol.*, **18**, 25–29 (1992).
168. J. F. Griffith et al., "Dose-response Studies with Chemical Irritants in the Albino Rabbit Eye as a Basis for Selecting Optimum Testing Conditions for Predicting Hazard to the Human Eye," *Toxicol. Appl. Pharmacol.*, **55**, 501–513 (1980).
169. K. Inoue et al., "Mutagenicity Tests and in Vitro Transformation Assays on Triethanolamine," *Mutat. Res.*, **101**, 305–313 (1982).
170. A. Maekawa, H. Onodera, H. Tanigawa, K. Furuta, J. Kanno, T. Matsuoka, Y. Ogui, and Y. Hayashi, "Lack of Carcinogenicity of Triethanolamine in F344 Rats," *J. Toxicol. Environ. Health*, **19**, 345–357 (1986), in Y. Konishi et al., *Fundam. Appl. Toxicol.*, **18**, 25–29 (1992).
171. Y. Konishi et al., "Chronic Toxicity Carcinogenicity Studies of Triethanolamine in B6C3F1 Mice," *Fundam. Appl. Toxicol.*, **18**, 25–29 (1992).
172. M. Schaper et al., "Cutting Machine or Machining Fluids," *Fundam. Appl. Toxicol.*, **16**, 309–319 (1991).
173. C. Artom, "Methylation of Phosphatidyl Monomethylethanolamine in Liver Preparations," *Biochem. Biophys. Res. Commun.*, **15**, 201–206 (1964), in Reference 153.
174. W. J. Welstead, Jr., "Stimulants," in *Kirk-Othmer Encyclopedia of Chemical Technology*, 3rd ed., Vol. 21, H. F. Mark, D. F. Othmer, C. G. Overberger, and G. T. Seaborg, Eds., Wiley, New York, 1983, pp. 747–761.
175. G. Pepeu et al., *J. Pharmacol. Exp. Therap.*, **129**, 291 (1960), through J. S. Bindra, "Memory-enhancing Agents and Antiaging Drugs," in *Kirk-Othmer Encyclopedia of Chemical Technology*, 3rd ed., Vol. 15, H. F. Mark, D. F. Othmer, C. G. Overberger, and G. T. Seaborg, Eds., Wiley, New York, 1981, pp. 132–143.
176. S. Malitz et al., "A Pilot Evaluation of Deanol in the Treatment of Anxiety," *Curr. Therap. Res.*, **9**, 261–264 (1967), in Reference 153.
177. D. R. Haurbrich et al., "Deanol Affects Choline Metabolism in Peripheral Tissues of Mice," *J. Neurochem.*, **37**, 476 (1981), Reference 62.
178. D. R. Klonne et al., Union Carbide, Society of Toxicology Annual Meeting, 1987, Abstract 691.
179. (a) F. Stenback et al., "Hydroxylamine Effects on Cryptogenic Neoplasm Development in Mice," *Cancer Lett.*, **38**, 73–85 (1987), in Reference 171. (b) F. Stenback et al., "Effects of Lifetime Administration of Dimethylaminoethanol on Longevity, Aging Changes, and Cryptogenic Neoplasms in C3H Mice," *Mech. Ageing Dev.*, **42**, 129–138 (1988), in Reference 171.
180. O. E. Bell et al., "Ethylated Ethanolamines in Phospholipids," *Fed. Proc.*, **23**, 222 (1964), in Reference 153.
181. F. Hauschild, "Beitrag zur Frage der pharmakologischen Wirkung einiger aliphatischer Alkyl- und Alkanolamine," *Arch. Exptl. Pathol. Pharmakol.*, **201**, 569–579 (1943), in Reference 153.

182. H. Smyth Jr. et al., "Range Finding Toxicity Data list V," *AMA Arch. Ind. Health*, **10**, 61 (1954), through Reference 152.
183. B. B. Brodie et al., "The Fate of Procaine in Man Following its Intravenous Administration and Methods for the Estimation of Procaine and Diethylaminoethanol," *J. Pharm. Exp. Ther.*, **93**, 359 (1948), through Reference 152.
184. C. Kraatz et al., "Observation on Diethylaminoethanol," *J. Pharm. Exp. Ther.*, **98**, 111 (1950), in Reference 152.
185. B. Rosenberg et al., "Studies on Diethylaminoethanol, I. Physiological Disposition and Action on Cardiac Arrhythmias," *J. Pharm. Exp. Ther.*, **95**, 18 (1949), in Reference 152.
186. K. H. Kilburn, "Particles Causing Lung Disease," *Environ. Health Perspect.*, **55**, 97–109 (1984).

CHAPTER EIGHTEEN

# Fluorine-Containing Organic Compounds

Domingo M. Aviado, M.D.

## 1 INTRODUCTION

During the third millennium AD, medical practice will most likely be eclipsed by events relating to fluorine-containing organic compounds reviewed in this chapter. This sweeping prediction is based on two developments prior to the end of this century or this millennium. First is that essential refrigerants, solvents, and plastic blowing agents presently in use continue to deplete the stratospheric ozone layer, causing a rising incidence of diseases such as skin cancer, eye cataracts, and immunologic disorders. Second is that fluorine-containing organics are potentially useful blood substitutes and diagnostic aids that will revolutionize the management of cancer and heart disease, presently the leading causes of death in the United States, most developed countries, and some developing countries. These developments can either increase or decrease mortality during the next millenium, depending on a cooperative effort among chemical manufacturers, medical researchers, and multinational groups (governmental or otherwise) interested in the environment.

### 1.1 Stratospheric Ozone Depletion

In 1974, Molina and Rowland drew attention to a potential biologic hazard resulting from depletion of the ozone layer owing to the release of chlorofluorocarbons into

---

*Patty's Industrial Hygiene and Toxicology, Fourth Edition, Volume 2, Part B*, Edited by George D. Clayton and Florence E. Clayton.
ISBN 0-471-54725-5 © 1994 John Wiley & Sons, Inc.

the atmosphere (1). The resulting "ozone war" among chemical manufacturers, scientists, and nonscientists recounted by Dotto and Schiff (2) have been settled in favor of the general premise that photodissociation of chlorofluorocarbons in the stratosphere produces significant amounts of chlorine atoms and leads to the destruction of atmospheric ozone. A reduction in the ozone allows more ultraviolet light to reach the earth's surface and is anticipated to increase the incidence of eye cataracts, immunologic diseases, malignant melanomas (a serious form of skin cancer frequently causing death), and basal and squamous-cell carcinomas of the skin (less serious but much more prevalent). The effects of increased ultraviolet radiation on plants and animals due to ozone depletion are of unknown magnitude.

The chlorofluorocarbons do not photodissociate in the atmosphere. They diffuse into the stratosphere where they undergo photodissociation, resulting in the release of chlorine atoms. The consequent chemical reactions resulting in ozone ($O_3$) depletion are as follows:

$$Cl + O_3 \rightarrow ClO + O_2$$

$$\underline{ClO + O \rightarrow Cl + O_2}$$

$$\text{Net } O + O_3 \rightarrow 2O_2$$

The above sequence depends on persistence of chlorofluorocarbons that have a half-life of 60 to 200 years. It has been estimated that the chlorofluorocarbons that have been manufactured during the past 50 years will continue to deplete ozone layer in the stratosphere.

### 1.2 Classification of Fluorine-containing Organic Compounds

At the outset, it is necessary to review the terminology of hydrocarbons substituted with fluorine, chlorine, or hydrogen, also known as halogenated hydrocarbons. Most compounds are identified by two, three, or four digits, depending on the number of carbon, chlorine, fluorine, and hydrogen in the molecule. In the past, all these compounds were referred to by the term "fluorocarbons," with individual FC numerical notation for each compound. Recent events have forced a distinction among the four subgroups. For the purpose of this chapter, the halogenated hydrocarbons can be grouped into the following classes:

*Chlorofluorocarbon (CFC).* Chlorinated and fluorinated methanes and ethanes, such as trichlorofluoromethane ($CCl_3F$), also known by a numerical notation, such as compound CFC 11. The use of "C" to represent either chlorine or carbon is confusing, and purists in chemical symbols might prefer ClFC to CFC.

*Hydrogenated Chlorofluorocarbon (HCFC).* Hydrogenated halogenated compounds such as dichlorofluoromethane ($CCHCl_2F$), also known as compound HCFC 21.

*Hydrogenated Fluorocarbon (HFC)*. Fluorine-substituted nonchlorinated hydrocarbons such as difluoroethane or HFC 152a.

*Fluororcarbon (FC)*. Nonchlorinated and nonhydrogenated, such as octafluorobutane or FC C-318.

The above chemical groupings relate to their activity in terms of ozone depletion. All chlorine-containing compounds, that is, CFC and HCFC, are capable of depleting ozone and all commercially available compounds belong to these two groups. The chemical industry has utilized the fact that hydrogenated compounds (HCFC) do not deplete the ozone layer as much as nonhydrogenated compounds (CFC) and are in the process of developing HCFC compounds as new alternatives. Because ozone depletion is caused by presence of chorine in the molecule, the nonchlorinated compounds (FC and HFC) are the only acceptable developmental alternatives for the long term.

This chapter starts with a review of animal toxicologic studies of a widely used refrigerant and aerosol propellant, specifically of the prototype chlorofluorocarbon 11 or CFC 11 (Section 2), and then continuing to other chlorofluorocarbons (CFC), hydrogenated chlorofluorocarbons (HCFC), fluorocarbons (FC), and hydrogenated fluorocarbons (HFC). The subjects of animal toxicologic studies (Section 3) and human toxicologic studies (Section 4) emphasize the fact that all fluorine-containing organic compounds are known to cause similar adverse reactions on the cardiopulmonary, nervous, and dermato-mucosal systems. The continuing epidemic of asthmatic deaths probably caused by fluorine-containing aerosol propellants are reviewed (Section 5). The fluorine-containing therapeutic and diagnostic agents are highlighted (Section 6) because of their unique contribution to future medical practice.

## 2 PROTOTYPE CHLOROFLUOROCARBON

When chlorofluorocarbons (CFC) were introduced in the 1940s, they were regarded as "inert" refrigerants compared to sulfur dioxide, ammonia, carbon tetrachloride, and chloroform, which were then in use. Chemical manufacturers introduced aerosols containing household products, cosmetics, and medication with CFC used as propellants. The use and abuse of aerosols led to accidental deaths by cardiac arrest. The most widely used propellant, trichlorofluoromethane (CFC 11), was the most widely investigated compound and proved to be most cardiotoxic. My colleagues and I completed comparative studies of cardiopulmonary effects of most commercially available fluorine-containing organic compounds during the 1970s and 1980s, including some developmental alternative compounds (3–7). These studies were initiated following reported fatalities from the use, misuse, and abuse of aerosol products in general, and of bronchodilator aerosols in particular. The experimental procedures developed for the toxicologic evaluation of CFC is discussed below because it may prove to be useful in the risk assessment of new developmental alternatives. It has become apparent that selected animal species are useful in anticipating the potential toxicity of inhalants in humans.

## 2.1 Cardiotoxicity in Animal Studies

There are no published reports on estimation of the $LC_{50}$ for CFC 11 by inhalation. All available observations relate to the concentration that would be lethal to exposed animals. The most resistant species requiring the highest lethal concentration is the guinea pig. The rat is the most sensitive, and the mouse and cat are between the two extremes. Studies on cardiac arrhythmias indicate that dogs and monkeys are more sensitive to CFC 11 than rodents.

*Guinea Pig.* Nuckolls (8) reported the first investigation of the acute toxicity of CFC 11. Twelve guinea pigs divided into four groups of three each were exposed for 5 min, 30 min, 1 hr, and 2 hr, respectively. Exposure of 2.5 percent for 30 min caused occasional tremors and the rate of respiration became irregular. Exposure to 10 percent for 1 hr resulted in coma. The guinea pigs exposed to this concentration for 2 hr were sacrificed 8 days later. Whereas their lungs were found to contain mottled areas of congestion, other organs showed no pathological changes. Scholz (9) reported that exposure to a concentration of 20 percent for 1 hr was lethal. According to Caujolle (10), inhalation of a concentration of 25 percent for 30 min was lethal in half the guinea pigs tested. A concentration of 3 percent inhaled for 2 hr, although not fatal, caused unconsciousness.

*Rat.* Lester and Greenberg (11) exposed rats to CFC 11 in concentrations ranging from 5 to 50 percent for 30 min. Whereas a concentration of 5 percent produced no symptoms of intoxication, concentrations of 6 and 7 percent caused a loss of postural reflex, 8 percent a loss of righting reflex, and 9 percent complete unconsciousness. The following concentrations and times were lethal: 10 percent inhaled for 20 to 30 min; 15 percent for 8 min; 20 to 30 percent for 4 min; and 50 percent for 1 min. Scholz (9) reported a lethal concentration of 10 percent after 90 min. Friedman et al. (12) exposed anesthetized rats to increasing concentrations of CFC 11 and observed that apnea occurred 5 min after the concentration reached 20 percent.

*Cat and Mouse.* Scholz (9) reported that inhalation of 10 percent CFC 11 for 1 hr was lethal to the cat. Caujolle (10) determined the lethal concentration in mice. In an atmosphere containing 15 percent CFC 11, mice succumbed in a few minutes.

### 2.1.1 Cardiac Arrhythmia

The dog and the monkey require a lower concentration of CFC 11 to provoke cardiac arrhythmia than do the rat or the mouse. In each species, the sensitivity to the proarrhythmic action of CFC 11 can be altered by special procedures that simulate disease.

*Mouse.* Interest in the cardiotoxicity of CFC was initiated with the observations of Taylor and Harris (13), using aerosols released from a bronchodilator pressure unit containing three propellants, including CFC 11. Experiments relating to the inhalation of CFC 11 in gaseous form have been reported by Aviado and Belez (14). Mice under pentobarbital anesthesia did not show any cardiac arrhythmia following inhalation of 2 or 5 percent CFC 11. However, inhalation of 10 percent CFC 11 produced second-degree atrioventricular block, and inhalation of 5 percent CFC 11 also caused the appearance of atrioventricular block, following a concurrent intravenous injection of epinephrine. Mice that had experimental bronchitis developed arrhythmia during inhalation of CFC 11 even without injection of epinephrine.

*Dog.* Reinhardt et al. (15) reported sensitization of the heart to epinephrine in the unanesthetized dog. The inhalation of 0.35 to 0.61 percent CFC 11 for 5 min caused ventricular fibrillation and cardiac arrest following the injection of epinephrine. Although inhalation of lower concentrations (0.09 to 0.13 percent) did not sensitize the heart, administration of higher concentrations (0.96 to 1.21 percent) resulted in a greater frequency of cardiac arrhythmia. Exercising on a treadmill, known to effect release of endogenous epinephrine, did not induce arrhythmia in dogs inhaling 0.5 to 1.0 percent CFC 11 (16). Clark and Tinston (17) investigated the interaction between CFC 11 and two sympathomimetic drugs in unanesthetized dogs. Inhalation of 1.25 percent CFC 11 sensitized the heart to epinephrine but not to isoproterenol.

*Monkey.* The minimal concentration that elicited cardiac arrhythmia in the anesthetized monkey was 5 percent CFC 11 inhaled for less than 5 min (18). In a group of seven monkeys, two developed ventricular premature beats and atrioventricular block. The sensitivity of the heart to arrhythmia was increased by infusion of epinephrine or by coronary arterial occlusion, which reduced the threshold proarrhythmic concentrations inhaled to 2.5 and 1.25 percent, respectively. Combination of the two procedures further reduced the threshold concentration to 0.5 percent CFC 11, which marks a tenfold increase in sensitivity of the heart as a result of experimental infarction.

*Rat.* In the unanesthetized rat, the minimal concentration that produced arrhythmia (atrial fibrillation, ventricular extrasystole, and widening of the T-wave), was 2.5 or 5 percent (19). The induction of pentobarbital anesthesia reduced the incidence of arrhythmia and increased the threshold concentration to 10 percent CFC 11 (20). Rats that developed cardiac necrosis elicited by isoproterenol injections showed a reduction in threshold concentration to 5 percent. Likewise, those that developed pulmonary arterial thrombosis showed a similar increase in the proarrhythmic activity of CFC 11. The induction of pulmonary emphysema did not increase cardiac sensitivity (19). Adrenalectomy or injection of drugs that block cardiac adrenergic receptors protected the heart from CFC 11-induced arrhythmia (20).

## 2.1.2 Cardiac Rate

Three animal species show tachycardia in response to the inhalation of CFC 11. The threshold levels were as follows: 1.0 percent in the anesthetized dog (21); 2.5 percent in the anesthetized monkey with open chest (18) or closed chest (22); and 2.5 percent in the unanesthetized rat (19). The induction of anesthesia in the rat caused a conversion of the tachycardiac response to bradycardia, and a similar influence probably occurs in a fourth species, the mouse. Only the anesthetized mouse has been used; it responds with bradycardia during inhalation of 10 percent CFC 11 (23). It is useful to recall that bradycardia is the usual response in human subjects inhaling low concentrations of CFC. A similar bradycardia is encountered in the dog when the administration of CFC is limited to the upper respiratory tract, that is, oropharyngeal and nasal areas (24). It is reasonable to suggest that bradycardia in humans originates from irritation of the upper respiratory tract, and that cardiac effects can be initiated prior to bronchopulmonary absorption of CFC 11.

## 2.1.3 Myocardial Contractility

Three techniques in different species have been used to investigate the CFC 11 effects on myocardial contractility. In the canine heart–lung preparation, inhalation of 2.5 percent CFC 11 depressed ventricular function curve (25). In the anesthetized monkey with myocardial ischemia, inhalation of 0.5 percent CFC 11 caused depression of the force of contraction recorded by a strain gauge sutured to the ventricular surface (18). The third technique is an innervated canine heart in situ, which was developed specifically for evaluation of cardiac function during single and repeated exposure to new alternative refrigerants (6).

Taylor and Drew (26) have reported that the hamster with inherited cardiomyopathy is predisposed to CFC 11 cardiotoxicity, as compared to the normal hamster. After exposure to 2 percent CFC 11, the cardiomyopathic animal died of frank congestive heart failure. Balazs et al. (27) observed focal myocardial necrosis in the dog exposed for two consecutive days to aerosols containing CFC 11 as propellant, and isoproterenol as the bronchodilator drug. There were no reported cardiac function studies on the animal exposed to low levels of CFC 11 daily for more than a few days. This type of animal experiment is needed to prove or disprove that occupational exposure to CFC causes heart disease (Section 4).

## 2.1.4 Summary of Prototype Chlorofluorocarbon Cardiotoxicity

The above CFC 11 threshold concentrations that influence each animal species permit the following generalization. There is a striking similarity in threshold concentrations between the mouse and the rat on one hand, and the dog and the monkey on the other. The mouse and rat require a CFC concentration of 2.5 to 5.0 percent in order to affect the circulatory system. When anesthetized, these two species respond with bradycardia. The importance of the parasympathetic nervous system has been demonstrated by the use of atropine, which blocks the response. In the unanesthetized rat, only tachycardia results from inhalation of the prototype

CFC. The circulatory system of the monkey and dog can be influenced by a concentration of 0.5 percent CFC 11. This level causes cardiac arrhythmia in the unanesthetized dog, and in the anesthetized monkey with coronary arterial occlusion and epinephrine infusion.

The most serious sign of toxicity to CFC 11 acute inhalation is cardiac arrhythmia, which can be elicited in all four animal species. This may account for sudden deaths associated with the use, misuse, and abuse of aerosols. There are three procedures that increase the sensitivity of the heart to arrhythmia: (a) the injection of epinephrine; (b) coronary ischemia or cardiac necrosis; and (c) experimental bronchitis or pulmonary thrombosis. A common feature of all three procedures is an increase in cardiac irritability caused by epinephrine either directly by initiation of ectopic foci in areas of the heart, or indirectly through hypoxemia and hypercapnia resulting from lung lesions. On the other hand, certain procedures depress the cardiac sensitivity to proarrhythmic action of CFC 11, including adrenalectomy, adrenergic blockade, and general anesthesia. There is one preparation that has not been completed for testing CFC 11 and related compounds, namely, the unanesthetized monkey with myocardial ischemia. The preparation would resemble more closely the heart disease patient who is liable to be exposed to bronchodilator aerosols containing fluorinated propellants. The threshold level of 0.5 percent in the monkey may be further reduced by omission of the anesthesia. This possibility is likely because the dog and the rat have demonstrated a threshold decrease in the unanesthetized state.

Finally, the mechanism of CFC 11 cardiotoxicity requires additional comments. The adverse effects originate from irritation of the respiratory tract which in turn reflexly influences the heart rate even prior to absorption of the inhalant, followed by direct depression of the heart after absorption. Because the heart is sensitized to sympathomimetic amines, the combination of CFC with a sympathomimetic bronchodilator is potentially dangerous for the treatment of bronchial asthma. For the same reason, sympathomimetic drugs are contraindicated in cardiac resuscitation of patients suffering from CFC poisoning. A cardiotonic drug that is free of proarrhythmic activity is available in Europe (28, 29).

## 2.2 Bronchopulmonary and Respiratory Toxicity in Animal Studies

Although there has been a nearly comprehensive comparison between human studies and animal experiment on CFC 11 toxicity on the cardiovascular system, the information relating to respiratory or bronchopulmonary area is less extensive. Animal studies on CFC 11 are useful in interpreting human observations in the following respects: (a) sublethal concentration causes reduction in respiratory movements in animals that correspond to the general anesthetic properties of fluorothane, a widely used fluorine-containing compound in medicine (Section 6.1); (b) bronchospasm has been encountered in humans exposed to CFC 11, a response that can be reproduced in animal experiments; and (c) animal studies show a reduction in pulmonary compliance that may relate to the reduction in pulmonary surfactant postulated to occur in asthmatic deaths from aerosol propellants (Section 5).

### 2.2.1 Increased or Decreased Airway Resistance

The effects of CFC 11 on the airways are variable. Both bronchodilation and bronchoconstriction have been observed in animal species so far examined.

*Monkey and Dog.* The minimal concentration that reduced resistance is 5.0 percent CFC 11 in the anesthetized monkey (22) and 2.5 percent in the anesthetized dog (21). In the latter species the reduction in resistance is blocked by pretreatment with a sympathetic blocking drug, suggesting that the effect is mediated through adrenergic receptors.

*Rat and Mouse.* The anesthetized rat shows an increase in airway resistance while exposed to increasing concentrations of the prototype CFC (12). The minimal dose is about 2.5 percent concentration in the anesthetized rat (19). However, in the emphysematous rat with an elevated airway resistance, the inhalation of 2.5 percent CFC 11 has no influence, suggesting that the response appears only in the nonemphysematous state. The minimal concentration that increases resistance in the anesthetized mouse is 1 percent, which is blocked by pretreatment with atropine (23). These observations indicate that the bronchoconstriction observed in this animal species is precipitated by vagal innervation to the lungs.

### 2.2.2 Decreased Pulmonary Compliance

There are no changes in pulmonary compliance in the monkey (22) and in the dog (14) during inhalation of 1.0 to 5.0 percent CFC 11. In the rat a reduction in compliance occurs following inhalation of 2.5 percent CFC 11, which is more intense in the emphysematous rat (19). The mouse is more sensitive, because 1 percent causes a significant reduction of compliance that is not mediated through the vagus (23). The cause of changes in elasticity of the lung has not been examined. One possibility is the formation of acute pulmonary edema. In another species, the dog, inhalation of 2.5 percent CFC 11 does not cause an elevation of pulmonary arterial pressure (30). Because it is technically impossible to measure pulmonary arterial pressure in the mouse, the cause of the reduction in compliance must be identified by another means. A reduction in tracheobronchial clearance as a cause of decreased compliance has been excluded in experimental observations in the donkey (31).

### 2.2.3 Decreased Respiratory Minute Volume

Unlike fluorine-containing general anesthetics, CFC 11 causes only depression of respiratory minute volume that is not preceded by stimulation of breathing. There is ultimate cessation of respiration, which is a manifestation of generalized depression of the central nervous system by CFC 11.

*Monkey.* The inspired concentration that causes a significant reduction in respiratory minute volume is 5 percent. In the same group of monkeys, circulation is influenced by 2.5 percent of the inhalant (18). The respiratory effect is brought about by a combination of reduced respiratory rate and tidal volume.

*Dog.* The respective threshold doses for the dog are lower than those for the monkey. Inhalation of 1 percent CFC 11 influences heart rate and blood pressure but not respiration. The minimal dose that depresses respiratory minute volume is 2.5 percent CFC 11 (21).

*Rat.* With administration of increasing concentrations of CFC 11, a 40 percent depression of respiratory minute volume occurs at 10 percent concentration (12). In the emphysematous rat, the reduction in response to inhalation of 2.5 percent CFC 11 is less than the control, indicating that pulmonary lesions cause a decreased response (19).

*Mouse.* The only published dose tested for respiratory effect in the mouse is 2.5 percent CFC 11. This concentration causes a 65 percent depression of respiratory minute volume, supported by a reduction in rate and tidal volume (23).

## 2.3 Extrathoracic Toxicity in Animal Studies

Some toxicologists have raised questions on the health hazards of exposure to CFC and related compounds, with regard to hepatotoxicity, reproductive abnormalities, mutagenicity, and carcinogenicity. Because CFC 11 is structurally similar to tetrachloroethane it was thought that the toxicity profiles may be alike. This argument was silenced by Clayton, who contrasted the difference between chlorine and fluorine substitution to hydrocarbons. In his reviews, Clayton concluded that fluorine stabilizes the adjacent $C-C_1$ bonds and reduces biologic activity (32–34). Human exposures to lethal and sublethal concentrations of CFC 11 show involvement only of the cardiopulmonary and central nervous systems, without involvement of other viscera (Section 4). This section discusses the animal and in vitro studies that support the general proposition that although fluorine-containing hydrocarbons are less toxic than chlorinated hydrocarbons, the possibility that CFC 11 causes heart disease, reproductive disorders, and cancer has not been excluded.

### 2.3.1 Hepatotoxicity

Repeated exposure to CFC 11 does not produce pathological lesions of the liver and other visceral organs. The studies were conducted as follows: 1.25 or 2.5 percent CFC 11 for 3.5 hr each on 20 consecutive days using guinea pigs, rats, and cats (8); and 0.1 percent CFC 11 continuously for 90 days, or 1.025 percent CFC 11 for 8 hr for 30 days, in guinea pigs, rats, dogs, and monkeys (35). Slater (36) compared the oral toxicity of CFC 11 with tetrachloromethane. He concluded that the latter compound is hepatotoxic, whereas CFC 11 is not. The most likely explanation is that tetrachloromethane undergoes biochemical metabolism to a potent hepatotoxin but such a transformation appears insignificant in the case of CFC 11. In a latter study, Slater suggested that the most likely route of metabolism for tetrachloroethane, but not for CFC 11, is the formation of free-radical products resulting in lipid peroxidation, which is requisite for liver necrosis (37). Cox et al. (38) have challenged Slater's conclusion by performing in vitro studies that show binding of CFC 11 with liver microsomal enzymes. Coincidentally, Blake and Mergner (39) showed that carbon-14-labeled CFC 11 is refractory to biotransformation and is rapidly exhaled in its unaltered form in beagle dogs, similar to the results of human studies (Section 5.2).

### 2.3.2 Tetratogenicity

There is no published study of the reproductive effects of CFC 11 inhalation. Paulet et al. (40) used a propellant mixture of 10 percent CFC 11 and 90 percent dichlorodifluoromethane or CFC 12; 20 percent of the mixture in air was administered, for 2 hr daily, to rats from day 4 to day 16 of gestation and to rabbits from day 5 to day 20. Adverse effects on the offsprings of the exposed pregnant animals were not observed.

### 2.3.2 Mutagenicity

Mutagenicity testing gave negative results in the following: Salmonella assay (41, 42) cell transformation assay (42), and mammalian cell testing (43). However, testing of other CFC compounds shows positive mutagenicity (Section 3.4).

### 2.3.4 Carcinogenicity

A National Cancer Institute sponsored bioassay on carcinogenicity has been completed. The oral gavage administration of 488 and 977 mg/kg/day to male rats, 538 and 1077 mg/kg/day to female rats, and 1962 and 3925 mg/kg/day to mice of both sexes did not show carcinogenicity (44). This study is significant because it differentiates the noncarcinogenicity of CFC 11 from the carcinogenicity of carbon tetrachloride in a similar oral bioassay study. The negative results temporarily have refuted the potential carcinogenicity of fluorocarbons in general, and CFC 11 in particular, suggested by injection studies performed by Epstein et al. (45). However, the question of carcinogenicity has been raised owing to suspicious cancer seen in exposed animals (Section 3.4) and the higher incidence of cancer among medical personnel exposed to fluorine-containing general anesthetics (Section 6.1). The only means of excluding carcinogenicity in bioassay is by inhalation route.

## 3. ANIMAL TOXICOLOGIC STUDIES

The next logical step after recounting toxicologic features of prototype CFC 11 is to review other fluorine-containing organic compounds. I have decided not to do so because of the inevitable stoppage of manufacture of currently available compounds. At the time of writing this chapter, the CFC compounds and the hydrogenated alternatives (HCFC) continue to be under suspicion as causing ozone depletion of the stratosphere (Section 1). The only promising group is the nonchlorinated compounds, that is, the hydrogenated fluorocarbons (HFC). The general plan in this section is to refrain from a compound by compound description and instead emphasize general toxicologic features and the means of obtaining relevant information. It is not possible at this time to anticipate which fluorine-containing compounds will be in commercial production during the remainder of this century.

### 3.1 Chemical Structure of Chlorofluorocarbons and Fluorocarbons

Because production of all chlorofluorocarbons will be phased out during the mid-1990s, there is no justification to review animal toxicologic studies, compound by compound, as was done in the 1982 edition of this chapter, and in reviews written by me (3–7). Interested readers can consult reviews sponsored by the American Conference of Governmental Industrial Hygienists (46), National Research Council (47), World Health Organization (48, 49), and the U.S. Environmental Protection Agency (50). The U.S. National Institute of Occupational Safety and Health has a Registry of Toxic Effects of Chemical Substances (RTECS) which is constantly revised as TOXLINE database, or printed in hard copy with microfiche (51). Chemical manufacturers have a Material Safety Data Sheet for each commercially available compound but information contained therein is generally brief, with incomplete references.

#### 3.1.1 Completely Halogenated Compounds (CFC)

There are 10 completely halogenated compounds containing both fluorine and chlorine; thus the group is called chlorofluorocarbons (CFC). The prototype compound, CFC 11, belongs to this group consisting of the following:

| Compound Number | Chemical Formula | Chemical Name |
|---|---|---|
| CFC 11 | $CCl_3F$ | Trichlorofluoromethane |
| CFC 12 | $CCl_2F_2$ | Dichlorodifluormethane |
| CFC 13 | $CClF_3$ | Chlorotrifluoromethane |
| CFC 112 | $Cl_2FC-CCl_2F$ | 1,1,2,2-Tetrachloro-1,2-difluoroethane |
| CFC 112a | $Cl_3C-CClF_2$ | 1,1,1,2-Tetrachloro-2,2-difluoroethane |
| CFC 113 | $Cl_2FC-CClF_2$ | 1,1,2-Trichloro-1,2,2-trifluoroethane |
| CFC 113a | $Cl_3C-CF_3$ | 1,1,1-Trichloro-2,2,2-trifluoroethane |
| CFC 114 | $ClF_2C-CClF_2$ | 1,2-Dichloro-1,1,2,2-tetrafluorethane |
| CFC 114a | $Cl_2FC-CF_3$ | 1,1-Dichloro-1,2,2,2-tetrafluoroethane |
| CFC 115 | $ClF_2C-CF_3$ | 1-Chloro-1,1,2,2,2-pentafluoroethane |

The above list of 10 CFCs are presently used as refrigerants, aerosol propellants, cleansing solvents, and plastic blowing agents. All the above compounds have been condemned as causing ozone depletion in the stratosphere.

### 3.1.2  Hydrogenated Chlorofluorocarbon Compounds (HCFC)

The substitution of hydrogen for halogen in methane and ethane reduces the duration of persistence in stratosphere and therefore causes less depletion of ozone, compared to completely halogenated compounds (CFC).

| Compound Number | Chemical Formula | Chemical Name |
|---|---|---|
| HCFC 21 | $CHCl_2F$ | Dichlorofluoromethane |
| HCFC 22 | $CHClF_2$ | Chlorodifluoromethane |
| HCFC 123 | $HCl_2C-CF_3$ | 2,2-Dichloro-1,1,1-trifluoroethane |
| HCFC 124 | $HClFC-CF_3$ | 1-Chloro-1,2,2,2-tetrafluoroethane |
| HCFC 132b | $H_2ClC-CClF_2$ | 1,2-Dichloro-1,1-difluoroethane |
| HCFC 133a | $H_2ClC-CF_3$ | 2-Chloro-1,1,1-trifluoroethane |
| HCFC 141b | $H_3C-CCl_2F$ | 1,1-Dichloro-1-fluoroethane |
| HCFC 142b | $H_3C-CClF_2$ | 1-Chloro-1,1-difluoroethane |
| HFC 152a | $H_3C-CHF_2$ | 1,1-Difluoroethane |

The above compounds continued to be of interest to the chemical manufacturers through the early 1990s. However, because these alternative compounds contain chlorine, their ozone depleting potential is significant although less than that of currently available chlorofluorocarbons.

### 3.1.3  Nonchlorinated Hydrogenated Fluorocarbons (HFC)

Until 1992, the only hope of developing ozone-friendly refrigerants was the chemical industry. An article in the *Wall Street Journal* (52) revealed to scientists outside of the chemical industry for the first time that the Air and Energy Engineering Research Laboratory (AEERL of U.S. Environmental Protection Agency) has been developing new chemical alternatives, and that 34 compounds have been prepared of sufficient stability and in sufficient yields and purity to obtain measurements of physicochemical properties. Eleven fluorocarbons have been selected for evaluation as replacement for four commercially available CFC compounds, because of similarities in boiling temperature, vapor pressure, and other thermophysical properties. The following list of promising alternatives includes two hydrogenated fluoroethers (HFE):

| Compound Number | Chemical Formula | Chemical Name |
|---|---|---|
| For Replacing CFC 11: | | |
| HFC 245ca | $F_2HC-CF_2-CFH_2$ | 1,1,2,2,3-Pentafluoropropane |
| HFC 245fa | $F_3C-CH_2-CF_2H$ | 1,1,3,3-Pentafluoropropane |

# FLUORINE-CONTAINING ORGANIC COMPOUNDS

For Replacing CFC 12:

| | | |
|---|---|---|
| HFC 227ea | $F_3C$—CHF—$CF_3$ | 1,1,1,2,3,3,3-Heptafluoropropane |
| HFC 227ca | $F_3C$—$CF_2$—$CF_2H$ | 1,1,1,2,2,3,3-Heptafluoropropane |
| HFC 245cb | $F_3C$—$CF_2$—$CH_3$ | 1,1,1,2,2-Pentafluoropropane |
| HFE 143a | $F_3C$—O—$CH_3$ | 1,1,1-Trifluorodimethyl ether |

For Replacing CFC 114:

| | | |
|---|---|---|
| HFC 236ea | $F_3C$—CHF—$CF_2H$ | 1,1,1,2,3,3-Hexafluoropropane |
| HFC 236fa | $F_3C$—$CH_2$—$CF_3$ | 1,1,1,3,3,3-Hexafluoropropane |
| HFC 236cb | $F_3C$—$CF_2$—$CFH_2$ | 1,1,1,2,2,3-Hexafluoropropane |
| HFC 254cb | $F_2HC$—$CF_2$—$CH_3$ | 1,1,2,2-Tetrafluoropropane |

For Replacing CFC 115:

| | | |
|---|---|---|
| HFE 125 | $CF_3$—O—$CF_2H$ | 1,1,1,2,2-Pentafluorodimethyl ether |

The above listed hydrogenated fluorocarbon compounds (HFC) were synthesized by scientists of AEERL, the Electric Power Research Institute, and Chemistry Departments of Clemenson University and University of Tennessee (53–55). Since the announcement of these HFC compounds, the manufacturers of currently available CFC have announced their HFC candidates.

| Compound Number | Chemical Formula | Chemical Name |
|---|---|---|
| HFC 125 | $HF_2C$—$CF_3$ | 1,1,2,2,2-Pentafluoroethane |
| HFC 134a | $H_2FC$—$CF_3$ | 1,2,2,2-Tetrafluoroethane |
| HFC 152a | $H_3C$—$CHF_2$ | 1,1-Difluoroethane |
| HFC 32 | $CH_2F_2$ | Difluoromethane |

I have not been successful in obtaining results of toxicologic studies that are discussed within industry group and scientists receiving industrial contracts.

## 3.2 Animal Models for Evaluation of Developmental Alternatives

At present, an overall assessment of relative lethality and cardiotoxicity of fluorine-containing organic compounds permits grouping into three classes. CFC 11 has the highest level of toxicity and is used as a reference standard of comparison for the other compounds.

| Compound Number | Chemical Name | Chemical Formula |
|---|---|---|
| *High Level of Toxicity* | | |
| CFC 11 | Trichlorofluoromethane | $CCl_3F$ |
| CFC 113 | Trichlorotrifluoroethane | $Cl_2FC$—$CClF_2$ |
| HCFC 21 | Dichlorofluoromethane | $CHCl_2F$ |

*Intermediate Level of Toxicity*

| | | |
|---|---|---|
| CFC 114 | Dichlorotetrafluoroethane | $ClF_2C-CClF_2$ |
| CFC 12 | Dichlorodifluoromethane | $CCl_2F_2$ |
| HCFC 22 | Chlorodifluoromethane | $CHClF_2$ |
| HCFC 142b | Chlorodifluoroethane | $H_3C-CClF_2$ |
| CFC 115 | Chloropentafluoroethane | $ClF_2C-CF_3$ |

*Low Level of Toxicity*

| | | |
|---|---|---|
| HFC 152a | Difluoroethane | $H_3C-CHF_2$ |
| FC C-318 | Octafluorocyclobutane | $C_4F_8$ |

Coincidentally, the above grouping shows a direct relationship between the number of chlorine atoms and the level of toxicity. The two examples of least toxic fluorocarbons do not contain chlorine atoms. The second nonchlorinated compound (FC C-318) was tested in the course of searching for a low level of cardiotoxicity for use as an aerosol propellant (56).

Although nonchlorine-containing hydrogenated fluorocarbons (HFC) are preferred from the environmental standpoint, they will have to be examined to prove that they have low levels of toxicity similar to those of FC C-318 and HFC 152a. After application of various techniques to measure responses in four animal species, it has become apparent that certain generalizations can be made regarding the sensitivity of each animal model. The following discussion points out the specific application of the mouse, rat, dog, and monkey in determination of the potential hazard of an inhalant to the circulatory and respiratory systems. Comparison with prototype CFC 11 is essential but unfortunately it has been overlooked by recent preliminary studies on developmental alternatives.

### 3.2.1 Mouse

The proarrhythmic activity of fluorine-containing compounds has been examined by administration in various concentrations while recording the electrocardiogram (14). The inhalant was also administered while injecting epinephrine to determine sensitization of the heart to developing arrhythmia. As a rule, fluorine-containing compounds that provoke spontaneous arrhythmia also sensitize the heart to epinephrine. There are also compounds that do not induce spontaneous arrhythmia but cause sensitization. Lastly, there are compounds that do not exert spontaneous or epinephrine-associated arrhythmia provided that a lack of activity is confirmed in one other species.

There is a wide range in concentration (5 to 40 percent) for the various fluorine compounds to cause arrhythmia. The sensitivity of the mouse can be increased by experimental induction of bronchitis (23), indicating that the minimal concentration can be reduced by disease. Compared to the other species, the mouse is least sensitive for determination of threshold concentration that would provoke cardiac arrhythmia.

In the course of testing CFC, a technique for measuring airway resistance and pulmonary compliance was developed for application to the mouse (23). This was not hitherto possible so that for the first time the sensitivities of the respiratory and circulatory systems have been compared in this species. The minimal concentration of CFC 11 is 1 percent to produce bronchoconstriction, and 2.5 percent to depress respiration. For the same CFC, 10 percent is needed to produce arrhythmia, indicating that the airways are more sensitive than the heart in the mouse.

### 3.2.2 Rat

A comparison of the rat with and without general anesthesia indicates that the unanesthetized state is more suitable for the investigation of cardiotoxicity (19). Anesthesia blocks the cardioaccelerator response, or even converts it to bradycardia. Although the unanesthetized rat is more sensitive than the mouse for demonstration of cardiotoxicity, the dog and monkey show effects when exposed to even lower concentration of the inhalant.

In the rat, the cardiotoxicity of fluorine compounds is reduced by adrenalectomy or prior treatment with adrenergic blocking drugs (20). This observation if significant because it demonstrates that the sympathetic nervous system participates in the cardiac effects of inhalants.

The rat is less sensitive than the mouse in manifesting respiratory toxicity (11). However, there are forms of experimentally induced diseases applicable to the rat. In addition to pulmonary emphysema, thrombosis of the pulmonary artery and myocardial necrosis have been induced in the rat. In the emphysematous animal, it has been demonstrated that lungs become more sensitive to a reduction in pulmonary compliance (19). Rats with cardiac necrosis or pulmonary arterial thrombosis show an increase in sensitivity to cardiotoxicity of fluorine-containing compounds (20).

### 3.2.3 Dog

The dog is the most sensitive animal for eliciting hypotension, depression of myocardial contraction, tachycardia, and cardiac arrythmia (21). For the last-mentioned effect, the unanesthetized dog with epinephrine injection is 5 to 10 times more sensitive to CFC and HCFC than the anesthetized dog without injection of epinephrine. This preparation is sufficiently sensitive that it can demonstrate the proarrhythmic activity of one inhalant (HFC 152a) even though the three other animal species fail to do so. The heart–lung preparation can be used to demonstrate direct depression of ventricular contractility, without participation of adrenal glands or autonomic innervation to the heart (18). As a rule, the CFC concentrations that influence respiration are higher than those for circulation. The dog appears to be the least sensitive animal for demonstrating respiratory toxicity. However, in the investigation of mode of action, there are several techniques applicable to the dog. The bronchodilation induced by some CFC is mediated by adrenergic receptors (21). In addition, CFC stimulates upper and lower respiratory tract receptors, which in turn influence respiration, bronchomotor tone, and heart rate.

### 3.2.4 Monkey

The advantages and disadvantages of the monkey have not been fully appreciated owing to its limited application. As a rule, the circulatory system is more responsive than the respiratory system (18, 22). However, there are instances in which the monkey's response is opposite to that of the other species, and the relevance to humans is unsettled. There are fluorinated inhalants that produce cardiotoxicity in all three species but not in the monkey, some that cause myocardial depression and hypotension in the dog but not in the monkey, and some that cause early respiratory depression, bronchoconstriction, or decreased compliance in the dog, rat, and mouse, but not in the monkey. Opposite results have also been encountered; for example, some compounds are toxic to the monkey but not to the other species. It has been generally assumed that the monkey's response closely resembles the human's, but there has been no definitive comparison. When parallel human and monkey studies on fluorine-containing organic compounds become available, it will be possible to decide the significance of the various animal models specifically for the prediction of human toxicity.

### 3.3 In vitro Studies Relating to Mutagenesis

In 1968 Landry and Fuerst (56) reported on the mutagenic activity of some 20 gaseous compounds using *Escherichia coli*. They found that compounds containing fluorine atoms were mutagenic. In 1974, Foltz and Fuerst reported on mutation studies performed on four fluorinated hydrocarbons using *Drosophila melanogaster* (57). All four fluorocarbons studied [octafluorocyclobutane (FC C-318), fluoroform (FC 23), 1,1-difluoroethane (HFC 152a), and perfluorobutene-2] significantly increased the recessive lethal mutation rates of progeny over control, with FC 23 being the most mutagenic. Four years later, using protracted incubation times in gaseous environment, Longstaff and McGregor reported that chlorodifluoromethane (HCFC 22) was mutagenic in *Salmonella typhimurium* (58). However, HCFC 22 has been shown to be nonmutagenic when tested in other submammalian test systems by Loprieno et al. (59). A closely related compound, trichlorofluoromethane (CFC 11), was also nonmutagenic when tested in *Salmonella typhimurium* strain TA1535 (41).

More recently, Longstaff and colleagues reported on results of genotoxicity studies on selected chlorofluorocarbons using two short-term in vitro tests for potential carcinogenicity (60, 61). These tests included the *Salmonella* reverse mutation test and the BHK21 cell transformation test. Compounds studied include almost all CFC, HCFC, and HFC listed in Section 3.1. The results emphasize a lack of structure–mutagenesis relationship.

### 3.4 Animal Studies on Mutagenesis and Carcinogenesis

To validate results of in vitro mutagenicity testing further, chronic in vivo studies were performed using Wistar-derived male and female rats that were administered

HCFC 22, HCFC 31, HCFC 133a, or HFC 134a, at a dose of 300 mg/kg body weight (3 percent solution in corn oil) for 52 weeks (61). Both male and female rats dosed with HCFC 31 had a high incidence of malignant stomach tumors, mainly squamous cell carcinomas of the nonglandular stomach and fibrosarcomas. Female cats that received HCFC 133a had a markedly increased incidence of malignant uterine adenocarcinomas. Males, on the other hand, showed a high incidence of benign testicular interstitial cell tumors and marked tubular atrophy of the testes even when tumors were not present. The remaining compounds showed no significant increase in the tumor incidence in any tissue. These results confirmed the short-term in vitro tests in case of HCFC 31, but surprisingly showed that HCFC 133a (negative in both in vitro assays), was found to be a carcinogen in in vivo assay.

Compound HCFC 22, which was shown to be positive in the Ames test, failed to induce point mutation or DNA repair in in vitro assays (60). Likewise, in vivo mammalian studies with HCFC 22 showed no mutagenic or carcinogenic activity. There is incomplete testing on carcinogenicity of HCFC 22 (62).

When tested without metabolic activation in *S. typhimurium* TA1538 strain, dichlorodifluoromethane (CFC 12), dichlorofluoromethane (HCFC 21), and chlorodifluoromethane (HCFC 22) were not mutagenic. Long-term carcinogenicity studies reported recently by Maltoni et al. showed that rats and mice exposed to 1000 or 5000 ppm of CFC 11, CFC 12, or HCFC 22, 4 hr/day, 5 days/week for 104 weeks (rats) or 78 weeks (mice), exhibited no carcinogenic effect (63). Similarly, male and female rats exposed to up to 10,000 ppm of 1,1,2-trichloro-1,2,2-trifluoroethane (CFC 113), 6 hr/day, 5 days/week for 24 months, showed no increase in tumor incidence over the control group. However, females exposed to 20,000 ppm showed a statistically significant increase in pancreatic islet cell adenomas (64). Nonetheless, the investigators maintained that the incidence of 5 out of 86 is within the normal historical background levels. Additional studies published by Mahurim and Bernstein showed that although CFC 113 per se is not mutagenic, it may pose a carcinogenic risk by enhancing carcinogenic activation (65). This suggestion was based on the finding that liver microsomal extract from mice exposed to CFC 113 at 20,000 ppm for 8 hr had enhanced ability to activate aminofluorene as a mutagen.

Compound HCFC 142b (1-chloro-1,1-difluoroethane) was tested by exposing rats to up to 20,000 ppm, 6 hr/day, 5 days/week for 104 weeks (66). Seckar et al. maintained that no evidence for mutagenic potential was observed in either the dominant lethal or cytogenic assay, despite a transient increase in week 7 of dead uterine implants, and an increase in the incidence of chromosome breaks. No genotoxicity or oncogenicity was observed.

The above results are preliminary in nature and need verification for positive oncogenicity. At the proper time, submission for approval of manufacturing under provisions of the Toxic Substances Control Act (TSCA) will identify potential health hazards.

### 3.5 Animal Studies on Reproduction

Although low-molecular-weight fluorine-containing compounds are generally considered to be "safe," a few studies performed and reported under section 8(e) of

TSCA indicate that these compounds may cause developmental and reproductive toxicity (67). For example, when pregnant rats were exposed to 2000 or 5000 ppm of HCFC 132b (1,2-dichloro-1-1-difluoroethane) on gestation day 7 to day 16, the mean number of early resorptions per litter was 1.2 and 1.7, respectively, compared to the control group (0.6). Further, exposure of male rats to 2000 or 5000 ppm of HCFC 132b, 6 hr/day, 5 days/week for 90 days, resulted in a disruption of spermatogenesis as manifested by disorganization of the seminiferous tubular epithelium and degeneration of germ cells leading to aspermatogenesis (8 EHQ-0388-0676) (67).

In another study, male mice were exposed to 2-chloro-1,1,1-trifluoroethane (HCFC 133a) at a concentration of 10,000 or 20,000 ppm for 6 hr/day for 5 consecutive days (after 2 days, the high dose group received only 5000 ppm owing to excessive toxicity). Two days after final exposure, animals were mated with untreated females for 5 days. The pattern of mating continued for 8 weeks. Fertility, as measured by percentage of pregnant females, the number of successful matings per male, or total number of implantations, was significantly reduced. In addition, a significant increase in the percentage of early death per total implants was noted. Further, a significant reduction in testis and epidydimis weights was also observed (8e-267) (67).

Compound HCFC 133a also proved to be a developmental toxin. When pregnant rats were exposed to 517, 2,034, 5200, or 19,600 ppm for 6 hr/day on day 6 to day 15 of gestation, postimplantation loss was significantly increased and litter size was significantly decreased at doses as low as 2034 ppm. Furthermore, surviving fetuses showed a significantly dose-related increase in the incidence of hydronephrosis and other malformations (8e-267) (67).

The above examples illustrate difficulties in obtaining definitive information on developmental alternatives. I have not been successful in obtaining additional information relating to above TSCA section 8(e) submissions by directly contacting manufacturers. There are no comparable data on prototype CFC 11 and a known reproductive toxin conducted by the same testing firm.

## 4 OCCUPATIONAL EXPOSURE

Manufacturing processes use hydrofluoric acid from fluorospar in the production of most fluorine-containing organic compounds. Some processes use carbon tetrachloride from carbon disulfide or as a co-product of perchloroethylene and chlorination of propylene, or chloroform from chlorination of methanol. At the outset, it is important to emphasize that most fluorine-containing compounds (CFC, HCFC, and HFC) are less toxic than any of the process materials used in their manufacture. The major hazards relate primarily to the inadvertent release of hydrofluoric acid or carbon tetrachloride, rather than to the manufactured final product.

## 4.1 Inhalation Hazards

Chlorofluorocarbon vapors are four to five times heavier than air. Thus high concentrations tend to accumulate in low-lying areas, resulting in the hazard of inhalation of concentrated vapors, which may be fatal. Under certain conditions, vapors may decompose on contact with flames or hot surfaces, creating the potential hazard of inhalation of toxic decomposition products (68).

There are several sources of exposure to CFC in manufacturing. Exposure in chemical plant operations and production is generally low but highly variable, and may be high in areas without adequate ventilation. Cylinder packers and shippers have occasional high exposure. Exposure during tank farm operations, and tank and drum filling, may exceed the threshold limit value. Tank truck and tank car fillers have potentially high exposure, which may be intermittent with the occurrence of accidents. Maintenance operators, laboratory analysts, and supervisory personnel have low exposure. The highest exposure in the plant occurs with venting of gases from returnable cylinders. The formation of high-temperature thermal decomposition products may occur during storage.

Administrative controls may be applied to limit occupational exposure to CFC, HCFC, and HFC during manufacture, packaging, and use. Enclosure of process materials and isolation of reaction vessels and proper design and operation of filling heads for packaging and shipping are such measures. Inhalation of vapors should be avoided. Forced air ventilation at the level of vapor concentration together with the use of individual breathing devices with independent air supply will minimize the risk of inhalation. Life lines should be worn when entering tanks or other confined spaces. Filling areas should be monitored to ensure that the ambient CFC, HCFC, and HFC concentrations do not exceed prevailing work standards.

## 4.2 Threshold Limit Values

The present work standards for CFC and HCFC are derived entirely from recommendations of the American Conference of Governmental Industrial Hygienists (46).

| CFC No. | Chemical Name | TLV ppm | TLV mg/m³ |
|---|---|---|---|
| CFC 11 | Trichlorofluoromethane | 1000 | 5600 |
| CFC 12 | Dichlorodifluoromethane | 1000 | 4950 |
| HCFC 21 | Dichlorofluoromethane | 10 | 40 |
| HCFC 22 | Chlorodifluoromethane | 1000 | 3500 |
| CFC 112 | 1,1,2,2-Tetrachloro-1,2-difluoroethane | 500 | 4170 |
| CFC 112a | 1,1,1,2-Tetrachloro-2,2-difluoroethane | 500 | 4170 |
| CFC 113 | 1,1,2-Trichloro-1,2,2-trifluoroethane | 1000 | 1250 |
| CFC 114 | 1,2-Dichloro-1,1,2,2-tetrafluoroethane | 1000 | 7000 |
| CFC 115 | 1-Chloro-1,1,2,2,2-pentafluoroethane | 1000 | 6320 |

As stated earlier elsewhere, there is no uniformity of comparative toxicity for cardiotoxicity of CFC and HCFC (Section 3.4). If any of the above nine compounds were to be reintroduced under TSCA, additional animal studies would be needed relating to carcinogenesis and reproductive effects. There is a hundredfold difference between the most toxic to the least toxic inhalant, that is, 10 ppm for HCFC 21 compared to 500 ppm for CFC 112 and CFC 112a, and 1000 ppm for the prototype CFC 11 and five other compounds. The difference is in the higher potency as hepatotoxin for HCFC 21 similar to that of carbon tetrachloride.

Most developed and some developing countries have adapted the same American work standards (69, 70). Some countries have stricter levels (i.e., CFC 11 of 500 ppm in Sweden) and other countries have included additional compounds [i.e., dichlorofluoroethane (CFC 141, 1000 ppm in the former Soviet Union].

### 4.3 Occupational Poisoning

Deaths associated with exposure to CFC and HCFC been encountered in the occupational setting, including refrigeration, air conditioning, solvent applications, and plastic foam blowing. The literature on poisoning from presently available CFC and HCFC is reviewed below because it may apply to developmental alternatives, specifically, non-ozone-depleting HFC.

#### *4.3.1 Refrigeration*

The widely used refrigerants include the halogenated fluorocarbons (CFC 11, CFC 12, CFC 113, CFC 114, CFC 115) and the hydrogenated chlorofluorocarbons (HCFC 22, HCFC 142b, HFC 152a). Mechanical vapor compression systems use CFC or HCFC for refrigeration and air conditioning, and account for the vast majority of refrigeration capability in the United States. The occupational hazard is almost entirely in the exposure to the refrigerants during manufacturing of refrigeration equipment, whereas product safety hazard is negligible because these compounds are held in sealed units.

Prior to 1960, the reported fatalities resulted from acute exposure to refrigerants such as methyl chloride alone (71–73), CFC or HCFC alone (74, 75), and a combination of CFC thermal decomposition products and sulfur dioxide (76). Dalhamn (68) reported two cases of phosgene poisoning from disintegration of CFC 11 at an open flame in an enclosure. Mendeloff (77) reported the case of a refrigeration equipment repair mechanic who claimed chronic exposure to refrigerant gases including CFC, HCFC, methyl chloride, ethanol, and sulfur dioxide. The illness began insidiously with malaise, chills, fever, and nausea; during the following days the patient had abdominal pain with further nausea and vomiting, backache, headache, swollen abdomen, and mild episodes of epistaxis. He was admitted to the hospital on the sixth day of the illness with an initial temperature of 98.6°F that rose to 100°F. The X-ray findings and postmortem lesions in the lungs resembled those of viral pneumonia, yet Mendeloff concluded that this case represented the first reported death from chronic exposure to fluorine-containing refrigerants. Leinoff (78) reported workers exposed to refrigerants who developed coronary thrombosis. However, there is no experimental work to prove or disprove that chronic exposure to fluorine-containing refrigerants leads to coronary heart disease.

In spite of the widespread publicity regarding fatalities from occupational exposure to refrigerants, numerous episodes of refrigerant poisoning were reported in scientific journals and lay press. The following examples represent some fatal incidents: refrigeration repairing (79–81) car air-conditioning repairing (82), and shipboard air-conditioning maintenance (83). The occurrence of cardiac arrhythmia in such cases of refrigerant poisoning is supported by results of animal studies (Section 3). However, the occurrence of hypertension reported in the literature cannot be supported by adequate animal studies (84).

### 4.3.2 Solvent Applications

The use of CFC and HCFC as solvents is more hazardous to human health than is their use of refrigerants, because solvents are generally employed in open containers. These fluorine-containing organic compounds find wide application as cleaning and drying solvents for defluxing and electronics cleaning, degreasing, displacement drying, and miscellaneous specialty applications. As nonpolar solvents, CFC and HCFC will clean metal, glass, plastic, and electroplated surfaces of manufactured goods.

Compound CFC 113 is the most widely used cleansing solvent. Fatalities have been reported in the following setting: degreasing tank (85, 86), lens factory (87), and electronic equipment factory (88, 89). Although cardiac arrhythmia and arrest can be reasonably ascribed to CFC 113 exposure, it is not known whether the following diseases can be ascribed to such occupational exposure: motor dysfunction (90), psycho-organic syndrome (91), and brain cancer (92). It is reasonable to suspect nervous system diseases because of brain tissue uptake of fluorine-containing organic compounds.

### 4.3.3 Foam Blowing Application

Nearly 17 percent of the use of CFC and HCFC in the United States is in the fabrication of urethane and nonurethane closed-cell plastic foams, and flexible urethane foams (for 1985, ref. 48). Suppliers of basic raw materials and foam systems, fabricators and suppliers of flexible foams, and producers and fabricators of rigid foams make use of fluorine-containing blowing agents. The plastic foams are manufactured for application in thermal insulation and packaging. Rigid urethane foams require gas for foaming, provided by carbon dioxide released during polymerization, or from volatile fluorine-containing liquids, especially CFC 11. Volatilization of the blowing agent is caused by the heat of reaction between the isocyanide and hydroxyl components. A substantial portion of the blowing agent is trapped within the closed plastic cells and release is nil. In the production of flexible urethane foams, the flowing agent augments the effect of carbon dioxide released by the reaction of free isocyanide groups with water. Virtually all of the fluorine-containing blowing agents are released soon after manufacturing.

## 4.4 Dermal Contact Hazards

The degreasing effect of CFC and HCFC may cause dermatologic problems, although there is negligible dermal absorption. Some fluorine-containing liquids remove natural oils from the dermis, causing irritation and development of dry, sensitive skin. These lower-boiling liquids may also be splashed onto the skin or into the eyes, causing freezing, temporary irritation, or serious damage. The freezing effect is produced upon evaporation of CFC or HCFC from the skin surface, and is a manifestation of cooling by evaporation, whereby molecules with high kinetic energy escape from the system. Thus heat energy is taken with the escaping molecules, leaving behind the low kinetic energy (low temperature) molecules. Because the temperature of a fluid is based on the concentration and average kinetic energy of its molecules, a rapid cooling effect is thus produced. Frostbite may be a complication of this freezing effect (93). If frostbite occurs, the exposed area should be soaked in lukewarm water within 20 to 30 min after exposure. Ice cold or hot water should not be used; body temperature is ideal. Soaking should be eliminated if treatment begins more than 30 min after exposure. A light coating of bland ointment, such as petroleum jelly, should be applied together with a light bandage. If the frostbitten area is large or severely affected, administration of an anticoagulant or vasodilator drug may be necessary in order to avoid the onset of gangrene. The eyes may also be damaged by freezing. If such contact occurs, they should be flushed with water for several minutes. Neoprene gloves, protective clothing, and eye protection minimize the risk of topical contact. The degreasing effect on the skin can be treated with lanolin ointment. Other dermal contact hazards include erythema multiforme (94) and allergic dermatitis (95). The most extreme form of hand injury is the accidental injection of CFC mixture from a high-pressure injection pump used for cleaning and maintenance of computers and computer parts (96). The irreversible injury of deep tissue is unrelated to skin refrigeration, which is usually reversible, based on supportive studies on cryosthesia of human skin (97–99) and animal skin (100, 101).

## 4.5 Fire Extinguishing Application

Although most CFC and HCFC are nonflammable, they have not been used to extinguish fires because of the availability of bromine-containing fluorocarbons, specifically bromotrifluoromethane (Halon 1301) and bromochlorodifluoromethane (Halon 1211). The bromine-containing fluorocarbons operate by chemical interruption of the combustion chain, and are used in total flooding systems for computer rooms and telephone facilities, as well as aircraft and portable fire extinguisher, including use on the U.S. Air Force P-13 rapid intervention crash trucks. Halon emerges from the fire extinguisher nozzle as a mixture of 85 percent liquid and 15 percent vapor and is discharged over long distances. It completely vaporizes upon contact with fire. There is some exposure in charging and refilling, testing, and leakage of the fire extinguishing systems. There can be accidental in-service discharges. During use in fires, thermal decomposition products present toxicity hazards. Exposure to bromofluorocarbons in fire extinguishing equipment has led to paresthesia, tinnitus, anxiety reactions, electroencephalographic changes, slurred speech, and decreased performance on psychological tests (102, 103). Like CFC 11, the Halons have been abused by inhalation, causing cardiac arrhythmia and cardiac arrest (104–107). The results of animal studies on Halons are essentially similar to those of CFC exposure (108–109). A controlled exposure to Halon 1211 indicates the potential cardiotoxicity during muscular exercise (110). The approach used by Kaufman et al., reported in 1992, is an ideal one for testing developmental alternative HCFC and HFC, particularly if they are used as aerosol propellants, discussed in Section 5.

### 4.5 Pyrolysis of Polymers

Fluorine-containing monomers are polymerized as gas-phase reactions. The major hazard associated with monomers arises primarily from inhalation, with the kidney as the major toxicologic target organ (111–113). Tetrafluoroethylene and its polymer (Teflon) are the most widely investigated compounds. Its carcinogenicity following tissue implantation was evaluated by IARC in 1979 (114). Additional studies have not been reported to resolve questions raised by IARC. In 1965, an epidemic of "polymer-fume fever" was reported in one factory handling polytetrafluoroethylene (115). Fever and chills occurred several hours after exposure to products of pyrolysis. Reports of polymer-fume fever continued through the 1980s (116). Although the syndrome has been known since 1951, the identity of pyrolytic products was initiated 40 years later (117).

## 5 PROPELLANTS FOR AEROSOLS

When fluorine-containing refrigerants were introduced in the 1940s, they were regarded as "inert" compounds compared to sulfur dioxide, ammonia, carbon tetrachloride, and chloroform that were then in use. The first aerosol application of "inert" refrigerants was an insecticide bomb used in World War II to protect troops in tropical areas against malaria and other vector-borne diseases (118). As other forms of aerosols were introduced, it became apparent that two compounds were needed to propel the active ingredient, one at high pressure (such as dichlorodifluoromethane, or CFC 12) and one at low pressure. Of the latter, trichlorofluoromethane (CFC 11) proved to be the most popular because of its vapor depressant effect and also its solvent and flame retardant actions. Following the reports of fatalities from the misuse and abuse of aerosol products, toxicologic studies were conducted; it became apparent that CFC compounds are not "inert" and that the most widely used propellant (CFC 11) is also the most toxic (Section 2).

### 5.1 Aerosol Use

In the mid-1980s, the annual worldwide production of CFCs exceeded 1 million tons. Approximately 300,000 tons are produced in the United States, allocated to the following: 39 percent refrigerants; 17 percent foam blowing; 14 percent solvents; 14 percent manufacture of plastics and resins; and 16 percent miscellaneous applications, including 2 percent as aerosol propellants (48).

Although cosmetic products and household aerosols no longer contain CFC 11, it is important to review retrospectively the toxicologic information, in the event that nonchlorinated fluorocarbons (HFC or FC) are introduced in the future. The cosmetic aerosol products include deodorants, antiperspirants, breath fresheners, women's personal hygiene products, skin lotions, foot hygiene preparations, hair shampoos, and hair sprays. In the past, the major problem was in identification of the role of CFC 11 in the overall toxicity of the hair spray aerosol products. The observation of Zuskin and Bouhuys (119) that placebo aerosol-containing propellants cause a reduction in mean expiratory flow rates when inhaled, signifies a bronchoconstrictor action. However, the effect is less than that elicited by the hair spray aerosol. These authors suggested that the hair spray particles and aerosol propellants release histamine because pretreatment with atropine (vagal blockade) or chlorpheniramine (antihistaminic) protected the subjects. Valic et al. (120) observed a reduction in forced expiratory volume following exposure of human subjects to hair sprays. They did not test the effect of propellants alone but postulated that the interaction between CFC 11 and alcohol would form aldehydes and hydrochloric acid, which are irritating to the airways.

The only published long-term study of toxicity of the propellant relative to hair sprays was completed by Giovacchini et al. (121), who exposed beagle dogs for a 10-sec period twice daily, with the dogs remaining in the chamber for 15 min after each spray period. There were no pathological lesions after 1 and 2 years of exposure. In an editorial that commented on the study (122), it was pointed out that because beagle dogs do not develop pulmonary granulomatous lesions, this animal model may not be useful in elaboration of the pulmonary lesions seen among beauticians exposed to hair sprays. It was also pointed out that guinea pigs would have been more suitable for the inhalation study, a suggestion that has not been answered by experimentation. Other investigators prefer the use of rats for testing spray can ingredients (123).

## 5.2 Human Studies on Cardiopulmonary Effects

As early as 1967, Plaut (124) reported one subject who developed bronchospasm after inhalation of CFC (unspecified) with no bronchodilator drug, from an aerosol unit. Subsequently, Fabel et al. (125) tested patients with chronic obstructive pulmonary disease given 10 to 16 actuations of the propellant, each dose comprising 12.6 ml of mixture of 4 parts CFC 11 and 6 parts dichlorofluoromethane (HCFC 21). A third study by Brooks et al. (126) showed that in 5 of 13 asthmatic patients there was a measurable increase in airway resistance following five inhalations of aerosol propellant. The publication specifies dichlorodifluoroethane, which was not commercially available, although a typographical error could have meant dichlorodifluoromethane (CFC 12), which is commonly used in bronchodilator aerosol units.

The bronchospasm provoked by inhaling CFC was unequivocally demonstrated by Valic et al. (127). The propellant gases were generated from commercial aerosol units and applied to the subject from a distance of 50 cm for periods of 15 to 60 sec. At a measured concentration of 95,000 mg/m$^3$ (1700 ppm), there was a biphasic change in ventilatory capacity, the first reduction occurring within a few minutes after exposure, and the second delayed until 13 to 30 min after exposure. Most subjects developed bradycardia and inversion of the T-wave. A 10 to 90 percent mixture of CFC 11 and CFC 12, respectively, caused more severe respiratory effects than either fluorocarbon inhaled singly.

The threshold for the cardiopulmonary effects of CFC 11 was not determined by Valic et al. (127). However, from the studies of Stewart et al. (128), it is certain that human exposure to 1000 ppm 8 hr/day, 5 days/week for a total of 18 exposures had no untoward subjective effects, and there were no changes in the electrocardiogram or pulmonary function tests. The venous blood levels of CFC 11 after 8 hr were as high as 4.69 µg/ml. The gradual attainment of this level represents a low uptake of the gas, similar to that of halothane, a fluorine-containing compound widely used to induce a general anesthesia (Section 6).

### 5.2.1  Respiratory Uptake and Elimination

The human exposure studies by Stewart et al. (128), described in the preceding paragraph, include post-exposure breath data that reveal a predictable excretion pattern. The rate of CFC 11 excretion in the expired air is a function of the duration of exposure, although there is no significant CFC 11 accumulation in the body following 8-hr exposures to 1000 ppm, repeated every 24 hr. The post-exposure breath decay curves serve to refine those reported earlier by Paulet et al. (129, 130).

Radioactive tracer techniques were used by Morgan et al. (131, 132) to measure the partition coefficient of CFC and HCFC, including CFC 11. As a group, CFC compounds have low lipid solubility compared to aliphatic chlorinated hydrocarbons. Chlorine-38-labeled CFC was poorly absorbed in the lung, with much of the inhaled vapor exhaled. After 30 min, the amount retained in the lungs was about 23 percent of the total unexpired CFC 11. Presumably, the compound remained in the lung tissue; after 5 min only a small fraction of the retained material was present in the blood.

Using fluorine-18-labeled CFC 11, Williams et al. (133) derived the same partition coefficient (olive oil/air partition = 27) as that derived by Morgan et al. (131). The former group of investigators characterized the subsequent fate of CFC 11 following its inhalation. The fall in pulmonary concentration was consistent with rapid uptake into the tissues followed by slow elimination into expired air (133).

### 5.2.2  Blood Levels of Exposed Individuals

Marier et al. (134) failed to detect CFC by random sampling of blood from household aerosols users. The conclusion that aerosol use poses no health hazard has been criticized (135). There are patients inhaling bronchodilator aerosols who show a significant CFC concentration in their blood. Paterson et al. (136) observed peak concentrations ranging from 0.13 to 2.60 μg/ml venous blood in a group of nine subjects. In a group of asthmatics, Dollery et al. (137, 138) detected peak arterial levels as follows:

|  | Level in Arterial Blood (μg/ml) | |
| --- | --- | --- |
| Actuations | CFC 11 | CFC 12 |
| Two, 30 sec apart | 0.53 to 3.1 | 0.2 to 3.13 |
| One | 0.26 to 2.0 | 0 to 2.03 |

Dollery and co-workers (137–139) have concluded that the blood levels are not sufficiently high to exert cardiotoxicity. However, they have not considered the possibility that CFC may induce reflexes from the respiratory tract, which in turn influence the heart. In other words, the effects on the heart are partially triggered by the inhaled CFC prior to its absorption.

Angerer et al. (140) exposed three volunteers in a room containing CFC 11 vapor at a mean concentration of 65 ml/m$^3$. The average pulmonary retention value was 18.2 percent with CFC levels of 537 ml/m$^3$ alveolar air and 2.8 mg/l in blood. After termination of exposure, CFC 11 concentrations in alveolar air and blood were excreted with biologic half-lives of 7 and 11 min, respectively, during the first phase of elimination (pulmonary) and 1.8 and 1.0 hr, respectively, during the second phase of elimination (urinary). Alveolar air analysis is recommended by the investigators as the preferred parameter for biologic monitoring.

### 5.3 Aerosol Use and Abuse

Although there have been hundreds of deaths from inhaling aerosols containing CFC 11 in a plastic bag, it is difficult to obtain an accurate estimation. The first documented report was by Baselt and Cravey (141) in 1968, who described a 15-year-old boy found dead with a plastic bag and a 9-oz aerosol can of a spray-on coating for frying pans lying adjacent to him. Both CFC 11 and CFC 12 used as propellants were detected in the tissues removed at the autopsy:

|  | CFC 11 (μl/100 g) | CFC 12 (μl/100 g) |
| --- | --- | --- |
| Blood | 0.86 | 0.05 |
| Kidney | 1.65 | 0.05 |
| Brain | 1.33 | 0.67 |
| Liver | 0.83 | 0.67 |
| Stomach contents | 5.78 | 6.92 |

The investigators concluded that death was due to asphyxia, because the prevalent opinion at that time was that the propellants were "inert" gases.

By 1975, the explanation for the death of a teenager due to inhalation of CFC-containing aerosol was cardiotoxicity instead of simple asphyxia. Poklis (142) reported the distribution of propellants in postmortem tissues as follows:

|         | CFC 11 (mg/100 g) | CFC 12 (mg/100 g) |
|---------|-------------------|-------------------|
| Blood   | 3.2               | 0.32              |
| Brain   | 6.1               | 0.45              |
| Liver   | 4.5               | 0.39              |
| Lung    | 3.2               | 0.32              |
| Kidney  | 2.5               | 0.18              |
| Trachea | 2.1               | 0.16              |
| Bile    | 0.6               | —                 |

The above results show concentrations in tissues higher than those reported for the 1968 case (141). There is a significant accumulation of propellants in the brain, liver, and lungs compared to blood levels, signifying propellant tissue distribution similar to that of chloroform. The accumulation in parenchymal organs has been confirmed in a 1991 aerosol abuse case report (143).

The change in explanation for causation of death from aerosol propellant abuse between 1968 and 1975 was triggered by Bass (144), who in 1970 reviewed the case histories of 110 American youths who died from the "sudden sniffing death syndrome." His eyewitness reports included one case of a 17-year old boy who died from inhaling a plastic bag filled with the contents of a spray can of hair shampoo; another case was that of a 15-year-old boy who inhaled an antitussive aerosol from a plastic bag; and the third case was a 15-year-old girl who sprayed frying pan aerosol into a plastic bag. The total number of deaths is of special significance because Bass postulated the cause of death as severe cardiac arrhythmia, a phenomenon proven by subsequent human investigation and animal experimentation.

Reports of deaths from the abuse of aerosol products continued to appear in the literature through the 1990s, mostly in the lay press. Chapel and Thomas (145) reported six deaths in Missouri from the abuse of aerosol glass-chill or from aerosol disinfectant spray. The aerosol was either sprayed directly into the oral cavity or inhaled after collection in a plastic bag.

Kramer and Pierpaoli (146) reported a 15-year-old girl from Connecticut who used deodorant aerosol to produce a "flashback," in which the individual experiences the symptoms of a hallucinogenic "trip" without actually having taken an active hallucinogenic agent. Crooke (147) described a 17-year-old boy from California who inhaled a can of hair shampoo from a plastic bag. The immediate effect was a compulsion to run, and after 150 yards, he collapsed and was pronounced dead. Kamm (148) reported a 16-year-old male from Louisiana who inhaled aerosol deodorant and subsequently died of ventricular fibrillation. The only findings at autopsy were cerebral edema, pulmonary edema, and generalized visceral congestion.

Another CFC propellant poisoning case is a Canadian youth who continually inhaled a lipid aerosol containing soybean extract, used to prevent food adherence to cooking pans (149). The addict developed acute adult-type respiratory distress syndrome, rather than sudden cardiac death. After performing laboratory tests, Fagan et al. concluded that CFC 11 and CFC 12 had damaged lung surfactant function. During the 1980s, CFC-containing aerosols continued to be available in Europe. Cases of sudden cardiac death (150), mouth frostbite (151), and pulmonary alveolitis (152) have been reported.

### 5.4 Asthmatic Deaths from Bronchodilator Aerosols

The treatment of bronchial asthma was much improved by the introduction of bronchodilator aerosols, first marketed commercially in the United States in 1956. A form whereby the sympathomimetic bronchodilators (epinephrine, isoproterenol, and salbutamol) could, aided by propellants, be dispersed from a pressurized container small enough to be conveniently carried in a purse or pocket seemed to have much to recommend it. The pressurized unit had a valve specifically designed to release a measured and reproducible dose of the drug for inhalation by the patient. The bronchodilation produced was reported to be prompt and less likely to be accompanied by tachycardia than when the drug was administered by subcutaneous injection.

The novel bronchodilator aerosols have been a commercial success both in the United States and abroad. The assumptions that underlay this modality—that the propellants were "inert" and that the dose of the drug inhaled was unlikely to do any harm—remained uncontested, and the pressure units were made available for over-the-counter sale in most countries including the United States. It became immediately apparent to certain physicians that the aerosols were being misused by some patients. The first such warnings were made in 1958 by Harris (153) in the United States. In 1961, Pavlik (155) reported that in 46 patients tested, only 18 gained full benefit. More important were his findings that 15 patients, although experiencing partial relief, were incorrectly using the aerosols, and that 13 showed no response at all. Yet these articles were overshadowed by the large number of publications and advertisements attesting the efficacy of the aerosol bronchodilators. It was against this background that the asthmatic deaths in Great Britain, Australia, and continental Europe, and the appearance of the locked-lung syndrome in the United States, were reported. Asthmatic deaths continue to be reported through the 1990s, so it is necessary to review past events.

#### 5.4.1 Asthmatic Deaths in Great Britain

In 1966, Smith (156) reported an increase in the annual number of deaths from bronchial asthma in England and Wales. Between 1959 and 1964, the overall death rate increased 42 percent, from 2.7 per 100,000 persons per year. The increase over the same period for the age group 5 to 34 years was 165 percent, and for those aged 5 to 14 years it was 330 percent. However, the rate of deaths at young ages from other respiratory conditions was stable. Together with the fact that there was no evidence to indicate that greater numbers of persons were suffering from bronchial asthma, this finding suggested that the increase might be due not to vicissitudes in the criteria for diagnosis of asthma, but rather to changes in the assignment of this disease entity as cause of death. The possibility that the increase was drug-induced (iatrogenic) found partial confirmation in the coincidental introduction of bronchodilator aerosols at a time when the mortality rate among asthmatics started to increase.

Pickvance (157) and Speizer et al. (158) examined the annual number of deaths attributed to bronchial asthma in England and Wales from 1960 to 1965. The increase was more pronounced at ages 5 to 34 years than at older ages, and was greater among those aged 10 to 14, in which group fatalities increased nearly eightfold in 7 years. Corticosteroids had been used increasingly since 1953, and pressurized aerosols containing sympathomimetics had enjoyed great popularity beginning in 1960.

Speizer et al. (159) and Fraser et al. (160) examined copies of the death certificates of 171 persons aged 5 to 34 years who, according to those records, had succumbed to bronchial asthma during a 6-month period from 1966 to 1967. Signs of severe asthma were found in 91 percent of the necropsies for which data were available. Death was sudden and unexpected in 81 percent of the patients. Although two-thirds had received corticosteroids before the terminal episode, such information about their use that was available provided no suggestion that excessive use was responsible for a large proportion of the deaths. However, pressurized aerosol bronchodilators had been used by 84 percent of the patients, and several instances of aerosol abuse were described. A similar increase in mortality rates occurred in other parts of Great Britain as well. In Eire, Linehan (161) reported a 130 percent increase in deaths of those aged 5 to 34 years during the years 1960 to 1968.

In December 1968, the pressurized aerosol bronchodilators ceased to be available to the public over the counter, and became obtainable only by prescription in Great Britain. Although the high mortality rate of asthma was reduced, it became more difficult to evaluate further what factors were involved, or to establish a direct correlation between the death rate from asthma and changes in the distribution and frequency of use of bronchodilator aerosols. Statistics collected by Inman and Adelstein (162) yielded a mortality curve for the period 1961 to 1967 that closely resembled the curves for sales of pressurized aerosols during these same years, that is, a period of steady increase followed by a period of leveling, with a terminal sharp decline.

The first deaths attributable to the use of pressurized bronchodilator aerosols were reported in 1967 by Greenberg and Pines (163). Four asthmatic patients were found unexpectedly dead, with empty pressurized aerosols either by their sides or in their hands. Eight other patients who had been hospitalized for treatment of varying degrees of severe bronchial asthma had also died suddenly. These deaths were also without obvious cause and remained unexplained after postmortem examination. The report by Greenberg and Pines cited ventricular fibrillation resulting from excessive use of aerosols as the cause of death, and was followed by reports of other deaths under similar circumstances: two adolescent asthmatics (164), a 32-year-old asthmatic (165), a 46-year-old asthmatic (166), a 50-year-old female asthmatic, a 68-year-old coal miner with chronic bronchitis (167), and three adolescent asthmatics (168).

### 5.4.2 Asthmatic Deaths in the United States

Until the 1960s, a consistent finding reported among asthmatic deaths in the United States was mucus plugs in the airways. Shapiro and Tate (169) ascribed other cases of sudden death to cardiac arrhythmia, and Van Metre (170) reported 17 deaths associated with acute asthmatic attack. In none of these deaths had the individual used bronchodilator aerosol excessively. In a subsequent report, however, Herxheimer (171) reported that nine patients whose deaths were attributable primarily to asthma had used bronchodilator aerosols well beyond the point of diminishing returns, and 29 patients who became resistant to the drug actually improved when they stopped inhaling the medicated aerosols.

In a review of 20 deaths over a 9-year period, Ghanman et al. (172) concluded that no one factor could reliably predict a fatal outcome in patients known to use medicated aerosols extensively. In their review of hospital admissions at a pediatric hospital from 1935 to 1968, Palm et al. (173) noted that there was an increase in the number of patients hospitalized for bronchial asthma during the preceding 10 years and there was a concomitant decline in mortality rate. In elderly patients, death was usually associated with carbon dioxide retention (174). Most of the subsequent reports in the United States related to the ineffectiveness of aerosol bronchodilators in asthmatic patients and, by illustrating both the limitations and hazards of these agents, contributed to limiting their distribution and the number of fatalities attributable to their use.

### 5.4.3 Explanations for Asthmatic Deaths

One outcome of the publicity surrounding the reports of deaths and locked-lung syndrome associated with the use of bronchodilators was the prohibition of their over-the-counter sale in some countries such as Great Britain. Physicians, once informed of the situation by national publications, tended to prescribed bronchodilator aerosols more cautiously, and also helped protect the public from the possible dangers inherent in their discriminate use. Although the causes for deaths and cases of locked-lung syndrome observed have not been identified, such explanations as have been proposed tend to emphasize one or the other of two groups of hypotheses. The first hypothesis related to the toxicity of the bronchodilators when inhaled. Two opposite suggestions were made as to the manner in which the sympathomimetics contribute to death among asthmatics. Herxheimer (175) emphasized the ineffectiveness of the drug, whereas Stolley (176) attributed the phenomenon to the high concentrations of the drug available in certain countries, including Great Britain. The second hypothesis emphasized the toxicity of CFC 11 and other fluorine-containing propellants. Although fatalities associated with the use of aerosols had been reported, the tendency was to underestimate the hazards of the propellants. The two reasons for this misconception were that (*a*) the propellants were claimed to be "inert" and nontoxic; and (*b*) the amount contained in the inspired aerosol was small, and little was thought to be absorbed into the blood. Phillips (177) questioned the nontoxicity of the propellants, because the only available information was based on topical testing on the skin of animals. The first experimental evidence offering unequivocal proof of the cardiotoxicity of the propellants themselves was supplied by Taylor and Harris (13) and Taylor et al. (178) (Section 2).

### 5.4.4 Global Epidemic of Asthmatic Deaths

During the 1980s, bronchodilator drugs were administered not only by metered-dose inhaler using CFC propellants, but also by nebulizer and dry-powder inhalers. The aggressive marketing of metered-dose inhaler has been responsible for its overwhelming popularity resulting in abuse and fatalities among asthmatic children (179–182). During the 1980s, there was a significant increase in mortality rate of childhood and adult asthma that has not been fully explained. Unsupervised excessive medication, particularly of beta-2 sympathomimetic bronchodilators by metered-dose inhaler, has been suggested as a causative agent in the following countries: United States (183, 184), Canada (185), Great Britain (186), continental Europe (187–192), Australia (193–195), and New Zealand (196). It is my opinion that the continued use of CFC as aerosol propellant is a significant contribution to the increase in asthmatic deaths, as was proposed in an earlier publication (197). A retrospective study of causation of death would be helpful in determining the role of bronchodilator drugs and CFC propellants. It is disappointing that a recent consensus conference on aerosol delivery completely overlooked the potential cardiotoxicity of CFC and HCFC (198–200).

## 6 PHARMACEUTICAL PRODUCTS

The last group of fluorine-containing organic compounds to be discussed are pharmaceuticals used in the treatment or diagnosis of diseases. The target organs discussed from the CFC toxicologic standpoint are also the same from the therapeutic standpoint, namely, the circulatory, respiratory, and central nervous systems. Only the last mentioned system has been sufficiently investigated to lead to introduction of useful general anesthetics. Drugs for the circulatory and respiratory systems need further refinement.

### 6.1 General Anesthetics

Although some CFC compounds have general anesthetic properties, none is in current use because of the availability of fluorine-containing anesthetics such as the following: Halothane (2-bromo-2-chloro-1,1,1-trifluoroethane); Methoxyflurane (2,2-dichloro-1,1-difluoroethyl methyl ether); Enflurane (2-chloro-1,1,2-trifluoroethyl difluoromethyl ether); and Isoflurane (1-chloro-2,2,2-trifluoroethyl difluoromethyl ether).

The use of fluorine-containing inhalation anesthetics involves significant exposure in the handling and processing of waste anesthetic gases and vapors, as well as in their initial use. Guidelines have been established for regulating exposure to these anesthetic wastes. The NIOSH criteria document on waste anesthetic gases and vapors has insisted upon a level of 5 ppm as the occupational standard for exposure to anesthetic vapors including halothane (201). Some scientists are raising the question that 5 ppm should logically be applied to CFC and HCFC in general, because some are mutagens and carcinogens (Sections 3.4 and 3.5). Epidemiologic studies in female hospital personnel and nurse anesthetists have shown associations between exposure to anesthetic vapors and the occurrence of cancers, spontaneous abortions, and congenital anomalies (201). It is possible that the 5 ppm standard will be extended to lowering the limits for exposure to all CFC and HCFC from the current standards of 1000 ppm (Section 4.2). In addition to inhalation anesthetic exposure in operating rooms, hospital personnel are also exposed to other fluorine-containing compounds. Gas sterilization units may employ CFC compounds. Clinical pathologists exposed to CFC in the preparation of frozen tissue sections have been seen to develop coronary heart disease. Speizer et al. (202) have undertaken an epidemiologic study of hospital personnel exposed to HCFC 22 for tissue freezing. They reported a 3.5-fold excessive incidence of palpitation in the exposed individuals compared to nonexposed hospital personnel. Although there are animal models to determine the influence of HCFC 22 on the pathogenesis of coronary heart disease, such a study has not been executed. It is more reasonable to explain the palpitation as a form of cardiac arrhythmia from noncoronary vascular etiology similar to that of CFC 11.

## 6.2 Synthetic Oxygen Transport Fluids

Perfluorochemicals are cyclic or straight-chain hydrocarbons in which hydrogen atoms are replaced with fluorine. Although they were initially proposed as oxygen-carrying resuscitation fluids or blood substitutes, their uses in clinical medicine and biomedical research are limited to specific examples, such as microcirculatory support in ischemic myocardium, coronary angioplasty, cancer therapy, contrast media and diagnostic imaging, surfactant in respiratory distress, decompression sickness, and ophthalmic surgery (203–210). There are at least three perfluorochemicals that have been tested in humans with encouraging results: Fluorosol DA, perfluorooctyl bromide, and perfluorotributylamine. Additional clinical trials are necessary prior to their commercial introduction.

## 7 CONCLUSION

This chapter began with the statement that during the next millenium, the ozone layer will probably be depleted by the persistence of CFC and HCFC in the stratosphere. The developmental alternatives consist of nonchlorinated fluorocarbons that are under toxicologic evaluation. It is very likely that new alternative HFC compounds, such as existing CFC and HCFC, are potentially toxic to the cardiovascular and bronchopulmonary systems. Deaths from the use, misuse, and abuse of aerosol products have been extensively documented. There are isolated reports of poisoning from exposure to refrigerants and solvents, and some studies showing a higher incidence of coronary heart disease among hospital personnel are required to establish causal relationship between fluorine-containing organic compounds, and cardiovascular and bronchopulmonary diseases among exposed workers. The high incidence of cancer among hospital personnel repeatedly exposed to fluorine-containing general anesthetics raises a fundamental need to examine other chlorofluorocarbon-exposed workers for similar effects.

This chapter has been written as a guide to toxicologic evaluation of nonchlorinated fluorocarbons (HFC and FC) that will be introduced during the 1990s through the next millennium. All existing CFC and HCFC (both chlorinated and fluorinated), are banned from production by the mid-1990s, and developmental alternatives are presently undergoing toxicologic evaluation to conform with TSCA. The industrial chemical manufacturers have not released new information until governmental approval for their manufacture is forthcoming. There has been no open scientific debate on the selection process, and my awareness of the process was made possible only by lay press coverage (Section 3.2.3). A selection process for the most desirable perfluorinated blood substitute is undoubtedly being undertaken, and the first one that will be commercially introduced will have several medical applications. Fluorine-containing pharmaceuticals include general anesthetics that are presently used. The new pharmaceuticals will assist in the treatment of diseases of the circulatory system and the eye and of cancer. So far the potential use of perfluorinated compounds in treatment of eye disease and of cancer does not include the cataract and skin cancer resulting from ozone depletion. The diseases caused by ozone depletion are unlikely to be treated by fluorine-containing organic compounds, although the possibility of this circumstance during the next millenum cannot be entirely dismissed.

## REFERENCES

1. F. S. Rowland and M. J. Molina, "The Role of Halocarbons in Stratospheric Ozone Depletion," in *Ozone Depletion, Greenhouse Gases and Climate Change*, Proceedings of Joint Symposium, March 23, 1988, National Academy Press, Washington, DC, 1989, pp. 33–55.
2. L. Dotto and H. Schiff, *The Ozone War*, Doubleday, Garden City, NY, 1978, 348 pp.
3. D. M. Aviado, "Toxicity of Propellants," in *Progress in Drug Research*, E. Jucker, Ed., Birkhauser Verlag, Basel, 1974, pp. 365–398.
4. D. M. Aviado, "Toxicity of Aerosol Propellants in the Respiratory and Circulatory Systems. X. Proposed Classification," *Toxicology*, **3**, 321–332 (1975).
5. D. M. Aviado, "Comparative Cardiotoxicity of Fluorocarbons," Chapter 14 in *Cardiac Toxicity*, T. Balazs, Ed., CRC Press, Boca Raton, FL, 1979, pp. 213–222.

6. S. Zakhari and D. M. Aviado, "Cardiovascular Toxicology of Aerosol Propellants, Refrigerants and Related Solvents," in *Cardiovascular Toxicology*, E. W. Van Stee, Ed., Raven Press, New York, 1982, pp. 281–327.
7. S. Zakhari and H. Salem, "Cardiac Toxicology Solvents," in *Principles of Cardiac Toxicology*, S. I. Baskin, Ed., CRC Press, Boca Raton, FL, 1991, pp. 465–501.
8. A. H. Nuckolls, *The Comparative Life, Fire and Explosion Hazards of Refrigerants*, Underwriters Laboratory, Chicago, 1959, approx. 113 pp.
9. J. Scholz, "New Toxicologic Investigations of Freons Used as Propellants for Aerosols and Sprays," *Fortschr. Biol. Aerosol-Forsch.*, **4**, 420–429 (1962).
10. F. Caujolle, "Toxcite Comparee des Fluids Frigorigenes," *Bull. Inst. Int. Froid*, **1**, 21–55 (1964).
11. D. Lester and L. A. Greenberg, "Acute and Chronic Toxicity of Some Halogenated Derivatives of Methane and Ethane," *Arch. Ind. Hyg.*, **2**, 335–344 (1950).
12. S. A. Friedman, M. Cammarato, and D. M. Aviado, "Toxicity of Aerosol Propellants on the Respiratory and Circulatory Systems. II. Respiratory and Bronchopulmonary Effects in the Rat," *Toxicology*, **1**, 345–355 (1973).
13. G. J. Taylor, IV, and W. S. Harris, "Cardiac Toxicity of Aerosol Propellants," *J. Am. Med. Assoc.*, **214**, 81–85 (1970).
14. D. M. Aviado and M. A. Belez, "Toxicity of Aerosol Propellants on the Respiratory and Circulatory Systems. I. Cardiac Arrhythmia in the Mouse," *Toxicology*, **2**, 31–42 (1974).
15. C. F. Reinhardt, A. Azar, M. E. Maxfield, P. E. Smith, Jr., and L. S. Mullin, "Cardiac Arrhythmias and Aerosol 'Sniffing,' " *Arch. Environ. Health*, **22**, 265–279 (1971).
16. L. S. Mullin, A. Azar, C. F. Reinhardt, P. E. Smith, Jr., and E. F. Fabrvka, "Halogenated Hydrocarbon-Induced Cardiac Arrhythmias Associated with Release of Endogenous Epinephrine," *Am. Ind. Hyg. Assoc. J.*, **33**, 389–396 (1972).
17. D. G. Clark and D. J. Tinston, "Cardiac Effects of Isoproterenol, Hypoxia, Hypercapnia, and Fluorocarbon Propellants and Their Use in Asthma Inhalers," *Ann. Allergy*, **30**, 536–541 (1972).
18. M. A. Belej, D. G. Smith, and D. M. Aviado, "Toxicity of Aerosol Propellants in the Respiratory and Circulatory Systems. IV. Cardiotoxicity in the Monkey," *Toxicology*, **2**, 381–395 (1974).
19. T. Watanabe and D. M. Aviado, "Toxicity of Aerosol Propellants in the Respiratory and Circulatory Systems. VII. Influence of Pulmonary Emphysema and Anesthesia in the Rat," *Toxicology*, **3**, 225–240 (1975).
20. R. E. Doherty and D. M. Aviado, "Toxicity of Aerosol Propellants in the Respiratory and Circulatory Systems. VI. Influence of Cardiac and Pulmonary Vascular Lesions in the Rat, *Toxicology*, **3**, 213–244 (1975).
21. M. A. Belez and D. M. Aviado, "Cardiopulmonary Toxicity of Propellants for Aerosols," *J. Clin. Pharmacol.*, **15**, 105–115 (1975).
22. D. M. Aviado and D. G. Smith, "Toxicity of Aerosol Propellants in the Respiratory and Circulatory Systems. VIII. Respiration and Circulation in Primates," *Toxicology*, **3**, 241–252 (1975).
23. R. S. Brody, T. Watanabe, and D. M. Aviado, "Toxicity of Aerosol Propellants on the Respiratory and Circulatory Systems. III. Influence of Bronchopulmonary Lesion on Cardiopulmonary Toxicity in the Mouse," *Toxicology*, **2**, 173–184 (1974).

24. D. M. Aviado and J. Drimal, "Five Fluorocarbons for Administration of Aerosol Bronchodilators," *J. Clin. Pharmacol.*, **15**, 116–128 (1975).
25. D. M. Aviado and M. A. Belez, "Toxicity of Aerosol Propellants on the Respiratory and Circulatory Systems. V. Ventricular Function in the Dog," *Toxicology*, **3**, 79–86 (1975).
26. G. J. Taylor and R. T. Drew, "Cardiomyopathy Predisposes Hamsters to Trichlorofluoromethane Toxicity," *Toxicol. Appl. Pharmacol.*, **32**, 177–183 (1975).
27. T. Balazs, F. L. Earl, G. W. Bierbower, and M. A. Weinberger, "The Cardiotoxic Effects of Pressurized Aerosol Isoproterenol in the Dog," *Toxicol. Appl. Pharmacol.*, **26**, 407–417 (1973).
28. D. M. Aviado, "Effects of Fluorocarbons, Chlorinated Solvents, and Inosine on the Cardiopulmonary System," *Environ. Health Persp.*, **26**, 207–215 (1978).
29. D. M. Aviado, "Inosine: A Naturally Occurring Cardiotonic Agent," *J. Pharmacol. (Paris)*, **14**(Suppl. 3), 47–71 (1983).
30. J. A. Simaan and D. M. Aviado, "Hemodynamic Effects of Aerosol Propellants. II. Pulmonary Circulation in the Dog," *Toxicology*, **5**, 139–146 (1975).
31. D. E. Bohning, R. E. Albert, M. N. Lippmann, and V. R. Cohen, "Effects of Fluorocarbon 11 and 12 on Tracheobronchial Particle Deposition and Clearance in Donkeys," *Am. Ind. Hyg. Assoc. J.*, **36**, 902–908 (1975).
32. J. W. Clayton, Jr., "The Mammalian Toxicology of Organic Compounds Containing Fluorine," *Handb. Exp. Pharmacol.*, **20**, 459–500 (1966).
33. J. W. Clayton, Jr., "Fluorocarbon Toxicology and Biological Action," *Fluorine Chem. Rev.*, **1**, 197–252 (1967).
34. J. W. Clayton, Jr., "Fluorocarbon Toxicity: Past, Present, Future," *J. Soc. Cosmet. Chem.*, **18**, 333–350 (1967).
35. L. J. Jenkins, Jr., R. A. Jones, R. A. Coon, and J. Siegel, "Repeated and Continuous Exposures of Laboratory Animals to Trichlorofluoromethane," *Toxicol. Appl. Pharmacol.*, **16**, 133–142 (1970).
36. T. S. Slater, "A Note on the Relative Toxic Activities of Tetrachloromethane and Trichlorofluoromethane on the Rat," *Biochem. Pharmacol.*, **14**, 178–181 (1965).
37. T. F. Slater and B. C. Sawyer, "The Stimulatory Effects of Carbon Tetrachloride and Other Halogenoalkanes on Peroxidative Reactions on Rat Liver Fractions *in vitro*; General Features of the Systems Used," *Biochem. J.*, **123**, 805–814 (1971).
38. P. J. Cox, L. J. King, and D. V. Parke, "Comparison of the Interactions of Trichlorofluoromethane and Carbon Tetrachloride With Hepatic Cytochrome P-450," *Biochem. J.*, **130**, 87P (1976); *Xenobiotica*, **6**, 363 (1976).
39. D. A. Blake and G. W. Mergner, "Inhalation Studies on the Biotransformation and Elimination of $^{14}$C-Trichlorofluoromethane and $^{14}$C-Dichlorodifluoromethane in Beagles," *Toxicol. Appl. Pharmacol.*, **30**, 396–407 (1974).
40. G. Paulet, S. Desbrousses, E. Vidal, "Absence d'effet teratogene des fluororcarbones chez le rat et le lapin," *Arch. Mal. Prof. Med. Trav. Secur. Soc.*, **35**, 658–661 (1974).
41. H. Uehleke, T. Werner, H. Greim, and M. Kramer, "Metabolic Activation of Haloalkanes and Tests *in Vitro* for Mutagenicity," *Xenobiotica*, **7**, 393–400 (1977).
42. E. Longstaff, M. Robinson, C. Bradbrook, J. A. Styles, and I. F. H. Purchase, "Genotoxicity and Carcinogenicity of Fluorocarbons: Assessment by Short-term *in Vitro* tests and Chronic Exposure in Rats," *Toxicol. Appl. Pharmacol.*, **72**, 15–31 (1984).

43. D. F. Krahn, F. C. Barsky, and K. T. McCooey, "CHO/HGPRT Mutation Assay: Evaluation of Gases and Volatile Liquids," *Environ. Sci. Res.*, **25**, 91–103 (1982).
44. National Cancer Institute Carcinogenesis Technical Report Series No. 106, *Bioassay of Trichlorofluoromethane for Possible Carcinogenicity*, 1978, 46 pp.
45. S. S. Epstein, S. Joshi, J. Andrea, P. Clapp, H. Falk, and N. Mantel, "Synergistic Toxicity and Carcinogenicity of 'Freons' and Piperonyl Butoxide," *Nature*, **214**, 526–528 (1967).
46. American Conference of Governmental Industrial Hygienists Inc., *Documentation of the Threshold Limit Values and Biological Exposure Indices*, 5th ed., ACGIH, Cincinnati, 1986, 696+ pp.
47. National Research Council, *Emergency and Continuous Exposure Limits for Selected Airborne Contaminants*, Vol. 2, Committee on Toxicology, National Academy Press, Washington, DC, 1984, 123 pp.
48. World Health Organization, "Fully Halogenated Chlorofluorocarbons," in *International Programme on Chemical Safety, Environmental Health Criteria 113*, World Health Organization, Geneva, 1990, 164 pp.
49. World Health Organization, "Partially Halogenated Chlorofluorocarbons (Methane Derivatives)," in *International Programme on Chemical Safety, Environmental Health Criteria 126*, World Health Organization, Geneva, 1991, 97 pp.
50. U.S. Environmental Protection Agency, *Hydrofluorocarbons and Hydrochlorofluorocarbons, Interim Report*, USEPA, Washington, DC, 1990, approx. 100 pp.
51. National Institute for Occupational Safety and Health, *Registry of Toxic Effects of Chemical Substances*, 1985–1986 ed., D. V. Sweet, Ed., 6 volumes; also TOXLINE database updated on October 14, 1992.
52. A. K. Naj, "EPA is Creating Safer Substitutes for Refrigerants," *Wall Street Journal*, B6 (April 7, 1992).
53. N. D. Smith, "New Chemical Alternatives for CFCs and HCFCs," EPA-600/F-92-012, March 24, 1992, 2 pp.
54. A. L. Beyerlein, D. D. DesMarteau, S. H. Hwang, N. D. Smith, and P. Joyner, "Physical Property Data on Fluorinated Propanes and Butanes as CFC and HCFC Alternatives," in *Proceedings of International CFC and Halon Alternative Conferences*, Alliance for Responsible CFC Policy, Baltimore, December 3–5, 1991, pp. 1–10.
55. D. D. DesMarteau, A. L. Beyerlein, S. H. Hwang, Y-C. Shen, S-W. Li, R. Mendonca, K. N. Naik, N. D. Smith, and P. Joyner, "Selection and Synthesis of Fluorinated Propanes and Butanes as CFC and HCFC Alternatives," ASHRAE Society Meeting, Chicago, January 23–27, 1993, pp. 1–9.
56. M. Landry, and R. Fuerst, "Gas Ecology of Bacteria," *Dev. Ind. Microbiol.*, **9**, 370–380 (1968).
57. V. C. Foltz, and R. Fuerst, "Mutation Studies with *Drosophila melanogaster* Exposed to Four Fluorinated Hydrocarbon Gases," *Environ. Res.*, **7**, 275–285 (1974).
58. E. Longstaff, and D. B. McGregor, "Mutagenicity of a Halocarbon Refrigerant Monochlorodifluoromethane (R22) in *Salmonella typhimurium*," *Toxicol. Lett.*, **2**, 1–4, (1978).
59. N. Loprieno, R. Barale, S. Presciuttini, A. M. Rossi, I. Sbrana, G. Stretti, L. Zaccaro, A. Abbondandolo, S. Bonatti, and R. Fiorio, "Comparative Data with Different Test Systems Using Microorganisms and Mammalian Cells on References and Environmental Mutagens," *Mutat. Res.*, **64**, 119 (1979).

60. E. Longstaff, M. Robinson, C. Bradbrook, J. A. Styles, and I. F. Purchase, "Genotoxicity and Carcinogenicity of Fluorocarbons: Assessment by Short-Term *in vitro* Test and Chronic Exposure in Rats," *Toxicol. Appl. Pharmacol.*, **72**, 15–31 (1984).
61. E. Longstaff, "Carcinogenic and Mutagenic Potential of Several Fluorocarbons," *Ann. N.Y. Acad. Sci.*, **534**, 283–298 (1988).
62. International Agency for Research on Cancer, "Chlorodifluoromethane," *IARC Monographs on the Evaluation of the Carcinogenic Risk of Chemicals to Humans*, **41**, 237–252 (1986).
63. C. Maltoni, G. Lefemine, D. Tovoli, and G. Perino, "Long-term Carcinogenicity Bioassays on Three Chlorofluorocarbons (Trichlorofluoromethane, FC-11; Dichlorodifluoromethane, F-12; Chlorodifluoromethane, FC-22) Administered by Inhalation to Sprague-Dawley Rats and Swiss Mice," *Ann. N.Y. Acad. Sci.*, **534**, 261–282 (1988).
64. H. J. Trochimowicz, G. M. Rusch, T. Chiu, and C. K. Wood, "Chronic Inhalation Toxicity/Carcinogenicity Study in Rats Exposed to Fluorocarbon 113," *Fundam. Appl. Toxicol.*, **11**, 68–75 (1988).
65. R. G. Mahurin and R. L. Bernstein, "Fluorocarbon-Enhanced Mutagenesis of Polyaromatic Hydrocarbons," *Environ. Res.*, **45**, 101–107, (1988).
66. J. S. Seckar, H. J. Trochimowicz, and G. K. Hogan, "Toxicological Evaluation of Hydrochlorofluorocarbon 142b," *Food Chem. Toxicol.*, **24**, 237–240 (1986).
67. U.S. Environmental Protection Agency, *Toxic Substance Control Act*, Section 8(e) submissions.
68. T. Dalhamn, "Freon som orsak till forgiftningsfall," *Nord. Hyg. Tidskr.*, **39**, 165–169 (1958).
69. W. A. Cook, *Occupational Exposure Limits—Worldwide*, American Industrial Hygiene Association, 1987, 308 pp.
70. International Labour Office, *Occupational Exposure Limits for Airborne Toxic Substances*, 3rd ed., Occupational Safety and Health Series No. 37, International Labour Office, Geneva, 1991, 455 pp.
71. A. H. Kegel, W. D. McNally, and A. S. Pope, "Methyl Chloride Poisoning from Domestic Refrigerators," *J. Am. Med. Assoc.*, **93**, 353–358 (1929).
72. H. M. Baker, "Industrial Methyl Chloride Poisoning," *Am. J. Publ. Health*, **20**, 294–295 (1930).
73. W. D. McNally, "Eight Cases of Methyl Chloride Poisoning with Three Deaths," *J. Ind. Hyg. Toxicol.*, **28**, 94–97 (1946).
74. J. Cheymol, "Un refrigerant industriel recent: le dichlorodifluoromethane," *Presse Med.*, **44**, 1123–1124 (1936).
75. E. H. Schiotz, "Freezing Agent Dichlorodifluoromethane from Point of View of Industrial Hygiene," *Nord. Hyg. Tidskr.*, **27**, 230–244 (1946).
76. T. Marti, "Note sur l'intoxication par le freon," *Ann. Med. Leg. Crim. Police Sci. Toxicol.*, **28**, 147–150 (1948).
77. J. Mendeloff, "Death After Repeated Exposures to Refrigerant Cases; Report of Case," *Arch. Ind. Hyg. Occup. Med.*, **6**, 518–524 (1952).
78. H. D. Leinoff, "Acute Coronary Thrombosis in Industry; Indirect Injuries from Toxic Gases and Other Physical Agents," *Am. Heart J.*, **24**, 187–195 (1942).
79. C. Edling, C. G. Ohlson, G. Ljungkvist, A. Oliv, and B. Soderholm, "Cardiac Ar-

rhythmia in Refrigerator Repairmen Exposed to Fluorocarbons," *Brit. J. Ind. Med.*, **47**, 207–212 (1990).
80. M. Antti-Poika, A. J. Heikkila, and L. Saarinen, "Cardiac Arrhythmias During Occupational Exposure to Fluorinated Hydrocarbons," *Brit. J. Ind. Med.*, **47**, 138–140 (1990).
81. D. D. Campbell, J. E. Lockey, J. Petajan, B. J. Gunter, and W. N. Rom, "Health Effects Among Refrigeration Repair Workers Exposed to Fluorocarbons," *Brit. J. Ind. Med.*, **43**, 107–111 (1986).
82. T. Astrom, A. Jonsson, and B. Jarvholm, "Exposure to Fluorocarbons During the Filling and Repair of Air-Conditioning Systems in Cars—A Case Report," *Scand. J. Work Environ. Health.*, **13**, 527–528 (1987).
83. M. A. Clark, J. W. Jones, J. J. Robinson, and J. T. Lord, "Multiple Deaths Resulting from Shipboard Exposure to Trichlorotrifluoroethane," *J. Forens. Sci.*, **30**, 1256–1259 (1985).
84. J. F. Burris and S. L. Schwartz, "Occupational Exposure to Freon: An Association With Severe Hypertension?," *J. Am. Med. Assoc.*, **267**, 569 (1992).
85. D. C. May and M. J. Blotzer, "A Report of Occupational Deaths Attributed to Fluorocarbon-113," *Arch. Environ. Health*, **39**, 352–354 (1984).
86. K. Yonemitsu, S. Tsunenari, and M. Kanda, "An Industrial Accidental Death Due to Freon 113 Poisoning—Toxicological Analysis of Its Cause of Death," *Jap. J. Legal Med.*, **37**, 428–433 (1983).
87. M. B. McGee, R. F. Meyer, and S. G. Jejurikar, "A Death Resulting from Trichlorotrifluoroethane Poisoning," *J. Forens. Sci.*, **35**, 1453–1460 (1990).
88. Y. Hoshika, T. Tsuchiya, and N. Murayama, "Two Cases of Acute Poisoning of 1,1,2-Trichloro-1,2,2-Trifluoroethane (Freon-113)," *Jap. J. Ind. Health.*, **31**, 248–249 (1989).
89. T. Kawakami, T. Takano, and R. Araki, "Enhanced Arrhythmogenicity of Freon 113 by Hypoxia in the Perfused Rat Heart," *Toxicol. Ind. Health.*, **6**, 493–498 (1990).
90. R. Sandyk and M. A. Gillman, "Motor Dysfunction Following Chronic Exposure to a Fluoroalkane Solvent Mixture Containing Nitromethane," *Eur. Neurol.*, **23**, 479–481 (1984).
91. K. Rasmussen, H. J. Jeppesen, and P. Arlien-Soborg, "Psychoorganic Syndrome from Exposure to Fluorocarbon 113—an Occupational Disease?," *Eur. Neurol.*, **28**, 205–207 (1988).
92. R. M. Park, M. A. Silverstein, M. A. Green, and F. E. Mirer, "Brain Cancer Mortality at a Manufacturer of Aerospace Electromechanical Systems," *Am. J. Ind. Med.*, **17**, 537–552 (1990).
93. E. E. Wegener, K. R. Barraza, and S. K. Das, "Severe Frostbite Caused by Freon Gas," *South. Med. J.*, **84**, 1143–1146 (1991).
94. M. Jeanmougin, D. Bonvalet, J. Civatte, A.-A. Ramelet, and C. Vilmer, "Toxidermies non Medicamenteuses par Voie Respiratoire ou Percutanee Probable," *Ann. Dermatol. Venereol.*, **11**, 437–444 (1984).
95. R. Valdivieso, J. Pola, C. Zapata, J. Cuesta, J. Puyana, C. Martin, and E. Losada, "Contact Allergic Dermatitis Caused by Freon 12 in Deodorants," *Cont. Dermat.*, **17**, 243–245 (1987).
96. E. V. Craig, "A New High-Pressure Injection Injury of the Hand," *J. Hand Surg.*, **9A**, 240–242 (1984).

97. M. C. Zimmerman, "Skin Refrigerants," *J. Dermatol. Surg. Oncol.*, **10**, 506 (1984).
98. Anon., "Skin Refrigerants [letter]," *J. Dermatol. Surg. Oncol.*, **11**, 8–9 (1985).
99. C. W. Hanke, H. H. Roenigk, Jr., and J. B. Pinksi, "Complications of Dermabrasion Resulting from Excessively Cold Skin Refrigeration," *J. Dermatol. Surg. Oncol.*, **11**, 896–900 (1985).
100. L. M. Dzubow, "Histologic and Temperature Alterations Induced by Skin Refrigerants," *J. Am. Acad. Dermatol.*, **12**, 796–810 (1985).
101. C. W. Hanke, and J. J. O'Brian, "A Histologic Evaluation of the Effects of Skin Refrigerants in an Animal Model," *J. Dermatol. Surg. Oncol.*, **13**, 664–669 (1987).
102. J. N. Harrison, D. J. Smith, R. Strong, M. Scott, J. Davey, and C. Morgan, "The Use of Halon 1301 for Firefighting in Confined Spaces," *J. Soc. Occup. Med.*, **32**, 37–43 (1982).
103. A. M. Sass-Kortsak, D. L. Holness, and G. J. Stopps, "An Accidental Discharge of a Halon 1301 Total Flooding Fire Extinguishing System," *Am. Ind. Hyg. Assoc. J.*, **46**, 670–673 (1985).
104. C. Steadman, L. C. Dorrington, P. Kay, and H. Stephens, "Abuse of a Fire-Extinguishing Agent and Sudden Death in Adolescents," *Med. J. Aust.*, **141**, 115–117 (1984).
105. W. M. I. Smeeton and M. S. Clark, "Sudden Death Resulting from Inhalation of Fire Extinguishers Containing Bromochlorodifluoromethane," *Med. Sci. Law*, **25**, 258–262 (1985).
106. Y. Lerman, E. Winkler, M. S. Tirosh, Y. Danon, and S. Almog, "Fatal Accidental Inhalation of Bromochlorodifluoromethane (Halon 1211)," *Human Exp. Toxicol.*, **10**, 125–128 (1991).
107. L. S. Mullin, C. F. Reinhardt, and R. E. Hemingway, "Cardiac Arrhythmias and Blood Levels Associated With Inhalation of Halon 1301," *Am. Ind. Hyg. Assoc. J.*, **40**, 653–658 (1979).
108. G. A. de S. Wickramaratne, D. J. Tinston, D. L. Kinsey, and J. E. Doe, "Assessment of the Reproductive Toxicology of Bromochlorodifluoromethane (BCF, Halon 1211) in the Rat," *Brit. J. Ind. Med.*, **45**, 744–760 (1988).
109. J. A. Styles, C. R. Richardson, R. D. Callander, M. F. Cross, I. P. Bennett, and E. Longstaff, "Activity of Bromochlorodifluoromethane (BCF) in the Three Mutation Tests," *Mutat. Res.*, **142**, 187–192 (1985).
110. J. D. Kaufman, M. S. Morgan, M. L. Marks, H. L. Green, and L. Rosenstock, "A Study of the Cardiac Effects of Bromochlorodifluoromethane (Halon 1211) Exposure During Exercise," *Am. J. Ind. Med.*, **21**, 223–233 (1992).
111. G. L. Kennedy, Jr., "Toxicology of Fluorine-Containing Monomers," *Crit. Rev. Toxicol.*, **21**, 149–170 (1990).
112. J. Odum and T. Green, "The Metabolism and Nephrotoxicity of Tetrafluorethylene in the Rat," *Toxicol. Appl. Pharmacol.*, **76**, 306–318 (1984).
113. C. L. Potter, A. J. Gandolfi, R. Nagle, and J. W. Clayton, "Effects of Inhaled Chlorotrifluoroethylene and Hexafluoropropene on the Rat Kidney," *Toxicol. Appl. Pharmacol.*, **59**, 431–440 (1981).
114. International Agency for Research on Cancer, *Tetrafluoroethylene and Polytetrafluoroethylene*, IARC Monographs on the Evaluation of the Carcinogenic Risk of Chemicals to Humans, Vol. 19, 1979, pp. 185–301.

115. C. E. Lewis and G. R. Kerby, "An Epidemic of Polymer-Fume Fever," *J. Am. Med. Assoc.*, **191**, 375–378 (1965).
116. Anon., "Polymer-Fume Fever Associated With Cigarette Smoking and the Use of Tetrafluoroethylene—Mississippi," *Morbid. Mortal. Week. Rep.*, **36**, 515–516, 521 (1987).
117. D. B. Warheit, W. C. Seidel, M. C. Carakostas, and M. A. Hartsky, "Attenuation of Perfluoropolymer Fume Pulmonary Toxicity: Effect of Filters, Combustion Method, and Aerosol Age," *Exp. Mol. Pathol.*, **52**, 309–329 (1990).
118. W. N. Sullivan, "The Coupling of Science and Technology in the Early Development of the World War II Aerosol Bomb," *Mil. Med.*, **136**, 157–158 (1971).
119. E. Zuskin and A. Bouhuys, "Acute Airway Responses to Hair-Spray Preparations," *N. Engl. J. Med.*, **290**, 660–663 (1974).
120. F. Valic, E. Zuskin, Z. Skuric, and M. Denich, "Effects of Aerosols of Common Use on the Ventilatory Lung Capacity. 1. Change in $FEV_{1.0}$ in Exposure to Hair Sprays," *Acta Med. Yugosl.*, **28**, 231–246 (1974).
121. R. P. Giovacchini, G. H. Becker, M. J. Brunner, and F. E. Dunlap, "Pulmonary Disease and Hair-Spray Polymers, Effects of Long-Term Exposure in Dogs," *J. Am. Med. Assoc.*, **193**, 298–299 (1965).
122. Anon., "Hair Sprays; The Pot Remains in the Boil," *Food Cosmet. Toxicol.*, **4**, 73–74 (1966).
123. J. Pauluhn, L. Machemer, G. Kimmerle, and A. Eben, "Methodological Aspects of the Determination of the Acute Inhalation Toxicity of Spray-Can Ingredients," *J. Appl. Toxicol.*, **8**, 431–438 (1988).
124. G. S. Plaut, "Bronchodilator Aerosols," *Lancet*, **2**, 721 (1967).
125. H. Fabel, R. Wettengel, and W. Hartmann, "Myokardischamie und Arrhythmien durch den Gebrauch oon Dosteraerosolen bein Menschen," *Dtsch. Med. Wochenschr.*, **97**, 428–431 (1972).
126. S. M. Brooks, S. Mintz, and E. Weiss, "Changes Occurring after Freon Inhalation," *Am. Rev. Respir. Dis.*, **105**(4), 640–643 (1972).
127. F. Valic, Z. Skuric, Z. Bantic, M. Rudar, and M. Hecej, "Effects of Fluorocarbon Propellants on Respiratory Flow and ECG," *Brit. J. Ind. Med.*, **34**, 130–136 (1977).
128. R. D. Stewart, P. E. Newton, E. D. Baretta, A. A. Herrmann, H. V. Forst, and R. J. Soto, "Physiological Response to Aerosol Propellants," *Environ. Health Perspect.*, **26**, 275–285 (1978).
129. G. Paulet and R. Chevrier, "Modalites de l'Elimation par l'Air Expire du Fluorane 11 Inhale. Etude Chez l'Homme et Chez l'Animal," *Arch. Mal. Prof. Med. Trav. Secur. Soc.*, **30**, 251–256 (1969).
130. G. Paulet, R. Chevrier, J. Paulet, M. Duchene, and J. Chappet, "De la Retention des Freons par les Poumons et les Voies Aeriennes. Etude Faite Chez l'Homme et l'Animal," *Arch. Mal. Prof.*, **30**, 101–120 (1969).
131. A. Morgan, A. Black, M. Walsh, and D. R. Belcher, "The Absorption and Retention of Inhaled Fluorinated Hydrocarbon Vapors," *Int. J. Appl. Radiat. Isot.*, **23**, 285–291 (1972).
132. A. Morgan, A. Black, and D. R. Belcher, "Studies on the Absorption of Halogenated Hydrocarbons and their Excretion in Breath Using $^{38}Cl$ Tracer Techniques," *Ann. Occup. Hyg.*, **15**, 273–283 (1972).

133. F. M. Williams, G. H. Draffan, C. T. Dollery, J. C. Clark, and A. J. Palmer, "Use of 18F-Labeled Fluorocarbon 11 to Investigate the Fate of Inhaled Fluorocarbons in Man and in the Rat," *Thorax*, **29**, 99–103 (1974).

134. G. Marier, H. MacFarland, G. S. Wiberg, H. Buchwald, and P. Dassault, "Blood Fluorocarbon Levels Following Exposure to Household Aerosols," *Canad. Med. Assoc. J.*, **111**, 39–42 (1974).

135. D. B. Rix, and T. King, "Household Aerosols [letter]," *Canad. Med. Assoc. J.*, **111**, 645 (1974).

136. J. W. Paterson, M. F. Sudlow, and S. R. Walker, "Blood Levels of Fluorinated Hydrocarbons in Asthmatic Patients After Inhalation of Pressurized Aerosols," *Lancet*, **2**, 565–568 (1971).

137. C. T. Dollery, G. H. Draffan, D. S. Davies, F. M. Williams, and M. E. Conolly, "Blood Concentrations in Man of Fluorinated Hydrocarbons After Inhalation of Pressurized Aerosols," *Lancet*, **2**, 1164–1166 (1970).

138. C. T. Dollery, F. M. Williams, G. H. Draffan, G. Wise, H. Sahyoun, J. W. Paterson, and S. R. Walker, "Arterial Blood Levels of Fluorocarbons in Asthmatic Patients Following Use of Pressurized Aerosols," *Clin. Pharmacol. Ther.*, **15**, 59–66 (1974).

139. G. H. Draffan, C. T. Dolery, F. M. Williams, and R. A. Clare, "Alveolar Gas Concentrations of Fluorocarbons-11 and -12 in Man After Use of Pressurized Aerosols," *Thorax*, **29**, 95–98 (1974).

140. J. Angerer, B. Schroder, and R. Heinrich, "Exposure to Fluorotrichloromethane (R-11)," *Int. Arch. Occup. Environ. Health*, **56**, 67–72 (1985).

141. R. C. Baselt and R. H. Cravey, II, "A Fatal Case Involving Trichloromonofluoromethane and Dichlorodifluoromethane," *J. Foren. Sci.*, **13**, 407–410 (1968).

142. A. Poklis, "Determination of Fluorocarbon 12 in Postmortem Tissues; A Case Report," *Forensic Sci.*, **5**, 53–59 (1975).

143. J. Hamill and T. G. Kee, "The Detection of Aerosol Propellants in Body Fluids and Tissue by Gas Chromatography-Mass Spectrometry," *J. Foren. Sci. Soc.*, **31**, 301–307 (1991).

144. M. Bass, "Sudden Sniffing Death," *J. Am. Med. Assoc.*, **212**, 2075–2079 (1970).

145. J. L. Chapel and G. Thomas, "Aerosol Inhalation for 'Kicks,'" *Mo. Med.*, **67**, 378–380 (1970).

146. R. A. Kramer and P. Pierpaoli, "Hallucinogenic Effect of Propellant Components of Deodorant Sprays," *Pediatrics*, **48**, 322–323 (1971).

147. S. T. Crooke, "Solvent Inhalation," *Tex. Med.*, **68**, 67–69 (1972).

148. R. C. Kamm, "Fatal Arrhythmia Following Deodorant Inhalation; Case Report," *Forensic Sci.*, **5**, 91–93 (1975).

149. D. G. Fagan, J. B. Forrest, G. Enhorning, M. Lamprey, and J. Guy, "Acute Pulmonary Toxicity of a Commercial Fluorocarbon-Lipid Aerosol," *Histopathology*, **1**, 208–223 (1977).

150. R. J. Goldsmith, "Death by Freon [letter]," *J. Clin. Psychiatr.*, **50**, 36–37 (1989).

151. D. C. Elliott, "Frostbite of the Mouth: A Case Report," *Mil. Med.*, **156**, 18–19 (1991).

152. G. M. Wright and A. Lee, "Alveolitis After Use of a Leather Impregnation Spray," *Brit. Med. J.*, **292**, 727–728 (1986).

153. H. C. Harris, "The Use and Abuse of Pocket Nebulizers in the Treatment of Asthma," *Postgrad. Med.*, **23**, 170–173 (1958).

154. I. Pavlik, "Nehoby a Nebezpeci Pri Acrosollove Inhalacni Lecbe," *Cas. Lek. Cesk.*, **100**, 275–279 (1961).
155. K. B. Saunders, "Misuse of Inhaled Bronchodilator Agents," *Brit. Med. J.*, **1**, 1037–1038 (1965).
156. J. M. Smith, "Death from Asthma [letter]," *Lancet*, **1**, 1042 (1966).
157. W. Pickvance, "Pressurized Aerosols in Asthma [letter]," *Brit. Med. J.*, **1**, 756 (1967).
158. F. E. Speizer, R. Doll, and P. Heaf, "Observations on Recent Increase in Mortality from Asthma," *Brit. Med. J.*, **1**, 335–338 (1968).
159. F. E. Speizer, R. Doll, P. Heaf, and L. B. Strang, "Investigation into Use of Drugs Preceeding Death from Asthma," *Brit. Med. J.*, **1**, 339–343 (1968).
160. P. M. Fraser, F. E. Speizer, S. D. M. Waters, R. Doll, and N. M. Mann, "The Circumstances Preceding Death from Asthma in Young People in 1968 to 1969," *Brit. J. Dis. Chest*, **65**, 71–84 (1971).
161. W. D. Linehan, "Asthma Deaths in Eire," *Brit. Med. J.*, **4**, 172–173 (1969).
162. W. H. W. Inman and A. M. Adelstein, "Rise and Fall of Asthma Mortality in England and Wales in Relation to Use of Pressurized Aerosols," *Lancet*, **2**, 279–285 (1969).
163. M. J. Greenberg and A. Pines, "Pressurized Aerosols in Asthma," *Brit. Med. J.*, **1**, 563 (1967).
164. A. Koivikko and T. Peltonen, "Death From Asthma in Children and Adolescents in Finland in 1952–1965," *Acta Paediat. Scand.*, **177**, 109 (1967).
165. E. M. Douglas, T. Hillier, and I. C. Johnson, "Pressurized Aerosol in Asthma [letter]," *Brit. Med. J.*, **2**, 53 (1967).
166. P. D. Exon, "Pressurized Aerosols in Asthma [letter]," *Brit. Med. J.*, **2**, 178 (1967).
167. G. S. Graham, "Two Sudden Unexplained Deaths in Asthmatics," *Scot. Med. J.*, **13**, 282–283 (1968).
168. A. P. Norman and S. Sanders, "Mortality in Asthma in Childhood," *Practitioner*, **201**, 909–914 (1968).
169. J. B. Shapiro and C. F. Tate, "Death in Status Asthmaticus: A Clinical Analysis of Eighteen Cases," *Dis. Chest*, **48**, 484–489 (1965).
170. T. E. Van Metre, Jr., "Death In Asthmatics," *Trans. Am. Clin. Climatol. Assoc.*, **78**, 58–69 (1966).
171. H. Herxheimer, "Asthma Mortality and Pressurized Aerosols [letter]," *Lancet*, **2**, 242–243 (1969).
172. R. D. Ghanman, L. Schreirer, and N. A. Vanselow, "Fatal Bronchial Asthma: An Analyses of Terminal Treatment in Twenty Cases," *Ann. Allergy*, **26**, 194–205 (1968).
173. C. R. Palm, M. A. Murcek, T. R. Roberts, H. C. Mansmann, Jr., and P. Fireman, "A Review of Asthma Admissions and Deaths at Children's Hospital of Pittsburgh from 1935 to 1968," *J. Allergy*, **46**, 257–269 (1970).
174. W. C. Tabb and J. L. Guerrant, "Life Threatening Allergy," *J. Allergy*, **42**, 249–260 (1968).
175. H. Herxheimer, "Asthma Deaths [letter]," *Brit. Med. J.*, **4**, 795 (1972).
176. P. D. Stolley, "Asthma Mortality. Why the United States was Spared an Epidemic of Deaths Due to Asthma," *Am. Rev. Resp. Dis.*, **105**, 883–890 (1972).
177. M. A. Phillips, "Bronchodilator Aerosols," *Lancet*, **2**, 677 (1967).
178. G. J. Taylor, IV., W. S. Harris, and M. D. Bodgonoff, "Ventricular Arrhythmias

Induced in Monkeys by the Inhalation of Aerosol Propellants," *J. Clin. Invest.*, **50**, 1546–1550 (1971).
179. H. F. Pratt, "Abuse of Salbutamol Inhalers in Young People," *Clin. Allergy*, **12**, 203–208 (1982).
180. I. M. Slessor, "Addiction to Aerosol Treatment [letter]," *Brit. Med. J.*, **288**, 485 (1984).
181. J. A. Kuzemko, "Near-Miss Asthma Deaths in Children [letter]," *Lancet*, **1**, 49 (1985).
182. C. O'Callaghan and A. D. Milner, "Aerosol Treatment Abuse," *Arch. Dis. Child.*, **63**, 70 (1988).
183. W. M. Wardell, "Nonprescription Sales of Metaproterenol Aerosols [letter]," *N. Engl. J. Med.*, **311**, 405–406 (1984).
184. S. F. Lanes and A. M. Walker, "Do Pressurized Bronchodilator Aerosols Cause Death Among Asthmatics?" *Am. J. Epidemiol.*, **125**, 755–760 (1987).
185. M. Spino, "Bioequivalence of Generic Aerosol Bronchodilators: What Are the Issues?" *Can. Med. Assoc. J.*, **141**, 883–887 (1989).
186. H. L. J. Markowe, "Asthma Mortality in England and Wales [letter]," *Lancet*, **2**, 636–637 (1986).
187. L. Andersson and G. Boethius, "Effect of Domiciliary Treatment with Nebulized Beta 2-Stimulants—A Retrospective Follow-up of 102 Patients," *Eur. J. Resp. Dis.*, **136**, 205–212 (1984).
188. P. V. Wichert, "Asthma Therapy—Too Much or Too Little? [editorial]," *Dtsch Med. Wochenschr.*, **113**, 799–800 (1988).
189. M. Moulin, F. C. Hugues, and M. C. Bigot, "Medicaments Adrenergigues et Pharmacovigilance: les Aerosols de Beta-2-Sympathomimetiques dans l Asthme," *Allerg. Immunol.*, **20**, 306, 309, 311 (1988).
190. Y. Maria and D. A. Moneret-Vautrin, "Les Beta$_2$-Mimetiques dans le Traitement de l'asthme: Mythes et Realities," *Rev. Pneumol. Clin.*, **45**, 185–193 (1989).
191. R. Estopa Miro, "Estimulantes Betaadrenergicos Topicos y Paro Respiratorio," *Med. Clin.*, **92**, 24–25 (1989).
192. M. Rissee, A. Kloppel, and G. Weiler, Todesursache im Asthma-Anfall bei Selbstmedikation mit Dosier-Aerosolem [letter]," *Dtsch. Med. Wochenschr.*, **115**, 1165 (1990).
193. A. J. Woolcock, "Therapies to Control the Airway Inflammation of Asthma," *Eur. J. Resp. Dis.*, **147**, 166–174 (1986).
194. S. Birdseye, L. D. Renouf, and R. M. Ford, "Death Rates in Asthmatic Patients [letters]," *Med. J. Aust.*, **147**, 469 (1987).
195. R. Pierce, "Asthma Aerosols and CFCs [editorial]," *Aust. N.Z. J. Med.*, **20**, 6–7 (1990).
196. A. McNeill, "Asthma, Inhalers, and Technique [letter]," *N.Z. Med J.*, **103**, 55 (1990).
197. D. M. Aviado, "Toxicity of Aerosols," *J. Clin. Pharmacol.*, **15**, 86–104 (1975).
198. J. R. Balmes et al., "Aerosol Consensus Statement," *Chest*, **100**, 1106–1109 (1991).
199. J. R. Balmes, "The Environmental Impact of Chlorofluorocarbon Use in Metered Dose Inhalers," *Chest*, **100**, 1101–1102 (1991).
200. D. J. Pierson, "Toward International Consensus on Clinical Aerosol Administration," *Chest*, **100**, 1100–1101 (1991).
201. National Institute for Occupational Safety and Health, *Criteria for a Recommended*

*Standard—Occupational Exposure to Waste Anesthetic Gases and Vapors*, 1977, approx. 255 pp.
202. F. E. Speizer, D. H. Wegman, and A. Ramirez, "Palpitation Rates Associated With Fluorocarbon Exposure in Hospital Setting," *N. Engl. J. Med.*, **292**, 624–626 (1975).
203. G. P. Biro and P. Glais, "Perfluorocarbon Blood Substitutes," *Crit. Rev. Oncol. Hematol.*, **6**, 311–374 (1987).
204. D. M. Long, D. C. Long, R. F. Mattrey, R. A. Long, A. R. Burgan, W. C. Herrick, and D. F. Shellhamer, "An Overview of Perfluoroctylbromide—Application as a Synthetic Oxygen Carrier and Imaging Agent for X-ray, Ultrasound and Nuclear Magnetic Resonance," *Biomat. Art. Cells Art. Org.*, **16**, 411–420 (1988).
205. M. Nabih, G. A. Peyman, L. C. Clark, Jr., R. E. Hoffman, M. Miceli, M Abou-Steit, M. Tawakol, and K. R. Liu, "Experimental Evaluation of Perfluorophenanthrene as a High Specific Gravity Vitreous Substitute: A Prelimianry Report," *Ophthalmol. Surg.*, **20m**, 286–293 (1989).
206. L. Zarif, J. Greiner, S. Pace, and J. G. Riess, "Synthesis of Perfluoroalkylated Xylitol Ethers and Esters: New Surfactants for Biomedical Uses," *J. Med. Chem.*, **33**, 1262–1269 (1990).
207. R. K. Spence, S. McCoy, J. Costabile, E. D. Norcross, M. J. Pello, J. B. Alexander, C. Wisdom, and R. C. Camishion, "Fluosol DA-20 in the Treatment of Severe Anemia: Randomized, Controlled Study of 46 Patients," *Crit. Care Med.*, **18**, 1227–1230 (1990).
208. J. C. Garrelts, "Fluosol: An Oxygen-Delivery Fluid for Use in Percutaneous Transluminal Coronary Angioplasty," *Ann. Pharmacother.*, **24**, 1105–1112 (1990).
209. H. E. Rice, R. Virmani, C. L. Hart, F. D. Kolodgie, and A. Farb, "Dose-Dependent Reduction of Myocardial Infarct Size With the Perfluorochemical Fluosol-DA," *Am. Heart J.*, **120**, 1039–1046 (1990).
210. K. C. Lowe, "Synthetic Oxygen Transport Fluids Based on Perfluorochemicals: Applications in Medicine and Biology," *Vox Sanguinis*, **60**, 129–140 (1991).

CHAPTER NINETEEN

# Aliphatic Hydrocarbons

Finis Cavender, Ph.D., D.A.B.T., C.I.H.

## 1 SATURATED HYDROCARBONS, ALKANES, PARAFFINS

### 1.1 General

*1.1.1 Occurrence and Use*

Saturated aliphatic hydrocarbons occur naturally as constituents of natural gas and swamp gas and in the paraffin fraction of crude oil. Some also occur in coal, the natural resins of plants, and the fats of animals. Shorter alkanes are produced during the combustion of fuels, by the catalytic cracking of crude petroleum oil, or by other specialized petrochemical processes. They can be released to the environment in the exhaust of gasoline and diesel engines, in the flue gas of municipal waste incinerators, and from vulcanization and extrusion operations. Paraffin mixtures are used extensively as fuels, solvents, lubricants, degreasers, protective coatings, propellants, refrigerants, intermediates in the synthesis of organic chemicals, food additives, and vehicles in the application of pesticides.

Petroleum distillates are produced by fractionation of crude petroleum oil. Although all petroleum distillates are hydrocarbons, not all hydrocarbons are petroleum distillates. These distillates consist of straight- and branched-chain hydrocarbons and are commonly divided into the fractions shown in Table 19.1.

*1.1.2 Physical Characteristics*

The boiling ranges of major alkane mixtures are given in Table 19.2. The toxicity of the alkanes is generally related to vapor pressure, viscosity, surface tension, and lipid solubility.

---

*Patty's Industrial Hygiene and Toxicology, Fourth Edition, Volume 2, Part B*, Edited by George D. Clayton and Florence E. Clayton.
ISBN 0-471-54725-5   © 1994 John Wiley & Sons, Inc.

**Table 19.1.** Petroleum Distillates and Their Uses

| Fraction | Synonym | Use |
|---|---|---|
| Petroleum ether | Benzine | Dry cleaning, thinner, solvent |
| Naphtha | Ligroin | Thinner, polish |
| Gasoline | Petroleum spirits | Fuel, thinner |
| Mineral spirits | Stoddard solvent, white spirits | Solvent, degreasing dry cleaning |
| Kerosene | Coal oil | Fuel, solvent, lighter fluid |
| Fuel oil | Diesel fuel | Fuel |
| Lubricating oil | Oil | Lubrication |
| Petrolatum | Petroleum jelly | Laxative, ointment |
| Paraffin wax | Paraffin | Sealant, polish |
| Asphalt | Tar | Sealant, construction |

*1.1.2.1 Volatility.* Volatility is the tendency of a liquid to evaporate and form a vapor. In general, the higher the volatility, the higher the vapor concentration in air. The most common industrial exposures are the inhalation of gases and vapors. Volatility correlates with the vapor pressure and evaporation rate. Vapor density relates to the dispersion of a vapor and is also an important consideration in industrial hygiene practice.

*1.1.2.2 Solubility.* Dermal contact with solvents and other liquid hydrocarbons occurs frequently in industrial operations. All alkanes are lipid soluble and can cause dermatitis and burns. A single exposure may result in mild skin irritation, whereas chronic exposure may cause severe dermatitis characterized by acne, folliculitis, or photosensitivity. Contact burns are the result of the defatting properties of lipid-soluble alkanes. Lipid solubility also relates to the general anesthetic properties of alkanes and to their potential for respiratory and dermal absorption.

**Table 19.2.** Boiling Range of Major Alkanes

| Principal Alkane (carbon number) | Commercial Product | Boiling Range (°C) |
|---|---|---|
| $C_1, C_2$ | Natural gas | Gas at 25°C |
| $C_3, C_4$ | Liquid, bottled gas | Gas at 25°C |
| $C_4-C_7$ | Petroleum ether | 20–90 |
| $C_6-C_8$ | Naphtha | 65–120 |
| $C_5-C_{10}$ | Gasoline | 36–210 |
| $C_7-C_9$ | Mineral spirits | 150–210 |
| $C_9-C_{16}$ | Kerosene | 170–300 |
| $C_5-C_{16}$ | Jet fuels | 40–300 |
| $>C_{16}$ | Lubricating oils | 300–700 |

## ALIPHATIC HYDROCARBONS

**Table 19.3.** Hazard Properties of Selected Alkanes

| Alkane | Vapor Pressure (mm Hg) (25°C) | Viscosity (cP) | Octanol/Water Coefficient (Log $K_{ow}$) | Flash Point (°F) | Lower Flammability Limit (%) | Vapor Density (air = 1) |
|---|---|---|---|---|---|---|
| Methane | Gas | 0.0109 | 1.09 | — | 5 | 0.55 |
| Ethane | Gas | 0.009 | 1.81 | — | 3 | 1.04 |
| n-Propane | Gas | 0.008 | 2.36 | −156 | 2.1 | 1.56 |
| n-Butane | Gas | 0.0074 | 2.89 | −76 | 1.6 | 2.07 |
| n-Pentane | 500 | 0.24 | 3.39 | −58 | 1.5 | 2.49 |
| n-Hexane | 150 | 0.33 | 4.11 | −10 | 1.3 | 2.97 |
| n-Heptane | 40 | 0.41 | 4.66 | 25 | 1.1 | 3.43 |
| n-Octane | 14.1 | 0.54 | 5.18 | 56 | 1.0 | 3.86 |
| n-Nonane | 4.45 | 0.71 | 5.46 | 88 | 0.8 | 4.41 |
| n-Decane | 1.43 | 0.92 | 5.98 | 115 | 0.8 | 4.90 |

*1.1.2.3 Flammability.* Flammability and explosiveness are measures of the potential for an alkane to ignite and burn. Some alkanes are flammable enough to be used as fuels, whereas others are relatively nonflammable. Although these properties are not direct measures of the toxicity of a substance, they do relate to the hazard in handling alkanes in industrial operations.

The volatility, lipid solubility, and flammability properties for selected alkanes are given in Table 19.3.

### 1.1.3 Toxicity

The aliphatic hydrocarbons are practically nontoxic for single exposures below the lower flammability limit. Virtually all paraffins will cause nausea, vomiting, abdominal pain, and occasionally diarrhea when ingested (1–3). Serious toxic effects of aliphatic hydrocarbons include asphyxia and chemical pneumonitis for many paraffins, axonal neuropathy for n-hexane, and cancer for butadiene. Dermatitis, central nervous system (CNS) depression, anesthesia, and cardiac sensitization have also been noted for many paraffins. Acutely, the most common toxic effects are CNS depression and asphyxia following inhalation and chemical pneumonitis after the aspiration of ingested alkanes. Asphyxia occurs when the oxygen in air is displaced by high concentrations of a gas or vapor. When the oxygen concentration is lowered from ambient levels to 10 percent or less, hypoxia results and the body is starved for oxygen. At this level of oxygen deprivation, death occurs swiftly.

Chemical or aspiration pneumonitis occurs following the aspiration of ingested hydrocarbons. Aspiration is often a problem in children, who may ingest hydrocarbons stored in familiar beverage containers. Viscosity is the most important determinant in evaluating the aspiration potential of hydrocarbons (4). The lower the viscosity, the more easily it spreads as a monolayer on surfaces including migration from the throat into the tracheobronchial tree. Material with a viscosity below 60 Saybolt Universal Standard (SUS) units has a very high risk of inhalation and aspiration. Viscosity also determines how deep the hydrocarbons will penetrate into the airways. Those paraffins with low viscosity will penetrate to the alveoli causing irritation and edema, that is, chemical pneumonitis consisting of burning of the oral mucosa and tracheobronchial tree, coughing, choking, and gasping. As the material spreads to the alveoli, bronchospasm, perfusion abnormalities, edema, hemorrhaging, and hypoxia may occur (2, 4). Emesis is contraindicated following the ingestion of hydrocarbons because of the threat of aspiration. The capacity of hydrocarbons to induce spontaneous vomiting must be carefully considered in persons who have ingested hydrocarbons.

Dermal irritation and CNS depression are common problems with liquid aliphatic hydrocarbons in chronic exposures. Dermal irritation occurs in workers repeatedly exposed to liquid hydrocarbons as solvents. The paraffins are lipid solvents and dissolve or extract the fats from the skin, resulting in painful drying and cracking of the skin, that is, chronic eczematoid dermatitis, with itching and inflammation.

CNS depression occurs as the inhaled vapor or gas crosses the alveolar-capillary membrane to be absorbed into the bloodstream. At levels that cause CNS depression, the lung itself is spared injury (1–3). The CNS depressant properties of some alkanes have led to substance abuse in the form of "glue sniffing," usually toluene or *n*-hexane. Other abusers have utilized gasoline, paints containing solvents such as xylene, methyl ethyl ketone, acetone, ethyl acetate, ethyl benzene, and isobutyl acetate, typewriter correction fluids, aerosol can propellants including propane and isobutane, and exhaust emissions. Abusers often exhibit a drunken appearance and suffer from learning or memory impairment, personality disorders, seizures, neuropsychological disorders, and tachycardia (1–3).

Chronic abuse and occupational exposure to $C_6$ alkanes have resulted in sensorimotor distal axonopathy. *n*-Hexane is used in shoe factories where worker exposure to high concentrations of *n*-hexane occurs daily (5, 6). Many of the workers develop axonopathy, especially in Japan, Italy, and China. Workers exposed to methyl *n*-butyl ketone in fabric print plants also develop this axonopathy (7). Further study revealed that 2,5-hexanedione is a key intermediate in producing axonopathy (8). The entire axonopathy can be reproduced in rats and other mammals (9). Interestingly, rats exposed to 10,000 ppm 6 hr/day, 5 days/week for 13 weeks exhibited minimal evidence of axonopathy (10), whereas continuous exposure to as little as 250 ppm for 4 months produced severe axonopathy (11, 12). Similar results are seen in mice (13). The pharmacokinetics of hexane metabolism show that the concentration of 2,5-hexanedione in the blood lags some 8 to 10 hr behind the hexane concentration in blood (14). Thus by the time 2,5-hexanedione is starting to build up, the daily exposure is over and the 2,5-hexanedione is excreted prior to the next exposure. However, with continuous exposure, the 2,5-hexanedione concentration continues to build to steady-state levels. This is quite important in industrial hygiene considerations. Whereas most workers in the United States work an 8-hr shift and then leave the plant, workers in China and Japan may commute to the factory to work and stay at the factory all week, even eating and sleeping in the factory between work shifts until the work week is over. Thus these workers are at much greater risk in developing axonopathy from hexane exposure.

Other hexane isomers do not produce axonopathy (15). Also other diketones such as 3,4-hexanedione (16) and 2,4-pentanedione from methyl ethyl ketone (17) do not produce axonopathy. It has been shown that the toxic chemical species is not 2,5-hexanedione, but a pyrrole formed from 2,5-hexanedione reacting with free amino groups in proteins (18). This was established by showing that 3,3-dimethyl-2,5-hexanedione, which cannot form a pyrrole, is not neurotoxic (19) and that 3,4-dimethyl-2,5-hexanedione, which forms the pyrrole faster, is more neurotoxic than 2,5-hexanedione (20). The pyrroles undergo oxidation leading to covalent cross-linking with proteins (19–21). Recently, a hexanedione derivative that is not oxidized as a pyrrole was synthesized and tested, and it is not neurotoxic (22). Thus only those alkanes that can be metabolized through the 2,5-hexanedione moiety to form pyrroles produce this axonopathy (23, 24).

Several $C_5$ or lower-carbon-number alkanes have been shown to be cardiac sensitizers, but are much less effective than chlorofluorocarbons.

Chronic exposure to petroleum or its products can result in petroleum toxicity, characterized by the formation of oil droplets in the liver and other tissues (25).

In general, branched-chain derivatives are less toxic than the corresponding parent straight-chain alkanes. Odorant properties increase whereas analgesic properties decrease with increasing chain length. Both dermal and pulmonary irritant properties increase with increasing chain length up to $C_{14}$ derivatives.

### 1.1.4 General Microbiological Characteristics

Of the aliphatic alkanes and alkenes, $C_1$ to $C_4$, some protect *Neurospora crassa* against gamma irradiation, but others enhance the radiation damage (26). Most $C_2$ to $C_{10}$ paraffins prevent germination of *Bacillus megaterium* spores, and the $C_3$ to $C_9$ members show total inhibition (27). Almost all hydrocarbons are susceptible to microbial oxidation. In the Atlantic Ocean, hydrocarbon-oxidizing bacteria are most abundant in the coastal areas (28). *Pseudomonas aeruginosa* has been shown to oxidize *n*-paraffins and to epoxidize α-olefins (29). Petroleum fractions containing $C_7$ and $C_{12}$ to $C_{20}$ hydrocarbons are actively degraded by such oxidizing bacteria (30). A variety of hydrocarbons, $C_6$ to $C_{19}$, are readily oxidized by the enzyme ω-hydroxylase of *Pseudomonas oleovorans* (31). A decrease in a $C_{10}$ to $C_{18}$ crude oil fraction and the $C_{20}$ to $C_{25}$ fraction is documented to be actively degraded by selected paraffin-utilizing bacteria (32). Others up to $C_{44}$ have been observed to be metabolized by a variety of microorganisms (33). However, degradation above $C_{20}$ is considerably less than for $C_6$ to $C_{20}$.

### 1.1.5 Industrial Hygiene

To collect alkanes for analysis, the lower-molecular-weight members of the alkane series are directly collected as gas, those from $C_5$ to $C_{16}$ are absorbed on charcoal, and the $C_{17}$ members and above are collected on filters as particulate matter (34). The individual compounds are quantified using gas chromatography and other methods (35). In the field, portable devices such as combustible gas indicators or organic vapor analyzers can also be used. According to the classification by the Occupational Safety and Health Administration (OSHA) (36), the lower alkanes

Table 19.4. Physicochemical Properties of Alkanes (40–56)

| Compound | B.P. (°C) | CAS Registry No. | Density (20/4°C) | Emp. Formula | Flammability Limits (%) | Flash Pt. (°C) | Freezing Pt. (°C) |
|---|---|---|---|---|---|---|---|
| Methane | −161.49 | 74-82-8 | | $CH_4$ | 5.3–14 | −187.78 | −182.48 |
| Ethane | −88.63 | 74-84-0 | | $C_2H_6$ | 3.2–12.45 | −135 | −183.27 |
| Propane | −42.07 | 74-98-6 | 0.5005 | $C_3H_8$ | 2.37–9.5 | −104.44 | −189.69 |
| Butane | −0.50 | 106-97-8 | 0.5788 | $C_4H_{10}$ | 1.86–8.41 | −60.0 | −138.35 |
| Isobutane | −11.73 | 75-28-5 | 0.5572 | $C_4H_{10}$ | 1.8–8.44 | −82.78 | −159.6 |
| Pentane | 36.07 | 109-66-0 | 0.6262 | $C_5H_{12}$ | 1.42–7.80 | −49.0 | −129.72 |
| Isopentane | 27.85 | 78-78-4 | 0.6197 | $C_5H_{12}$ | 1.32–8.3 | 51.0 | −159.9 |
| Neopentane | 9.50 | 463-82-1 | 0.5910 | $C_5H_{12}$ | 1.4–8.3 | −6.67 | −16.55 |
| Hexane | 68.74 | 110-54-3 | 0.6594 | $C_6H_{14}$ | 1.18–7.80 | −22 | −95.348 |
| Isohexane | 60.27 | 107-83-5 | 0.6532 | $C_6H_{14}$ | 1.2–7.7 | −23.33 | −153.67 |
| 3-Methylpentane | 63.28 | 96-14-0 | 0.6643 | $C_6H_{14}$ | 1.2–7.0 | −20 | −118 |
| Neohexane | 49.74 | 75-83-2 | 0.6492 | $C_6H_{14}$ | 1.2–7.0 | −47.8 | −99.87 |
| 2,3-Dimethylbutane | 57.99 | 79-29-8 | 0.6616 | $C_6H_{14}$ | 1.2–7.7 | −28.89 | −128.53 |
| Heptane | 98.43 | 142-82-5 | 0.6838 | $C_7H_{16}$ | 1.2–6.7 | −4.4 | −90.61 |
| Isoheptane | 90.0 | 591-76-4 | 0.6786 | $C_7H_{16}$ | 1.0–7.0 | −27.8 | −118.286 |
| Neoheptane (2,2-dimethylpentane) | 79.20 | 590-35-2 | 0.6739 | $C_7H_{16}$ | | | −123.82 |
| 2,3-Dimethylpentane | 89.78 | 565-59-3 | 0.6951 | $C_7H_{16}$ | 1.1–6.7 | −28.9 | −135 |
| 2,4-Dimethylpentane | 80.50 | 108-08-7 | 0.6727 | $C_7H_{16}$ | | −28.9 | −119.14 |
| Octane | 125.67 | 111-65-9 | 0.7025 | $C_8H_{18}$ | 0.96–4.66 | 13.0 | −56.79 |
| 2,2,4-Trimethylpentane | 99.24 | 540-84-1 | 0.6919 | $C_8H_{18}$ | 1.1–6.0 | −12 | −116 |
| 2,3,4-Trimethylpentane | 113.47 | 565-75-3 | 0.7191 | $C_8H_{18}$ | | 5 | −109.0 |
| Nonane | 150.80 | 111-84-2 | 0.7176 | $C_9H_{20}$ | 0.87–2.9 | 31.1 | −53.52 |
| 2,2,5-Trimethylhexane | 124.08 | 3522-94-9 | 0.7072 | $C_9H_{20}$ | | 12.8 | −105.79 |
| Decane | 174.1 | 124-18-5 | 0.7301 | $C_{10}H_{22}$ | 0.78–2.6 | 46.1 | −29.662 |
| 2,7-Dimethyloctane | 159.87 | | 0.7242 | $C_{10}H_{22}$ | | | −54.0 |
| Undecane | 195.89 | 1120-21-4 | 0.7402 | $C_{11}H_{21}$ | | 65 | −25.59 |
| Dodecane | 216.28 | 112-40-3 | 0.7487 | $C_{12}H_{26}$ | 0.6 | 73.89 | −9.55 |
| Tridecane | 235.44 | 629-50-5 | 0.7564 | $C_{13}H_{28}$ | | 79.44 | −5.39 |
| Tetradecane | 253.57 | 629-59-4 | 0.7628 | $C_{14}H_{30}$ | 0.5 | 100 | 5.86 |
| Pentadecane | 270.63 | 629-62-9 | 0.7685 | $C_{15}H_{35}$ | | | 9.93 |
| Hexadecane | 286.79 | 544-76-3 | 0.7734 | $C_{16}H_{32}$ | | 135 | 18.7 |
| Heptadecane | 301.82 | 629-78-7 | 0.7780 | $C_{17}H_{36}$ | | 148.89 | 21.98 |
| Octadecane | 316.12 | 593-45-3 | 0.7819 | $C_{18}H_{38}$ | | 165.56 | 28.18 |
| Nonadecane | 329.7 | 629-92-5 | 0.7855 | $C_{19}H_{40}$ | | 168.33 | 32.1 |
| Pristane | 296.0 | 1921-70-6 | 0.7827 | $C_{19}H_{40}$ | | | |
| Eicosane | 342.7 | 112-95-8 | 0.7887 | $C_{20}H_{42}$ | | 182.22 | 36.8 |

[a]Solubility in water/alcohol/ether: v = very soluble; s = soluble; d = slightly soluble; i = insoluble.
[b]At 760 mm Hg, 25°C.
[c]At atmospheric pressure.

are simple asphyxiants, and air concentrations are recommended to be held below 1000 ppm. The American Conference of Governmental Industrial Hygienists (ACGIH) threshold limit values (TLVs) are 1000 for the $C_1$ to $C_3$ hydrocarbons and 600 ppm for $C_4$ hydrocarbons (37).

## 1.2 Methane

Methane, $CH_4$, is a flammable gas (38) that occurs in natural gas at a concentration of 60 to 80 percent (39). It collects in coal mines or geologically similar earth deposit sites, evolves as marsh gas, and forms during certain fermentation and sludge

| Mol. Wt. | Refractive Index, $n_D^{20}$ | Solubility[a] w/al/et | Sp. Gr. (25°C) | Density (air = 1) | Vapor Pressure[b] (mm Hg) (°C) | Viscosity (SUS) | W/V Conversion (mg/m³ = 1 ppm) |
|---|---|---|---|---|---|---|---|
| 16.042 | — | s/s/s | 0.42 | 0.55 | 40[c] (−86.3) | N.A. | 0.66 |
| 30.069 | 1.0377 | i/d/— | 0.374 | 1.05 | 40[c] (23.6) | N.A. | 1.23 |
| 44.096 | 1.2898 | s/s/v | 0.5077 | 1.551 | 10[c] (26.9) | N.A. | 1.80 |
| 58.123 | 1.3326 | s/v/v | 0.5844 | 2.07 | 2[c] (18.8) | N.A. | 2.38 |
| 58.123 | — | s/v/v | 0.5631 | 2.07 | 2[c] (7.5) | N.A. | 2.38 |
| 72.150 | 1.3575 | d/v/v | 0.6312 | 2.49 | 400 (18.5) | <32 | 2.95 |
| 72.150 | 1.3537 | i/v/v | 0.2648 | 2.5 | 595 (21.1) | <32 | 2.95 |
| 72.150 | 1.342 | i/s/s | 0.5967 | 2.48 | 1100 (21.8) | <32 | 2.95 |
| 86.177 | 1.3749 | i/v/s | 0.6640 | 2.97 | 100 (15.8) | <32 | 3.52 |
| 86.177 | 1.3715 | i/s/s | 0.6579 | 3.0 | | <32 | 3.52 |
| 86.177 | 1.3749 | i/s/v | 0.6690 | 2.97 | 400 (10.5) | <32 | 3.52 |
| 86.177 | 1.3686 | i/s/s | 0.6540 | 3.0 | 400 (31.0) | <32 | 3.52 |
| 86.177 | 1.3750 | i/s/s | 0.6664 | 3.0 | 400 (39.0) | <32 | 3.52 |
| 100.203 | 1.3876 | i/v/v | 0.6882 | 3.52 | 40 (22.3) | <32 | 4.10 |
| 100.203 | 1.38495 | i/s/v | 0.6830 | — | 40 (14.9) | <32 | 4.1 |
| 100.203 | 1.3822 | i/s/s | | | | <32 | 4.10 |
| 100.203 | 1.3920 | i/s/s | 0.6994 | 3.45 | 40 (13.9) | <32 | 4.10 |
| 100.203 | 1.3815 | i/s/s | 0.6722 | 3.48 | 8.2 (21.0) | <32 | 4.10 |
| 114.230 | 1.3974 | i/v/s | 0.7068 | 3.94 | 10 (19.2) | <32 | 4.67 |
| 114.230 | 1.3915 | i/v/s | 0.6962 | 3.93 | 40.6 (21) | <32 | 4.67 |
| 114.230 | 1.4042 | i/v/v | 0.7233 | | 0.970 | <32 | 4.67 |
| 128.257 | 1.4054 | i/v/v | 0.7217 | 4.41 | 10 (38) | <32 | 5.25 |
| 128.257 | 1.3997 | i/v/v | 0.7174 | 4.7 | 12.9 (21) | | 5.25 |
| 142.284 | 1.4119 | i/v/s | 0.7341 | 4.9 | 1 (16.5) | <32 | 5.82 |
| 142.284 | 1.4086 | —/—/s | | | | | 5.82 |
| 156.306 | 1.4172 | i/v/v | 0.7441 | 5.4 | | <32 | 6.39 |
| 170.337 | 1.4216 | i/v/v | 0.7526 | 5.96 | 1 (47.8) | <32 | 6.97 |
| 184.362 | 1.4256 | i/v/v | 0.7603 | | 1 (59.4) | 33.8 | 7.54 |
| 198.392 | 1.4289 | i/v/v | 0.7667 | 6.83 | 1 (76.4) | 35.4 | 8.11 |
| 212.418 | 1.4319 | i/v/v | 0.7724 | | 1 (91.6) | 37.0 | 8.69 |
| 226.444 | 1.4345 | i/d/v | 0.7773 | 7.8 | 1 (105.3) | 39.1 | 9.26 |
| 240.471 | 1.4369 | i/d/v | 0.7818 | | 1 (115) | 41.9 | 9.84 |
| 254.498 | 1,4390 | i/d/v | 0.7858 | | 1 (119) | | 10.41 |
| 268.525 | 1.4409 | i/d/s | 0.7893 | | 1 (133.2) | | 10.98 |
| 268.525 | 1.4385 | i/—/s | | | | 39.1 | 10.98 |
| 282.552 | 1.4426 | i/—/s | 0.7925 | | Solid | | 11.56 |

degradation processes. It is often accompanied by other low-molecular-weight hydrocarbons. Methane is odorless and colorless (40), but has practically no physiological effects below the lower flammability limit. Selected physical data are presented in Table 19.4 (40–56).

Methane is practically nontoxic, but acts as a simple asphyxiant at very high concentrations (57). A concentration of 87 percent has caused asphyxiation and 90 percent respiratory arrest in mice (58). However, methane is a rather inefficient anesthetic agent, owing to its low protein-binding properties (59). Methane is readily absorbed (60, 61) and metabolized by mammals (62). When inhaled, the majority of the absorbed dose is exhaled unchanged. A small amount of methane is converted to methanol and ultimately to carbon dioxide (46). Uptake in humans is less rapid than in the rat (63). The liquefied gas may cause frostbite on skin contact.

Methane inhibits spore germination of *Bacillus megaterium*; however, in combination with other gases, it appears to promote germination in *Escherichia coli* and *Neurospora crassa* (64), but not spore formation (27). Also, the utilization of methane as a carbon source by some cultures, such as *Methylococcus* and others (28, 65), emphasizes that methane is biodegradable.

The 96-hr aquatic toxicity rating (TLm96) is greater than 1000 ppm (66). ACGIH (37) recommends that methane be treated as a simple asphyxiant, that is, an inert gas or vapor that acts as a simple asphyxiant without other physiological effects when present at high concentrations in air. For simple asphyxiants, a time-weighted average (TWA) of 1000 ppm is suggested. Methane is on the Environmental Protection Agency Toxic Substances Control Act (EPA TSCA) Chemical Inventory and the Test Submission Data Base (67).

### 1.3. Ethane

Ethane, $CH_3CH_3$, is a flammable gas (33) that occurs in natural gas at a concentration of 5 to 9 percent (39). Ethane is practically nontoxic and selected physical properties are listed in Table 19.4. Small quantities of ethane, along with other short-chain alkanes and alkenes, have been detected on mined coal samples (68). It is released in the exhaust of diesel and gasoline engines, from municipal incinerators, and from the combustion of natural gas, gasoline, and polypropylene. Additionally, ethane is produced as a catabolic product of lipid peroxidation in rats (69). The odor of ethane can be detected between 185 and 1106 $mg/m^3$ (70). Guinea pigs exposed to 2.2 to 5.5 percent for 2 hr show slight signs of irregular respiration, which is readily reversible on cessation of the exposure (71). At high concentrations, ethane causes CNS depression (49). At higher concentrations, ethane acts as a simple asphyxiant by displacing oxygen from the air (57). Mixtures of oxygen containing concentrations of 15 to 19 percent ethane act as a weak cardiac sensitizer (72). The liquid causes severe frostbite (46). *Neurospora crassa* can use ethane as the sole carbon source (26). It also promotes growth of *Mycobacterium vaccae* (73), and is documented to serve as a medium for petroprotein production (74). However, ethane prevents the germination of *Bacillus megaterium* spores (27). Industrially, ethane is handled similarly to methane, and an occupational exposure limit of 1000 ppm is recommended. Ethane is on the EPA TSCA Chemical Inventory and the Test Submission Data Base (67). A chromatographic method to separate and identify ethane in gaseous samples of $C_2$ to $C_5$ hydrocarbons has been developed (75).

### 1.4 Propane

Propane, $CH_3CH_2CH_3$, occurs in natural gas at a concentration of about 3 to 18 percent (39), and traces of propane have been measured in human expired air (76).

It is emitted into the atmosphere from furnaces, automobile exhausts, and natural gas sources and from the combustion of polyethylene and phenolic resins. Measurements in a medium-sized U.S. city in 1972 have shown urban air concentrations of approximately 50 ppb (77).

Propane is odorless, highly flammable (38), and explosive. The odor of propane can be detected between 1800 and 36,000 mg/m$^3$ (70). Selected physical data are presented in Table 19.4. Propane is used as an aerosol propellant and as a fuel source. With sufficient oxygen, it burns to carbon dioxide and water, but to carbon monoxide and water when oxygen is deficient (78). At 650°C, it decomposes to ethylene and ethane. Propane is an anesthetic (57) and is nonirritating to the skin and eyes. Direct skin or mucous membrane contact with liquefied propane causes burns and frostbite. At air concentration levels below 1000 ppm, propane exerts very little physiological action (79). It is on the FDA's GRAS list, substances "generally regarded as safe" (80). At very high levels, propane has CNS depressant and asphyxiating properties. Twenty cases of "sudden death" have been reported (81) in which propane and propylene were quantified in blood, urine, and cerebrospinal fluid. Among operators of propylene-fueled forklifts, 17 cases of carbon monoxide poisoning were noted (82). Inhalation studies in cats (83) indicate a gas concentration of 89 percent is not anesthetic but will depress the blood pressure. Guinea pigs showed irregular breathing, tremors, stupor, and CNS depression when exposed to 2.2 to 5.5 percent for 5 to 120 min, which was rapidly reversible upon cessation of exposure (71). In dogs (84) 1 percent causes hemodynamic changes, whereas 3.3 percent decreases inotropism of the heart, a decrease in mean aortic pressure, stroke volume and cardiac output, and increase in pulmonary vascular resistance. In primates, 10 percent propane induces some myocardial effects, whereas exposure to 20 percent causes aggravation of these parameters and respiratory depression (85, 86). Ten percent propane in the mouse (85) and 15 percent in the dog (72) produced no arrhythmia but weak cardiac sensitization. Propane is metabolized by *Microbacterium vaccae* (73), and is readily degraded by soil bacteria (87). *Mycobacterium phlei* is capable of growing on propane as the sole carbon source (88). Propane is metabolized by the various microorganisms via the malonyl succinate pathway (87). Propane reduces the viable cell count of *Escherichia coli* and produces biochemical mutations (89) but is not mutagenic in the Ames system. The aquatic TLm96 is greater than 100 ppm (66).

The OSHA permissible exposure limit (PEL) is 1000 ppm (1800 mg/m$^3$) (36), and ACGIH treats propane as a simple asphyxiant (37). Propane is on the EPA TSCA Chemical Inventory and Test Submission Data Base (67). For monitoring purposes, dual range explosive gas meters are used, properly calibrated for propane, in the range of 481 to 2016 ppm at 20.5°C and 760 mm Hg (90). A chromatographic method to quantify propane in gaseous mixtures of $C_2$ to $C_5$ hydrocarbons has been developed (75).

## 1.5 Butanes

### 1.5.1 n-*Butane*

Butane, $CH_3(CH_2)_2CH_3$, occurs in natural gas and in the ambient urban air, in small concentrations. It has been measured in the exhaust of diesel engines at 22 ppm (91) and detected in the exhaust of gasoline engines and in air above landfills and disposal sites. The pure or mixed-isomer material is used as an aerosol propellant, refrigerant, food additive, fuel source, or chemical feedstock for special chemicals in the solvent, rubber, and plastics industries (52). It is a flammable and explosive gas (38) and the odor of butane can be detected between 2.9 and 14.6 mg/m$^3$ (70) and in water at 6.2 ppm (92). Selected physical properties are listed in Table 19.4.

Upon direct contact, the liquefied product may cause burns or frostbite to the eyes, skin, or mucous membranes. The inhalation of 10,000 ppm for 10 min may result in CNS depression but produces no systemic effects (79). It can cause blurred vision and can be aspirated resulting in pneumonitis. The 4-hr $LC_{50}$ for the rat is 658 mg/l (93). The 2-hr $LC_{50}$ for the mouse is 680 mg/l (93). $n$-Butane is anesthetic to mice at 13 percent in 25 min and at 22 percent in 1 min. In the dog, 25 percent is required for anesthesia (94). In guinea pigs, concentrations of 2.1 to 5.6 percent cause sniffing and chewing movements with rapid breathing; however, the animals quicky recovered after cessation of exposure. Mixing butane and isobutylene has an additive CNS depressant effect in rats (95). The mechanism concerning the anesthetic properties of butane is similar to that of ethane and propane.

Some parallelism between the toxicity and effective cerebral concentrations have been observed (96). Butane is a weak cardiac sensitizer in the dog (97). Concentrations of 5000 ppm in the anesthetized dog may cause hemodynamic changes, such as a decrease in cardiac output, left ventricular pressure and stroke volume, a decrease in myocardial contractility, and aortic pressure (84). Butane is partially absorbed in the rat lung and is translocated to the brain, kidney, liver, spleen, and adipose tissue (93). Microsomal enzyme systems oxidize butane to its parent alcohol (98). Conversely, butane promotes lipid synthesis in *Mycobacterium vaccae* (73). Butane inhibits the growth of some bacteria and fungi spores and some plant seeds (27). It also inhibits enzymatic lysis of bacterial spores (99). *Mycobacterium crassa* and *phlei* can grow using butane as the sole carbon source (88). In combination with various concentrations of oxygen, butane supports the growth of *Neurospora crassa* (64), as well as the germination of *N. ascrospores* and growth of *Escherichia coli* strains B and Sd4 (79, 89). Butane is not mutagenic in the Ames system (46). The aquatic TLm96 is above 1000 ppm (66). The ACGIH threshold limit value (TLV) (37) and the OSHA PEL (36) are 800 ppm (1900 mg/m$^3$). Butane is on the EPA TSCA Chemical Inventory and the Test Submission Data Base (67). For detection and quantification in air, color detector tubes are available (100). A chromatographic method to determine butane in gaseous mixtures of $C_2$ to $C_5$ hydrocarbons has been developed (75, 101).

### 1.5.2 2-Methylpropane, Isobutane

Isobutane, 2-methylpropane, $CH(CH_3)_3$, a flammable gas (38), occurs in small quantities in natural gas and crude oil. It has been detected in urban atmospheres (102) at concentrations of 44 to 74 ppb (103). It also evolves from natural sources and has been measured in diesel exhaust at 1.4 to 11 ppm (91, 104) and in cigarette smoke at 10 to 60 ppm (105). Isobutane is produced in refining processes and used as a raw material for petrochemicals (106), an aerosol propellant (107), an industrial carrier gas, and a fuel source. It represents one of the basic raw materials used in the chemical industry for the production of propylene glycols and oxides and polyurethane foams and resins. Selected physical properties are listed in Table 19.4.

Toxicologically, the vapor exerts no effect on skin and eyes. The odor of methylpropane is recognized at a concentration of 45 mg/m$^3$ (70). Direct contact with the liquid produces chemical burns. Human volunteers exposed to 250 to 1000 ppm for 1 min to 8 hr and 500 ppm for 1 to 8 hr/day for 10 days showed no adverse effects (108). The 1-hr $LC_{50}$ for the mouse is 52 mg/l (84). At concentrations in the range of the $LC_{50}$, mice exhibit CNS depression, rapid and shallow respiration, and apnea. Isobutane is a weak cardiac sensitizer (84, 97). At high concentrations, a decrease in pulmonary compliance and tidal volume was noted in the rat (109). No effects were noted in anesthetized dogs at concentrations up to 2 percent, but decreased myocardial contractility was noted at 2.5 percent, exaggerated effects at 5 percent, with a decrease in ventricular and aortic pressure, and at 10 percent decreased left ventricular pressure, mean arterial flow, and stroke volume, with increased pulmonary vascular resistance. Isobutane is a CNS depressant in the mouse at 15 percent in 60 min, and at 23 percent in 26 min (94). At 22 to 27 percent, it is anesthetic in the mouse in 8.7 min, but causes respiratory arrest in 15 min (109). In the dog, isobutane is anesthetic at 45 percent in 10 min (94). Isobutane is oxidatively metabolized by rat liver microsomes to its parent alcohol (98). The vapor inhibits the growth of some bacteria (110) but supports the growth of *Mycobacterium phlei* (88).

The ACGIH TLV (37) and the OSHA PEL (36) is 800 ppm (1900 mg/m$^3$). Isobutane can be isolated and quantified by gas chromatography.

### 1.6 Pentanes

#### 1.6.1 n-Pentane

Pentane, $CH_3(CH_2)_3CH_3$, is a flammable liquid (38) and has been detected in the urban atmosphere from combustion exhausts and natural sources (101). Selected physical properties are given in Table 19.4. It is used as an aerosol propellant, a solvent, an important component of engine fuel, and a raw material for the production of chlorinated pentanes and pentanols (51). Pentane is a CNS depressant, but is not as effective as the $C_1$ to $C_4$ gases. The intensity of CNS depression appears generally to decrease with increasing molecular weight, but increases for the highly symmetrical compounds (111). For humans, pentane's lethal concentration is 130,000 ppm, and toxic effects are noted at 90,000 ppm (112). It is a dermal irritant and is easily aspirated (43). It is a weak cardiac sensitizer of the dog heart to epinephrine (72, 97). Only a small increment in dose separates CNS depression and lethality. Inhalation of 5000 ppm (0.5 percent) pentane for 10 min was not irritating to mucous membranes and did not produce local or systemic effects. Concentrations of 200 to 300 mg/l caused incoordination and inhibition of the righting reflex of mice (113). Pentane is anesthetic to mice at 7 percent concentration in 10 min and 9 percent concentration in 1.3 min (94). A concentration of 128,000 ppm caused deep anesthesia in mice (114), whereas 9 to 12 percent concentration for 5 to 60 min resulted in CNS depression and 12.8 percent for 37 min resulted in death. The intravenous $LD_{50}$ is 446 mg/kg and the 2-hr $LC_{50}$ is 325 mg/m$^3$ (67). Air concentrations of 10.4, 50.9, and 94.7 mg/m$^3$ showed histological changes in the developing cerebral cortex of the rat (115). Doses of 40 percent resulted in death of mice, which showed collapsed lungs on autopsy (116). Chronic exposure has resulted in anoxia (57). $n$-Pentane had a slow inhibitory action on the myelin sheath of the peripheral nerve tissue in the rat (117). A nerve impulse blockage has been demonstrated in the squid axon and in the frog sciatic nerve (118) that could be verified for pentane and hexane (119). Similarly, a solvent mixture of 80 percent pentane, 5 percent hexane, and 14 percent heptane caused polyneuritis (120). Pure pentane does not cause axonopathy (121). Some isolated systemic effects indicate local affinity and destruction of the myelin sheath of peripheral nerve tissue (117). This may be due to hexane contamination rather than pentane itself. Pentane is metabolized by hydroxylation to pentanol (98), conjugated with glucuronate, and subsequently excreted (122). Pentanes and higher alkanes can be used as a sole carbon source by *Pseudomonas fluorescens* and *Corynebcaterium* (123), *Mycobacterium vaccae* (73), *M. phlei* (88), and *Micrococcus lysodeikticus*. Pentane is not mutagenic in the Ames system (124). The aquatic TLm96 ranges between 10 to 100 ppm (66). The OSHA PEL (36) is 600 ppm (1800 mg/m$^3$) and the ACGIH TLV (37) is 600 ppm (1770 mg/m$^3$). The NIOSH recommended exposure limit (REL) for pentane is 120 ppm (350 mg/m$^3$) with a ceiling concentration of 610 ppm (1800 mg/m$^3$) for 15 min (125). Saturated air contains 66 percent (vol./vol.) pentane, at 25°C/760 mm Hg. The odor threshold for pentane is between 6.6 mg/m$^3$ (70) and it exhibits a moderate odor intensity at 5000 ppm (79). It is on the EPA TSCA Chemical Inventory and the Test Submission Data Base (67). An air monitoring procedure for pentane has been established by NIOSH (68). Samples are collected on charcoal, desorbed with carbon disulfide, and quantified using a flame ionization detector. A chromatographic method to isolate and quantify pentane in air mixtures of $C_2$ to $C_5$ hydrocarbons has been developed (75).

### 1.6.2  2-Methylbutane, Isopentane

Isopentane, 2-methylbutane, $(CH_3)_2CHCH_2CH_3$, is a flammable liquid (38) and exhibits physical properties very similar to those of pentane. Isopentanes have been detected in urban air (102–104). Isopentanes are used for the production of amylnaphthalenes and isoprene (51). Selected physical data are given in Table 19.4. As an anesthetic, isopentane is less potent (94) than the shorter-chain alkanes; however, metabolically it appears more active. High vapor concentrations are irritating to the skin and eyes. Inhalation of up to 500 ppm appears to have no effect on humans (79). The 1-hr $LC_{50}$ in the mouse is estimated to be 1000 mg/l. Methylbutane causes CNS depression between 270 and 400 mg/l (58), and is a weak cardiac sensitizer (72). Isopentane is metabolized by hydroxylation to the parent alcohol

(98). The aquatic TLm96 is between 100 to 1000 ppm (66). *Mycobacterium phlei* is capable of growth on isopentane as the sole carbon source (88).

OSHA and ACGIH have not established exposure values; however, ceiling values established for pentane are recommended. Monitoring and analytical methods are similar to those for pentane.

### 1.6.3  2,2-Dimethylpropane, Neopentane

Neopentane, 2,2-dimethylpropane, $C(CH_3)_4$, is a flammable liquid (38) and is physically similar to butane. Neopentane, like pentane, is an important component of petroleum fuel mixtures. Limited toxicity data are available (52). However, based on structure–activity relationship extrapolations from physical, chemical, and biologic data, the toxicity characteristics of neopentane are lower than those of isopentane and pentane. This has been confirmed in experimental work (94). Neopentane is hydroxylated by rat liver microsomes to the parent alcohol (98).

Industrial hygiene information for neopentane is similar to that for isopentane.

### 1.7  Hexanes

#### 1.7.1  n-Hexane

*n*-Hexane, $CH_3(CH_2)_4CH_3$, is a flammable liquid (38) isolated from natural gas and crude oil (51). It is widely used pure or as commercial-grade solvent, which may be *n*-hexane mixed with isohexanes and cyclopentane. Selected physical properties are given in Table 19.4.

Hexane may be the most highly toxic member of the alkanes. It is an anesthetic (118, 119, 126). When ingested, it causes nausea, vertigo, bronchial and general intestinal irritation, and CNS depression; it also presents an acute aspiration hazard (4). About 50 g may be fatal to humans. Acute inhalation effects are euphoria, dizziness, and numbness of limbs (127). An exposure of 880 ppm for 15 min can cause eye and upper respiratory tract irritation in humans (128). Exposure to 2000 ppm was without effect, whereas exposure to 5000 ppm for 10 min caused marked vertigo and nausea (79). The odor of hexane is irritating at 1800 $mg/m^3$ (70). Concentrations of 100 ppm affect the righting reflex in the mouse (113). Chronic industrial exposure has produced motor polyneuropathy in workers (129, 130). Similarly, among 1264 workmen, 32 cases of neuropathy were observed in the shoe manufacturing industry, utilizing hexane–cyclohexane solvent mixtures (131). Among 1662 Japanese workers handling *n*-hexane, 93 cases of polyneuropathy were observed (132). The prognosis for total recovery is good for most patients with hexane-induced axonopathy (133, 134). Also, hexane caused eye lesions in the macula of 11 of 15 Finnish workers exposed 5 to 21 years in an adhesive bandage factory (135). Hexane is an irritant to the skin. Following subcutaneous injections, hexane produces pathological changes in mice similar to those seen in humans (136). The lowest lethal concentration ($LC_{LO}$) for mice is 120 mg/l (113) and the lowest lethal dose ($LD_{LO}$) for rats is 9100 mg/kg (137). Intramuscular injection of hexane in rabbits causes edema and hemorrhaging of the lungs and tissues, with polymorphonuclear leukocytic reactions. The oral $LD_{50}$ in rats is 28.7 mg/kg (138). Dermal application of 2 to 5 ml/kg for 4 hr to rabbits resulted in ataxia and restlessness (139). No deaths occurred at 2 ml/kg, but lethality was noted at 5 ml/kg. Inhalation of 1000 to 64,000 ppm for 5 min in the mouse resulted in irritation of the respiratory tract and anesthesia (114), 30,000 ppm produced CNS depression (4), and 34,000 to 42,000 ppm was lethal (140).

Subchronic exposure of rats for 14 weeks resulted in peripheral distal axonopathy (140). Subchronic exposure of pigeons to 3000 ppm, 5 hr/day, for 82 days over 17 weeks produced no nerve tissue damage (141). However, exposure of rats to 400 to 600 ppm, 24 hr/day, 5 days/week, has resulted in peripheral neuropathy in 45 days (11). Continuous exposure of mice to 100 ppm hexane produced no evidence of peripheral neuropathy in 7 months; however, 250 ppm resulted in peripheral nerve injury (142). Exposure of rats to 850 ppm for 143 days showed loss in weight and degeneration of myelin and axis cylinders in the sciatic nerve (143–145). Degeneration of axons and nerve terminals has also been observed as a result of "glue sniffing" (146). Chronic effects from glue sniffing over a period of 5 to 15 months have been described as distal symmetrical motor sensory polyneuropathy (12, 144, 145).

Hexane is actively absorbed by mammals and accumulates in tissues proportional to the lipid content (147). It is metabolized to hydroxy derivatives (31) by a cytochrome P450-containing mixed function oxidase system (148, 149) before being converted to keto forms. The metabolites 5-hydroxy-2-hexanone and 2,5-hexanedione, also common to other neurotoxic hexacarbons, have been identified (8). Hexane activates various enzyme systems, including UDP-glucuronyl transferase (150). Some normal hydroxylated intermediates are excreted as the glucuronides (122). Hexane is inactive as a tumor-promoting agent (151). The straight-chain hexacarbons have an affinity to nerve tissue, where they affect blockage of nerve impulses in the frog (118) and structural changes of the myelin sheath in the rat (117). Hexane is a reproductive toxin in rats (152) but not in mice (153). It is also a fetotoxin in rats and mice (67). The aquatic TLm96 is greater than 1000 ppm (66). A variety of microorganisms degrade hexane by oxidation mechanisms similar to the lower homologues (31). Hexane is also utilized by the microorganisms *Mycobacterium vaccae* (73) and *M. phlei* (88) and is not mutagenic in the Ames system (154). The OSHA PEL (36) is 50 ppm (180 mg/m$^3$), the ACGIH TLV (37) is a TWA of 50 ppm (176 mg/m$^3$), and the NIOSH REL (125) is 500 ppm (1800 mg/m$^3$). The OSHA PEL, ACGIH TLV, and NIOSH REL for other hexane isomers are 500 ppm with a ceiling of 1000 ppm (36, 37, 125). Hexane is on the EPA TSCA Chemical Inventory and the Test Submission Data Base (67). The concentration in saturated air is 19.94 percent at 25°C and 760 mm Hg. Detailed protection information is available (125).

Air samples are collected on charcoal tubes, the materials are desorbed with carbon disulfide, and the hexane or its isomers are quantified by gas chromatography, using flame ionization detectors (90).

## 1.7.2 2-Methylpentane, Isohexane

Little information is available on isohexane. Selected physical properties are listed in Table 19.4. No physiological data are available. However, the isohexanes are expected to be skin, eye, and mucous membrane irritants and to have a low oral toxicity, but to be aspirated into the lungs and absorbed through the skin. Isohexanes are predicted to be CNS depressants and are cardiac sensitizers (72), but are not expected to have neurotoxic properties. The TLm96 is greater than 1000 ppm (66). The OSHA PEL (36) is 500 ppm (1800 mg/m$^3$), the ACGIH TLV (37) is 500 ppm (1760 mg/m$^3$), and the NIOSH REL is 100 ppm (350 mg/m$^3$) for hexane isomers other than $n$-hexane (125).

## 1.7.3 Other Hexane Isomers

The selected physical data for 2-methylpentane also applies to 3-methylpentane, as listed in Table 19.4. Very little information is available on 2,2-dimethylbutane, neohexane. For toxicity, see 2-methylpentane. Neohexane is a flammable liquid (38); selected physical data are given in Table 19.4.

Selected data as given in Table 19.4 on 2-methylpentane also apply to 2,3-dimethylbutane.

## 1.8 Heptanes

### 1.8.1 n-Heptane

$n$-Heptane, $CH_3(CH_2)_5CH_3$, a flammable liquid (38), occurs in natural gas, crude oil, and pine extracts, and is produced in refining processes. It is used as an industrial solvent, automotive starter fluid, and paraffinic naphtha, and as a gasoline knock-testing standard. Selected physical properties are given in Table 19.4. Humans exposed to 0.1 percent heptane exhibited slight vertigo in 6 min; to 0.2 percent, vertigo in 4 min; and to 0.5 percent, CNS depression in 7 min (79). The lingering taste of gasoline for several hours after exposure has also been reported (79). The odor threshold for heptane is between 200 and 1280 mg/m$^3$ (70). Concentrations of 4.8 percent caused respiratory arrest within 3.0 min (114). Survivors show marked vertigo and incoordination requiring 30 min for recovery. They also exhibited mucous membrane irritation, slight nausea, and lassitude (79). A concentration of 40 mg/l affected the righting reflexes of mice, and 70 mg/l was lethal (113). The 2-hr $LC_{50}$ in mice is 75 mg/l (67). Direct skin contact may cause pain, burning, and itching. The time for symptom reversal is longer than for pentane or hexane (125). CNS depression in mice occurs at 1.0 to 1.5 percent within 30 to 60 min. However, tetanic convulsions occasionally occur (155). Heptane is a weak cardiac sensitizer (97). A narrow margin exists between the onset of CNS depression or convulsions and cardiac sensitization and recovery or death. Heptane is metabolized to its parent alcohol (121), conjugated by glucuronates, and subsequently excreted. Heptane is metabolized at relatively high rates to hydroxy derivatives by a cytochrome

P450-containing mixed function oxidase system before being converted to the corresponding keto forms. Heptane is not neurotoxic, because it forms only small amounts of 2,5-heptanedione (156). Heptane is a lipid solvent, is readily absorbed through the skin, and exhibits systemic effects other than neurotoxicity similar to those of hexane.

Petroleum-adapted organisms biodegrade heptane from 19 to 55 percent in 92 hr, depending on the cultures used (30). *Mycobacterium phlei* is capable of utilizing heptane as the sole carbon source (88), and promotes lipid synthesis by *M. vaccae* (75). The odor threshold is about 200 ppm (79). The aquatic TLm96 is greater than 1000 ppm (66). The OSHA PEL (36) and ACGIH TLV (37) are 400 ppm (1600 mg/m$^3$) and the NIOSH REL (125) is 85 ppm (350 mg/m$^3$). Heptane is on the EPA TSCA Chemical Inventory and the Test Submission Data Base (67). For monitoring and chemical analysis, heptane is quantified using gas chromatography (90).

### 1.8.2  2-Methylhexane, Isoheptane

Isoheptane, $(CH_3)_2CH(CH_2)_3CH_3$, is a flammable liquid (38) often used in fuel mixtures. It is also used as raw material for the production of plasticizers. Selected physical characteristics are given in Table 19.4. Physiologically, isoheptane is somewhat less active than hexane. It is a CNS depressant at 0.8 to 1.1 percent (113). For isoheptane, the loss of the righting reflex in mice occurred at concentrations of 50 mg/l, whereas the loss was observed at 40 mg/l for heptane (113). Thus the isomer appears somewhat less toxic and is not a neurotoxin.

Industrial hygiene information for isoheptane is similar to that for heptane.

### 1.8.3  Other Heptane Isomers

The other branched isomers, such as 3-methylhexane, are expected to have toxic properties similar to those of 2-methylhexane. Selected physical data on neoheptane, 2,2-dimethylpentane, are given in Table 19.4.

## 1.9  Octanes

### 1.9.1  n-Octane

*n*-Octane, $CH_3(CH_2)_6CH_3$, is a flammable liquid (38), and is a component of natural gas and crude oil. It is used as a solvent, a chemical raw material, and an important chemical agent in the petroleum industry. Octane offers desirable blending values that achieve certain antiknock and combustion qualities for high compression engine fuels. Selected physical properties are given in Table 19.4.

Orally, octane may be more toxic than its lower homologues. The odor threshold for octane is between 725 and 1208 mg/m$^3$ and is irritating at 1450 mg/m$^3$ (55, 70, 90). If the material is aspirated into the lungs, it may cause rapid death due to cardiac arrest, respiratory paralysis, and asphyxia (4). It is a dermal irritant (157). The CNS depressant potential of octane is approximately that of heptane, but does not appear to exhibit other CNS effects seen in lower homologues. Octane does not cause axonopathy (155). A concentration of 35 mg/l resulted in the loss of righting reflexes in mice and 50 mg/l caused a total loss of reflexes (113). A concentration of 9.5 percent causes loss of reflexes in mice in 125 min; however, up to 1.9 percent is easily tolerated for 143 min, with the effects being reversible (155). Octane is metabolized to hydroxy derivatives via a cytochrome P450 oxidase system (148). *Pseudomonas oleovarans* (31) and other microorganisms degrade octane (73) by oxidation or hydroxylation to octanol.

The OSHA PEL (36) is 300 ppm (1450 mg/m$^3$), the ACGIH TLV (37) is 300 ppm (1400 mg/m$^3$), and the NIOSH REL (125) is 75 ppm (350 mg/m$^3$). Octane is on the EPA TSCA Chemical Inventory and the Test Submission Data Base (67). Vapor samples are collected on charcoal, desorbed with carbon disulfide, and analyzed by flame ionization gas chromatography (90).

### 1.9.2 Octane Isomers

Isooctanes are blended with lower alkane homologues to form preignition additives for high compression engine fuel (51). Selected physical data for trimethyl isomers are given in Table 19.4. Orally, the isooctanes are more toxic than the lower homologues and are readily aspirated into the lungs of rats (4). 2,5-Dimethylhexane causes CNS depression at 70 to 80 mg/l for the mouse, but is not as effective as octane (113). Following intramuscular injection into rabbits, isooctane produces hemorrhage, edema, and polymorphonuclear leukocytic reactions in the pulmonary tract (158). Specifically, angitis, interstitial pneumonitis, abscess formation, thrombosis, and fibrosis were noted (158). A concentration of 16,000 ppm of isooctane causes respiratory arrest in mice (114). The compounds are metabolized via a cytochrome P450 oxidase system to form hydroxylate isooctanes (31).

2,2,4-Trimethylpentane, 2,3,4-trimethylpentane, and mixed dimethylhexanes are widely used in the chemical industry. Selected physical properties are given in Table 19.4; however, very few toxicity data are presently available. By structure–activity extrapolation, these compounds can be compared to 2,5-dimethylhexane. None of the branched octanes are expected to have neurotoxic properties. *Pseudomonas fluorescens*, *Corynebacterium* (123), and *P. oleovorans* use 2,5-dimethylhexane or 2,2,4-trimethylpentane as their sole carbon source (31). Industrial hygiene practice will be similar to the methodology used for octane.

### 1.10 Nonanes

#### 1.10.1 n-Nonane

Nonane, $CH_3(CH_2)_7CH_3$, is an important component of gasoline (51) and has been detected in urban atmospheres (104). Concentrations of 1.6 to 4.4 ppb of *n*-nonane have been identified in polluted air. Selected physical data are given in Table 19.4.

Few toxicologic data are available. The odor threshold for nonane is 3412 mg/m$^3$ (70). It can be aspirated into the lung following ingestion (43). Nonane is a CNS depressant (40) and a primary skin irritant (43). The 4-hr $LC_{50}$ in rats is 3200 ppm (17 mg/l) (159). No-effect levels of 1.9 and 3.2 mg/l were noted for rats exposed

6 hr/day, 5 days/week for 13 weeks (159). A concentration of 8.1 mg/l (1500 ppm) resulted in mild tremors, slight coordination loss, and low irritation of the eyes and extremities (159). Chronic inhalation of nonane vapors may cause altered neutrophils but no pulmonary lesions were noted (160). It does not cause axonopathy (157). Nonane is used as a carbon source by some bacteria, but prevents spore formation in *Bacillus megaterium* (27). Nonane is metabolized in the rat to hydroxyl derivatives prior to conversion into the corresponding keto form, using a cytochrome P450-containing mixed function oxidase system (148). The OSHA PEL (36) and the ACGIH TLV (37) are 200 ppm (1050 mg/m$^3$). Nonane is on the EPA TSCA Chemical Inventory and the Test Submission Data Base (67). Monitoring and sampling methods are identical to those recommended for octane.

### 1.10.2  2,2,5-Trimethylhexane

Trimethylhexane or neononane, $(CH_3)_3C(CH_2)_2CH(CH_3)_2$, is a flammable liquid (38) and is a component of gasoline. Selected physical data are given in Table 19.4. Little toxicologic information is available. By extrapolation, neononane is expected to be less toxic than nonane. The estimated $TC_{LO}$ is greater than 4000 ppm and the $LD_{LO}$ is 4 g/kg. Industrial hygiene methods are identical to those for octane.

## 1.11  Decanes

### 1.11.1  n-Decane

Decane, $CH_3(CH_2)_8CH_3$, is a flammable liquid (38) and is a component of engine fuel (51). Selected physical and chemical properties are given in Table 19.4. Decane has been detected in urban atmospheres (104), and concentrations of 1 to 2.7 ppb have been identified in polluted air (158). Decane is a simple asphyxiant and causes CNS depression (40). Decane is considered relatively nontoxic (111). The odor threshold for decane is 11.3 mg/m$^3$ (45). It can be aspirated and causes edema and hemorrhaging (157). Rats exposed to 540 ppm of *n*-decane 18 hr/day, 7 days/week for 57 days stimulated weight gains and decreased the total white blood count, but no bone marrow changes or other organ changes were noted. The lowest dermal toxic dose ($TD_{LO}$) for mice is 25 g/kg for 52 weeks (151, 161). The 2-hr $LC_{50}$ in mice is 73.3 mg/l (67). Decane is highly lipid soluble and causes pulmonary pneumonitis when aspirated (4). Decane is readily hydroxylated to decanol by *Pseudomonas desmolytica S11* (162) and *P. oleovorans* (31). Decane is degraded by *Candida tropicalis* via the 1,2-dehydrogenated and 1-hydroxylated intermediates (145). Minor alternate pathways proceed via the 1,2-diol and include terminal oxidation (163). Decane is metabolized to hydroxy derivatives (31) before being converted to the respective keto form (148), using a cytochrome P450-containing mixed function and microsomal oxidase systems or aerobic spore-forming bacteria (164). Neither OSHA nor ACGIH has set exposure standards for decane. It is on the EPA TSCA Chemical Inventory and the Test Submission Data Bank (67).

Industrial hygiene information for decane will be identical to that discussed for octane. Analytical methodology for decane and isodecane is available (35, 165).

## 1.11.2  2,7-Dimethyloctane

2,7-Dimethyloctane, or $(CH_3)_2CH(CH_2)_4CH(CH_3)_2$, diisoamyl, did not cause CNS depression (113). Selected chemical and physical data are given in Table 19.4.

## 1.12  Undecane

### 1.12.1  n-Undecane

*n*-Undecane or hendecane, $CH_3(CH_2(_9CH_3$, is a flammable liquid (38) and is a component of gasoline (51). Physical data are given in Table 19.4. Undecane can be aspirated into the lung following ingestion (4) and does not cause axonopathy (157). Undecane inhibited bacterial spore germination on certain growth media.

## 1.13  Dodecane

Dodecane, $CH_3(CH_2)_{10}CH_3$, is a flammable liquid (38) and is a component of gasoline (51). Selected physical and chemical properties are given in Table 19.4.

Dodecane is not highly toxic but has been shown to be a promoter of skin carcinogenesis for benzo(*a*)pyrene dissolved in decalin applied to the mouse skin (166–168). The odor threshold is 37 mg/m³ (45). The lowest toxic dose ($TD_{LO}$) for mice is 11 g/kg for 22 weeks for the above mixture (162). A mixture of 50 percent dodecane and 50 percent decalin provides a suitable carrier for strongly carcinogenic agents for the skin of mice (169). Dodecane can be used as the sole carbon source by *Pseudomonas fluorescens*, *Corynebacterium* (123), and *P. oleovorans* (31). In *Candida* species, it may inhibit carbohydrate transport (170). Neither OSHA nor ACGIH has set exposure standards for dodecane.

## 1.14  Tridecane and Higher Homologues

Hexadecanes and other higher homologues are used as stock for hydrocracking processes (106). Selected physical and chemical properties are given in Table 19.4 for *n*-tridecane ($C_{13}$), *n*-tetradecane ($C_{14}$), *n*-pentadecane ($C_{15}$), *n*-hexadecane ($C_{16}$), *n*-heptadecane ($C_{17}$), *n*-octadecane ($C_{18}$), *n*-nonadecane ($C_{19}$), pristane ($C_{19}$), and *n*-eicosane ($C_{20}$) (49). Pristane is used as a lubricant and transformer oil, and is oxidatively degraded by *Pseudomonas oleovorans* (31). The lowest toxic dose ($TD_{LO}$) of tetradecane for mice is 9600 mg/kg for 20 weeks (151). The $C_{13}$ to $C_{16}$ alkanes, when aspirated into the lungs, are asphyxiants similar to the $C_6$ to $C_{10}$ members (4). These alkanes cause death more slowly and can cause chemical pneumonitis (4).

Topical applications of hexadecane to the guinea pig skin caused marked hyperkeratosis and increased acid phosphatase (171). Conversely, hexadecane in combination with 2-butanone or cyclohexane potentiated local anesthetics (172). Exposure to high vapor or mist concentrations of squalene ($C_{30}$) results in lung changes (116). Hexadecane is readily degraded through mineralization by marine bacteria (173). Following oral gavage of 1 g to rats, heptadecane was found in concentrations of 0.7, 1.4, 1.2, and 0.2 percent in the intestinal wall, liver, intestinal content, and feces, respectively (174). In rats fed a diet of 25 percent *Spirulina algea*, heptadecane accumulated in adipose tissue at 80.2 and 272.0 µg/g in males and females and in lung and muscle at ~10 to 20 µg/kg, respectively (175). Dietary concentrations of 52 ppm fed to pigs for 12 months resulted in the excretion of some heptadecane in milk during lactation (175). The higher homologues of the alkanes have been detected as oil droplets in a variety of human tissues (176), whereas $C_{20}$ and higher alkanes occur naturally in a variety of waxes (39). Varying with the season, $C_{14}$ to $C_{35}$ alkanes have been isolated from milk fat (177).

*Candida lipolytica* (178) and *Nocardia corralina* (179) can use alkanes as their sole carbon source. Tetra- and hexadecanes can be utilized by *Pseudomonas* strains for the synthesis of phenazine carboxylic acids (180) and a *Candida* strain is capable of using tetradecane as the sole carbon source (181). In *Candida tropicalis*, alkane, alcohol, and aldehyde dehydrogenase-dependent synthetases have been detected (181). Cocarcinogenic activity may be common to many $C_{12}$ to $C_{30}$ aliphatic hydrocarbons (169). Tetradecane and hexadecane can be metabolized by a cytochrome P450 mixed function oxidase system or hydroxylated by an enzyme from *Pseudomonas oleovorans* (31) with the $C_{12}$, $C_{14}$, and $C_{16}$ *n*-paraffins hydroxylated by *P. aeruginosa* (31). Hexadecane is degraded via mineralization by a variety of oceanic bacteria (182). Using radioactive tracers, 2-methylhexadecane, 2,2-dimethylhexadecane, and pristane have been demonstrated to undergo microbial oxidation readily (183). Heptadecane is readily taken up into mussel (184). The alkanes to $C_{17}$ can be collected on charcoal and the higher members as particulate matter on filters, desorbed with carbon disulfide, and quantified using gas chromatography (35).

## 2 ALKENES (OLEFINS)

### 2.1 General Correlations

#### 2.1.1 Occurrence and Use

Olefins that are utilized in industry are primarily produced by petroleum cracking processes or by pyrolysis from paraffins (106). These alkenes serve for the preparation of a wide variety of petrochemicals (106) that are extensively utilized in the pharmaceutical, cosmetic, chemical, and rubber processing industries. Small quantities of the lower alkenes have been detected in urban atmospheres (102). In a comprehensive environmental sampling and identification study (158), only isobutene was found to occur in higher quantities among a variety of 98 alkyl and aromatic hydrocarbon derivatives. Considerable quantities of lower alkenes were detected at an experimental diesel engine source; however, they had almost vanished at a 6-ft distance from the source (91). Lower alkenes were also detected in diesel fume exhaust (185).

## ALIPHATIC HYDROCARBONS

### 2.1.2 Physical and Chemical Properties

Alkenes are chemically more reactive than alkanes (39), primarily through addition reactions across the double bonds. When heated or in the presence of catalysts, most olefins will polymerize. They have higher boiling points than the parent paraffins as shown in Table 19.5 (40–56). They are slightly more toxic than the alkanes, and readily react with ozone and other photochemical species to form chemiluminescent and other oxidation products (186).

### 2.1.3 General Toxicity

Alkenes are only slightly more toxic than alkanes. Ethylene, propylene, butylene, isobutylene, and butadiene are weak anesthetics and simple asphyxiants. Pentene has been used for surgical anesthesia. Because of increasing mucous membrane irritancy and cardiac effects with increasing chain length, the hexylenes and higher members are unsuitable as anesthetic agents. The higher members may cause CNS depression (187), but are not sufficiently volatile to be considered vapor hazards at room temperature. Branching decreases the toxicity of $C_3$ alkenes, does not appreciably affect the $C_4$ and $C_5$ alkenes, and increases the toxicity of $C_6$ to $C_{18}$ alkenes. Unlike the hexanes, the olefins do not produce axonopathy. Repeated exposure to high concentrations of the lower members of the alkenes results in hepatic damage and hyperplasia of the bone marrow in animals. However, no corresponding effects have been noted in humans. Alpha olefins are more reactive and toxic than beta isomers. The alkadienes are more irritant and generally more toxic than the corresponding alkanes (188).

Alkenes can be aspirated into rat lungs to cause death due to chemical pneumonitis (4). The rate of death appears practically parallel between alkanes and alkenes of equal carbon chain length. Some unsaturated compounds can undergo addition or oxidation reactions to form adducts causing plant leaf damage, which is mild for the ethylenes to butylenes, and decene to tetradecene, but more serious for isobutylene and the pentenes to nonenes (189).

A variety of selective mycobacteria have been found to biodegrade alkenes.

### 2.1.4 Industrial Hygiene

Industrial hygiene procedures for alkenes and chemical analyses are similar to those for the alkanes. Specific procedures and chromatographic methods are available for quantifying selected alkenes.

### 2.2 Ethene (Ethylene)

Ethene or ethylene, $CH_2:CH_2$, is a colorless, flammable (38), and explosive gas, with a faintly sweet odor, and is strikingly reactive. Selected physical properties are given in Table 19.5. It occurs up to 4 percent in illuminating gas and is used as a plant hormone in ripening fruit (49). It has been detected in urban atmospheres at very low levels (190), but is more prevalent in large metropolitan areas (77).

Table 19.5. Physiocochemical Properties of Alkenes (40–56)

| Compound | B.P. (°C) | CAS Registry No. | Density (20/4°C) | Emp. Formula | Flammability Limits (%) | Flash Pt. (°C) | Freezing Pt. (°C) |
|---|---|---|---|---|---|---|---|
| Ethene | −103.71 | 74-85-1 | 0.5674 | $C_2H_1$ | 2.7–36 | −136.11 | −169.15 |
| Propene | −47.4 | 115-07-1 | 0.5139 | $C_3H_6$ | 2.0–11.0 | −108 | −185.25 |
| Butene-1 | −6.26 | 25167-67-3 | 0.5951 | $C_3H_8$ | 1.6–10 | −112 | −185.35 |
| cis-Butene-2 | 3.72 | 390-18-1 | 0.6213 | $C_4H_8$ | 1.7–9.0 | −73.3 | −138.91 |
| trans-Butene-2 | 0.88 | 624-64-6 | 0.6042 | $C_4H_8$ | 1.8–9.7 | −73.3 | −105.550 |
| Isobutene | −6.90 | 115-11-7 | 0.5942 | $C_4H_8$ | 1.8–9.6 | −76.11 | −140.3 |
| Pentene-1 | 29.97 | 109-67-1 | 0.6405 | $C_5H_{10}$ | 1.5–8.7 | −51.11 | −165.22 |
| cis-Pentene-2 | 36.94 | 627-20-3 | 0.6556 | $C_5H_{10}$ | | | −151.39 |
| trans-Pentene-2 | 36.35 | 646-04-8 | 0.6482 | $C_5H_{10}$ | | | −140.244 |
| Hexene-1 | 63.485 | 592-41-6 | 0.6732 | $C_6H_{12}$ | 1.2–6.9 | −26.11 | −139.82 |
| cis-Hexene-2 | 68.84 | 592-43-8 | 0.6869 | $C_6H_{12}$ | | −20.6 | −141.13 |
| trans-Hexene-2 | 67.87 | 592-43-8 | 0.6784 | $C_6H_{12}$ | | −17.8 | −132.97 |
| Isohexane-2 | 67.29 | | 0.6863 | $C_6H_{12}$ | | | −135.07 |
| Heptane-1 | 93.64 | 592-76-7 | 0.6930 | $C_7H_{11}$ | | −3.89 | −119.03 |
| cis-Heptene-2 | 97.95 | | 0.7012 | $C_7H_{11}$ | | −2.2 | −109.48 |
| trans-Heptene-3 | 95.67 | | 0.6981 | $C_7H_{11}$ | | −6.1 | −136.63 |
| Octene | 121.28 | 111-66-0 | 0.7149 | $C_8H_{16}$ | | 21.11 | −101.74 |
| Nonene | 146.67 | 124-11-8 | | $C_9H_{18}$ | | 26 | −81.11 |
| Decene | 170.56 | 872-05-9 | 0.7408 | $C_{10}H_{20}$ | | 48.89 | −66.3 |
| Dodecene | 213.4 | 25378-22-7 | 0.7584 | $C_{12}H_{21}$ | | 100 | −35.23 |
| Hexadecene | 284.4 | 629-73-2 | 0.7811 | $C_{16}H_{32}$ | | | 4.1 |
| Octadecene | 179 (15) | 112-88-9 | 0.7891 | $C_{18}H_{36}$ | | 100 | 17.5 |
| Propadiene | −34.5 | 463-49-0 | 1.787 | $C_3H_4$ | 2.1-lel | −136.6 | |
| 1,2-Butadiene | 10.85 | | 0.652 | $C_4H_6$ | 2–12 | | −139.90 |
| 1,3-Butadiene | −4.413 | 106-99-0 | 0.6211 | $C_4H_6$ | 2.0–11.5 | | −108.92 |
| Isoprene | 34.07 | 78-79-5 | 0.6810 | $C_5H_8$ | 1.5–8.9 | −54 | −145.94 |
| 1,3-Hexadiene | 73.0 | | 0.7050 | $C_6H_{10}$ | 2.0–6.1 | −21 | |
| 1,5-Hexadiene | 59.5 | | 0.6923 | $C_6H_{10}$ | 2.0–6.1 | −21 | −140.68 |
| 1,4-Heptadiene | 93 | | 0.7270 | $C_7H_{12}$ | | | |
| 1,7-Octadiene | | | | $C_8H_{14}$ | | | |
| Squalene | 280 | 111-02-4 | 0.8584 | $C_{30}H_{50}$ | | | −20 |
| Lycopene | | | | $C_{40}H_{56}$ | | | 175 |
| β-Carotene | | 432-70-2 | 1.00 | $C_{40}H_{56}$ | | | 188 |

[a]Solubility in water/alcohol/ether: v = very soluble; s = soluble; d = slightly soluble; i = insoluble.
[b]25°C.
[c]Atmospheric pressure.

Ethene has been detected in cigarette smoke at levels between 0.093 and 0.097 percent (105). Ethene is prepared by cracking of ethane and propane and other petroleum gases, and by catalytic dehydration of ethanol (106). Ethene is one of the most important industrial raw materials for a variety of chemicals, petrochemicals, polymers, and resins. It is used as a fruit and vegetable ripening agent and has been used as a surgical anesthetic (191).

The toxicity of ethene is not remarkable. Concentrations of less than 2.5 percent are systemically inert. However, very high concentrations may cause CNS depression, unconsciousness, and asphyxia due to oxygen displacement (192).

Ethene is a useful anesthetic agent. Its advantages over comparable human anesthetics are rapid onset and recovery time after exposure termination (191) with little or no effect on cardiac and pulmonary functions. Respiration, blood pressure, and pulse rates are rarely changed, even under anesthetic conditions. Cardiac

| Mol. Wt. | Refractive Index, $n_D^{20}$ | Solubility[a] w/al/et | Sp. Gr. (25°C) | Vapor Density | Vapor Pressure[b] (mm) (°C) | Viscosity (SUS) | W/V Conversion (mg/m³ = 1 ppm) |
|---|---|---|---|---|---|---|---|
| 28.054 | — | i/d/s | — | 0.987 | 40[c] (1.5) | N.A. | 1.15 |
| 42.08 | — | i/v/v | 0.522 | 1.46 | 10[c] (19.8) | N.A. | 1.72 |
| 56.11 | 1.3962 | i/v/v | 0.6013 | 1.93 | 3480 (21) | N.A. | 2.30 |
| 56.11 | 1.3931[b] | i/v/v | 0.6272 | 1.9 | 760 (3.7) | N.A. | 2.30 |
| 56.11 | 1.3848[b] | — | 0.6100 | 1.9 | 760 (0.9) | N.A. | 2.30 |
| 56.11 | 1.3926[b] | i/v/v | 0.6002 | 2.01 | 400 (21.6) | | 2.30 |
| 70.134 | 1.3715 | i/v/v | 0.6457 | 2.42 | 400 (12.8) | <32.6 | 2.87 |
| 70.134 | 1.3830 | i/v/v | 0.668 | 2.4 | | | 2.87 |
| 70.134 | 1.3793 | i/v/v | 0.6533 | 2.4 | | | 2.87 |
| 84.161 | 1.3879 | i/s/s | 0.6780 | 3.0 | 310 (35) | <32.6 | 3.44 |
| 84.161 | 1.3977 | i/s/s | 0.6916 | 2.9 | | | 3.44 |
| 84.161 | 1.3935 | i/s/s | 0.6832 | 3.0 | | | 3.44 |
| 84.161 | 1.4004 | i/s/— | | | | | 3.44 |
| 98.188 | 1.3988 | i/s/s | 0.7015 | 3.39 | 101.4 (21.1) | <32.6 | 4.02 |
| 98.188 | 1.4045 | i/s/s | 0.7057 | 3.34 | | | 4.02 |
| 98.188 | 1.4043 | i/s/s | 0.7075 | 3.38 | | | 4.02 |
| 112.21 | 1.4087 | i/v/s | 0.719 | 3.9 | 36 (38) | <32.6 | 4.59 |
| 126.24 | | | 0.733 | 4.35 | 12 (38) | <32.6 | 5.16 |
| 140.268 | 1.4215 | i/v/v | 0.745 | 4.84 | 1 (95.7) | <32.6 | 5.74 |
| 168.312 | 1.4300 | i/s/s | 0.70 | 5.81 | 1 (47.2) | <32.6 | 6.88 |
| 224.429 | 1.4412 | i/s/s | | | | 35.5 | 9.18 |
| 252.482 | 1.4448 | i/—/— | 0.79 | 0.71 | | Solid | 10.33 |
| 40.0646 | 1.4168 | i/d/v | 0.657 | 1.40 | 10 (33) | | 1.66 |
| 54.0914 | 1.4205 | i/v/v | 0.658 | 1.9 | 760 (18.5) | | 2.21 |
| 54.0914 | 1.4292 | i/s/s | 0.6272 | 1.87 | 1840 (21) | | 2.21 |
| 68.118 | 1.4219 | i/v/v | 0.6861 | 2.35 | 2 (15.3) | | 2.79 |
| 82.145 | 1.4380 | i/s/s | 0.710 | 2.8 | | | 3.36 |
| 82.145 | 1.4042 | i/s/s | 0.6970 | 2.8 | | | 3.36 |
| 96.172 | 1.4370 | i/—/s | | | | | 3.93 |
| 110.199 | | | | | | | 4.51 |
| 410.725 | 1.4990 | i/d/s | | | | <32.6 | 16.80 |
| 536.882 | | i/d/s | | | | | 21.96 |
| 536.882 | | i/d/s | | | | | 21.96 |

arrhythmias occur infrequently and ethene has little effect on renal and hepatic functions (192); however, cyanosis often occurs (193). Mice repeatedly exposed at concentrations causing CNS depression showed no histopathological changes in kidneys, adrenals, hearts, or lungs (194). However, the disadvantage as an anesthetic is its explosion and flammability properties (195).

Humans exposed to ethene may experience subtle signs of intoxication, resulting in prolonged reaction time (196). Exposure at 37.5 percent for 15 min resulted in marked memory disturbances (192). Humans exposed to as much as 50 percent ethene in air, whereby the oxygen availability is decreased to 10 percent, experience loss of consciousness (197). Prolonged inhalation of 85 percent ethene in air is slightly toxic, whereas 94 percent in oxygen is fatal (43). Death is certain at 8 percent oxygen (197). Therefore, ethene used as an anesthetic agent should be supplemented with the appropriate oxygen concentration, which, however, increases the potential for combustion. Dogs exposed to 1.4 percent ethene were rapidly anesthetized (191). Ethene equilibrated in alveolar, arterial, brain, and muscle tissue in 2 to 8.2 min, even more rapidly than ethyl ether (191).

Male rats exposed to 10,000, 25,000, or 57,000 ppm for 4 hr exhibited increased serum pyruvate levels and liver weights (198). The liver mitochondrial volume increased in rats treated with ethene (199). Rats exposed to 90 percent ethene and 10 percent oxygen are anesthetized in about 20 min and exhibit slight nervous symptoms (200). Deep anesthesia occurs in a few seconds with 95 percent ethene and 5 percent oxygen, accompanied by marked cyanosis and depression with a slow fall in blood pressure (198). Ethene is not a cardiac sensitizer in the dog (72). It is nonirritating to the skin and eyes (58).

In a chronic study, 1-day-old and adult rats continuously exposed to 3 $mg/m^3$/day for 90 days exhibited hypertension, disruption of the subordination chronaxy, and decreased cholinesterase activity (201). Conversely, ethene has speeded up the wound healing process in muscle injuries of mice (202). In rats exposed to 0, 300, 1000, or 3000 ppm 6 hr/day, 5 days/week for up to 24 months, no lesions attributable to ethene were noted (203). Metabolically, ethene may alter carbohydrate metabolism and has caused temporary hypoglycemia (204). Additionally, a reduction of inorganic phosphates has been noted (58). The aquatic TLm96 is between 100 to 1000 ppm (66). Ethene was not mutagenic in *E. coli* or several *Bacillus* species (89). It is a plant hormone, effective at concentrations as low as 60 $mg/m^3$ (205, 206). At higher concentrations, it inhibits plant metabolism (207).

A TLV of 1000 ppm, as for other simple asphyxiants, has been recommended by ACGIH (37). The odor threshold for ethene is 299 $mg/m^3$ (70). Ethene is on the EPA TSCA Chemical Inventory and the Test Submission Data Base (67). NIOSH (90) has published analytic methods for industrial hygiene monitoring and sampling. Ethene can be analytically determined using chemiluminescent reactions with ozone (1985). Owing to the highly flammable and explosive properties, extreme caution is advised when handling ethene. Also, skin and eye contact with the compressed gas should be prevented.

### 2.3 Propene (Propylene)

Propene, propylene, or methylethylene, $CH_2:CHCH_3$, is a colorless, flammable, practically odorless gas (38). Commercially, it is available in liquefied form with trace impurities of lower alkanes and alkenes (208). Selected physical properties are given in Table 19.5. Propene occurs in refined petroleum products (106) and is produced in the petroleum cracking process (51). Propene has been detected in cigarette smoke (105) and is released in diesel engine exhaust fumes (91). It occurs in ambient air of metropolitan areas, and has been detected in smog samples of Los Angeles (188), St. Louis (77), and Tokyo (209). It occurs naturally in bananas and apples (210) but also occurs in ocean sediments as a microbial degradation product (211).

Propene is highly reactive and is utilized as raw material for a variety of processes, including plastics manufacture (51, 208). It is a common feedstock for the production of gasoline (51) and synthetic rubber (39) and is used as an aerosol propellant (212). Early investigations indicated its possible use as an anesthetic (213), and it was found twice as potent as ethene (214). Propene is of generally low toxicity and acts as a simple asphyxiant and mild anesthetic (49, 214). The vapor is nonirritating to the skin, but the liquefied product may cause burns from direct contact (192). Propene has been used in dental surgery as a temporary anesthetic (215). At a concentration of 6.4 percent for 2.25 min, mild intoxication, paresthesias, and inability to concentrate were noted (58). At 12.8 percent in 1 min, symptoms were markedly accentuated and at 24 and 33 percent, unconsciousness followed in 3 min (58, 216). Human exposure to 23 percent propene for 3 to 4 min did not result in unconsciousness (217). Two subjects exposed to 35 and 40 percent propene vomited during or after the exposure, and one complained of vertigo (214). Exposure to 40, 50, or 75 percent for a few minutes caused initial reddening of the eyelids, flushing of the face, lacrimation, coughing, and sometimes flexing of the legs (215). No variations in respiratory or pulse rates or electrocardiograms were noted (215). A concentration of 50 percent prompted anesthesia in 2 min, followed by complete recovery (213).

Cats do not exhibit any toxic signs when anesthesia is induced with propene concentrations of 20 to 31 percent (213). Subtle effects occur from exposure to 40 to 50 percent, blood pressure decrease and rapid pulse at 70 percent, and the unusual ventricular ectopic beat from 50 to 80 percent (213). A concentration of 40 percent produced light anesthesia in rats, but no toxic symptoms within 6 hr (200). Exposure to 55 percent for 3 to 6 min, 65 percent for 2 to 5 min, or 70 percent for 1 to 3 min resulted in deep anesthesia with no CNS signs or symptoms (200). However, propene was found to be a cardiac sensitizer in the dog (72). Chronic exposure of mice to concentrations causing CNS depression resulted in moderate to very slight fatty degeneration of the liver, but somewhat less than caused by ethylene (218). Exposure of rats and mice to 0, 200, 1000, or 5000 ppm 7 hr/day, 5 days/week for 18 to 24 months did not reveal any carcinogenic effects in either species (43). In another study with exposures to up to 10,000 ppm, rats exhibited non-neoplastic lesions in the nasal cavity (219). Propene is metabolized via hydration to the alcohol and is excreted as the conjugated alcohol or propionic acid. These reactions may progress in a similar manner as recorded in *Pseudomonas oleovorans*; propene is hydroxylated but does not form the epoxide (220). Propene is not mutagenic in *E. coli* and protects against mutation (89). Propene exhibits a stimulating effect on plant growth at low concentrations but is inhibitory at high levels (189). ACGIH (37) recommends a TLV of 1000 ppm as for propene and other simple asphyxiants. Propene is on the EPA TSCA Chemical Inventory and the Test Submission Data Base (67). The odor threshold for propene is 39.6 mg/m$^3$ and the odor is described as aromatic (70). A light sensitivity to the eyes following exposure to approximately 1 mg/m$^3$ has been reported (221). Analytical methods for propene are available (35).

Because propene is highly flammable and explosive, the compressed gas or the liquid should be handled with all necessary precautions.

## 2.4 Butenes

The mono-unsaturated butenes occur in various forms, saturated in either position 1 or 2, that is, alpha or beta, and as linear or branched derivatives. The β-unsaturated compound can occur in cis or trans configuration, the trans being the more stable form. All derivatives are chemically more reactive but the α-butenes to a greater extent than the comparable butane. Selected physical data are given in Table 19.5. The butenes exhibit toxicity properties similar to butane. They are simple asphyxiants and can be used as anesthetics (196).

### 2.4.1 1-Butene

1-Butene, butylene, α-butylene, or ethylethylene, $CH_2:CHCH_2CH_3$, is a colorless, flammable gas (38) with a slightly aromatic odor (52). Selected physical properties are given in Table 19.5. Butene occurs as a petroleum refining by-product (51) and has been detected in diesel exhaust (91). It is not detected in the urban atmosphere, probably owing to its chemical reactivity (91). When thermally reacted, butene pyrolyzes to $C_1$ and $C_2$ alkanes, toluene, and several $C_5$ and $C_6$ cyclanes, or cyclenes (106). Butylene is produced by cracking of petroleum products. It is used in the production of a wide variety of chemicals in gasoline and rubber processing industries (39). It is very reactive and readily undergoes addition reactions. It is recovered from catalytically cracked petroleum oil (51) and is mainly used for the production of polymers (222). Butene is liberated from tetra- and tributyllead or tin during oxidative dealkylation by hepatic microsomes (223) and has been detected in human expired air (76). 1-Butene is a simple asphyxiant and classified as nontoxic (38). At concentrations above the flammability range, it is an anesthetic (224). It has a low acute toxicity and is mildly irritating to the eye. On direct eye and skin contact liquid butene can cause burns and frostbite. As an anesthetic, it is 4.5 times more potent than ethylene (58). Exposure of mice to concentrations of 15 percent butene resulted in reversible signs of incoordination, confusion, and hyperexcitability; at 20 percent deep anesthesia in 8 to 15 min, with subsequent respiratory failure in 2 hr; and at 30 percent in 2 to 4 min and 40 min, respectively (200). A concentration of 40 percent resulted in profound anesthesia in 30 sec, with no CNS symptoms but with death in 10 to 15 min (200). Butene is metabolized slowly through its 1-hydroxy derivative (225). In plants, 1-butene has been recorded to cause slight leaf damage (189). Butene supports spore germination in *Neurospora crassa* (64) and growth of some *E. coli* strains (26) but inhibits spore germination of *Bacillus megaterium* (27).

Threshold concentrations of 0.4 percent (4000 ppm) have been suggested to protect adequately for the lower explosive level (226). Neither OSHA nor ACGIH has set exposure levels for butene, but butene is on the EPA TSCA Chemical Inventory and the Test Submission Data Base (67). Odor thresholds in air of 55 mg/m$^3$ (92), 15.4 mg/m$^3$ (221), and 0.93 to 1.3 ppm (92) have been reported. Chemical identification by spectral methods is available (35). Butene should be handled with care, owing to its flammability and reactivity with other chemicals and polymerization potential (39, 206).

## 2.4.2 2-Butene

2-Butene, β-butylene, butylene, or dimethylethylene, $CH_3CH:CHCH_3$, can occur in trans or cis conformation, and is also known as pseudobutylene (52). It is a colorless, flammable gas (80). Selected physical properties are given in Table 19.5. It has been recovered from refining gases or produced by petroleum cracking (51). It is a component in the production of gasolines, butadiene, and a variety of other chemicals (51). 2-Butene has been detected in diesel exhaust (91). The more highly reactive *trans*-2-butene occurs at much lower concentrations in the atmosphere than other comparable hydrocarbons (227). Concentrations of 13 to 13.5 percent (300 to 400 mg/l) causes deep CNS depression in mice and about 19 percent (120 to 420 mg/l) is fatal (58). It is a mild mucous membrane irritant (58). It is a cardiac sensitizer in dogs (72). 2-Butene has been detected in exhaled air. In the majority of subjects, concentrations of the cis form were greater than trans isomer (76).

Neither OSHA nor ACGIH have set exposure levels for butene; however, butene is on the EPA TSCA Chemical Inventory and the Test Submission Data Base (67). The odor threshold for 2-butene is generally 0.05 to 0.059 mg/l, and specifically is 4.8 mg/m$^3$ for the *trans*-2 isomer (92). Analytical methods for butene-2 have been described (35).

## 2.4.3 Isobutene

Isobutene, α-butylene, isobutylene, or 2-methylpropene, $CH_2:C(CH_3)_2$, is a highly volatile, flammable liquid (38, 226). Selected physical properties are given in Table 19.5. It has been detected in the urban atmosphere at low concentrations (103). Isobutene is used as a monomer or the formation of a copolymer for the production of synthetic rubber (51) and various plastics (49). Isobutene is produced in refinery streams by absorption on 65 percent $H_2SO_4$ at about 15°C, or by reacting with an aliphatic primary alcohol and then hydrolyzing the resulting ether (49). Approximately 95 percent of the commercial isobutene is used to produce dimers, trimers, butyl rubber, and other polymers. The other 5 percent is used to produce antioxidants for food or food packaging and for plastics (49).

The toxicity of isobutene is similar to that of other lower alkenes. It is a simple asphyxiant and causes CNS depression at higher concentrations. At 30 percent, it produces no CNS depression in mice, and excitement and CNS depression in 7 to 8 min at 40 percent, but immediate CNS depression in 2 to 2.25 min at 50 percent, or in 50 to 60 sec at 60 to 70 percent (58). The 2-hr $LC_{50}$ in the mouse is 415 mg/l and the 4-hr $LC_{50}$ in the rat is 620 mg/l (95).

In metabolic studies on rats and mice, inhaled isobutene levels in the brain and parenchymatous organs were similar, but the level in the fatty tissue was significantly higher than in the brain, liver, kidneys, or spleen. There was a linear relationship between the degree of CNS depression and the cerebral concentrations (93, 96). The effects of butane and butene were additive (95).

Isobutene is nonhazardous below the lower flammability limit. Except for asphyxia at high concentrations for prolonged periods, no untoward effects of isobutene on the health of workers have been reported. The industrial hygiene measures for isobutene are similar to those for butene. Exposure standards have not been established for isobutene (37). A colorimetric method for the determination of isobutene without interference by ethene or propene has been developed.

## 2.5 Pentenes (Amylenes)

### 2.5.1 1-Pentene

1-Pentene, α-$n$-amylene, 1-amylene, or propylethylene, $CH_2:CH(CH_2)_2CH_3$, is a flammable liquid (38) that occurs in coal tar (49) and in petroleum cracking mixtures (106). It has a highly disagreeable odor and polymerizes on extended periods of storage. Selected physical properties are given in Table 19.5.

The anesthetic action of pentene appeared about 15 times more potent than that of ethene (217). However, at CNS depression levels, it causes more severe primary excitement (217). It produces anesthesia at 6 percent in 15 to 20 min, but is more cardiotoxic than the lower homologues (58). At one time it was used without success as a human anesthetic but produced better results in dentistry (196). In animals, pentene causes respiratory and cardiac depression and the primary excitation observed in humans (58). Pentene is oxidized at the double bond and excreted as the alcohol or its conjugate (58). 1-Pentene, but not 2-pentene, causes specific smoglike damage to agricultural plants (189).

Exposure standards have not been established for pentenes (36, 37). The threshold odor of pentene is 0.54 to 6.6 mg/m$^3$ and is 0.19 ppm for 1-pentene (92). 1-Pentene is on the EPA TSCA Chemical Inventory (67).

### 2.5.2 2-Pentene

2-Pentene, β-$n$-amylene, 2-amylene, or ethylmethyl ethylene, $CH_3CH:CHCH_2CH_3$, is similar to 1-pentene. 2-Pentene is produced by petroleum cracking and is a component of high octane fuel (106). The odor threshold is between 0.54 and 6.6 mg/m$^3$ (92) and 2-pentene is on the EPA TSCA Chemical Inventory (67).

### 2.5.3 Isopentenes

1-Isopentene is also called α-isoamylene and 3-methyl-1-butene, $CH_2:CHCH(CH_3)_2$. 2-Isopentene is β-isoamylene or 2-methyl-2-butene, $(CH_3)_2C:CHCH_3$. Selected properties are given in Table 19.5. More toxicity data are available for 2-isopentene.

2-Isopentene at 5.4 percent in mice produced incoordination, marked hyperexcitability, and light anesthesia in 20 min, with a tendency toward convulsions. At 6.12 percent, it produced deep anesthesia in 6 to 15 min, with respiratory failure 1 hr post-exposure (200). At 6.12 percent for 30 to 45 min, convulsions and death resulted (200).

CNS depression is observed at concentrations of 112 mg/l and the minimal lethal concentration is 212 mg/l for mice (113). β-Amylenes have been identified as cilia toxins, which may play a role towards their respiratory paralytic action (228).

Exposure standards have not been established for isopentene. The odor threshold is between 0.54 and 6.6 mg/m$^3$ (70).

## 2.6 Hexenes

### 2.6.1 1-Hexene

1-Hexene, butylethylene, α-hexylene, or 1-hexylene, $CH_2:CH(CH)_3CH_3$, is a colorless liquid, which is highly volatile and flammable (38). Selected physical properties are given in Table 19.5. Hexene is produced by olefin cracking (106) and is used primarily for organic syntheses and in fuels. Hexene is a component of coffee aroma (229). Hexene is a low to moderate irritant to the skin and eyes. When ingested it presents a moderate aspiration hazard (4). When inhaled, it produces CNS depression in humans at a concentration of about 0.1 percent, with accompanying mucous membrane irritation, vertigo, vomiting, and cyanosis. The minimal concentration causing CNS depression in mice was 2.9 percent or 29,100 ppm, and the minimal fatal concentration was 4.08 percent or 40,800 ppm (113). In pollution simulation research, 1-hexene has been shown to cause smoglike damage to various agricultural plants (189).

Exposure standards have not been set. Hexene is on the EPA TSCA Chemical Inventory and the Test Submission Data Base (67).

### 2.6.2 2-Hexene

2-Hexene, β-hexylene, 2-hexylene, or methylpropylethylene, $CH_3CH:CH(CH_2)_2CH_3$, is a highly flammable liquid (38). Selected physical properties are given in Table 19.5. Its toxicity resembles that of 1-hexene.

### 2.6.3 Isohexene

Isohexene, β-isohexylene, or 2-methylpent-2-ene, $(CH_3)_2C:CHCH_2CH_3$, is a component of cigarette smoke and is classified as a cilia toxin to the respiratory tract (228). Selected physical properties are given in Table 19.5. Analytic methods for quantifying all hexene isomers are available (35).

## 2.7 Heptenes

1-Heptene, α-heptylene, or pentylethylene, $CH_2:CH(CH_2)_4CH_3$, is a colorless, flammable liquid (38). Selected physical properties are given in Table 19.5. The 2- and 3-heptenes occur in the cis and trans configurations. 2-Heptene is β-heptylene and 3-heptene is γ-heptylene. The general toxicity of the heptenes is somewhat lower than that of the hexenes, except that its aspiration hazard is increased (4). Heptene is a liver toxin and destroys cytochrome P450 (230). In mice, concentrations of 60 mg/l cause loss of the righting reflex (113). CNS depression and lethality have been noted at 60 and above 200 mg/l, respectively (58). Heptene is not mutagenic in the Ames system (231). A plant physiological study showed that the 3-heptene caused greater damage on leaves than 1-heptene (189).

## 2.8 Octenes and Higher Alkenes

1-Octene or α-octylene, $CH_2:CH(CH_2)_5CH_3$; 2-octene, or β-octylene, $CH_3CH:CH(CH_2)_4CH_3$; isooctene or 2-methyl-2-heptene, $(CH_3)_2C:CH(CH_2)_3CH_3$;

the trimethylpentenes; and 2-ethyl-1-hexene, $CH_2:C(C_2H_5)(CH_2)_3CH_3$, are flammable liquids (38). Selected physical properties are given in Table 19.5. The trimethylpentenes have a gasoline-like odor (44). In humans, they cause headache, inability to pay sustained attention, vertigo, nausea, and CNS depression (232). Octenes may be more irritant to mucous membranes, skin, and eyes than the lower homologues. Octenes, when ingested, may rapidly be aspirated into the lungs (40, 232) and may act as a simple asphyxiant (44). The 4-hr $LC_{50}$ in the rat is 4000 ppm for 2-ethyl-1-hexene (233). All octenes are metabolically hydroxylated, conjugated, and excreted from the mammalian system (58). The aquatic TLm96 range is between 100 and 1000 ppm (66). Octene is less damaging to plant leaves than hexene and heptene (189). The odor threshold for 1-octene is 2.0 ppm (92). Exposure standards have not been set for the higher alkenes. Various analytic determination procedures are available.

Nonene is an aspiration hazard when ingested (4). It is irritating to the skin, eyes, nose, and throat, and at high concentrations, it is an anesthetic (44). The aquatic TLm96 is greater than 1000 ppm (66). The isomer 4-nonene is less active toward plant leaves than the lower homologues, whereas the higher members do not cause any damage (189). 1-Decene or $n$-decylene, $CH_2:CH(CH_2)_7CH_3$, is a colorless, flammable liquid (38). It is irritating to the eyes and respiratory tract (44) and has CNS depressant properties (52). Decene and dodecene also may present an aspiration hazard when ingested (4). The odor of decene is pleasant (44) and the odor threshold of 1-decene in air is 7 ppm (92). Selected physical properties are given in Table 19.5.

Analytic methods for the determination of decene and dodecene are available. Hexa- and octadecene have been found in land and ocean natural deposits and sediment cores (180). The long-chain olefins, 9-tricosene, and related compounds have insect attractant and psychedelic propeties (234).

### 2.9 Propadiene

Propadiene, allene, or dimethylenemethane, $CH_2:C:CH_2$, is a colorless, flammable (38), and unstable gas with a sweetish odor. Selected physical properties are given in Table 19.5. Propadiene is produced in small quantities in petroleum cracking processes. It is of low general toxicity, but may cause CNS depression at very high concentrations (226). In mice, concentrations of 20 percent caused restlessness and CNS depression in 11 min, 30 percent caused CNS depression in 3 min, and 40 percent in 1 to 2 min (58). For industrial use, adequate precautions in regard to inhalation and skin exposure are required (226).

### 2.10 Butadiene

Butadiene, biethylene, 1,3-butadiene, α-butadiene, divinyl, erythrene, 1-methylallene, or vinylethylene, $CH_2:CHCH:CH_2$, is a colorless, flammable gas (38). It is potentially explosive when mixed with air and has a characteristic, aromatic odor (208). Selected physical properties are given in Table 19.5. Butadiene is formed in tobacco smoke (235). It has been produced from ethanol or butane, but is primarily formed from butene or by catalytic cracking of light oil or naphtha (49). It does not occur naturally. Several enzyme systems biosynthesize acetoin, which can be converted to butadiene by chemical action (236). Butadiene is used for the preparation of a variety of chemicals, but is the major component for the production of synthetic rubber, often in combination with styrene (237). Butadiene is highly reactive, dimerizes to form 4-vinylcyclohexene, and polymerizes easily (106). Therefore polymerization inhibitors have to be added for its storage and transport (208).

The acute and subchronic toxicity of butadiene is of low order and is not cumulative. It has anesthetic action and at very high concentrations causes CNS depression, respiratory paralysis, and death. Human signs and symptoms from overexposure are blurred vision, nausea, and prickling and dryness of the mouth, throat, and nose, followed by fatigue, headache, vertigo, nausea, decreased blood pressure and pulse rate, unconsciousness, and respiratory paralysis. No hematologic disorders have been attributable to butadiene (238).

At lower concentrations, it causes slight irritation to the skin and eyes. Dermatitis among butadiene workers appears to represent a secondary effect of additives, accelerators, or inhibitors (239, 240). On direct dermal contact, the liquefied product causes burns and frostbite. Humans exposed to 1000 ppm showed no irritant effects (241), whereas 2000 to 4000 ppm resulted in mild irritation of the eyes and difficulty in focusing on instrument scales (241). Voluneers exposed to 8000 ppm for 8 hr also showed no effects other than mild irritation of the eyes and upper respiratory tract, and 10,000 ppm for 5 min showed no effects on blood pressure or respiration (208).

The oral $LD_{50}$ in the rat is 5480 mg/kg (242), and in the mouse is 3210 mg/kg (242). The lowest lethal concentration in the rabbit is 250,000 ppm, which caused death in 23 min (241). The 4-hr $LC_{50}$ in rats is 285 mg/l (67) and 250 mg/l in mice (67). In mice exposed to concentrations of 10 percent, butadiene caused no symptoms, of 15 percent light CNS depression, of 20 percent some excitement and CNS depression in 6 to 12 min, and of 30 to 40 percent excitement with twitching in 1 to 1.2 min and 40 to 60 sec, respectively (58). As a CNS depressant, it is more potent than propadiene, but only half as potent as butene. Rabbits exposed to 25 percent (250,000 ppm) butadiene exhibited light, relaxed anesthesia in 1.6 min, followed by loss of various reflexes, CNS effects, and death in 23 min (241). Rats gavaged with 100 mg/kg butadiene for 2.5 months showed some lymphohistocytic infiltration in the heart, liver, and kidney (243). Rats, guinea pigs, and a dog exposed to butadiene at 600, 2300, and 6700 ppm for 7.5 hr/day, 6 days/week for 8 months showed slightly retarded growth at the highest dose, but no blood dyscrasias were found (241). Butadiene causes thyroid tumors in rats and mice (244) and is classified as a 2B carcinogen by IARC (245). In distribution studies, butadiene was found at higher concentrations in the medulla oblongata than in the cerebellum or cerebral cortex (93, 95).

Most literature sources reflect that toxicologic tests were conducted on butadiene as commonly used in mixtures or copolymers, such as styrene. Some combinations have shown added irritancy. In children playing with butadiene-mineral or clay mixtures, dermatitis or specific rashes developed within a few days post-exposure (224). In pollution studies, butadiene produces plant-leaf damage similar to that produced by $C_5$ to $C_8$ linear and branched monoolefins (189).

Currently, the OSHA PEL (36) is 1000 ppm (2200 mg/m$^3$) for butadiene, the ACGIH TLV (37) is 10 ppm (22 mg/m$^3$), and the NIOSH REL (125) is the lowest feasible concentration due to carcinogenic concerns. Butadiene is on the EPA TSCA Chemical Inventory and the Test Submission Data Base (67). NIOSH has published a monitoring procedure (90) and various analytic methods are available for environmental and clinical samples (241). Handling procedures should be strictly followed, because its boiling point is 23.54°F. The greatest danger exists when it is mixed with air to the explosive limit of 2 percent. The use of protective garments and equipment is required.

### 2.11 Isoprene

Isoprene or 2-methyl-1,3-butadiene, $CH_2:C(CH_3)CH:CH_2$, is a colorless, volatile, flammable liquid (38). It is highly reactive, usually occurs as its dimer (246), and unless inhibited undergoes explosive polymerization (54). Selected physical properties are given in Table 19.5. Industrially, isoprene is produced by dehydrogenation in the oxo process from isopentene or by high-temperature thermal cracking of petroleum oil (51), gas oils, and naphthas (39). It is used in the manufacture of butyl and synthetic rubber, plastics, and a variety of other chemicals (51).

Isoprene also has been detected in the vapor phase of tobacco smoke (247) and has been detected in a number of biologic systems, mammalian tissues, and plant bacteria, probably as a phosphate or pyrophosphate derivative (246) or as a radical. Isoprene possesses two reactive centers and can combine linearly or three-dimensionally to form polyenes, cyclics, aromatics, and diverse polymers. Much literature is available on its di- and polymerization to farnesene ($C_{10}$), squalene ($C_{30}$), and a variety of $C_{40}$ compounds, such as the lycopenes and carotenes. Therefore the isoprene unit is the most important building block for lipids, steroids, terpenoids, and a wide variety of natural products, including latex, the raw material for natural rubber (236, 248). In tobacco smoke, isoprene has been determined to be the precursor of a number of polycyclic aromatics, as demonstrated by thermal condensations in the range of 450 to 700°C (235).

The toxicity of isoprene is similar to that of butadiene and butylene (239), although it is more irritating than comparable alkenes or alkanes of similar volatility (188). At low concentrations, except as a result of cigarette smoking and in some occupational connections, few toxic effects have been documented, although at high concentrations it is a CNS depressant and asphyxiant (224). Isoprene has been detected in human expired air at 15 to 390 µg/hr in smokers and from 40 to 250 µg/hr in nonsmokers (76). Twenty percent of inhaled isoprene was absorbed in the upper respiratory tract and a total of 70 to 99 percent was retained in the lungs (249, 250). In human volunteers, the average odor threshold occurred at 10 mg/m$^3$, and slight irritation of the upper respiratory mucosa, larynx, and pharynx was noted at 160 mg/m$^3$ (251). Isoprene is also effective in reducing the tracheal mucous flow (252).

In rabbits 0.19 to 0.75 mg/l caused an increased respiratory rate (251). At 109 mg/l for 2 hr, rats lost the righting reflex (251). The 2-hr LC$_{50}$ values are 148 mg/l for the female and 139 mg/l for the male mouse (67). At 24 hr post-exposure survivors in both groups exhibited improved swimming time (251). Enlarged lungs were found in mice that died during the study (251). The 4-hr LC$_{50}$ in rats is 180 mg/l (67). In mice, the no-effect level was 20,000 ppm, deep CNS depression was seen at 35,000 to 45,000 ppm, and death was noted at 50,000 ppm (253). Similar results showed loss of the righting and total reflexes in the mouse at 120 mg/l but no deaths at 200 mg/l (113). Mice and rabbits repeatedly exposed to 2.2 to 4.9 mg/l, 4 hr/day for 4 months, and rats for 5 months, showed no weight differential from the control. However, after the third month, oxygen consumption in rats decreased. In rabbits there were increased numbers of leukocytes, decreased numbers of erythrocytes, and some increased organ weights (251).

Metabolic studies with isoprene (96) suggest that concentrations in the rat brain indicate the relative toxic potency for a given exposure. When applied to mouse skin, isoprene reduced the number of papillomas, but at a much lower rate than retinyl acetate (254). Occupational exposures to concentrations above the maximum permissible concentrations have allegedly resulted in CNS and cardiac alterations (255) and subtle immunologic changes (256). However, these signs and symptoms may be due to a variety of other materials utilized and produced in the isoprene rubber industry (257). The aquatic TLm96 is between 10 and 100 ppm (66).

No exposure standards have been established, although a maximum allowable concentration of 40 mg/m$^3$ has been suggested (251). Isoprene is on the EPA TSCA Chemical Inventory and the Test Submission Data Base (67). The odor threshold is 5 mg/m$^3$ (92). Environmental samples are collected on charcoal and desorbed with carbon disulfide and quantified by coulometric titration (35).

## 2.12 Hexadiene and Higher Alkadienes

1,3-Hexadiene or ethylbutadiene, $CH_2:CHCH:CHCH_2CH_3$, is a flammable liquid (38) with an odor threshold of 2.0 ppm (92). 1,4-Hexadiene or allyl allene, $CH_2:CHCH_2CH:CHCH_3$, and 1,5-hexadiene or diallyl, $(CH_2:CHCH_2)_2$, are colorless, flammable liquids (38). Selected physical properties are given in Table 19.5. Analytic methods are available.

The odor threshold of 1,4-heptadiene, $CH_2:CHCH_2CH:CHCH_2CH_3$, is 9.0 ppm (92) and that of 1,3-, 1,4-, and 2,4-octadiene is 2.0, 1.5, and 1.2 ppm, respectively (92). For 1,7-octadiene, the oral LD$_{50}$ in the rat is 19.7 ml/kg (258). The dermal LD$_{50}$ is 14.1 ml/kg and the saturated vapor caused death in 15 min (258). It was moderately irritating to the rabbit skin and produced low corneal injury (258).

## 2.13 Alkatrienes and Polyenes (Tri- and Polyolefins)

The alkatrienes or triolefins are also named cumulenes, after cumulene, which is the simplest compound of the series. Higher homologues of this class are squalene

($C_{30}H_{48}$), lycopene, and carotenes ($C_{40}H_{60}$). These occur naturally in animals, plants, and lower organisms. Therefore their toxicity at low concentrations is negligible. The oral $LD_{50}$ in mice is 5 g/kg (67). Squalene is on the EPA TSCA Chemical Inventory (67). Selected physical properties are given in Table 19.5.

## 3 ALKYNES

### 3.1 General

The alkynes do not exert any acute local toxicity. The lower members are anesthetics and cause CNS depression. They are practically nonirritating to the skin, but cause pulmonary irritation and edema at very high concentrations. The higher-molecular-weight members can be aspirated into the lungs when ingested.

### 3.2 Acetylene

Acetylene, ethine, ethyne, or narcylene, CH:CH, is a colorless, highly flammable (38), and explosive gas. Selected physical properties are given in Table 19.6 (40–56). Acetylene is prepared from calcium carbide and occasionally contains phosphine and arsine as impurities up to 95 and 3 ppm, respectively (259, 260). These impurities account for the etheral to garlic-like odor and its secondary toxic effects (187, 208). Acetylene is highly reactive and forms explosive mixtures with oxygen, chlorine, and fluorine (197). When heated, it undergoes explosive, exothermic reactions (197). Acetylene has been detected at concentrations of 0.03 ppm in U.S. urban smog (102). In Japan, acetylene has been detected in the air near industrial and suburban areas (261). Various plant and bacterial systems can reduce and inactivate acetylene through their nitrogen-fixing mechanisms (262, 263). An important industrial raw material, acetylene is used to produce solvents and alkenes, which in turn serve as monomers in plastic production (51). Acetylene is also utilized in brazing, cutting, flame scarfing, and metallurgical heating and hardening, and

**Table 19.6.** Physicochemical Properties of Alkynes (40–56)

| Compound | B.P. (°C) | CAS Registry No. | Density (20/4°C) | Emp. Formula | Flammability Limits (%) | Flash Pt. (°C) | Freezing Pt. (°C) |
|---|---|---|---|---|---|---|---|
| Acetylene | −84.00 | 74-86-2 | 0.6208 | $C_2H_2$ | 2.5–100 | | −81.0 |
| Propyne | −23.22 | 74-99-7 | 0.7062 | $C_3H_4$ | 2.4–11.7 | | −102.7 |
| 1-Butyne | 8.07 | 107-00-6 | 0.650 | $C_4H_6$ | | −28.9 | −125.72 |
| 2-Butyne | 27.0 | 503-17-3 | 0.6910 | $C_4H_6$ | 1.4 lel | | −32.26 |
| 1-Pentyne | 40.18 | 627-19-0 | 0.6901 | $C_5H_8$ | | | −105.7 |
| Isopentyne | 26.35 | | | $C_5H_8$ | | | −89.7 |
| 1-Decyne | 174.10 | 764-43-2 | 0.7655 | $C_{10}H_{18}$ | | | −44.0 |
| Butadiyne | 10.3 | 463-12-8 | 0.7364[b] | $C_4H_2$ | | | |
| 1,6-Heptadiyne | 112.0 | 2396-63-6 | 0.8164 | $C_7H_8$ | | | −85.0 |

[a]Solubility in water/alcohol/ether: v = very soluble; s = soluble; d = slightly soluble; i = insoluble.
[b]0/4°C.
[c]Atmospheric pressure.

in the glass industry (187). In optometry, it is a component in contact lens coatings (264). Like ethylene, acetylene is used to ripen fruit (265–267) and mature rubber trees (268) or flowers (266).

Acetylene is nontoxic below its lower explosive limit of 2.5 percent. At higher concentrations, it has anesthetic action and at higher levels is a simple asphyxiant (57). In the 1920s, acetylene was used as an anesthetic, because it afforded immediate recovery without after effects (196). However, owing to its explosive characteristics, it now finds only limited application as an anesthetic. It causes CNS depression at slightly higher concentrations than for ethylene (196). Humans can tolerate an exposure of 100 mg/l for 30 to 60 min (208). Marked intoxication occurs at 20 percent, incoordination at 30 percent, and unconsciousness at 35 percent in 5 min. There is no evidence that tolerable levels have any deleterious effects on health (253, 269), although two deaths (270, 271) and a near fatality at 40 percent (224) have been recorded, which occurred during acetylene manufacturing using calcium carbide. The intoxications have been attributed to the phosphine and arsine impurities in crude acetylene (40, 43). Mammals have shown tolerance at 10 percent, intoxication at 25 percent, and death at 50 percent in 5 to 10 min (272). Rodents exposed to 25, 50, or 80 percent acetylene in oxygen for 1 to 2 hr daily, up to 93 hr, showed no organ weight changes or cellular injuries (269). A rise in blood pressure was noted in cats exposed to an 80 percent acetylene–oxygen mixture.

The NIOSH REL for acetylene is 2500 ppm (2662 mg/m³) as a ceiling (274), whereas ACGIH (37) treats acetylene as a simple asphyxiant. Acetylene is on the EPA TSCA Chemical Inventory and the Test Submission Data Base (67). The odor threshold for acetylene is 657 mg/m³ (70). Acetylene detection tubes are available, and acetylene can be quantified by gas chromatography to the parts per billion level. Various scrubber systems for acetylene have been designed. The handling of acetylene warrants special attention in view of its wide-range flammability and capability of forming explosive mixtures (239). Separate storage in a cool, well-ventilated area is advised (43).

| Mol. Wt. | Refractive Index, $n_D^{20}$ | Solubility[a] w/al/et | Sp. Gr. (25°C) | Vapor Density | Vapor Pressure[b] (mm) (°C) | Viscosity (SUS) | W/V Conversion (mg/m³ = 1 ppm) |
|---|---|---|---|---|---|---|---|
| 26.038 | 1.00051 | d/d/— | 0.65 | 0.91[c] | 40[c] (16.8) | N.A. | 1.07 |
| 40.065 | 1.3863 | d/v/— | | 1.38 | 3876[c] (20) | N.A. | 1.64 |
| 54.091 | 1.3962 | i/s/s | | | 760 (2.4) | N.A. | 2.21 |
| 54.091 | 1.3921 | i/s/s | 0.69 | 1.86 | 760 (8.7) | | 2.21 |
| 68.118 | 1.3852 | i/v/v | | | | | 2.79 |
| 68.11 | 1.3723 | i/v/v | | | | N.A. | 2.79 |
| 138.252 | 1.4265 | i/s/s | | | | | 5.65 |
| 50.057 | 1.4189 | v/s/v | | | | | 2.05 |
| 92.140 | 1.451 | i/—/— | | | | | 3.77 |

## 3.3 Propyne

Propyne or methylacetylene, $CH_3C:CH$, is a colorless, highly flammable (38), and explosive gas. Selected physical data are given in Table 19.6. Like acetylene, propyne is used as a welding torch fuel (275). It is a highly reactive material, but its toxic properties are minimal. Propyne has been detected in exhaled air at 0.81 μg/hr in nonsmokers, and at 1.1 and 2.3 μg/hr in smokers (76). It is 18 times as potent as acetylene in causing CNS depression (196). It is a simple anesthetic and in high concentrations is an asphyxiant (40).

The OSHA PEL is 1000 ppm (1650 mg/m$^3$) (36) and the ACGIH TLV is 1000 ppm (1640 mg/m$^3$) (37). The gas should be handled with caution, as described for acetylene.

## 3.4 Higher Alkynes

Practically no toxicity data are available on butynes and above. Selected physical properties are given in Table 19.6 for 1-butyne, 2-butyne, 1-pentyne, isopentyne, 1-decyne, butadiyne, and 1,6-heptadiyne. Some symmetrical butadiynes have fungistatic effect (276). Isopentyne or 3-methylbutyne, $(CH_3)_2CHC:CH$, has caused loss of righting reflexes in mice at 150 mg/l and was fatal at 250 mg/l (58, 113). The $C_6$ to $C_{10}$ alkynes, when ingested, may present aspiration hazards (4).

Oral $LD_{LO}$ values of 350 and 639 mg/kg (277) have been recorded for 1-buten-3-yne. For 1,6-heptadiyne, $CH:C(CH_2)C:CH$, the oral $LD_{50}$ is 2300 mg/kg in the rat, 2620 mg/kg in the rabbit, and 3830 mg/kg in the dog (278). The taste threshold of 1-decyne is 0.1 ppm, and odor recognition occurs at 4 ppm in air (92). Some analytic procedures are available for these high alkynes.

## REFERENCES

1. J. Rosenberg, "Solvents," *Occupational Medicine*, J. LaDou, Ed., Appelton & Lange, Norwalk, CT, 1990, pp. 359–386.
2. L. R. Goldfrank, A. G. Kulberg, and E. A. Bresnitz, "Hydrocarbons," *Goldfrank's Toxicologic Emergencies*, 4th ed., L. R. Goldfrank, N. E. Flourenbaum, N. A. Lewin, R. S. Weisman, and M. A. Howland, Eds., Appleton & Lange, Norwalk, CT, 1990, pp. 759–780.
3. P. D. Bryson, *Comprehensive Review of Toxicology*, 2nd ed., Aspen Publishers, Inc. Rockville, MD, 1989, pp. 553–563.
4. H. Gerarde, *Arch. Environ. Health*, **6**, 329 (1963).
5. Y. Yamamura, "*n*-Hexane polyneuropathy," *Folia Psychiatr. Neurol.*, **23**, 45–57 (1969).
6. S. Sanagi, Y. Seki, K. Siigemoto, and M. Hirata, *Int. Arch. Occup. Environ. Health*, **47**, 69–79 (1980).
7. N. Allen, J. R. Mendell, J. Billmaier, R. E. Fontaine, and J. O'Neill, "Toxic Polyneuropathy due to Methyl *n*-Butyl Ketone: an Industrial Outlook," *Arch. Neurol.*, **32**, 209–222 (1980).

8. G. D. DiVincenzo, C. J. Kaplan, and J. Dedinas, *Toxicol. Appl. Pharmacol.*, **36**, 511–522 (1976).
9. P. S. Spencer, H. H. Schaumburg, M. I. Sabri, and B. Veronesi, *CRC Crit. Rev. Toxicol.*, **7**, 279–356 (1976).
10. F. L. Cavender, H. W. Casey, H. Salem, D. G. Graham, J. A. Swenberg, and E. J. Gralla, *Fundam. Appl. Toxicol.*, **4**, 191–201 (1984).
11. H. H. Schaumburg and P. S. Spencer, *Brain*, **99**, 182 (1976).
12. H. Altenkirsch, J. Mager, G. Stoltenburg, and J. Helmbrecht, *J. Neurol.*, **214**(2), 137 (1977).
13. J. K. Dunnick, D. G. Graham, R. S. Yang, S. B. Haber, and H. R. Brown, *Toxicology*, **57**, 163–172 (1989).
14. J. S. Bus, E. L. White, P. J. Gillies, and C. S. Barrow, *Drug Metab. Dispos.*, (1981).
15. G. Egan, P. Spencer, H. Schaumburg, K. J. Murray, M. Bischoff, and R. Scale, *Neurotoxicology*, **1**, 515–524 (1980).
16. W. J. Krasavage, J. L. O'Donoghue, G. D. DiVincenzo, and C. J. Terhaar, *Toxicol. Appl. Pharmacol.*, **52**, 433–441 (1980).
17. F. L. Cavender, H. W. Casey, H. Salem, J. A. Swenberg, and E. J. Gralla, *Fundam. Appl. Toxicol.*, **3**, 264–270 (1983).
18. D. G. Graham, D. C. Anthony, K. Boekelheide, N. A. Maschmann, R. G. Richards, J. W. Wolfram, and B. R. Shaw, *Toxicol. Appl. Pharmacol.*, **64**, 415–422 (1982).
19. L. M. Sayre, C. M. Shearson, T. Wongmongkolrit, R. Medori, and P. Gambetti, *Toxicol. Appl. Pharmacol.*, **84**, 36–44 (1986).
20. M. B. Genter, Gy. Szakal-Quin, D. W. Anderson, D. C. Anthony, and D. G. Graham, *Toxicol. Appl. Pharmacol.*, **87**, 351–362 (1987).
21. A. P. DeCaprio, N. L. Strominger, and P. Weber, *Toxicol. Appl. Pharmacol.*, **68**, 297–307 (1983).
22. M. B. G. St.Clair, V. Amarnath, M. A. Moody, D. C. Anthony, C. W. Anderson, and D. G. Graham, *Chem. Res. Toxicol.*, **1**, 179–185 (1988).
23. J. M. Batterskill, H. P. A. Illing, R. O. Shillaker, and A. M. Smith, Her Majesty's Stationery Office, U.K., 1987.
24. D. C. Anthony and D. G. Graham, "Toxic Response of the Nervous System," *Casarett and Doull's Toxicology*, M. O. Amdur, J. Doull, and C. D. Klassen, Eds., 4th ed., Pergamon Press, New York, 1991, pp. 407–429.
25. K. Pennington and R. Fuerst, *Arch. Environ. Health*, **22**, 476 (1971).
26. R. Fuerst and S. Stephens, *Dev. Ind. Microbiol.*, **53**, 32 (1965).
27. L. J. Rode and J. W. Foster, *Microbiology*, **53**, 32 (1965).
28. C. E. ZoBell, *API Proc.—Joint Conf. Prev. Control Oil Spills*, 1969, p. 317.
29. A. C. Van Der Linden and R. Juybregtse, *Antonie van Leeuwenhoek J. Microbiol. Serol.*, **33**(4), 382 (1967).
30. H. I. Kator, C. H. Oppenheimer, and R. J. Miget, "Microbial Degradation," *API Proc.—Joint Conf. Prev. Control Oil Spills*, 1971, p. 287.
31. E. J. McKenna and M. J. Coon, *J. Soil Chem.*, **245**(15), 3882 (1970).
32. R. J. Miget, C. H. Oppenheimer, H. I. Kator, and P. A. LaRock, *API Proc.—Joint Conf. Prev Control Oil Spills*, 1969, p. 327.
33. J. R. Haines and M. Alexander, *Appl. Microbiol.*, **28**(6), 1084 (1974).

34. T. Nagata, T. Kojima, and S. Makisumi, *Jap. J. Legal Med.*, **25**, 439 (1971).
35. L. Meites, *Handbook of Analytical Chemistry*, 1st ed., McGraw-Hill, New York, 1963.
36. "Occupational Health and Environmental Control," Subpart G, Sec. 1910.93, "Air Contaminants," *Fed. Reg.*, **39**(125), 23540 (1974).
37. *Threshold Limit Values for Chemical Substances and Physical Agents and Biological Exposure Indices for 1991–92*, American Conference of Governmental Industrial Hygienists, Cincinnati, OH, 1991.
38. "Hazardous Materials Regulations and Miscellaneous Amendments," Title 49, Chapter 1. Materials Transportation, Bureau of Transportation, *Fed. Reg.*, **41**(252), 57018 (1976).
39. C. R. Noller, *Chemistry of Organic Compounds*, 3rd ed., W. B. Saunders, Philadelphia, 1966.
40. N. I. Sax and R. J. Lewis, Sr., Eds., *Hawley's Condensed Chemical Dictionary*, 11th ed., New York, Van Nostrand Reinhold, 1987.
41. E. W. Flick, *Industrial Solvents Handbook*, 3rd ed., Noyes Publications, Park Ridge, NJ, 1985.
42. J. H. Kuney, Ed., *Chemcyclopedia 90*, American Chemical Society, Washington, DC, 1990.
43. International Labor Office, *Encyclopedia of Occupational Health and Safety*, Vols. I & II, Inernational Labor Office, Geneva, 1983.
44. U.S. Coast Guard, Department of Transportation, *CHRIS—Hazardous Chemical Data*, Vol. II, U.S. Government Printing Office, Washington, DC, 1984–1985.
45. K. Verschueren, *Handbook of Environmental Data of Organic Chemicals*, 2nd ed., Van Nostrand Reinhold, New York, 1983.
46. Hazardous Substance Data Bank, TOXNET System, National Library of Medicine, Bethesda, MD, 1992.
47. D. R. Lide, Ed., *CRC Handbook of Chemistry and Physics*, 71st ed., Boca Raton, FL, CRC Press, 1990–1991.
48. *Laboratory Waste Disposal Manual*, 2nd ed., revised, Manufacturing Chemists Association, Inc. Washington, DC, 1974.
49. S. Budavari, Ed., *The Merck Index*, 11th ed., Merck and Company, Rahway, NJ, 1989.
50. *Registry of Toxic Effects of Chemical Sustances*, U.S. Dept. Health, Education, and Welfare, Cincinnati, OH, September, 1991.
51. V. B. Guthrie, Ed., *Petroleum Products Handbook*, McGraw-Hill, New York, 1960.
52. N. Sax, *Dangerous Properties of Industrial Materials*, 7th ed., Van Nostrand Reinhold, New York, 1987.
53. *Toxic and Hazardous Industrial Chemicals Safety Manual for Handling and Disposal with Toxicity Data*, International Technical Information Institute, Tokyo, Japan, 1988.
54. *Fire Protection Guide on Hazardous Chemicals*, 10th ed., National Fire Protection Association, Quincy, MA, 1991.
55. *Handbook of Organic Industrial Solvents*, 4th ed., American Mutual Insurance Alliance, Chicago, 1972.
56. F. Rossini, K. Pitzer, R. Arnett, R. Braun, and G. Pimentel, *Selected Values of Physical Thermodynamic Properties of Hydrocarbons and Related Compounds*, Carnegie Press, Pittsburgh, PA, 1953.

57. Y. Henderson and H. W. Haggard, *Noxious Gases*, 2nd ed., Reinhold, New York, 1943.
58. W. R. von Oettingen, *Toxicity and Poential Dangers of Aliphatic and Aromatic Hydrocarbons*, Publ. Health Bull. No. 255, 1940.
59. D. Balasubramanian and D. B. Wetlaufer, *Physiology*, **55**, 762 (1956).
60. E. A. Wahrenbrock, E. I. Eger, R. G. Laravuso, and G. Maruschak, *Anesthesiology*, **40**(1), 19 (1974).
61. A. C. Carles, T. Kawashire, and S. Pueper, *Arch. Ges. Physiol.*, **359**, 209 (1975).
62. R. W. Dougherty, J. J. O'Toole, and M. S. Allison, *Proc. Soc. Exp. Biol. Med.*, **124**, 1155 (1967).
63. S. S. Kappus, S. Qureshi, and R. Fuerst, *Dev. Ind. Microbiol.*, **15**, 397 (1974).
64. S. Stephens, C. DeSha, and R. Fuest, *Dev. Ind. Microbiol.*, **12**, 356 (1971).
65. A. J. Lawrence, M. B. Kemp, and J. R. Quayle, *Biochem. J.*, **116**, 631 (1970).
66. K. W. Hann and P. A. Jensen, *Water Quality Characteristics of Hazardous Materials*, Vols. 1–4, Environmental Engineering Division, Civil Engineering Department, Texas A&M University, 1974.
67. Registry of Toxic Effects of Chemical Substances, On-Line, TOXNET System, National Library of Medicine, Bethesda, MD, 1992.
68. A. G. Kim and L. J. Douglas, *U.S. Natl. Tech. Inf. Serv. PB Rep. No. 22157514*, 1973, 13 pp.
69. U. Koester, D. Albrecht, and H. Kappus, *Toxicol. Appl. Pharmacol.*, **41**, 639 (1977).
70. J. H. Ruth, *Am. Ind. Hyg. Assoc. J.*, **47**, A-142–A-151 (1986).
71. A. H. Nuckolls, *Underwriters Laboratory Report No. 2375*, Nov. 13, 1933.
72. J. C. Krnatz, Jr., C. J. Carr, and J. F. Vitcha, *J. Pharmacol. Exp. Therap.*, **94**, 315 (1948).
73. J. R. Vestal and J. J. Perry, *Can. J. Microbiol.*, **17**, 445 (1970).
74. B. Voleski and J. E. Zajic, *Appl. Microbiol.*, **21**(4), 614 (1971).
75. H. H. Westberg, R. A. Rasmussen, and M. Holdren, *Anal. Chem.*, **46**(12), 1852 (1974).
76. J. P. Conkle, B. J. Camp, and B. E. Welsh, *Arch. Environ. Health*, **30**(6), 290 (1975).
77. S. L. Kopczynski, W. A. Lonneman, T. Winfield, and R. Seila, *J. Air, Pollut. Control Assoc.*, **25**(3), 251 (1975).
78. H. Eyer, *Muench. Med. Wochenschr.*, **114**(13), 10 (1972).
79. F. A. Patty and W. P. Yant, *Odor Intensity and Symptoms Produced by Commercial Propane, Butane, Pentane, Hexane and Heptane Vapor*, reprinted from the U.S. Bureau of Mines Report of Investigation No. 2979 (1929).
80. FDA's GRAS List—*Substances Generally Accepted as Safe*—CFR Title 21, 121 (1991).
81. T. Ikoma, *Nichidai Igaku Zasshi*, **31**(2), 71 (1972).
82. T. A. Fawcett, R. E. Moon, P. J. Frocica, G. Y. Mebrane, D. R. Theil, and C. A. Piantadosi, *J. Occup. Med.*, **34**, 12–15 (1992).
83. W. E. Brown and V. E. Henderson, *J. Pharmacol. Exp. Ther.*, **27**, 1 (1925).
84. D. M. Aviado, S. Zakhari, and T. Watanabe, *Non-fluorinated Propellants and Solvents for Aerosols*, CRC Press, Cleveland, OH, pp. 49–81 (1977).
85. D. M. Aviado, *Toxicology*, **3**, 321 (1975).
86. D. M. Aviado and D. G. Smith, *Toxicology*, **3**, 241 (1975).

87. J. R. Vestal and J. J. Perry, *J. Bacteriol.*, **99**(1), 216 (1969).
88. W. E. O'Brien and L. R. Brown, *Dev. Ind. Microbiol.*, **9**, 389 (1967).
89. M. M. Landry and R. Fuerst, *Dev. Ind. Microbiol.*, **9**, 370 (1968).
90. *NIOSH Manual of Analytical Methods*, 3rd ed., Vols. 1 and 2, U.S. Dept. HEW, U.S. Printing Office, Washington, DC, 1985.
91. M. C. Battigelli, *J. Occup. Med.*, **5**(1), 54 (1963).
92. W. H. Stahl, Ed., *Compilation of Odor and Taste Threshold Values Data*, American Society for Testing and Materials, Philadelphia, 1973.
93. B. B. Shugaev, *Arch. Environ. Health*, **18**, 878 (1969).
94. R. W. Stoughton and P. D. Lamson, *J. Pharmacol. Exp. Ther.*, **58**, 74 (1936).
95. B. B. Shugaev, *Farmakol. Toksikol.*, **30**, 102 (1967).
96. B. B. Shugaev, *Farmakol. Toksikol.*, **31**, 360 (1968).
97. C. F. Reinhardt, A. Azar, M. E. Maxfield, P. E. Smith, and L. S. Mullin, *Arch. Environ. Health*, **22**, 265 (1971).
98. U. Frommer, V. Ulrich, and H. J. Staudinger, *Z. Physiol. Chem.*, **351**, 913 (1970).
99. K. Watanabe and S. Takesue, *Enzymologia*, **41**, 99 (1971).
100. I. Sunshine, Ed., *CRC Handbook of Analytical Toxicology*, The Chemical Rubber Company, Cleveland, Ohio, 1969.
101. G. Holzer, H. Shanfield, A. Zlatkis, W. Bertsch, P. Juarez, H. Mayfield, and H. M. Liebich, *J. Chromatogr.*, **142**, 755 (1977).
102. A. P. Altshuller and T. A. Bellar, *J. Air Pollut. Control Assoc.*, **13**(2), 81 (1963).
103. R. J. Gordon, H. Mayrsohn, and R. M. Ingels, *Environ. Sci. Technol.*, **2**(12), 1117 (1968).
104. A. P. Altshuller, W. A. Lonneman, F. D. Sutterfield, and S. L. Kopczynski, *Environ. Sci. Technol.*, **5**(10), 1009 (1971).
105. H. W. Patton and G. D. Touey, *Anal. Chem.*, **28**, 1865 (1956).
106. R. F. Gould, Ed., *Refining Petroleum for Chemicals*, American Chemical Society Publication, Washington, DC, 1970.
107. R. E. Gosselin, R. P. Smith, and H. C. Hodge, *Clinical Toxicology of Commercial Products*, 5th ed., Williams and Wilkins, Baltimore, 1984.
108. R. D. Stewart, A. A. Herrmann, E. D. Baretta, H. V. Forster, J. J. Sikora, P. E. Newton, and R. J. Soto, *Scand. J. Work Environ. Health*, **3**(4), 234 (1977).
109. S. A. Friedmann, M. Cammarato, and D. M. Aviado, *Toxicology*, **1**, 345 (1973).
110. K. Watanabe and S. Takesue, *Agric. Biol. Chem.*, **36**, 825 (1972).
111. D. J. Crisp, A. O. Cristie, and A. F. A. Ghobashy, *Comp. Biochem. Physiol.*, **22**, 629 (1967).
112. *Documentation of the Threshold Limit Values for Substances in Workroom Air*, 3rd ed., American Conference of Governmental Industrial Hygiene, Cincinnati, 1984.
113. N. W. Lazarew, *Arch. Exp. Pathol. Pharmacol.*, **143**, 223 (1929).
114. H. E. Swann, Jr., B. K. Kwon, G. H. Hogan, and W. M. Snellings, *Am. Ind. Hyg. Assoc. J.*, **35**(9), 311 (1974).
115. T. I. Bonashevskaya and D. P. Partsef, *Gig. Sanit.*, **36**(9), 11 (1971).
116. R. E. Pattle, C. Schock, and J. Battensby, *Brit. J. Anesth.*, **44**(11), 1119 (1972).
117. M. G. Rumsby and J. B. Finean, *J. Neurochem.*, **13**, 1513 (1966).

118. D. A. Haydon, B. M. Hendry, S. R. Levinson, and J. Requena, *Biochem. Biophys. Acta*, **470**, 17 (1977).
119. D. A. Haydon, B. M. Hendry, and S. R. Levinson, *Nature*, **268**, 356 (1977).
120. M. Gaultier, G. Rancural, C. Piva, and M. L. Efthymion, *J. Eur. Toxicol.*, **6**(6), 294 (1973).
121. Y. Takeuchi, Y. Ono, N. Hisanaga, J. Kotoh, and Y. Suigiura, *Brit. J. Ind. Med.*, **37**, 241–247 (1980).
122. W. R. F. Notten and P. W. Henderson, *Biochem. Pharm.*, **24**, 1093 (1975).
123. K. M. Fredericks, *Nature*, **209**, 1047 (1966).
124. C. J. Kirwin, W. C. Thomas, and V. F. Simmon, *J. Soc. Cosmet. Chem.*, **31**, 367–370 (1980).
125. National Institute for Occupational Safety and Health, *NIOSH Recommendations for Occupational Safety and Health Standards NIOSH/CDC*, Atlanta, GA, 1988.
126. T. DiPaolo, *J. Pharm. Sci.*, **67**(4), 566 (1978).
127. K. W. Nelson, J. F. Ege, M. Ross, L. E. Woodman, and L. Silverman, *J. Ind. Hyg. Toxicol.*, **25**, 282 (1943).
128. R. Korobkin, A. Ashbury, and S. Neilson, *Arch. Neurol.*, **32**, 158 (1975).
129. N. Battistini, G. L. Lenji, E. Zanette, C. Fieschi, F. Battista, A. Franzinelli, and E. Sartorelli, *Riv. Pat. Nerv. Ment.*, **95**, 871 (1974).
130. G. W. Paulson and G. W. Waylonis, *Arch. Intern. Med.*, **136**, 880 (1976).
131. C. Carapella, *Ann. 1st Super Sanita*, **13**(1–2), 353 (1977).
132. M. Iida, Y. Yamamura, and I. Sobue, *Electromyography*, **9**, 247 (1969).
133. Y. C. Chang, *Brit. J. Ind. Med.*, **47**, 485–489 (1990).
134. L. S. Andrews and R. Snyder, "Toxic Effects of Solvents and Vapors," *Casarett and Doull's Toxicology*, 4th ed., M. O. Amdur, J. Doull, and C. D. Klassen, Eds., Pergamon Press, New York, 1991, pp. 681–722.
135. A. M. Seppalainen and C. Raitha, *Proc. 2nd Finnish-Estonian Symposium on Early Effects of Toxic Substances*, 1981, pp. 180–187.
136. N. Ishii, A. Herskowitz, and H. Schaumburg, *J. Neuropathol. Exp. Neurol.*, **31**, 198 (1972).
137. M. L. Keplinger, G. E. Lanier, and W. B. Deichmann, *Toxicol. Appl. Pharmacol.*, **1**, 156 (1959).
138. E. T. Kimura, D. M. Elery, and P. W. Dodge, *Toxicol. Appl. Pharmacol.*, **19**, 699 (1971).
139. C. H. Hine and H. H. Zuidema, *Ind. Med.*, **39**(5), 215 (1970).
140. P. S. Spencer, M. C. Bischoff, and H. H. Schaumburg, *Toxicol. Appl. Pharmacol.*, **44**, 17 (1978).
141. V. Foa, R. Gilioli, C. Bugheroni, M. Maroni, and G. Chiappino, *Med. Lav.*, **67**(2), 136 (1976).
142. T. Inoue, S. Yamada, H. Miyagaki, and Y. Takeuchi, *Proc. XVI Int. Congr. Occup. Health (Tokyo)*, 522 (1969).
143. K. Kurita, *Jap. J. Ind. Health Exp.*, **9**, 672 (1974).
144. A. K. Ashbury, S. L. Nielsen, and R. Telfer, *J. Neuropathol. Exp. Neurol.*, **33**, 191 (1974).

145. E. G. Gonzales and J. A. Downey, *Arch. Phys. Med. Rehab.*, **53**, 333 (1972).
146. J. Towfighi, N. K. Gonatas, D. Pleasure, H. S. Cooper, and L. McCree, *Neurology*, **26**(3), 238 (1977).
147. P. Bohlen, U. P. Schlunegger, and E. Lauppi, *Toxicol. Appl. Pharmacol.*, **25**, 242 (1973).
148. A. Y. H. Lu, H. W. Strobel, and M. J. Coon, *Mol. Pharm.*, **6**, 213 (1970).
149. H. Kramer, H. Staudinger, and V. Ullrich, *Chem-Biol. Interactions*, **8**, 11 (1974).
150. H. Vainio, *Acta Pharmacol. Toxicol.*, **34**(3), 152 (1974).
151. J. Sice, *Toxicol. Appl. Pharmacol.*, **9**(1), 70 (1966).
152. C. DeMartino, W. Malorni, M. C. Amantini, P. S. Barcellona, and N. Frontali, *Exp. Mol. Pathol.*, **46**, 199–216 (1987).
153. T. J. Mast, P. L. Hakett, J. R. Decker, R. B. Westerberg, and L. B. Sasser, *Govt Rep. Announcements Index*, **89**, 916, 572 1989.
154. K. Mortelmans, S. Haworth, T. Lawlor, W. Speck, B. Tainer, and E. Zeiger, *Environ. Mutagen.*, **8**(suppl. 7), 1–119 (1986).
155. H. Fuehner, *Biochem. Z.*, **115**, 235 (1929).
156. J. Bahima, A. Cert, and M. Merendez-Gallego, *Toxicol. Appl. Pharmacol.*, **76**, 473–482 (1984).
157. R. Snyder, Ed., *Ethel Browning's Toxicity and Metabolism of Industrial Solvents*, 2nd ed., Vol. 1, *Hydrocarbons*, Elsevier, Amsterdam, 1987.
158. M. Aaira, *Tokyo Ika Daigaku Zasshi*, **33**, 4 (1975).
159. C. P. Carpenter, D. L. Geary, Jr., R. C. Myers, D. J. Nachreiner, L. J. Sullivan, and J. M. King, *Toxicol. Appl. Pharmacol.*, **44**, 53 (1978).
160. G. I. Vinogradov, I. A. Chernichenko, and E. M. Makarenko, *Gig. Sanit.*, **8**, 10 (1974).
161. C. A. Nau, J. Neal, and M. Thornton, *Arch. Environ. Health*, **12**, 382 (1966).
162. D. L. Liu and B. J. Dutka, *J. Water Pollut. Control Fed.*, **45**(2), 232 (1973).
163. J. M. Lebeault, B. Roche, Z. Duvnjak, and E. Azoulay, *Arch. Microbiol.*, **72**, 140 (1970).
164. E. I. Kvasnikov, E. F. Solomko, A. M. Zhuravel, and V. M. Romanenko, *Mikrobiologiya*, **40**(5), 858 (1971).
165. Y. A. Guzhova and V. S. Fadeev, *Zh. Anal. Khim.*, **34**(1), 184 (1979).
166. K. Adachi and S. Yamasawa, *Nature*, **222**, 191 (1969).
167. E. Bingham and H. L. Falk, *Arch. Environ. Health*, **19**, 779 (1969).
168. A. W. Horton, D. N. Eshleman, A. R. Schuff, and W. H. Perman, *J. Natl. Cancer Inst.*, **56**(2), 387 (1976).
169. A. W. Horton and G. M. Christian, *J. Natl. Cancer Inst.*, **53**(4), 1017 (1974).
170. C. O. Gill and C. Ratledge, *J. Gen. Microbiol.*, **75**(1), 11 (1973).
171. T. Maruta, H. Koga, M. Kinoshita, K. Takei, and R. Ogura, *Kurume Med. J.*, **22**(3), 183 (1975).
172. J. Ziegenmeyer, N. Reuter, and F. Meyer, *Arch. Int. Pharmacol. Therap.*, **224**(2), 238 (1976).
173. J. D. Walker and R. R. Colwell, *Appl. Environ. Microbiol.*, **31**(2), 198 (1976).
174. M. Popovic, N. Aliaga, and N. Gerencevic, *Acta Pharmacol. Jugosl.*, **24**(1), 17 (1974).

175. J. Tulliez, G. Bories, C. Boudene, and C. Fevrier, *Ann. Nutr. Aliment.*, **29**(6), 563 (1976).
176. J. K. Boitnott and S. Margolis, *Johns Hopkins Med. J.*, **127**, 65 (1970).
177. R. Ristow and H. Weiner, *Fette Seifen, Anstrichm.*, **70**, 273 (1968).
178. A. Tanaka and S. Fukui, *J. Ferment. Technol.*, **48**, 137 (1970).
179. V. W. Jamison, R. L. Raymond, and J. O. Hudson, *Appl. Microbiol.*, **17**(6), 853 (1969).
180. T. Higashihara and A. Sato, *Agric. Biol. Chem.*, **33**(12), 1802 (1969).
181. Z. Duvnjak, B. Roche, and E. Azoulay, *Arch. Mikrobiol.*, **72**(2), 135 (1970).
182. G. J. Mulkins-Phillips and J. E. Steward, *Canad. J. Microbiol.*, **20**, 955 (1974).
183. D. F. Jones and R. Howe, *J. Chem. Soc.*, **22**, 2809 (1968).
184. R. F. Lee, R. Sauerheber, and A. A. Benson, *Science*, **177**, 344 (1972).
185. R. H. Linnell and W. E. Scott, *Arch. Environ. Health*, **5**, 616 (1962).
186. J. N. Pitts, Jr., B. J. Finlayson, H. Akimoto, W. A. Kummer, and R. J. Steer, International Symposium on Identification and Measurement of Environmental Pollutants, Ottawa, Ontario, Canada, June 14–17, 1971.
187. M. M. Kay, A. F. Henschel, J. Butler, R. N. Ligo, and I. R. Tabershaw, *Occupational Diseases, A Guide to their Recognition*, U.S. Department of Health, Education, and Welfare, Washington, DC, 1977.
188. H. B. Elkins, *The Chemistry of Industrial Toxicology*, 2nd ed., Wiley, New York, 1959.
189. A. J. Haagen-Smit, E. F. Darley, M. Zaitlin, H. Hull, and W. Noble, *Plant Physiol.*, **27**, 18 (1952).
190. E. R. Stephens, *U.S. Natl. Tech. Inf. Serv. PB Rep. No 23099318GA*, 1973.
191. A. L. Cowles, H. H. Borgstedt, and A. J. Gillies, *Anesthesiology*, **36**(6), 558 (1972).
192. Manufacturing Chemists Assoc., Inc., Chemical Safety Data Sheets, Washington, DC, 1991
193. C. Thienes and T. J. Haley, *Clinical Toxicology*, 5th ed., Lea and Febiger, Philadelphia, 1972.
194. C. Reynolds, *Anesth. Analg.*, **6**, 121 (1927).
195. A. G. Gilman, T. W. Rall, A. S. Nies, and P. Taylor, Eds., *The Pharmacological Basis of Therapeutics*, 8th ed., Pergamom, New York, 1990.
196. L. R. Riggs, *Proc. Soc. Exp. Biol. Med.*, **22**, 269 (1924–1925).
197. *Encyclopedia of Occupational Health and Safety*, Vols. I and II, McGraw-Hill, New York, 1972.
198. R. B. Conolly, R. J. Jaeger, and S. Szabo, *Exp. Mol. Pathol.*, **28**, 25 (1978).
199. A. O. Olson and M. Spencer, *Can. J. Biochem.*, **46**, 283 (1968).
200. L. K. Riggs, *J. Am. Pharm. Assoc.*, **14**, 380 (1925).
201. M. L. Krasovitskaya and L. Malyarova, *Gig. Sanit.*, **33**(5), 7 (1968).
202. P. Pietsch and M. Chenoweth, *Proc. Soc. Exp. Biol. Med.*, **130**, 714 (1968).
203. T. E. Hamm, Jr., D. Guest, and J. G. Dent, *Fundam. Appl. Toxicol.* **4**, 473–478 (1984).
204. P. Cazzamali, *Clin. Chirurg.*, **34**, 477 (1931).
205. H. K. Pratt and J. D. Goeschl, *Ann. Rev. Plant Physiol.*, **20**, 542 (1969).

206. R. E. Holm and J. L. Key, *Plant Physiol.*, **44**, 1295 (1969).
207. G. D. Clayton and T. S. Platt, *Am Ind. Hyg. Assoc. J.*, **28**, 151 (1967).
208. W. Braker and A. Mossman, *Matheson Gas Data Book*, 5th ed., Matheson Gas Products, East Rutherford, NJ, 1971.
209. I. Watanabe, T. Okita, and S. Seino, *Koshu Eiseiin Kenkyu Hokoku*, **20**(2), 107 (1971).
210. D. F. Meigh, *Nature*, **184**(Suppl. 14), 1072 (1959).
211. J. W. Swinnerton and V. J. Linnenbom, *Science*, **156**, 1119 (1967).
212. Serta Ltd., Brit. Patent 1,026,685 (CI.A OIN), April 20, 1956.
213. W. E. Brown, *J. Pharmacol. Exp. Therap.*, **23**, 485 (1924).
214. J. T. Halsey, C. Reynolds, and W. A. Prout, *J. Pharmacol. Exp. Therap.*, **26**, 479 (1926).
215. M. H. Kahn and L. K. Riggs, *Ann. Int. Med.*, **5**, 651 (1932).
216. B. M. Davidson, *J. Pharmacol. Exp. Therap.*, **26**, 33 (1926).
217. L. K. Riggs and H. D. Goulden, *Anesth. Analg.*, **4**, 209 (1925).
218. C. Reynolds, *J. Pharmacol. Exp. Therap.*, **27**, 93 (1926).
219. J. A. Quest et al. *Toxicol. Appl. Pharmacol.* **76**, 288–295 (1984).
220. S. W. May, R. A. Schwartz, B. J. Abbott, and O. R. Zaborsky, *Biochim. Biophys. Acta*, **403**, 245 (1975).
221. M. L. Krasovitskaya and L. K. Malyarova, *Biol. Deistvie Gig. Znachenie Atm. Zagryaz.*, 74 (1966).
222. M. W. Ranney, *Synthetic Lubricants*, Noyes Data Corp., Park Ridge, NJ, 1972.
223. J. E. Casida, E. C. Kimmel, B. Holm, and G. Widmark, *Acta Chem. Scand.*, **25**(4), 1497 (1971).
224. W. B. Deichmann and H. W. Gerarde, *Toxicology of Drugs and Chemicals*, Academic Press, New York, 1969.
225. B. Testa and D. Mihailova, *J. Med. Chem.*, **21**(7), 683 (1978).
226. W. Braker, A. L. Mossman, and D. Siegel, *Effects of Exposure to Toxic Gases—First Aid and Medical Treatment*, 2nd ed., Matheson, Lyndhurst, NJ, 1977.
227. R. Gould, *Photochemical Smog and Ozone Rections*, American Chemical Society, Washington, DC, 1972.
228. C. E. Searle, *Chemical Carcinogens*, American Chemical Society, Washington, DC, 1976.
229. *Fenaroli's Handbook of Flavor Ingredients*, 2nd ed., Vol. 2, edited, translated, and revised by T. E. Furia and N. Bellanca, The Chemical Rubber Co., Cleveland, OH, 1975.
230. P. R. Ortiz De Montellano and B. A. Mico, *Mol. Pharmacol.* **18**, 128–136 (1980).
231. P. R. Ortiz De Montellano and A. S. Boparti, *Biochim. Biophys, Acta*, **544**, 504–509 (1978).
232. *International Maritime Dangerous Goods Code*, International Maritime Ory., London, 1988.
233. C. P. Carpenter, H. F. Smyth, Jr., and U. C. Pozzani, *J. Ind. Hyg. Toxicol.*, **32**(6), 344 (1949).
234. I. Richter, H. Krain, and H. K. Mangold, *Experientia*, **32**(2), 186 (1976).
235. E. Gil-Av and J. Shabtai, *Nature*, **197**, 1065 (1963).

236. J. S. Fruton and S. Simmons, *General Biochemistry*, 2nd ed., Wiley, New York, 1958.
237. M. W. Ranney, *Lubricant Additives*, Noyes Data Corporation, Park Ridge, NJ, London, 1973.
238. F. Benini, V. Colamussi, and M. L. Zannoni, *Arcisp. S. Anna Ferrara*, **23**(6), 511 (1970).
239. A. Hamilton and H. L. Hardy, *Industrial Toxicology*, 3rd ed., PSG Publishing Company, Inc., Littleton, MA, 1974.
240. I. M. Mirzoyan, *Gig. Tr. Prof. Zabol.*, **11**, 38 (1972).
241. C. P. Carpenter, C. B. Shaffer, C. S. Weil, and H. F. Smyth, Jr., *J. Ind. Hyg. Toxicol.*, **26**, 69 (1944).
242. *Gig. Tru. Prof. Zab.*, **13**, 42, (1969).
243. E. I. Donetskaya and F. S. Shvartsapel, *Tr. Permsk. Med. Inst.*, **82**, 223 (1970).
244. R. Melnick and M. G. Bird, Symposium On The Toxicity, Carcinogenesis, and Human Health Aspects of 1,3-Butadiene, *Environ. Health Perspect.*, **86**, 1–171 (1990).
245. *International Agency for Research on Cancer*, Suppl. 7, 1987, pp. 136–137.
246. P. deMayo, *Mono- and Sesquiterpenoids*, Vol. II, Interscience, New York, 1959.
247. T. Dalhamn and R. Rylander, *Arch. Environ. Health*, **20** (1970).
248. D. P. Gough and F. W. Hemming, *Biochem. J.*, **118**, 163 (1970).
249. T. Dalhamn, M. Edfors, and R. Rylander, *Arch. Environ. Health*, **17**, 252 (1968).
250. J. L. Egle, Jr., and B. J. Gochberg, *Am. Ind. Hyg. Assoc. J.*, **36**(5), 369 (1975).
251. V. D. Gostinskii, *Gig. Tr. Prof. Zab.*, **9**(1), 36 (1965).
252. L. Weissbecker, R. M. Creamer, and R. D. Carpenter, *Am. Rev. Respir. Dis.*, **104**(2), 182 (1971).
253. F. A. Patty, *Industrial Hygiene and Toxicology*, 2nd rev. ed., Vol. II, Wiley, New York, 1963.
254. R. J. Shamberger, *J. Natl. Cancer Inst.*, **47**(3), 667 (1971).
255. S. A. Pigolev, *Gig. Tr. Prof. Zabol.*, **15**(2), 49 (1971).
256. A. A. Nikultseva, *Gig. Tr. Prof. Zabol*, **11**(12) 41 (1976).
257. A. G. Pestova and O. G. Petrovskaya, *Vrach. Delo*, **4**, 135 (1973).
258. H. F. Smyth, Jr., C. P. Carpenter, C. S. Weil, W. C. Pozzani, J. A. Streigel, and J. S. Nycum, *Am. Ind., Hyg. Assoc. J.*, **30**, 470 (1969).
259. R. N. Harger and L. W. Spolyar, *Arch. Ind. Health*, **18**, 497 (1958).
260. C. H. Theines and T. J. Haley, *Clinical Toxicology*, 5th ed., Lea & Febiger, Philadelphia, 1972.
261. M. Yoshioka, S. Tsujimoto, N. Oki, M. Tamaki, Y. Torihashi, and N. Takada, *Hyogo-ken Kogai Keneyusho Kenkyu Hokoku*, **8**, 48 (1976).
262. M. J. Dilworth, *Biochim. Biophys. Acta*, **127**, 285 (1966).
263. M. Kelly, *Biochem. J.*, **107**, 1 (1968).
264. H. Yasuda, M O. Bumgarner, H. C. Marsh, B. S. Yamanashi, D. P. DeVito, M. L. Wolbarsht, J. W. Reed, M. Bessler, M. B. Landers III, D. M. Hercules, and J. Carver, *J. Biomed. Mater. Res.*, **9**(6), 629 (1975).
265. E. M. J. Gifford, *Am. J. Bot.*, **56**(8), 892 (1969).
266. W. W. Aldrich and H. Y. Nakasone, *J. Am. Soc. Hort. Sci.*, **100**(4), 410 (1975).
267. T. W. Speitel and S. M. Seigel, *Plant Cell Physiol.*, **16**(2), 383 (1975).

268. E. Yip, W. A. Southorn, and J. B. Gomez, *J. Rubber Res. Inst. Malays.*, **24**(2), 103 (1974).
269. B. M. Davidson, *J. Pharmacol.*, **25**, 119 (1925).
270. A. T. Jones, *Arch. Environ. Health*, **5**, 417 (1960).
271. D. S. Ross, *Ann. Occup. Hyg.*, **16**, 85 (1973).
272. F. Flury, *Arch. Exp. Pathol. Pharmakol.*, **138**, 65 (1925).
273. H. Franken and L. Miklos, *Zentralbl. Gynaekol.*, **42**, 2493 (1933).
274. National Institute for Occupational Safety and Health, *Criteria for a Recommended Standard, Occupational Exposure to Acetylene*, U.S. Department of Health, Education, and Welfare, Washington, DC, 1976.
275. T. R. Norton, "Metabolism of Toxic Substances," Chapter 4 in *Toxicology, The Basic Science of Poisons*, L. J. Casarett and J. Doull, Eds., Macmillan, New York, 1975.
276. J. Reisch. W. Spitzner, and K. E. Schulte, *Arzneim.-Forsch.*, **17**(7), 816 (1967).
277. Shell Chemical Company, unpublished report, 1961.
278. F. Sperling, *Fed. Proc.*, **19**(1), 389 (1960).

CHAPTER TWENTY

# Alicyclic Hydrocarbons

Finis Cavender, Ph.D., D.A.B.T., C.I.H.

## 1 CYCLOALKANES OR CYCLOPARAFFINS

The alicyclic hydrocarbons include the cycloalkanes, also called cycloparaffins, cyclanes, and naphthenes; the cycloalkenes, also known as cycloolefins and cyclenes; and a variety of their substituted derivatives. Some naturally occurring alkanyl- or alkenyl-substituted mono-, di-, or polycyclenes are commonly known as terpenes, which occur in various plants. Other compounds are isolated from crude petroleum refinery distillates or catalytically cracked petroleum products. Cycloalkanes are extensively usd to produce re-formed aromatics (1) and some are administered as anesthetics. They are synthesized in pure form by the reduction of dihalogenated propane precursors. Some physical properties of the cycloalkanes are given in Table 20.1 (2–17) and toxicity data in Tables 20.2 and 20.3 (18–38).

### 1.1 Toxicity

The lower cycloalkanes are gases and have been used as anesthetics, especially cyclopropane. The $C_5$ and higher members are liquids and cause central nervous system (CNS) depression. However, from $C_6$ on, the margin of safety between CNS depression and death is very narrow and symptomatically barely recognizable. The alicyclics, in general, are CNS depressants with low acute and chronic toxicity. They are exhaled in unchanged form or are rapidly metabolized into water-soluble metabolites, usually as glucuronides. Exposure of humans and laboratory animals to high concentrations may cause excitement, loss of equilibrium, stupor, and coma,

---

*Patty's Industrial Hygiene and Toxicology, Fourth Edition, Volume 2, Part B*, Edited by George D. Clayton and Florence E. Clayton.
ISBN 0-471-54725-5 © 1994 John Wiley & Sons, Inc.

Table 20.1. Physical and Chemical Data for Cycloalkanes (2–17)

| Compound | B.P. (°C) (mm Hg) | CAS Registry No. | Density (g/ml, 20°C) | Empirical Formula | Flammability Limits (%) | Flash Pt. (°C) | Freezing Pt. (°C) |
|---|---|---|---|---|---|---|---|
| Cyclopropane | −32.7 | 75-19-4 | 0.6769$^b$ | $C_3H_6$ | 2.4–10.4 | — | −127.6 |
| Cyclobutane | 12.0 | 287-23-0 | 0.720$^c$ | $C_4H_8$ | | <10 | −50.0 |
| Cyclopentane | 49.26 | 287-92-3 | 0.7454 | $C_5H_{10}$ | | −7 | −93.88 |
| Methylcyclopentane | 71.81 | 96-37-7 | 0.7486 | $C_6H_{12}$ | 1.2–8.4 | −28.9 | −142.465 |
| Ethylcyclopentane | 103.47 | 1640-89-7 | 0.7665 | $C_7H_{14}$ | 1.1–6.7 | | −138.44 |
| Cyclohexane | 80.74 | 110-82-7 | 0.7786 | $C_6H_{12}$ | 1.33–8.4 | −20 | 6.55 |
| Methylcyclohexane | 100.93 | 108-87-2 | 0.7694 | $C_7H_{14}$ | 1.2–6.7 | −4 | −126.59 |
| Ethylcyclohexane | 131.78 | 1678-91-7 | 0.7880 | $C_8H_{16}$ | 0.9–6.6 | 35 | −111.32 |
| Dimethylcyclohexane | 119.54 | 2207-64-7 | 0.7584 | $C_8H_{16}$ | | 11.1 | −33.50 |
| Cycloheptane | 118.48 | 291-64-5 | 0.8098 | $C_7H_{14}$ | | <37.8 | 12.0 |
| Cyclooctane | 149.0 (749) | 292-64-8 | 0.8439 | $C_8H_{16}$ | | 14.3 | 14.3 |
| Cyclononane | 353.1 | | | $C_9H_{18}$ | | 11.1 | 11.1 |

$^a$Solubility in water/alcohol/ether: v = very soluble; s = soluble; i = insoluble.
$^b$−30°C.
$^c$5°C.
$^d$−42°C.
$^e$SUS = Saybolt universal standard.

but rarely death. Oral adminsitration in animals has resulted in severe diarrhea, vascular collapse, and heart, lung, liver, and brain degeneration. Cycloparaffins are dermal irritants, they defat the skin to cause morphological changes and hypothermia, and they are CNS depressants. The liquid members up to $C_8$, and to a lesser exent $C_{12}$, are aspiration hazards.

A number of alicyclic hydrocarbons are nephrotoxic in male rats. Repeated dose oral (39) or inhalation (40) studies result in tubular degeneration characterized by hyaline droplet formation, necrosis, intratubular casts, and medullary mineralization. These effects are associated with the presence of α-2-microglobulin in a number of rat strains including Fischer 344, Sprague–Dawley, Wistar, Buffalo, and Norway Brown. It is not produced in the NCI–Black–Reiter strain (41) or in female rats (42), and nephropathy is not produced by alicyclic hydrocarbons in these animals. The hydrocarbon nephropathy has been produced by cyclohexane (43, 44), 2,2,4-trimethylpentane (45), JP-5 jet fuel (45), JP-10 jet fuel (46), d-limonene (47), decalin (42, 48), and tetralin (49). The nephropathy has been studied extensively using a short-term decalin inhalation model. (40). The nephropathy is not seen in mice or dogs (44).

This protein was of primary interest in the carcinogenesis studies of unleaded gasoline (50). Kidney tumors were produced only in males and were associated with α-2-microglobulin. Further study revealed that humans do not produce the α-2-microglobulin. On this basis, the National Toxicology Program stated that the kidney tumors seen in male rats were not relevant to carcinogenesis in man (51). Unleaded gasoline did not produce kidney tumors in male NCI–Black–Reiter rats (52).

# ALICYCLIC HYDROCARBONS

| Mol. Wt. | Refractive Index, | Solubility[a] w/al/et | Sp. Gr. | Vapor Density | Vapor Pressure[b] (mm Hg) (°C) | Viscosity (SUS)[c] | Wt./Vol. Conversion (mg/m³ = 1 ppm) |
|---|---|---|---|---|---|---|---|
| 42.08 | 1.3700[d] | s/v/v | 0.720 | 1.45 | | N.A. | 1.72 |
| 56.11 | 1.426 | i/v/v | | 1.93 | | N.A. | 2.30 |
| 70.14 | 1.4065 | i/v/v | 0.751 | 2.42 | 400 (31.0) | <32.6 | 2.87 |
| 84.16 | 1.4097 | i/v/v | 0.754 | 2.90 | 100 (17.9) | <32.6 | 3.44 |
| 98.19 | 1.4198 | i/v/v | 0.771 | 3.4 | | <32.6 | 4.02 |
| 84.16 | 1.4262 | i/v/v | 0.779 | 2.90 | 100 (60.8) | <32.6 | 3.44 |
| 98.19 | 1.4231 | i/s/s | 0.774 | 3.39 | 43 (25) | <32.6 | 4.02 |
| 112.21 | 1.4330 | i/s/s | 0.792 | 3.9 | | <32.6 | 4.59 |
| 112.21 | 1.4290 | i/s/s | 0.767 | 3.86 | 10 (10.2) | | 4.59 |
| 98.19 | 1.4436 | i/v/v | 0.783 | 3.3 | | | 4.02 |
| 112.21 | 1.4586 | i/—/— | | | | | 4.59 |
| 123.23 | | i/—/— | 0.854 | | | | 5.16 |

## 1.2 Cyclopropane

Cyclopropane or trimethylene, $\overline{CH_2CH_2CH_2}$, the simplest cycloalkane, is a colorless, flammable (53), and explosive gas. Its mild, sweet odor resembles that of petroleum ether (4, 16). Selected physical data are given in Table 20.1 (2–17). Reagent-grade cyclopropane is prepared by the reduction of 1,2-dibromocyclopropane with zinc and alcohol. It is stable at room temperature but undergoes ring opening to propene when heated (54). In anesthesiology, it is the preferred agent for a wide range of purposes (30, 55). It is readily absorbed in the respiratory tract and is a potent CNS depressant (27, 56). There is a wide margin between anesthetic effect and toxicity and it is a cardiac sensitizer (57, 58). It is minimally irritating to the pacemaker and conduction tissue and the respiratory tract (27). Mice exhibited anesthesia in 3 min at 5.8 mmol/l without death and recovery in 1.5 min (28), and a concentration of 18 percent was lethal in 39 min. Experimentally, cyclopropane acts rapidly in the dog and serves as a model for predicting anesthetic concentrations for human application (59). In chickens, exposure for 6 hr produced embryonic abnormalities at 10 to 20 percent and death at 20 to 30 percent in 12-hr exposures (60). The 2-hr $LC_{LO}$ in mice is 282 mg/l (22). In general, however, cyclopropane should not present an industrial hazard. At present, exposure standards have not been established, although as a simple asphyxiant, a TLV of 400 ppm is recommended (61). Cyclopropane is on the Environmental Protection Agency's Toxic Substances Control Act (EPA TSCA) Chemical Inventory and Test Submission Data Base (22). Analytic determination procedures are available.

## 1.3 Cyclobutane

Cyclobutane or tetramethylene, $\overline{CH_2(CH_2)_2CH_2}$, is a colorless, flammable (53), explosive gas. Selected physical properties are given in Table 20.1. Cyclobutane is

**Table 20.2. Inhalation Toxicity of Some Cycloalkanes**

| Compound | Species | Concentration ppm | Concentration mg/l | Time | Findings | Ref. |
|---|---|---|---|---|---|---|
| Cyclopropane | Mouse | 142,000 | 5.8 | 3 min | Anesthesia in 3 min, recovery in 1.5 min | 28 |
| | | 18,000 | 309.6 | 39 min | Lethal | 28 |
| Cyclopentane | Human | 10–15 | 0.029–0.043 | | Tolerable | 29 |
| | Mouse | 38.3 | 0.110 | | Minimal narcotic concentration, loss of reflexes and lethal | 30 |
| | Rat | 112–1139 | 0.039–0.397 | 6 hr/day × 3 weeks | No effects in female and male rats | 29 |
| | | 8110 | 2.328 | 6 hr/day × 12 weeks | Decreased body weight gains in females | 29 |
| Cyclohexane | Mouse | 18,000 | 61.9 | 5 min | Trembling | 31 |
| | | | | 15 min | Disturbed equilibrium | 31 |
| | | | | 25 min | Recumbent | 31 |
| | Rabbit | 427 | 1.47 | | No effect | 32 |
| | | 440 | 1.51 | | No effect | 32 |
| | | 780–3300 | 2.68–11.3 | | Some micropathological changes | 32 |
| | | 3300 | 11.46 | | No visible effects | 31 |
| | | <7500 | <25.8 | | Some deaths | 33 |
| | | 14,535 | 50 | | Loss of righting reflex | 34 |
| | | 18,000 | 61.9 | 6 min | Trembling | 31 |
| | | | | 15 min | Disturbed equilibrium | 31 |
| | | | | 30 min | Recumbent | 31 |
| | | 18,500 | 63.4 | 8 hr | Not lethal | 31 |
| | | 26,600 | 89.4 | 1 hr | Lethal | 31 |

| Compound | Species | | | Effect | Ref |
|---|---|---|---|---|---|
| | | 17,400–20,350 | 60–70 | Lethal | 18 |
| | | 26,050 | 89.6 | Before death, changes occur in hemoglobin and erythrocyte quantities, leukocytes increase | 32 |
| | Guinea pig | 18,000 | 61.9 | Slight trembling | 31 |
| | | 18,000 | 61.9 | Disturbed equilibrium | 31 |
| | Cat | | | Recumbent | 31 |
| Cyclohexane | Primate | 1243 | 4.28 | 6 hr/day × 50 weeks | No effect | 31 |
| | Rabbit | 443 | 1.49 | 8 hr/day × 26 weeks | No effect, no pathological effects | 33 |
| | | 786 | 2.70 | 6 hr/day × 50 weeks | Minor microscopic changes in liver and kidney | 31 |
| | | 3300 | 11.46 | 6 hr/day × 50 weeks | No deaths, no signs of injury | 31 |
| | | 7,400–18,500 | 25.5–63.4 | 6 hr/day × 10 weeks | Some fatalities | 31 |
| Methylcyclohexane | Rabbit | 1050 | 4.22 | | No effect | 32 |
| | | 1162 | 4.67 | | No effect | 32 |
| | | 2830 | 11.35 | | Very subtle cellular injury to kidney and liver | 32 |
| | Mouse | 7463–9950 | 30–40 | | Loss of righting reflex | 34 |
| | | 7500–10,800 | 30.2–43.4 | | Lethal | 32 |
| | | 9950–12,440 | 40–50 | | Lethal | 18 |
| | Primate, rabbit | 363 | 1.46 | 6 hr/day × 50 weeks | No effect | 32 |
| | | 9850 | 39.6 | 6 hr/day × 50 weeks | Tendency to light convulsions | 32 |
| | | 14,900 | 59.9 | 6 hr/day × 50 weeks | Convulsions | 32 |
| Dimethylcyclohexane | Mouse | 4360–5450 | 20–25 | | Loss of righting reflex | 34 |
| | | 5450–6540 | 25–30 | | Death | 34 |
| Ethylcyclohexane | Mouse | 3270 | 15 | | Loss of righting reflex | 34 |
| | | 7625 | 35 | | Death | 34 |

**Table 20.3.** General Toxicity of Some Cycloalkanes

| Compound | Route | Species | Parameter | Result | Ref. |
|---|---|---|---|---|---|
| Cyclohexane | Oral | Rat | $LD_{50}$ | 8.0–39.0 ml/kg | 35 |
| | | | $LD_{50}$ | 29.82 g/kg | 36 |
| | | Mouse | $LD_{50}$ | 1.30 g/kg | 22 |
| | | Rabbit | $LD_{LO}$ | 5.5–6.0 g/kg | 33 |
| | | | $TD_{LO}$ | 1.5–5 g/kg | 33 |
| | Dermal | Rabbit | $LD_{100}$ | >180.2 g/kg | 33 |
| | Intravenous | Rabbit | $LD_{LO}$ | 77 mg/kg | 33 |
| | Toxicity | Aquatic | TLm96 | 100–1000 ppm | 37 |
| Methylcyclo-hexane | Oral | Rabbit | $TD_{LO}$ | 1.0–4.0 g/kg | 33 |
| | | | $LD_{LO}$ | 4.0–4.5 g/kg | 33 |
| | | | $LD_{100}$ | 4.5–10.0 g/kg | 33 |
| | Dermal | Rabbit | $LD_{50}$ | >86.7 g/kg | 33 |
| Cyclododecane | Subcutaneous | Mouse | $LD_{50}$ | 10 g/kg | 38 |

produced from cyclobutene and is reutilized in catalytic cracking processes (1) It is an anesthetic, a weak cardiac sensitizer (62), and a simple asphyxiant (7). The 2-hr $LC_{LO}$ in mice is 282 mg/l (22). Industrially, it is similar to cyclopropane and it is not expected to be a health hazard. Cyclobutane is on the EPA TSCA Chemical Inventory and Test Submission Data Base (22).

### 1.4 Cyclopentane

Cyclopentane or pentamethylene, $\overline{CH_2(CH_2)_3CH_2}$, is the first liquid member of the cycloalkane series. It is a constituent of the cycloalkane fraction of crude oil (4). It is flammable (26) and its vapors are explosive. Selected physical properties are given in Table 20.1 (2–17). Cyclopentane is not sufficiently stable to occur naturally in large quantities. It is produced in petroleum refining processes, and is found as an impurity in technical-grade hexane (58). It is commonly used for cracking aromatics (1). Commercially, cyclopentane is used to produce a variety of analgesics, sedatives, hypnotics, antitumor agents, CNS depressants, prostaglandins, insecticides, and many other products (63).

Cyclopentane is a CNS depressant and lipid solvent (56). In mice there is no safety margin between the minimal CNS depressant concentration, loss of reflexes, and lethality, all occurring at 110 mg/l (30). When ingested, cyclopentane presented a low to moderate aspiration hazard in mice (64). Cyclopentane applied to guinea pig skin produced slight erythema (65). The aquatic TLm96 is greater than 1000 ppm (24). The Occupational Safety and Health Administration (OSHA) permissible exposure limit (PEL) (66) and American Conference of Governmental Industrial Hygienists (ACGIH) threshold limit valve (TLV) (67) are 600 ppm (1720 mg/m³). Protective clothing and barrier creams are recommended when handling the cyclopentane (7). The National Institute for Occupational Safety and Health (NIOSH) method of analysis involves absorption on charcoal, desorption with carbon disulfide, and quantification using a gas chromatograph with a flame ionization detector (FID) (68). Cyclopentane is on the EPA TSCA Chemical Inventory and Test Submission Data Base (22).

### 1.4.1 Methylcyclopentane

Methylcyclopentane, $CH_3\overline{CH(CH_2)_3CH_2}$, is a colorless, flammable (26) liquid and has a sweetish odor. Selected physical properties are given in Table 20.1. Methylcyclopentane has been detected in the expired air in smokers (69). Its toxicity resembles that of cyclopentane.

Methylcyclopentane is used as an solvent extractant for essential oils from plants (70). Chemically, the substituted cyclopentane ring plays an integral part in the synthesis of natural products, such as prostaglandins (63, 71). It is also a feedstock for aromatization to benzene (1). There is no safety margin between the onset of CNS depression and death.

### 1.4.2 Ethylcyclopentane

In addition to causing CNS depression and anesthesia, methyl- and ethylcyclopentane cause convulsions (30). The toxicity to the mouse increases from the cyclopentane to the methyl to the ethyl derivative with CNS depression seen at 110 mg/l (30) and a lethality at 30 to 50 mg/l (32). The aquatic TLm96 is greater than 1000 ppm (37). Selected physical properties are given in Table 20.1.

## 1.5 Cyclohexane

Cyclohexane, hexahydrobenzene, or hexamethylene, $\overline{CH_2(CH_2)_4CH_2}$, is a colorless, flammable (53) liquid with a sweetish odor similar to chloroform (16). Saturated air at 760 mm and 26.3°C contains 13.66 percent cyclohexane with a vapor density of 1.23 (31). It is fractionated from crude oil, and is prepared synthetically from benzene (54) or by hydrocracking of cyclopentane (1). Selected physical properties are given in Table 20.1 and toxicity data in Tables 20.2 and 20.3. Cyclohexane is used as a solvent for resins, fats, waxes, rubber, and adhesives (72); as a raw material for a number of chemicals including adipic acid; and as the precursor of nylon-66 (73). Cyclohexane can be converted to benzene or when acidified to methylcyclopentane by petroleum reforming processes (1).

Cyclohexane was shown to accelerate the penetration of local anesthetics through intact guinea skin (74). In a comparative study with humans and animals, cyclohexane was therapeutic for reducing granulocytes (75). Cyclohexane promotes drug action and potentiates the effect of tri-*o*-cresyl phosphate (76).

Cyclohexane is a CNS depressant (56) and causes dizziness, nausea, and unconsciousness (30). It is a lipid solvent and a skin irritant (4), and on repeated contact defats the skin. The vapor is irritating to the skin, eyes, and respiratory tract. The oral $LD_{50}$ values in rats range between 8.0 and 39.0 ml/kg, varying with the age of the animal (35). Selected toxicity data are given in Tables 20.2 and 20.3. The no-effect level has been estimated to be 0.016 ml/kg or 1.5 ml per 60 kg weight (35). At lethal doses, rabbits exhibit, within 1 to 1.5 hr, diarrhea and widespread vascular damage and collapse, hepatocellular degeneration, and toxic glomerulonephritis (33). Cyclohexane is absorbed by the skin, and massive applications to rabbit skin result in microscopic changes in the liver and kidneys (33). Cyclohexane is of low acute toxicity, and exhibits a narrow margin of safety between CNS depression, loss of reflexes, and death. Chronic studies have demonstrated that the mouse is somewhat more sensitive than the rabbit. Cyclohexane is absorbed by the respiratory tract but much of this is exhaled unchanged. Another portion is excreted in the urine and the balance is metabolized via the hepatic and nephric systems. Cyclohexane induces liver microsomal hydroxylases, which, in turn, oxidize it to cyclohexanol (77–80). Cyclohexanol then appears to be excreted mainly as the sulfate or glucuronide conjugate (32). However, the action of cyclohexane differs from that of benzene, inasmuch as the cytochrome P450 monooxygenase system is not affected (81). Microbial systems also oxidize cyclohexane to cyclohexanol (79). Cyclohexane undergoes photodecomposition and mineralization (82). The ACGIH (67) TLV is 300 ppm (1030 mg/m$^3$) and the OSHA (66) PEL is 300 ppm (1050 mg/m$^3$). Few industrial morbidity reports have been published. For biologic monitoring exhaled breath and urine should be analyzed for cyclohexane or its metabolites. Any decrease in the ratio of inorganic to total sulfates is roughly proportional to concentration of inhaled cyclohexane (31). In general, cyclohexane is a safe solvent when handled according to good industrial hygiene practice (72). Cyclohexane is on the EPA TSCA Chemical Inventory and the Test Submission Data Base (22). NIOSH recommends absorption on charcoal, desorption with carbon disulfide, and quantification using FID–GC (gas chromatography) (68).

### 1.5.1 Methylcyclohexane

Methylcyclohexane or hexahydrotoluene, $CH_3CH(CH_2)_4CH_2$, is a colorless, flammable (53) liquid. Selected physical properties are given in Table 20.1. Saturated air at 760 mm and 25°C contains 5.65 percent methylcyclohexane, and its vapor density is 1.14 (31). It occurs in crude petroleum oils and is separated by distillation. It is also produced by hydrogenation of toluene or by acidic hydrocracking of polycyclic aromatics (1). Methylcyclohexane is used as a solvent for cellulose ethers and as raw material in a variety of synthetic processes (83).

Exposure of rabbits to methylcyclohexane resulted in diarrhea within 1 to 1.5 hr of the exposure (33). Selected toxicity data are given in Tables 20.2 and 20.3. When topically applied to the skin of rabbits, methylcyclohexane caused defatting of the skin followed by hardening of the keratin layer, some cellular injury, and slight hypothermia (17). For inhalation exposures, the no-effect level is 1200 ppm for the rabbit and 300 ppm for the primate (32). Exposure to lethal concentrations in the primate results in mucous secretion, lacrimation, salivation, labored breathing, and diarrhea (32). Methylcyclohexane as a mist represents a greater aspiration hazard than cyclohexane or cycloheptane (64).

When inhaled, methylcyclohexane is absorbed, distributed to the liver and kidneys, hydroxylated to the *trans*-4-ol by hepatic microsomal enzymes (84), and finally excreted as the sulfate or the glucuronide (32). Microbial ω-hydroxylases are also capable of hydroxylating methylcyclohexane to methylcyclohexanol (79). In plant pathology, methylcyclohexane has been found active in tumor destruction (85).

The OSHA (66) PEL is 400 ppm (1600 mg/m$^3$). The ACGIH (67) TLV is 400 ppm (1610 mg/m$^3$). NIOSH recommends collection on charcoal, desorption with carbon disulfide, and quantification using FID–GC (68).

### 1.5.2 Other Alkylcyclohexanes

Dimethylcyclohexane, $CH_3\overline{CH(CH_2)_4CH}CH_3$, is used as a solvent. Selected physical properties are given in Table 20.1. Dimethylcyclohexane appears to be more toxic than the monomethyl derivative. Exposures of mice to 20 to 25 mg/l resulted in the loss of the righting reflex and 25 to 30 mg/l caused death (34). Ethylcyclohexane is slightly more toxic than dimethylcyclohexane. The loss of righting reflex occurred at a concentration of 15 mg/l and death at 35 mg/l (34). Dimethyl- and ethylcyclohexane are used in catalytic reforming to produce $C_8$ aromatic compounds (1). Selected toxicity data are given in Tables 20.2 and 20.3.

## 1.6 Cycloheptane

Cycloheptane or heptamethylene, $\overline{CH_2(CH_2)_5CH_2}$, is a colorless, flammable (53) liquid. Selected physical properties are given in Table 20.1. The toxicity of cycloheptane resembles that of methylcyclohexane. It is a CNS depressant (7) and is an aspiration hazard (64). Heptamethylene causes morphological changes in guinea pig skin and increases arginase activity (63).

For industrial hygiene information, see the procedures recommended for cyclohexane. Mass spectral data are available for its analytic determination (86).

## 1.7 Cyclooctane and Higher Cycloalkanes

Cyclooctane or octamethylene, $\overline{CH_2(CH_2)_6CH_2}$, is a flammable liquid (53). Selected physical properties are given in Table 20.1. When ingested, it is an aspiration hazard (64). When applied to the guinea pig skin, cyclooctane causes morphological changes and increases arginase activity (65). Cyclooctane and cyclononane can be quantified using gas chromatography (86). Cyclododecane is used as a mothproofing agent (87). The $LD_{50}$ for subcutaneous injection in mice is 10 mg/kg (38). Selected toxicity data are given in Table 20.3. Cyclododecane is not a skin irritant (38).

## 2 CYCLOALKENES OR CYCLOOLEFINS

The cyclic olefins are more highly reactive than their paraffin counterparts. They contribute to photochemical smog by reacting with ozone and other small molecular or ionic species (88). Selected physical properties are given in Table 20.4 (2–17) and toxicity data in Tables 20.5 to 20.7.

In $C_4$ to $C_7$ cycloalkenes, toxicity increases with molecular weight. Cycladienes

**Table 20.4.** Physical and Chemical Data for Cycloalkenes and Dicyclic Hydrocarbons (2–17)

| Compound | B.P. (°C) (mm Hg) | CAS Registry No. | Density (g/ml, 20°C) | Empirical Formula | Flammability Limits (%) | Flash Pt. (°C) | Freezing Pt. (°C) |
|---|---|---|---|---|---|---|---|
| Cyclopentene | 44.24 | 142-29-0 | 0.7720 | $C_5H_8$ | | −37.22 | −135.08 |
| Cyclohexene | 82.98 | 110-83-8 | 0.8110 | $C_6H_{10}$ | | −6 | −103.51 |
| Cycloheptene | 115.0 | | 0.8228 | $C_7H_{12}$ | | | −56.0 |
| 1-Vinylcyclohexene | 145.0 | 2622-21-1 | 0.8623 | $C_8H_{12}$ | | | |
| 4-Vinylcyclohexene | 128.9 | 100-40-3 | 0.8299 | $C_8H_{12}$ | | 21.11 | |
| Limonene | 178.0 | 138-86-3 | 0.8411 | $C_{10}H_{16}$ | | | |
| 1,3-Cyclopentadiene | 40.0 | 542-92-7 | 0.8021 | $C_5H_6$ | | | |
| Methylcyclohexadiene-1,4 | 101.5 | 1489-57-2 | 0.8354 | $C_6H_8$ | | | |
| 1,3,5-Cycloheptatriene | 117.0 | 544-25-2 | 0.8875 | $C_7H_8$ | | | |
| Cycloocta-1,5-diene | 150.8 | 111-78-4 | 0.8818 | $C_8H_{12}$ | | | |
| Cyclooctatetraene | 140.56 | | 0.9206 | $C_8H_8$ | | | −4.68 |
| cis-Decalin | 195.65 | 91-17-8 | 0.8965 | $C_{10}H_{18}$ | 0.7–4.9[c] | 57 | −43.01 |
| trans-Decalin | 186.7 | | 0.8699 | | | 57 | −30.40 |
| Tetralin | 207.57 | 119-64-2 | 0.9702 | $C_{10}H_{12}$ | 0.8–5[c] | 71 | −35.79 |
| Dicyclopentadiene | 170.0[d] | 77-73-6 | 0.9302[b] | $C_{10}H_{12}$ | | | 35 (95) |
| α-Pinene | 156.2 | 80-56-8 | 0.8582 | $C_{10}H_{16}$ | | 33 | |
| Turpentine | 153–175 | 8006-64-2 | 0.854–0.868 | | 0.8 (lel)[e] | 35–39 | −50 to −60 (−45.6 to −51.1) |
| Camphene | 160.2 | 79-92-5 | 0.8450 | $C_{10}H_{16}$ | | | |
| α-Caryophyllene | 123 (10) | 6753-98-6 | 0.8905 | $C_{15}H_{24}$ | | | |

[a]Solubility in water/alcohol/ether: v = very soluble; s = soluble; d = slightly soluble; i = insoluble.
[b]At 35°C.
[c]At 212°C.
[d]Slight decomposition.
[e]Lower explosive limit.

and polyenes appear to possess increasingly irritant, toxic, and sensitizing properties, peaking somewhat higher for cycladienes than for monocyclic alkenes. The aspiration hazard also appears higher for the cycloalkenes than the cycloalkanes (64).

## 2.1 CYCLOPENTENE

Cyclopentene, $\overline{CH:CH(CH_2)_2CH_2}$, is a highly flammable liquid (69) with a low flash point (7). It reacts readily with oxidizing agents (7). Selected physical properties are given in Table 20.4 and toxicity data are given in Tables 20.5 and 20.6. The oral $LD_{50}$ in the rat is 1656 ml/kg and the dermal $LD_{50}$ in the rabbit is 1231 ml/kg. Inhalation of the concentrated vapor was lethal to rats in 5 min, and a 4-hr exposure to 16,000 ppm was lethal to four of six rats (89). Skin irritation and corneal injury were moderate to severe in rabbits (89). Short-term exposure of cyclopentene to humans revealed a tolerable level of only 10 to 15 ppm (90). Chronic exposure of rats to 112 to 1139 ppm for 12 weeks showed no effects, whereas 8110 ppm 6 hr/day, 5 days/week for 3 weeks resulted in decreased body weight gains of female rats (29). Cyclopentene is on the EPA TSCA Chemical Inventory and the Test Submission Data Base (22).

| M. P. (°C) | Mol. Wt. | Refractive Index, $n_D^{20}$ | Solubility[a] w/al/et | Sp. Gr. | Vapor Density (Air = 1) | Vapor Pressure[b] (mm Hg) (°C) | Wt./Vol. Conversion (mg/m³ = 1 ppm) |
|---|---|---|---|---|---|---|---|
| −93.3 | 68.12 | 1.4225 | i/s/s | 0.778 | | | 2.79 |
| −103.51 | 82.15 | 1.4465 | i/v/v | 0.816 | 2.8 | 160 (38) | 3.36 |
| −56.0 | 96.17 | 1.4552 | i/s/s | | | | 3.93 |
| | 108.18 | 1.4915 | i/—/s | | | | 4.43 |
| | 108.18 | 1.4639 | i/—/s | 0.834 | 3.76 | 25.8 (38) | 4.43 |
| −74.35 | 136.24 | 1.4730 | i/v/v | | | | 5.57 |
| −97.2 | 66.10 | 1.440 | i/v/v | 0.80 | | | 2.70 |
| | 94.16 | 1.4763 | i/v/s | | | | 3.85 |
| −79.49 | 92.154 | 1.5343 | i/s/s | | | | 3.77 |
| −69 to −70 | 108.18 | 1.4905 | i/—/— | | | | 4.43 |
| −7 | 104.54 | 1.5381 | —/s/s | | | 7.9 (25) | 4.26 |
| −43.01 | 138.25 | 1.4810 | i/v/v | 0.874 | 4.76 | 1 (22.5) | 5.65 |
| −30.7 | 138.25 | 1.4695 | i/v/v | | 4.76 | 10 (47.2) | 5.65 |
| −35.79 | 132.20 | 1.5414 | i/v/v | 0.975 | 4.55 | | 5.41 |
| 32 | 132.20 | 1.5050 | i/s/v | 0.93 | 4.55 | 10 (47.6) | 5.41 |
| −55.0 | 136.24 | 1.4658 | i/v/v | 0.86 | 4.7 | 10 (37.3) | 5.57 |
| | 134.5 | | i/s/s | 0.857 | 4.6–4.84 | | 5.50 |
| 52.0 | 143.24 | 1.4750 | i/d/s | | | | 5.49 |
| | 204.36 | 1.53038 | i/—/— | | | | 8.36 |

## 2.2 Cyclohexene

Cyclohexene or 1,2,3,4-tetrahydrobenzene, $\overline{CH:CH(CH_2)_3CH_2}$, is a flammable (53) liquid that occurs in coal tar (4). Selected physical properties are given in Table 20.4. Cyclohexene is prepared by dehydration of cyclohexanol (4) by thermal reaction of a mixture of ethylene–propylene–butadiene (1). It is used as stabilizer for high-octane gasoline, as an alkylation component, and in the chemical synthesis of adipic, maleic, and hexahydrobenzoic acids and aldehydes (4).

The toxicity of cyclohexene, as for the comparable cyclenes, is low. Selected toxicity data are given in Table 20.6. It is an irritant, and it defats the skin on direct contact. It is a anesthetic and CNS depressant. When ingested, it represents a low to moderate pulmonary aspiration hazard (64). Concentrations of 30 mg/l cause a loss of righting reflex in the mouse, and 45 to 50 mg/l is lethal. Dogs exhibit tremors and staggering gait following cyclohexene exposure (112). Rats, guinea pigs, and rabbits were exposed to 75, 150, 300, or 600 ppm cyclohexene 6 hr/day, 5 days/week for 6 months. Rats exposed to 600 ppm exhibited a lower weight gain and a significant increase in alkaline phosphatase (108). It is hydroxylated, conjugated, and eliminated in the urine (113). Microsomal oxidases readily hydroxylate cyclohexene to the corresponding dihydroxy derivatives (114).

**Table 20.5** General Toxicity of Some Cycloalkenes and Dicyclic Hydrocarbons

| Compound | Route | Species | Parameter | Result Time or Dose | Findings | Ref. |
|---|---|---|---|---|---|---|
| Cyclopentene | Oral | Rat | $LD_{50}$ | 2.14 ml/kg | Skin irritation moderate to severe | 88 |
| | Dermal | Rabbit | $LD_{50}$ | 1.59 ml/kg | Corneal injury moderate to severe | 88 |
| | Eye | Rabbit | Draize eye test | 0.1 ml | | 88 |
| Vinylcyclohexene-4 | Oral | Rat | $LD_{50}$ | 3.08 g/kg | | 88 |
| | Dermal | Rabbit | $LD_{50}$ | 20 ml/kg | | 88 |
| | | Mouse | Dermal | 145 g/kg × 54 weeks | Light neoplastic signs | 89 |
| Limonene | | | | | | |
| Acute | Oral | Rat | $LD_{50}$ | 5.0 g/kg | | 90 |
| | | Mouse | $LD_{50}$ | 5.6–6.6 g/kg | | 91 |
| | IP | Mouse | $LD_{50}$ | 1.3 | | 91 |
| | SC | Mouse | $LD_{50}$ | 25.6 ml/kg | | 92 |
| | | Aquatic | TLm96 | 1000 ppm | | 37 |
| Subchronic | Oral | Mouse | | 277–2770 mg/(kg)(day), 1/day × 1 month | Slight decrease in body weight and food consumption | 91 |
| | | Dog | | 1.2–3.6 ml/(kg)(day), 1/day × 6 months | Frequent vomiting, decrease in body weight, blood sugar, and cholesterol, some kidney effects | 93 |
| Teratogenic | Oral | Rat | | 2.87 g/(kg)(day), days 9–15 of gestation | Weak teratogen | 94 |
| | | Mouse | | 2.36 g/(kg)(day), days 7–12 of gestation | Weak teratogen | 95 |
| | | | | 0.250 g/kg | Inhibits tumorigenesis | 96 |

| | | | | | |
|---|---|---|---|---|---|
| Cyclopentadiene | | | | | |
| Dimer | Oral | Rat | $LD_{50}$ | 0.82 g/kg | 97 |
| | Dermal | Rabbit | $LC_{50}$ | 6.72 ml/kg | 97 |
| Monomer | SC | Rabbit | $LD_{00}$ | 0.5–1.0 cm³ | No effects | 30 |
| | | | $LD_{100}$ | 3.0 cm³ | Narcosis with fatal convulsions | 30 |
| Cycloheptatriene | Oral | Rat | $LD_{50}$ | 57 mg/kg | | 98 |
| | | Mouse | $LD_{50}$ | 171 mg/kg | | 98 |
| | SC | Rat | $LD_{50}$ | 442–884 mg/kg | Severe dermal irritant, not a sensitizer | 98 |
| Cyclooctadiene | SC | Rat | | | Skin sensitizer | 99 |
| Cyclododecatriene | SC | Rat | | | Skin sensitizer | 99 |
| Decalin | Oral | Rat | $LD_{50}$ | 4.17 g/kg | | 100 |
| | Dermal | Rat | $LD_{50}$ | 5.90 ml/kg | | 100 |
| Tetralin | Oral | Human | $LT_{LO}$ | 1–1.5 ml/kg | Transient effects | 101 |
| | Oral | Rat | $LD_{50}$ | 2.86 g/kg | | 100 |
| | Dermal | Rat | $LD_{50}$ | 17.3 g/kg | | 100 |
| Dicyclopentadiene | Oral | Rat (M) | $LD_{50}$ | 520 mg/kg | | 102 |
| | | Rat (F) | $LD_{50}$ | 378 mg/kg | | 102 |
| | | | | 0.353 ml/kg | | 103 |
| | | Mouse (M) | $LD_{50}$ | 190 mg/kg | | 102 |
| | | Mouse (F) | | 250 mg/kg | | 102 |
| | IP | Rat | $LD_{50}$ | 0.31 ml/kg | | 103 |
| | Dermal | Rabbit | $LT_{LO}$ | 2.0 g/kg | Irritant, not a sensitizer | 102 |
| | Dermal | Rabbit | $LD_{50}$ | 5.08 ml/kg | | 103 |
| | Eye | Rabbit | | 0.1 ml | Temporary irritation | 102 |
| Turpentine | SC | Rat | | 1.0 ml/kg × 14 exp. | 96 hr post-exposure leukocytosis | 31 |
| | Oral | Man | $LD_{LO}$ | 60–100 g | Fatal, but also recovery from 120 g | 104 |
| | Oral | Rat | | 1.8 mg/(kg)(day) × 3 days | Stimulation of liver microsomes | 105, 106 |

**Table 20.6. Inhalation Toxicity of Some Cycloalkenes**

| Compound | Species | Concentration ppm | Concentration mg/l | Time | Findings | Ref. |
|---|---|---|---|---|---|---|
| Cyclopentene | Human | 10–15 | 28–42 | 5 min | Tolerable | 88 |
| | Rat | Concd. vapor | | 4 hr | Lethal | 88 |
| | Rat | 16,000 | 44.6 | | Lethal to 4 of 6 | 29 |
| | Rat | 112–1139 | | 6 hr/day × 12 weeks | No effects | 29 |
| | Rat | | 8110 | 6 hr/day × 5 weeks | Decreased body weight gain of females | 29 |
| Cyclohexene | Mouse | 8830 | 30 | | Loss of righting reflexes | 34 |
| | | 13,400–14,900 | 45–50 | | Fatal | 34 |
| | Rat, guinea pig, rabbit | 75 | 0.25 | 6 hr/day × 5 days/week | Increased alkaline phosphatase | 107 |
| | | 150 | 0.540 | | Increased alkaline phosphatase | 107 |
| | | 300 | 1.008 | | Increased alkaline phosphatase | 107 |
| | | 600 | 2.016 | | Lower weight gain, increased alkaline phosphatase | 107 |
| Vinylcyclohex-1-ene | Mouse | | 7.5 | $LC_{50}$ | Narcosis | 118 |
| | | | 13.7 | $LC_{50}$ | | 118 |
| Vinylcyclohex-4-ene | Mouse | 6095 | 27.0 | $LC_{100}$ | | 108 |
| | | Satd. vapor | | | Time to death, 15 min | 88 |

Cyclohexene also forms sulfur-containing metabolites in the rabbit (115). Incubation of 1-methyl-1-cyclohexene with *Aspergillus niger* results in the formation of stereospecific (+)-1-methyl-1-cyclohexen-6-ol (116).

For cyclohexene, the OSHA (66) PEL is 300 ppm (1015 mg/m$^3$) and the ACGIH (67) TLV is 300 ppm (or 1010 mg/m$^3$). NIOSH suggests collection of environmental samples on charcoal (68). Cyclohexene can be quantified using infrared and ultraviolet spectroscopy, coulometric titration, nuclear magnetic resonance spectroscopy, and gas chromatography (68, 86). Cyclohexene is on the EPA TSCA Chemical Inventory and the Test Submission Data Base (22).

## 2.3 Cycloheptene

Cycloheptene, suberene, or suberylene, $\overline{CH:CH(CH_2)_4CH_2}$, is a flammable liquid (53). Cycloheptene is metabolized to form the sulfur-containing metabolites *N*-acetyl-*S*-cycloheptyl-L-cysteine and *N*-acetyl-*S*-hydroxycycloheptyl-L-cysteine in the rabbit (115). Its toxicity is similar to that of cyclohexene.

Cycloheptene can be quantified using gas chromatography (86).

## 2.4 Cyclooctene

Cyclooctene, $\overline{CH:CH(CH_2)_5CH_2}$, is a flammable liquid (53). It is primarily used as raw material in the chemical industry. In rats and rabbits, cyclooctene is metabolized to 2- and 3-hydroxycyclooctylmercapturic acid (117).

## 3 ALKENYL CYCLOALKENES

### 3.1 Vinylcyclohexenes

#### 3.1.1 1-Vinylcyclohex-1-ene
Vinylcyclohex-1-ene, cyclohexenylene, 1-ethenylcyclohexene, or tetrahydrostyrene, $CH_2:CHC:CH(CH_2)_3CH_2$, is a flammable liquid (53) and serves as an important chemical intermediate. Vinylcyclohexene is a common component of tobacco smoke and is thought to be formed by dimerization from butadiene (118). Selected physical properties are given in Table 20.4 and toxicity data in Table 20.6.

In mice exposed to 1-vinylcyclohexene, CNS depression was noted at a concentration of 7.5 mg/l. The 4-hr $LC_{50}$ in mice is 13.7 mg/l (119).

#### 3.1.2 4-Vinylcyclohex-1-ene

4-Vinylcyclohex-1-ene or 4-ethenylcyclohexene, $CH_2:CHCH(CH_2)_2\overline{CH:CHCH_2}$, is a colorless liquid. Selected physical properties are given in Table 20.4 and toxicity data in Tables 20.5 and 20.6. Methylated cyclohexene is used in perfume formulations (120). It is an irritant and a CNS depressant at high concentrations. It has a low degree of toxicity following ingestion or dermal absorption.

The oral $LD_{50}$ of 4-vinylcyclohex-1-ene in the rat is 3080 mg/kg, and the dermal $LD_{50}$ in the rabbit is 20 ml/kg (89). Time to death for rats was 15 min in the saturated vapor (89) and a concentration of 8000 ppm for 4 hr proved lethal to four of six rabbits (89). The 4-hr $LC_{50}$ for 4-vinylcyclohex-1-ene in mice is 27 mg/l (109). A dose of 145 g/kg of 4-vinylcyclohex-1-ene applied to mouse skin for 54 weeks provided weak evidence of carcinogenicity (90).

Exposure standards have not been established. However, it has been suggested that the vapor concentration be kept below 100 ppm in any workroom environment.

## 3.2 Terpenes

Further side-chain substitution of cyclohexene, with a methyl, isopropenyl, ethenyl, or other group, forms a class of chemicals named terpenes. Limonene is a $C_{10}$ cyclic olefin and α-pinenes are common monoterpenes.

### 3.2.1 Limonene

Limonene, carvene, dipentene, p-mentha-1,8-diene, or 1-methyl-4-isopropenylcyclohex-1-ene, $CH_2C(CH_3):CHCH_2CH[C(CH_3):CH_2]CH_2$, occurs in the oil of many plants (121). Selected physical properties are given in Table 20.4. Limonene has a highly fragrant citrus odor and is the main constituent (up to 86 percent) of the terpenoid fraction of fruit, flowers, leaves, bark, and pulp from shrubs, annuals, or trees including anise, mint, caraway, polystachya, pine, lime, and orange oil (91, 122). Limonene is found in the gas phase of tobacco smoke (123) and has been detected in urban atmospheres (124). It is used to add fragrance and taste to fruit and essence to flowers and leaves. Limonene has antimicrobial, antiviral (125), antifungal (126), antilarval (127), and insect attractant (128) and repellent (129) properties. In Japan, it has been used to dissolve gallstones (130) and in wound healing (131). It is used as an odorant (132), a solvent (133), an aerosol stabilizer (134), and a wetting and dispersing agent (4). Polylimonene is used as a flavor fixative (12). Limonene is of low acute toxicity; selected oral and dermal toxicity data are given in Table 20.5. Concentrations of 1.2 to 3.6 ml/(kg)(day) for 6 months produced mild symptoms in dogs (94). Limonene is absorbed 100 times more rapidly than water and 10,000 times faster than sodium and chloride ions (135). However, limonene is a dermal irritant and may be a skin sensitizer (4). When dosed with 2869 mg/kg during days 9 to 15 of gestation, rats exhibited a decrease in body weight gain and the fetuses exhibited prolonged ossification of the metacarpal bone and proximal phalanx and slightly decreased spleen and ovarian weights (95). In mice given 2363 mg/kg limonene orally on days 7 to 12 of gestation, the fetuses exhibited a decrease in weight gain and increase in the incidence of abnormal bone formation (96). Rabbits showed decreased body weight gain and six deaths of 21 animals administered 1000 mg/kg during gestation. A dose of 250 mg/kg resulted in no teratogenic effects (136). Limonene and its hydroperoxide injected subcutaneously into C57BL/6 mice decreased the incidence of dibenzopyrene-induced tumors appreciably (97). Metabolic studies in humans and animals demonstrated that 75 to 95 percent of the material was excreted in the urine and up to 10 percent in the feces (137). Limonene administration results in the exacerbation of hyaline droplet formation in the kidneys of male rats (47). The major metabolite, limonene 1,2-oxide, was associated with α-2-microglobulin in the male rat kidney. Limonene enhances hepatic functions in rat liver (138). In another study, 25 percent of the administered limonene was excreted in the bile within 48 hr (139). A metabolite isolated from rabbit urine was identified as *p*-mentha-1,8-dien-10-ol (140). Limonene is readily degraded in soil (141, 142).

The odor threshold for limonene in water is 10 ppb (143). It is quantified using gas chromatography. With proper handling precautions, this material presents no health hazard. Limonene is on the EPS TSCA Chemical Inventory and Test Submission Data Base (22).

## 4 CYCLOOLEFINS AND POLYENES

### 4.1 Cyclopentadiene

Cyclopentadiene, *p*-pentine, pentole, or pyropentylene, $\overline{CH:CHCH_2CH:CH}$, occurs in the $C_6$ to $C_8$ petroleum distillation fraction (1) or from coke-oven light oil fractions (4). Selected physical properties are given in Table 20.4 and toxicity data in Table 20.5. Cyclopentadiene polymerizes easily in the presence of peroxides and trichloroacetic acid (4). It is used as an intermediate in chemical syntheses, especially for Diels–Alder reactions (4).

Cyclopentadiene is irritating to the eyes, respiratory tract, and nose. The 1-hr $LC_{50}$ is 39 mg/l in the rat and 15 mg/l in the mouse (22). When injected subcutaneously into the rabbit, 3.0 ml caused CNS depression with fatal convulsions (30). Signs and symptoms during CNS depression include primary motor unrest and decreased, intermittent respiration rate prior to death. Its vapors produce CNS depression in the frog in 10 min, but recovery is complete in 70 min (30).

The OSHA (66) PEL is 75 ppm (200 mg/m³) and the ACGIH (67) TLV is 75 ppm (203 mg/m³). Cyclopentadiene is on the EPA TSCA Chemical Inventory and the Test Submission Data Base (22).

#### 4.1.1 Methylcyclopentadiene

Methylcyclopentadiene, $CH_3\overline{C:CHCH_2CH:CH}$, is a flammable liquid, isolated in the $C_6$ to $C_8$ petroleum distillation fraction (1).

#### 4.1.2 Methylcyclohexadiene

Methylcyclohexa-1,3-diene, $\overline{CH_2C(CH_3):CHCH_2CH:CH}$, is steam distilled from natural lime oil and terpenes and is used as a fragrance. Selected physical properties are given in Table 20.4.

#### 4.1.3 Cyclooctadiene

Cycloocta-1,5-diene, $\overline{CH_2CH:CH(CH_2)_2CH:CHCH_2}$, is a flammable, highly reactive liquid; selected physical properties are given in Table 20.4. It is produced from petroleum distillation fractions and is used as an intermediate in the plastics industry, as a synthetic lubricant, and in numerous other applications. Cyclooctadiene is corrosive to the skin, with necrosis of the epidermis and ulceration and marked inflammation of the dermis (100). Nonoccluded applications of cyclooctadiene to the skin of rabbits, guinea pigs, and hairless mice produced an immediate erythematous reaction. It is also a skin sensitizer (100).

Cycloocta-1,3-diene, $\overline{CH:CH(CH_2)_3CH:CHCH_2}$, exhibits the same properties as the 1,5 isomer but is more reactive. It has been shown to reduce glutathione concentration in the rat liver (117).

In the rat and rabbit both the 1,3 and 1,5 isomers are metabolized to dihydroxycyclooctylmercapturic acids and to sulfate and glucuronide conjugates (117). Cyclooctadiene applied to the guinea pig skin on three alternate days was irritating to the skin, causing erythema, dry appearance, slight dermal weight increase, and increased arginase activity (65).

### 4.2 Cycloheptatriene and Cyclododecatriene

Cyclohepta-1,3,5-triene or tropilidene, $\overline{CH:CHCH:CHCH:CHCH_2}$, and cyclododecatriene or cyclododeca-1,5,9-triene, $\overline{CH_2[CH:CH(CH_2)_2]_2CH:CHCH_2}$, are flammable liquids. Selected physical properties for cyclohepta-1,3,5-triene are given in Table 20.4. The oral $LD_{50}$ of cyclohepta-1,3,5-triene is 57 mg/kg in the rat and 171 mg/kg in the mouse (99). The dermal $LD_{50}$ in the rat is 442 mg/kg (99). It is a severe dermal irritant but not a sensitizer. Both compounds, when applied to the guinea pig skin on three alternate days, caused erythema, thickening, and increased weight of the epidermal layer and increased dermal arginase activity. All dermal effects were less remarkable than for the cyclooctadienes (65); however, cyclododecatriene is a more potent skin sensitizer (100). Both the $C_8$ and $C_{12}$ compounds were immediately irritant to the rabbit eye, producing mild conjunctivitis which cleared within 48 hr. The eyelids became swollen and exuded a discharge. Blephanitis resolved somewhat faster for the $C_8$, but was still apparent for the $C_{12}$ homologue 1 week after the application.

Analytic procedures are available for quantifying a number of the cycloalkenes. Protective garments should be worn to prevent contact with the skin and eyes. Cycloheptatriene is on the EPA TSCA Test Submission Data Base (22).

## 5 DICYCLIC ALKANES

### 5.1 Decalin

Of the dicyclic alkanes, decalin, bicyclo[4.4.0]decane, decahydronaphthalene, napthalane, or naphthane is the most important member industrially. It is a flammable liquid. Selected physical properties are given in Table 20.4. Decalin occurs naturally in crude oil and is produced commercially by the catalytic hydrogenation of naphthalene. It is also a product of combustion and is released from natural fires (12). It is widely used as a solvent for naphthalene, fats, oils, resins, and waxes, as an alternate for turpentine in lacquers, shoe polish, and floor waxes, as a component in motor fuels and lubricants, and as a fuel for stoves (4). Decalin has a slight odor resembling menthol (4). Decalin has been utilized as a vehicle in long-term skin painting (144, 145) and somatic mutation studies (146).

Decalin is irritating to the eyes, skin, and mucous membranes. Toxicity data are given in Tables 20.5 and 20.7. Dermatitis without serious systemic poisoning has been reported in painters (31). When used as a cleaning agent, it has caused eczema, pruritis, and skin sensitiziation (25). The lowest vapor concentration to affect humans was 100 ppm (10).

The oral $LD_{50}$ is 4.17 g/kg in the rat and the dermal $LD_{50}$ is 5.9 g/kg in the rabbit (101). Exposure to the saturated vapor was lethal to rats in 2 hr (101). The 4-hr $LC_{50}$ is 500 ppm for the rat and 993 ppm for the mouse (22). Exposure to 500 ppm for 4 hr was lethal to four out of six rats (101). Of three guinea pigs exposed to 319 ppm (1.8 mg/l) for 8 hr/day, one died on day 1, the second on day 21, and the third on day 23 (31). Gross and microscopic evaluation revealed lung congestion, kidney, and liver injury. Decalin applied to the skin of guinea pigs on two successive days resulted in death within 10 days of exposure. The systemic tissue injury was identical to injury from inhaled decalin (31).

The aquatic TLm96 was between 10 to 100 ppm (37). Both *cis*- and *trans*-decalin gave rise in the rabbit to racemic decanols (147), which were excreted in the urine, conjugated with glucuronic acid. Guinea pigs dosed orally with decalin exhibited a brownish green urine, an occurrence also reported in workers exposed to a mixture of decalin and tetralin (31). This was not seen in inhalation or dermal studies. Marine organisms can utilize decalin as their sole carbon source (148). *Pseudomonas fluorescens* and *Corynbacterium* degraded decalin more slowly than alkanes and alkenes (149). Decalin may be collected on charcoal, desorbed with carbon disulfide, and quantified by gas chromatography or mass spectroscopy. Decalin is on the EPA TSCA Chemical Inventory and the Test Submission Data Base (22).

## 6 DICYCLIC ALKENES

### 6.1 Tetralin

Tetralin or 1,2,3,4-tetrahydronaphthalene is a flammable liquid. Selected physical properties are given in Table 20.4. The odor of tetralin resembles that of benzene and menthol (4). On standing or storage, peroxides form that can cause explosions during the distillation of tetralin (4). It is prepared by the catalytic hydrogenation of naphthalene or during acidic, catalytic hydrocracking of phenanthrene (1). At 700°C, tetralin yields tars that contain appreciable quantities of 3,4-benzopyrene (150). Tetralin is used widely as a solvent for fats and oils, as an alternative to turpentine in polishes and paints (102), and as a pesticide (12).

Tetralin is irritating to the skin, eyes, and mucous membranes and causes CNS depression at high concentrations (4). Toxicity data are given in Table 20.5. Fol-

**Table 20.7. Inhalation Toxicity of Some Dicyclic Hydrocarbons**

| Compound | Species | No. | Concentration ppm | Concentration mg/l | Time | Parameter | Result | Ref. |
|---|---|---|---|---|---|---|---|---|
| Decalin | Rat | 4 | 500 | 2.83 | | $LC_{65}$ | Lethal to 4 of 6 animals | 100 |
| | | 2 | Satd. vapor | | | $LC_{100}$ | Lethal | 100 |
| | Guinea pig | 8 | 319 | 1.8 | 8 hr/day × 23 days | | 1 of 3 lethal on day 1, 1 on day 21, and 34 on day 23 | 31 |
| Dicyclopentadiene | Rat | 6M, | 359.4 | 1.90 | 4 hr | $LC_{50}$ | | 103 |
| | | 6F | 385.2 | 2.08 | | $LC_{50}$ | | 109 |
| | Rat | 2M, | 1000 | 5.40 | 4 hr | $LC_{100}$ | Eye, nose | 109 |
| | | 2F | | | | $LC_{100}$ | Dyspnea, muscular incoordination, tremors; autopsy: lung, liver congestion | |
| | | 2M, | 2500 | 13.53 | 1 hr | $LC_{100}$ | Eye, nose irritation, dyspnea narcosis; autopsy: congestion of lung liver, kidney | 109 |
| | | 2F | | | | $LC_{100}$ | | |
| | Mouse | 6 | 145.5 | 0.70 | | $LC_{50}$ | | 103 |
| | Guinea pig | 6 | 770.5 | 3.721 | | $LC_{50}$ | | 103 |
| | Rabbit | | 771.0 | 3.723 | | | | 103 |
| | Rat | | 19.7 | 0.093 | 7 hr × 89 days | | No effects | 103 |
| | | | 35.2 | 0.170 | 7 hr × 89 days | | Kidney lesions in male rats | 103 |

| | | | | | | |
|---|---|---|---|---|---|---|
| | | 73.8 | 0.356 | 7 hr × 89 days | | Kidney lesions in male rats | 103 |
| | 4M | 100 | 0.483 | 6 hr × 15 dys | LC$_{00}$ | No toxic signs; organs normal | 103 |
| | 2M | 250 | 1.207 | 6 hr × 10 days | LC$_{25}$ | Weight loss, nose irritation, dyspnea tremors, hypersensitivity; hematology: organs normal | 109 |
| Turpentine | | | | | | | |
| Human | | 720–1100 | 3.96–6.05 | | LT$_{Lo}$ | Complaint of eye irritation, headache, dizziness, nausea, chest pain, and visual disturbances | 31 |
| Cat | | 746–782 | 4.1–4.3 | 3.5–4 hr | LT$_{Lo}$ | Lethargy, incoordination, nausea | 110 |
| | | 1090 | 6.0 | 6 hr | | Collapse, later recovery | 110 |
| | | 1455 | 8.0 | 0.5–1 hr | | Incoordination, tonic convulsions | 110 |
| | | 2909–4364 | 16–24 | 0.3–1.5 hr | LC$_{80}$ | Lethal to 80% of test animals | 110 |
| Dog | | 54.5–112 | 0.3–1 | 1 hr/day × 8 days | LD$_{00}$ | No effects | 110 |
| | | 818 | 4.5 | 3.5–4.5 hr | LT$_{Lo}$ | Nausea, incoordination, weakness, light paralysis | 110 |

lowing the ingestion of 250 ml of an ectoparasiticide containing tetralin, nausea, vomiting, intragastric discomfort, transient liver and kidney damage, green-gray urine, and some clinical and enzymatic changes were noted (102). All signs, symptoms, and effects subsided within 2 weeks (102). The oral $LD_{50}$ is 2.86 g/kg in the rat and the dermal $LD_{50}$ is 17.3 g/kg in the rabbit (101). The exposure of rats to the saturated vapor for 8 hr was not lethal (101). The 8-hr $LC_{50}$ in guinea pigs is 275 ppm (22). Tetralin is moderately irritating to the skin and mildly irritating to the eye (101). Tetralin causes cataracts in some animals (25). The aquatic TLm96 is between 10 and 100 ppm (37). Tetralin is cytotoxic to ascites tumor cells when incubated in vitro for 5 hr (151).

Of the administered dose, 87 to 99 percent was excreted in the urine as glucuronide and 0.6 to 1.8 percent in the feces (152). Tetralin hydroperoxide is produced in rat liver (153). Tetralin was not metabolized by the mussel and up to 80 percent was excreted in unchanged form when the organism was transferred to fresh seawater (154). Tetralin has been successfully used to destroy crown gall and olive knot neoplasms (85).

Occupational exposure occurs in the paint, solvent, and varnish industries. No official exposure standards have been established. The odor threshold in water is 18.0 ppm (143). Tetralin can be quantified by gas chromatography and mass spectrometry. Tetralin is on the EPA TSCA Chemical Inventory and the Test Submission Data Base (22).

### 6.2 Dicyclopentadiene

Dicyclopentadiene or 3a,4,7,7a-tetrahydro-4,7-methanoindene is a colorless, crystalline combustible solid (7, 110). Selected physical properties are listed in Table 20.4. It is produced by thermal cracking of petrochemical feedstocks or as a byproduct of the coke-oven industry (156). It is also formed by spontaneous dimerization of cyclopentadiene. Cyclopentadienes are used in the synthesis of chlorinated hydrocarbon pesticides; in paint, varnish, and resin manufacture (157); and in elastomers used as water pond liners (155). It is moderately to highly toxic, as shown in Tables 20.5 and 20.7. Dicyclopentadiene has a disagreeable odor similar to camphor (4). Occupational exposure resulted in headaches for up to 2 months (18). To the eyes it represents a temporary hazard, because immediate washing with water does not shorten injury time (103). The oral $LD_{50}$ is 0.35 g/kg in the rat, 0.19 g/kg in the mouse, and 1.2 g/kg in cattle, and the dermal $LD_{50}$ is 6.72 g/kg in the rabbit. The 4-hr $LC_{50}$ is 145 ppm in the mouse, 770 ppm in the guinea pig, and 771 ppm in the rabbit (104, 110). Following oral doses in rats, general congestion, hyperemia, and focal hemorrhage of the kidney, intestine, stomach, bladder, and the lung occurred. Leukocytosis was noted in rats following a single 5-ml or repeated 1-ml subcutaneous injections. Rats exposed to 55 or 74 ppm 7 hr/day for 89 days exhibited kidney and lung damage (18). Some of the inhaled dicyclopentadiene is exhaled in unchanged form; the balance is hydroxylated, conjugated, and excreted in the urine as glucuronide. Dicyclopentadiene was not mutagenic in the Ames test (158). The acquatic TLm96 is between 1 and 10 ppm (37).

In 1976, limits of 1.3 ppm in food and water for drinking, irrigation, recreation, and aquatic life were recommended (159). The OSHA PEL (66) is 5 ppm (30 mg/m$^3$) and the ACGIH TLV (67) is 5 ppm (27 mg/m$^3$). A variety of analytic determination methods using color detection indicators, thin-layer gas chromatography, and ultraviolet spectral procedures are available.

The odor threshold is 0.03 mg/m$^3$ and is irritating at 2.7 mg/m$^3$ (160). Protective clothing including boots, gloves, and goggles as well as respiratory protection is recommended (16). Dicyclopentadiene is on the EPA TSCA Chemical Inventory and the Test Submission Data Base (22).

## 6.3 Pinenes

α-Pinene, 2-pinene, pinene, or 2,6,6-trimethylbicyclo[3.1.1]hept-2-ene is a fragrant, flammable liquid. Selected physical properties are given in Table 20.4. β-Pinene and γ-pinene are position isomers and exhibit similar characteristics. They occur naturally in a variety of trees and shrubs, and air concentrations near pine forests may reach 500 to 1200 μg/m$^3$ (135). Pinene is obtained from wood turpentine or pine oil by distillation and is widely used in the manufacture of camphor as an insect attractant and repellent, insecticides, perfume bases, plasticizers, solvents, and synthetic pine oil (4). The pinenes have been utilized as antibacterial agents.

Pinene is readily absorbed through the respiratory tract, skin, and intestine (4). It is irritating to the skin and mucous membranes and causes dermal eruption and an occasional benign tumor following chronic contact (4). Large doses result in delirium, ataxia, and kidney damage. Prolonged exposure has resulted in palpitation, dizziness, nervous disturbances, chest pain, bronchitis, and nephritis. The human oral fatal dose is estimated to be 180 g (4). Subjects with cardiac diseases may experience increased olfactory sensitivity toward pinene (162). Pinene is lethal in conifer needle-chewing insects (163), causes leukemic changes in fowl (164), and causes deviations in avian plasma proteins with accompanying erythroblastosis (165). Pinene exhibits choleretic action in male rats (166). It increases the microsomal protein content (167), cytochrome P450, and the activities of some hydroxylases (168). Pinene is degraded by microbiological organisms in soil (169).

The odor threshold of 2-pinene in water is 60 ppb (143). Occupationally, workers should be protected from inhaling pinene vapors and from direct skin contact. Analytically, it can be quantified using gas chromatographic procedures.

## 6.4 Turpentines

Gum turpentine is the steam-volatile fraction of pine tree pitch. Chemically, it consists of 58 to 65 percent γ-pinene along with β-pinene and other isomeric terpenes. Wood turpentine, obtained from waste wood chips or sawdust, contains 80 percent γ-pinene, 15 percent monocyclic terpenes, 1.5 percent terpene alcohols, and other terpenes. Sulfate turpentine is a by-product in paper manufacture. The turpentines are used in surface coatings and as solvents for oils, fats, waxes and resins, lacquers, and polishes. Turpentine is also used therapeutically as a human ointment and counterirritant (4) and in veterinary practice as an expectorant, rubifacient, and antiseptic (4), owing to its antimicrobial properties. Selected physical properties are given in Table 20.4.

Turpentine presents a moderate health hazard, because it is readily absorbed through the respiratory tract or skin. Toxicity data are given in Tables 20.5 and 20.7. It is irritating to the skin, eyes, nose, and mucous membranes, and major systemic effects include kidney and bladder injury (58). The fatal oral dose may be as low as 110 g; however, recovery from a 120 g has been reported (105). Symptoms of intoxication consist of gastroenteric pain, nausea, vomiting, toxic nephritis with hematuria, albuminuria, and oliguria (4). At high concentrations, coma may be followed by death. At lower concentrations, pronounced anemia occurs occasionally. When ingested, turpentine is an aspiration hazard (23). When excreted in the urine, turpentine or its volatile components can be recognized by the very specific odor of violets (105). Skin contact can cause eczema (58). Occupational contact dermatoses are common among workers in the chemical, rubber, and welding industries (170–172) and also in the home (172). In humans, chronic inhalation of turpentine has caused extensive glomerulonephritis (174). Turpentine can be corrosive to the eye (58). Chronic dermal contact produces inflammation, with an effect on the collagen of the dermis (175). Turpentine is considered to be a skin sensitizer (176). Of the many cases of dermatitis due to turpentine, approximately 14.8 percent have resulted in eczema (177). Hypersensitivity to wood turpentine may be due to impurities such as formic acid, formaldehyde, and phenols (178). Chronic dermal contact may cause allergic erythema, headaches, coughing, and sleeplessness (179). In workers, acute concentrations of 0.01 to 5.2 mg/l caused irritation (172). Repeated injections of turpentine and insulin caused necrotizing fasciitis (180). The 1-hr $LC_{50}$ in rats is 20 mg/l, the 6-hr $LC_{50}$ in rats is 12 mg/l, and the 2-hr $LC_{50}$ in mice is 29 mg/l (181). Signs and symptoms included CNS depression, increased respiration rate, and decreased tidal volume, but no pulmonary lesions (181). The highest concentrations in organs were in the brain and spleen (181). Exposure to 5 to 10 mg/l for 1 to 2 hr was lethal to several test animals (174). In cats exposed to 4.1 to 4.3 mg/l of turpentine for 3.5 to 4 hr, lethargy, incoordination, and nausea were noted and at 6.0 mg/l for 3 hr, prostration was noted with recovery in 20 min. Exposure to 8 mg/l for 1.0 to 1.5 hr resulted in incoordination, whereas exposure to 16 to 24 mg/l for 40 min to 1.5 hr was lethal to four of five cats (111). In dogs, exposure to 0.3 to 1 mg/l daily for 8 days yielded no effects; 4.5 mg/l for 3.5 to 4.5 hr caused nausea; and 6.0 mg/l for 3 hr caused nausea, incoordination, weakness, and light paralysis, with fairly rapid recovery (111). No effects were noted in guinea pigs exposed to 715 ppm (182). In rabbits, intradermal injection of turpentine in peanut oil produced erythematosis and granulocytes in connective tissue, which did not clear in 48 hr. After 9 days, anastomosis, round cell infiltration, and evidence of connective tissue remodeling were noted (183). Intradermal abscesses are easily produced (184) in the rat but are not seen in the mouse (185). In avian species, erythroblastosis has been observed (165).

A portion of the turpentine absorbed in industrial exposures is exhaled unchanged in expired air. The remainder is metabolized, conjugated with glucuronic acid, and excreted in the urine (15).

Metabolic studies showed that certain protein synthesis inhibitors suppressed turpentine-induced inflammation (186). Oral doses of 1.8 mg/kg/day for 3 days in the rat stimulated microsomal enzymes and reduced the toxicity toward parathion (81, 82). In the guinea pig, turpentine protected against the hypersensitivity to 6-mercaptopurine (187). When applied to the skin, turpentine supported the growth of tumors in the rabbit, but not in the mouse (188).

The OSHA PEL (66) is 100 ppm (560 mg/m$^3$) and the ACGIH TLV (67) is 100 ppm (556 mg/m$^3$). No effects were noted at 100 ppm, but throat irritation was seen at 125 ppm and eye and nose irritation at 175 ppm (189). NIOSH recommends the absorption of turpentine on charcoal, desorption with carbon disulfide, and quantification using gas chromatography (68). Procedures are available for the removal of turpentine from air and water (190). Occupationally, turpentine is a hazard on direct skin contact or when vapors are inhaled. Protective measures include the use of skin barriers and ointments (191, 192). Turpentine is on the EPA TSCA Chemical Inventory and the Test Submission Data Base (22).

## 7 OTHER CYCLIC OLEFINS

### 7.1 Ethylidenenorbornene

Ethylidenenorbornene or 5-ethylidenebicyclo[2.2.1]hept-2-ene, a compound closely related to pinene, has an oral $LD_{50}$ of 3.2 g/kg in rats (193). The 4-hr $LC_{50}$ is 732 ppm for the female mouse and 3100 ppm for the male rabbit. Based on exposures of 7 hr/day for 3 months in dogs, a TLV of 5 ppm for an 8-hr workday has been suggested (194).

### 7.2 Camphene

Camphene or 3,3-dimethyl-2-methylene norcamphone can be classified as a terpene or as diisoprene. It occurs naturally in the oils of several plants and can be prepared from α-pinene. It is isolated as cubic crystals (4). It forms flammable vapors (7) and is used in tablet form for mothproofing (195). It is also used in the manufacture of camphor and in the cosmetic, perfume, and food flavoring industries (196). Selected physical properties are given in Table 20.4.

Camphene was not irritating when applied at 4 percent in petrolatum (197). The oral $LD_{50}$ value in the rat is >5 g/kg, and the dermal $LD_{50}$ is >2.5 g/kg in the rabbit (198). It inhibits plant and fungal growth and is a natural protector against insects and mites (197). Camphene can be exhaled unchanged following dermal application or intravenous injection and is metabolized to the glycol, conjugated, and excreted in the urine (197). Camphene is analyzed by gas chromatography or mass spectrometry. Protective clothing including gloves and face shield are recommended (16). Camphene is on the EPA TSCA Chemical Inventory and the Test Submission Data Base (22).

### 7.3 Caryophyllene

Caryophyllene, a dicyclic sesquiterpenoid, is a fragrant liquid that occurs naturally in many plants (4, 121). It occurs in the alpha, beta, and iso forms (4). It is widely used in the perfume industry. Selected physical properties for α-caryophyllene are given in Table 20.4.

In the boll weevil, caryophyllene suppresses intestinal microbial flora (189). It is a skin irritant in the rabbit (22) and can be quantified using gas chromatography. Caryophyllene is on the EPA TSCA Chemical Inventory (22).

## REFERENCES

1. R. F. Gould, Ed., *Refining Petroleum for Chemicals*, American Chemical Society, Washington, DC, 1970.
2. D. R. Lide, Ed., *CRC Handbook of Chemistry and Physics*, 71st ed., CRC Press, Boca Raton, FL, 1990–1991.
3. *Laboratory Waste Disposal Manual*, 2nd ed., revised, Manufacturing Chemists Association, Inc., Washington, DC, 1974.
4. S. Budavari, Ed., *The Merck Index*, 11th ed., Merck and Company, Rahway, NJ, 1989.
5. *Registry of Toxic Effects of Chemical Substances*, U.S. Dept. Health, Education, and Welfare, Cincinnati, OH, September, 1991.
6. V. B. Guthrie, Ed., *Petroleum Products Handbook*, McGraw-Hill, Inc., New York, 1960.
7. N. Sax, *Dangerous Properties of Industrial Materials*, 7th ed., Van Nostrand Reinhold, New York, 1987.
8. *Toxic and Hazardous Industrial Chemicals Safety Manual for Handling and Disposal with Toxicity Data*, International Technical Information Institute, Tokyo, 1988.
9. *Fire Protection Guide on Hazardous Chemicals*, 10th ed., National Fire Protection Association, Quincy, MA, 1991.
10. *Handbook of Organic Industrial Solvents*, 4th ed., American Mutual Insurance Alliance, Chicago, 1972.
11. F. R. Rossini, K. S. Pitzer, R. L. Arnett, R. M. Braun, and G. C. Pimentel, *Selected Values of Physical Thermodynamic Properties of Hydrocarbons and Related Compounds*, Carnegie Press, Pittsburgh, PA, 1953.
12. Hazardous Substance Data Bank, TOXNET System, National Library of Medicine, Bethesda, MD, 1992.
13. N. I. Sax and R. J. Lewis, Sr., Eds., *Hawley's Condensed Chemical Dictionary*, 11th ed., Van Nostrand Reinhold Co., New York, 1987.
14. E. W. Flick, *Industrial Solvents Handbook*, 3rd ed., Noyes Publications, Park Ridge, NJ, 1985.

15. J. H. Kuney, Ed., *Chemcyclopedia 90*, American Chemical Society, Washington, DC, 1990.
16. U.S. Coast Guard, Department of Transportation, *CHRIS-Hazardous Chemical Data*, Vol. II, U.S. Government Printing Office, Washington, DC, 1984–1985.
17. K. Verschueren, *Handbook of Environmental Data of Organic Chemicals*, 2nd ed., Van Nostrand Reinhold, New York, 1983.
18. *Documentation of the Threshold Limit Values for Substances in Workroom Air*, 3rd ed., American Conference of Governmental Industrial Hygienists, Cincinnati, 1984.
19. J. Rosenberg, "Solvents," *in Occupational Medicine*, J. LaDou, Ed., Appelton & Lange, Norwalk, CT, 1990, pp. 359–386.
20. L. R. Goldfrank, A. G. Kulberg, and E. A. Bresnitz, "Hydrocarbons" *in Goldfrank's Toxicologic Emergencies*, 4th ed., L. R. Goldfrank, N. E. Flourenbaum, N. A. Lewin, R. S. Weisman, and M. A. Howland, Eds., Appelton & Lange, Norwalk, CT, 1990, pp. 759–780.
21. P. D. Bryson, *Comprehensive Review of Toxicology*, 2nd ed., Aspen Publishers, Rockville, MD, 1989, pp. 553–563.
22. Registry of Toxic Effects of Chemical Substances, On-Line, TOXNET System, National Library of Medicine, Bethesda, MD, 1992.
23. R. E. Gosselin, R. P. Smith, and H. C. Hodge, *Clinical Toxicology of Commercial Products*, 5th ed., Williams and Wilkins, Baltimore, 1984.
24. L. S. Andrews and R. Snyder, "Toxic Effects of Solvents and Vapors," in M. O. Amdur, J. Doull, and C. D. Klassen, Eds., *Casarett and Doull's Toxicology*, 4th ed., Pergamon Press, New York, 1991, pp. 681–722.
25. R. Snyder, Ed., *Ethel Browning's Toxicity and Metabolism of Industrial Solvents*, 2nd ed., Vol. 1, *Hydrocarbons*, Elsevier, Amsterdam, 1987.
26. Manufacturing Chemists Association, Inc., Chemical Safety Data Sheets, Washington, DC, 1991.
27. A. G. Gilman, T. W. Rall, A. S. Nies, and P. Taylor, Eds., *The Pharmacological Basis of Therapeutics*, 8th ed., Pergamon, New York, 1990.
28. R. W. Stoughton and P. D. Lamson, *J. Pharmacol. Exp. Ther.*, **58**, 74 (1936).
29. G. Kimmerle and J. Thyssen, *Int. Arch. Arbeitsmed.*, **34**(3), 177 (1975).
30. W. F. Von Oettingen, *Toxicity and Potential Dangers of Aliphatic and Aromatic Hydrocarbons*, Publ. Health Bull. No. 255 (1940).
31. F. A. Patty, *Industrial Hygiene and Toxicology*, Vol. II, Wiley, New York, 1963.
32. J. F. Treon, W. E. Crutchfield, and K. V. Kitzmiller, *J. Ind. Hyg. Toxicol.*, **25**(8), 323 (1943).
33. J. F. Treon, W. E. Crutchfield, and K. V. Kitzmiller, *J. Ind. Hyg. Toxicol.*, **25**(6), 199 (1943).
34. N. W. Lazarew, *Arch. Exp. Path. Pharmakol.*, **143**, 223 (1929).
35. E. T. Kimura, D. M. Ebert, and P. W. Dodge, *Toxicol. Appl. Pharmacol.*, **19**, 699 (1971).
36. W. B. Deichmann and T. J. LeBlanc, *J. Ind. Health Toxicol.*, **25**, 415, (1943).
37. R. W. Hann and P. A. Jensen, *Water Quality Characteristics of Hazardous Materials*, Vols. 1–4, Environmental Engineering Division, Civil Engineering Department, Texas A & M University, 1974.

38. D. Irie, T. Sasaki, and R. Ito, *Toho Igakkai Zasshi*, **20**(5–6), 772 (1973).
39. D. R. Webb, G. M. Ridder, and C. L. Alden, *Food Chem. Toxicol.*, **27**, 639–649 (1989).
40. L. C. Stone, M. S. McCracken, R. L. Kanerva, and C. L. Alden, *Food Chem. Toxicol.*, **25**, 35–41 (1987).
41. G. M. Ridder, E. C. von Bargen, C. L. Alden, and R. D. Parker, *Fundam. Appl. Toxicol.*, **15**, 732–743 (1990).
42. C. L. Alden, R. L. Kanerva, G. M. Ridder, and L. C. Stone, *Renal Effects of Petroleum Hydrocarbons*, Princeton Scientific, Princeton, 1984, pp. 107–120.
43. A. M. Bernard, R. de Russis, J. C. Normand, and R. R. Lauwerys, *Toxicol. Letters*, **45**, 271–280 (1989).
44. C. L. Gaworski, C. C. Haun, J. D. MacEwen, E. H. Vernot, and R. H. Bruner, *Fundam. Appl. Toxicol.*, **5**, 785–793 (1985).
45. T. E. Eurell, *Govt. Rep. Announcements Index*, **91**, 540–541 (1991).
46. M. P. Serve, *Govt. Rep. Annoncements Index*, **86**, 72–79 (1986).
47. R. L. Kanerva and C. L. Alden, *Food Chem. Toxicol.*, **25**, 355–358 (1987).
48. R. L. Kanerva, M. S. McCracken, C. L. Alden, and L. C. Stone, *Food Chem. Toxicol.*, **25**, 53–61 (1987).
49. M. P. Serve, *Govt. Rep. Announcements Index*, **89**, 588–594 (1989).
50. C. A. Halder, C. E. Holdsworth, B. Y. Cockerell, and V. J. Piccirillo, *Toxicol. Indust. Health*, **1**, 67–68 (1985).
51. J. A. Swenberg, B. Short, S. Borghoff, J. Strasser, and M. Charbonneau, *Toxicol. Appl. Pharmcol.*, **97**, 35–47 (1989).
52. D. R. Dietrick and J. A. Swenberg, *Fundam. Appl. Toxicol.*, **16**, 749–762 (1991).
53. "Hazardous Materials Regulations and Miscellaneous Amendments," Title 49, Chapter 1, Materials Transportation, Bureau of Transportation, *Fed. Reg.*, **4**(252), 57018 (1976).
54. J. C. Arcos, M. F. Argus, and G. Wolf, *Chemical Induction of Cancer*, Vol. I, Academic Press, New York, 1958.
55. E. A. Wahrenbrock, E. I. Eger, R. B. Laravuso, and G. Maruschak, *Anesthesiology*, **40**(1), 19 (1974).
56. W. B. Deichmann and H. W. Gerarde, *Toxicology of Drugs and Chemicals*, Academic Press, New York, 1969.
57. G. W. Seuffert and K. F. Urbach, *Anesth. Analg. Curr. Res.*, **46**(2), 267 (1967).
58. M. M. Key, A. F. Henschel, J. Butler, R. N. Ligo, and I. R. Tabershaw, *Occupational Diseases, A Guide to their Recognition*, rev. ed., U.S. Department of Health, Education, and Welfare, Washington, DC, June, 1977.
59. A. L. Cowles, H. H. Borgstadt, and A. J. Gillies, *Anesthesiology*, **36**(6), 588 (1972).
60. N. B. Anderson, *Anesthesia*, **29**(1), 113 (1968).
61. *International Labor Office. Encyclopedia of Occupational Health and Safety*, Vols. I and II, Geneva, Switzerland, 1983.
62. J. C. Krantz, Jr., C. J. Carr, and J. F. Vitcha, *J. Pharmcol. Exp. Ther.*, **94**, 315 (1948).
63. J. K. Sugden and B. K. Razdan, *Pharmacol. Acta Helv.*, **47**(5), 257 (1972).
64. H. W. Gerarde, *Arch. Environ. Health*, **6**, 329 (1963).
65. V. K. H. Brown and V. L. Box, *Brit. J. Dermatol.*, **85**, 432 (1971).

66. "Occupational Health and Environmental Control," Subpart G, Sec. 1910.93 Air Contaminants, *Fed. Reg.*, **39**(125), 235–240 (1974).
67. *Threshold Limit Values for Chemical Substances and Physical Agents and Biological Exposure Indices for 1991–1992*, American Conference of Governmental Industrial Hygienists, Cincinnati, 1991.
68. J. P. Conkle, B. J. Camp, and B. E. Welch, *Arch. Environ. Health*, **30**(6), 290 (1975).
69. A. M. Aliev, G. Y. Alikishi-Zade, and V. A. Shlyapnikov, *Otkrytiya, Izobret, Prom. Tovarnye Znaki*, **51**(44), 59 (1974).
70. P. Foss, G. Takeguchi, H. Tai, and C. Sik, *Ann. N.Y. Acad. Sci.*, **180**, 126 (1971).
71. H. B. Elkins, *The Chemistry of Industrial Toxicology*, 2nd ed., Wiley, New York, 1959.
72. C. R. Noller, *Chemistry of Organic Compounds*, 3rd ed., W. B. Saunders, Phildelphia, 1966.
73. Z. Ziegenmeyer, M. Reuter, and F. Meyer, *Arch. Int. Pharmacodyn. Ther.*, **224**(2), 338 (1976).
74. L. Braier, *Haematologica*, **58**(7–8), 491 (1973).
75. I. Franchini, A. Cavatorta, M. D'Errico, M. DeSantis, G. Romito, R. Gatti, G. Juvarra, and G. Palla, *Experientia*, **34**(2), 250 (1978).
76. V. Ullrich, *Z. Physiol. Chem.*, **350**(3), 357 (1969).
77. W. Diehl, J. Schaedelin, and V. Ullrich, *Z. Physiol. Chem.*, **351**(11), 1359 (1970).
78. E. J. McKenna and M. Coon, *J. Biol. Chem.*, **245**(15), 3882 (1970).
79. G. Mohn, *Xenobiotica*, **7**(1–2), 96 (1977).
80. A. Kraemer, H. Staudinger, and V. Ullrich, *Chem.-Biol. Interactions*, **8**, 11 (1974).
81. J. D. Walker and R. R. Colwell, *Appl. Environ. Microbiol.*, **31**(2), 198 (1976).
82. T. R. Norton, "Metabolism of Toxic Substances," Chapter 4 in *Toxicology, the Basic Science of Poisons*, L. J. Casarett and J. Doull, Eds., Macmillan, New York, 1975.
83. U. Frommer, V. Ullrich, and H. Staudinger, *Z. Physiol. Chem.*, **351**, 913 (1970).
84. M. N. Schroth and D. C. Hildebrand, *Phytopathology*, **58**, 848 (1968).
85. L. Meites, *Handbook of Analytical Chemistry*, 1st ed., McGraw-Hill, New York, 1963.
86. K. Masui and H. Kawauchi, Jap. Pat. 78:08777, March 31, 1978.
87. R. G. Gould, *Photochemical Smog and Ozone Reactions*, American Chemical Society, Washington, DC, 1972.
88. H. F. Smyth, Jr., C. C. Carpenter, C. S. Weil, U. C. Pozzani, J. A. Striegel, and J. S. Nycum, *Am. Ind. Hyg. Assoc. J.*, **30**, 470 (1969).
89. B. L. Van Duuren, N. Nelson, L. Orris, E. D. Palmes, and F. L. Schmitt, *J. Natl. Cancer Inst.*, **31**, 41 (1963).
90. D. J. Opdyke, *Food Cosmet. Toxicol.*, **13**, 733 (1975).
91. M. Tsuji, Y. Fujisaki, Y. Arikawa, S. Masuda, S. Kinoshita, A. Okubo, K. Noda, H. Ide, and Y. Iwanaga, *Oyo Yakuri*, **9**(3), 387 (1975).
92. M. Tsuji, Y. Fujisaka, K. Yamachika, K. Nakagami, F. Fujisaki, M. Mito, T. Aoki, S. Kinoshita, A. Okubo, and I. Watanabe, *Oyo Yakuri*, **8**(10), 1439 (1974).
93. M. Tsuji, Y. Fujisaki, Y. Arikawa, S. Masuda, T. Tanaka, K. Sato, K. Noda, H. Ide, and M. Kikuchi, *Oyo Yakuri*, **9**(5), 775 (1975).
94. M. Tsuji, Y. Fujisaki, A. Okubo, Y. Arikawa, K. Noda, H. Ide, and T. Ikeda, *Oyo Yakuri*, **10**(2), 179 (1975).

95. R. Kodama, A. Okubo, E. Araki, K. Noda, H. Ide, and T. Ikeda, *Oyo Yakuri*, **13**(6), 863 (1977).
96. F. Homburger, A. Treger, and E. Boger, *Oncology*, **25**(1), 1 (1971).
97. H. F. Smyth, Jr., C. P. Carpenter, C. S. Weil, and U. C. Pozzani, *Arch. Ind. Hyg. Occup. Med.*, **10**, 61 (1954).
98. V. K. H. Brown, L. W. Ferrigan, and D. E. Stevenson, *Ann. Occup. Hyg.*, **10**, 123 (1967).
99. V. K. H. Brown and C. G. Hunter, *Brit. J. Ind. Med.*, **25**, 75 (1968).
100. H. F. Smyth, Jr., C. P. Carpenter, and C. S. Weil, *Arch. Ind. Hyg. Occup. Med.*, **4**, 199 (1951).
101. D. E. Drayer and M. M. Reidenberg, *Drug Metab. Dispos.*, **1**(3), 577 (1973).
102. E. R. Hart and J. C. Dacre, *Proc. First Int. Congr. Toxicol., Toronto*, G. L. Plaa and W. A. M. Duncan, Eds., 1977.
103. E. R. Kinkead, U. C. Pozzani, D. L. Geary, and C. P. Carpenter, *Toxicol. Appl. Pharmacol.*, **20**, 552 (1971).
104. S. Moeschlin, *Poisoning, Diagnosis and Treatment*, 1st Am. ed., Grune & Stratton, New York, 1965.
105. F. Sperling, H. K. U. Ewenike, and T. Farber, *Environ. Res.*, **5**(2), 164 (1972).
106. R. Y. Omirov and A. Y. Aberkulov, *Med. Zh. Uzb.*, **1**, 50 (1972).
107. S. Laham, *Toxicol. Appl. Pharmacol.*, **37**(1), 155 (1976).
108. *Monographs on the Evaluation of Carcinogenic Risks of Chemicals to Man*, Vol. 11, 1976, p. 277.
109. J. C. Gage, *Brit. J. Ind. Med.*, **27**(1), 1 (1970).
110. K. B. Lehmann, *Arch. Hyg.*, **83**, 239 (1914).
111. J. Pohl, *Zentralbl. Gewerbehyg. Unfallverhuet.*, **12**, 91 (1925).
112. W. J. Canady, D. A. Robinson, and H. D. Colby, *Biochem. Pharmacol.*, **23**(21), 3075 (1974).
113. K. C. Leibman and E. Ortiz, *J. Pharmacol. Exp. Therp.*, **173**(2), 242 (1970).
114. S. P. James, D. J. Jeffery, R. H. Waring, and D. A. White, *Biochem. Pharmacol.*, **20**, 897 (1971).
115. K. Ganapathy, K. S. Khanchandani, and P. K. Bhattacharyya, *Indian J. Biochem.*, **3**, 66 (1966).
116. R. H. Waring, *Xenobiotica*, **1**(3), 303 (1971).
117. E. Gil-Av and J. Shabtai, *Nature*, **197**, 1065 (1963).
118. M. F. Savchenkov, *Gig. Sanit.*, **30**(7), 28 (1965).
119. D. Helmlinger and P. Naegeli, Ger. Pat. 2622611, January 13, 1977.
120. P. deMayo, *Mono- and Sesquiterpenoids*, Vol. II, Interscience, New York, 1959.
121. D. J. Opdyke, *Food Cosmet. Toxicol.*, **13**, 731 (1975).
122. H. Elmenhorst and H. P. Harke, *Z. Naturforsch.*, **23b**, 1271 (1968).
123. W. Bertsch, R. C. Chang, and A. Alatkis, *J. Chromatogr. Sci.*, **12**, 175 (1974).
124. A. S. Bondarenko, B. E. Aizenman, L. A. Bakena, and I. S. Kozhina, *Rastit. Resur.*, **10**(4), 583 (1974).
125. H. T. Brodrick, *Phytophylactica*, **3**(2), 69 (1971).
126. K. L. Stevens and L. Jurd, U.S. Pat. 3954991, May 4, 1976.

127. G. O. Osborne and J. F. Boyd, *N. Z. J. Zool.*, **1**(3), 371 (1974).
128. R. P. Bordasch and A. A. Berryman, *Can. Entomol.*, **109**(1), 95 (1977).
129. H. Igimi, T. Hisatsugu, and M. Nishimura, *Am. J. Dig. Dis.*, **21**(11), 926 (1976).
130. M. O. Karryev, *Mater. Ybileinoi Resp. Nauchin. Konf. Farm.*, **62** (1972).
131. T. Yoshino, Jap. Pat. 77:72825, June 17, 1977.
132. P. R. Perez, P. Carmona, B. Lafuente, J. Bellanato, and A. Hidalgo, *Rev. Agroquim. Technol. Aliment.*, **17**(1), 59 (1977).
133. S. Iscowitz, U.S. Pat. 3977826, August 31, 1976.
134. H. Roemmelt, A. Zuber, K. Dirnagl, and H. Drexel, *Muench. Med. Wochenschr.*, **116**(11), 537 (1974).
135. R. Kodama, A. Okubo, K. Sato, E. Araki, K. Noda, H. Ide, and T. Ikeda, *Oyo Yakuri*, **13**(6), 885 (1977).
136. R. Kodama, T. Yano, K. Furkawa, K. Noda, and H. Ide, *Xenobiotica*, **6**(6), 377 (1976).
137. T. Ariyoshi, M. Arakaki, K. Ideguchi, Y. Ishizuka, K. Node, and H. Ide, *Xenobiotica*, **5**(1), 33 (1975).
138. H. Igimi, M. Nishimura, R. Kodama, and H. Ide, *Xenobiotica*, **4**(2), 77 (1974).
139. R. Kodama, K. Noda, and H. Ide, *Xenobiotica*, **4**(2), 85 (1974).
140. N. R. Ballal, P. K. Bhattacharyya, and P. M. Rangachari, *Biochem. Biophys. Res. Commun.*, **23**(4), 473 (1966).
141. R. S. Dhavalikar, P. M. Rangachari, and P. K. Bhattacharyya, *Indian J. Biochem.*, **3**(3), 258 (1966).
142. W. H. Stahl, *Compilation of Odor and Taste Threshold Values Data*, American Society for Testing and Materials, Philadelphia, 1973.
143. A. W. Horton and G. M. Christian, *J. Natl. Cancer Inst.*, **53**(4), 1017 (1974).
144. E. Bingham, *Arch. Environ. Health*, **19**, 779 (1969).
145. K. Adachi, S. Yamasawa, R. R. Suskind, and G. Christian, *Nature*, **22**, 191 (1969).
146. T. H. Elliott, J. S. Robertson, and R. T. Williams, *Biochem. J.*, **100**, 403 (1966).
147. G. J. Mulkins-Phillips and J. E. Stewart, *Can. J. Microbiol.*, **20**, 955 (1974).
148. K. M. Fredericks, *Nature*, **209**, 1047 (1966).
149. G. M. Badger and J. Novotny, *Nature*, **198**, 1086 (1963).
150. H. Holmberg and T. Malmfors, *Environ. Res.*, **7**, 183 (1974).
151 T. H. Elliott and J. Hanam, *Biochem J.*, **108**, 551 (1968).
152. C-C Lin and C. Chen, *Biochim. Biophys. Acta*. **192**(1), 133 (1969).
153. R. F. Lee, K. Sauerheber, and A. A. Benson, *Science*, **177**, 344 (1972).
154. T. A. Sullivan and W. C. McBee, *Proc. Miner. Waste Util. Symp.*, **4**, 245 (1974).
155. Kirk-Othmer *Encyclopedia of Chemical Technology*, 3rd ed., Wiley, New York, 1984.
156. M. Sittig, *Handbook of Toxic and Hazardous Chemicals and Carcinogens*, 2nd ed., Noyes Data Corporation, Park Ridge, NJ, 1985.
157. K. Mortelmaus, S. Haworth, T. Lawlor, W. Speck, B. Tainer, and E. Zeiger, *Environ. Mutagen*, **8**(Suppl. 7), 1–19 (1986).
158. W. D. Burrows, *Joint Conf. Sens. Environ. Pollut.* [Conf. Proc.], **4**, 80 (1978).
159. J. H. Ruth, *Am. Ind. Hyg. Assoc. J.*, **47**, A-142–A-151 (1986).

160. Y. I. Taradin, G. V. Buravlev, G. S. Bokareva, N. Y. Kuchmina, and L. N. Shavrikova, *Toksikol. Gig. Prod. Neftekhim. Proizvod.*, **197** (1972).
161. L. Z. Geikhman and Z. A. Dubrovskii, *Vrach. Delo.*, **2**, 141 (1970).
162. A. K. Oshkaev, *Izhv. Vyssh. Uchebn. Zaved. Lesn. Zh.*, **20**(5), 28 (1977).
163. C. Q. Darcel, R. W. Bide, and M. Merriman, *Can. J. Biochem.*, **46**(5), 503 (1968).
164. M. Merriman and C. Q. Darcel, *Can. J. Biochem.*, **43**(10), 1667 (1965).
165. K. Moersdorf, *Chim. Ther.*, **7**, 442 (1966).
166. A. Pap and F. Szarvas, *Acta Med. Acad. Sci. Hung.*, **33**(4), 379 (1976).
167. J. Lesznyak, S. Benko, R. Szabo, and E. Muller, *Kiserl. Orvostud.*, **24**(6), 571 (1972).
168. O. P. Shukla, M. N. Moholay, and P. K. Bhattacharyya, *Indian J. Biochem.*, **5**(3), 79 (1968).
169. G. P. Elizarov, *Gig. Tr. Prof. Zabol.*, **1**(2), 32 (1967).
170. H. Duengeman, S. Borelli, and J. W. Wittmann, *Arbeitsmed. Sozialmed. Arbeitshyg.*, **7**(4), 85 (1972).
171. W. Schneider, *Berufsdermatosen*, **21**(2), 45 (1973).
172. C. Eberhartinger, *Wien. Med. Wochenschr.*, **123**(26–27), 449 (1973).
173. E. M. Chapman, *J. Ind. Hyg. Toxicol.*, **23**(7), 277 (1941).
174. A. J. Bailey, T. J. Sims, M. LeLous, and S. Bazin, *Biochem. Biophy. Res. Commun.*, **66**(4), 1160 (1975).
175. A. A. Fisher, *Dermatitis*, 2nd ed., Lea and Febiger, Philadelphia, 1973.
176. R. Brun, *Dermatologica*, **150**, 193 (1975).
177. C. P. McCord, *J. Am. Med. Assoc.*, **86**, 1978 (1926).
178. P. Mikhailov, N. Berova, A. Tsutsulova, *Allerg. Asthma*, **16**(4/5), 201 (1970).
179. C. Oh, F. Ginsberg-Fellner, and H. Dolger, *Diabetes*, **24**(9), 856 (1975).
180. F. Sperling, W. L. Marcus, and C. Collins, *Toxicol. Appl. Pharmacol.*, **10**(1), 8 (1967).
181. H. F. Smyth and H. F. Smyth, Jr., *J. Ind. Hyg.*, **10**(8), 261 (1928).
182. G. S. Lazarus, *J. Invest. Dermatol.*, **62**, 367 (1974).
183. R. C. Hays and G. L. Mandell, *Proc. Soc. Exp. Biol. Med.*, **147**, 29 (1974).
184. T. B. Wellington and J. V. Jones, *Immunology*, **27**, 125 (1974).
185. R. W. Schayer and M. Reilly, *Am. J. Physiol.*, **215**(2), 472 (1968).
186. S. M. Phillips and B. Zweiman, *J. Exp. Med.*, **137**(6), 1494 (1973).
187. F. Homburger and E. Boger, *Cancer Res.*, **28**, 2372 (1968).
188. K. W. Nelson, J. F. Ege, M. Ross, L. E. Woodman, and L. Silverman, *J. Ind. Hyg. Toxicol.*, **25**, 282 (1943).
189. M. Sittig, *How to Remove Pollutants and Toxic Materials from Air and Water*, Noyes Data Corporation, Park Ridge, NJ, 1977.
190. R. Schuppli, *Z. Haut-Geschlechtskr.*, **46**(20), 751 (1971).
191. C. Eberhartinger, *Wien Med. Wochenschr.*, **12**(25–26), 513 (1971).
192. V. V. Dobrynina and E. I. Lyublina, *Gig. Tr. Prof. Zabol.*, **10**, 52 (1974).
193. E. R. Kinkead, U. C. Pozzani, D. L. Geary, and C. P. Carpenter, *Toxicol. Appl. Pharmacol.*, **20**(2), 250 (1971).

194. V. N. Suchkov, S. N. Sakharova, E. A. Pogodina, and A. G. Shalatilova, *Otkrytiya Izobret. Prom. Obraztaz Tovarnye Znaki*, **50**(17), 11 (1973).
195. D. L. J. Opdyke, *Food Cosmet. Toxicol.*, **13**, 735 (1975).
196. D. L. J. Opdyke, *Monograph on Fragrance Raw Materials*, 1979.
197. J. Lesznyak and G. Lusztig, *Gefaesswand Blutplasma, Symp.*, **4**, 253 (974).
198. P. A. Hedin, O. H. Lindig, P. P. Sikorowski, and M. Wyatt, *J. Econ. Entomol.*, **71**(3), 394 (1978).

CHAPTER TWENTY-ONE

# Aromatic Hydrocarbons

Finis Cavender, Ph.D., D.A.B.T., C.I.H.

## 1 MONOCYCLIC AROMATIC COMPOUNDS (ARENES)

### 1.1 General Considerations

#### 1.1.1 Introduction

The aromatic hydrocarbons are the class of chemicals that deal with benzene, its derivatives, and homologues. The aromatic hydrocarbons are of considerable economic importance as industrial raw materials, solvents, and components of innumerable commercial and consumer products. However, the aromatics differ vastly in chemical, physical, and biologic characteristics from the aliphatic and alicyclic hydrocarbons. In addition, the aromatics are more toxic to humans and other mammals. Of prime importance are (1) the hematopoietic toxicity of benzene resulting in aplastic anemia in humans and other mammalian species; (2) benzene-induced leukemia in humans; (3) the cerebellar lesions and loss of central nervous system (CNS) integrative functions in "glue sniffers" exposed to high levels of toluene; and (4) the carcinogenicity of styrene and the polycyclic aromatic hydrocarbons.

Chemically, the aromatic hydrocarbons can be divided into three groups: (*a*) the alkyl-, aryl-, and alicyclic-substituted benzene derivatives, (*b*) the di- and polyphenyls, and (*c*) the polycyclic compounds composed of two or more fused benzene ring systems. The basic chemical entity is the benzene nucleus, which occurs alone, substituted, joined, or fused.

The simplest compound is benzene, the nonsubstituted ring system. When one

---

*Patty's Industrial Hygiene and Toxicology, Fourth Edition, Volume 2, Part B*, Edited by George D. Clayton and Florence E. Clayton.
ISBN 0-471-54725-5 © 1994 John Wiley & Sons, Inc.

methyl group is attached to the ring, toluene is formed, and with two attached methyl groups, xylene is formed. Xylene occurs in three isomeric forms. The hemimellitines and mesitylenes possess three methyl groups, durene four, and the penta- and hexamethylbenzenes, five and six methyl groups, respectively. Other industrially important compounds are ethylbenzene, isopropylbenzene or cumene, and styrene or vinylbenzene.

The aromatics are moderately reactive and undergo photochemical degradation in the atmosphere. Synthetically, they are of prime value as chemical raw materials.

### 1.1.2 Physical Properties

The aromatic compounds occur in liquid, vapor, or solid form. The lower-molecular-weight derivatives possess higher vapor pressures, volatility, absorbability, and solubility in aqueous media than do the comparable aliphatic or alicyclic compounds. Selected physical data of the benzenes and naphthalenes are given in Table 21.1 (1–22). These properties contribute to their biologic activities. They are characterized also by miscibility or conversion to compounds soluble in aqueous body fluids, high lipid solubility, and donor–acceptor and polar interactions (23). Because of their low surface tension and viscosity, the aromatics may be aspirated into the lungs during ingestion, where they cause chemical pneumonitis.

### 1.1.3 Origin and Other Sources

Benzene and its alkyl derivatives, the polyphenyls, and polycyclic aromatics (PAHs) are obtained as products or by-products in petroleum or coal refining, burning, or pyrolysis processes. From coke-oven operations, the aromatics are recovered from the gases and the coal tars. From crude oil distillation, they are produced by fractionated distillation, solvent extraction, naphthenic dehydrogenation, alkylation of benzene or alkenes, or from alkanes by catalytic cyclization or aromatizations. Benzene has been detected in cigarette smoke at 47 and 64 ppm (24, 25). Benzene and toluene have been detected in rainwater at concentrations of 0.1 to 0.5 µg/liter, and up to 1.0 µg/liter in the ambient air (26).

### 1.1.4 Utilization

The aromatic hydrocarbons are used widely as chemical raw materials, intermediates, solvents, in oil and rosin extractions, as components of multipurpose additives, and extensively in the glue and veneer industries owing to their rapid drying characteristics. Aromatics serve in the dry-cleaning industry, in the printing and metal processing industries, and for many other similar applications. They serve as important constituents of aviation and automotive gasolines and represent important raw materials in the preparation of pharmaceutical products.

### 1.1.5 General Toxicity

The aromatics are primary skin irritants, and repeated or prolonged skin contact may cause dermatitis, dehydrating, and defatting the skin. Eye contact with aromatic liquids may cause lacrimation, irritation, and on prolonged contact, severe burns. Conjunctivitis and corneal burns have been reported for the $C_6$ to $C_8$ members. Naphthalene causes cataracts in the eyes of experimental animals. Its vapors are respiratory and mucous membrane irritants and may cause severe systemic injury. Direct aerosol deposition or contact from ingestion and subsequent aspiration can cause severe pulmonary edema, pneumonitis, and hemorrhage (27). The alkylbenzenes, with side chains $C_1$ to $C_4$, are readily aspirated and can produce instant death, via cardiac arrest and respiratory paralysis. For example, in hexylbenzene exposure, death occurred in 18 min; during which time extensive pulmonary edema occurred (28), resulting in a considerable increase in lung weight. The higher alkylbenzenes showed few or no effects. The unique effects of benzene on bone marrow and blood-forming mechanisms are of major importance. For the alkylbenzenes in general, the acute toxicity is higher for toluene than for benzene, and decreases further with increasing chain length of the substituent, except for highly branched $C_8$ to $C_{18}$ derivatives. The toxicity increases again for the vinyl derivatives. Pharmacologically, the alkylbenzenes are CNS depressants, for they exhibit a particular affinity to nerve tissues.

*1.1.5.1 Acute Toxicity.* In comparison, benzene is more toxic than any of the substituted benzene derivatives, except for toluene and styrene. Aromatic hydrocarbons cause local irritation and changes in endothelial cell permeability, and are absorbed rapidly. Secondary effects have been observed in the liver, kidney, spleen, bladder, thymus, brain, and spinal cord in animals (29). The aromatic hydrocarbons, even from a single dose, exhibit a special affinity to nerve tissue. Animals dosed with alkylbenzenes exhibit signs of CNS depression, sluggishness, stupor, anesthesia, and coma. This is in sharp contrast with benzene, which is a neuroconvulsant, producing tremors and convulsions. The CNS depressant potency of the alkylbenzenes depends on branching or side chain length. It diminishes with increasing numbers of substituents or side-chain carbon number up to dodecylbenzene, which has practically no CNS depressant activity (29).

*1.1.5.2 Chronic Toxicity.* Because of its affinity to blood-forming tissue and myelotoxic activity, chronic exposure to benzene is considered more serious than for all other alkylbenzenes. All substituted benzene derivatives tested are devoid of this myelotoxicity. This has been clearly demonstrated with toluene, which does not alter leukocyte count or bone marrow nucleation in the rat (29). Workers with long-term exposure to some aromatic hydrocarbons have shown changes in leukocyte alkaline phosphatase (30). In lifetime studies, benzene is carcinogenic in rodents.

*1.1.5.3 Metabolism.* Benzene and its derivatives are readily hydroxylated and alkyl side chains oxidized to carboxylic acids. Benzene may also be metabolized by ring opening. Benzene immediately increases the urinary excretion of organic sulfate. Of the alkylbenzenes, only *m*-xylene and mesitylene follow this trend.

**Table 21.1.** Physical Properties for Some Aromatic Hydrocarbons (1–22)

| Compound | B.P. (°C) | CAS Registry No. | Density (20/4°C) | Empirical Formula | Flammability Limits[a] (%) | Flash Pt. [°C (°F)] | Freezing Pt. (°C) |
|---|---|---|---|---|---|---|---|
| Benzene | 80.10 | 71-43-2 | 0.8787 | $C_6H_6$ | 1.4–7.9 | −11(12) | 5.53 |
| Toluene | 110.62 | 108-88-3 | 0.8869 | $C_7H_8$ | 1.4–6.7 | 4.4(40) | −94.99 |
| o-Xylene | 144.41 | 95-47-6 | 0.8802 | $C_8H_{10}$ | 1.0–6.0 | 32(90) | |
| m-Xylene | 139.10 | 108-38-3 | 0.8642 | $C_8H_{10}$ | 1.1–7.0 | 29(84) | |
| p-Xylene | 138.35 | 106-42-3 | 0.8611 | $C_8H_{10}$ | 1.1–7.0 | 27(81) | |
| Xylenes, mixed | 138.3 | 1330-20-7 | 0.864 | $C_8H_{10}$ | 1.0–7.0 | 37.6(100) | |
| Trimethylbenzene | | | | | | | |
| 1,2,3- | 176.1 | 526-73-8 | 0.8944 | $C_9H_{12}$ | 0.88 lel | | −25.4 |
| 1,2,4- | 169.35 | 96-63-6 | 0.8758 | $C_9H_{12}$ | 0.88 lel | 54.5(130) | −42.2 |
| 1,3,5- | 164.7 | 108-67-8 | 0.8652 | $C_9H_{12}$ | 0.88 lel | | |
| Tetramethylbenzene | | | | | | | |
| 1,2,3,4- | 205.0 | 488-23-3 | 0.9052 | $C_{10}H_{14}$ | | | |
| 1,2,3,5- | 198.0 | 527-53-7 | 0.8670 | $C_{10}H_{14}$ | | | |
| 1,2,4,5- | 196.0 | 95-93-2 | 0.8875 | $C_{10}H_{14}$ | | | |
| Ethylbenzene | 136.2 | 100-41-4 | 0.8670 | $C_8H_{10}$ | 1.6–7 | 12.8(55) | |
| Methylethylbenzene | 161.3 | 620-24-4 | 0.8645 | $C_9H_{12}$ | | | |
| 1,2-Diethylbenzene | 183.4 | 135-01-3 | 0.8800 | $C_{10}H_{14}$ | | | |
| 1,3-Diethylbenzene | 181.0 | 141-93-5 | 0.8620 | $C_{10}H_{14}$ | | | |
| 1,4-Diethylbenzene | 183.8 | 105-05-5 | 0.8620 | $C_{10}H_{15}$ | | | |
| Diethylbenzene, mixed | 183.8 | 25340-17-4 | 0.868 | $C_{10}H_{14}$ | | 55.6(132) | |
| n-Propylbenzene | 159.2 | 103-65-1 | 0.8620 | $C_9H_{12}$ | 0.8–6 | 30(86) | |
| Cumene | 152.4 | 98-82-8 | 0.8618 | $C_9H_{12}$ | 0.9–6.5 | 36(96) | |
| o-Cymene | 178.15 | 527-84-4 | 0.8766 | $C_{10}H_{14}$ | | | |
| m-Cymene | 175.14 | 535-77-3 | 0.8610 | $C_{10}H_{14}$ | | | |
| p-Cymene | 177.1 | 99-87-6 | 0.8573 | $C_{10}H_{14}$ | 0.7–5.6 | 47(117)[d] | |
| n-Butylbenzene | 183 | 104-51-8 | 0.8601 | $C_{10}H_{14}$ | 0.8–5.8 | 71(160) | |
| sec-Butylbenzene | 173.0 | 135-98-8 | 0.8621 | $C_{10}H_{14}$ | 0.8–6.9 | 52.2(126) | |
| Isobutylbenzene | 172.8 | 538-93-2 | 0.8532 | $C_{10}H_{14}$ | 0.8–6.0 | 52.2(126) | |
| tert-Butylbenzene | 169 | 98-06-6 | 0.8655 | $C_{10}H_{14}$ | 0.7–5.7[c] | 60(140) | |
| tert-Butyltoluene | 192.8 | 98-51-1 | 0.8575 | $C_{11}H_{14}$ | | 68.3(155) | |
| Dodecylbenzene | 290–410 | 123-01-3 | 0.9 | $C_{10}H_{20}$ | | 140.6(285) | |
| Styrene | 145.2 | 100-42-5 | 0.9060 | $C_6H_6$ | 1.1–6.1 | 32(90) | |
| ∝-Methylstyrene | 165.4 | 98-83-0 | 0.9062 | $C_9H_{10}$ | 1.9–6.1 | 54(129) | |
| p-Methylstyrene | 171 | 622-97-9 | | $C_9H_{10}$ | 0.8–11.0 | 52.8(127) | |
| o-Divinylbenzene | 178.5 | 98-83-9 | 0.934 | $C_{10}H_{10}$ | | | |
| m-Divinylbenzene | 199.5 | 108-57-6 | 0.9289 | $C_{10}H_{10}$ | 0.3 | 73.9(165) | |
| p-Divinylbenzene | 83.6 | 105-06-6 | 0.913 | $C_{10}H_{10}$ | | | |
| Allylbenzene | 156 | | 0.8920 | $C_9H_{10}$ | | | |
| 1-Phenylbutene-2 | 175 | 1560-06-1 | 0.888 | $C_{10}H_{12}$ | | 71.1(160) | |
| Phenylacetylene | 142.4 | 536-74-3 | 0.9300 | $C_8H_6$ | | | |
| Diphenyl | 255.9 | 95-52-4 | 0.8660 | $C_{12}H_{10}$ | | | |
| Diphenylmethane | 265.5 | 101-81-5 | 1.0060 | $C_{13}H_{12}$ | | 130(266) | |
| cis-Stilbene | 135 | 103-30-0 | | $C_{14}H_{12}$ | | | |
| trans-Stilbene | 305 | 645-49-8 | 0.9707 | $C_{14}H_{12}$ | | | |
| o-Terphenyl | 322 | 84-15-1 | 1.14 | $C_{18}H_{14}$ | | 163(325)[c] | |
| m-Terphenyl | 365 | 92-06-8 | | $C_{18}H_{14}$ | | 135(675)[c] | |
| p-Terphenyl | 405 | 92-94-1 | 1.236 | $C_{18}H_{14}$ | | 240(465)[c] | |
| Naphthalene | 217.9 | 91-94-1 | 1.0253 | $C_{10}H_6$ | | 87.8(190)[c] | |
| 1-Methylnaphthalene | 244–64 | 90-12-0 | 1.0202 | $C_{11}H_{10}$ | | | |
| 2-Methylnaphthalene | 241.05 | 91-57-6 | 1.0058 | $C_{11}H_{10}$ | | | |
| 1,4-Dimethyl-naphthalene | 268 | | 1.0166 | $C_{12}H_{12}$ | | | |

[a]Lower limit (lel) to upper.
[b]Solubility in water/alcohol/ether: v = very soluble; s = soluble; d = slightly soluble; i = insoluble
[c]−100°C.
[d]Closed cup.
[e]Open cup.

# AROMATIC HYDROCARBONS

| M.P. (°C) | Mol. Wt. | Refractive Index ($n_D^{20}$) | Solubility[b] w/al/et | Sp. Gr. (25°C) | Vapor Density (Air −1) | Vapor Pressure [mm Hg (°C)] | Viscosity (SUS) | Wt./Vol. Conversion (mg/m³ ≃1 ppm) |
|---|---|---|---|---|---|---|---|---|
| 5.5 | 78.11 | 1.5011 | d/v/v | 0.880 | 2.8 | 100 | <32.6 | 3.19 |
| −95 | 92.14 | 1.4961 | i/v/v | 0.87 | 3.1 | 36.7 | <32.6 | 3.77 |
| −25.18 | 106.17 | 1.5055 | i/v/v | 0.90 | 1.1 | 6.8 | <32.6 | 4.34 |
| −47.87 | 106.17 | 1.4972 | i/v/v | 0.87 | 1.03 | 8.3 | <32.6 | 4.34 |
| 13.26 | 106.17 | 1.4958 | i/v/v | 0.86 | 1.03 | 8.9 | <32.6 | 4.34 |
|  | 106.17 |  | i/v/v | 0.9 |  | 6–16 | <32.6 | 4.34 |
| −25.37 | 120.19 | 1.5139 | i/v/v | 0.899 |  |  | <32.6 | 4.92 |
| −43.8 | 120.19 | 1.5067 | i/s/s | 0.880 | 4.1 | 341 | <32.6 | 4.92 |
| −44.7 | 120.19 | 1.4994 | i/v/v | 0.870 | 4.1 | 1.82 | <32.6 | 4.92 |
| −6.25 | 134.22 | 1.5203 | i/v/v |  |  |  | <32.6 | 5.49 |
| −23.68 | 134.22 | 1.5130 | i/v/v |  |  |  | <32.6 | 5.49 |
| 79.24 | 134.22 | 1.5116 | i/v/v |  |  |  | <32.6 | 5.49 |
| −94.97 | 106.17 | 1.4959 | i/v/v | 0.867 | 3.7 | 10 | <32.6 | 4.34 |
| −95.55 | 120.19 | 1.4966 | i/v/v |  |  |  | <32.6 | 4.92 |
| −31.2 | 134.22 | 1.5035 | i/v/v |  |  |  | <32.6 | 5.49 |
| −83.89 | 134.22 | 1.4955 | i/v/v |  |  |  | <32.6 | 5.49 |
| −42.85 | 134.22 | 1.4967 | i/v/v |  |  |  | <32.6 | 5.49 |
|  | 134.22 |  | i/v/v | 0.88 | 4.62 | 1 | <32.6 | 5.49 |
| −99.5 | 120.19 | 1.4920 | i/v/v | 0.86 | 4.14 | 10 | <32.6 | 4.92 |
| −96 | 120.19 | 1.4915 | i/v/v | 0.86 | 4.1 | 10 | <32.6 | 4.92 |
| −71.54 | 134.22 | 1.5006 | i/v/v |  |  |  | <32.6 | 5.49 |
| −63.75 | 134.22 | 1.4930 | i/v/v |  |  |  | <32.6 | 5.49 |
| −67.94 | 134.22 | 1.4909 | i/v/v | 0.857 | 4.62 | 1 | <32.6 | 5.49 |
| −88.0 | 134.22 | 1.4898 | i/v/v | 0.8656 | 4.6 | 2.4 | <32.6 | 5.49 |
| −75 | 134.22 | 1.4895 | i/v/v | 0.8664 | 4.62 | 4.02 |  | 5.49 |
| −51.5 | 134.22 | 1.4366 | i/v/v | 0.8576 | 4.62 | 1.0 |  | 5.49 |
| −57.85 | 134.22 | 1.4927 | i/v/v | 0.8710 | 4.62 | 5.7 |  | 5.49 |
| −52.4 | 148.25 | 1.4921 | i/s/s |  | 4.62 | 0.65 |  | 5.98 |
|  | 246.44 | — | i/s/s |  | 8.47 | — |  | 10.08 |
| −30.63 | 140.15 | 1.5168 | i/s/s | 0.905 | 3.60 | 4.5 |  | 4.26 |
| −23 | 118.18 | 1.5386 | i/s/s | 0.91 | 4.08 |  |  | 4.83 |
|  | 118.18 |  | i/s/s | 0.90 |  |  |  | 4.83 |
|  | 130.19 |  | i/d/d |  |  |  |  | 5.33 |
| −66.9 | 130.19 |  | i/d/d | 0.93 | 4.48 | 1 |  | 5.33 |
| 31 | 130.19 |  | i/d/d |  |  |  |  | 5.33 |
| −40 | 118.18 | 1.5131 | i/s/s |  |  |  |  | 4.83 |
|  | 132.21 |  |  |  |  |  |  | 5.41 |
| −44.8 | 102.14 | 1.5489 | i/s/s | 0.933 |  |  |  | 4.18 |
| 71 | 154.21 | 1.558 | i/s/s |  |  |  |  | 6.31 |
| 23.35 | 168.24 | 1.5723 | i/s/s |  |  |  |  | 6.88 |
| −5 | 180.25 | 1.6264 | i/s/s |  |  |  |  | 7.37 |
| 124.5–124.8 | 180.25 | 1.6264 | i/d/s |  |  |  |  | 7.37 |
| 58 | 230.31 |  | i/s/s | 1.14 |  | 7.9 |  | 9.42 |
| 89 | 230.31 |  | i/s/s | 1.16 |  | 7.9 |  | 9.42 |
| 213 | 230.31 |  | i/s/s | 1.24 |  |  |  | 9.42 |
| 80.55 | 128.17 | 1.4003 | i/s/v | 1.15 | 4.42 |  |  | 5.2 |
| −22 | 142.20 | 1.6170 | i/v/v |  |  |  |  | 5.82 |
| 34.48 | 142.20 | 1.6019 | i/v/v |  |  |  |  | 5.82 |
| 7.66 | 156.23 | 1.6127 | i/v/v |  |  |  |  | 6.39 |

*1.1.5.4 Marine Species.* Aromatic hydrocarbons appear to accumulate in marine animals to a greater extent and are retained longer than alkanes (31). In all species tested, the accumulation of aromatic hydrocarbons depended primarily on the octanol/water partition coefficient. Once absorbed, higher-molecular-weight hydrocarbons are released more slowly (31). The concentration to produce deep CNS depression in barnacle larvae after immersion for 15 min was 3.1 percent for benzene and 4.5 percent for toluene (32).

*1.1.5.5 Microbial Effects.* To some bacteria, as little as 0.01 percent of toluene, xylene, mesitylene, phenol, or cresol may be bacteristatic or bactericidal (33), whereas other microorganisms tolerate concentrations up to 0.5 percent hydrocarbon. Benzene is least susceptible to bacterial oxidation; increasing substitution and chain length, especially to even numbers, promotes the ease of oxidation (33).

### 1.1.6 Industrial Hygiene

From an industrial hygiene standpoint, the aromatics require close monitoring and evaluation; particularly, benzene, toluene, xylene, ethylbenzene, cumene, *p-tert*-butyltoluene, and styrene. Within the past several years, threshold limit values have been lowered incrementally for some of the aromatic compounds. This has been a consequence of the development of better sampling and analytic techniques and more extensive toxicity testing.

### 1.1.7 Medical Aspects

Industrial monitoring programs should be continually evaluated. Where excursion values are found, biologic monitoring should be carried out in addition to regular medical surveillance programs.

## 1.2 Benzene

### 1.2.1 General Aspects

Benzene is the simplest aromatic compound. In commerce, benzene is one of the most important industrial chemicals. However, currently as an analytic agent and in many household and industrial products, benzene has been replaced by other solvents, such as toluene.

### 1.2.2 Chemical and Physical Data

Benzene, benzol, or phene is a clear, colorless liquid with a characteristic sweet odor at low concentrations, disagreeable and irritating at high levels. Selected physical data are summarized in Table 21.1. Benzene forms a highly flammable (34) and explosive mixture with air at 1.4 to 8.0 percent. It is an excellent solvent. Chemically it is fairly stable, but it readily undergoes substitution reactions to form halogen, nitrate, sulfonate, and alkyl derivatives. Commercial benzene has three standard grades and usually contains varying concentrations of toluene, xylene, and phenol, and traces of carbon disulfide, thiophene, alkenes, naphthalene, and related compounds.

The photochemical formation of nitrobenzenes and nitrophenols from benzene has been observed in the presence of nitrogen oxides. Benzene also combines photochemically with halogens to produce eye and mucous membrane irritants (35). Ozone reacts 10 to 20 times more slowly with benzene than with toluene or other methyl-substituted derivatives (36).

### 1.2.3 Occurrence, Industrial Sources, and Preparation

Benzene occurs in coal tar and petroleum naphtha, from which it is commercially prepared (5). Benzene occurs in thermal degradation gases from high-density polystyrene (37) and in solid waste gasification products (38). Also, it has been identified in condensates of tobacco smoke (24). Benzene is a constituent of gasoline. Even recently, European gasolines, often called benzin, contained up to 5 percent benzene, being produced from higher aromatic re-formates (39). The benzene concentration in the gasoline vapor phase is lower than that in the liquid, but depends somewhat on the concentration of other hydrocarbon and metal additives (39). Some consumer products have contained 50 percent benzene or more (40); however, current products with a benzene content above 0.1 percent must be adequately labeled with warnings to the consumer (41).

Benzene has been detected in the expired air in nonsmokers and smokers (42). It occurs as a degradation product in the rat after administration of triphenyllead acetate (43) or *p*-toluic acid phenylhydrazide (44).

The benzene concentration in air samples from urban areas over a period of 15 months ranged from 1.3 to 15 ppb (45). Benzene in other urban samples ranged from 2 to 98 ppb (46, 47).

Benzene is produced in billion gallon quantities each year, mainly by fractionated distillation from crude oil, solvent extractions or crystallization, catalytic dehydrogenation, and re-forming of light naphthas, alkanes, and alkenes (48). It is also produced as a by-product in the coke-oven industry.

### 1.2.4 Utilization

The extensive use of benzene in industry stems from its availability at a relatively low cost. It is used mainly in chemical processes as a raw material and a solvent. Some important processes include the manufacture of ethylbenzene, styrene, cumene, phenolic resins, ketones, adipic acid, caprolactam, hylon (5), and various dyes (49). However, it has been replaced in most solvent applications for consumer products, unless so labeled (41). Benzene is a high-energy component of aviation and automotive fuels, especially as a replacement for alkyl lead compounds.

### 1.2.5 Toxicity

*1.2.5.1 Acute Human Exposure.* The primary toxic effects of benzene are hematopoietic effects, leukemia, and CNS depression. Selected toxicity data are summarized in Tables 21.2 to 21.5 (11–13, 21, 27–29, 50–64).

*Oral Effects.* Oral ingestion of 9 to 12 g benzene (50) has caused signs of staggering gait, vomiting, somnolence, tachycardia, loss of consciousness, and delirium, with subsequent chemical pneumonitis, collapse involving initial stimulation, then abrupt CNS depression. At moderate concentrations, symptoms are dizziness, excitation, and pallor, followed by flushing, weakness, headache, breathlessness, constriction of the chest, and fear of impending death. Visual disturbances and convulsions are frequent. At higher concentrations, the clinical signs are excitement, euphoria, and hilarity, then suddenly change to weariness, fatigue, and sleepiness, followed by coma and death (65).

*Dermal Effects.* Dermal contact is a route of absorption, but the rate is much lower than through the respiratory system. Benzene is irritating to the skin (50) and, by defatting the keratin layer, causes erythema, vesiculation, and dry and scaly dermatitis (66).

*Inhalation Effects.* When benzene first was obtained in pure form, attempts were made to use it as an anesthetic agent. However, as a result of unpleasant side effects and aftereffects, the practice was abandoned (50).

Because of benzene's high volatility, inhalation is the most prevalent route of exposure (67). Acute human exposure has occurred mainly to vapors in accidental spills, drying of soiled clothing in poorly ventilated areas in dry-cleaning businesses, in the home, or in confined spaces where benzene was used as a solvent or product component.

In exposure to 25 ppm for 2 hr, benzene appeared quickly in the blood and cleared with 300 min (68). Absorption of 79.8 fo 84.8 percent occurred in exposures to 10 to 16 mg/l (50). Some benzene is exhaled in unchanged form (69). Exposure to 0.340 mg/l for 5 hr resulted in 33 to 65 percent retention; 3.8 to 27.8 percent was exhaled unchanged and 0.1 to 0.2 percent excreted. Of the absorbed benzene, 9.7 to 42 percent is metabolized to urinary phenol, 0 to 5.4 percent to pyrocatechol, and 0.1 to 3.3 percent to hydroquinone (70), mainly as conjugated sulfates (70). Benzene concentration in tissues is high, because of its high octanol/water partition coefficient (71).

Benzene intoxication resembles that of gasoline (52) in the early acute stages, causing primarily CNS effects. The rate of recovery depends on the initial concentration and exposure time, but symptoms may persist for several weeks. The main signs of intoxication are drowsiness, dizziness, headache, vertigo, and delirium, and may proceed to loss of consciousness. Exposure to 16 mg/l may give a feeling of warmth (50). An acute first-time moderate exposure to benzene may produce CNS depression, with headache, giddiness, and sometimes transient mild irritation of the respiratory and alimentary tracts. A workman exposed to benzene experienced severe vomiting 6 hr after work; however, the symptoms subsided rapidly with aftereffects (52). Acute, high exposure may cause dyspnea, inebriation with euphoria, and tinnitus, which rapidly leads to typical deep anesthesia. If the victim is not treated at this stage, respiratory arrest rapidly ensues, often associated with muscular twitching and convulsions (52). When inhaled, benzene has no effect

**Table 21.2.** Human Acute Benzene Exposure

| Route | Dose or Concentration | Results, Signs, or Symptoms | Ref. |
|---|---|---|---|
| Oral | 9–12 g | Staggering gait, vomiting, somnolence, shallow rapid pulse, loss of consciousness, delirium, death | 50 |
| | 10 ml | May be approximate fatal dose | 51 |
| | 30 g | May be approximate fatal dose | 52 |
| Inhalation | 1.5 ppm (5 mg/m$^3$) | Olfactory threshold | 27 |
| | 25 ppm (0.08 mg/l) × 480 min | No effect, detectable in blood | 29 |
| | 50–150 ppm (0.16–0.48 mg/l) × 300 min | Headache, lassitude, weariness | 29 |
| | 500 ppm (1.6 mg/l) × 60 min | Headache | 29 |
| | 1500 ppm (4.8 mg/l) × 60 min | Symptoms of illness | 29 |
| | 3000 ppm (9.6 mg/l) × 30 min | May be tolerated for 0.5–1 hr | 29, 50 |
| | 7500 ppm (24.0 mg/l) × 60 min | Signs of toxicity in 0.5–1 hr | 50 |
| | 3100–5000 ppm (10–16 mg/l) × 30 min | Subtle signs of intoxication, absorbed 79.8–84.8% | 50 |
| | 19,000–20,000 ppm (61–64 mg/l) × 5–10 min | May be fatal in 5 to 10 min | 29 |

at 25 ppm, but at 50 to 150 ppm produces headache, lassitude, and weariness, and at 500 ppm causes more exaggerated symptoms; 3000 ppm may be tolerated for 0.5 to 1.0 hr, 7500 ppm may result in toxic signs in 0.5 to 1.0 hr, and 20,000 ppm may be fatal in 5 to 10 min (50). In most industrial exposures, mixtures of benzene and other hydrocarbons, mainly toluene, are involved. A case of sudden death has been described (72), where a tank started to overflow in a light oil loading area. A combination of high concentrations of benzene and toluene, excitement due to the mishap, and running through the tank farm appeared to have contributed to the death of a worker. The benzene content was determined to be 0.38 mg percent in the blood, 1.38 percent in the brain, and 0.26 percent in the liver. A sudden death occurred in a 16-year-old boy as a result of "sniffing" rubber cement containing benzene as a solvent (73, 74). A blood sample revealed 94 μg of benzene per 100 ml and kidney tissue contained 0.55 mg/100 g. Many such "glue-sniffing" cases have been recorded. In addition, a laborer installing plastic tiles with liquid adhesives was found dead in the basement at the workplace and another worker unconscious nearby. The causes of the intoxication and death were attributed to vapors of benzene and toluene (75). The cause of the deaths may have been ventricular fibrillation, probably a result of myocardial sensitization to endogenous epinephrine (66). It is believed that the metabolism of benzene in humans proceeds similarly to that observed in the rabbit, but at a more rapid rate than in the dog (50).

*Cellular Effects.* Acute erythromyelosis was diagnosed in a worker following exposure to moderate benzene concentrations over a period of 30 years (76). In

**Table 21.3.** Acute Animal Experiments with Benzene

| Acute | Species | Dose or Concentration | Results, Signs, or Symptoms | Ref. |
|---|---|---|---|---|
| Oral | Rat | 3.4 g/kg | $LD_{50}$ | 83 |
| | Rat | 5.6 g/kg | $LD_{50}$ | 84 |
| | Rat | <1.0–5 g/kg | $LD_{50}$, variation, age and strain dependent | 83 |
| | Rabbit | 150–300 mg/kg (2.5–11.0 μci) | Use of cold and radioactive benzene, no effect; 48.5% exhaled unchanged, 51.5% urinary excretion: 18.2–21.2% as phenol, 4.8% as quinol, 4.4% catechol | |
| | Mouse | 4.70 g/kg | $LD_{50}$ | 54 |
| | Dog | 4.70 g/kg | Lowest reported lethal dose | 54 |
| Intrapulmonary instillation | Rat | 0.25 ml | $LD_{100}$, cardiac arrest, death | 28 |
| Eye | Rabbit | 0.10 ml | Irritancy, moderate conjunctival irritant, causes transient corneal injury | 84 |
| Subcutaneous | Mouse | 0.088 g/kg | No effect | 54 |
| | | 0.44 g/kg | 27% inhibition of circulatory erythrocytes | 54 |
| | | 2.20 g/kg | 50% inhibition of circulatory erythrocytes | 54 |
| | | 2.70 g/kg | Found to be a teratogen | 54 |
| | Frog | 1.40 g/kg | Lethal dose | 54 |
| | Rat | 1.15 g/kg | Lethal dose | 54 |
| | Mouse | 0.468 g/kg | $LD_{50}$ | 54 |
| | Guinea pig | 0.527 g/kg | Lethal dose | 54 |

| Route | Species | Dose | Effect | Ref |
|---|---|---|---|---|
| Inhalation | Rat | 10,000 ppm (31.9 mg/l) × 7 hr | $LC_{50}$ | 29 |
| | Mouse | 2,195 ppm (7.0 mg/l) | CNS depression | 50 |
| | Rabbit | 4,000 ppm (12.8 mg/l) | CNS depression | 91 |
| | | 10,000 ppm (31.9 mg/l) | Death | 91 |
| | | 35,000–45,000 ppm (111.6–144 mg/l) | | |
| | | 4–71 min | Slight anesthesia in 4 min and death in 22–71 min | 90 |
| | | 3.7 min | Light anesthesia, relaxed | 90 |
| | | 5.0 min | Excitation, tremors, running movements | 90 |
| | | 6.5 min | Loss of pupil reflex | 90 |
| | | 11.4 min | Loss of blinking reflex | 90 |
| | | 12.0 min | Pupillary contraction | 90 |
| | | 15.6 min | Involuntary blinking | 90 |
| | | 36.2 min | Death | 90 |
| | Guinea pig | 6,270 ppm (20 mg/.)—30 min peak | Benzene blood level 3.2 mg/100 ml, 1.7 mg/m³ phenol and clearing | 94 |
| | | 15,675 ppm (50 mg/l)—30 min peak | Benzene blood level 8.0 mg/100 ml, 1.8 mg/m³ phenol and clearing | 94 |
| | Dog | 45,800 pm (146 mg/l) | Lethal dose | 54 |
| | Cat | 53,300 ppm (170 mg/l) | Lethal dose | 54 |
| Aquatic | Striped bass | 100–10 ppm | TLm96 | 101 |
| | *Tigriopus californicus* | 10.9 µl/l | $LC_{50}$ 96-hr toxicity similar to that of crude oil | 101 |
| | | 0.1 ml/l | Lethal and more toxic than other petroleum products | 102 |

Table 21.4. Human Chronic Exposure to Benzene

| Years of Exposure | Year Published | No. of Cases | Type of Exposure | Exposure Concn. | Findings | Ref. |
|---|---|---|---|---|---|---|
| 3–54 years | 1939 | 286/332 | Rotogravure ink solvent application and personal dry-cleaning agent | 11–1060 ppm 4–7 days/wk | Lowest concentration: fatigue and dizziness<br>Medium concentration: fatigue, dryness of mucous membranes, hemorrhaging, nausea or vomiting, lethargy<br>Highest concentration: weakness, fatigue, epistaxis, dryness of mucous membranes, loss of appetite, nausea or vomiting, shortness of breath, dizziness, insomnia, and lethargy | 103 |
| 1946–1956 | 1956 | 107/147 | Shoe manufacturing | ~400 ppm | Hematopathy and thrombocytopenia most common | 113 |
| Prior to 1964 | 1964 | 6/47 | Paint thinners, printing inks, adhesives | | Six myeloid, hemocytoblastic, or lymphatic effects, some reversible, of 47 cases with hemopathy | 109 |
| 1945–1965 | 1965 | 20 | Benzene vapor | — | Aplastic anemia | 76 |
| Prior to 1966 | 1966 | 3 | | | Aplastic anemia with osmotic fragility | 104 |
| Prior to 1955 | 1966 | 125/147 | Shoe adhesive solvents | ~400 ppm 1955–1966 | Nine-year follow-up of 125–147 cases with some decreased thrombocytosis from previous benzene exposure of 100/147 with abnormal hematology | 97 |

| Year | N | Industry/Use | Exposure | Effects | Ref |
|---|---|---|---|---|---|
| 1938–1968 | 1 | Occupational varnishing | — | Rare case of acute erythro-myelosis | 76 |
| Prior to 1971 3 months–17 years | 51/217 | Shoe adhesive solvents | 30–210 ppm | Leukopenia 9.7%, pancytopenia 2.8%, eosinophilia 2.3%, thrombocytopenia 1.8%, basophilia 0.5%, giant platelets 0.5%; anemia was reversible | 111 |
| Benzene prior to 1953 then toluene | 34 | Rotogravure benzene and toluene | 125–>525 ppm | Some higher chromosome aberration in peripheral blood lymphocytes | 107 |
| Prior to 1961 5–18 years | 5/216 | Watch industry | — | Ten-year follow-up: 3 thrombocytopenias, 1 thrombocytopenia and anemia, 1 death of aplastic anemia | 106 |
| 4 months–15 years | 32 | Shoe adhesive and solvents | 150–650 ppm | Reversible pancytopenia, 28 thrombocytopenia, 14 macrocytic anemia, 3 megaloblastic erythropoiesis | 108 |
| Prior to 1973 | 299/1000 | — | — | Absolute monocytosis in an average of 14%, 19 reticuloendothelial hyperplasia | 112 |
| 1 year | 40 | Shoe industry benzene and other solvents | — | Rate of leukemia-like effects of occupationally exposed workers 13.0/100,000 vs. 6.0/100,000 in general population | 97 |
| Prior to 1950 and later | 350/594 | U.S. industrial benzene production and use | — | Various hematopathies and several cases with leukemia-like effects | 97 |

Table 21.5. Chronic Benzene Studies

| Route | Species | Dose or Concentration | Time | Results or Findings | Ref. |
|---|---|---|---|---|---|
| Oral | Rat (F) | 1 mg/kg/day | 187 days | No effect | 84 |
| | | 10 mg/kg/day | 187 days | Very slight leukopenia | 84 |
| | | 50 mg/kg/day | 187 days | Leukopenia and erythrocytopenia | 84 |
| | | 100 mg/l$^z$ | 187 days | Leukopenia and erythrocytopenia | 84 |
| | | | 4 months | Development of leukopenia during 1st month, CNS disruption after 4 months | 97 |
| | | 250 mg/l$^a$ | 40–50 days | Temporary changes in leukocyte counts and cholinesterase activity | 97 |
| | | 250 mg/l$^a$ + 10 mg/m | 20 days | Temporary changes in leukocyte counts and cholinesterase activity | 97 |
| | | 1.6 ml/kg | 3 days | Increased liver weight, decreased protein weight and enzyme activities | 97 |
| | | 1.5 mg/kt + 0.51 mg/m$^3$ | 100 days | Sensitizing effect in 30–40 days | 154 |
| | | 9.0 mg/kg + 6.63 mg/m$^3$ | 100 days | Very rapid sensitizing and intoxicating effect | 154 |
| | Rabbit | 1 mg/kg | 6 months | Mild leukopenic, splenic, and testicular degeneration | 84 |
| Dermal | Rat | 0.6 g/kg | 4 hr/day × 4 months | Plasmic cell increase in bone marrow, disruption of erythropoietic element maturation | 155 |
| | Mouse (hairless) | To cover dorsal skin | 2 years | Only slightly above spontaneous frequency of several tumor types, but some skin papillomas | 172 |
| | | 2 microl/appln. | 3 ×/day × 3 day | Epidermal hyperplasia with neural invasion | 170 |
| | Rabbit | 25% in Vaseline | 2 ×/week | Tremors, excitement, cachexia | 50 |

| | Animal | Dose | Duration | Effect | Ref |
|---|---|---|---|---|---|
| | Rabbit (ear) | Undiluted | 10–20 appln. | Slight to moderate irritation, moderate necrosis | 84 |
| Subcutaneous | Rat | 1 mg/kg | 12 days | Chromosomal damage in 50.94% of bone marrow cells | 243 |
| | | 2.5 mg/kg | 30 days | Prolonged generation and transit time of granulocytes and reticulocytes, and decreased maturation time of neutrophilic granulocytes and erythroblasts | 97 |
| | | 0.2 g/kg | 12 days | Lymphopenia, with toluene increased neutrophils and bacillonuclear neutrophils | 97 |
| | | 1 g/kg | 12 days | Chromosomal analysis at metaphase; chromatid breaks 50.9%, gaps 44.7%, isochromatid breaks 4.34%, somewhat higher than from toluene | 243 |
| | | 2 ml/kg | 21 days | Cellular injury, mitochondrial swelling, severe dilation of membrane system, rough and smooth endoplasmic reticulum/ severely disintegrated hematopoietic cells | 97 |
| | | 1 ml/kg/day | 14 days | Leukopenia, no significant change in hematocrit | 97 |
| | | 1 ml/kg/day | 5 weeks | Rapid decrease in femoral marrow nucleated cell count and DNA phosphorus % (of bone marrow dry wt) | 97 |
| | | 2 ml/kg/day | 3 weeks | | |
| | Mouse | 0.025–0.1 ml | 54 weeks, 104 weeks | Not a carcinogen | 186 |
| | | 0.5 ml/kg | 2 appln./day × 1–10 days | Dose-dependent binding to liver, residue increasing with time; initial binding to bone marrow, disassociation after day 6 | 97 |

**Table 21.5.** (Continued)

| Route | Species | Dose or Concentration | Time | Results or Findings | Ref. |
|---|---|---|---|---|---|
| Subcutaneous | Rabbit | 0.1 ml/kg/day × 13 months | | Increased pseudoeosinopoiesis | 97 |
| | | 0.5 ml/kg/day × 10–15 days | | Initial stimulation of pseudoeosinopoiesis, bone marrow promyelocyte, and myelocyte count | 97 |
| | | 0.5 ml/kg/day × 3–4 weeks | | Decrease of blood leukocytes | 97 |
| | | 0.3 ml/kg/day × 1–9 weeks | | Pancytopenia, hypoplasia of bone marrow, with severe inhibition of the DNA synthesis | 181 |
| | | 0.5 ml/kg/day × 2–3 weeks | | Decrease in peripheral blood leukocyte count, decreases mitochondrial respiration | 97 |
| | | 1.0 ml/kg/day × 12 days | | Increased erythrocyte permeability, induction of porphyrin biosynthesis in brain gray matter | 176 |
| Inhalation | Rat | 4.8 ppm (15 mg/m$^3$) | Continuous | Hyperplasia and vacuolization of smooth reticulum, selective destruction of olfactory basal cells | 97 |
| | Rat, guinea pig, dog, primate | 17 ppm (56 mg/m$^3$) 30 ppm (98 mg/m$^3$) 256 ppm (817 mg/m$^3$) | 8 hr/day × 5 days/week cont. 9–127 days | Histopathology of all organs essentially negative, some slight weight reduction at the highest level | 238 |
| | Rat | 94 ppm (350 mg/m$^3$) | 5.5 hr/day × 30 days | Increased α-aminobutyric acid level decreased in cerebellum but not mesencephalon, aspartic and gluramic acid decreased at both sites | 97 |
| | | 158 ppm (500 mg/m$^3$) | 4 hr/day interm. | Leukopenia and muscle antagonistic chromaxy | 97 |

| | | | | | |
|---|---|---|---|---|---|
| | | 400 ppm (1260 mg/m³) | 13 wk | Increased thrombocytes, leukopenia, decreased segmented granulocytes | 97 |
| | | 158–1580 ppm (50–5000 mg/m³) | Continuous | Leukopenia, leukocytosis, reduced blood cholinesterase, alteration of reversible chromaxy ratio of muscle antagonists | 156 |
| Inhalation | Rat | 88 ppm (0.28 mg/l) | 7–8 hr/day × 5 days/week × 6 months for all concentrations | Slight splenomegaly | 84 |
| | | 2200 ppm (7.00 mg/l) | | Narcosis and growth reduction | |
| | | 4400 ppm (14.00 mg/l) | | Above symptoms amplified | |
| | | 6600 ppm (21.00 mg/l) | | Liver histopathology and bone marrow changes | |
| | | 9400 ppm (30.00 mg/l) | | Liver histopathology and bone marrow changes | |
| | | 1650 ppm (5.22 mg/l) | 6 hr/day × 5 days/week × 12 weeks | Leukopenia, partially reversed by phenobarbital | 97 |
| | | 8500 ppm (27.00 mg/l) | 6 hr/day × 10 days | Enzymatic changes in peripheral blood leukocytes, reduction of lymphocytes, granular lymphocytes | 97 |
| | Mouse | 4680 ppm (14.80 mg/l) | 8 hr/day | Depletion of bone marrow colony-forming cells | 97 |
| | Rabbit | 12,000 ppm (38.60 mg/l) | 1 hr/day × several days | Development of grayish white cornea, transient but persistent irritation | 50 |
| | Guinea pig | Vapor | 6 hr/day × 5 weeks | Increased blood serum aspartate and alanine aminotransferase activity, slightly decreased cholinesterase activity first week, decreased serum protein, increased prothrombin time | 97 |
| | Dog | 500–800 ppm (1.6–2.55 mg/l) | 2–8 hr/day × 24–410 days | Base-line benzene in blood increased from 5–6.5 times post-exposure | 97 |

addition, effects on precursor cell hematopoiesis (77), peripheral blood, and bone marrow changes (78), as well as autopsy reports of hemorrhaging in the brain, pericardium, urinary tract, mucous membranes, and skin (66) have been described.

*1.2.5.2 Acute Animal Studies.* Exposures of animals to benzene produces CNS effects, hematopoietic effects, and cardiac sensitization. Fasted animals show different pathways of metabolism, hydroxylation, and conjugation (79). High benzene concentrations increased the serotonin level in the rat brain (80). The incorporation of iron citrate into red blood cells is destroyed by benzene (81). The combination of benzene and lead inhibits both intact reticulocyte heme and protein synthesis (82).

*Oral Administration.* Oral $LD_{50}$ values in the rat range from 930 to 4900 mg/kg, depending on the age and strain of the rat (83). The $LD_{50}$ in mice is 4700 mg/kg and the $LD_{LO}$ in dogs is 2 g/kg (54). Secondary to ingestion, aspiration into the lungs may occur (28). Direct instillation into the lungs of rats causes pulmonary edema and hemorrhage at the site of contact (28). Cardiac arrest was observed when 0.25 ml of benzene was instilled into rat lungs (28).

*Eye Toxicity.* In rabbits, benzene is a moderate eye irritant, causing conjunctival irritation and transient corneal injury (84).

*Dermal Effects and Permeability.* When applied to the guinea pig skin, benzene elicits increased dermal permeability (85). It is absorbed across intact skin (64, 21), but does not produce systemic toxicity. Benzene appears to be a dermal sensitizer in the female guinea pig (86).

*Inhalation of Benzene.* In acute exposures, adult rats and mice were more resistant to the effects of benzene than were young animals (87). The effects vary with the rate of respiration and the fraction of the inhaled benzene retained in the body. The retention was 63 to 81 percent for the dog (88) and was independent of ventilation rates. The highest retention occurred at the lowest ventilatory rate. The 4-hr $LC_{50}$ in rats is 13,700 ppm (89). In rabbits, benzene retention was 37 to 54.4 percent (50). Rabbits exposed to concentrations of 35,000 to 45,000 ppm showed slight anesthesia in about 4 min and death in 22 to 71 min (90). The exposure to saturated benzene vapor resulted in ventricular extrasystole in the cat and the primate, with periods of ventricular tachycardia that occasionally terminated in ventricular fibrillation. In the mouse, 7 mg/l produced CNS depression in 295 minutes, 15 mg/l in 51 min, and 38 mg/l in 8 min, with recovery. At 38 mg/l, death occurred in 38 to 295 min, and at 77 mg/l in about 50 min (50). In the rabbit, 4000 ppm produced CNS depression, and more than 10,000 ppm proved fatal (91). At 35,000 to 45,000 ppm the effects were mainly on the CNS; anesthesia occurred in 3.7 min, with excitation and tremors in 5.0 min and loss of pupil reflexes in 6.5 min (90). In the rabbit, sudden death from ventricular fibrillation has also been observed. Pharmacokinetic events include rapid release of adrenal hormones, epinephrine and norepinephrine, which then may sensitize the cardiac system, especially the myocardium, to the action of benzene (92). In male rats, benzene-induced respiratory paralysis occurred, followed by ventricular fibrillation (93). In guinea pigs exposed to benzene, blood concentrations were monitored (94). Different absorption and clearance kinetics were seen for 20- and 50-mg/l exposure levels. In 20 min, the blood level increased to 3.2 and 8.0 mg benzene/100 ml, respectively, and cleared to 1.9 and 5.1 mg/100 ml 5 min post-exposure. During a 20-min period following benzene exposure, the blood phenol level increased from 1.6 to 1.8 mg/100 ml, but appeared to be independent of exposure concentration. These data support the CNS effects being due to benzene, rather than phenol. Histochemical studies revealed increased alkaline phosphatase in the spinal cord of the mouse, indicating disturbed neuronal transport characteristics (95). It appears that the central nervous system is more susceptible than kidney tissue to the effects of benzene (96). The cellular effects in acute exposures are secondary to CNS effects. In mice, an 8-hr exposure to 4680 ppm and an in vitro cell culture study showed significant depletion of bone marrow colony-forming cells (77).

*1.2.5.3 Acute Metabolism.* In humans and animals, 25 to 80 percent of absorbed benzene is exhaled unchanged. The retained benzene is oxidized to phenol, and in minor quantities to 1,2-di-, 1,4-di-, or 1,2,3-trihydroxybenzenes, cresols, and mercapturic or muconic acids conjugated to the sulfate or glucuronide. The key intermediate is the formation of benzene oxide (58, 97). These pathways depend on the nutritional state (98), route of entry (50), dose, or possible metabolic pathway overload involved (58). Young rabbits excrete mainly nonconjugated compounds (79). Conjugation reactions take place mainly in the liver and lung (99) and depend on the induction level of microsomal cytochrome P450 (58). Phenobarbital pretreatment of rats increased the urinary excretion of benzene and its metabolites but decreased the fraction exhaled (100). Ascorbic acid appears to protect from intoxication by the facilitation of metabolic reactions (50).

*1.2.5.4 Aquatic Studies.* Some TLm and $LC_{50}$ studies listed in Table 21.3 reveal similar values for benzene and crude oil (101). Benzene was lethal to *Tigriopus californicus*, a tide pool copepod, within 2 days of exposure (102).

*1.2.5.5 Chronic Human Exposure.* Formerly, benzene was used extensively as a fast-drying solvent in a variety of trades. When the hematopoietic potential of benzene was realized, it was controlled more strictly or subsequently replaced wherever feasible. Benzene intoxication results from its use in improperly ventilated areas. A summary of high exposures and chemical effects is given in Table 21.4.

*Human Inhalation Exposure.* The chronic exposure to low levels of benzene is related to a number of pathological conditions. Symptoms include headaches, dizziness, fatigue, anorexia, dyspnea, visual disturbances, and vague symptoms not connected usually with benzene poisoning. Signs also include fatigue, vertigo, pallor, visual disturbances, and loss of consciousness (70). Findings following occupational exposures include membrane effects (103), hyperbilirubinemia (104), spleno-

and adrenomegaly (105), blood dyscrasias with hemolytic effects (103), anemia (106), aplastic anemia (107, 108), hemocytoblastic and lymphatic involvement (109, 110) and reticulocytosis (104), leukopenia, pancytopenia, eosinophilia, basophilia, and thrombocytopenia (111), monocytosis (112), and hyperplastic bone marrow effects (104, 58). Also, an increase in chromosomal aberrations has been observed (107). There are three stages of involvement: (1) low, intermittent exposure resulting in very subtle hematopoietic changes; (2) moderate to high exposures affecting enzyme synthesis and causing sensitizing effects and anemias; and (3) high exposures producing irreversible blood dyscrasias (113).

*Metabolism during Chronic Exposure.* Benzene poisoning stems almost exclusively from inhalation exposure. In repeated exposures, different metabolic reactions occur from those observed in acute exposures. Alternate pathways are activated as the result of the depletion of a variety of protective enzyme systems, including those active in iron incorporation or membrane transport. Other affected systems include agglutination mechanisms, cell mitosis and tubular aggregation, and bone marrow stem cells. The activity of granulocytic alkaline phosphatase is reduced by benzene exposure (114). The solubility of benzene in the body fluids is limited but it rapidly accumulates in any type of fatty tissue. It is absorbed through membranes, because of its high octanol/water partition coefficient. For example, the blood saturation equilibrium is reached at 2.1 mg/l for benzene concentrations of 100 ppm in the inhaled air (115). This is a function of the fluid–air partition coefficient, which is 11.7 for blood and 5.5 for plasma, and the tissue–blood partition coefficient, which is 16.2 for bone marrow and 58.5 for fat (91). Benzene from lipid storage is released much more slowly in the female than in the male. A portion of ingested or inhaled benzene is exhaled unchanged or excreted in the urine. However, the majority is metabolized through a variety of pathways. The primary site of metabolism is the liver, where benzene is oxidized to phenol (hydroxybenzene), catechol (1,2-dihydroxybenzene), or quinol (1,4-dihydroxybenzene). Phenol is subsequently conjugated with inorganic sulfate to phenylsulfate and excreted in the urine (116). Peak phenylsulfate excretion occurs about 4 to 8 hr after exposure to benzene (117–119). Minor pathways include further oxidation of catechol to 1,2,3-trihydroxybenzene, catabolism to *cis,cis-* or *trans,trans-*muconic acids (ring opening), and phenol conjugation with glucuronic acid to form glucuronides or with cysteine to produce 2-phenylmercapturic acid (116). If the metabolism is overwhelmed from excessive exposure, the excretion of the glucuronides becomes the primary metabolic pathway (120).

*Other Effects.* In chronic exposures, workers exhibit signs of CNS lesions, abnormal caloric labyrinth irritability, and impairment of hearing (121). In combination with other solvents, workers exhibit neurological syndromes of asthenoneurotic or asthenovegetative polyneuritis with occasional neuronal progression, even after cessation of the exposure (122).

It is possible that benzene is a dermal sensitizer (86). Repeated skin contact with benzene results in defatting of the skin, which leads to erythema, dry scaling, and in some cases the formation of vesicular papules. Prolonged exposure may produce lesions resembling first- or second-degree burns. Systemic intoxication by dermal absorption is unlikely, because absorption is much more likely via inhalation (70).

Intoxication at extremely high levels produces cardiac sensitization (123). Benzene causes dyspnea and tachycardia (124). Benzene decreases arterial pressure and peripheral resistance, and causes diffuse-dystrophic myocardial alterations, which are reversible following termination of the exposure. The increased cardiac output, accelerated circulation rate, and the greater capacity of the precapillary beds are apparently compensatory–adaptive measures, to promote tissue oxygenation (125).

Benzene is fetotoxic but these effects may be due to maternal toxicity (97). The concentration of benzene is about the same in maternal and cord blood (126), indicating that benzene does cross the placenta. Some effects on testes in exposed animals have been reported (97). No other reproductive effects for benzene are known.

*Hematologic Effects.* Benzene's effect on the hematopoietic system and its unique myelotoxicity have been known for many years and occur in three stages. Initially there are bloodclotting defects caused by functional, morphological, and quantitative platelet alteration (thrombocytosis), as well as reduced numbers of blood components (pancytopenia and aplastic anemia). At this stage, if diagnosed and treated, the effects are readily reversible. At a more advanced stage, the bone marrow becomes hyperplastic, then hypoplastic, iron metabolism is disturbed, and internal hemorrhaging occurs. Sometimes the effects are associated with monocytosis or absolute lymphocytosis (112). At this stage, diagnosis and treatment should be prompt and intense, and workers protected from benzene exposure. Indicative clinical findings are erythrocyte counts below 3.5 million, leukocytes below 4500, decreased platelet numbers, increased iron, and decreased transferrin. In the third phase, bone marrow aplasia becomes progressive (52). It is possible that bone marrow regeneration is excessively stressed through augmented destruction of peripheral red cells, leading to final regenerative exhaustion, although individual variations in hematologic absolute values and continuous changes are extreme. For example, leukocyte counts may vary more during a life cycle than can be achieved through exposure to benzene (91). Exposure to benzene and other chemical agents may cause hyporegenerative anemias or pancytopenias, which in some cases may evolve into myeloblastic (127, 128), granulocytic (129), or hemocytoblastic leukemia (108, 130). Benzene exposure, possibly in combination with other solvents, has been linked to 50 cases of leukemia (131).

Four shoemakers developed acute leukemia (132) from exposures to very high concentrations of benzene and other adhesive components and solvents; one case of lymphoblastic and one of myeloblastic leukemia were observed (133). In 1967–1973, 31 of 28,500 workers were diagnosed with acute leukemia or preleukemia (134, 135). These incidents represent 13 cases per 100,000 workers and more than six cases of leukemia per 100,000 of the general population (134, 135). The number of cases decreased in 1974–1975 and no new cases were reported in 1976 following the gradual replacement of benzene in gasoline (136). Apparently, genetic and other extraneous factors play an etiologic role, as observed in two families with the occurrence of leukemia in pairs of identical twins (137). Also, a case of familial chronic lymphocytic leukemia occurred in a family that had worked in the drycleaning business for 20 to 30 years (138). Conversely, among 38,000 workers in the petroleum and petrochemical industries, the incidence of leukemia did not exceed the expected incidence (139).

*Subcellular Effects.* Chromosomal aberrations were found in peripheral lymphocytes of benzene workers (140) but not in personnel handling toluene (107). Both chromatid and chromosomal aberrations occur in bone marrow preparations (141). Aberrations in some subjects were detected several years after cessation of the benzene exposure (142). One female exposed to benzene during pregnancy showed an increase in chromosomal aberrations, but delivered a normal, healthy boy and later a girl (142). In one case of granulocytic leukemia after intermittent benzene exposure over 7 years, the bone marrow exhibited cells lacking one or more G-group chromosomes and others with supernumerary C-group chromosomes (129). Another case with erythroleukemia showed inconsistent chromosomal changes (143).

Current analyses are complicated by preexisting familial chromosomal anomalies and the natural changes that occur during normal life cycles (144). Benzene interferes with DNA replication, inhibits hepatic polyribosomes, and inhibits protein synthesis (145). More specifically, experiments with rabbits demonstrate that benzene interferes with the thymidine incorporation into bone marrow DNA (146).

*Immunologic Response.* It has long been suspected that benzene exposure may alter the immune response and general functions (91). Bone marrow functions appear to be closely related to immune mechanisms. Workers exposed to benzene, isopropylbenzene, hydroperoxide, and phenol exhibit decreased phagocytic activity, diminished bactericidal activity of the skin, and lowered lysozyme and blood cholinesterase activity. In addition, they exhibit a tendency to hypotension, decreased pulse rate, and CNS changes (147). In 62 of 79 workers exposed to benzene, toluene, and xylene, a decrease serum complement concentration was noted (148).

*Clinical Determinations.* Elevated phenol excretion above 20 mg/l can indicate that benzene exposure has occurred in the preceding 8 to 10 hr. This is not definitive, however, because salicylate or other drugs and certain foods may more than double the phenol level to 75 mg/l (149). An exposure of 25 to 30 ppm may elevate phenol excretion to 100 or 200 mg/l (51, 148). Conversely, with up to 5 ppm, no changes may occur. Ratio changes of urinary inorganic to organic sulfate excretion can also serve as a biomarker. Exposures of 10 to 40 ppm benzene may lower the ratio to 72 percent, 40 to 75 ppm to 61 percent, 75 to 100 ppm to 43 percent, and 100 to 700 ppm to 38 percent (29). Early signs of benzene intoxication include urinary phenol and sulfate concentrations and direct blood changes. Better indicators are sequential hematologic and cytological changes. It should be noted that other chemical agents or drugs may induce similar signs and symptoms (150, 151).

*Prophylaxis and Follow-up.* It appears that nutrition, genetics, and immunocompetence may predispose some individuals' sensitivity toward benzene effects. Vitamins aid in prophylaxis and treatment (152). Recovery from moderate benzene exposure may take 1 to 4 weeks, whereby unsteady gait, nervous irritability, and breathlessness may persist for 2 weeks, and cardiac distress and the peculiar yellow color of the skin as long as a month (70). Recovery from chronic high exposure may take from 2 months to several years (70). In 186 workers, nervous system changes returned to normal and urinary phenol levels were found at the base lines for 76 percent of the employees over a 5-year period (153).

### 1.2.5.6 Chronic Bioassays

*Oral Administration.* Oral concentrations of 1 mg/kg in the rat were without effect, whereas higher concentrations produced leukocyte changes. Very high doses affected rat liver and protein fraction weights and decreased hepatic aminopyrine $N$-demethylase, but increased acetanilide hydroxylase activity. Simultaneous ingestion of 9.0 mg/kg and exposure to 6.63 mg/m$^3$ demonstrated sensitizing effects in the rat (154). In rabbits, "exhausted" bone marrow is characterized with initial increases, followed by compensatory decreases of eosinophils, neutrophils, and myelocytes.

*Dermal Application.* Dermal absorption of benzene occurs to a moderate degree. Some papillomas and systemic hematopoietic effects have been noted, although the rate of false-negative studies in mouse skin tests is high (97). Benzene is negative in skin painting studies (21).

*Inhalation Exposure.* Six-week old rats exposed to 1.6 ppm (5 mg/m$^3$) benzene exhibited a lower resistance than 4.5-month old animals. At 9.5 ppm (0.03 mg/l), tolerance to benzene was independent of age (155). In general, continuous inhalation exposures exert a more pronounced effect on the animals than intermittent exposures (156). The enzymatic activity of liver microsomes increased in rats exposed to benzene vapors for several weeks (157). Continuous inhalation of 4.8 ppm of benzene caused changes in the rat endoplasmic reticulum, associated with inflammation. Histopathological, hematologic, cytological, and biochemical changes were induced. Exposure of mice to 300 ppm 6 hr/day, 5 days/week, for 16 weeks is associated with leukemia, thymic and nonthymic lymphoma, Zymbal gland tumors, and lung adenomas (97).

*Metabolism.* Absorption, distribution, and metabolism of benzene in laboratory animals are similar to that observed for humans, except the relative absorption appears higher in humans (50). In female and male rats with a large body fat content, benzene was eliminated more slowly and stored longer than in lean animals

(158). Dietary fats and paraffin oils stimulated intestinal absorption of benzene in the rat (159). Benzene concentration in the rabbit was highest in the adipose tissue, high for bone marrow, and lower for brain, heart, kidney, lung, and muscle (160), although direct binding was higher in the liver than in bone marrow (161). In the mouse, benzene is metabolized to free phenol and phenylsulfate (161). Metabolic changes in rabbits have been observed in the brain cortex and erythrocytes, in δ-aminolevulinic acid, prophobilinogen, and protoporphyrin (162). In the liver, benzene oxidation to phenol requires microsomal cytochrome P450 (163). Hepatic microsomes are the main site for benzene metabolism (164). A 4-hr benzene exposure to 11.9 mg/l enhanced hepatic microsomal cytochrome P450 and NADPH oxidase activities in rats, whereas 35 mg/l (the $LC_{10}$) inhibited the enzymes (165). This was accompanied by a decreased rate of hepatic hydroxylations, decreased concentration of free SH groups, and reduced tissue oxygen tension (165). In some cases, pretreatment with phenobarbital protected the organism; in others it did not influence the rate of benzene detoxification (89, 166–168). In mouse blood, benzene inhibits iron uptake into red cells (146).

*Cytotoxic Effects.* As seen in cats, the primary action of benzene is its capability to change vessel membrane permeability and secondarily to affect the nutritional state at the cellular level (169). In the hairless mouse, signs of hyperplasia of the regenerative epithelium were noted (170). Histochemical determinations indicate that benzene affects neuron enzymes of the spinal cord in mice (95). Benzene causes morphological changes in the liver of rats and rabbits following injections of 0.05 g or 1 g/kg for 20 to 70 days (171). Degenerative muscle fibers were detected in the heart, bronchi, blood vessels, and bile duct in the rat, and degenerative and dystrophic changes in the central nervous system, cortex, thalamus, brain, and brainstem in the rabbit (171). Liver injury is restricted mainly to the endoplasmic reticulum (172, 173), as is true also in the kidney, along with decreased numbers of mitochondria (174).

*Hematologic and Subcellular Effects.* Hematologic effects appear in female rats more rapidly than in the male (175). Benzene induces the biosynthesis of porphyrins in rabbit erythrocytes (176). In the guinea pig, benzene initiates the destruction of erythrocytes in the spleen and lymph nodes, Kupffer cells in the liver, and enhanced phagocytic and hemolytic action (50). In rats, chronic exposure to 1 $g/m^3$, 4 hr daily for 6 months caused an inhibition of the phagocytic activity of leukocytes (177).

Subtle effects of decreased mitochondrial and lysosomal enzyme activities were observed in rabbits exposed to 16 mg/l, 6 hr/day for 14 days (178). No effect was detected at 10 mg/l (178). The myelotoxic effect caused by chronic benzene exposure resembles that of ionizing radiation or exposure to radiomimetic substances (179). Benzene affects erythropoiesis and leukopoiesis of different cell lines (179). Myelotoxic effects include a depression of hematopoiesis in the stem cells of the bone marrow and spleen and a reduction of the number of hematopoietic stem cells (180). This is parallel to the reduction of erythrocytes, thrombocytes, and leukocytes in peripheral blood (180). Subcutaneous injection of 0.3 ml/kg to rabbits for 1 to 9 weeks was accompanied by pancytopenia and hypoplasia of the bone marrow (181).

In vitro incubation of rabbit or human reticulocytes with iron transferrin at nonhemolytic concentrations of 56 to 113 m$M$ resulted in cellular protein synthesis inhibition (182).

*Immunologic Response.* In rabbits, benzene affects autoimmune processes leading to hemopathy (183).

*Sex Differences, Teratology, and Reproduction.* In general, females or species with large quantities of adipose tissue are more susceptible to the effects of benzene. Benzene alone exerted no effect on fertility, conception, or embryonic development in rats exposed on days 1 to 19 of gestation (97).

*Mutagenicity.* Benzene injected subcutaneously into rats at 200 mg/kg once a day for 12 days produced some chromosomal aberrations, breaks, and gaps (184). In a host-mediated assay, benzene was not mutagenic (185).

*Carcinogenicity.* Benzene, when injected subcutaneously at 0.05 to 0.2 ml twice a week for 44 weeks, then once a week for 10 weeks into weanling male C57B1/6N mice produced no carcinogenic effects during the 50-week observation period (186). Skin of white mice painted with benzene for 6 weeks showed ultrastructural changes in the dermis and epidermis, indicating nonspecific irritation of these tissues (187, 188). Conversely, benzene as a solvent or carrier, potentiates the action of carcinogens (189).

*1.2.5.7 Biodegradation.* *Pseudomonas* strains can use benzene as their sole carbon source (190). Ring opening occurs in *Pseudomonas* and *Moraxella* species (191). Benzene is metabolized by the avocado fruit (192) and grapes (193) to carbon dioxide.

### 1.2.6 Industrial Hygiene

*1.2.6.1 Threshold Air Concentrations.* The Occupational Safety and Health Administration (OSHA) permissible exposure limit (PEL) (148) is 1 ppm (5 mg/m$^3$) with a short-term exposure limit (STEL) of 5 ppm (15 mg/m$^3$) and the American Conference of Governmental Industrial Hygienists (ACGIH) threshold limit value (TLV) (194) is 10 ppm (32 mg/m$^3$); both carry a cancer notation. The odor of benzene can be detected in the air at concentrations of 1.5 to 5 ppm (27, 51, 55). The odor is irritating at 9000 ppm (55). The National Institute for Occupational Safety and Health (NIOSH) (195) recommends absorption of benzene on charcoal, desorption with carbon disulfide, and quantification by flame ionization gas-chromatography.

*1.2.6.2 Biologic Monitoring.* In biologic monitoring, urinary phenol determinations are a valuable guide. Benzene in exhaled air is proportional to recent benzene exposure levels. Urinary phenol levels of 100 mg/l were measured following a 1-hr exposure to 200 ppm or an 8-hr to 25 ppm. Base-line levels are less than 25 mg/l. Decreases in $^{59}$Fe-incorporation into erythrocytes is a measure of benzene-induced depression of erythropoietic function (58). Frequent determination of sulfate is advised (196). Medical surveillance should include blood pressure check, pulmonary function, blood chemistry, hematology, urinalysis, and skin examinations (196).

*1.2.6.3 Precautionary Measures.* A variety of procedures and devices are available for the removal of benzene from circulating air. Protective clothing including neoprene gloves, face shields, and NIOSH-approved respirators are required (197, 198, 21). Local exhaust ventilation should be applied (21). Benzene is on the Environmental Protection Agency (EPA) Toxic Substances Control Act (TSCA) Chemical Inventory and the Test Submission Data Base (54).

## 1.3  Toluene

### *1.3.1  Properties, Sources, and Use*

Toluene is the lowest-molecular-weight alkylbenzene. Alkylbenzenes possess properties similar to those described for benzene, except they are not hematotoxic. Alkyl substitution changes physical characteristics and, in turn, the absorption, partition, and metabolism properties.

The time required for hydroxylation, excretion, and sulfate ratio normalization increases with increasing doses and molecular weight of the side chain. The possibility of ring hydroxylation rises with increasing side-chain size (199).

Toluene or methylbenzene is a clear, colorless, noncorrosive, flammable liquid (34) with a sweet, pungent, benzene-like odor. Selected physical properties are given in Table 21.1. (1–22, 200).

Toluene concentrations in urban air ranged from 0.01 to 0.05 ppm (198). It is released from manufacturing plants, automobile and coke-oven emissions, gasoline evaporation (198), and cigarette smoke (24). Toluene is a component of high-flash aromatic naphthas, which are produced from crude oil by primary distillation and as by-products in the coal tar industry. Presently, toluene is produced from petroleum, and specifically from methylcyclohexane containing naphthas by catalytic reforming processes (5). Re-forming of *n*-heptane at 977°C yields about 62 percent toluene (48). Toluene is a pyrolysis product of thermal cracking (48) and is also produced from coal tar (200).

Toluene is used extensively as a solvent in the chemical, rubber, paint, and drug industries (66), as a thinner for inks (103), perfumes, and dyes, and as a nonclinical thermometer liquid (201).

### *1.3.2  Toxicity*

Toluene toxicity resembles that of benzene except it is devoid of benzene's chronic hematopoietic effects. Selected toxicity data are given in Tables 21.6 and 21.7.

**Table 21.6.** Acute Toluene Toxicity

| Route | Species | Dose or Concentration | Results or Findings | Ref. |
|---|---|---|---|---|
| Inhalation | Human | 100 ppm (0.38 mg/l) | Psychological effects, transient irritation | 54 |
| | | 200 ppm (0.76 mg/l) | Central nervous system effects | 54 |
| | | 400 ppm (1.52 mg/l) | Mild eye irritation, lacrimation, hilarity | 90 |
| | | 600 ppm (2.3 mg/l) | Lassitude, hilarity, slight nausea | 90 |
| | | 800 ppm (3.03 mg/l) | Metallic taste, headache, lassitude, slight nausea | 217 |
| | | 100–1000 ppm (0.38–4.1 mg/l) | Absence of illness, decreased erythrocytes (evaluated for 1940–1941) | |
| Oral | Rat | 2.5 g/kg | Lethal to ~30% of test animals | 27 |
| | Rat (14 day old) | 3.0 ml/kg | $LD_{50}$ | 83 |
| | Rat (young adult) | 6.4 ml/kg | $LD_{50}$ | 83 |
| | | 7.0 g/kg | $LD_{50}$ | 84 |
| | | 7.4 g/kg | $LD_{50}$ | 83 |
| | | 7.53 ml/kg | $LD_{50}$ | 54 |
| Inhalation | Rat | 1700 ppm (6.4 mg/l) × 4 hr | Dose was tolerated | 239 |
| | | 4000 ppm (15.2 mg/l) × 4 hr | $LC_{LO}$ | 54 |
| | | 8000 ppm (30.4 mg/l) × 4 hr | $LC_{50}$ | 54 |
| | | 8800 ppm (35.0 mg/l) × 4 hr | $LC_{50}$ | 239 |
| | | 15,000–25,000 ppm (40.0–66.5 mg/l) × 15–35 min | Lethal to 4 of 5 rats; blood, liver, and brain toluene concentration, 0.27, 0.64, and 0.87 mg/g, respectively | 230 |
| | | 15,000–25,000 ppm (40.0–66.5 mg/l) + $O_2$ × 80–130 min | Lethal to 7 of 10 rats with $O_2$ supplied | 230 |
| | | 45,000–70,000 ppm (170–265 mg/l) × 2.9 min | Light anesthesia, relaxed | 54 |
| | | 45,000–70,000 ppm (170–265 mg/l) × 9.5 min | Pupillary contraction | 54 |
| | | 45,000–70,000 ppm (170–265 mg/l) × 14.8 min | Loss of blink reflex | 54 |
| | | 45,000–70,000 ppm (170–265 mg/l) × 16.1 min | Excitation, tremors, running movements | 54 |
| | Mouse | 8520 ppm (32.1 mg/l) × 8 hr | Lethal to 87.5% | 271 |
| | Cat | 7800 ppm (31.0 mg/l) × 6 hr | CNS effect, mydriasis, mild tremors, prostration in 80 min, light anesthesia in 2 hr | 239 |
| Dermal | Dog | 760 ppm (3.0 mg/l) × 6 hr | No signs of discomfort | 239 |
| | Rabbit | 14 g/kg | $LD_{50}$ | 54 |

**Table 21.7.** Chronic Toluene Studies

| Route | Species | Dose or Concentration | Results or Findings | Ref. |
|---|---|---|---|---|
| Oral | Rat (F) | 118 mg/kg/day × 193 days | No effect | 84 |
| | | 353 mg/kg/day × 193 days | No effect | 84 |
| | | 590 mg/kg/day × 193 days | No effect | 84 |
| Dermal | Rabbit | Undiluted, 10–20 applications | Slight to moderate irritation, slight necrosis | 84 |
| Inhalation | Rat | 7.7–255 ppm (0.03–1.0 mg/l) | Reduced blood cholinesterase level | 156 |
| | Rat, guinea pig, dog, primate | 107 ppm (0.39 mg/l) × 8 hr/day × 90–127 days | No histopathological or organ effects | 238 |
| | | 1085 ppm (4.10 mg/l) × 8 hr/day × 90–127 days | No histopathological or organ effects | 238 |
| | Rat, dog | 245 ppm (0.95 mg/l) × 6 hr/day × 13 weeks | No significantly different effects from the controls | 239 |
| | | 490 ppm (1.9 mg/l) × 6 hr/day × 13 weeks | No significantly different effects from the controls | 239 |
| | | 1515 ppm (3.9 mg/l) × 6 hr/day × 13 weeks | No significantly different effects from the controls | 239 |
| | Rat | 390 ppm (1.0 mg/l) × 4 hr/day × 6 months | Produced some inhibition of the phagocytic activity of leukocytes | 177 |
| | Rat (M) | 1000 ppm (3.8 mg/l) × 8 hr/day × 4 weeks | Increased adrenal weight and plasma hydrocorticoids; decreased eosinophiles | 54 |
| | Rat | 6450 ppm (24.28 mg/l) × 5 hr/day × 4 months | Decreased serum albumin, increased β- and γ-globulin and lipoprotein levels | 54 |

Older reports of positive hematopoietic effects may have been caused by benzene impurities in toluene. It is a CNS depressant and a skin and mucous membrane irritant. Severe dermatitis may result from its drying and defatting action. Toluene is readily absorbed by inhalation, ingestion, and somewhat through skin contact. If ingested, toluene is an aspiration hazard (28). In guinea pigs, the 4-hr no-effect level is below 1250 ppm (202).

*1.3.2.1 Human Studies*

*Human Acute Inhalation.* The most rapid route of entry is through the pulmonary system. Toluene in exhaled air is a reflection of toluene exposure (69). Of the inhaled toluene, 86 to 96 percent is retained in the body (203). The no observable effect level (NOEL) for 6-hr exposures is 40 ppm and mild intoxication occurs at 100 ppm (200). Volunteers exposed to low toluene concentrations exhibited transitory mild upper respiratory tract irritation at 200 ppm; mild eye irritation, lacrimation, and hilarity at 400 ppm; lassitude, hilarity, and slight nausea at 600 ppm; and rapid irritation, nasal mucous secretion, metallic taste, drowsiness, and impaired balance at 800 ppm (90).

High concentrations may result in paresthesia, disturbance of vision, dizziness, nausea, CNS depression, and collapse (204). An employee was found unconscious after an exposure to high vapor concentrations for 18 hr. Tests indicated hepatic and renal involvement with myoglobinuria, with all effects reversible within 6 months (204). Some instant deaths have been recorded (73, 123) from "glue sniffing." A glue-soaked cloth in a paper bag may create a toluene concentration from 200 to 5000 ppm (205). During exercise, the pulmonary toluene concentrations increase up to twofold (206, 207), whereas the mental productivity and reaction time decrease (208). Liver effects have been noted in workers exposed to 324 ppm for 2 months or more (200).

*Human Chronic Exposure.* Experience in the varnish and paint industries has shown toluene to be a moderate toxicant. Toluene is a cardiac sensitizer and fatal cardiotoxin (209). It also causes hepatomegaly and is hepatotoxic and nephrotoxic (51, 200, 210). In several cases of habitual "glue sniffing," renal, neural, and cerebellar dystrophy occurred (208, 211, 212). Isolated cases with some cytochemical lymphocytic changes have been recorded (213, 214). Workers in a pharmaceutical plant exposed to toluene fumes developed leukopenia, and especially neutropenia. Within 6 months, those affected showed an increase in clotting time and a decrease in the prothrombin level, with urinary hippuric acid excretion of 4 g/l (215). Periodontal effects were also seen (216). A survey of 106 paint workers repeatedly exposed to 100 to 1100 ppm toluene exhibited no signs of illness. Subtle clinical findings included some cases of enlarged liver and decreased erythrocytes, with other hematologic parameters normal (217).

*Oral and Dermal Exposure.* One death occurred following the ingestion of 625 mg/kg (200). Prolonged contact of toluene with the skin may cause drying and defatting, leading to fissured dermatitis (66, 218). However, toluene is not a dermal sensitizer (218).

*Sensitization.* Toluene is a cardiac sensitizer, which caused "sudden death" from habitual sniffing of glue (123). The cardiotoxic effect of toluene appears somewhat lower than that of benzene (93).

*Human Metabolism.* In humans, 93 percent of the inhaled toluene is retained. Absorption of toluene in the oral cavity is 29 percent (203). Toluene is absorbed into the vascular system (219), followed by distribution to various tissues and metabolism (220). The uptake of toluene is linear, resulting in 0.55 mg/100 ml from 200 ppm up to 2.23 mg/100 ml from an exposure to 800 ppm (27, 219).

Toluene is subsequently oxidized to benzoic acid and, in turn, conjugated with glycine to form hippuric acid or with glucuronic acid to yield benzoylglucuronates (116). Attempts have been made to correlate toluene blood levels with hippuric acid excretion rates (221–225). Urinary hippuric acid is an indicator of toluene exposure (226) and becomes increasingly reliable in exposures to 800 ppm and above (219). A biologic threshold concentration of 1000 to 1100 mg/l has been proposed (224, 227). In painters exposed to combinations of toluene and xylene, urinary excretion of glucuronates and hippuric acid is increased to five times the normal level (117).

Another metabolite, toluene oxide, is formed by microsomal enzymes (228). Methylbenzenes have been shown to induce cytochrome P450 in the Southern armyworm (229). Conversely, toluene inhibits mitochondrial oxidative phosphorylation (230).

### 1.3.2.2 Animal Studies

*Acute Animal Studies.* With respect to acute oral effects, toluene appears to be the least toxic of the alkylbenzenes (84), but is somewhat more acutely toxic than benzene (50).

*Oral Effects.* The $LD_{50}$ in the rat ranges between 636 and 7300 mg/kg (200).

*Inhalation.* In the dog, 91 to 94 percent of the inhaled toluene is retained with no direct respiratory rate dependence (88). In rats, exposure to 1500 to 2500 ppm for 15 to 35 min resulted in the recovery of 0.27 mg in the blood, 0.64 mg in the liver, and 0.87 mg in the brain (231). When combined with oxygen, the above values increased to 0.43, 1.07, and 1.27 mg, respectively (231, 232). The 1-hr $LC_{50}$ in the rat is more than 26,700 ppm (54), the 4-hr $LC_{50}$ in rats is 8800 ppm (200), and the 24-hr $LC_{50}$ in the mouse is 400 ppm (54).

*Dermal and Eye.* Toluene is a primary skin irritant. The dermal $LD_{50}$ in the rabbit is 12 to 14 g/kg (54). In the eye, it causes rapid and intense turbidity of the cornea and inflammation of the conjunctiva (50).

*Acute Metabolism.* In rabbits, toluene is absorbed into body fluids. The partition coefficient for adipose tissue was higher than for any other organ tested (160). In mice, the highest concentration was measured in adipose tissue with less in liver, kidney, and cerebrum (233). Metabolites from toluene in the rat included *o*- and *p*- but not *m*-cresol (234), benzyl alcohol, and hippuric acid (234).

*Toxicity to Marine Life.* An aquatic threshold toxicity in goldfish was 23 ppm (235). Toluene was less toxic to coho salmon (236) than benzene and xylene. Toluene was absorbed by the mussel, but to a lesser degree than the comparable $C_7$ paraffins (237).

*Chronic Oral Studies.* Repeated ingestion of toluene in female rats at 118, 354, and 590 mg/(kg)(day) for 193 days produced no marked effects (84).

*Dermal Effects.* Repeated application of undiluted toluene to rabbit skin caused slight to moderate irritation and slight necrosis (84). Subcutaneous injection produced transient granulocytopenia in the rabbit (238).

*Inhalation.* Long-term exposure to 100 to 1000 ppm toluene produced no effects (239, 240). Behavioral symptoms consisting of circling by rats have been reported (241), and decreased learning capabilities have been reported (242).

*Metabolism.* The metabolic pattern following chronic exposure is similar to that seen in acute exposures. However, tissue distribution appears more extensive and the highest quantities are found in liver, brain, bone marrow, cerebellum, and adrenals (27). Rats exposed to 186.6 mg/m$^3$ for 2 weeks to 6 months exhibited increased peroxidase action and decreased catalase activity (243). Cellular and subcellular changes appear to be reversible contrary to those caused by benzene, except for chromosome gaps and chromatid and isochromatid breaks, noted in rat bone marrow cells (244).

### 1.3.2.3 Effects on Plants.
Toluene is metabolized by a variety of plants (245), such as grapes (190) and avocados (193).

### 1.3.2.4 Microbial Effects.
Toluene exerts limited bacteriostatic, but pronounced fungistatic effects toward some species and can reduce drastically *Actinomycetes* populations (246). It also possesses anthelmintic properties (247). Conversely, bacterial strains isolated from soil that closely resemble *Pseudomonas desmolytica* can utilize toluene as their sole carbon source (248), as can *Nocardia* cultures (249), *P. aeruginosa*, and *P. oleovorans* (250). The latter microorganisms require supplemental protein for maximal hydroxylation (250, 251). Toluene is biodegraded primarily through side-chain hydroxylation (252).

## 1.3.3 Industrial Hygiene and Occupational Medical Aspects

### 1.3.3.1 Threshold Limits.
The OSHA PEL (148) is 100 ppm (375 mg/m$^3$) with a STEL of 150 ppm (560 mg/m$^3$), and the ACGIH TLV (194) is 100 ppm (377 mg/

m³) with a STEL of 150 ppm (565 mg/m³). A TWA based on human experimental data suggests air concentrations up to 480 ppm (1.9 mg/l) represent no-effect levels (240). The odor threshold ranges between 2.5 and 8 ppm (240, 55). The odor is irritating at 750 ppm (55).

*1.3.3.2 Analytic Determinations.* NIOSH recommends collection on charcoal, desorption using carbon disulfide, and quantification using flame ionization gas chromatography (253). Selected methods are available for the determination of toluene in water (254), blood (255, 256), other body fluids (256), and tissues (257).

*1.3.3.3 Medical Aspects.* Comprehensive preplacement and biennial medical examinations are recommended for all workers exposed to toluene. Laboratory tests should include hematology and urine analyses (258). The average urinary biologic threshold of hippuric acid is 1000 to 1100 mg/l (227). A high protein diet is advised as a prophylactic measure (259).

*1.3.3.4 Precautionary Measures.* Protective clothing should include gloves, barrier creams, eye goggles or face shields, and a cartridge-type or self-contained breathing apparatus (260, 21, 261). Toluene is on the EPA TSCA Chemical Inventory and the Test Submission Data Base (54).

## 1.4 Xylenes

### 1.4.1 General Aspects and Physical Properties

The xylenes or dimethylbenzenes occur in three isomeric forms, that is, the *o-*, *m-*, or *p*-xylene, or the 1,2-, the 1,3-, and the 1,4-dimethylbenzene, respectively.

Xylene occurs in many petroleum products, in coal naphthas, and as an impurity in petrochemicals, such as benzene and toluene. The three isomers have been identified among the volatile products in tobacco smoke (24). Also, m- and p-xylene have been detected in particulate samples of urban air (45).

The three isomers possess similar properties. They are commercially available separated or mixed as colorless, flammable liquids (34, 29, 60). Xylene readily dissolves fats, oils, and waxes. Selected physical properties are given in Table 21.1 and toxicity data in Tables 21.8 and 21.9.

### 1.4.2 Production and Use

Xylene is produced by catalytic reforming and, depending on the feedstock, yields of above 85 percent can be achieved (48). Commercially, xylene is recovered also from coal tar, yielding a typical mixture of about 10 to 20 percent ortho, 40 to 70 percent meta, and 10 to 25 percent para isomer. Impurities include ethylbenzene, benzene, toluene, phenol, thiophene, and pyridine (29, 60).

The xylenes are used widely as thinners (103), as solvents for inks, rubber, gums, resins, adhesives, and lacquers, as paint removers, in the paper coating industry (224), as solvents and emulsifiers for agricultural products (229, 262), as fuel com-

**Table 21.8.** Xylene Toxicity in Humans

| Route of Entry | Species Number | Dose or Concentration | Effects or Results | Ref. |
|---|---|---|---|---|
| Eye | Human | 460 ppm (1980 mg/m$^3$) | Irritant to 4 of 6 subjects | 266 |
| Dermal | Human | Undiluted | Burning effect, also drying and defatting of the skin | 66, 268 |
| Inhalation (acute) | Human 6 | 1 ppm (4.3 mg/m$^3$) | Odor threshold | 266 |
| | 6 | 40 ppm (17.2 mg/m$^3$) | Identification threshold | 266 |
| | 10 | 100 ppm (43.10 mg/m$^3$) | Satisfactory for occupational 8-hr exposure | 313 |
| | 10 | 200 ppm (860 mg/m$^3$) × 3–5 min | Irritant to eyes, nose, throat | 313 |
| | | 110–460 ppm (472–1980 mg/m$^3$) | Irritant to eyes, nose, throat | 266 |
| | 8 | Unknown | Respiratory irritation to 6 or 8 with clinical signs | 272 |
| | 3 | ~10,000 ppm (~43.1 mg/l) 18.5 hr | One death, lung congestion, brain hemorrhage, 2 workers unconscious 19–24 hr, retrograde amnesia and some renal effects, one hypothermia and lung congestion | 273 |
| Inhalation (chronic) | Human | 59 ppm (0.254 mg/l) | Hippuric acid 1998 + 1197 mg/l | 278 |
| | | 93 ppm (0.40 mg/l) | Hippuric acid 1812 + 1275 mg/l | 278 |
| | | 256 ppm (1.10 mg/l) | Hippuric acid 3821 + 1113 mg/l | 278 |
| | | 398 ppm (1.71 mg/l) | Hippuric acid 5500 + 1690 mg/l | 278 |

ponents (49, 263, 264), and commonly in the chemical industry as intermediates (265). Xylene is utilized widely to replace benzene, especially as a solvent. Specifically, the *o*-xylene serves as a raw material for the production of plasticizers, alkyd resins, and glass-reinforced polyesters; the para derivative for polyester fibers and films; and the meta isomer to produce isophthalic acid, polyester, and alkyd resins (48).

### 1.4.3 Toxic Effects

Most animal studies indicate that xylene is more toxic than toluene. Threshold effects occur at lower doses for toluene; however, at high doses, xylene, is more toxic. Moreover, the fact that benzene occurs as an impurity may render older toxicity reports unreliable (115).

**Table 21.9.** Acute Xylene Studies

| Material | Route of Entry | Species | Dose or Concentration | Results or Effects | Ref. |
|---|---|---|---|---|---|
| o-Xylene | Oral | Rat | 2.5 ml/kg | Lethal to 7 of 10 animals | 27 |
| m-Xylene | Oral | Rat | 2.5 ml/kg | Lethal to 3 of 10 animals | 27 |
| p-Xylene | Oral | Rat | 2.5 ml/kg | Lethal to 6 of 10 animals | 27 |
| Xylene | Oral | Rat | 4.3 g/kg | $LD_{50}$ | 84 |
|  |  | Rat | 10.0 ml/kg | $LD_{50}$ | 750 |
|  | Eye | Rabbit | 13.8 mg | Turbidity and irrigation of the conjunctiva, lacrimation, edema | 50 |
|  |  | Cat | Undiluted | Vacuoles in the cornea resembling "polishers' keratitis" | 29 |
| o-Xylene | Inhalation | Mouse | 1.80 ml/kg | $LD_{50}$ | 313 |
|  |  | Rat | 6350 ppm (27.4 mg/l) × 4 hr | $LC_{50}$ | 750 |
|  |  |  | 6700 ppm (29 mg/l) × 4 hr | $LC_{50}$ | 266 |
|  |  | Cat | 9500 ppm (41 mg/l) × 2 hr | $LC_{100}$, with typical CNS effects | 325 |
|  |  | Mouse | 3500–4600 ppm (15–20 mg/l) | Produces narcosis | 313 |
|  |  | Rat | 6125 ppm (26.3 mg/l) × 12 hr | Lethal dose | 115 |
|  |  | Mouse | 6920 ppm (30 mg/l) | Lethal dose | 313 |
| m-Xylene | Inhalation (acute) | Rat | 8000 ppm (34.5 mg/l) × 4 hr | Lethal dose | 313 |
|  |  | Mouse | 2010 ppm (8.7 mg/l) × 2 hr | Lethal dose | 115 |
|  |  |  | 2300–3500 ppm (10–25 mg/l) | $LT_{LO}$, narcosis | 115, 325 |
|  |  |  | 11,500 ppm (50 mg/l) | $LC_{100}$ |  |
| p-Xylene |  | Rat | 4912 ppm (21.1 mg/l) × 24 hr | Lethal | 313 |
|  |  | Mouse | 2300 ppm (10 mg/l) | Produces narcosis | 325 |
|  |  |  | 3500–8100 ppm (15–35 mg/l) | $LC_{100}$ | 115, 325 |
| Xylene | Eye (chronic) | Cat | Undiluted | Corneal vacuoles | 313 |
|  |  | Rabbit | Undiluted | No effect | 266 |
| Xylene | Dermal | Rabbit | Undiluted | Moderate to marked irritation, moderate necrosis | 84 |
|  | Subcutaneous | Rabbit | 300 mg/kg/day × 6 weeks | No myelotoxic effects | 313 |
|  |  |  | 700 mg/kg/day × 9 weeks | No myelotoxic effects | 313 |

| | | | | | |
|---|---|---|---|---|---|
| | | Guinea pig | 1–2 ml/kg/day × 10 days | No effects except slight reduction in red blood cells without affecting white blood cells | 50 |
| | Inhalation | Mouse | 11.5 ppm (0.05 mg/l) 4 hr/day × 12 months | Hematologic and immunologic changes with eventual decomposition | 295 |
| | | | 46.4 ppm (0.20 mg/l) 2 hr/day × 12 months | Hematologic and immunologic changes with eventual decomposition | 295 |
| | | Rat, guinea pig, primate | 78 ppm (0.337 mg/l) 8 hr/day × 5 days/week × 90 exp. | Essentially no hematologic effects | 238 |
| Xylene | Inhalation | Rat, beagle | 180 ppm (0.77 mg/l) 6 hr/day × 5 days/week × 13 weeks | No statistically significant effects | 60 |
| | | Guinea pig | 300 ppm (1.3 mg/l) 4 hr/day × 6 days/week × 64 exp. | At necropsy some liver degeneration and inflammation of the lungs | 201 |
| | | Rat, guinea pig, primate | 780 ppm (3.36 mg/l) 8 hr/day × 5 days/week × 30 exp. | Essentially no hematologic effects | 238 |
| | | Rat, beagle | 460 ppm (2.0 mg/l) 6 hr/day × 5 days/week × 13 weeks | No statistically significant effects | 60 |
| | | Rat, rabbit | 700 ppm (3.0 mg/l) 8 hr/day × 6 days/week × 130 days | No significant erythro- and thrombocyte changes, slight reduction of leukocytes | 293 |
| | | Rat, beagle | 310 ppm (3.5 mg/l) 6 hr/day × 5 days/week × 13 weeks | No statistically significant effects | 60 |
| | | Rabbit | 1150 ppm (5.0 mg/l) 8 hr/day × 6 days/week × 55 days | Subtle reduction of erythro-, leuko-, and thrombocytes, some bone marrow hyperplasia without structural changes | 293 |
| | | | 2300 ppm (10.0 mg/l) 6 hr/day × 32 days (180 hr) | Subtle reduction of erythro- and lymphocytes, increase in leukocytes death due to pneumonia | 294 |
| | | Cat | 2300 ppm (10.0 mg/l) 6 hr/day × 9 days (46.5 hr) | Subtle reduction of erythro- and leukocytes, also monocytes, death due to pneumonia | 294 |

### 1.4.3.1 Human Studies

*Acute Human Exposure.* Ingestion of xylene causes severe gastrointestinal distress. Aspiration into the lung causes chemical pneumonitis, pulmonary edema, and hemorrhage (29). Ingestion of small quantities of xylene can produce urinary dextrose and urobilinogen excretion, with toxic hepatitis, which is reversible in 20 days (266).

A concentration of 460 ppm was irritating to the eyes of a human panel (267). Conjunctivitis and corneal burns have been reported following direct contact of the eyes with xylene (29). Xylene is an irritant and causes defatting, which may lead to dryness, cracking, blistering, or dermatitis (50, 66). When inhaled at high concentrations, signs include a flushing and reddening of the face and a feeling of increased body heat due to the dilation of superficial blood vessels (29). In addition, disturbed vision, dizziness, tremors, salivation, cardiac stress, CNS depression, confusion, and coma (268), as well as respiratory difficulties have been noted (269). Xylene is a cilia toxin and mucous coagulating agent (270). One death has been ascribed to the misuse of a shampoo–solvent mixture (271). Xylene can cause instant death (271) owing to sensitization of the myocardium to epinephrine (123), so that even endogenous hormones precipitate sudden and fatal ventricular fibrillation (92) or respiratory arrest and consequent asphyxia (272).

Exposure to vapors from epoxy resin–concrete disposal containing xylene caused upper respiratory irritation in six of eight workers (273). Clinical findings were temporary albuminuria, microhematuria, and pyuria (273). While painting in a tank, three workers were exposed to xylene thinner and traces of toluene. One death was noted with lung congestion, focal intra-alveolar hemorrhage, acute pulmonary edema, and hepatic, anoxic, and neuronal damage. The other two workers were found unconscious after 18.5 hr, but regained consciousness 1 and 5 hr following emergency treatment. Clinical findings included hepatic and renal impairment, with significant hypothermia in one case (274). Female workers appear to be more susceptible to the effects of xylene (275).

*Chronic Human Exposure.* Signs and symptoms from chronic exposure resemble those from acute exposures, but are generally more severe.

Repeated, prolonged exposure to vapors may produce conjunctivitis of the eye and dryness of the nose, throat, and skin (66). Direct dermal contact may result in flaky or moderate dermatitis.

Exposure to xylene vapors may cause CNS excitation, followed by CNS depression, characterized by paresthesia, tremors, apprehension, impaired memory, weakness, nervous irritation, vertigo, headache, anorexia, nausea, and flatulence (276, 277), and may lead to anemia and mucosal hemorrhage (262). Clinically, no bone marrow aplasia, but hyperplasia (50), moderate liver enlargement, necrosis, and nephrosis may occur (50, 277). In a hematology laboratory, air concentrations of xylene measured 21 to 120 ppm (278). Hippuric acid, phenol, glucuronic acid, and uric acid determinations in groups of ship painters were correlated with xylene exposure concentrations. Uric acid was decreased, the other parameters elevated

(279). In the intaglio printing industry, xylene exposure is related to the occurrence of anemia (280). During several days of painting a water tank with an agent containing 65 percent xylene and 35 percent benzene, half of the work force experienced nausea and excreted red to coffee-brown urine. Almost all complained of headaches, loss of appetite, and extreme fatigue, and one death occurred (281). The most dangerous exposures industrially, commercially, or in the home, involve spray painting. Five of the nine mothers of stillborn infants with caudel regression syndrome had been exposed during pregnancy to xylene vapors (282). Based on 11 paired cord blood samples, xylene crosses the placenta (126).

*Absorption and Metabolism.* Xylene, when ingested, is quickly absorbed. Absorption through the intact abraided skin occurs readily. Experimentally, *m*-xylene was absorbed by healthy human subjects with one or both hands immersed, at an approximate rate of 2 $\mu g/cm^2/min$ (283, 284). The amount absorbed during the immersion of both hands in *m*-xylene for 15 min equals the amount absorbed during exposure to 100 ppm (283). Absorption during dermal exposure to 600 ppm xylene vapor for 3.5 hr corresponded to the amount retained during a 5.5 hr exposure to 10 ppm (285).

Xylene is absorbed mainly through the mucous membranes and pulmonary system (50). In subjects exposed to xylene vapor, 64 percent was retained (286). Absorbed xylene is distributed through the vascular system. Generally, the xylenes are metabolized to the corresponding *o*-, *m*-, or *p*-toluic acids (116) and excreted in urine primarily conjugated with glycine as methylhippuric acid. Normally, urinary phenol is not elevated (277). Of the xylene absorbed through the skin, 80 to 90 percent was eliminated as methylhippuric acid (283, 284). Application of barrier creams did not significantly influence the rate of absorption (284). Exposure to *m*- and *p*-xylene resulted in the excretion of *m*- and *p*-methylhippuric acid (223). A linear relationship was found between atmospheric xylene concentration and excreted toluic acid (287). Of the 64 percent retained by human volunteers, 95 percent was metabolized, and 5 percent exhaled unchanged through the lung (288). In volunteers exposed to 100, 300, and 600 ppm, the retention of the vapor tended to decrease at the end of an exposure (289). Alanine aminotransferase increased and serum cholinesterase activity decreased in workers occupationally exposed to xylene and other organic solvent mixtures (290).

### 1.4.3.2 Animal Studies

*Acute Animal Toxicity.* The oral $LD_{50}$ in the rat is 4.3 mg/kg (54). Visual disturbances including rotary and positional nystagmus were observed in rabbits (291). In the rabbit eye, 13.8 mg xylene produced marked turbidity and severe irritation of the conjunctiva, with lacrimation and edema (50, 84). Xylene has been found to increase the dermal permeability to water (85, 292).

When inhaled, 450 ppm xylene was very toxic to the guinea pig, but 300 ppm was tolerated for some time without harm (50). The 4-hr $LC_{50}$ in the rat is 5000 ppm and the $LC_{LO}$ in the guinea pig is 450 ppm (54).

In fish, the aquatic TLm96 value is between 10 and 100 ppm for mixed *o*- and *p*-xylene (293) and the 1-hr $LC_{50}$ is 17 ppm (235). Rainbow trout exposed to 7.1 ppm xylene survived for 2 hr, but all died at 16.1 ppm.

*Subacute and Chronic Toxicity.* Corneal vacuoles were seen in the cat (268) but not in the rabbit following repeated exposure to xylene vapor (267). Repeated application of undiluted xylene to the rabbit skin produced moderate to marked irritation and moderate necrosis (84). Subtle changes in the erythrocytes but not in leukocytes were noted in the rats and rabbits exposed to 3 mg/l. Slight reductions in erythro-, leuko-, and thrombocytes were observed, and some bone marrow hyperplasia was noted at 5 mg/l (294, 295). In mice exposed to 0.05 mg/l, 4 hr/day, a three-phase effect was noted. There were subcellular changes with decreased immunobiologic response in phase 1, from 1 to 3 months; decompensation in phase 2; and normalization in phase 3 (296). In rats exposed continuously to 158 to 1580 ppm, reduced blood cholinesterase level altered the reversible chromaxy ratio of muscle antagonists, and leukopenia or leukocytosis was noted (156). Exposure of rats to 1 mg/l of *m*-xylene 4 hr/day for 6 months caused an inhibition of the phagocytic activity of leukocytes (177). Xylene exerts little or no teratogenic action on the rat or chicken (297).

*Metabolism.* In the rabbit, xylene is absorbed rapidly and partitioned between the blood and tissue with the highest fat content (160). In the dog, xylene is excreted as the toluic acid–glycine conjugate, methylhippuric acid (298). In rabbits, oral gavage of 0.4 ml/kg/day for 1 week resulted in the excretion of *o*-toluic acid glucuronide (299). However, a further metabolite, 3,4-dimethylphenol glururonide, was identified by ether hydrolysis and infrared analysis (299). The same metabolites were isolated and identified in the rat and guinea pig. In all three species, *m*-xylene was excreted as *m*-toluic acid derivatives. *p*-Xylene was excreted as *p*-toluic acid derivative, but also a 2,5-dimethylphenol glucuronide was isolated from all three species (234). *m*-Xylene is converted to 2,4-dimethylphenol in the rat (234). The ortho, meta, and para isomers, in decreasing order, are demethylated to phenol (300). In the rabbits exposed to 1 mg/l 4 hr/day for 32 days, *m*-xylene produced 57.3 to 63.9 mg of *m*-methylhippuric acid (300). The metabolite was cleared within 24 hr following exposure.

In vitro xylene hydroxylation by rat microsomal lung and liver preparations has demonstrated that the oxidation to toluic acid is mediated by hepatic but not the pulmonary microsomal enzyme systems (302). In vitro activation experiments demonstrated that also rabbit hepatic but not pulmonary microsomal enzyme systems were affected by phenobarbital pretreatment (303). Furthermore, phenobarbital, 3-methylcholanthrene, and chlorpromazine increase the $LC_{50}$ of inhaled *p*-xylene (304).

*1.4.3.3 Microbial Studies.* Xylene can serve as the sole carbon source for *Pseudomonas desmolytica* (248) and other *Pseudomonas* strains that produce large quantities of toluic acid (305, 306). *P. aeruginosa* converts *p*-xylene into *p*-methylbenzyle alcohol and possibly further to methylbenzoic acid (252). However, *m*-xylene can be metabolized also to methylsalicyclic acid and further to 3-methylcatechol (307). Another pathway for *p*-xylene is the conversion of *p*-methylbenzyl alcohol to *p*-methylbenzoic acid, *p*-toluic acid, *p*-cresol, *p*-hydroxybenzoic alcohol, *p*-hydroxybenzaldehyde, *p*-hydroxybenzoic acid, and 3,4-dihydroxybenzoic acid (308). A similar sequence for *Pseudomonas* and *p*-xylene has been proposed except for the last few steps, whereby *p*-toluic acid may convert to 4-methylcatechol and then to 2-hydroxy-5-methylmuconic semialdehyde (309). *P. putida* converts *p*-xylene to *cis*-3,6-methyl-3,5-cyclohexadiene-1,2-diol (310). A *Pseudomonas* strain, *P. pxy*, can grow on *m*- and *p*-xylene and utilize them as the sole source of energy. A mutant *P. pxy*-82 can transform *m*-xylene to 3-methylcatechol and 3-methylsalicyclic acid (311). From *Nocardia* cultures, xylene is converted to 2,3-dihydroxy-*p*-toluic acid and 3,6-dimethylpyrocatechol (249, 312).

### 1.4.4 Industrial Hygiene and Biologic Monitoring

***1.4.4.1 Threshold Limit Values.*** The OSHA PEL (148) with a STEL of 150 ppm (655 mg/m$^3$) is 100 ppm (435 mg/m$^3$) and the ACGIH TLV (194) is 100 ppm (434 mg/m$^3$) with a STEL of 150 ppm (65 mg/m$^3$). The NIOSH recommended exposure limit (REL) (298) is 100 ppm for a 10-hr/day, 40-hr work week, with a ceiling of 200 ppm (868 mg/m$^3$) as determined by a 10-min sampling period. All standards apply to the mixed, the *m*-, the *o*-, and the *p*-xylene.

The odor threshold in air is 0.35 to 1.0 ppm for xylene (313, 267, 55). The odor threshold for xylene in water is 2.12 ppm (313).

***1.4.4.2 Air Sampling.*** NIOSH (195) recommends charcoal tube collection, desorption with carbon disulfide, and analysis by flame ionization gas chromatography. Photoionization and electron capture detectors as well as mass spectral analysis are also recommended (314).

***1.4.4.3 Biologic Monitoring and Medical Aspects.*** Xylene is quantified in blood by gas chromatography (315). Several methods have been published on the quantification of methylhippuric acid in urine (222, 316–321) using chromatographic techniques. Employees should undergo comprehensive preplacement (66) and biennial medical checkups (298). Air and biologic monitoring programs should be established and evaluated regularly.

***1.4.4.4 Industrial Hygiene Considerations.*** Occupational exposure to xylene vapors and direct liquid contact is possible in xylene manufacture and use. Protective clothing should include gloves and face shield (54). Local ventilation and other precautionary measures are warranted. Labeling, warning precautions, and personnel protection have been described (298).

Methods for xylene waste gas treatment (322), waste purification (323), and solid waste xylene recycling (324, 325) are available. Xylene is on the EPA TSCA Chemical Inventory and Test Submission Data Base (54).

## 1.5 Tri- and Polymethylbenzenes

The properties of tri- and polymethylbenzenes resemble those of their lower homologous. Toxicity data for tri- and tetramethylbenzenes are given in Table 21.10.

### 1.5.1 Trimethylbenzenes

The trimethylbenzenes are colorless, flammable liquids, which occur in three isomeric forms: 1,2,3-trimethylbenzene or hemimellitine; 1,2,4-trimethylbenzene or pseudocumene; 1,3,5-trimethylbenzene or mesitylene. Selected physical data are given in Table 21.1. All isomers occur in refined petroleum and coal tars (3). They are used in industry as chemical raw materials, paint thinners, solvents, and as motor fuel components (29). They have been detected in trace quantities in urban air (45). Trimethylbenzenes have been identified in exhaled air (42). Trimethylbenzenes cross the placenta (126). In nature, the 1, 2, 4 isomer is an insect attractant.

*1.5.1.1 Toxic Effects.* The loss of righting response for 1,2,4-trimethylbenzene in the mouse occurs at 40 mg/l (8130 ppm) and loss of reflexes at 40 to 45 mg/l (8130 to 9140 ppm); for the 1,3,5 isomer these losses were 25 to 35 mg/l (7110 to 9140 ppm) and 35 to 45 mg/l (7110 to 9140 ppm), respectively (326).

Exposure to 1 mg/l of trimethylbenzene 4 hr/day for 6 months caused an inhibition of phagocytic activity of the leukocytes of rats (177).

The aquatic TLm for the three isomers is between 100 and 1000 ppm (293). The 96-hr $LC_{50}$ is 13 ppm for the mixed product (235).

*1.5.1.2 Metabolism.* The trimethylbenzenes are readily absorbed into the vascular system. When 1.2 g/kg of 1,2,3-trimethylbenzene was orally administered to the rat, it was excreted as various urinary metabolites with minor quantities as trimethylphenol, whereas the 1,2,4-isomer was excreted as the corresponding 3,4-dimethyl derivative (327). The excretion of the free trimethylphenols in minor quantities was determined for the 1,2,4 and 1,3,5 isomers (234).

Exposure to 1.5, 3.0, and 6.0 mg/l of 1,3,5-trimethylbenzene for 6 hr resulted in increases in granulocytes, but a decrease in lymphocytes (328). Exposures to 3.0 mg/l 6 hr/day for 5 weeks did not significantly affect the hematologic picture (328). However, it elevated the serum alkaline phosphatase activity in acute exposures, and the serum glutamic–oxalacetic transaminase activity in chronic exposures (329). Apparently, 3,5-dimethylbenzoic acid is formed in humans and in the rat (330).

*1.5.1.3 Industrial Hygiene.* The OSHA PEL (148) is 25 ppm (125 mg/m$^3$) and the ACGIH TLV is 25 ppm (123 mg/m$^3$). Protective clothing and precautionary measures are the same as recommended for xylene.

Sampling procedures and analytic methods are similar to those recommended for xylene (42). A biologic microdetermination method for mesitylenic acid, a metabolite of 1,3,5-trimethylbenzene, has been published (330).

**Table 21.10. Toxicity of Tri- and Tetramethylbenzenes**

| Material | Route of Entry | Species | Dose or Concentration | Results or Effects | Ref. |
|---|---|---|---|---|---|
| Trimethylbenzene | | | | | |
| 1,2,3- | Oral (subacute) | Rat | 1.2 g/kg | Urinary metabolites: <br> 2,3-dimethylhippuric acid, 17.3% <br> 2,3-dimethylbenzoic glucuronide, 19.4% <br> 2,3-dimethylbenzoic sulfate, 19.9% | 326 |
| 1,2,4- | Aquatic | Fish | 1000–100 ppm | TLm96 | 292 |
| | IP (acute) | Guinea pig | 1.788 g/kg | Minimum fatal dose ($LD_{LO}$) | 297 |
| | Inhalation | Mouse | 8130 ppm (40 mg/l) | Loss of righting response (prostration) | 230 |
| | | | 8130–9140 ppm (40–45 mg/l) | Loss of reflexes | 230 |
| | Subcutaneous (subacute) | Rabbit | 2–3 g/kg/day | Local infiltration and necrosis | 297 |
| | Oral | Rat | 1.2 g/kg/day | Metabolites: <br> 3,4-dimethylhippuric acid, 43.2% <br> 3,4-dimethylbenzoic glucuronide, 6.6% <br> 3,4-dimethylbenzoic sulfate, 12.9% <br> 3,4-dimethylbenzoic acid, – | 326 |
| 1,3,5- | Aquatic | Fish | 1000–100 ppm | TLm96 | 292 |
| | Oral (acute) | Rat | 23 g/kg | Lethal to 7 of 10 test animals | 27 |
| | Intraperitoneal | Rat | 1.5–2.0 g/kg | $LD_{100}$, minimal fatal dose | 54 |
| | Inhalation | Mouse | 7110–9140 ppm (25–35 mg/l) | Loss of righting response (prostration) | 325 |
| | | | 8130 ppm (40 mg/l) | Loss of reflexes | 325 |
| | Inhalation (subacute) | Rabbit | 0.12 mg/kg/day | Moderate thrombocytosis | 54 |
| | | Rat | 0.2 ml/kg | Primary thrombopenia | 54 |
| | | | 1 mg/l × 4 hr/day × 6 months | Inhibition of phagocytic actions | 177 |
| | Oral (subacute) | Rat | 1.2 g/kg/day | Metabolites: <br> 3,5-dimethylhippuric acid, 78.0% <br> 3,5-dimethylbenzoic glucuronide, 7.6% <br> 3,5-dimethylbenzoic sulfate, 1.2% | 326 |
| Durene | Aquatic | Fish | 1000–100 ppm | TLm96 | 292 |
| | | Goldfish | 13 ppm | 96-hr $LC_{50}$ | 234 |
| | Oral (acute) | Rat | >5 g/kg | $LD_{50}$ | 27 |

### 1.5.2 Tetramethylbenzene

Tetramethylbenzene occurs as the 1,2,3,4 isomer or prehnitine, the 1,2,3,5 isomer or isodurene, and the 1,2,4,5 isomer or durene. Durene is of greatest commercial importance.

Durene is a white, odorless solid. Selected physical properties are given in Table 21.1. Durene is produced by methylation of low-boiling aromatics, principally pseudocumene (29). It is used as a solvent, a constituent of motor fuels, and a chemical raw material. Durene appears to exhibit a toxicity similar to that of the trimethylbenzenes. It is metabolized similarly to other alkylbenzenes. A saturated solution exhibits fungistatic properties (29, 331, 332).

Exposure limits have not been established; however, the limits for trimethylbenzene are recommended. Analytic methods include gas chromatography and mass spectrometry (333).

### 1.5.3 Pentamethylbenzene

Pentamethylbenzene is a white solid at ambient temperatures. It enhances microsomal drug-metabolizing enzymes (334) by induction of mixed function oxidases (335). It also has fungistatic properties (332). Analytic methods are available.

### 1.5.4 Hexamethylbenzene

Hexamethylbenzene is a solid material produced and used in refining petroleum and in chemical syntheses.

The hexamethylated ring appears far more toxic than the lower anklyated benzenes. It was lethal to nine of 10 rats, versus none of 10 for durene when 2.5 ml was administered orally (27). Hexamethylbenzene is not carcinogenic in long-term skin studies in mice (336).

Analytic methods are available.

## 1.6 Ethylbenzenes

Toxicity data for the ethylbenzenes are given in Table 21.11.

### 1.6.1 Ethylbenzene

*1.6.1.1 General.* Ethylbenzene or phenylethane is a colorless, flammable liquid (34) with a pungent odor. Being heavier than air, its vapors may travel a considerable distance and ignite and backflash (8). It evaporates about 94 times more slowly than ether (9). Selected physical data are presented in Table 21.1 (1–22, 337).

Traces of ethylbenzene have been detected in exhaled air (42). It also occurs in the gas phase of smoke condensate (24) and has been detected at 3.1 to 4.5 ppb in urban air (45).

Ethylbenzene is a petrochemical and is prepared by dehydrogenation of naphthenes or from catalytic cyclization and aromatization (115), but mainly by alkylation of benzene. It is used in the production of styrene and synthetic polymers (48), but also as a solvent (115, 338) and component for automotive and aviation fuels (29).

*1.6.1.2 Toxicity.* Ethylbenzene appears more irritating than its lower homologues. Exposure to 200 ppm is irritating to the eyes and 5000 ppm is intolerable (21). Exposure to high vapor concentrations causes irritation and CNS effects. Dermal contact causes erythema and inflammation (6, 21). It is absorbed also through the skin. Ethylbenzene has been detected in subcutaneous adipose tissue samples of workers 3 days after exposure to rubber manufacturing components (339). Prolonged exposure results in functional disorders, irritation to the upper respiratory tract, and hepatobiliary complaints (21, 337). It is an aspiration hazard (21, 337).

The oral $LD_{50}$ in rats ranges from 3.5 to 5.5 g/kg (3, 21, 54) and the dermal $LD_{50}$ ranges from 15.5 to 17.8 g/kg (54, 337). The 4-hr $LC_{LO}$ in rats is 4000 ppm (54). Ethylbenzene crosses the placenta and has been detected in cord blood samples (126) but is not fetotoxic in rats, mice, or rabbits (337). Reproductive effects have been studied in rats, rabbits, and monkeys, but these studies are not conclusive as to the effects of ethylbenzene (337).

*1.6.1.3 Metabolism.* In humans, the major metabolites are mandelic acid (70 percent) and phenylglyoxylic acid (25 percent) (337). In the rabbit (29, 340), it is metabolized to a number of oxidation products and subsequently excreted. The major urinary metabolite is hippuric acid. The oxidation products are benzoic acid, phenylacetic acid, and mandelic acid, excreted as the glycine conjugate, and also methylphenylcarbinol, 1-phenylethanol, excreted as the glucuronide. In another study, urinary metabolites included about 0.3 percent *p*-ethylphenol and smaller quantities of 1- and 2-phenylethanol (234). When absorbed through the skin, mandelic acid was excreted at 4.6 percent, whereas following exposure, the majority of the ethylbenzene was excreted as mandelic acid conjugated with glycine (341).

Ethylbenzene stimulates microsomal enzyme synthesis and phenobarbital enhances its metabolic hydroxylation (342). *Pseudomonas putida* is capable of oxidizing ethylbenzene to (+)-*cis*-3-ethyl-3, 5-cyclohexadiene-1, 2-diol and related compounds (343).

*1.6.1.4 Industrial Hygiene.* The OSHA PEL (148) is 100 ppm (435 mg/m$^3$) with a STEL of 125 ppm (435 mg/m$^3$) and the ACGIH TLV (194) is 100 ppm (434 mg/m$^3$) with a STEL of 125 ppm (543 mg/m$^3$). The odor threshold for ethylbenzene is 8.7 ppm and is irritating at 870 ppm (55). The NIOSH sampling procedure includes collection in charcoal tubes, desorption with disulfides, and quantification by flame ionization gas chromatography (195). Other analytic techniques include ultraviolet spectrometry (255).

Protective clothing and equipment should include boots, rubber overclothing, goggles, and self-contained breathing apparatus (21). Urinary mandelic acid and exhaled ethylbenzene serve as biologic monitoring procedures (194, 21, 337). Ethyl-

**Table 21.11.** Toxicity of Ethylbenzenes

| Material | Route of Entry | Species | Dose or Concentration | Results or Effects | Ref. |
|---|---|---|---|---|---|
| Ethylbenzene | Dermal | Human | 22–33 mg/m³ hr | Absorption rate high on hands and forearms | 340 |
| | Inhalation | Human | 1000–2000 ppm (4.92–9.84 mg/l) × 6 min | Severe eye irritation, lacrimation, gradual response, fatigue, but increasing vertigo, chest constriction, and dizziness when leaving | 336 |
| | Inhalation (chronic) | Human | 5000 ppm (24.6 mg/l) | Unacceptable concentration | 336 |
| | | | 100 ppm (0.492 mg/l) | Irritant | 336 |
| | Oral (acute) | Rat | 2.7 g/kg | Lethal to 7 of 10 test animals | 27 |
| | | | 3.5 g/kg | $LD_{50}$ | 84 |
| | | | 5/46 ml/kg | $LD_{50}$ | 336 |
| | Eye | Rabbit | Undiluted | Slight conjunctival irritation | 84 |
| | | | | Corneal injury in some rabbits—classification 2/10 | 336 |
| | Dermal | Rabbit | Undiluted | Irritant, classification 4/10 | 336 |
| | | | 5.0 g/kg | $LD_{50}$ | 336 |
| | | | 17.8 ml/kg | $LD_{50}$ | 336 |
| | Intraperitoneal | Guinea pig | 539 mg/kg | Slightly more toxic and irritant than benzene | 50 |
| | Inhalation (acute) | Rat | 4000 ppm (19.7 mg/l) × 4 hr | ~$LC_{50}$, 3 of 6 survived | 336 |
| | | Mouse | 3050 ppm (15 mg/l) | Loss of righting response | 325 |
| | | | 9150 ppm (45 mg/l) | Death in 2 hr | 325 |
| | | Guinea pig | 1000 ppm (4.92 mg/l) × 3 min | Slight nasal irritation | 336 |
| | | | 1000 ppm (4.92 mg/l) × 8 min | Eye irritation | 336 |
| | | | 2000 ppm (9.84 mg/l) × 1 min | Moderate eye and nasal irritation | 336 |
| | | | 2000 ppm (9.84 mg/l) × 345 min | One animal unconscious | 336 |

| Compound | Route | Dose | Effect | Ref. |
|---|---|---|---|---|
| | | 2000 ppm (9.84 mg/l) × 390 min | Apparent vertigo | 336 |
| | | 2000 ppm (9.84 mg/l) × 480 min | Static and motor ataxia | 336 |
| | | 5000–10,000 ppm (24.6–49.2 mg/l) | Immediate, intense irritation to conjunctivas and nasal mucous membrane, lacrimation, staggering gait. On pathological examination, intense cerebral congestion, lung edema and congestion, blood cyanotic | 336 |
| | Aquatic | Fish | 100–10 ppm | TLm96 | 292 |
| | Oral (subchronic) | Rat | 13.6–136 mg/kg/day × 182 days | No effects | 84 |
| | Dermal | Rabbit | 408–680 mg/kg/day × 182 days | Liver and kidney weight increases, slight pathological signs | 84 |
| | | | Undiluted, repeated | Moderate irritation with slight necrosis | 84 |
| | Inhalation | Rat | 400–2200 ppm (1.7–9.5 mg/l) × 7 hr/day × 144–214 days | Slight liver and kidney weight increases, slight pathological changes at two highest doses | 84 |
| | | Rabbit, guinea pig, primate | 400–600 ppm (1.7–2.6 mg/l) × 7 hr/day × 186–214 days | Little or no effect | 84 |
| Methylethyl-benzene | Inhalation | Mouse | 3000 ppm (15 mg/l) | Loss of righting reflex, no deaths | 230 |
| Diethyl-benzene | | | | | |
| o-, m-, p- | Oral (acute) | Rat | 1.2 g/kg | $LD_{50}$ | 84 |
| | Oral (acute) | Rat | 5.0 g/kg | Lethal to all 10 animals | 29 |
| | Oral (acute) | Rat | 5.0 g/kg | Lethal to 8 of 10 animals | 29 |
| | Inhalation | Mouse | >5500 ppm (>30 mg/l) | Loss of righting reflex, no deaths | 230 |
| | Aquatic | Fish | 100–10 ppm | TLm96 | 292 |
| | Oral | Rat, rabbit | 0.0025 mg/kg/day | No effect | 344 |
| | | | 0.25 mg/kg/day | No effect | 344 |
| | | | 2.5 mg/kg/day | Some adrenal gland weight decrease | 344 |
| | Dermal | Rabbit | Undiluted | Moderate irritation, slight necrosis | 84 |

benzene is on the EPA TSCA Chemical Inventory and the Test Submission Data Base (54).

### 1.6.2 Methylethylbenzenes

Methylethylbenzene is a clear, flammable liquid. Selected physical data are given in Table 21.1 and toxicity data in Table 21.11. It has been detected in the urban air at 1.5 to 4.0 ppb (45). The toxicity of this material resembles that of ethylbenzene. Little information appears available on dimethylethylbenzene. However, it has been documented to cross the placenta (126).

### 1.6.3 Diethylbenzene

Diethylbenzene is a colorless, mobile, flammable liquid. Selected physical data are given in Table 21.1 and toxicity data in Table 21.11.

Diethylbenzene is slightly more toxic than the ethylbenzene. The mixed compound was tested as 25 percent ortho, 40 percent meta, and 35 percent para isomer (29). In rats, 120 mg/kg caused slight hemorrhages and dystrophic changes in the liver, gastric mucosa, duodenum, spleen, and kidneys, and also hepatic decreases in protein and glycogen. Several *Pseudomonas* strains have been found capable of using *m*- and *p*-diethylbenzene as the sole carbon source (248). The metabolism proceeds via side-chain oxidation, rather than by dehydrogenation and subsequent hydration, yielding *p*-ethylphenylacetic acid (344).

No threshold limit values have been established for occupational exposure. However, the limits for ethylbenzene may apply. The odor and taste threshold in water is 0.04 to 0.05 mg/l (345).

For handling diethylbenzene and worker protection, similar measures may apply as recommended for ethylbenzene.

## 1.7 Propylbenzenes

Some of the propylbenzene isomers, such as cumene and cymene, are of considerable industrial and commercial importance.

### 1.7.1 Propylbenzene

Propylbenzene or *n*-propylbenzene is a colorless, flammable liquid. Selected physical data are given in Table 21.1 and toxicity data in Table 21.12. *n*-Propylbenzene in the mouse produces a loss of righting response at 10 to 15 mg/l, loss of reflexes at 15 mg/l, and death at 20 mg/l (29, 326). The oral $LD_{50}$ is 6040 mg/kg (3, 21, 54) and the 2-hr $LC_{50}$ in rats is 65,000 ppm (54). It is metabolized in mammals similarly to other alkylbenzenes (199, 346). An analytic determination procedure is available (255, 21). Propylbenzene is on the EPA TSCA Chemical Inventory and the Test Submission Data Base (54).

### 1.7.2 Isopropylbenzene (Cumene)

*1.7.2.1 General.* Isopropylbenzene or cumene is a colorless, flammable liquid with a sharp, aromatic or gasoline-like odor (19, 21). Selected physical data are given in Table 21.1. Occupational information is presented in a NIOSH/OSHA standard (347).

Cumene occurs in a variety of petroleum distillates and commercial solvents. It has been detected also in the exhaled air of human volunteers, and higher quantities were noted in smokers than in nonsmokers (42). Hemolytic effects may be produced whenever isopropylbenzene is permitted to oxidize to the peroxide (348).

Cumene is produced commercially by alkylation of benzene with propylene and recovered from petroleum by fractional distillation (115). Most of the commercially available material is used as a thinner for paints and enamels, as a constituent of some petroleum-based solvents, in the synthesis of phenol (29), and in the perfume industry (349).

*1.7.2.2 Toxicity.* Cumene appears slightly less toxic than its *n*-propyl isomer, but more so than benzene or toluene. Like its lower homologues, it may be irritant to the eyes and skin. It is a CNS depressant characterized by slow induction and long duration of effects (350). Exposure to vapor concentrations may cause dizziness, slight incoordination, and unconsciousness. Prolonged skin contact may result in skin rashes. In acute exposures, animals exhibit damage to the spleen and fatty changes in the liver, but no renal or pulmonary effects (115). Subacute exposures showed no significant changes in peripheral blood, but some liver, kidney, and lung effects. Selected toxicity data are given in Table 21.12. The oral $LD_{50}$ in rats is 1.4 to 2.91 g/kg, the oral $LD_{50}$ in mice is 12.8 g/kg, the 4-hr $LC_{50}$ in rats is 8000 ppm, the 7-hr $LC_{50}$ in mice is 2000 ppm, and the dermal $LD_{50}$ in the rabbit is 12.3 g/kg (3, 20, 21, 54).

*1.7.2.3 Metabolism.* Cumene is absorbed readily in mammals and is oxidized at the side chain, forming dimethylphenylcarbinol glucuronide (351, 352). In rats, no phenolic metabolites were detected following ingestion (234), and less than 5 percent cumene was exhaled unchanged in the rabbit (27).

*Pseudomonas desmolytica* (248), *P. convexa*, and *P. ovalis* are capable of growing on cumene (248, 353). Oxidation products were identified as 3-isopropylcatechol and (+)-2-hydroxy-7-methyl-6-oxooctanoic acid (354). In soil, cumene inhibits ammonification and nitrification mechanisms (355).

*1.7.2.4 Industrial Hygiene.* The OSHA STEL (148) is 50 ppm (245 mg/m$^3$) with a skin notation and the ACGIH STEL (194) is 50 ppm (246 mg/m$^3$). The odor threshold is 0.039 ppm and is irritating at 32 ppm. A NIOSH sampling procedure (195) involves the collection of vapor on charcoal and analysis by flame ionization gas chromatography. A gas chromatographic method for metabolites is available (351).

Exposure prevention includes proper eye, skin, and face protection and a cartridge-type of self-contained breathing apparatus (261, 21). Appropriate medical surveillance programs include a screen for kidney disease, chronic respiratory disease, liver disease, and skin disease (261, 21). Cumene is on the EPA TSCA Chemical Inventory and the Test Submission Data Base (54).

Table 21.12. Toxicity of Propylbenzenes and Higher Alkylbenzenes

| Material | Route of Entry | Species | Dose or Concentration | Results or Effects | Ref. |
|---|---|---|---|---|---|
| n-Propylbenzene | Oral | Rat | 4.830 g/kg | $LD_{LO}$ | 54 |
| | | | 5.0 g/kg | 2 deaths of 10 | 29 |
| | Inhalation | Mouse | 2000–3000 ppm (10–15 mg/l) | Loss of righting response | 230 |
| | | | 3000 ppm (15 mg/l) | Loss of reflexes | 230 |
| | | | 4100 ppm (20 mg/l) | Death | 230 |
| Cumene | Oral (acute) | Rat | 1.4 g/kg | $LD_{50}$ | 84 |
| | | | 5.0 g/kg | 6 deaths in 10 test animals | 27 |
| | Eye | Rabbit | Undiluted | Slight conjunctival irritation | 84 |
| | Inhalation | Human | 200 ppm (1.0 mg/l) | Irritant | 9 |
| | | Rat | 8000 ppm (30 mg/l) × 4 hr | 50% mortality | 29 |
| | | Mouse | 2000 ppm (10 mg/l) × 7 hr | Lethal dose | 115 |
| | | | 4100 ppm (20 mg/l) | Loss of righting response | 325 |
| | | | 5100 ppm (25 mg/l) | Loss of reflexes, no deaths | 325 |
| | Aquatic | Fish | 100–10 ppm | TLm96 | 292 |
| | Oral (subchronic) | Rat (F) | 154 mg/kg/day × 194 days | No effect | 84 |
| | | | 462 mg/kg/day × 194 days | Slight kidney weight increase | 84 |

| Compound | Route | Species | Dose | Effect | Ref. |
|---|---|---|---|---|---|
| | Skin | Rabbit | 769 mg/kg/day × 194 days | Moderate kidney weight increase | 84 |
| | Inhalation | Rat, rabbit | Undiluted | Moderate irritation, slight necrosis | 84 |
| | | | 500 ppm (2.5 mg/l) 8 hr/day × 6 days/week × 150 days | Hyperemia and congestion of lungs, liver, and kidney | 115 |
| Cymene | Oral | Rat | 4750 mg/kg | $LD_{50}$ | 54 |
| | Inhalation | Rat | 5000 ppm (27.5 mg/l)/45 min | $LC_{LO}$ | 54 |
| n-Butylbenzene | Oral | Rat | 5.00 g/kg | 2 deaths of 10 | 29 |
| sec-Butylbenzene | Oral | Rat | 2.240 g/kg | $LD_{50}$ | 54 |
| | | | 5.0 g/kg | Lethal to 8 of 10 animals | 29 |
| tert-Butylbenzene | Oral | Rat | 5.00 g/kg | Lethal to 7 of 10 animals | 27 |
| tert-Butyltoluene | Inhalation (acute) | Human | 10 ppm (0.06 mg/l) × 3 min | Irritant | 54 |
| | Oral | Rat | 20 ppm (0.12 mg/l) × 5 min | CNS | 54 |
| | | | 1.6 g/kg | Lethal to 50% of colony | 29 |
| | | Mouse | 1.8 ml/kg | $LD_{50}$ | 54 |
| | | Rabbit | 0.9 ml/kg | $LD_{50}$ | 54 |
| | | Rat, mouse | 2.0 ml/kg | $LD_{50}$ | 54 |
| | Inhalation | Rat, mouse | 248 ppm (1.5 mg/l)/4 hr | $LC_{50}$ | 54 |
| | Inhalation (subacute) | Rat (F) | 25–50 ppm (0.15–0.3 mg/l) × 1–7 hr | Hemoglobin increase, erythrocyte, and leukocyte decreases | 54 |
| Dodecylbenzene | Oral | Rat | 5 g/kg | No deaths | 29 |

### 1.7.3 Isopropyltoluene (Cymene)

Isopropyltoluene or cymene is a fragrant, flammable liquid. Cymene occurs as the ortho, meta, and para isomer in a wide variety of essential oils. It is produced by catalytic cracking of petroleum in the wood pulp sulfite process (356). The para isomer is used for the synthesis of *p*-cresol (356). The toxicity of cymene equals that of cumene and its lower homologues. It is readily absorbed in mammals and distributes in tissues as do similar solvents (357). The oral $LD_{50}$ in the rat is 4.75 g/kg (3). Toxicity data are given in Table 21.12.

*p*-Cymene is capable of serving as a carbon source for several strains of *Pseudomonas* (248). Isopropylbenzoic acid, *p*-isopropylbenzyl alcohol, and the aldehyde were also identified (358).

Solvent-resistant gloves, splash-proof goggles, and a self-contained breathing apparatus are recommended (19). Cymene is on the EPA TSCA Chemical Inventory and Test Submission Data Base (54).

### 1.7.4 Diisopropylbenzenes

Diisopropylbenzene occurs as the ortho, the meta, and the para isomers, all flammable liquids. However, they are of very low toxicity orally. Of 10 rats dosed with 5.0 mg/kg, no deaths for the meta, and one death for the ortho and the para isomers occurred (29). Rats and rabbits exposed to 0.2 to 1.0 mg/l (30 to 150 ppm) for 90 min daily for 5 weeks caused vascular hyperemia, hemorrhaging in most major organs, fatty and protein dystrophy in the liver, kidney, and heart, and hyperplasia of the bone marrow (359). *m*-Diisopropylbenzene caused decreased fertility in rats and mice exposed to 1 to 3 mg/l for 30 days (360).

## 1.8 Butylbenzenes

### 1.8.1 Monobutylbenzenes

Butylbenzene occurs in four isomeric forms, the *n*-butylbenzene or 1-phenylbutane, the *sec*-butylbenzene or 2-phenylbutane, the isobutylbenzene or 2-methyl-1-phenylpropane, and *tert*-butylbenzene or 2-methyl-2-phenylpropane. They are odorous, flammable liquids. Selected physical data in Table 21.1 and selected toxicity data are given in Table 21.12. The neurotoxicity of *tert*-butylbenzene is believed to be a consequence of hemorrhage in the spinal cord due to vascular injury. A single oral dose of 0.075 ml produced an irreversible foreleg paralysis in the rat (29). The toxicity of butylbenzenes is similar to that for isopropylbenzene. They are metabolized by side-chain hydroxylation and conjugation for urinary excretion.

Isobutylbenzene is hydroxylated to isobutylcatechol by *Pseudomonas desmolytica* (354).

### 1.8.2 Polysubstituted Butylbenzenes

Tertiary butyltoluene or *p*-methyl-*tert*-butylbenzene is the isomer of commercial importance. It is a clear, colorless, combustible liquid, with a distinct aromatic odor (29). Saturated air contains about 800 ppm vapor. Selected physical data are given in Table 21.1. *tert*-Butyltoluene is produced by the alkylation of toluene with isobutylene (29, 347). It is used as a solvent in resin preparation and as a raw material in the chemical and pharmaceutical industries (29, 347). Selected toxicity data are given in Table 21.12. Workers handling *tert*-butyltoluene experience nasal irritation, nausea, malaise, headache, and weakness (29). *tert*-Butyltoluene may also decrease blood pressure, increase pulse rate, and cause CNS and hematopoietic effects (29).

The OSHA PEL (148) is 10 ppm (60 mg/m$^3$) with a STEL of 20 ppm (120 mg/m$^3$). The ACGIH TLV (194) is 10 ppm (61 mg/m$^3$) with a STEL of 20 ppm (121 mg/m$^3$). NIOSH recommends a sampling procedure and medical surveillance programs (347).

### 1.9 Other Alkylbenzenes

Very little information exists on amylbenzene or hexylbenzene. Dodecylbenzene is used as a vehicle and solvent for polynuclear aromatics in carcinogenesis studies. It is a flammable, odorous liquid. Selected physical data are given in Table 21.1. It is produced by the alkylation of benzene with propylene tetramer (5), and is used to produce arylakyl sulfonates in the soap and detergent industries (5).

It has been used as a chemical vehicle in carcinogenesis studies; however, it may act as a promotor for materials such as dimethylbenzanthracene (359). The oral toxicity of dodecylbenzene is very low; an oral dose of 5 g/kg caused no deaths in rats (29). It is a mild skin irritant and a skin sensitizer (21). The odor is a weak oily odor (19).

Industrially, dodecylbenzene vaporizes slowly and can be contained easily. Also, it can be sulfonated in wastes and recycled (360). Goggles or a face shield and rubber gloves are recommended for those handling dodecylbenzene (21). It is on the EPA TSCA Chemical Inventory and the Test Submission Data Base (54).

### 1.10 Styrene

Styrene is the simplest member of the alkenylbenzenes. At one time, its toxicity was equated to that of vinyl chloride (66). However, no parallelism has been found to date.

#### 1.10.1 General

Styrene, cinnamene, styrol, or vinylbenzene (29) is a colorless to yellow flammable liquid with a sweet, floral odor at low, but disagreeable at high concentrations (21, 55, 361). Selected physical data are given in Table 21.1 and toxicity data in Tables 21.13 and 21.14. It occurs naturally in the sap of the styracaceous tree (115) and has been detected in the urban atmospheres (45).

Styrene is produced by alkylation of benzene with ethylene, followed by catalytic dehydrogenation (5) or by demethylation of cumene (347). Chemically, styrene is highly reactive and polymerizes readily, sometimes accompanied by violent explosions. Because the reactions occur rapidly at elevated temperatures, it is necessary to add polymerization inhibitors for transport and storage (362). Styrene monomer is one of the world's major organic chemicals. It is used in plastics and resins and as a dental filling component, chemical intermediate, component in agricultural products (363), and stabilizing agent. Styrene is present in the oily fraction of cigarette smoke (24). Various polymeric combinations contain styrene (364, 365).

### 1.10.2 Toxicity

Styrene is irritating to the eyes, skin, mucous membranes, and respiratory system. High exposure levels may cause anesthesia and systemic effects.

*1.10.2.1 Acute Human Exposure.* Human volunteers experienced no effect when exposed to 10 ppm or less, but strong irritation at 600 ppm. Selected toxicity data are given in Table 21.13. Healthy volunteers exposed to 50 or 150 ppm of styrene at rest or light physical exercise demonstrated that styrene in the alveolar air varied with the level of exercise. The concentration of styrene in the arterial and venous blood increased sharply, and about 50 percent of the uptake was excreted as mandelic acid (366). When exposed to 350 ppm, a statistically significant impairment of the volunteers' reaction time was observed (367). Volunteers exposed to 376 ppm showed signs of transient neurological effects (368). In addition, accidental poisonings have been recorded (369). Styrene vapor may be absorbed by the skin (367).

*1.10.2.2 Acute Animal Studies.* Oral administration of styrene to rats showed rather low toxicity. Selected toxicity data are given in Tables 21.13 and 21.14. Styrene causes moderate conjunctival irritation and slight, transient corneal injury in rabbits. Nystagmus was demonstrated in rabbits, and during styrene exposure the directions of the rotary nystagmus reversed. The blood levels indicated possible CNS involvement.

Under ambient conditions, the vaporization of styrene is too low to be lethal to laboratory animals in a few minutes. The highest concentration to cause no serious systemic disturbances in 8 hr was 1300 ppm. Animals exposed to 2500, 5000, or 10,000 ppm showed eye and nasal irritation. Those exposed to 2500 ppm exhibited varying degrees of weakness and stupor, followed by incoordination, tremors, and unconsciousness. Unconsciousness occurred at 2500 ppm in 10 hr, at 5000 ppm in 1 hr, and at 10,000 ppm in a few minutes (370). The oral $LD_{50}$ in rats is 1 to 5 g/kg (20, 54) and the oral $LD_{50}$ in mice is 316 mg/kg (54). The 4-hr $LC_{50}$ in rats is 6000 ppm (370) and the 4-hr $LC_{50}$ in mice is 9500 mg/m$^3$ (54). Styrene is a sensory irritant of the upper airways in the mouse (371).

*1.10.2.3 Chronic Human Exposure.* Signs and symptoms experienced by a group of employees handling styrene in a manufacturing plant included nausea, vomiting, loss of appetite, and general weakness (372).

Table 21.13. Toxicity of Styrene

| Route of Entry | Species | Dose or Concentration | Results or Effects | Ref. |
|---|---|---|---|---|
| Inhalation (acute) | Human | >10 ppm (0.04 mg/l) | Odor not detectable | 84 |
| | | 60 ppm (0.26 mg/l) | Detectable, but nonirritant | 84 |
| | | 100 ppm (0.43 mg/l) | Strong odor, but without excessive discomfort | 84 |
| | | 200–400 ppm (0.85–1.7 mg/l) | Objectionable strong odor | 84 |
| | | 376 ppm (1.6 mg/l) × 1 hr | Neurological impairment | 367 |
| | | 600 ppm (2.6 mg/l) | Very strong odor, strong eye and nasal irritant | 84 |
| | | 800 ppm (3.4 mg/l) × 3 hr | Immediate eye and throat irritation, increased nasal mucous secretion, metallic taste, drowsiness, vertigo; after test termination, slight muscular weakness, accompanied by inertia and depression | 90 |
| Oral | Rat | 5 g/kg | $LD_{50}$ | 84 |
| | | 5 ml/kg | One mortality in 10 test animals | 29 |
| Eye | Rabbit | Undiluted | Moderate conjunctival irritation and slight, transient corneal injury | 84 |
| Inhalation | Rat | 6000 ppm (26 mg/l) × 4 hr | Approximate $LC_{50}$ | 369 |
| | Guinea pig | 5200 ppm (22 mg/l) × 4 hr | Approximate $LC_{50}$ | 369 |
| Aquatic | Fish | 100–10 ppm | TLm96 | 292 |

**Table 21.14. Subacute and Chronic Exposure to Styrene**

| Route of Entry | Species | Dose or Concentration | Results or Effects | Ref. |
|---|---|---|---|---|
| Oral | Rat | 66.7 mg/kg/day × 5 days/week × 185 days | No effect | 84 |
| | | 100 mg/kg/day × 5 days/week × 28 days | No effect | 369 |
| | | 133 mg/kg/day × 5 days/week × 185 days | No effect | 84 |
| | | 400 mg/kg/day × 5 days/week × 185 days | Growth, liver and kidney weight deviations | 84 |
| | | 500 mg/kg/day × 5 days/week × 28 days | Poor weight gain, no significant pathology | 369 |
| | | 667 mg/kg/day × 5 days/week × 185 days | Kidney weight, moderate growth and liver weight deviations | 84 |
| | | 1000 mg/kg/day × 5 days/week × 28 days | Irritant to esophagus and GI tract; some deaths before 28 days | 369 |
| | | 2000 mg/kg/day × few days | Highly irritant to esophagus and GI tract, resulting in rapid death | 369 |
| | Rabbit | 600 mg/kg/day × 3–10 days | Increased serum cholinesterase, carboxyl- and arylesterase activity | 381 |
| Dermal | Rabbit | Undiluted × 20 appln. × 4 weeks | Moderate irritant, with blistering and hair loss | 369 |
| | | Undiluted × 10–20 appln. × 2–4 weeks | Moderate irritant, slight necrosis | 84 |
| Subcutaneous | Rabbit | 600 mg/kg/day × 3–10 days | Increased cholinesterase activity | 403 |
| | Rat | 2.5 g/kg/day × 15–20 days | Decreased serotonin level in blood, lungs, intestine, and brain | 403 |

| | | | | |
|---|---|---|---|---|
| Inhalation | Rat | 6.5 ppm (35 mg/m³) × 4 hr/day × 5 days/week × 1–4 months | Decreased neutrophil phagocytic activity, increased susceptibility towards staphylococcal infection | 401 |
| | | 300 ppm (7.9 mg/m³) × 6 hr/day × 2–11 weeks | Enzymatic adaptive changes | 401 |
| | | 1300 ppm (6.3 mg/l) × 7 hr/day × 139 exp × 7 months | Eye and nasal irritation | 84 |
| | | 2000 ppm (9.3 mg/l)/day × 105 exp. × 5 months | Eye and nasal irritation, moderate growth depression | 84 |
| | Mouse | 6.5 ppm (35 mg/m³) × 4 hr/day × 3 months | Susceptibility towards staphylococcal infection initially decreased, then increased, then lowered again for 1 month each | 401 |
| | Rabbit | 6 ppm (29 mg/m³) × 4 hr/day × 4 months | Decreased phagocytic activity of | 401 |
| | | 1300 ppm (6.3 mg/l) × 7 hr/day × 264 exp./12 months | No effect | 84 |
| | | 2000 ppm (9.3 mg/l) × 7 hr/day × 126 exp./5 months | No effect | 84 |
| | Guinea pig | 650 ppm (3.0 mg/l) × 7 hr/day × 130 exp./6 months | No effect | 84 |
| | | 1300 ppm (6.3 mg/l) × 7 hr/day × 139 exp./7 months | Eye and nasal irritation, slight | 84 |
| | | 2000 ppm (9.3 mg/l) × 7 hr/day × 98 exp./5 months | Eye and nasal irritation, moderate growth depression | 84 |
| | Rhesus monkey | 1300 ppm (6.3 mg/l) × 7 hr/day × 264 exp./12 months | No effect | 84 |

An ocular examination of 345 workers in a styrene plant revealed conjunctival irritation in 22 percent, but no retrobulbar neuritis or central retinal vein occlusion (373).

Repeated or prolonged skin contact causes dermatitis, marked by rough, dry, and fissured skin (29). In general, fair-skinned individuals appear to be less resistant than dark-skinned persons to the defatting and dehydrating action of styrene (372).

Industrial exposure occurs during the synthesis and handling of styrene and in the production of polystyrene and its copolymers (374). Exposure of 150 ppm or higher resulted in prolonged reaction time (375). In 494 production workers, acute lower respiratory symptoms and a very low percentage with $FEV_1/FVC$ less than 75 percent and FVC <80 percent were noted; however, liver function was normal even at the high exposure level (376).

One study of 50 petroleum workers producing synthetic rubber indicated one-half of the work force experienced gastric acidity reduction, liver detoxification, and pancreatic changes (377, 378). In addition, moderate anemia, leukopenia, reticulocytosis, increased clotting time, and a rise in capillary permeability were noted (379). Clinical analyses indicated changes in blood protein composition (380, 381) and increased cholinesterase activities (382).

A group of 98 workers exposed to styrene showed psychological function changes parallel to air concentrations as determined by mandelic acid excretion (383, 384), abnormal brain waves (385), and some peripheral nerve lesions (385).

Styrene has been observed to cross the placenta (126), and some CNS defects were observed in children whose mothers has been exposed to styrene during pregnancy (386).

*1.10.2.4 Subacute and Chronic Animal Studies.* In two oral studies (84, 370), the no-effect level was 133 to 667 mg/(kg)(day). Low growth, and organ weight changes were noted at 1 g/kg/day, which resulted eventually in some deaths. Additionally, 2 g/kg/day proved so highly irritant to the esophagus and stomach that death ensued quickly (370). In the rabbit, styrene causes increase in serum cholinesterase (383).

Styrene is moderately irritating to the skin, even when applied undiluted (84, 370). Styrene increased cholinesterase, carboxylesterase, and arylesterase activities in the rabbit (382). Rats and guinea pigs exposed to 1300 ppm of styrene for 7 to 8 hr/day exhibited eye and nasal irritation and appeared unkempt (84, 115), whereas the rabbit and the rhesus monkey exhibited no adverse signs at this concentration.

In rats, styrene accumulates in tissues, increases cholinesterase activity, and decreases spleen vitamin C content (387).

Exposure of female rats to styrene vapor prolonged the estrus cycle (388) and was embryotoxic (389). Malformations were shown in chick embryos (390). Styrene is positive in some mutagenicity assays and negative in others (391, 392). Male rats exposed to 300 ppm, 6 hr/day, did show some increase in bone marrow chromosomal aberrations (393), from 8 to 12 percent versus 1 to 6 percent for the controls. In long-term feeding tests, styrene was noncarcinogenic in the Fischer 344 rat and the $B_6C_3F_1$ mouse (393). In inhalation studies, styrene was carcinogenic in rodents and

IARC has classified styrene as a 2B carcinogen, that is, a possible human carcinogen (64).

### 1.10.3 Absorption and Metabolism

The absorption of styrene in humans proceeds by all routes, but mainly through the respiratory tract. The partition coefficients for air and water and air and blood have been determined (395), and the ratio between atmospheric concentration, length of exposure, and initial blood levels for humans are proportional (396). Following absorption, styrene is readily metabolized (397). In humans, the two majority urinary metabolites of styrene are mandelic acid and phenylglyoxylic acid (396) excreted as 85 and 10 percent, respectively, of the retained dosage (396), with about 2 percent exhaled in unchanged form (29). Below 250 ppm, the excretion of mandelic acid can be related directly to exposure concentration (398). Exposure to 500 ppm increases the excretion of hippuric acid. Minor metabolites are 17-ketosteroids (399). Nonexcreted styrene appears to accumulate in adipose tissue (339). Usually a weekend is not sufficient time to excrete all of the metabolites (397).

Various pathways have been proposed for the metabolism of styrene in humans and animals. Experiments have shown that the excreted metabolites in the rat and the rabbit differ with the route and dose of administration, ranging from 9 to 32 percent for mandelic acid, 0 to 11 percent for phenylglyoxylic acid, 10 to 40 percent for hippuric acid, 6 to 8 percent for glucuronides, 5 to 9 percent for sulfur compounds, and traces for 1- and 2-phenylethanol (396) and 4-vinylphenol (234, 396).

The 4-hr $LC_{50}$ in the rat is 11.8 mg/l and the 2-hr $LC_{50}$ in the mouse is 21 mg/l (396). Following exposure, 25.0 mg/100 g was observed in the brain tissue, 20.0 in the liver, 14.7 in the kidney, 19.1 in the spleen, and 133 mg/100 g in perirenal fat (396). Mouse brain contained 18.0 mg/100 g styrene (396).

When administered orally to the rat, styrene is converted to benzoic acid and excreted as hippuric acid (400); the beta carbon is exhaled primarily as carbon dioxide (401). The minor metabolites are mandelic acid and the glucuronide of phenylglycol (400). The metabolism of styrene has been summarized (29, 396, 402–404).

### 1.10.4 Industrial Hygiene

*1.10.4.1 Threshold Values.* The OSHA PEL (148) is 50 ppm (215 mg/m$^3$) with a STEL of 100 ppm (425 mg/m$^3$) and the ACGIH TLV (194) is 50 ppm (213 mg/m$^3$) with a STEL of 100 ppm (426 mg/m$^3$). The human no-effect level is estimated to be 650 ppm (84). The odor threshold is 0.2 ppm (55), is recognized at 50 ppm, and is irritating at 430 ppm (55).

*1.10.4.2 Atmospheric Monitoring.* NIOSH suggests the collection of styrene on charcoal, desorption with carbon disulfide, and quantification by flame ionization gas chromatography (195). An air sampling method applicable to the rubber vulcanization industry also is available (405). Methods for styrene determination in emission gases are also available (21, 406).

***1.10.4.3 Biologic Monitoring.*** OSHA has published medical surveillance programs (347). Biologic monitoring includes the quantification of urinary creatinine, mandelic acid, and phenylglyoxylic acid (407–409). Because styrene alters the metabolism of amino acids, amino acid determination may present a direct indicator of the exposure (410).

***1.10.4.4 Handling Practices.*** Protective equipment should include neoprene gloves, apron, safety goggles, and a cartridge-type respiratory breathing apparatus (21). Styrene is on the EPA TSCA Chemical Inventory and Test Submission Data Base (54).

## 1.11 Vinyltoluene

### *1.11.1 General*

Vinyltoluene or methylstyrene is a colorless, combustible liquid with a strong, disagreeable odor (8). It usually occurs as a mixture of the meta and the para isomers at 50 to 70 and 30 to 45 percent, respectively (84). At elevated temperatures, vapors may form explosive mixtures with air, and polymerization may occur under explosive expansion. Selected physical data are given in Table 21.1.

Vinyltoluene is produced by the dehydrogenation of *m*- and *p*-ethyltoluene and by catalytic re-forming. It is used in the plastics industry, in resin production, as a block-packaging component for radioactive waste (411), and as an insecticide component (412).

### *1.11.2 Toxicity*

Vinyltoluene causes eye, skin, and upper respiratory tract irritation. At high concentrations, it exhibits anesthetic and systemic effects similar to that of styrene. Selected toxicity data are given in Table 21.15. Exposures of 580 ppm are well tolerated by most laboratory animals (84).

Vinyltoluene is metabolized by similar oxidative mechanisms as seen for styrene.

### *1.11.3 Industrial Hygiene*

The OSHA PEL (148) is 50 ppm (240 mg/m$^3$) with a STEL of 100 ppm (485 mg/m$^3$) and the ACGIH TLV (194) is 50 ppm (242 mg/m$^3$) with a STEL of 100 ppm (483 mg/m$^3$). The odor is irritating at 240 ppm (55). NIOSH suggests collection of vinyltoluene on charcoal, desorption with carbon disulfide, and quantification by flame ionization gas chromatography (195). Air and medical monitoring procedures has been issued (347).

Eye, skin, and respiratory protective equipment should be utilized when handling vinyltoluene.

Table 21.15. Toxicity of Vinyltoluene and Other Alkenylbenzenes

| Material | Route of Entry | Species | Dose or Concentration | Results or Effects | Ref. |
|---|---|---|---|---|---|
| Vinyltoluene | Inhalation (acute) | Human | <10 ppm (< 0.05 mg/l) | Odor not detectable | 84 |
| | | | 50 ppm (0.24 mg/l) | Detectable odor, but no irritation | 84 |
| | | | 200 ppm (1.0 mg/l) | Strong odor, but tolerated without discomfort | 84 |
| | | | 300 ppm (1.5 mg/l) | Objectionable, strong odor | 84 |
| | | | >400 ppm (>2.0 mg/l) | Very potent odor, strong eye and nasal irritant | 84 |
| | Oral | Rat | 2.5 ml$^a$ | Lethal to 4 or 10 animals | 27 |
| | | | 4.0 g/kg | $LD_{50}$ | 84 |
| | | | 4.9 g/kg$^b$ | $LD_{16}$ | 401 |
| | | | 5.7 g/kg$^b$ | $LD_{50}$ | 401 |
| | | Mouse | 3.16 g/kg$^b$ | $LD_{50}$ | 401 |
| | Eye | Rabbit | Undiluted | Slight conjunctival irritation, no corneal injury | 84 |
| | Inhalation (acute) | Mouse | 62 ppm (300 mg/l)$^c$ | $LC_{50}$ | 401 |
| | Inhalation (subchronic) | Guinea pig | 6 ppm (20 mg/m³) × 4 months | Teratogenic effects | 401 |
| | | Rat | 580 ppm (2.8 mg/l) × 7 hr/day | No effect | 84 |
| | | | 1130 ppm (5.5 mg/l) | Moderate growth depression | 84 |
| | | | 1350 ppm (6.5 mg/l) | Moderate growth and liver weight depression; increased mortality | 84 |
| | | Mouse | 6200 ppm (30 mg/l) × 1 month | Slight weight reduction | 403 |
| | | Rabbit, rhesus monkey | 580–1350 ppm (2.8–6.5 mg/l)/7 hr/day$^c$ | No significant effects | 84 |

**Table 21.15.** (Continued)

| Material | Route of Entry | Species | Dose or Concentration | Results or Effects | Ref. |
|---|---|---|---|---|---|
| | | Guinea pig | 580 ppm (2.8 mg/l) × 7 hr/day[c] | No effect | 84 |
| | | | 1130 ppm (5.5 mg/l) | Slight growth depression | 84 |
| | | | 1350 ppm (6.5 mg/l) | Slight growth depression, slight pathological effect | 84 |
| | | | 6200 ppm (30 mg/m³) × 1 month | Some teratogenic effects | 403 |
| Divinylbenzene | Oral | Rat | 2.5 ml[a] | 5 deaths of 10 | 27 |
| | | | 4.040 g/kg | $LD_{50}$ | 403 |
| Allylbenzene | Oral | Rat | 3.60 g/kg | $LD_{50}$ | 54 |
| | | | 4.620 g/kg | Lethal effects | 54 |
| | | Mouse | 2.90 g/kg | $LD_{50}$ | 377 |
| α-Methyl-styrene | Inhalation | Human | <10 ppm (<0.05 mg/l) | Odor not detectable | 84 |
| | | | 50 ppm (0.25 mg/l) | Detectable odor, but no irritation | 84 |
| | | | 100 ppm (0.5 mg/l) | Strong odor, but tolerated without discomfort | 84 |
| | | | 200 ppm (1.0 mg/l) | Objectionable, strong odor | 84 |
| | | | >600 ppm (>2.9 mg/l) | Very potent odor, strong eye and nasal irritant | 84 |
| | Oral | Rat | 4.9 g/kg | $LD_{50}$ | 84 |
| | Eye | Rabbit | Undiluted | Slight conjunctival irritation | 84 |
| | Inhalation | Rat | 3000 ppm (14.5 mg/l) | Lethal effects | 29 |
| | | Guinea pig | 3000 ppm (14.5 mg/l) | Lethal effects | 29 |
| | Aquatic | Fish | 100–10 ppm | TLm96 | 292 |
| | Dermal | Rabbit | Undiluted | Moderate to marked irritation, slight necrosis | 84 |
| | Inhalation (subchronic) | Rat | 200 ppm (0.97 mg/l) × 7 hr/day × 139 exp. | No effect | 84 |

| Compound | Route/Species | Dose | Effects | Ref. |
|---|---|---|---|---|
| | | 600–800 ppm (2.90–3.9 mg/l) × 7 hr/day × 28–149 exp. | Slight kidney and liver depression, also growth depression at 800 ppm | 84 |
| | | 3000 ppm (14.9 mg/l) × 7 hr/day × 3–4 exp. | High mortality | 84 |
| | Rabbit, rhesus monkey | 200–600 ppm (0.97–2.9 mg/l) × 7 hr/day × 139–152 | No effect except some growth depression and slightly increased mortality in the rabbit at 600 ppm | 84 |
| | Guinea pig | 200–600 ppm (0.97–2.9 mg/l) × 7 hr/day × 139–144 exp. | No effect except some liver weight depression at 600 ppm | 84 |
| | | 800 ppm (3.9 mg/l) × 7 hr/day × 27 exp. | Slight growth, liver and kidney weight depression | 84 |
| | | 3000 ppm (14.5 mg/l) × 7 hr/day × 3–4 exp. | High mortality | 84 |
| 1-Phenyl-butene-2 | Oral | 2.5 ml[a] | Lethal to 8 of 10 animals | 27 |
| 4-Phenyl-butene-1 | Oral | 2.5 ml[a] | Lethal to all 10 rats | 27 |
| Phenyl- | Oral | 5.00 g/kg | Lethal effects | 4 |

[a] Per animal 1:1 in olive oil.
[b] Ortho and para isomers 28:72; other test for meta and para derivatives.
[c] 92–100 exposures.

## 1.12 Divinylbenzene

### 1.12.1 General

Divinylbenzene or vinylstyrene is a water-white, combustible liquid, which occurs as the ortho, meta, and para isomers. Selected physical data are given in Table 21.1. Divinylbenzene is prepared by the dehydrogenation of diethylbenzene. Inhibitors are added to divinylbenzene to prevent autopolymerization, which occurs at elevated temperatures for the meta and the para isomers (8). The monomer is utilized as an insecticide stabilizer, as an ion-exchange resin, as a cross-linking agent in water purification, as a sustained release agent, and as a dental filling component. It has been used as an experimental clotting agent for sustained life research (413).

### 1.12.2 Toxicity

Divinylbenzene is moderately irritating to the eye, skin, and respiratory tract.

The toxicity of divinylbenzene resembles that of styrene. An intravenous injection of polystyrene–divinylbenzene copolymer particles was without effect in rats.

Selected toxicity data are given in Table 21.15.

### 1.12.3 Industrial Hygiene

The OSHA PEL (148) is 10 ppm (50 mg/m$^3$) and the ACGIH TLV (194) is 10 ppm (53 mg/m$^3$). Eyes, skin, and respiratory tract should be protected.

## 1.13 Propenylbenzenes

Propenylbenzene occurs as the 1-propenyl or 2-propenyl derivative. The 1-propenylbenzene is also named β-methylstyrene, $CH_3CH:CHC_6H_5$; the 2-propenylbenzene is also called allylbenzene, $CH:CHCH_2C_6H_5$. Allylbenzene occurs naturally in the essential oils of a variety of *Aniba* species (414).

The acute toxicity of 1-propenylbenzene is slightly below that of the vinylbenzenes. Selected toxicity data for 2-propenylbenzene are given in Table 21.15.

Acidic and neutral metabolites of allylbenzene have been identified as 1′-hydroxyallylbenzene and cinnamyl alcohol (415).

Industrial hygiene precautionary measures include eye, skin, and respiratory protection when handling the liquid or vapor.

## 1.14 α-Methylstyrene

### 1.14.1 General

α-Methylstyrene or 2-phenylpropylene is a colorless, combustible liquid with a sweet, aromatic odor (55). Selected physical data are given in Table 21.1. Methylstyrene is synthesized by the catalytic alkylation of benzene with propylene in hydrofluoric and sulfuric acids (29) and by the dehydrogenation of cumene (5). It is used as a chemical raw material and as an intermediate in plastic and resin manufacture.

### 1.14.2 Toxicity

α-Methylstyrene is irritating to the eyes, skin, and upper respiratory tract. Prolonged skin contact may cause dermatitis, and repeated inhalation may result in CNS depression.

At 100 ppm, the odor is tolerable (84). Selected toxicity data are given in Table 21.15. Methylstyrene may cause hepatic dysfunction, enzyme and immunologic changes, and vitamin $B_{12}$ deficiency in workers (416–418).

The oral $LD_{50}$ in rats is 4.9 g/kg. In rats, exposure to a mixture of butadiene and α-methylstyrene at 99.8 and 5.2 mg/m$^3$, respectively, resulted in a decrease in the number of leukocytes, phagocytic rate, respiration rate, and vitamin B and C levels of the blood and other organs (419). This was seen for α-methylstyrene alone in the rat and the rabbit (420). Single and repeated exposures increased the activity of cholinesterases in several rat organs (421–423).

α-Methylstyrene is a moderate to marked dermal irritant with slightly necrotic effects (84). It induced changes in connective tissue and neuron fibers in rat skin, although the effects were reversible in 30 to 60 days (424). α-Methylstyrene caused inflammation, hyperemia, edema, and hyperkeratosis after 20 applications to the rabbit skin (425).

In subchronic exposures, no effects were observed up to 600 ppm in several animal species.

### 1.14.3 Industrial Hygiene

The OSHA PEL (148) is 50 ppm (240 mg/m$^3$) with a STEL of 100 ppm (485 mg/m$^3$) and the ACGIH TLV (194) is 50 ppm (242 mg/m$^3$) with a STEL of 100 ppm (483 mg/m$^3$). NIOSH recommended absorption on charcoal, desorption with carbon disulfide, and quantification by flame ionization gas chromatography (195). Methods for the quantification of methylstyrene in waste water and biomaterials are also available.

α-Methylstyrene appears to be somewhat less toxic than styrene or vinyltoluene. However, industrial hygiene and biologic monitoring recommendations have issued (347).

## 1.15 Other Alkenylbenzenes

Elongation of the side-chain increases toxicity (27). Selected physical data are given in Table 21.1 and toxicity data in Table 21.15. Phenylbutene is used as a raw material and intermediate in the chemical and pharmaceutical industries.

## 1.16 Alkynylbenzenes

### 1.16.1 Phenylacetylene

Phenylacetylene is a highly odorous, flammable liquid. Selected physical data are given in Table 21.1. It is prepared from ω-bromostyrene and potassium hydroxide (356) and used as a chemical raw material. The oral $LD_{50}$ in rats is 5.00 g/kg (5).

## 2 DI-, TRI-, AND TETRAPHENYLIC COMPOUNDS

### 2.1 Diphenyls

### 2.2.1 Diphenyl

*2.2.1.1 General.* Diphenyl or phenylbenzene is a colorless solid crystallized in leaflets, with a pleasant, peculiar odor (3). Selected physical data are given in Table 21.1 and toxicity data in Table 21.16.

Diphenyl is prepared by heating phenyldiazonium chloride with copper (356), by dehydrogenation of benzene (29), by passing benzene through an iron tube at 650 to 800°C (29), or by passing benzene over ferrous–ferric oxide and heating to 1000°C (49).

Diphenyl is thermally stable and is used as heat transfer fluid (29, 426). It is also used as a raw material in the chemical industry. In the citrus fruit and vegetable industries, it has been applied to fruit and vegetables or used in food packaging materials as a preservative (332), and fungistat (427, 428).

The residue tolerance for diphenyl on citrus fruit skin is 110 ppm (29).

*2.2.1.2 Toxicity.* In acute exposures, diphenyl causes eye and skin irritation, and may affect the central and peripheral nervous systems. Chronic human exposure is characterized by fatigue, headache, tremor, insomnia, sensory impairment, and mood changes, accompanied by clinical findings of cardiac or hepatic impairment and irregularities of the peripheral and central nervous systems. Selected toxicity data are given in Table 21.16. One death has been recorded (429). In a follow-up study, some neurophysiological abnormalities were still observed 1 and 2 years after the initial investigation (430). Prolonged skin contact may produce sensitization or dermatitis (66).

In the rat and the rabbit, slight irritation of the gastrointestinal tract and hepatic and renal effects were noted (431). A mild paralysis of the hind legs was observed in some animals (29).

In subchronic studies, hepatic, cardiac, and renal cell degeneration, with occasional spleen and other tissue involvement, were noted.

*2.2.1.3 Metabolism.* Diphenyl is absorbed through the skin, the mucous membranes, and the pulmonary system. It is metabolized in the liver to water-soluble hydroxy derivatives. Diphenyl is excreted unchanged by the biliary system of the rat following phenobarbital intraperitoneal injections of 70 mg/kg for 4 days (432).

The 2-hydroxydiphenyl was identified in the urine of dogs after oral administration of diphenyl (433), in the rabbit (434), and in vitro in the rat (435, 436), hamster (437, 436), mouse, cat, coypu, and frog (436), as well as in various marine species (438). The 3-hydroxylation product has been identified in the urine of the rabbit after oral administration (433). The 4-hydroxylase was more prevalent than 2-hydroxylase in 11 laboratory animals tested (436), in rat microsomal preparations (435), hamster microsomes (437), the mouse (439), rabbit urine (433), and various marine animals (438).

In rabbit urine, the 3,4- and 4,4'-dihydroxydiphenyl derivatives were identified (433, 434, 440). In the rat, 3,4- and 4,4'-dihydroxyphenyl, along with diphenylmercapturic acid and diphenylglucuronide, were excreted (411). The related derivatives, 3-methoxy-4-hydroxy- and 3-hydroxy-4-methoxydiphenyl, also have been identified as urinary metabolites (432). The 2-hydroxylase is prevalent in the young rat but diminishes with aging, whereas the 4-hydroxylase increases with age (441).

*2.2.1.4 Microbial Studies.* Several organisms have been found that can utilize diphenyl as the sole carbon source. These include *Pseudomonas desmolyticum*, *P. putida*, and *Acinetobacter* species (442) and 258 strains originating from Japanese natural resources (443). Several bacterial degradation products have been identified, including 2,3-dihydroxydiphenyl (444), 2,3-dihydro-2,3-dihydroxydiphenyl, 3-hydroxydiphenyl and benzoic acid (445), 2-hydroxy-6-oxo-6-phenylhexa-2,4-dienoic acid (446), and γ-benzoylbutyric acid (443).

*2.2.1.5 Industrial Hygiene.* The OSHA STEL (148) is 0.2 ppm (1 mg/m$^3$) and the ACGIH TLV (194) is 0.2 ppm (1.3 mg/m$^3$). Prolonged exposure to 0.75 ppm (0.005 mg/l) is considered a human health hazard (431). The odor threshold is 0.0062 ppm and is irritating at 7.5 ppm (55).

Sampling techniques have been published (429). Spectrophotometric (29), thin-layer (434), and gas and liquid chromatographic (440, 447) procedures have been developed for the quantification of the above diphenyl and its metabolites.

Medical and industrial monitoring programs should include protection of the eyes, skin, and respiratory tract.

## 2.1.2 Alkyldiphenyls

Methyldiphenyl, *o*-benzyltoluene, occurs naturally as a component of the essential oil of the flower from *Astragalus sinicus* (448).

Alkyldiphenyls, such as 4-*sec*-butyldiphenyl, the 4,4'-di-*sec* derivative, and related derivatives are used as nonspreading lubricants (449).

## 2.2.3 Diphenylalkanes

Diphenylmethane or 1,1'-methylenebisbenzene occurs in orthorhombic needles and possesses the odor of oranges (3). Selected physical data are given in Table 21.1.

**Table 21.16.** Toxicity of Diphenyl

| Material | Route of Entry | Species | Dose or Concentration | Results or Effects | Ref. |
|---|---|---|---|---|---|
| Diphenyl | Inhalation (acute) | Human | 3–4 ppm (19–25 mg/m$^3$) | Irritation to eyes and mucous membranes | 29 |
| Diphenyl | Inhalation (chronic) | 47 workers | <1.6 ppm (<1 mg/m$^3$) | No symptoms and no deviations of cardiac or hepatic functions | 425 |
| Workers | Unknown | 33 workers | 4.4–128 ppm (28–800 mg/l) | Transient nausea, vomiting, bronchitis Abdominal pain, headache, cardiac, hepatic, renal effects, peripheral and CNS abnormalities, 1 death | 21 428, 429 |
| Diphenyl | Oral (acute) | Rat | 3.28 g/kg | LD$_{50}$ | 430 |
|  |  | Rabbit | 2.41 g/kg | LD$_{50}$ | 430 |
|  | Inhalation | Rat | 47.5 ppm (0.3 mg/l) | No effects, no deaths in liver | 430 |
| Diphenyl (dust) | Inhalation | Rat | 6.5–47.5 ppm (0.04–0.3 mg/l) × 7 hr/day × 5 days/week for 64 days | Irritation of nasal mucosa, respiration difficulties, some deaths, some bronchopulmonary lesions; hepatic and renal tissue effects | 430 |
|  |  | Mouse | 0.8 ppm (0.0005 mg/l) × 7 hr × 62 exp. | Same effect as rat | 430 |

| | | | | |
|---|---|---|---|---|
| | Rabbit | 6.5–47.5 ppm (0.04–0.3 mg/l) × 7 hr/day × 5 days/week × 64 days | No effects | 430 |
| | | 50–100 mg × 13 months | Above changes enhanced, with affected thyroid and parathyroid functions, some papilloma and squamous cell carcinoma of the forestomach | 29 |
| | Rabbit | 1% in the diet | Growth inhibition in weanlings | 29 |
| Diphenyl | Mouse | 1 g × 2–3 times/week | Cumulative; total fatal dose 10–50 g | 29 |
| | | Dilution (with croton oil and acetone) | No significant effects | 459 |
| Dermal (subacute) | Mouse | 46 mg/kg | $TD_{LO}$, some neoplastic signs | 21 |
| Subcutaneous | Rabbit | 0.5 g/kg × 2 hr/day × 5 days | Some deaths, growth depression, slight cardiac, hepatic, and renal tissue changes, follicular atrophy, necrosis and leukocytic infiltration of the spleen | 524 |
| Oral (chronic) | Rat | 50–100 mg × 2 months 50–100 mg × 13 months | Moderate degenerate changes in liver | 21 |

**Table 21.17.** Toxicity of Terphenyls

| Material | Route of Entry | Species | Dose or Concentration | Results or Effects | Ref. |
|---|---|---|---|---|---|
| Terphenyl | Inhalation (acute) | Human | Spills with short-term exposure | Headaches and sore throat, reversible in 24 hr | 425 |
| | Inhalation (chronic) | Human | 0.01–0.94 ppm (0.0094–0.89 g/m$^3$) | No effect on blood pressure, pulmonary function even improved in exposed group; isocitric dehydrogenase borderline but not statistically elevated | 425 |
| | Skin (chronic) | Human | 0.01–0.94 ppm (0.094–0.89 mg/m$^3$) | Six of 200 workers developed nonspecific, readily reversible skin rashes; not a skin sensitizer | 425 |
| Terphenyl (non-irradiated) | Oral (acute) | Rat | 17.5 g/kg | LD$_{50}$ | 54 |
| | | Mouse | 12.5 g/kg | LD$_{50}$ | 54 |
| | | Rat | 6.0 g/kg | LD$_{50}$ | 54 |
| | | Mouse | 6.0 g/kg | LD$_{50}$ | 54 |
| o-Terphenyl | | Rat | 1.90 g/kg | LD$_{50}$ | 54 |
| m-Terphenyl | | Rat | 2.40 g/kg | LD$_{50}$ | 54 |
| p-Terphenyl | | Rat | >10.0 g/kg | LD$_{50}$ | 54 |
| Hydroterphenyl | | Rat | 6.6 g/kg | LD$_{50}$ | 461 |
| | | Mouse | 4.2 g/kg | LD$_{50}$ | 461 |

| Compound | Route | Species | Dose | Effect | Reference |
|---|---|---|---|---|---|
| Terphenyl | Inhalation (acute) | Rat | 100 ppm (0.94 mg/l) (0.94 g/m$^3$) | No mortality, but pulmonary pathology after a 1-hr exposure | 54 |
| | | | 320 ppm (3.0 mg/l) | 4/8 deaths, with early asphyxial death due to crystalline plugs in the trachea | 54 |
| o-Terphenyl | Oral (chronic) | Rat | 0.25–0.5 g/kg/day × 30 days | Increased liver and kidney to body weight ratios | 425 |
| m-Terphenyl | | | 0.25–0.5 g/kg/day × 30 days | Increased liver to body weight ratios | 425 |
| p-Terphenyl | | | 2.5–5.0 g/kg/day × 1 month | Insignificant weight decreases, intensification of antitoxic functions of the liver | 54 |
| mixed Terphenyl Nonirradiated | | Mouse | 0.25 g/kg/day × 8 weeks | Changes in cytoplasm of hepatocytes, no lesions | 461 |
| | | | 0.6 g/kg/day × 16 weeks | Severe chemical nephrosis | 54 |
| | | | 1.2 g/kg/day × 16 weeks | Intensified nephritis, especially affecting the proximal tubules | 54 |
| | | | 1.2 g/kg/day × 16 weeks | Lethal | 54 |
| Irradiated | Inhalation (chronic) | Laboratory animals | 0.3 ppm (3 mg/m$^3$) × 1 month | No effect | 54 |
| p-Terphenyl | | | 3.7 ppm (35 mg/m$^3$) × 1 month | Functional and morphological changes | 54 |
| | | | 212 ppm (2000 mg/m$^3$) × 4 hr/day × 5 days/week × 8 weeks | Cell debris in lungs, but rapidly cleared | 462 |

It is prepared from methylene chloride and benzene by the Friedel–Crafts reaction, with aluminum chloride as the catalyst (3). It is used as a fragrance in the cosmetics industry (450). Diphenylmethane may be one of the degradation products of DT (451) and can serve as the sole carbon source to some *Hydrogenomas* (452) and *P. putida* strains. However, it is one of the most persistent compounds in nature by forming a tetraphenyl ether (451).

### 2.1.4 Diphenylalkenes

Diphenylethylene or stilbene occurs in the trans or cis configuration; the trans form is more stable and more prevalent. Stilbene is a solid at room temperature. Selected physical data are given in Table 21.1. It is used as a nutritional aid in agriculture and as a chemical intermediate in the dye industry (356).

Stilbene is hydroxylated, primarily in the para position and secondarily in the meta position, and readily metabolized in several species (453–456).

Stilbene is covalently bound to rat liver microsomal protein. This binding is inhibited by some drug systems but accelerated by 3-methylcholanthrene pretreatment (457). In rabbits, the liver microsomes are capable of cleaving the ethylenic linkage to produce benzoic acid (458).

When handling diphenylethylene, gloves and other protective clothing should be worn.

## 2.2 Triphenyls

### 2.2.1 General

The open ring system of triphenyl is known also as terphenyl. It occurs naturally in petroleum oil. There are three chemical isomers, the ortho, meta, and para derivatives, of which the ortho and the para forms appear industrially most prevalent (459). All forms are solid at room temperature. Selected physical data are given in Table 21.1. The terphenyls are industrially important as chemical intermediates in the manufacture of nonspreading lubricants, as nuclear reactor coolants, and as heat storage and transfer agents (426). *p*-Terphenyl has been used as a sunscreen lotion component (460). The terphenyls have low vapor pressures and are not significant industrial hazards.

### 2.2.2 Toxicity

*2.2.2.1 Human Response.* The toxicity of the terphenyls to humans is relatively low. Selected toxicity data are given in Table 21.17. No adverse effects have been detected in a work force except for some reversible skin rashes.

*2.2.2.2 Animal Studies.* The rat and the mouse show variable responses depending on the isomer used and the irradiation state. Although the nonirradiated mixture is nontoxic, the irradiated *o*-terphenyl is highly toxic.

Acute exposure showed some effects in the guinea pig, depending on the particle size of the aerosol (461). Terphenyls are not skin sensitizers (426).

In skin painting studies, one isolated papilloma was obtained. There is a possibility that terphenyl may have cocarcinogenic potential, as shown with tars (459). In mice exposed to terphenyl, cell debris was noted in the lungs as nonspecific pulmonary membrane and cellular damage. This, however, cleared rapidly (462).

***2.2.2.3 Absorption, Distribution, and Metabolism.*** An intragastric dose of $^{14}$C-labeled *o*-terphenyl was rapidly absorbed and distributed. It was almost completely excreted within 48 hr, mainly in bile in the rat, mouse, and rabbit (463). Accumulation in the liver peaked at 4.5 hr in the mouse and was completely cleared in 1 week (464). Young rats when fed *o*- and *m*-terphenyl exhibited depressed body weight gain, and *m*-terphenyl induced liver hypertrophy (465). Following exposure, hydrogenated terphenyls were rapidly exhaled (464). Cholinesterase inhibition has been noted in exposures to 20 mg/m$^3$ of hydroterphenyl (466). *Pseudomonas desmoluticum* can grow on and degrade *m*-terphenyl (442).

### 2.2.3 Industrial Hygiene and Medical Surveillance

The OSHA (148) ceiling is 0.5 ppm (5 mg/m$^3$) and the ACGIH (194) ceiling is 0.5 ppm (4.7 mg/m$^3$). NIOSH recommends collection on a cellulose membrane filter, extraction with carbon disulfide, and quantification by flame ionization chromatography (195).

Industrial hygiene monitoring and handling procedures are described, and specified preemployment and periodic medical examinations are required (467).

## 3 DINUCLEAR SYSTEMS

### 3.1 Naphthalene

#### *3.1.1 General*

Naphthalene or moth flake is a white solid that exhibits a typical mothball or tar odor. Chemically, it is composed of two fused benzene rings.

Naphthalene occurs naturally in the essential oils of the roots of *Radix* and *Herba ononidis* (468). It is formed in cigarette smoke by pyrolysis (469, 470) and is a photodecompostition product of carbaryl (471). It is the most abundant component (11%) of coal tar (3). Naphthalene also occurs in crude oil and is isolated from cracked petroleum (5), from coke-oven emissions (115), or from high-temperature carbonization of bituminous coal (472). Naphthalene, in both solid and liquid forms, is flammable. The vapors or dusts, with air, can produce explosive mixtures. In contact with water, molten naphthalene (above 110°C), may undergo violent foaming (473). Selected physical data given in Table 21.1 and toxicity data in Table 21.18.

Naphthalene is used extensively as a raw material and intermediate in the chemical, plastics, and dye industries (115), as a moth repellent in the form of balls or disks, as an air freshener (3, 201), and as a surface-active agent. It is utilized in

**Table 21.18.** Toxicity of Naphthalene

| Material | Route of Entry | Species | Dose or Concentration | Results or Effects | Ref. |
|---|---|---|---|---|---|
| Naphthalene | Oral (acute) | Human (child) | 2–3 g | Lethal dose | 476, 54 |
| | | (adult) | 5–15 g | Lethal dose | 29 |
| | Oral | Rat | 2.6 g/kg | All metabolized in 2 days | 54 |
| | | Albino | 1 g/kg/day × 2 days | Slight eye effects | 501 |
| | | Pigmented | 1 g/kg/day × 2 days | More severe eye effects | 501 |
| | Subcutaneous | Mouse | 5.1 g/kg | $LD_{50}$ | 54 |
| | Intraperitoneal | Rat | 100 mg/kg | 20–30% excreted in urinary metabolism | 519 |
| | | | 1 g/kg | Lethal to 67.8% in 2–3 hr | 54 |
| | | Mouse | 150 mg/kg | $LD_{LO}$, lowest lethal dose published | 54 |
| | Aquatic | Fish | 1–10 ppm | TLm96 | 292 |
| | | Crab | 8–12 ppb | Lethal to all animals | 516 |
| | Oral (repeated) | Rabbit | 1 g/kg/day | Browning of lens and eye humor, degeneration of retina and cataract formation | 502 |
| | Subcutaneous | Rat | 820 mg/day | Not tumorigenic after >1000 days observation | 54 |
| Naphthalene | | | | | |
| Methyl- | Oral | Rat | 4.360 g/kg | $LD_{50}$ | 54 |
| 1-Methyl- | Oral | Rat | 5.00 ml/kg | Lethal to all animals | 29 |
| 2-Methyl- | Oral | Rat | 5.00 ml/kg | Lethal to all animals | 29 |
| 1,6-Dimethyl- | Oral | Rat | 5.00 ml/kg | Lethal to 7 of 10 animals | 29 |

the manufacture of insecticides, fungicides, lacquers, and varnishes (5) and to preserve wood and other materials (56, 115). In medicine, it has been applied as an antiseptic, anthelmintic, and dusting powder in skin diseases. In a microbial production unit, it is used to make salicyclic acid (21).

### 3.1.2 Human Toxicity

***3.1.2.1 Acute Oral Effects.*** Ingestion of naphthalene in the form of mothballs has resulted in no ill effects in some of the cases described. The ingested material was eliminated unchanged in the feces (52). Greater danger exists when it is ingested in combination with fats, which facilitate the absorption of naphthalene (52). In severe cases, ingestion caused gastroenteric distress, tremors, and convulsions (269). The most characteristic sign is acute intravascular hemolysis (56). Within 2 to 7 days moderate to severe anemia may develop (52). Heinz bodies appear, and the serum takes on a yellowish brown color. Peripheral blood may exhibit hypochromia and polynucleosis. The bone marrow may appear hyperplastic and show an increased proportion of nucleated erythrocytes (29). In some cases, hemoglobinuria, possible occlusion of the renal tubules, and altered renal functions may occur (52). Death may ensue due to respiratory failure (268).

A 1-year-old child who had accidentally ingested naphthalene showed increasing lethargy and anorexia 2 weeks later, followed by hemolytic anemia (474). One case showed early symptoms of toxic kidney attack, yellow skin, dark coloration of the urine, and sharp onset of pain (475). In a 6-year-old child, 2 g administered over a 2-day period caused death (268, 476). In most mortality cases, the lethal effects are acute hemolytic anemia, presence of Heinz bodies, and fragmented erythrocytes (477, 478). A 36-year-old pharmacist was given 5 g of unpurified naphthalene in an emulsion of castor oil. On awakening, he had intense bladder pain and was nearly blind (479).

Individuals with congenital erythrocyte glucose-6-phosphate dehydrogenase deficiency are particularly susceptible to hemolytic agents because they rapidly cause hemolytic anemia (268, 474, 480–483).

***3.1.2.2 Acute Eye and Dermal Effects.*** Naphthalene is irritating to the eye (66), and retinotoxicity has been recorded (484, 485).

Upon direct skin contact, naphthalene is a primary irritant (66). Diapers or clothing stored with mothballs and used directly on infants have caused skin rashes and systemic poisoning (51, 269). Dermal absorption is facilitated by the use of baby oil (51, 486, 487).

***3.1.2.3 Acute Inhalation Effects.*** Naphthalene may volatilize and sublime at room temperature and it can also be inhaled as dust particles. Signs and symptoms due to exposure to naphthalene vapors resemble those observed from ingestion or dermal absorption.

Naphthalene vapors may cause eye and respiratory tract irritation, headache, nausea, and profuse perspiration, depending on the duration and exposure concentration. Also, optic neuritis has been observed (29). Acute effects can be the result of dermal and inhalation exposure from clothing stored in moth balls (488).

*3.1.2.4 Chronic Oral Effects.* Oral intoxication in industry is unlikely; however, in the general population, 50 cases of severe chronic effects from repeated ingestion of a naphthalene–isopropyl alcohol "cocktail" have been recorded (489). The symptoms resembled those of ethanol intoxication and consisted of tremors, restlessness, extreme apprehension, and hallucinations. The effects subsided in a few days (489).

*3.1.2.5 Chronic Eye Effects.* Repeated exposure to naphthalene vapor or dust can cause corneal ulceration, cataracts (490), lenticular opacities (491), and general opacities (492).

*3.1.2.6 Chronic Dermal Effects.* On repeated contact, naphthalene may cause erythema and dermatitis, especially in hypersensitive individuals (66). Occasional allergic responses are rare.

*3.1.2.7 Chronic Inhalation Effects.* Repeated exposure to vapors may produce malaise, headache, and vomiting. Toxic effects have been observed in industry where electric vaporizing devices were used (478, 493–495).

*3.1.2.8 Other Human Effects.* Naphthalene can cross the placenta and is fetotoxic (496). In 21 newborns, a rather severe form of hemolytic anemia was produced from blankets stored in naphthalene (21).

*3.1.2.9 Human Metabolism.* When inhaled, naphthalene is rapidly absorbed but is more slowly absorbed through intact skin or when ingested. Conversion to the hemolytic agents, $\alpha$- and $\beta$-naphthol and $\alpha$- and $\beta$-naphthoquinone, is rapid in the adult and very slow in the newborn (29, 497–499). The naphthols are partially excreted as the glucuronides (482). Naphthalene per se is nonhemolytic (498). The most active metabolite is probably naphthalene 1,2-oxide (499, 500).

### 3.1.3 Toxicity in Animals

*3.1.3.1 Acute Oral Effects.* In the rat, mouse, and guinea pig, the oral $LD_{50}$ is 490 mg/kg, 533 mg/kg, and 1200 mg/kg, respectively (54). Selected toxicity data are given in Table 21.18. A dose of 1.0 g/kg was mildly toxic to the eye (501). Rabbits fed naphthalene exhibited browning of the lenses and eye humors (502), inhibition of the ciliary body and ascorbic acid transport, and the development of cataracts (503). Cataract formation was accompanied by glycolytic decreases and changes in the electrophoretic pattern of lens proteins (504) and the disappearance of oxidized glutathione (505). Dogs administered 3 g of naphthalene developed distemper-like attacks and moderate anemia (467).

***3.1.3.2 Acute Dermal Effects.*** The dermal $LD_{50}$ in rats is greater than 20 g/kg (54). A dose of 2.5 g/kg was not lethal in rats (499).

***3.1.3.3 Repeated Exposures.*** In rabbits, daily oral administration of 1 g/kg of naphthalene produced lenticular opacity and peripheral swelling of the lens, slightly visible after three doses, and causing marked changes after 20 doses (115, 502, 506). Naphthalene causes changes in amino acid, ascorbic acid, protein, and carbohydrate metabolism of the eye, producing calcium oxalate crystals (502, 507).

At 1.5 g/kg/day by stomach tube, white spots appeared in the rabbit eye periphery, but were distributed over the entire retina in young animals (508).

Rats injected intraperitoneally with 40 mg/kg naphthalene for 3 days produced arylhydroxylase inhibition (509).

In addition to its retinotoxic action, naphthalene vapor may also lead to the formation of cataracts (485).

***3.1.3.4 Toxicity to Insects.*** Naphthalene is an effective insecticide against fruit-piercing moths (510). Some resistant and susceptible strains of the housefly, *Musca domestica*, showed that single doses of naphthalene were excreted more rapidly by the male than the female (511). The resistance depended on microsomal activity (512).

***3.1.3.5 Toxicity to Marine Species.*** Naphthalene was absorbed by the common marine mussel, *Mytilus edulis*, and released again in unchanged form (237). In *G. mirabilis*, *O. maculosus*, and *C. stigmacus*, naphthalene was metabolized to 1,2-dihydro-1,2-dihydroxynaphthalene and excreted in the urine (513). The 96-hr TLm is between 1 and 10 ppm. In rainbow trout, an 8-hr exposure to 0.005 mg/l resulted in tissue concentrations 20 to 100 times that of the water levels (514). The highest retention was in liver tissue (515). The half-life in fat was less than 24 hr (514). Naphthalene and its combination with serum albumin at 8 to 12 ppb in flowing seawater produced 100 percent mortality (516). Naphthalene accumulated in marine animals, but when they were transferred to oil-free seawater, the naphthalene was excreted in 2 to 60 days (517). Naphthalene was extracted from No. 2 fuel oil by the polychaete *Neanthes arenaceodentata*, accumulated, and subsequently released within 400 hr by the male, but retained for 3 weeks by the female (518). Larvae of the exposed females contained up to 18 ppm naphthalene, which decreased to undetectable levels during development (518).

***3.1.3.6 Mutagenicity and Carcinogenicity.*** Naphthalene is not a mutagen and is carcinogenically inactive (270, 499, 519). No carcinomas were seen in hairless mice (520). The naphthalene has low DNA- (521, 522) and RNA-binding capacities (523). Its low carcinogenic potential may be due to specific fluorescence spectral characteristics (524) and the potent naphthalene 1,2-epoxide hydrase activity in mammals (500). Naphthalene inhibits the activity of Ehrlich carcinoma cells (525) and the tumorigenic potential of tobacco smoke condensate (526). It also inhibited the induction of skin tumors in mice receiving naphthalene and benzo(*a*)pyrene (499).

### 3.1.4 Absorption, Metabolism, and Excretion

Absorption of naphthalene readily occurs during inhalation exposures (115). In the rat, naphthalene is readily converted to 1- or 2-naphthol and 1,2-hydroxyglucuronide or the 1-sulfate (527). It is possible that naphthalene 1,2-epoxide is a short-lived intermediate (528). The 1,2-oxide then may spontaneously be isomerized to 1-naphthol (270, 529–531), also leading to naphthalene oxides and naphthoquinones (270, 532, 533), glutathione and mercapturic acid conjugates (270), and naphthalene dihydrodiol glucuronide as isolated from liver microsomes (534). At 100 mg/kg intraperitoneally, 20 to 30 percent was excreted in rat urine with 85 to 90 percent in the form of acid conjugates; 5 to 10 percent was excreted in the bile with 70 to 80 percent as acid conjugates (519). The major metabolite was naphthalene-1,2-dihydrodiol (519). In addition, the triol and tetrol metabolites may occur (519). A methylthio derivative has been identified also in rat urine (535). The 1,2-diol may be catalytically oxidized to 1,2-naphthoquinone, a reaction reversed by ascorbic acid (502). An interaction of the quinone with protein may be responsible for the browning of the eye preceding cataract formation (502).

Naphthalene 1,2-epoxidation progresses through a liver microsomal arene oxidase system (519, 536, 537) by forming a cytochrome P450 complex (164, 538) and utilizing an ascorbic acid–iron–oxygen coenzyme system (531). The absorption of naphthalene, microsomal oxidation, and similar types of metabolism have been observed also in the housefly (539).

### 3.1.5 Microbial Studies

Several *Pseudomonas* strains are capable of growing on naphthalene (540, 541), decomposing (542) and utilizing it in soil or liquid media (543, 544). *P. desmolyticum* was tested for the degradation of industrial naphthalene in waste water (442) and petroleum effluents (545). A *Pseudomonas* strain degraded naphthalene to 1,2-dihydro-1,2-dihydroxynaphthalene (546), and *P. aeruginosa* produced salicylate (547). Conversely, naphthalene and carbaryl inhibited microbial vitamin $B_{12}$ formation in soil (548). Another derivative was selectively destructive to tomato tumors (549). Naphthalene reduced the photosynthetic processes in *Nitzschia palea* (550) and inhibited growth and the photosynthetic capacity of the green algae, *Chlamydomonas angulosa* (551).

### 3.1.6 Industrial Hygiene and Biological Monitoring

The OSHA PEL (148) is 10 ppm (50 mg/m$^3$) with a STEL of 15 ppm (75 mg/m$^3$), and the ACGIH TLV (194) is 10 ppm (52 mg/m$^3$) with a STEL of 15 ppm (79 mg/m$^3$). Saturated air at 25°C contains about 100 ppm naphthalene. The odor threshold is 1.5 (55), is recognizable at about 25 ppm (115), and is irritating at 75 ppm (55).

NIOSH recommends collection on charcoal, desorption with carbon disulfide,

# AROMATIC HYDROCARBONS

and quantification by flame ionization gas chromatography. Other methods, including ultraviolet spectrophotometric procedures, have been described (21, 499, 510). Biologically, urine can be monitored for naphthalene, 1-hydroxynaphthalene, or related metabolites (29).

Workers handling naphthalene should be provided with impervious clothing, boots, gloves, and face shields to prevent contact and a cartridge-type respirator for vapor concentrations about 10 ppm (19, 21, 49). Naphthalene is on the EPA TSCA Chemical Inventory and the Test Submission Data Base (54).

## 3.2 Alkylnaphthalenes

Alkylnaphthalenes are formed as pyrolysis products in cigarette smoke (552). Some have been identified in commercial carbon paper (553). They also are the major components of the $C_{10}$ to $C_{13}$ alkylnaphthalene concentrate fraction, which distills at 400 to 550°F (48).

A $C_{11}$ to $C_{12}$ petroleum mixture of re-formates containing about 23 percent alkylnaphthalenes caused skin and eye effects (554). The alkylnaphthalenes appear more toxic to marine species than the alkylbenzenes (555). The toxicity and the bioaccumulation increase with molecular weight (517).

*Nocardia* cultures, isolated from soil, preferentially oxidized alkylnaphthalenes when methylated in the 2 position (249).

### 3.2.1 Methylnaphthalene

Methylnaphthalene can occur as the 1 or 2, the alpha or the beta isomer, $C_{11}H_{10}$. Selected physical data are given in Table 21.1 and toxicity data in Table 21.18. 1-Naphthalene, a flammable solid, also has been identified in the waste water of coking operations (556), in textile processing plants (557), and as a photodecomposition product of 1-naphthylacetic acid (558). Methylnaphthalene is used as a component in slow-release insecticides (559) and in mole repellents (560). Workplace exposures to 18 to 32 $\mu g/m^3$ for 2-methylnaphthalene have been reported (499).

Methylnaphthalene is not a human skin irritant or photosensitizer (29). Whole-body irradiation decreased the sensitivity to ingested methylnaphthalene in the rat (561). In rats, intraperitoneal injection of the 1-methyl isomer is one-fourth as lethal as naphthalene, and the 2-methyl derivative showed no effect (562). 2-Methylnaphthalene is an eye and skin irritant in rabbits (54).

Methylnaphthalene accumulates in marine species (518); however, biliary excretion of 2-methylnaphthalene in the trout was potentiated by 2,3-benzanthracene (563). Of the methyl derivatives, *Nocardia* cultures oxidized only the 2-methyl isomer (249).

Exposure limits have not been set for methylnaphthalene. It is on the EPA TSCA Chemical Inventory and the Test Submission Data Base (54).

### 3.2.2 Di and Polymethylnaphthalene

Dimethylnaphthalene can occur in various isomeric forms. The 1,2-dimethyl derivative has been used as a selective organic solvent (564). It also inhibits crown gall and olive knot neoplasms (549). It accumulates in shrimp (229), clams, and other marine species (565). Dimethyl derivatives with position 2 occupancy were oxidized to the acid by *Nocardia* species (249). The 2,3-dimethyl derivative acted as a weak accelerator of skin tumor induction (526). The oral toxicity in the rat appeared lower for the 1,6-dimethyl derivative than for the monomethyls (29).

Trimethylnaphthalene was active as a termiticide (566).

## 4 TRI AND POLYNUCLEAR RING SYSTEMS

### 4.1 General Aspects

Fused-ring systems occur in linear, staggered, or three-dimensional configurations. The linear three-ring system is anthracene and the staggered system is phenanthrene. Selected physical data are given in Table 21.19 and toxicity data in Table 21.20.

Polycyclic aromatic hydrocarbons or polyarylhydrocarbons (PAHs) are not synonymous with carcinogens. However, some of these compounds or their derivatives have carcinogenic or cocarcinogenic potential. This potential depends on the species, the tissue, and the immunologic and nutritional states involved.

#### 4.1.1 Occurrence

Two polycyclic aromatic hydrocarbons, benzo($a$)pyrene (BaP) and benz($a$)anthracene, were quantified at 0.2 ppb in meat and 98 ppb in coconut oil (567). Even higher concentrations occur in cooked food and secondary smoke (567, 568).

Polycyclic aromatic hydrocarbons are formed in cigarette smoke (568). Smoke from a single test cigarette contained 9.7 to 11.1 ng/m$^3$ (569). Environmental air concentrations of BaP varied with the season at 5 ng/m$^3$ in September and 68 ng/m$^3$ in March (569). Benzopyrene, dibenzopyrene, dibenzanthracene, and other PAHs have been identified in the exhaust gases of diesel engines (570). In peat deposit areas, anthracene, dibenzanthracene, benzopyrene, and methylcholanthrene have been observed (571). However, by far the greatest PAH quantities are emitted through energy production resulting in up to 6 g benzo($a$)pyrene per person per year in the United States (572).

#### 4.1.2 Toxicity

The polycyclic aromatic hydrocarbons are mainly solid materials, soluble in fats, oils, and organic solvents. Mutagenic or carcinogenic properties of PAHs have been linked to physicochemical properties, such as electronegativity (573) or $K$- and $L$-region reactivity indexes (574), electrophilic potency, dipole moment, in-

**Table 21.19. Physical Data for Polycyclic Aromatics (591)**

| Compound | B.P. (°C) | CAS Registry | Density | Empirical Formula | M.P. (°C) | Mol. Wt. | Refractive Index | Solubility[a] w/al/et | Sp. Gr. (25°C) | Vapor Density (Air = 1) | Vapor Pressure [mm Hg (°C)] | Wt./Vol. Conversion (mg/m³ = 1 ppm) |
|---|---|---|---|---|---|---|---|---|---|---|---|---|
| Acenaphthene | 279 | 83-32-9 | 1.189 | $C_{12}H_{10}$ | 96.2 | 154.21 | 1.6048 | i/s/s (95°C) | 1.02 | 5.32 | 10 (131.2) | 6.31 |
| Anthracene | 340 | 120-12-7 | 1.283 | $C_{14}H_{10}$ | 216.3 | 178.23 | | i/d/d | 1.25 | 6.15 | 1 (145) | 7.23 |
| Phenanthrene | 340 | 85-01-80 | 1.179 | $C_{15}H_{10}$ | 101 | 178.23 | | i/s/s | 1.18 | 6.14 | 1 (118.3) | 7.23 |
| 1,2-Benzanthracene | 435 | 56-55-3 | | $C_{18}H_{12}$ | 162 | 228.29 | | i/d/s | | | | 9.34 |
| 6-Methyl-1,2-benzanthracene | | 316-14-3 | | $C_{19}H_{14}$ | 151 | 242.32 | | i/s/s | | | | 9.91 |
| 7-Methyl-1,2-benzanthracene | | 2541-69-7 | | $C_{19}H_{14}$ | 183.3 | 242.32 | | i/s/— | | | | 9.91 |
| 10-Methyl-1,2-benzanthracene | | 2541-69-7 | | $C_{19}H_{14}$ | 141 | 242.32 | | i/s/s | | | | 9.91 |
| 9,10-Dimethyl-1,2-benzanthracene | | 57-97-6 | | $C_{20}H_{16}$ | 122.5 | 256.35 | | i/d/— | | | | 10.49 |
| 1,2-benzophenanthrene | 448 | 218-01-0 | 1.274 | $C_{18}H_{12}$ | 254 | 228.29 | | i/d/d | | | | 9.34 |
| 3,4-Benzophenanthrene | | 95-19-7 | | $C_{18}H_{12}$ | 68 | 228.29 | | i/d/— | | | | 9.34 |
| 3-Methylchrysene | | — | | $C_{19}H_{14}$ | 173 | 242.32 | | —/s/— | | | | 9.91 |
| Pyrene | 393 | 129-00-0 | | $C_{16}H_{10}$ | 156 | 202.26 | 1.59427 | i/s/s | | | | 8.27 |
| 3,4-Benzopyrene | 312 | 50-32-8 | 1.271 | $C_{20}H_{12}$ | 177 | 252.32 | | i/d/s | | | | 10.32 |
| Benzo(e)pyrene | | 192-97-2 | | $C_{20}H_{12}$ | 178.5 | 252.32 | | i/v/s | | | | 9.76 |
| Cyclopentapyrene | | | | $C_{19}H_{12}$ | 170 | 240.31 | | | | | | 9.76 |
| Methylcholanthrene | 280 | 56-49-5 | 1.28 | $C_{21}H_{16}$ | 180 | 268.36 | | i/s/— | | | | 10.98 |

[a] Solubility in water/alcohol/ether: v = very soluble; s = soluble; d = slightly soluble; i = insoluble.

Table 21.20. Toxicity of Polynuclear Aromatics

| Material | Route of Entry[a] | Species | Dose or Concentration[b] | Results or Effects | Ref. |
|---|---|---|---|---|---|
| Phenanthrene | Oral (acute) | Mouse | 700 mg/kg | $LD_{50}$ | 591 |
| Benzo(a)pyrene | Subcutaneous | Rat | 50 mg/kg | $LD_{50}$ | 577 |
|  | IP | Mouse | 500 mg/kg | Lethal dose | 591 |
| Anthracene | Oral (subchronic) | Rat | 4.5 g over >1000 days | No tumors | 591 |
|  | Dermal | Mouse | 10 μM/week × 25 days | Few papillomas in a small number of animals | 610 |
|  |  |  | 0.5% soln. 3/week × 25 days | Benign tumors not exceeding the control group | 591 |
|  | Subcutaneous | Mouse | 5 mg × 280 days | No tumors | 591 |
|  |  |  | 20 mg/week × 33 weeks | Sarcomas in 5 of 9 rats at injection site | 591 |
| Phenanthrene | Oral | Rat | 1 ml/s.d. × 310 days obs. | No mammary tumors | 591 |
|  | Dermal | Mouse | 0.5% × 6 week/total 20 appln. | Slight increase over control in benign tumors | 591 |
| Trimethylphenanthrene | Subcutaneous | Mouse | 5 mg/appln. × 372 days | No tumorigenic effects | 591 |
|  | Subcutaneous | Mouse | 0.5 mg in 0.25 cm³ lard/s.d., obs. 17 months | No tumors induced | 591 |
| Benz(a)anthracene | Oral | Rat | 200 mg/s.d. × 310 days obs. | No mammary tumors | 591 |
|  | SC | Mouse | 0.05 mg/isc × 22–28 months obs. | 11% with tumors after 315 days | 591 |
|  |  |  | 0.2 mg/isc × 22–28 months obs. | 24% with tumors after 346 days | 591 |
|  |  |  | 1.0 mg/isc × 22–28 months obs. | 34% with tumors after 298 days | 591 |
|  |  |  | 5.0 mg/isc × 22–28 months obs. | 55% with tumors after 299 days, dose-action related | 591 |
|  |  |  | 2.2 μM/week × 35 weeks | Multiple papillomas | 610 |
|  |  |  | 5 mg/s.d. × 15 months obs. | No tumors | 591 |
|  | IV | Rat | 2.0 mg (13 mg/kg)/appln. at 50, 55, 56 days × 7 months | No tumors observed | 620 |
|  |  | Mouse | 0.25 ml/s.d. × 20 weeks | Tumor incidence lower than for controls | 591 |

| Compound | Route | Dose | Result | Ref |
|---|---|---|---|---|
| 4-Methyl- | SC | Mouse | 2 mg/s.d. × 11 months obs. | Some tumors in 4 months, 75% of animals affected in 11 months | 591 |
| 7-Methyl- | SC | Mouse | 230 μg × 3/week × 12 months | 13 carcinomas/35 mice | 591 |
| | | | 100 mg/s.d. | 31% with mammary tumors | 591 |
| 8-Methyl- | SC | Mouse | 3 mg at 1st month/5 mg at 3 and 9 months | Lung tumors, 5.5 per mouse at 7 month observation time | 591 |
| 10-Methyl- | Dermal | Mouse | 300 μg/appln. × 2/week × 20 weeks | 72% with papillomas, dose-action related | 591 |
| 12 Methyl- | SC | Mouse | 100 μg × 3/week × 12 months | 17% with mammary tumors | 591 |
| 1′-Methyl- | SC | Mouse | 5 mg/s.d. × 15 months obs. | Not a tumorigen | 591 |
| 3,9-Dimethyl- | SC | Mouse | 2 mg/s.d. × 13 months obs. | Not a tumorigen | 591 |
| 4,9-Dimethyl- | SC | Mouse | 2 mg/s.d. × 13 months obs. | Very weak tumorigenic response | 591 |
| 7,12-Dimethyl- | Oral (gastric intubation) | Rat (F) | 1 ml/dose with 0.1 (0.5, 1.0, 5.0, 10, 50, 100) mg/rat in sesame oil | Lobular carcinoma | 591 |
| Chrysene | Dermal | Mouse | 0.2%, 2 × 1 drop/week | Tumor induction in 10 wk on isc skin | 591 |
| | SC | Mouse | 20 mg/s.d. × 12 months | 100% with mammary tumors | 591 |
| [Benzo(b)-phenanthrene] | Dermal | Mouse | 0.3–7.5% in benzene or mouse fat | 2 of 5 samples tested weekly tumorigenic when dissolved in benzene | 591 |
| 3,4-Benzophenanthrene | SC | Mouse | 5 mg/s.d. × 271 days obs. | 5 sarcomas in 23% of animals; weak carcinogen | 591 |
| 2-Methyl | IV | Mouse | 0.25 mg/s.d. × 20 weeks obs. | At 8 weeks, 1.5 lung tumors in 2 of 10 mice; at 14 weeks, 1.7 in 6 of 11 mice | 591 |
| 2,9-Dimethyl | SC | Mouse | 0.2 mg/s.d. × 18 months obs. | No tumors observed | 591 |
| | Dermal | Mouse | 8.3% in croton oil × 25 days obs. | Tumor incidence slightly above controls | 591 |
| Pyrene | | | 10 μM/week × 35 weeks | Few papillomas in a low number of animals | 610 |

Table 21.20. (Continued)

| Material | Route of Entry[a] | Species | Dose or Concentration[b] | Results or Effects | Ref. |
|---|---|---|---|---|---|
| Benzo(a)pyrene | Oral | Rat | 1 mg/g food during pregnancy | Teratogenic effects of stillbirths and reduced $F_1$ growth | 591 |
| | | | 100 mg/s.d. × >50 days obs. | Mammary tumors in 8 of 9 rats | 591 |
| | | Mouse | 0.15 mg/g food × 80–140 days | Gastric papilloma, squamous cell carcinomas, pulmonary adenomas, and leukemia | 591 |
| | Dermal | Mouse | 1% soln. 2×/weeks × 200 days | First tumor after 70 days; after 200 days all animals affected | 591 |
| | | Rabbit | 0.3% benzene soln. 2×/week × >400 days | At 400 days, 1 carcinoma, 10 of 12 animals exhibit skin tumors | 591 |
| | SC | Mouse | 0.09 mg/s.d. × 183 days obs. | Tumor yield within statistical range at 78% | 591 |
| | | | 4 mg + 0.2 ml carbowax days 11, 13, 15 | Pulmonary adenoma in progeny, general 2.36 adenomas/mouse | 591 |
| | | | 4 mg as above + 2 drops 1% croton oil in acetone dermally 1/day × 28 weeks | Skin papilloma in 23.6% of treated offspring | 591 |
| | | Primate | 10 mg in 0.2 ml olive oil/s.d. × 7 months obs. | One of 2 animals died within 24 hr, the second developed a palpable nodule in 6 months, at 7 months measured 30 × 40 × 21 nm at injection site, sarcoma on heart muscle | 591 |

| | | | | | |
|---|---|---|---|---|---|
| | IP | Rat (M) | 4 mg/s.d. × 9 months obs. | Tumors of spleen and pancreas, 2/30 developed mammary, 2/30 uterine adenocarcinomata | 591 |
| | | (F) | 10 mg/s.d. × 9 months obs. | | |
| | | Mouse | 2 mg/s.d. × 33 weeks obs. | Intra-abdominal tumors at 15 wk adhering to internal organs | 591 |
| | | | 2–4 mg/s.d. × 1 year obs. | 6/14 survivors showed lung adenomas | 591 |
| | IV | Rat (F) | 750 mg/kg/s.d. | Moderate mutagenic index | 591 |
| | | | 2 mg (13 mg/kg)/s.d. × 95 days obs. | Nine of 30 rats with mammary carcinomas at days 56–95 | 591 |
| | | Mouse | 0.25 mg/s.d. × 20 weeks obs. | At 8 weeks, 8/10 mice with 2.3 lung tumors per mouse; at 20 weeks, 10/10 with 3.7/mouse | 591 |
| Benzo(e)pyrene | Dermal | Mouse | 10 μM/week × 35 weeks | Multiple papillomas | 591 |
| | SC | Mouse | 1% in 0.20 cm³ lard/s.d. | One liposarcoma at injection site in 8 days | 696 |
| Methylcholanthrene 3-Methyl- | SC | Mouse | 0.02 mg/s.d. × 221 days obs. | 14 sarcomas per 27 mice | 591 |
| | IV | Rat | 2 mg at days 50, 53, 56 | Carcinomas at days 44–98 in 7 of 30 rats | 620 |
| Dibenz(a,h)anthracene | Dermal | Mouse | 2.5 μM/week × 35 weeks | Multiple papillomas | 591 |
| | SC | Mouse | 0.04 mg/s.d. | 6 sarcomas in 18 mice (33% with tumors), 195 days av. induction time | 591 |
| | IV | Mouse | 0.25 mg/s.d. | After 8 and 20 weeks, all mice showed lung tumors, av. 30.5/mouse at 20 weeks | 591 |

[a] IP = intraperitoneal; SC = subcutaneous; IV = intravenous.
[b] s.d. = single dose; isc = intrascapular.

tramolecular and subcellular binding (575), hydrophobicity, and others. However, these characteristics alone are inadequate for specific predictions.

For acute ingestion, the polynuclear aromatics are practically nontoxic. The oral $LD_{50}$ in the mouse ranged from 1 g/kg for phenanthrene to over 18 g/kg for anthracene. Acute dermal application produced little effect, and the subcutaneous $LD_{50}$ for benzo(a)pyrene was 50 mg/kg in the rat (576).

Repeated and chronic administration of some PAHs has produced carcinogenic and teratogenic effects. To date, compounds with linear ring structures have been neoplastically negative, whereas benzo(a)pyrene and some of its derivatives are active in animal studies (573, 577).

### 4.1.3 Metabolism

Polycyclic aromatic hydrocarbons are metabolized through epoxides and hydroxides and excreted as conjugates. Benz(a)anthracene forms the 5,6-epoxide by microsomal mixed function oxidases and NADPH when incubated with rat liver in the presence of DNA and protein. The epoxide undergoes spontaneous rearrangement to the 5-hydroxide, hydration, conjugation, or reaction with cellular constituents to form complexes. Most PAHs are metabolized through an epoxide (578). Also, direct hydroxylation to form diol and triol derivatives occurs with benzopyrene (579). Various hydroxide–epoxide (579, 580) or hydroxide–oxide combinations have been identified. Rat liver microsomes can also produce 3- or 6-hydroxymethyl metabolites, as shown with benzo(a)pyrene (581, 582).

The enzyme systems, such as aryl hydrocarbon hydroxylase (AHH), are present in almost all human (583) and animal cell tissues (584–587) and are inducible by noncarcinogenic (583, 588) and potentially carcinogenic hydrocarbons (581). The stability of cytochrome P450 epoxidase may depend on immunologic competence, as does the epoxide hydrase (589).

### 4.1.4 Teratogenicity and Mutagenicity

Benzo(a)pyrene has been shown to be both teratogenic and mutagenic in rodents (590, 591). Benzo(a)anthracene was also positive in one study (591).

### 4.1.5 Carcinogenicity

Five of the PAHs are carcinogenic in some studies. They are benzo(a)anthracene, benzofluoranthracene, benzo(a)pyrene, chrysene, and dibenzo(a,h)anthracene.

### 4.1.6 Industrial Hygiene

For coal tar volatiles that contain one or more of the PAHs, the OSHA and ACGIH ceiling is 0.2 ppm with a cancer notation.

Sampling techniques include collecting air particles using an absorbant glass sampler, desorption with pentane, and quantification using spectral analysis (592). Collection on acrylonitrile–PVC filters is also recommended (593). Analytic quantification is also achieved using gas chromatography–high-resolution mass spectrometry (589, 592) or chemiluminescence (591). Methods for cleanup from waste water are also available (591, 595).

## 4.2 Acenaphthene

Acenaphthene crystallizes in bipyramidal needles. It occurs in petroleum bottoms and is used as a dye intermediate, insecticide, and fungicide, and in the manufacture of plastics (3). Selected physical data are given in Table 21.19.

Toxic effects include irritation to the skin and mucous membranes (8). Treatment of *Allium cepa* root meristem cells with acenaphthalene vapor for 12 to 96 hr caused anomalies leading to random development of the cells (596). In *Allium cepa* and *Phloeum pratense*, acenaphthalene caused disorientation of microtubules, resulting in altered cellular expansion (597).

## 4.3 Anthracenes

### 4.3.1 Anthracene

Anthracene or green oil is the simplest tricyclic aromatic compound. Selected physical data are given in Table 21.19. It crystallizes as monoclinic plates but sublimes. The crystals are clear white with violet fluorescence when pure, and are yellow and fluorescent green with tetracene and naphthacene as impurities (3). Anthracene occurs in coal tar naphtha (66), from which it is isolated by sublimation (3). Anthracene is used for the preparation of anthraquinone, alizarin dyes, (3) and fluorescent dyes for the evaluation of pesticides applied to cattle (597). Saturated air contains 0.13 percent anthracene (29).

*4.3.1.1 Toxicity* Anthracene is phototoxic and photoallergic on the human skin (66, 269). Coal tar fumes containing mainly anthracene and phenanthrene used in field roofing operations condense as respirable particles (598). Flue dust from coal tar pitch also contains respirable particles (599). It is a skin and eye irritant in the mouse (591).

When added to tocopherol solutions, the antioxidative and antiradical properties were low for anthracene, but were intensified for 7,12-dimethylbenzanthracene and even more for 20-methylcholanthrene (600). Anthracene was negative in a mouse skin painting study (591), it is classified as noncarcinogenic (573), and it was negative in five of six short-term mutagenicity tests (601, 602).

*Daphnia pulex* accumulated 760 times the anthracene concentration found in water, reaching an equilibrium in 4 hr (603).

*4.3.1.2 Microbial Studies.* *Pseudomonas* and *Nocardia* species are capable of degrading anthracene in sediment cores and in shoreline waters in or near oil spills (604). In deep sediment cores, however, degradation was very slow (605).

*4.3.1.3 Industrial Hygiene.* No official monitoring methods have been recommended. However, analytic procedures are available using colorimetric, polarographic, ultraviolet, spectral, and gas chromatographic techniques (253, 591). An antipollution procedure is also available (606).

### 4.3.2 Derivatives of Anthracene

The methyl, anthryl, dimethyl, diprophyl, dinaphthyl, trimethyl, and tetramethyl derivatives of anthracene were found carcinogenically inactive with the exception of 9,10-dimethylanthracene, which may have contained impurities when tested (573).

In rabbits, the octahydro- and perhydroanthracenes were hydroxylated in a manner similar to unsaturated hydrocarbon (607).

## 4.4 Phenanthrenes

### 4.4.1 Phenanthrene

Phenanthrene, an isomer of anthracene, is a crystalline solid. Selected physical properties are given in Table 21.19. It occurs in coal tar (3) and can be isolated from several types of crude petroleum (608).

*4.4.1.2 Toxicity.* Phenanthrene is a mild allergen (8) and human dermal photosensitizer (3). It is low to moderately toxic and when dermally applied was weakly neoplastic (609). Phenanthrene produced one dicentric chromosome and one gap in chromatid aberrations (10) in vitro, but no sister chromatid exchanges in Chinese hamsters in vivo bone marrow (611). It was ineffective as an initiator (591). Phenanthrene is inactive in skin painting studies (573).

### 4.4.2 Derivatives of Phenanthrene

Little toxicologic information is available on substituted phenanthrenes. Carcinogenically, most derivatives are inactive, such as the 1-methyl, 3-isopropyl, 1-methyl-7-isopropyl, 1,9-dimethyl, and 1,2,3,4-tetramethyl derivatives (573).

Conversely, 1-8-octahydro- and 1-methylphenanthrene have been observed to reduce foliar wilt and vascular discoloration caused by *Fusarium oxysporum* in the tomato (612).

## 4.5 Benzanthracenes

### 4.5.1 Benzanthracene

Benzanthracene, benz(*a*)anthracene, or 1,2-benzanthracene is a crystalline solid at room temperature. Selected physical data are given in Table 21.19.

Benz(*a*)anthracene occurs in crude oil, coal tar, and flue dust (599), and as a pyrolysis product in tobacco smoke (613), and in coal-derived products (614). It is emitted in exhaust gas from gasoline and other petroleum products (613). In 1958, urban atmosphere in the United States contained 0.1 to 21.6 $\mu g/m^3$ or an average of 4.0 $\mu g/m^3$ (613). Common foods contain from 0.20 to 189 ppb (567).

*4.5.1.1 Toxicity.* In comparison to anthracene, benz(*a*)anthracene appears more highly toxic but less so than phenanthrene by the dermal or subcutaneous route. Selected toxicity data are given in Table 21.20. Benzanthracene is metabolized through a 3,4-epoxide (615) to a 3,4-diol, a 3,4-diol 1,2-epoxide (614), or 8,9-dihydroxybenz(*a*)anthracene 10,11-oxide (617). Benzanthracene induces aryl hydrocarbon hydroxylase by six to 12 times in the lung, four to nine times in the skin, and two to three times in the small intestine and the kidneys (229).

In long-term dermal studies in the mouse, signs of carcinogenic effects have been recorded, but not when administered intravenously or by similar parenteral routes. When toluene was used as a solvent for repeated dermal administration, the tumor induction was insignificant below a concentration of 0.2 percent, whereas benzanthracene in dodecane was still slightly tumorigenic at 0.0002 percent (618). When tested as an initiator with phorbol esters, it produces skin tumors (591). The *K*-region bond localization energy is relatively high and no hydroxides are formed at positions 5 and 6 (573). Short-term cell transformation tests are positive for 1,2-benzanthracene in human and rodent cell lines, but only when activated with S9 homogenates (602).

Microbial degradations of benzanthracene in freshwater sediments is higher than for naphthalene and anthracene (605).

### 4.5.2 Methylbenzanthracenes

Some confusion exists for methyl benzanthracenes because two types of numbering systems have been used. Selected physical data are given in Table 21.19 for 6-, 7-, and 10-methyl-1,2-benzanthracene (2). The acute toxicity depends on the position of methyl groups. Selected toxicity data for methylbenzanthracenes are given in Table 21.20.

*K*-Region bond localization energies predict the 5- and 6-methyl derivatives to be carcinogenically active and the 2- and 7-methyls moderately active (573). The sarcoma incidence was highest for the 6-, 7-, 8-, and 12-methyl compounds, and low or negative for the 1-, 4-, 5-, 9-, and 10-methylbenz(*a*)anthracenes (619, 620).

In rat liver and mouse skin preparations, 7-methylbenz(*a*)anthracene was metabolized to all five possible *trans*-dihydrodiols (621). One carcinogenically active intermediate, 3,4-dihydro-3,4-dihydroxy-7-methylbenz(*a*)anthracene 1,2-oxide, has been isolated (621, 622). In single 25-μg mouse skin applications, the 3,4-dihydrodiol was the most active compound tested (622). In mouse embryo cells, the 7-methyl compound formed a 7-methylbenz(*a*)anthracene 5,6-oxide, although it did not form the expected nucleic acid adducts (623, 624). A slight increase in the transformation rate of hamster embryo cells was noted (625). Of the non-*K*-region diols, the 3,4-diol was most active in transforming mouse fibroblasts and V79 Chinese hamster cells (626).

### 4.5.3 Dimethylbenzanthracenes

Many pairing combinations are possible for dimethylbenzanthracene; however, most of the investigations have been carried out with the 7,12-dimethylben-

(*a*)anthracene, or 7,12-DMBA. Selected physical data are given in Table 21.19. 7,12-DMBA crystallizes in platelets with a faint green-yellow tinge (3). It is produced by a number of synthetic routes (3).

*4.5.3.1 Toxicity.* Repeated doses have skin, sebaceous gland (188), and hepatic antioxidizing (627) effects, affecting the carbohydrate metabolism scheme (628) and the hematopoietic (629) and endocrine systems (630). Cellular (631), extracellular (632), and teratogenic effects (633) have been reported, including possible sterility in rodents (634). In oral administration in the rat, DMBA accumulated in adipose tissue and the mammary gland (635). Orally administered 7,12-DMBA at 20 mg to rats dissolved in the lipid fraction of the chylomicrons in the lymph (636). Following intraglandular injection, it was retained in the rat's mandibular gland (637). With liver homogenate, 7,12-DMBA metabolized to various water-soluble mono-, di-, and hydroxymethyl derivatives (638) with the formation of glutathione conjugates (639). Selected toxicity data are given in Table 21.20 for several dimethylbenzanthracenes.

*4.5.3.2 Tumorigenesis.* Methyl substitution transforms benzanthracene into compounds which, depending on the position of the methyl groups (620), have high carcinogenic potential (619). For example, 7,12-DMBA is one of the most powerful synthetic carcinogens (640). In humans, it may be an initiator when applied to the skin (641). It can produce tumors in situ (620), as shown with 100 mg 7,12-DMBA applied to the lip of rats, which provoked edema, muscle degeneration at the injection site, and fibrosarcoma of the lip 9 months after inoculation (642). DMBA may be transported also to various sites to cause the effect (643). Similarly, oral administration of 9,10-dimethyl-1, 2-benzanthracene at 15 mg/ml arachis oil to Wistar rats produced mammary tumors in 4 to 8 weeks (531). Weekly intravenous injection of 7,12-DMBA produced dermal melanocytomas and tumors of the forestomach, intestine, ovary, skin subcutis, and lymphoreticular tissue in the Syrian hamster (644).

*4.5.3.3 Mutagenicity.* For 7,12-dimethylbenzanthracene, low mutagenic effects were noted when tested in human cell cultures with S9-activation, but practically no effects when tested without activation (602, 645). In rodent cell tests, nonactivated exhibits weak and activated moderate mutagenic effects (602). It also causes sister chromatid exchanges and chromosome aberrations (646). For 9,10-dimethylbenzanthracene, mammalian cell transformation tests are slightly positive reaction when activated with S9 mixtures (602). Cellular effects of 7,8,12-trimethylbenz(*a*)anthracene are similar to those of 7,12-dimethylbenz(*a*)anthracene (646).

*4.5.3.4 Industrial Hygiene.* The benzanthracenes, especially the 4- to 10- and 12-methylated derivatives, warrant careful handling. Special clothing should protect the eyes and skin from contact with the solid or solutions.

## 4.6 Benzophenanthrenes

### 4.6.1 Chrysene

1,2-Benzophenanthrene or chrysene is a solid material. Selected physical data are given in Table 21.19. Chrysene can be isolated from crude petroleum and coal tar. It occurs in cigarette smoke and has been detected at 1.5 to 13.3 ng/m$^3$ in urban air (613).

Chrysene is mildly acutely toxic. Equivocal results have been presented for its carcinogenic potential (613, 573, 591); however, chrysene was positive in two skin painting studies (591). Toxicity data can be found in Table 21.20.

Industrially, chrysene should be handled with precaution; however, it presents a low risk unless contaminated or in solution. Antipollution methods are available (606). Several analytic methods are available (591).

### 4.6.2 3,4-Benzophenanthrene

3,4-Benzophenanthrene is also a solid material. Selected physical data are given in Table 21.19 and toxicity data in Table 21.20. Toxicity data are also given for 2-methyl- and 2,9-dimethyl-3,4-benzophenanthrene.

### 4.6.3 Methylchrysenes

The 1-, 2-, 3-, 5-, and 6-methylchrysenes occur in cigarette smoke (613). 1,2-Dimethylchrysene is a solid material. Selected physical data for 3-methylchrysene are given in Table 21.19. The 1,7-dimethylchrysene has been isolated from coal tar pitch.

## 4.7 Pyrenes

### 4.7.1 Pyrene

**4.7.1.1 General.** Pyrene or benzo($d,e,f$) phenanthrene is a colorless solid, soluble in organic solvents. Selected physical data are given in Table 21.19. It occurs in pyrolysis or cooking processes at the lower cooling temperatures (567) and has been detected in urban atmospheres in the United States (613).

**4.7.1.2 Toxicity.** Rats ingesting lethal doses die in 2 to 5 days and rats exposed to lethal concentrations die in 1 to 2 days. Dermal exposure to 10 g/kg was not lethal in mice. Selected toxicity data are given in Table 21.20. Inhalation also caused pathological changes in hepatic, pulmonary, and intragastric tissue and decrease in the number of neutrophils, leukocytes, and erythrocytes. Dermal applications for 10 days caused hyperemia, weight loss, and hematopoietic changes; applications for 30 days produced dermatitis; and chronic effects consisted of leukocytosis and lengthened chromaxia of the leg muscle flexors. Workers exposed to 3 to 5 mg/m$^3$ noted effects that disappeared at levels below 0.1 mg/m$^3$. Some teratogenic, but no blastomogenic or carcinogenic effects were noted (647, 648, 591), except for an occasional papilloma (649). Pyrene was inactive as a tumor initiator (591).

Rat liver microsomal systems metabolize pyrene to 1-hydroxy- and 4,5-dihydro-4,5-dihydroxypyrene, as well as 1,6- and 1,8-pyrenequinone (531).

### 4.7.2 Benzo(a)pyrene

Benzo($a$)pyrene or 3,4-benzopyrene crystallizes as yellow needles. Selected physical data are given in Table 21.19.

Most experimental work on benzopyrenes has been carried out with benzo-($a$)pyrene, because of the findings that it is an animal carcinogen and a suspected human carcinogen (230).

Benzo($a$)pyrene occurs naturally in crude oils, shale oils (650), and coal tars (632), and is emitted with gases and fly ash from active volcanoes (651). Cigarette smoke and tar contain up to 0.1 percent 3,4-benzo($a$)pyrene (519), pyrolyzed from isoprene and $C_6$ to $C_{10}$ alkylbenzene precursors. The gasoline engine emits up to 0.170 ng benzo($a$)pyrene/gal fuel but only 0.02 to 0.03 ng/gal in an emission-controlled vehicle (613, 652). The greatest emissions occur from residential energy production in coal and wood furnaces (613), mounting to tons of benzopyrene per year in the United States. Other sources represent industrial coke-oven emissions (653) and road abrasions. Atmospheric concentrations have been measured as 0.05 to 74 ng/m$^3$ in urban air worldwide (613, 649). Subsequently, benzopyrene may enter the food chain. It was quantified at 0.4 to 99 ppb in food products (654, 567). Benzopyrene also is produced when edible fats are superheated (655). Conversely, baking and irradiation decrease its content in foods (656). Soil contamination is generally proportional to prevailing air concentrations (657). From soil, migration into plants can occur (658), although benzopyrene is degradable by some soil microorganisms (658).

***4.7.2.1 Toxicity.*** the acute toxicity of benzo($a$)pyrene is low. Selected toxicity data are given in Table 21.20. Benzo($a$)pyrene is a potent skin carcinogen (591).

In an epidemiologic study (659), a relationship was observed between lung cancer, soot-borne benzo($a$)pyrene, soot per se, and U.S. per capita cigarette consumption versus death rates (660). These factors all increased, whereas utilization of coal and lignite declined (661).

Repeated oral ingestion caused hypoplastic anemia in mice (662); intratrachial instillation of 0.63 mg once weekly for life into Syrian hamsters resulted in the development of bronchogenic adenomas, growth of epithelial cords of cells into lung tissues, and tumor formation (663), with changes to hyperplastic, then squamous metaplastic epithelium, and papillomas (664). Subcutaneous injection of 0.5 mg into rats increased the antitoxidative activity of the pancreas with increases of the insular cell nucleus and nucleolus sizes (665). Benzopyrene rapidly crosses the placenta in mice (666). A dose of 12 mg administered to pregnant A and C57BL mice produced an incidence of lung tumors in 31.6 percent male and 9.1 percent female offspring versus 1.2 percent in control males (647). Pregnant rats administered 4 mg/kg during gestation produced no effects in the offspring, whereas 20 mg/kg resulted in a 20 percent tumor incidence (667). In a four-generation mouse study, the incidence of papillomas and carcinomas increased (668). There is increased benzo(a)pyrene hydroxylase activity in the early gestation period, and this especially in smokers (669) and increased sensitivity in fetal tissue (670).

*4.7.2.2 Absorption, Distribution, and Metabolism.* Benzo(a)pyrene was readily absorbed by marine species (237). Distribution to striated muscle cells was rapid in the cells of *Xenopus laevis*.

Benzo(a)pyrene appears to be metabolized readily by mammals or excreted in unchanged form (671). The rat hepatic microsomes convert benzopyrene into a variety of metabolites, of which seven have been identified as 9,10-, 4,5-, and 7,8-diol, 1,6- and 3,6-quinone, and 9- and 3-hydroxybenzo(a)pyrene (229, 671–673). A multiple pathway system (674), including possible epoxides, (572, 617, 672) has been reported. Also the 3- and 6-hydroxy and the 3- and 5-hydroxymethyl derivatives have been identified as metabolic products (675–678). A major metabolite in marine fish is 7,8-dihydro-7, 8-dihydroxybenzene (241). Metabolites produced by human placental microsomes were 3-hydroxy-, 4,5-, 7,8-, and 9,10-dihydrodihydroxybenzo(a)pyrenes and some quinones, as well as other unidentified diols (679).

In rat liver microsomes, the cytochrome P448, P450, and NADPH–cytochrome reductase c systems (680) are activated by 3,4-benzopyrene (681, 682). Their activity decreases in the order liver > intestine > lung > kidney (680). The mixed function oxidase systems metabolize benzopyrene to 7,8- and 9,10-diols (580, 683) and the 4,5- and 9,10-oxo derivatives (684, 685). One of the mixed function oxidase systems, aryl hydrocarbon hydroxylase, catalyzes the formation of the 3-hydroxy derivative (686). A variety of other hydroxylases (687) and monooxygenases are also activated by 3,4-benzopyrene. It is active as a tumor initiator (491).

*4.7.2.3 Mutagenicity.* In cell transformation studies, 3,4-benzopyrene was weakly positive in WI-38 test cells and was moderately active in the rodent system with activation (471). Conversely, the Ames test was negative (688).

*4.7.2.4 Microbial Studies.* Various soil types (689) and microorganisms are capable of degrading benzo(a)pyrene, including *Bacillus megaterium* (690–692) and *Pseudomonas aeruginosa* (693).

*4.7.2.5 Industrial Hygiene.* For benzo(a)pyrene, no official exposure limits have been established.

Sample collection by absorption on charcoal or silica may be used. A selection of analytic methods are available (491); the preferential techniques involved column chromatographic separation with gas chromatographic–mass spectral quantification. When benzo(a)pyrene-containing products are handled, protective garments should be worn, and adequate ventilation and respiratory equipment should be available.

### 4.7.3 Benzo(e)pyrene

Benzo(*e*)pyrene or 1,2-benzopyrene is a position isomer of benzo(*a*)pyrene. Selected physical data are given in Table 21.19 and toxicity data in Table 21.20.

Benzo(*e*)pyrene has been identified and can be isolated from coal tar (613). It has been found in the urban atmosphere in the United States (694).

1,2-Benzopyrene is inactive as an initiator (672).

Some older studies (695, 696) report neoplastic and carcinogenic effects in the mouse, but not in the guinea pig, possibly stemming from chemical impurities.

Analytic determination procedures are available (672).

### 4.7.4 Cyclopentapyrene

Cyclopenta(*c,d*)pyrene or acepyrene, $C_{17}H_{12}$, is a five-member fused ring system (697). Cyclopentapyrene has been isolated from carbon black (698) and identified by infrared, ultraviolet, and mass spectrometry (699).

In the Ames test, cyclopentapyrene causes frame shift mutations (700).

## 4.8 Cyclopenta- and Dibenzophenanthrenes

### 4.8.1 3-Methylcholanthrene

#### 4.8.1.1 General.
3-Methylcholanthrene or 20-methylcholanthrene is a solid that crystallizes from benzene and ether in pale yellow prisms (3). Selected physical data are given in Table 21.19.

#### 4.8.1.2 Toxicity.
3-Methylcholanthrene is moderately or, on repeated exposure, highly irritating to the skin. When administered orally to hamsters, 3-methylcholanthrene produced colon neoplasms (701). On repeated or chronic exposure, 3-methylcholanthrene was tumorigenic by almost all routes tested in the rat, mouse, hamster, guinea pig, rabbit, and dog. Toxicity data are given in Table 21.20.

#### 4.8.1.3 Metabolism.
In fetal rat liver, several metabolites including the 1- and 2-hydroxy-, the *cis* and *trans*-1,2-dihydroxy-, the 11,12-dihydroxy-11,12-dihydro-, and the 1- and 2-keto-3-cholanthrene were isolated (702). The 1- and 2- hydroxy derivatives were further metabolized to *trans*-9,10-dihydrodiols (579).

Methylcholanthrene is one of the prototypes of mixed function oxygenase inducers of cytochrome P448. Therefore, 3-methylcholanthrene is often used as an experimental positive control for aryl hydrocarbon hydroxylase induction. Methylcholanthrene also induces as arene and alkene oxide monooxygenases (703).

#### 4.8.1.4 Mutagenicity and Tumorigenesis.
3-Methylcholanthrene was negative in human cell mutagenicity tests but proved highly active in rodent transformation systems (602). It is weakly clastogenic in Chinese hamster cultures (610).

3-Methylcholanthrene appears to be a rapid, all-around neoplastic agent (704) and a potent liver tumorigen (694). It produced mammary tumors in hamsters by gavage (705) or via paraffin pellets (706).

### 4.9 Dibenz(a,h) anthracene

Dibenz($a,h$)anthracene or 1,2:5,6-dibenzanthracene has been isolated from coal tar pitch and occurs in coke-oven effluents (613). It has been detected in urban atmospheres and occurs in cigarette smoke (613). On topical application, it induces papillomas and carcinomas in mice (704, 491). It is also a tumor initiator (491). Toxicity data are given in Table 21.20.

## 5 PETROLEUM AS A HYDROCARBON SOURCE

Petroleum or crude oil recovery and transport normally is carried out using enclosed systems. Transfer operations represent points of exposure that constitute a relatively low human hazard. The health hazard is lower when handling crude oil than certain fractionated materials. A simplified list of crude oil fractions and their uses is presented in Table 21.21.

During refining, exposures vary greatly, being low for the alkane gases and moderate for kerosine, solvents, and light oil fractions. Repeated dermal contact or exposure to vapors are significant health hazards. Permissive threshold concentrations can be calculated if the specific composition is known. Thus, with proper precautions, the risks for toxic or tumorigenic effects are low.

A study of refinery workers indicated that deaths due to cancer and arterial diseases are slightly higher in refinery and petrochemical production workers than for others in the plant and the control population (707). In three work-year groups, the digestive cancer and respiratory system incidence appeared to increase with work-years, and the incidence of brain and other CNS tumors declined at the 10 to 19 work-year range (707).

A petroleum fraction, Iomex, administered orally to rats at 1 ml/kg for 4 weeks did not produce any toxic symptoms, but at higher doses the animals showed high mortality, hematologic alterations, and slight abnormalities in the lung, liver, and kidney (708).

Several marine species are able to absorb, metabolize, and release crude petroleum or fractions. In general, aromatic hydrocarbons are retained longer, and the highest molecular weight compounds released more slowly (31).

### 5.1 Crude Oil

Crude oil or petroleum is a flammable liquid and is a complex mixture of organic and some inorganic materials, varying with its geologic origin. Selected physical data are given in Table 21.22. It is utilized to manufacture gasoline and lubricating oils (709). Crude oil varies in distillate type from gasoline and accompanying products to highly viscous asphalts (5). The average composition includes paraffinic,

**Table 21.21.** Petroleum Fractionation (5, 115)

| Fraction | Organic Compound | Boiling Range [°C (°F)] | Use |
|---|---|---|---|
| Natural gas | $C_1$–$C_2$ | −164 to −88 (−263 to −126.4) | Fuel, chemical |
| Liquefied, or bottled, gas | $C_3$–$C_4$ | −44.4 to +1.0 (−48 to +34) | Fuel gas; for the synthesis of rubber components, petro-chemicals |
| Petroleum ether | $C_4$–$C_5$ | 20–60 (68–140) | Solvents |
| Gasolines | $C_5C_{10}$ | 32–149 (90–300) | Aviation fuel |
| | | 32–210 (90–410) | Motor gasoline |
| Naphthas | $C_5$–$C_{10}$ | 65–204.4 (149–400) | Cleaning fluids, solvents refining stock |
| Kerosines | $C_5$–$C_{16}$ | 40–300 (104–572) | Jet and turbofuels |
| Kerosine | | 350–550 (176.7–287.8) | Stove oil, tractor and gas turbine fuel |
| Gas oil | $C_9$–$C_{16}$ | 204–371 (400–700) | Furnace oil, Diesel oil |
| Lubricating stocks | $C_{17}$–higher | 204–400 (400–750) | White oils, Lubricating oils and greases |
| Waxes | $C_{20}$–higher | 204–400 (400–750) | Sealing wax, Food component |
| Bottoms | $C_{20}$–higher | 499–higher (750–higher) | Heavy fuel oil, Road oils, Asphalts |

naphthenic, aromatic, and sulfur-, nitrogen-, and oxygen-containing compounds, and a variety of metals, including cobalt, manganese (710), boron, chromium, nickel, sulfur, vanadium (711), and uranium (712).

### 5.1.2 Toxicity

The major acute effect of crude oil is CNS depression, although this is reversible even at high concentrations. On ingestion and aspiration or the inhalation of vapors, it will produce chemical pneumonitis. On prolonged dermal contact or inhalation, it is an irritant. Low and high sulfur-containing crude oils were administered at 37.0 to 123.0 ml/kg once a day for 5 days to cattle ranging in age from 6 months to 3.5 years. Toxic effects included vomiting, bloating, aspiration pneumonia, anorexia, weight loss, mild mental depression, and a decreased plasma glucose level (713). Southern U.S. crude oil was more toxic to annelids than Kuwait crude (714). Sublethal concentrations were reversibly toxic in the marine species (715). The fate of crude and fractionated oil in the ecosystem has been summarized (716, 717).

Crude oil affects the growth and photosynthetic action of microalgae (551, 718), but is degraded by a variety of microorganisms. Selected toxicity data are given in Table 21.23.

### 5.1.2 Industrial Hygiene

When handling crude oil, vapors or mists should not be inhaled. Gloves and glasses should be worn for dermal and eye protection.

## 5.2 Natural and Liquefied Gases

Natural gas is a colorless, odorless, flammable gas (34), which occurs naturally along with petroleum deposits in marshes, or from waste decomposition. It consists of 83 to 99 percent methane, as well as ethane, propane, and butane (8). The gas can be liquefied for transport and storage and is primarily used as fuel.

Petroleum gas, recovered during refining of crude oil, is a flammable gas (34) that is easily compressed to LPG, liquified petroleum or "bottled" gas. It consists of propane and butane, with minute quantities of mercaptans added for odorant warning properties (5). LPG is primarily used for fuel, as chemical raw material, and for refinery blending of a variety of materials.

Liquified gases are practically nontoxic below the explosive limits, cause CNS depression at high concentrations, and may cause asphyxia by oxygen displacement (269).

Generally, flammability and explosive hazards outweigh the toxic effects.

## 5.3 Gasoline

### 5.3.1 General

Gasoline or petrol is a flammable liquid (34) produced from the light distillates during petroleum fractionation. Selected physical data are given in Table 21.22. The distillation ranges are specified for the particular application, mainly the reciprocating, spark ignition, and internal combustion engines. The most critical property is the octane number, supplied with high-octane hydrocarbons and other compounds. The major components are primarily paraffins, olefins, naphthenes, and aromatics, and more recently 10 to 40 percent ethyl alcohol (5). The distillation from initial to final boiling point ranges from about 32 to 225°C (90 to 437°F) and the explosive limits, 1.3 to 6.0 percent (719).

### 5.3.2 Toxicity

Although gasoline grades vary with octane number and engine requirement, the general toxic effects do not differ appreciably, except with volatility and benzene content. When gasoline is inhaled at high concentrations, the additives exert only minor influence. Overall, few cases of intoxication have been reported in relation to gasoline quantities handled. Selected toxicity data are given in Table 21.23.

**Table 21.22.** Physical Properties of Petroleum Naphthas and Thinners

| Common Name | Alternate Name | B.P. (°C) | Flash Pt. (°C) | Mol. Wt. | Carbon Number | Class of Components | TLV [ppm (mg/m³)] | Ref. |
|---|---|---|---|---|---|---|---|---|
| Crude oil | Earth oil, petroleum | <0 to >1000 | −7 to −32 | N.A. | $C_1$–>$C_{50}$ | About 300 organic substances identified, some heavy metals | 5 mg/m³ for misting | 751 |
| Gasoline | Benzin, petrol, automotive or aviation fuel | 32–210 | 40–70 | ~100 | $C_4$–$C_{12}$ | n- and isoparaffins, olefins, aromatics | 300 (890) | 5, 39 |
| Petroleum ether | Ligroin, petroleum benzin | 30–60 | −57 to −46 | ~77 | $C_5$–$C_6$ | Paraffins (pentanes, hexanes, isohexanes) | 500 (1600) | 751 |
| Rubber solvent | | 45–125 | −13 (9)[a] | 84–97 | $C_5$–$C_7$ | Paraffins, monocycloparaffins, olefins (trace), benzene, alkylbenzenes | 400 (1590) | 715 |
| VM&P naphtha | Range of 80 thinner | 95–160 | −7 to 13 | 87–114 | $C_5$–$C_{11}$ | Paraffins, mono- and dicycloparaffins, benzene (trace), alkylbenzenes | 300 (1350) | 751 |
| Mineral spirits (petroleum spirits) | Refined petroleum solvent, white spirits | 150–200 | <0–35 | ~130 | $C_6$ | Paraffins, naphthenes, olefins, aromatics | 200 (350) | 751 |
| Stoddard solvent | White spirits | 160–210 | 38–43 | ~140 | $C_7$–$C_{12}$ | Paraffins, mono- and dicycloparaffins, benzene (trace), alkylbenzenes | 100 (25) | 751 |

| Name | | Boiling range | Flash point | Carbon range | Composition | Density |
|---|---|---|---|---|---|---|
| High flash naphtha | | 150–204 | | ~140 | $C_7$–$C_{12}$ | Paraffins, naphthenes, aromatics | 751 |
| 140° flash naphtha | | 185–207 | 59–60 | 154 | $C_5$–$C_{13}$ | Paraffins, mono- and dicycloparaffins, benzene (trace), alkylbenzenes | 751 |
| Aromatic petroleum naphtha | Aliphatic solvent naphtha | 93–315 | 2–38 | ~140 | $C_8$–$C_{13}$ | Paraffins, alkylbenzenes | 758, 195 |
| High aromatic naphtha | Coal tar naphtha | 184–206 | 62 (144)[a] | ~140 | $C_8$–$C_{13}$ | Paraffins, mono-, di-, and tricyclic naphthenes, alkylbenzenes and naphthalenes, olefins | 758 |
| Thinner | High aromatic solvent | | | | | | |
| 40 | | 186.7–230.6 | 49 (120)[a] | ~148 | $C_8$–$C_{13}$ | Paraffins, mono- and dicycloparaffins, mono- and diolefins | 751 |
| 50 | | 97.8–105 | 4.5 (40) | 97 | $C_6$–$C_8$ | Paraffins, olefins, naphthenes, aromatics | 751 |
| 60 | | 128.3– | 30 (86) | 120 | $C_5$–$C_{10}$ | Paraffins, monocycloparaffins, alkylbenzenes | 751 |
| 70 | | 157.2–210.6 | — | 132 | $C_5$–$C_{12}$ | Paraffins, monocycloparaffins, alkylbenzenes | 751 |
| 80 | | 96.7–142.2 | 30 (86)[b]<br>33 (38)[a] | 106 | $C_6$–$C_9$ | Paraffins, dicycloparaffins, alkylbenzenes | 751 |
| Kerosine | Stove oil | 163–288 | 49–52 | ~180 | $C_{10}$–$C_{16}$ | Aliphatics, mono- and dicycloparaffins, alkylbenzenes | 2, 5 |
| Deodorized kerosine | | 207.8–272.2 | 80 (176)[a]<br>88 (190)[b] | 179 | $C_6$–$C_{14}$ | Paraffins, mono- and dicycloparaffins, aromatics | 782 |

[a] Tagg closed cup.
[b] Tagg open cup.

Table 21.23. Toxicity of Petroleum Solvents

| Material | Route of Entry | Species | Dose or Concentration | Results or Effects | Ref. |
|---|---|---|---|---|---|
| Crude oil | Aquatic | Fish | >1000 ppm | TLm 96 | 292 |
| Gasoline (b.p. <230°F) | Oral | Human | 10–15 g | Lethal in children | 719 |
| | Inhalation (acute) | Human | 20–50 g | Toxic effects in adults | 52 |
| | | | 550 ppm (~2 mg/l) × 1 hr | No effects | 115 |
| | | | 900 ppm (~3.5 kg/l) × 1 hr | Slight dizziness, irritation of eyes, nose, throat | 115 978 |
| | | | 2000 ppm (~7.6 mg/l) × 1 hr | Dizziness, mucous membrane irritation, and anesthesia | 115 |
| | | | 10,000 ppm (~37 mg/l) × 1 hr | Nose and throat irritation in two minutes, dizziness in 4 mins, signs of intoxication in 4–10 min | 115 |
| | Inhalation (chronic) | Human | >500 ppm (~1.8 mg/l)/day | May cause vomiting, diarrhea, insomnia, headache, dizziness, anemia, muscle and neurological symptoms | 56 |
| | Inhalation | Mouse | 30,000 ppm (~110 mg/l) × 5 min | Lethal dose | 56 |
| Petroleum ether | Dermal | Human | Undiluted × 30 min | Disruption of horny layer, peeling | 753 |
| | | | Undiluted × 1 hr | Erythema, hyperemia, swelling pigmentation | 753 |
| | Inhalation | | Saturated | Cerebral edema | 753 |
| | | | 445–1250 ppm (1160–4400 mg/m³) | Blurred vision, cold sensation in extremities, fatiguability, headache; fatty degeneration of muscle fibers, demyelination and mild axonal degeneration | 753 |
| | | | 500–2500 ppm (1.5–7.8 mg/l)/day | Neurogenic atrophy | 753 |
| | | | 1000–2500 ppm (3.1–7.8 mg/l)/day | Polyneuropathy in 6–9 months | 753 |

| | | | | | |
|---|---|---|---|---|---|
| Aromatic petroleum naphtha | Inhalation | Human | 0.07 ppm (0.4 mg/m³) | Odor threshold | 760 |
| | | | 0.5–2.5 ppm (2.2–11 mg/m³) | Identification threshold | 665 |
| | | | 26 ppm (150 mg/m³) | Sensory threshold | 760 |
| | Inhalation | Rat | 66 ppm (0.38 mg/l) × 8 hr | No observable physiological effects | 760 |
| | | | 1500 ppm (8.7 mg/l) × 8 hr (2 exp.) | Lethal, erythrocyte fragility | 760 |
| B.P. 364–403°F | Inhalation of aerosol | Mouse | 550 ppm (3.1 mg/l) | Increased respiratory rate by 50% | 760 |
| B.P. 311–392°F | Inhalation (subchronic) | Cat | 150 ppm (8.2 mg/l) × 6 hr | CNS depression | 760 |
| | | Rat | 50 ppm (300 mg/m³) × 8 hr/day × 5 days/week × 90 days | No detectable changes | 554 |
| | | | 616 ppm (3.6 mg/l) × 18 hr/day × 7 days/week × 150 days | Decreased weight gain, lung congestion, hemorrhaging | 554 |
| | | | 1000 ppm (5.7 mg/l) × 18 hr/day × 7 days/week × 78 days | Congestive changes in lung, liver, spleen, and kidney, decreased white blood count | 554 |
| B.P. 392–480°F | | Rat | 50 ppm (300 mg/m³) × 8 hr/day × 5 days/week × 90 days | Slight bone marrow changes and DNA depression | 554 |
| | | | 200 ppm (1.4 mg/l) × 8 hr/day × 5 days/week × 90 days | Decreased white blood count and weight gain, one of 17 animals developed cataracts | 554 |
| | | | 500 ppm (2.8 mg/l) × 18 hr/day × 7 days/week | Lethal to 50% of the colony | 554 |
| B.P. 311–392°F | Inhalation | Primate | 50 ppm (300 mg/m³) × 7 hr/day × 5 days/week × 90 days | No effect | 554 |
| | | | 200 ppm (1.4 mg/l) × 7 hr/day × 5 days/week × 90 days | Equilibrium disturbances, decreased white blood count, initially tremors, loss of hair, dry skin, some myelocytic depression, erythrocytic changes | 554 |
| B.P. 392–480°F | Inhalation | Primate | 50 ppm (0.3 mg/l) × 7 hr/day × 5 days/week × 90 days | Diarrhea, increased erythrocyte activity, irritation of face and eyes, stimulation of bone marrow erythrocyte activity, decreased myelocytes | 554 |

Table 21.23. (Continued)

| Material | Route of Entry | Species | Dose or Concentration | Results or Effects | Ref. |
|---|---|---|---|---|---|
| Rubber solvent | Inhalation | Human | 200 ppm (1.4 mg/l) × 7 hr/day × 5 days/week × 90 days | Effects as for 50 ppm | 554 |
| | Inhalation | Human | 10 ppm (40 mg/m³) | Odor threshold concentration | 755 |
| | Inhalation | Rat | 2800 ppm (10 mg/l) | No effect | 755 |
| | Inhalation | Dog | 15,000 ppm (6.1 mg/l)/4 hr | $LC_{50}$ | 755 |
| | Inhalation | | 1500 ppm (6.1 mg/l) | No effect | 755 |
| VM&P naphtha | Inhalation | Human | 0.86 ppm (4 mg/m³) | Odor threshold | 757 |
| | Inhalation | Rat | 3400 ppm (16 mg/l) × 4 hr | $LC_{50}$; eye irritant in 30 min | 757 |
| | Inhalation | | 1200 ppm (5800 mg/m³) × 6 hr/day × 5 days/week × 40 days | At day 40, increased neutrophils, decreased lymphocytes, normal at day 65 | 757 |
| | Inhalation | Dog | 1200 ppm (5800 mg/m³) × 6 hr/day × 5 days/week × 40 days | Increased reticulocyte count, increased alkaline phosphatase, decreased GOT | 757 |
| Mineral spirits | Inhalation | Human | Unknown/day × 4 months | Aplastic anemia, lethal | 54 |
| | Inhalation | | 670–1670 ppm (1–2.5 mg/l) × 30 min | Nausea, vertigo during exercise or rest, one case of premature atrial beats, one case of t-wave inversion | 54 |
| | Inhalation | Dog | ~140–360 ppm (238–169 mg/m³)/day × 23.5 hr/day × 90 days | Normal hematology | 54 |
| | Inhalation | Guinea pig | ~200 ppm (363 mg/m³)/day × 7 days/week × 60–90 days | Lowest lethal effect | 54 |
| | Inhalation | | 360 ppm (619 mg/m³)/day × 7 days/week × 60–90 days | Increase in body weight | 54 |
| | Inhalation | Primate | ~320 ppm (555 mg/m³)/day × 7 days/week × 60–90 days | Decrease in body weight | 54 |
| | Inhalation | Human | (600 mg/m³) × 8 hr | Irritant | 54 |

| Substance | Route | Species | Dose/Exposure | Effect | Ref |
|---|---|---|---|---|---|
| | Intraperitoneal | Rat | 5680 mg/kg | Lethal | 54 |
| | | Mouse | (50,000 mg/m³) × 8 hr | | 54 |
| | Aquatic | Fish | >1000 ppm | TLm 96 | 292 |
| Stoddard solvent | Inhalation (acute) | Human | 9 ppm (50 mg/m³) | Odor threshold | 54 |
| | | | 400 ppm (2300 mg/m³) | No eye, nose, or throat irritation | 54 |
| | Dermal | | Undiluted 1–2/day × 6 months | Fatal when used for washing hands | 54 |
| | | | Undiluted × 2 yr | Fatal when used for removal of paint from hands | 54 |
| | Dermal + inhalation (chronic) | | Undiluted × 8 hr/day × 8 weeks | Follicular dermatitis from daily contact in dry cleaning | 54 |
| | Inhalation | Human | Fumes × 0.3 months | Hepatic involvement and sensitization, hepatic tests still elevated 1 year post-exposure | 54 |
| | | | Fumes 2–3×/month × 2 years | Lethal, moderate hypoplasia of bone marrow | 54 |
| | | | Fumes × 20 years | Worker survived after splenectomy | 54 |
| | | | Fumes × 17 years | Aplastic anemia, intracerebral hemorrhage, death | 54 |
| | Inhalation (acute) | Rat, dog | 410–1400 ppm (2.4–2.8 mg/l) | No significant difference between test and control | 54 |
| | | Cat | 1700 ppm (10 mg/l) | CNS depression | 54 |
| | Inhalation (chronic) | Rat | 84–330 ppm (0.48–1.9 mg/l) | Slight kidney pathological signs | 54 |
| | Inhalation | Dog | 84–330 ppm (0.48–1.9 mg/l) | No significant effects | 54 |
| 140°F Flash naphtha | | Human | 0.6 ppm (4 mg/m³) | Odor threshold level | 759 |
| | | | 17–49 ppm (110–310 mg/l) × 15 min/day × 2 days | Slight temporary dryness of eyes | 759 |
| | Inhalation (chronic) | | 37 ppm (230 mg/m³) | Suggested hygienic standard | 759 |
| | Inhalation | Rat, dog, cat | 33–300 ppm (0.21–1.9 mg/l) | Slight lacrimation of 1 dog, otherwise no significant clinical or observable effects | 759 |
| | | | 37 ppm (230 mg/m³) × 6 hr/day × 5 days/wk × 13 weeks | No effects | 759 |

**Table 21.23.** (Continued)

| Material | Route of Entry | Species | Dose or Concentration | Results or Effects | Ref. |
|---|---|---|---|---|---|
| Thinner 40 | Inhalation (aerosol) | Rat | 33 ppm (0.2 mg/l) × 7 hr | Dose tolerated | 54 |
| | | | ~140 ppm (8.3 mg/l) × 7 hr (1 μ diameter) | Irritation to extremities, loss of coordination | 54 |
| | | Dog | ~41 ppm (0.25 mg/l) × 8 hr | Dose tolerated | 54 |
| | (Aerosol) | Cat | 120 ppm (7.0 mg/l) × 6 hr (1 μ diameter) | No visible discomfort | 54 |
| Thinner 50 | Inhalation | Human | 2.5 ppm (10 mg/m³) | Odor threshold | 54 |
| | | | 430 ppm (1.7 mg/l) | 4 of 5 people willing to work for 8 hr | 54 |
| | | | 530 ppm (2.1 mg/l) | Lightheadedness, headache in 30 mins | 54 |
| | Inhalation | Rat | 1300 ppm (5.2 mg/l) | Dose tolerated | 54 |
| | | | 8300 ppm (33.0 mg/l) × 6 hr | $LC_{50}$ | 54 |
| | Inhalation | Dog | 600 ppm (2.4 mg/l) × 6 hr | No discomfort | 54 |
| | Inhalation | Cat | 7600 ppm (30.0 mg/l) × 6 hr | Signs of CNS effect, mydriasis, mild tremors, light anesthesia after 2–3 hr, but reversible | 54 |
| Thinner 60 | Inhalation | Human | 2 ppm (10 mg/m³) | Odor threshold | 54 |
| | | | 170 ppm (0.85 mg/l) | Tolerated | 54 |
| | | | 350 ppm (1.7 mg/l) | Tolerated | 54 |
| | | Rat | 170 ppm (0.85 mg/l) | No visible response | 54 |
| | | | 690 ppm (3.4 mg/l) | No visible response | 54 |
| | | | 2500 ppm (12.0 mg/l) | Slight effects, loss of coordination | 54 |
| | | | 4900 ppm (24.0 mg/l) × 4 hr | $LC_{50}$ | 54 |
| | | Beagle dog | 820 ppm (4.0 mg/l) | No visible effects | 54 |
| | | | 1900 ppm (9.5 mg/l) | Loss of coordination | 54 |
| | | Cat | 4100 ppm (20 mg/l) | Lethal in 4 hr | 54 |
| | | | 7700 ppm (38 mg/l) | Lethal in 150 min | 54 |

| Substance | Route | Species | Dose | Effect | Reference |
|---|---|---|---|---|---|
| Thinner 70 | Inhalation | Human | 0.7 ppm (4 mg/m$^3$) | Odor threshold | 54 |
| | | | 59 ppm (0.32 mg/l) | Sensory response, minimal | 54 |
| | | Rat | 180 ppm (0.95 mg/l) × 15 min | Ocular and nasal irritation | 54 |
| | | | 810 ppm (4.4 mg/l) × 8 hr | Lacrimation, loss of coordination, fine tremors | 54 |
| | | Beagle | 930 ppm (5.0 mg/l) × 4 hr | Convulsions in 2 hr | 54 |
| | | Cat | 370 ppm (2.0 mg/l) × 6 hr | CNS effects | 54 |
| | | Rat, dog | 200 ppm (1.0 mg/l) × 13 weeks | No ill effect level | 54 |
| | | | 410 ppm (2.2 mg/l) × 13 weeks | No recognizable effects except for slight reduction in weight gain | 54 |
| Thinner 80 | Inhalation | Human | 100 ppm (0.45 mg/l) | Odor threshold | 54 |
| | | | 150 ppm (0.65 mg/l) | Slight transitory eye irritation | 54 |
| | | | 230 ppm (1.0 mg/l) | Suggested hygienic standard | 54 |
| | | Rat | 800 ppm (3.5 mg/l) × 4 hr | Response variable | 54 |
| | | | 6200 ppm (27 mg/l) × 4 hr | Dose tolerated | 54 |
| | | Dog | 480 ppm (2.1 mg/l) × 4 hr | LC$_{50}$ | 54 |
| | | Cat | 5500 ppm (24.0 mg/l) × 4 hr | Dose tolerated | 54 |
| | | | | CNS effects, animals prostrate in 41–65 min, recovery in 14 days | 54 |
| | | Rat, dog | 390 ppm (1.7 mg/l) × 6 hr/day × 5 days/week × 14 weeks | No visible effects | 54 |
| Kerosine | Oral | Human | 0.5 oz | Lowest lethal dose | 730 |
| | | | 3–4 oz | Mean lethal dose | 730 |
| | | | 8 oz | Highest nonlethal dose | 730 |
| | Oral (with aspiration into lungs) | | <1 ml | May cause chemical pneumonitis systemic symptoms, including CNS effects | 268 |
| | Inhalation | | 14 ppm | No effect | 784 |
| | | | 20 ppm (140 mg/m$^3$) | Odor threshold | 784 |
| | Oral | Rabbit | 28 g/kg | Lethal to some animals | 54 |

Table 21.23. (Continued)

| Material | Route of Entry | Species | Dose or Concentration | Results or Effects | Ref. |
|---|---|---|---|---|---|
| Deodorized kerosine | Oral | Mouse | 64 ml/kg | Lethal to 4 of 10 animals | 784 |
| Kerosine | Oral | Mouse | 50 ml/kg | Intoxication in 12–15 min, with labored and rapid respiration, death 10 hr after second equal dose on 2nd day; congestion of renal tubules, lung surface hyperemia, liver yellow patches | 54 |
| Deodorized kerosine | | Rabbit | 64 ml/kg | Lethal to 1 of 10 animals | 784 |
| Kerosine | | Rabbit | 28 g/kg | $LD_{50}$ | 54 |
| | | Guinea pig | 20 g/kg | $LD_{50}$ | 54 |
| | Intratracheal | Rat | 800 mg/kg | Lowest lethal dose | 54 |
| | Eye | Rabbit | Undiluted | Practically innocuous | 54 |
| | Intraperitoneal | Rat | 10.7 g/kg | Lowest lethal dose | 54 |
| | | Rabbit | 6.6 g/kg | $LD_{50}$ | 54 |
| | | Dog | 50 ml/kg | Labored, rapid respiration, sedation | 54 |
| | Intravenous | Rabbit | 180 mg/kg | $LD_{50}$ | 54 |
| Deodorized kerosine | Inhalation | Rat | 14 ppm (0.10 mg/l) × 8 hr | No signs of distress | 784 |
| Kerosine | Dermal (subacute) | Rabbit | 3 ml (kg)(day) × 3 days | Hair loss, scaling, cracking of the epidermis, no systemic toxicity | 54 |
| | | Guinea pig | 0.5 ml/3 day × 2 weeks | Carrier for dinitrochlorobenzene, increases the reactivity, swelling, infiltrations | 54 |
| Deodorized kerosine | Inhalation (subchronic) | Rat, dog cat | 14 ppm (0.1 ml/l) × 8 hr/day × 13 weeks | No discomfort in saturated vapor | 784 |
| | Aerosol | Rat | 7.4 mg/l × 6 hr/day × 4 days | Skin irritation of extremities | 784 |
| | Aerosol | Cat | 6.4 mg/l × 6 hr/day × 4 days | No effect | 784 |

In children, death from accidental ingestion of as little as 10 to 15 g gasoline has been observed (52). In adults, ingestion of 20 to 50 g of gasoline may produce severe intoxication. Ingestion causes immediate severe burning of the pharynx and gastric region. With immediate gastric lavage, no general symptomatic effects were noted, except for transient liver damage (52). Symptoms in oral intoxication are mild excitation, loss of consciousness, occasionally convulsions, cyanosis, congestion, and capillary hemorrhaging of the lung and internal organs (52), followed by death due to circulatory failure (52); in milder cases, symptoms are inebriation, vomiting, vertigo, drowsiness, confusion, and fever (115). Unless prevented, aspiration into the lungs and secondary pneumonia may occur. Gasoline may cause hyperemia of the conjunctiva and is a skin irritant and a possible allergen. On acute inhalation, humans experience intense burning of the throat and respiratory system, and bronchopneumonia may develop. At extremely high concentrations where oxygen displacement is a factor, asphyxiation may occur. Severe intoxication is accompanied by CNS effects, coma, and convulsions. The deaths of two occupants of a light aircraft wreckage appear to be attributable to systemic fat embolism following massive, acute gasoline inhalation (720).

Repeated or chronic dermal contact may result in drying of the skin, lesions, and other dermatologic conditions (721). Inhalation of gasoline during bulk handling operations produced no physiological effects (722). In India, complaints from gasoline pump workers, possibly due to the warmer climate, included headache, fatigue, disturbance of sleep, and loss of memory (723). Urinary phenol levels above 40 mg/l could be directly related to quantities of gasoline handled per day (723). Workers chronically exposed to gasoline vapors showed a decrease in the phagocytic activity of peripheral blood granulocytes, globulin, and total protein levels (724). Occupational exposure to vapors of gasoline-powered equipment has been related to some nonlymphocytic leukemias (725). One case of acute hepatic and CNS effects due to high level gasoline inhalation has been reported (52).

"Gasoline sniffing" has produced morbidity and mortality cases owing to acute and chronic inhalation (726). One lethal case and one with signs of lead encephalopathy, elevated blood lead, and a marked decrease of δ-aminolevulinic acid dehydratase levels have been described (727). Gasoline can sensitize the myocardium and cause rapid CNS depression with respiratory failure (728). This could explain the sudden sniffing deaths (729) and lethal effects to workers cleaning storage tanks without proper respiratory protection (730). Intratracheal instillation of as little as 0.2 ml gasoline-type petroleum fractions have caused instant death in the rat (28). Exposure of rabbits to 310 mg/l for 2 hr resulted in decreased electrolytes, heart muscle alkaline phosphatase, and α-1 and α-2 globulins. Rabbits chronically exposed to gasoline vapors showed alterations in lipid metabolism and serum lipid changes (731) and lymphoid cell decreases (732). Rats exposed to 1 g/m$^3$, 5 hr/day, 5 days a week, resulted in reversible damage of the eye blood vessels at 3 months, but atrophy and necrosis after 6 to 9 months (733). Exposure of rats to 10 mg ethyl gasoline/l for 6 hr/day caused disseminated degenerative changes in the neurons of the central nervous system (734); at 49.7 mg/l, 4 hr/day for 27 weeks, it caused alterations in the ovarian and pituitary functions (735). When 0.25 ml/l was added to 8.5 ml rat brain homogenate, inhibition of monoamine oxidase occurred at a higher rate by leaded than unleaded gasoline (736). Limited experiments demonstrated that gasolines exhibited no gonadotrophic nor mutagenic action (737), nor did organic phosphate additives (738).

Considerable concern developed when unleaded gasoline was shown to cause kidney tumors in rats (739). Repeated oral doses (740) or inhalation studies (741) result in tubular degeneration characterized by hyaline droplet formation, necrosis, intratubular casts, and medullary mineralization. These effects are produced in males of a number of rat strains but are not produced in the NCI–Black–Reiter strain males (742), or in any female rats (743) or other mammalian species (744). The tumors develop in response to the accumulation of the protein, $\alpha$-2-microglobulin (58). It has been documented that female rats and other mammalian species do not accumulate $\alpha$-2-microglobulin. On this basis, the tumors are not considered relevant to humans (745).

### 5.3.3 Industrial Hygiene

The OSHA PEL (148) is 300 ppm (900 mg/m$^3$) with a STEL of 500 ppm (1500 mg/m$^3$) and the ACGIH TLV (194) is 300 ppm (890 mg/m$^3$) with a STEL of 500 ppm (1480 mg/m$^3$).

Gasoline can be collected on charcoal, desorbed with carbon disulfide, and quantified using flame ionization gas chromatography (722, 21). A combination of gas chromatography and mass spectrometry can be used for the analytic quantification of gasoline in the blood (746, 21). Oil adsorbents can be used for the removal of gasoline from aqueous effluents (747).

When working in an atmosphere where gasoline vapors may occur, precaution is advised. At no time should a facility be entered with levels above 500 ppm (115).

### 5.4 Petroleum Naphthas

#### 5.4.1 General

Petroleum naphthas or petroleum solvents are complex hydrocarbon mixtures that can be obtained from the petroleum light distillate or low-boiling fraction (5).

The light petroleum naphthas are composed mainly of alkanes, mono- and dicyclanes, alkenes, alkylbenzenes, naphthenes, and some benzenes, lending the mixture its specific physicochemical properties, such as boiling point range and flash point. Selected physical data are given in Table 21.22. It is used as rubber solvent, paint thinner, cleaning or degreasing agent, and petroleum refining stock. The hydrocarbon solvents are not used as food additives (480). Unfortunately, there has been wide abuse of glues and solvents, through the practice of inhalation or "glue sniffing" (204). Some have used naphthas, mineral spirits, or kerosine as a rubbing fluid, resulting in several deaths from misuse.

*5.4.1.1 Toxicity.* Petroleum benzin, mineral spirits, and naphthas are volatile and present a higher toxicity hazard than the higher-boiling fractions. Chronic exposure to petroleum distillate caused CNS damage (748). Aside from CNS depression, myocardial (208) and hematopoietic effects have been recorded. Myelotoxic effects and hypoplasia are ascribed to the benzene content (749). However, where possible, benzene now has been removed from most commercial materials.

Dermal contact causes erythema, blistering, and cellular damage, and naphthas present an allergenic potential (201). Repeated applications cause dermatitis and other lesions.

The oral $LD_{50}$ in rats ranges from 4.5 to >25 ml/kg for a series of petroleum distillates (750). With accidental ingestion, aspiration into the lungs may occur, causing endothelial injury, edema, and hemorrhage (269). In the rat, aspiration of 0.2 ml of a petroleum fraction containing gasoline, fuels, and naphthas, with a viscosity of 39 SUS (Saybolt Universal Standard) or less, resulted in eight of 10 deaths (28). $LC_{50}$ values range up to 73,680 ppm (750).

None of the petroleum solvents have carcinogenic potential, although some may act as carriers for carcinogenic agents by the solvent effect (704).

*5.4.1.2 Industrial Hygiene.* The threshold limits vary with the volatility of the solvent, from 100 to 1000 ppm. For sampling, activated charcoal collection has been recommended, and several analytic determination procedures are available (252, 751).

When petroleum hydrocarbon solvents are handled, gloves and respiratory protection are recommended. For special cases, barrier creams can be used (259).

### 5.4.2 Petroleum Ether

Petroleum ether, ligroin, or petroleum benzin is a flammable liquid (34) and low-boiling cut. Selected physical data are given in Table 21.22. Petroleum ether is used as a universal solvent and extractant for chemicals, fats, waxes, paints, varnishes, and furniture polishes, and is used as a detergent, in photography, and as fuel (3).

*5.4.2.1 Toxicity.* Petroleum ether consists principally of *n*-pentane and *n*-hexane. Thus the general effects of intoxication are peripheral nerve disorders, CNS depression, and skin and respiratory irritation, discussed in detail in Chapter 19 on alkanes. Ingestion has caused chemical pneumonia in children (752). On human skin, it causes erythema, edema, disruption of the horny layer, and peeling (753). Acute inhalation of petroleum ether, when it was mistakenly used as an anesthetic agent, caused reversible cerebral edema (753). Toxicity data are given in Table 21.23.

Numerous reports point to the neurotoxic effects on prolonged inhalation of petroleum ether in inadequately ventilated business establishments, where employees experienced polyneuropathy (753). Signs and symptoms included loss of appetite, muscle weakness, impairment of motor action, and paresthesia, similar to effects discussed for *n*-hexane.

*5.4.2.2 Industrial Hygiene.* The OSHA PEL (148) is 400 ppm (1600 mg/m$^3$).

## 5.4.3 Rubber Solvent

Rubber solvent is a clear, colorless, and flammable liquid, somewhat less volatile than petroleum ether. Selected physical data are given in Table 21.22 and toxicity data in Table 21.23. It is used as a solvent in the manufacture of adhesives, brake linings, rubber cements, tires, intaglio inks, paints, and lacquers, and is used in degreasing operations (5).

When inhaled in large concentrations, rubber solvent causes symptoms similar to benzene in toxicity. In six of eight recorded deaths, findings in rubber workers included a nonsignificant incidence of myeloid leukemia (754).

The 4-hr $LC_{50}$ in the rat is 15,000 ppm (61,000 mg/m$^3$) with a no-effect level of 2800 ppm and 1500 ppm in the rat and dog, respectively (755). Signs and symptoms included CNS depression and convulsions in the rat and the cat (755).

The human odor threshold is 10 ppm (755). The ACGIH TLV (194) is 400 ppm (1590 mg/m$^3$) and the NIOSH REL (753) is 350 mg/m$^3$.

## 5.5.4 Varnish Makers' and Painters' Naphtha (VM&P Naphtha)

VM&P naphtha or varnish makers' and painters' naphtha is a colorless to yellow, flammable and explosive liquid with an aromatic odor and of the boiling range 95 to 160°C (5). Selected physical data are given in Table 21.22 and toxicity data in Table 21.23. VM&P naphtha is used extensively as a solvent for lacquers, varnishes, and quick-evaporating paint thinner. It is a direct distillation product containing $C_5$ to $C_{11}$ hydrocarbons. This solvent is mildly irritating to the eye and nose. Exposure to heated VM&P naphtha caused labored breathing in 18 of 19 individuals; two were cyanotic with general excitation, tremors, nausea, and hyperactivity (756). The symptoms subsided in 30 min, except for one worker. In humans, 880 ppm (4.1 mg/l) produced eye and throat irritation with temporary olfactory fatigue (757). In animal studies, temporary hematologic effects were noted.

The OSHA PEL (148) is 300 ppm (1350 mg/m$^3$) with a ceiling of 400 ppm (1800 mg/m$^3$) and the ACGIH TLV (194) is 300 ppm (758 mg/m$^3$). The odor threshold is 0.86 ppm (4 mg/m$^3$) (755).

The NIOSH recommends collection on charcoal, desorption with carbon disulfide, and quantification by flame ionization gas chromatography (195).

## 5.4.5 Petroleum Spirits

Petroleum spirits or mineral spirits, compose a fraction slightly lower in boiling point than Stoddard solvent. There is some confusion on fraction names. Petroleum spirits are a fraction containing paraffins, naphthenes, and aromatics (115). The toxicity of petroleum spirits is comparable with that of heptanes and octanes. Selected physical data are given in Table 21.22 and toxicity data in Table 21.23.

Generally, mineral spirits are mildly irritating to the gastrointestinal tract on ingestion and to the skin on contact. Systemically, the central nervous (269) and cardiac (753) system may be affected. Aspiration has been reported from one ingestion of products containing mineral spirits (752). Aplastic anemia and

# AROMATIC HYDROCARBONS 1409

thrombocytopenia were diagnosed in a worker who had used white spirits for cleaning floors for 4 months. The exposure proved fatal 3 months later (753). Exposure to 1000 to 2500 mg/m$^3$ caused only slight effects (753). Animal studies show even milder effects.

The NIOSH REL (148) is 200 ppm (350 mg/m$^3$). The odor threshold for mineral spirits is 15.8 ppm (55). Industrial hygiene sampling may be carried out as recommended for petroleum distillate (195).

### 5.4.6 Stoddard Solvent

Stoddard solvent, also called white spirits, is a colorless, flammable fluid with a kerosine-like odor. Selected physical data are given in Table 21.22 and toxicity data in Table 21.23. Stoddard solvent is used widely in dry-cleaning processes, and as a general cleaning and universal solvent.

The toxicity of Stoddard solvent resembles that of gasoline. It produced no effect on the human eye (485). Major manifestations are defatting, drying, scaling of the skin on direct contact, and possible development of dermatitis (615). On ingestion, aspiration into the lungs may occur, causing pneumonitis, pulmonary edema, and hemorrhage. Acute effects from inhaling large concentrations include nausea, vomiting, cough, and pulmonary irritation (217). Chronic exposures in humans have also resulted in hepatic and hematopoietic changes similar to the effects of the lower naphthas.

The OSHA PEL (148) and ACGIH TLV (194) is 100 ppm (525 mg/m$^3$). The odor threshold is 5.3 ppm and is irritating at 2100 ppm (55).

NIOSH (195) recommends collection on activated charcoal, desorption with carbon disulfide, and quantification using flame ionization gas chromatography.

### 5.4.7 140° Flash Naphtha

140°F flash naphtha is a slightly higher-boiling petroleum spirit fraction. Selected physical data are given in Table 21.22 and toxicity data in Table 21.23.

The toxicity of 140°F flash naphtha resembles that of Stoddard or petroleum spirits. No significant effects were observed in the rat, dog, or cat in chronic inhalation experiments (759).

A TLV of 37 ppm was recommended for an 8-hr day, 40-hr week exposure (750).

### 5.4.8 Aromatic Petroleum Naphthas

Aromatic petroleum naphthas, coal tar or pyrolysis naphthas, are manufactured in three boiling ranges (762). Aromatic petroleum naphthas are processed from high-boiling distillate fractions, containing mainly alkylbenzenes, cumeme, toluene, and xylene (760). Selected physical data are given in Table 21.22 and toxicity data in Table 21.23. They are used as chemical raw materials, as degreasing agents, in varnishes, lacquers, synthetic enamels, and lithography inks, and in textile printing (760). Aromatic petroleum naphthas are also used as solvents for herbicides, fungicides, and insecticides.

The toxicity of aromatic naphthas resembles that of benzene and the lower alkylbenzenes. Acute signs of toxicity are eye, nose, and throat irritation, vertigo, nausea, dyspnea, CNS depression, and neurotoxicity if benzene is present. Chronic signs are CNS depression and slight to severe changes of the hematopoietic system, also depending on the benzene content. Upon ingestion, the naphthas present aspiration hazards.

A series of acute, subacute, and subchronic studies using rodents, canines, and primates showed decreased white blood count, bone marrow effects, lung congestion, CNS depression, and isolated cataracts.

The OSHA PEL (148) is 100 ppm (400 mg/m$^3$). The odor threshold is 0.07 ppm, with recognition between 0.5 to 2.5 ppm, and 26 ppm for sensory effects. NIOSH (195) recommends collection on activated charcoal, desorption using carbon disulfide, and quantification using flame ionization gas chromatography.

### 5.4.9 Thinners

A series of thinners or naphthas were tested for inhalation and odor recognition properties. The thinners are aliphatic-type naphthas of the petroleum spirits range. Selected physical data are given in Table 21.22. They are clear to yellow, flammable liquids, used in paints, glues, varnishes, and lacquers and as general solvents or degreasing agents.

The toxicity of thinners resembles that of the mineral spirits, hexane, and benzene. Neurotoxicity was seen in 18 juveniles who had sniffed glue thinner (761), and motor defects had not resolved 8 months later. The general toxic effects are CNS and occasional myelotoxic effects.

Odor thresholds appear too high to serve as warning properties. A TLV of 150 ppm is suggested for thinner 80, a value also applicable to other thinner grades.

### 5.4.10 Naphthenic Aromatic Solvents

Two naphthenic solvents with boiling ranges of 157 to 183 and 151 to 200°C have been tested in basic inhalation studies. Selected physical data are given in Table 21.22.

## 5.5 Middle Distillate Products

### 5.5.1 Kerosine

Kerosine or kerosene, coal oil, No. 1 fuel oil, or mineral seal oil (5), is a white to pale yellow, mobile, flammable, and combustible liquid. Kerosine is produced by direct fractionation from the "middle distillate fraction" (5). The individual kerosine composition varies widely, but consists mainly of linear and branched alkanes, alkenes, cyclanes, and aromatics in the $C_{10}$ to $C_{16}$ range. Selected physical data are given in Table 21.22 and toxicity data in Table 21.23.

Kerosine is used widely as illuminating, heating, and cooking fuel, as a cleaning, degreasing, and mold release agent, as a solvent in asphalt coating, and for enamels, paints, polishes, thinners, and varnishes.The deodorized product is utilized mainly for household sprays, herbicides, insecticides, and pesticides. Of the heavier kerosines, mineral seal oil has been used as railway coach and caboose lamp fuel (5). It is used also medicinally for veterinary decontamination (762).

**5.5.1.1 Toxicity.** The toxicity of kerosines varies according to their composition. The deodorized and refined kerosines are least toxic. Others may contain benzene or alkylbenzenes, which result in hematopoietic or similar manifestations. Human ingestion of kerosine results in rapid absorption from the gastrointestinal tract, systemic effects, and possible aspiration into the lungs (185). A comparative ratio of oral to aspirated lethal doses may constitute 500 ml versus 5 ml (763). Systemic effects are gastrointestinal irritation, vomiting, diarrhea, and in severe cases, drowsiness and CNS depression, progressing to coma and death (730).

Signs of lung involvement include increased rate of respiration, tachycardia, and cyanosis (268). Innumerable cases of accidental kerosine ingestion by children have been reported. In 1962, 28,000 nonfatal poisoning cases in the United States were attributed to petroleum distillates, mainly kerosine (730). Complications include bacterial pneumonia (730) and pneumatoctes (752). In 22 of 52 human cases, an increase of gastric fluid level has been observed (764). Owing to the aspiration hazard, inducing vomiting and lavage are contraindicated in cases of acute ingestion (268). Preferred antidotes are charcoal (765, 766) and milk (765). The administration of ipecac has aided in some cases (767). For mixtures containing 43 percent or more kerosine, the aspiration hazard is acute (28).

Kerosine even on single contact defats the skin, which may lead to irritation, infection, and dermatitis (269, 768). Several cases of blistering or diffuse redness with edema have been reported in children (769). The acute blistering effect was confirmed in patch tests (770–772). Kerosine is not irritant to the eye (485).

Kerosine, including most of the fuel oils, is not sufficiently volatile to constitute an acute inhalation hazard. When heated or emitted as an aerosol or mist, kerosine may cause mucous membrane irritation and chemical pneumonitis (115). Several such cases have been recorded, after kerosine was used to massage extremities, resulting in aplastic anemia and death (753, 773).

Animal experiments demonstrate the low oral toxicity to the rat, rabbit, and chicken, especially of deodorized kerosine. Aspiration into the lungs may increase the oral $LD_{50}$ by a factor of 1:140, as seen in the rat (767). Experiments with primates have shown that aspiration into the lungs causes cellular damage (774). Kerosine aerosols have varying effects, depending on droplet size and composition, causing mucous membrane irritation to polyemia (775). Absorption is practically negligible through the intact skin (776), but moderate through abraded skin. Kerosine causes moderate to severe injury in prolonged or repeated dermal contact. Sublethal doses injected intratracheally in rats resulted in an acute exudative inflammation representing the reaction of the alveolar capillaries, which reaches its maximum in 3 days and subsides in 7 days. This is followed by a chronic proliferative inflammation, which reaches its maximum in 10 days and subsides but is still present at the end of 1 month (777).

In marine species, kerosine is less toxic than diesel oil and lower fractions (102). Kerosine is readily absorbed through the gastrointestinal tract of the primate (778) and by fish (779) and is readily distributed to fatty tissues.

Kerosine is utilized as the sole carbon source by *Pseudomonas aeruginosa* (780) and *P. pseudomallei* (781) and mutants of *Candida lipolytica* (782), but it inhibits the growth of *Blakeslea trispora* (783).

*5.5.1.2 Industrial Hygiene.* The NIOSH REL (148) is 14 ppm (100 mg/m$^3$). The odor threshold is 0.09 ppm (0.6 mg/m$^3$) and the sensory threshold is 20 ppm (0.14 mg/l) (784). The odor is irritating at 123 ppm (55).

Vapors can be collected on charcoal, desorbed with carbon disulfide, and quantified using ultraviolet spectral (254) or gas chromatographic analyses.

When kerosine is handled, prolonged skin contact should be avoided, although the surface should be thoroughly washed in case of accidental contact. Kerosine should never be siphoned by mouth.

### 5.5.2 Jet Fuel

The term jet fuel encompasses the aircraft turbine engine and jet fuels. They are composed of hydrocarbons from the middle distillate fraction in the kerosine range, with some components from the light distillates. They are composed of $C_5$ to $C_{16}$ aliphatics, monocyclanes, aromatics, and alkenes (5). Selected physical data are given in Table 21.24.

The toxicity of jet fuels resembles that of kerosine. However, neurological effects have been recorded, indicating the presence of hexane-like constituents. In one acute case, a jet pilot became intoxicated and the cockpit concentration was estimated to have been 300 to 7000 ppm of JP4 (785). Long-term worker exposure in the aircraft manufacture using jet fuels caused symptoms of dizziness, headache, nausea, palpitation, and pressure in the chest (753). Concentrations of the solvent in the air were estimated later at 500 to 3000 ppm, based on a molecular weight of 170. Clinical findings included neurasthenia, psychasthenia, and polyneuropathy (753).

In animal studies, subacute exposures with the Fischer 344 rat, the C57B1/6 female mouse, and the beagle dog at 0.15 and 0.75 mg/l for 90 days produced increased blood urea nitrogen in the rat and decreased serum albumin in the dog (786). Jet fuel JP9 additives RJ-4 and RJ-5 in a 6-month inhalation study caused increased liver and kidney weights and some pulmonary irritation in the rat and the dog (787); however, they produced no effects on chronic vapor exposure with the dog and the primate. A concentration of >0.05 mg/l of RJ-5 decreased the hatchability of flagfish (788).

### 5.5.3 Diesel Fuel

Diesel fuel or fuel oil No. 2 is a gas oil fraction obtained from the middle distillate in petroleum separation (5). The composition of the various grades differ in ratios

| Common Name | Viscosity (SUS) | B.P. (°F) | Flash Pt. (°C) | Freezing Point | Mol. Wt. | Carbon Number | Class of Components | Ref. |
|---|---|---|---|---|---|---|---|---|
| Jet fuel | | | | | | | | |
| JP-1 | <32 | 410–572 | 35—63 | −76 | | $C_5$–$C_{16}$ | Aliphatics, mono- and dicycloparaffins, alkylbenzenes | 2, 5 |
| JP-3 | | 240–470 | ~23 to ~1–76 | | | | Aliphatics, mono- and dicycloparaffins, alkylbenzenes | 2, 5 |
| JP-4 | | <290–470 | | −76 | ~160 | $C_5$–$C_{16}$ | Aliphatics, olefins, mono- and dicycloparaffins, alkylbenzenes | 5 |
| JP-5 | | 400–550 | 35–63 | −40 | ~170 | $C_5$–$C_{16}$ | Aliphatics, olefins, mono- and dicycloparaffins, alkylbenzenes | 2, 5 |
| JP-6 | | 250–500 | | −65 | | $C_5 C_{16}$ | Aliphatics, olefins, mono- and dicycloparaffins, alkylbenzenes | 5 |
| Diesel fuel | <32–45 | 350–750 | 38–54 | | | $C_5$–>$C_{16}$ | Aliphatics, olefins, mono- and dicycloparaffins, alkylbenzenes | 5 |
| Distillate heating oils | | | | | | | | |
| No. 1 | ~35 | 420–625 | 38–74 | | | $C_{11}$–>$C_{16}$ | | 5 |
| No. 2 | 35–55 | 363–634 | 38 min. | | | $C_9$–>$C_{16}$ | | 5 |
| Motor oil | 60–530 | 690–1090 | | | | $C_7$–>$C_{20}$ | | 5 |
| White oil spray | 186.6 (37.8°C) | | | | | | Aliphatics | 3 |
| Petroleum wax | 50–92.2 (210°C) | 425–580 | | 490–655 | | $C_{20}$–$C_{32}$ | Aliphatics | 5 |

of alkanes, alkenes, cyclanes, and aromatics. They are slightly viscous, brown, flammable fluids. Selected physical data are given in Table 21.24.

Diesel fuel is used for diesel or semidiesel, high-speed engines requiring a type of fuel with low viscosity and moderate volatility (5). The heavier grades are used for railroad and marine diesel engines.

The toxicity of diesel fuel resembles that of kerosine. No carcinogenic compounds were formed in yeast cultivated on diesel fuel or in the unsaponifiable fraction of the muscle and liver of chickens and pigs that had consumed yeasts grown on the diesel media (789). Mosquito larvae exhibited chromatin clumping and loss of the granular matrix (790).

### 5.5.4 Heating Oils

Heating oils are available in six grades, the selection depending on the type of use. No. 1 fuel oil, stove oil, which is mainly kerosine, is available for home heating using pot burners and stoves (5). Selected physical data are given in Table 21.24. Fuel oil No. 2 resembles diesel fuel, and is used in furnaces, burners, and semidiesel engines. Fuel oils 4, 5, and 6 are heavier grades, produced from the residual distillate fraction (5).

The toxicity of No. 1 fuel oil is essentially that of kerosine. Fuel oil No. 2 resembles regular diesel fuel. Similarly, there is a low oral, moderate dermal, and high aspiration hazard. No. 2 oil was more toxic toward estuarine shrimp than crude oil (791). The fuel was incorporated rapidly into the mussel and still detectable after 35 days, of which 14 days were in clean seawater (792). Similarly, it was taken up into clams and retained for more than 2 weeks (793).

## 5.6 Products from Lubricating Stock Distillates

### 5.6.1 White Oils

The lubricating oils are vacuum distillation fractions. When further refined and treated, white oils or medicinal oils are obtained. White oil, liquid paraffin or mineral oil, is a mixture of middle aliphatic hydrocarbons. White oil is a flammable, oily, colorless liquid. Selected physical data are given in Table 21.23.

White oils are practically inert, and therefore can be applied internally as a laxative and externally as a protectant and lubricant.

The OSHA PEL (148) and ACGIH TLV (194) is 5 mg/m$^3$ as an oil mist. ACGIH also recommends a STEL of 10 mg/m$^3$.

### 5.6.2 Paraffins (Waxes)

Upon dilution with naphtha or cooling the lubricating oil fraction, the paraffins or waxes solidify and thus can be separated from the oil. There are two types of waxes, the paraffin wax, found in the low-boiling petroleum fractions, and the microcrystalline wax, found in the high-boiling fractions (5). Paraffin wax, or hard wax, is a mixture of solid hydrocarbons, mainly alkanes. The wax is white, somewhat translucent, odorless, and flammable. Selected physical data are given in Table 21.24. Paraffin wax can be added to medicinal agents. Petroleum wax and petrolatum are the only hydrocarbons permitted for use in food products. Paraffin wax is used as a household wax and extensively as a coating for food containers and wrappers.

Petroleum waxes are inert; wax fumes are mild eye, nose, and throat irritants (484). Paraffin wax is biodegradable (794).

The OSHA PEL (148) and ACGIH TLV (194) is 2 mg/m$^3$ as a fine dust or condensed fume. An analytic procedure is available (253).

### 5.6.3 Lubricating Oils (Automotive and Aviation Oils)

The lubricating oils are manufactured from the medium lube distillate and fall into several functional categories. These are the intermittently used automotive, aviation, and tractor oils, the continuously servicing turbine oils, and specially prepared insulating and hydraulic fluids (5).

*5.6.3.1 Toxicity.* The lube oils are composed of aliphatic, olefinic, naphthenic, and aromatic hydrocarbons. Lubricating oil additives include antioxidants, bearing protectors, wear resistors, dispersants, detergents, viscosity index improvers, pour-point depressors, and antifoaming and rust-resisting agents (5).

The overall viscosity ranges from 75 to 3000 SUS. Therefore, their oral and dermal toxicities are very low. The oral $LD_{50}$ in rodents is greater than 10 g/kg and the dermal $LD_{50}$ greater than 15 g/kg. Inhalation does not present a problem, except if misting occurs, although frequent and prolonged direct skin contact may produce skin irritation and dermatitis (115) owing to certain additives (481). Aspiration is not likely to occur, except for the lower-boiling, light oils.

Although the aromatic content of motor oil is relatively high, it does not include polycyclic aromatics. Scrotal tumors were noted in men exposed to coal tar and polycyclic aromatics containing oil (795). However, no analytic data were given

A variety of microorganisms, such as *Pseudomonas*, *Aeromonas*, *Aerobacter*, and *Xanthomas*, grow on lubricating oils (796).

*5.6.3.2 Industrial Hygiene.* The OSHA PEL (148) and ACGIH TLV (194) is 5.0 mg/m$^3$ and ACGIH recommends a STEL of 10 mg/m$^3$. NIOSH (195) recommends collection on a membrane filter, extraction using chloroform, and quantification using fluorescence spectroscopy. Emission photometric and gas chromatographic procedures also have been used (797, 798).

Handling of lubricant oils requires precautionary measures if misting occurs or vapors are formed.

### 5.6.4 Cutting Oils

Cutting oils are fluids that lubricate and cool the machining of metal. The toxicity of cutting oils resembles that of lubricating oils. They fall into three categories: (*a*) the water-insoluble petroleum blends (mineral oils), the straight cutting oils, and the concentrates; (b) the water-soluble petroleum sulfonates and the chemical emulsions; and (c) the synthetic fluids. They are composed of mineral and lard oil, sulfur compounds, and chlorine or chlorinated organic compounds (5)

*5.6.4.1 Toxicity.* Although the parent petroleum cut of the cutting oils is of low-order toxicity, the final products, with the additives, have caused contact dermatitis (799–801). This may be due to physical or physiological blocking of hair follicles progressing to folliculitis (201, 483). The problem may start on dorsal surfaces of hands and arms, later include forearms, thighs, and the abdomen, and include formation of perifollicular papules and pustules. Melanosis may develop later (483). The nickel and chromium salt additives may be the causative elements (802).

Separate from causing dermal effects, cutting oils are associated with skin (803) and scrotal cancer in Europe (804). However, the occurrence in the United States has been very low (801) and may be due to continuous improvements of additives (805) and decreasing the aromatic portions of the oils (806). Also, the term mineral oil in Great Britain has been used interchangeably with heavy or aromatic oil.

Repeated application of commercial cutting oil to mouse skin produced dysplasia or malignancy in 48 percent of the treated group and 8 percent in the controls (806).

*5.6.4.2 Industrial Hygiene.* Reviews on cutting oil degradation are available (807), including methods to reclaim the oil (808). When cutting oils are handled, cutaneous reactions largely can be prevented by good personal hygiene. This consists of minimized contact, prompt removal of oil from the skin with soap and water, and wearing clean work clothing (115) and protective shields (809). Protective skin creams may also be used (800, 810). If dermatitis should occur, prompt and expert medical attention should be sought (809). Prospective employees with a significant history of dermatitis or preexisting skin disorders should be protected from exposure where direct dermal contact with cutting oils, lubricants, or coolants is likely to occur (809).

### 5.6.5 Petrolatum

Petrolatum or petroleum jelly is the oldest marketed petroleum product. It is an odorless, tasteless, viscous, yellow to amber semisolid material.

Petrolatum is used widely in the pharmaceutical, medicinal, and household areas. Industrially, it is used in skin protective coatings. It is inert and is noncarcinogenic (811). It is nonallergenic and nonirritating and is thus utilized as a dermal test vehicle. Sample substances may remain allergenic in petrolatum for at least 1 year (201).

### 5.7 Residual Oils

Residual or heavy oils are the basis for the heavier fuel oils Nos. 4, 5, and 6, the bunker C, and railroad oils.

They are of low-order toxicity. Repeated or prolonged dermal contact may have systemic effects because of their aromatic content.

## REFERENCES

1. D. R. Lide, Ed., *CRC Handbook of Chemistry and Physics*, CRC Press, Cleveland, OH, 1991–1992.
2. *Laboratory Waste Disposal Manual*, 2nd rev. ed., Manufacturing Chemists Association, Inc., Washington, DC, 1974.
3. S. Budavari, Ed., *The Merck Index*, 11th ed., Merck and Company, Rahway, NJ, 1989.
4. *Registry of Toxic Effects of Clinical Substances*, U.S. Dept. Health, Education, and Welfare, Cincinnati, OH, September, 1991.
5. V. B. Guthrie, Ed., *Petroleum Products Handbook*, McGraw-Hill, New York, 1960.
6. N. Sax, *Dangerous Properties of Industrial Materials*, 7th ed., Van Nostrand Reinhold, New York, 1987.
7. *Toxicology and Hazardous Industrial Chemicals Safety Manual for Handling and Disposal with Toxicity Data*, International Technical Information Institute, publ. Japan, 1976.
8. *Fire Protection Guide on Hazardous Chemicals*, National Fire Protection Association, Quincy, MA, 1991.
9. *Handbook of Organic Industrial Solvents*, 4th ed., American Mutual Insurance Alliance, Chicago, 1972.
10. F. Rossini, K. Pitzer, R. Arnett, R. Braun, and G. Pimental, *Selected Values of Physical Thermodynamic Properties of Hydrocarbons and Related Compounds*, Carnegie Press, Pittsburgh, PA, 1953.
11. J. Rosenberg, "Solvents," in *Occupational Medicine*, J. LaDou, Ed., Appleton & Lange, Norwalk, CT, 1990, pp. 359–386.
12. L. R. Goldfrank, A. G. Kulberg, and E. A. Bresnitz, "Hydrocarbons," *Goldfrank's Toxicologic Emergencies*, 4th ed., in L. R. Goldfrank, N. E. Flourenbaum, N. A. Lewin, R. S. Weisman, and M. A. Howland, Eds., Appleton & Lange, Norwalk, CT, 1990, pp. 759–780.
13. P. D. Bryson, *Comprehensive Review of Toxicology*, 2nd ed., Aspen Publishers, Rockville, MD, 1989, 553–563.
14. C. R. Noller, *Chemistry of Organic Compounds*, 3rd ed., Saunders, Philadelphia, 1966.
15. N. I. Sax and R. J. Lewis, Sr., Eds., *Hawley's Condensed Chemical Dictionary*, 11th ed., Van Nostrand Reinhold, New York, 1987.
16. E. W. Flick, *Industrial Solvents Handbook*, 3rd ed., Noyes Publications, Park Ridge, NJ, 1985.
17. J. H. Kuney, Ed., *Chemcyclopedia 90*, American Chemical Society, Washington, DC, 1990.
18. International Labor Office, *Encyclopedia of Occupational Health and Safety*, Vols. I & II, International Labor Office, Geneva, 1983.
19. U.S. Coast Guard, Department of Transportation, *CHRIS—Hazardous Chemical Data*, Vol. II, U.S. Government Printing Office, Washington, DC, 1984–1985.
20. K. Verschueren, *Handbook of Environmental Data of Organic Chemicals*, 2nd ed., Van Nostrand Reinhold, New York, 1983.

21. *Hazardous Substance Data Bank*, TOXNET System, National Library of Medicine, Bethesda, MD, 1992.
22. F. Rossini, K. Pitzer, R. Arnett, R. Braun, and G. Pimentel, *Selected Values of Physical Thermodynamic Properties of Hydrocarbons and Related Compounds*, Carnegie Press, Pittsburgh, 1953.
23. L. Mitterhauszerova, K. Fralova, A. Sulkova, and L. Krasnec, *Acta Fac. Pharm. Univ. Comenianae*, **25**, 9(1974).
24. H. Elmenhorst and H. P. Harke, *Z. Naturforsch.*, **23b**, 1271 (1968).
25. J. R. Newsome, V. Norman, and C. H. Keith, *Tobacco Sci.*, **9**, 102 (1965).
26. E. Lahmann, B. Seifert, and D. Ulbrich, *Proc. Int. Clean Air Congr.*, *4th*, Tokyo, Japan, 595 (1977).
27. H. W. Gerarde, *Am. Med. Assoc. Arch. Ind. Health*, **19**, 403 (1959).
28. H. W. Gerarde, *Arch. Environ. Health*, **6**, 329 (1963).
29. H. W. Gerarde, *Toxicology and Biochemistry of Aromatic Hydrocarbons*, Elsevier, London, 1960.
30. R. Smolik, A. Lange, W. Zatonski, I. Juzwiak, and L. Andreasik, *Pol. Tyg. Lek.*, **28**(21), 769 (1973).
31. J. M. Neff, B. A. Cox, D. Dixit, and J. W. Anderson, *Mar. Biol*, **38**(3), 279 (1976).
32. D. J. Crisp, A. O. Christie, and A. F. A. Ghobasky, *Comp. Biochem. Physiol.*, **22**, 629 (1967).
33. C. E. ZoBell, *API Proc.—Joint Conf. Prev. Control Oil Spills*, 317 (1969).
34. "Hazardous Materials Regulations," Department of Transportation Bureau, *Fed. Reg.*, **41**, 57018 (1976).
35. I. Kesy-Dabrowska, *Rocz. Panstw. Zakl. Hig.*, **24**(3), 337 (1973).
36. C. T. Pate, B. Atkinson, and J. N. Pitts, Jr., *J. Environ. Sci. Health, Part A*, **A11**(1), 1 (1976)
37. E. G. Ivanyuk and A. I. Kobzar, *Gig. Sanit.*, **2**, 63 (1974).
38. W. F. Feldman, U.S. Pat. 4005994, February 1, 1977.
39. H. E. Runion, *Am. Ind. Hyg. Assoc. J.*, **36**, 338 (1975).
40. R. J. Young, R. A. Rinsky, P. F. Infante, and J. K. Wagoner, *Science*, **199**(4326), 248 (1978).
41. *Fed. Reg.*, **43**(98), 21838 (1978).
42. J. P. Conkle, B. J. Camp, and B. E. Welch, *Arch. Environ. Health*, **30**(6), 290 (1975).
43. B. Williams, L. G. Dring, and R. T. Williams, *Biochem. J.*, **127**(2), 24 (1972).
44. P. S. Jaglan, J. L. Nappier, R. E. Hornish, and A. R. Friedmann, *J. Agric. Food Chem.*, **25**(4), 963 (1977).
45. W. Bertch, R. C. Chang, and A. Zlatkis, *J. Chromatogr. Sci.*, **12**, 175 (1974).
46. S. Pilar and W. F. Graydon, *Environ. Sci. Technol.*, **7**(7), 628 (1973).
47. I. R. Tabershaw, F. Ottoboni, and W. C. Cooper, *Air Quality Monographs*, No. 69-5, February, 1969.
48. R. F. Gould, Ed., *Refining Petroleum for Chemicals*, American Chemical Society, Washington, DC, 1970.
49. *Encyclopaedia of Occupational Health and Safety*, Vols. I and II, International Labor Office, Geneva, 1983.

50. W. F. Von Oettingen, *Publ. Health Bull. No. 255*, U.S. Public Health Service, Washington, DC, 1940.
51. H. Thienes and T. J. Haley, *Clinical Toxicology*, Lea & Febiger, Philadelphia, 1972.
52. S. Moeschlin, Ed., *Poisoning, Diagnosis and Treatment*, 1st Am. ed., Grune and Stratton, New York, 1965.
53. D. C. Anthony and D. G. Graham, "Toxic Response of the Nervous System," in *Casarett and Doull's Toxicology*, 4th ed., M. O. Amdur, J. Doull, and C. D. Kalssen, Eds., Pergamon, New York, 1991, pp. 407–429.
54. *Registry of Toxic Effects of Chemical Substances*, On-Line, TOXNET System, National Library of Medicine, Bethesda, MD, 1992.
55. J. H. Ruth, *Am. Ind. Hyg. Assoc. J.*, **47**, A-142-A-151 (1986).
56. R. E. Gosselin, R. P. Smith, and H. C. Hodge, *Clinical Toxicology of Commercial Products*, 5th ed., Williams and Wilkins, Baltimore, 1984.
57. *Documentation of the Threshold Limit Values for Substances in Workroom Air*, 3rd ed., American Conference of Governmental Industrial Hygienists, Cincinnati, 1984.
58. L. S. Andrews and R. Snyder, "Toxic Effects of Solvents and Vapors," in *Casarett and Doull's Toxicology*, 4th ed., M. O. Amdur, J. Doull, and C. D. Kalssen, Eds., Pergamon, New York, 1991, pp. 681–722.
59. K. Mortelmans, S. Haworth, T. Lawlor, W. Speck, B. Tainer, and E. Zeiger, *Environ. Mutagen.*, **8**(Suppl. 7), 1–119 (1986).
60. R. Snyder, Ed., *Ethel Browning's Toxicity and Metabolism of Industrial Solvents*, 2nd ed., Vol. 1, *Hydrocarbons*, Elsevier, Amsterdam, 1987.
61. A. G. Gilman, T. W. Rall, A. S. Nies, and P. Taylor, Eds., *The Pharmacological Basis of Therapeutics*, 8th ed., Pergamon, New York, 1990.
62. W. B. Deichmann and H. W. Gerarde, *Toxicology of Drugs and Chemicals*, Academic, New York, 1969.
63. C. E. Searle, *Chemical Carcinogens*, American Chemical Society, Washington, DC, 1976.
64. International Agency for Research on Cancer, Suppl. 7, Geneva, 1987, pp. 136–137.
65. J. B. Lurie, *S. Afr. J. Clin. Sci.*, **3**, 212 (1952).
66. M. M. Key, A. F. Henschel, J. Butler, R. N. Ligo, and I. R. Tabershaw, Eds., *Occupational Diseases, a Guide to their Recognition*, NIOSH, Washington, DC, 1977.
67. H. H. Cornish, Chapter 19 in *Toxicology, the Basic Science of Poisoning*, L. J. Casarett and J. Doull, Eds., Macmillan, New York, 1975.
68. A. Sato and Y. Fugiwara, *Sangyo Igaku*, **14**(3), 114 (1972).
69. K. Nomiyama and H. Nomiyama, *Int. Arch. Arbeitsmed.*, **32**(1–2), 85 (1974).
70. *API Toxicological Review of Benzene*, American Petroleum Institute, New York, 1960.
71. A. Sato, T. Nakajima, Y. Fugiwara, and K. Hirosawa, *Int. Arch. Arbeitsmed.*, **33**(3), 169 (1974).
72. J. Tauber, *J. Occup. Med.*, **12**(3), 91 (1970).
73. C. L. Winek and W. D. Collom, *J. Occup. Med.*, **13**, 5 (1971).
74. C. L. Winek, W. D. Collom, and C. H. Wecht, *Lancet*, **1**, 683 (1967).
75. B. Block and G. F. Tadjer, *Harefuah*, **89**(2), 74 (1975).
76. C. Rozman, S. Woessner, and J. Saez-Serrania, *Acta Haematol.*, **40**(4), 234 (1968).
77. E. M. Uyeki, A. E. Ashkar, D. W. Shoeman, and T. U. Bisel, *Toxicol. Appl. Pharmacol.*, **40**(1), 49 (1977).

78. R. I. Volchkova, *Tr. Voronszh. Med. Inst.*, **87**, 29 (1972).
79. I. D. Gadaskina, A. Z. Buzina, and O. N. Dorofeeva, *Gig. Sanit.*, **3**, 30 (1973).
80. G. K. Kadyrow and M. I. Safarov, *Izv. Akad. Nauk, Azerb. SSR, Ser. Biol. Nauk.*, **3**, 109 (1972).
81. R. Snyder, in *Symposium on Toxicology of Benzene and Alkylbenzenes*, D. Braun, Ed., Industrial Health Foundation, Pittsburgh, 1974, pp. 44–53.
82. J. M. Wildman, M. L. Freedman, J. Roseman, and B. Goldstein, *Res. Commun. Chem. Pathol. Pharmacol.*, **13**(3), 473 (1976).
83. E. T. Kimura, D. H. Ebert, and P. W. Dodge, *Toxicol. Appl. Pharmacol.*, **19**, 699 (1971).
84. M. A. Wolf, V. K. Rowe, D. D. McCollister, R. C. Hollingsworth, and F. Oyen, *Am. Med. Assoc. Arch. Ind. Health*, **14**, 387 (1956).
85. R. H. Steele and D. Wilhelm, *J. Exp. Pathol.*, **47**(6), 612 (1966).
86. A. Bjornberg and H. Mobackin, *Berufsdermatosen*, **21**(6), 245 (1973).
87. Y. C. Manyashin, M. F. Savchenkov, and G. Sidnev, *Farmakol. Toksikol.*, **31**(2), 250 (1968).
88. J. L. Egle and B. J. Gochberg, *J. Toxicol. Environ. Health*, **1**(3), 531 (1976).
89. R. T. Drew and J. R. Fouts, *Toxicol. Appl. Pharmacol.*, **27**(1), 183 (1974).
90. C. P. Carpenter, C. B. Shaffer, C. S. Weil, and H. F. Smyth, Jr., *J. Ind. Hyg. Toxicol.*, **26**, 69 (1944).
91. B. K. Leong, *J. Toxicol. Environ. Health*, Suppl. 2, 45 (1977).
92. L. H. Nahum and H. E. Hoff, *J. Pharmacol. Exp. Ther.*, **50**, 336 (1934).
93. V. Morvai, A. Hudak, G. Ungvary, and B. Varga, *Acta Med. Acad. Sci. Hung.*, **33**(3), 275 (1976).
94. M. Peronet, *J. Pharm. Chim.*, **21**, 503 (1935).
95. J. Jonek, Z. Olkowski, and B. Zieleznik, *Acta Histochem,*. **20**, 286 (1965).
96. M. Kaminski, A. Karbowski, and J. Jonek, *Folia Histochem. Cytochem.* (Drakow), **8**(1), 63 (1970).
97. ATSDR, *Toxicological Profile for Benzene*, (Update), U.S. Dept. of Health and Human Services, Atlanta, GA, 1992.
98. H. H. Cornish and R. C. Ryan, *Toxicol. Appl. Pharmacol.*, **7**(6), 767 (1965).
99. C. Harper, R. T. Drew, and J. R. Fouts, *Drug Metab. Dispos.*, **3**(5), 381 (1973).
100. J. A. Timbrell and J. R. Mitchell, *Xenobiotica*, **7**(7), 415 (1977).
101. R. D. Meyerhoff, *J. Fish. Res. Board Can.*, **32**(10) 1864 (1975).
102. C. J. Barnett and J. E. Kontogiannis, *Environ. Pollut.*, **8**(1), 45 (1975).
103. L. Greenburg, M. Mayers, L. Goldwater, and A. Smith, *J. Ind. Hyg. Toxicol.*, **21**(8), 395 (1939).
104. M. Aksoy, S. Erdem, T. Akgun, O. Olur, and K. Dincol., *Blut*, **13**, 85 (1966).
105. G. P. Biscaldi, G. R. D. Cuna, and G. Pollini, *Haematol. Arch.*, **54**(8), 579 (1978).
106. E. Guberan and P. Kocher, *Schwiez. Med. Wochenschr.*, **101**, 1789 (1971).
107. A. Forni, E. Pacifico, and A. Limonta, *Arch. Environ. Health*, **22**, 373 (1971).
108. M. Aksoy, K. Dincol, S. Erdem, T. Akgun, and G. Dincol, *Brit. J. Ind. Med.*, **29**, 56 (1972).
109. E. C. Vigiliani and G. Saita, *N. Engl. J. Med.*, **271**(17), 872 (1964).

110. M. Askoy, S. Erdem, K. Dincol, T. Hepyuksel, and G. Dincol, *Blut*, **28**, 293 (1974).
111. M. Aksoy, K. Dincol, T. Akgun, S. Erdem, and G. Dincol., *Brit. J. Ind. Med.*, **28**, 296 (1971).
112. L. Roth, P. Turcanu, I. Dinu, and G. Moise, *Folia Haematol.*, **100**, 213 (1973).
113. B. D. Goldstein, *J. Toxicol. Environ. Health*, **2**, 69 (1977).
114. S. Pawelski, E. Wechrzycka, B. Modzewski, and S. Roszkowski, *Pol. Tyg. Lek.*, **19**(38), 1433 (1964).
115. F. A. Patty, Ed., *Industrial Hygiene and Toxicology*, Vol. II, Wiley-Interscience, New York, 1963.
116. S. Laham, *Ind. Med.*, **39**(5), 61 (1970).
117. P. Mikulski, R. Wiglusz, A. Bublewska, and J. Uselis, *Biul. Inst. Mid. Morsk Gdansk*, **23**(1/2), 67 (1972).
118. N. L. Kanner, *Gig. Tr. Prof. Zabol.*, **15**(10), 60 (1971).
119. *Review of the Health Effects of Benzene*, National Research Council, Commitee on Toxicology, National Academy of Sciences, Washington, DC, 1975.
120. G. M. Rusch, B. K. J. Leong, and S. Laskin, *J. Toxicol. Environ. Health Suppl.*, **2**, 23 (1977).
121. A. Brzecki, S. Misztel, and R. Kostolowski, *Ann. Acad. Med. Lodz.*, **14**(1), 55 (1973).
122. E. A. Drogichina, L. A. Zorina, and I. A. Gribova, *Gig. Tr. Prof. Zabol.*, **15**(5), 18 (1971).
123. C. F. Reinhardt, L. S. Mullins, and M. E. Maxfield, *J. Occup. Med.*, **15**(12), 953 (1973).
124. A. M. Monaenkova and K. V. Glotova, *Gig. Tr. Prof. Zabol.*, **13**(11), 32 (1969).
125. A. M. Monaenkova and L. A. Zorina, *Gig. Tr. Prof. Zabol.*, **4**, 30 (1975).
126. B. J. Dowty and J. L. Laseter, *Pediatr. Res.*, **10**, 696 (1976).
127. A. Forni and L. Moreo, *Eur. J. Cancer*, **3**, 251 (1967).
128. E. C. Vigliani and A. Forni, *J. Occup. Med.*, **11**(3), 148 (1969).
129. M. Sellyei and E. Kelemen, *Eur. J. Cancer*, **7**(1), 83 (1971).
130. E. C. Vigliani, *Ann. N.Y. Acad. Sci.*, **271**, 143 (1976).
131. P. A. Goguel, A. Cavigneaux, and J. Bernard, *Nouv. Rev. Fr. Hematol.*, **7**(4), 465 (1967).
132. M. Aksoy, K. Dincol, S. Erdem, and G. Dincol, *Am. J. Med.*, **21**, 160 (1970).
133. M. Aksoy, S. Erdem, G. Erdogan, and G. Dincol, *Human Heredity*, **24**, 70 (1974).
134. M. Aksoy, S. Erdem, and G. Dincol, *Blood*, **44**(6), 837 (1974).
135. M. Aksoy, S. Erdem, and G. Dincol, *Acta Haematol.*, **55**, 65 (1976).
136. M. Aksoy, *Lancet*, **1**, 441 (1978).
137. M. Aksoy, S. Erdem, G. Erdogan, and G. Dincol, *Human Heredity*, **26**, 149 (1976).
138. F. A. Walker, *Ann. Intern. Med.*, **85**(3), 404 (1975).
139. J. J. Thorpe, *J. Occup. Med.*, **16**(6), 375 (1974).
140. I. M. Tough, P. G. Smith, W. M. C. Brown, and D. G. Harnden, *Eur. J. Cancer*, **6**(1), 49 (1970).
141. H. Kahn and M. H. Khan, *Arch. Toxikol.*, **31**(1), 39 (1973).
142. A. M. Forni, A. Cappellini, E. Pacifico, and E. C. Vigliani, *Arch Environ. Health*, **23**, 358 (1971).
143. A. Forni and L. Moreo, *Eur. J. Cancer*, **5**(5), 459 (1969).

144. S. R. Wolman, *J. Toxicol. Environ Health Suppl.*, **2**, 63 (1977).
145. M. L. Freedman, *J. Toxicol. Environ. Health*, Suppl. 2, 37 (1977).
146. R. Snyder, E. W. Lee, J. J. Kocsis, and C. M. Witmer, *Life Sci.*, **21**(12), 1709 (1977).
147. R. V. Sapozhkov, *Tr. Sarat. Med. Inst.*, **71–88**, 18 (1970).
148. *Occupational Safety and Health Standards Subpart Z—Toxic and Hazardous Substances*, CFR, Title 29, Sec. 1910.93, 1992.
149. W. A. Fishbeck, R. R. Langner, and R. J. Kociba, *Am. Ind. Hyg. Assoc. J.*, **36**(11), 820 (1975).
150. P. A. Miescher, *Semin. Haematol.*, **10**(4), 311 (1973).
151. E. J. Gralla, *Vet. Clin. North Am.*, **5**(4), 699 (1975).
152. E. Browning, *J. Occup. Med.*, **7**, 554 (1965).
153. V. I. Boiko, L. M. Makareva, Z. G. Pedrez, M. Y. Burdygina, and V. A. Sukhanova, *Kazan. Med. Zh.*, **1**, 77 (1975).
154. L. A. Tepikina and M. S. Gordeeva, *Gig. Sanit.*, **3**, 23 (1978).
155. G. G. Avilova, I. P. Ulanova, E. E. Sarkisyants, and E. A. Karpukhina, *Gig. Sanit.*, **6**, 310774 (1974).
156. Y. E. Yakushevich, *Gig. Sanit.*, **38**(4), 6 (1973).
157. I. Gut, *Cesk. Hyg.*, **16**(6), 183 (1971).
158. A. Sato, T. Nakajima, Y. Fujiwara, and N. Nuragama, *Brit. J. Ind. Med.*, **32**, 321 (1975).
159. W. Laase, *Pharmazie*, **28**, 10 (1973).
160. A. Sato, Y. Fujiwara, and T. Nakajima, *Sangyo Igaku*, **16**(1), 30 (1974).
161. J. J. Kocsis, E. W. Lee, M. D'Souza, and R. Snyder, *Fed. Proc. Abstr.*, **37**(3), 505 (1978).
162. H. Kahn and I. Muzyka, *Work Environ. Health*, **10**(3), 140 (1973).
163. R. Snyder and M. Ideda, Int. Workshop on Toxicity of Benzene, Paris, November 9–11, 1976.
164. W. J. Canady, D. A. Robinson, and H. D. Colby, *Biochem. Pharmacol.*, **23**, 3075 (1974).
165. L. A. Tiunov, S. P. Nechiporenko, Z. I. Menshikova, N. M. Petushkov, V. A. Ivanova, T. S. Kolosova, and M. A. Akhmatova, *Farmakol. Toksikol.*, **40**(1), 97 (1977).
166. J. J. Jonek, M. Kaminski, H. Grzybek, P. Panz, and B. Gruszeczka, *Acta Histochem.* (Jena), **55**(1), 60 (1976).
167. I. Gut, *Arch. Toxicol.*, **35**(3), 195 (1975).
168. E. W. Lee, L. S. Andrews, C. M. Witmer, F. W. Deckert, J. J. Kocsis, and R. Snyder, *Fed. Proc.*, **34**(3), 227 (1975).
169. S. M. Gusman and V. A. Zhukov, *Strukt, Funkts. Gisto-Gematicheskikh Barerov, Mater. Soveshch. Probl. Gisto-Gematicheskikh Barerov*, 87 (1971).
170. M. J. T. Fitzgerald, J. C. Folan, and T. M. O'Brien, *J. Invest. Dermatol.*, **64**, 169 (1975).
171. V. N. Russkikh and A. V. Rodnikov, *Tr. Nauch. Konf. nauch-Issled. Inst. Gig. Vod. Transp.*, **2**, 137 (1972).
172. O. D. Laerum, *Acta Pathol. Microbiol. Scand.*, **81**(1), 57 (1973).
173. E. S. Reynolds, *Biochem. Pharmacol.*, **21**(19), 2555 (1972).
174. J. Jonek, J. Sroczynski, M. Kaminski, and H. Grzybek, *Arch. Mal. Prof. Med.*, **32**(9), 517 (1971).
175. F. Ito, *Showa Igakkai Zasshi*, **22**, 278 (1962).

176. V. Muzyka, *Vopr. Med. Khim.*, **15**(5), 521 (1969).
177. L. M. Bernshtien, *Vopr. Gig. Tr. Profzabol., Mater. Nauch. Konf.*, 53 (1972).
178. J. Sroczynski, K. Zapisz, G. Jonderko, and A. Wegiel, *Patol. Pol.*, **27**(1), 59 (1976).
179. M. Berlin, J. Gage, and E. Johnson, *Work Environ. Health*, **11**(1), 1 (1974).
180. V. N. Frash, B. K. Yushkov, and A. V. Karaulov, *Gig. Tr. Prof. Zabol.*, **12**, 44 (1976).
181. B. Speck, T. Schnider, U. Gerber, and S. Moeschlin, *Schwiez. Med. Wochenschr.*, **96**(38), 1274 (1966).
182. F. J. Fork, H. S. Cohen, J. Rosman, and M. C. Freedman, *Blood*, **47**(1), 145 (1976).
183. E. S. Tikhachek and V. N. Frash, *Gig. Tr. Prof. Zabol.*, **17**(8), 30 (1973).
184. V. B. Dobrokhotov and M. I. Enikeev, *Gig. Sanit.*, **1**, 32 (1977).
185. J. P. Lyon, *Diss. Abstr. Int. B.*, **36**(11), 5537 (1975).
186. J. H. Ward, J. H. Weisburger, R. J. Yamamoto, T. Benjamin, C. A. Brown, and E. K. Weisburger, *Arch. Environ. Health*, **30**, 22 (1975).
187. D. Tarin, *Int. J. Cancer*, **3**(6), 734 (1968).
188. D. K. Garibyan and S. A. Papoyan. *Nekot, Itogi. Izuch. Zagryazneniya Vnesh. Sredy Kanstserogen. Veshchestvami*, 112 (1972).
189. B. L. Van Duuren, N. Nelson, L. Orris, E. D. Palmer, and F. L. Schmitt, *J. Natl. Cancer Inst.*, **31**(1), 41 (1963).
190. D. T. Gibson, J. R. Koch, and R. E. Kallio, *Biochem.*, **7**(7), 2653 (1968).
191. W. C. Evans, *Nature*, **270**(5632), 17 (1977).
192. F. Jansen and A. Olsen, *Plant Physiol.*, **44**(5), 786 (1969).
193. P. Tkhelidze, *Soobsch. Akad. Nauk. Gruz. SSR*, **56**(3), 697 (1969).
194. *Threshold Limit Values for Chemical Substances and Physical Agents and Biological Exposure Indices for 1991–92*, American Conference of Governmental Industrial Hygienists, Cincinnati, OH, 1991.
195. *NIOSH Manual of Analytical Methods*, 3rd ed., Vols. 1 and 2, U.S. Dept. Health, Education, and Welfare, U.S Printing Office, Washington, DC, 1985.
196. *Criteria for Recommended Standard—Occupational Exposure to Benzene*, U.S. Dept. Health, Education, and Welfare, Cincinnati, OH, 1974.
197. H. M. D. Utidjian, *J. Occup. Med.*, **18**, 7 (1976).
198. P. Walker, *U.S. Natl. Tech. Inf. Serv. Publ. Bull. Rep. Issue PL-256735* (1976).
199. H. W. Gerarde and D. B. Ahlstrom, *Toxicol. Appl. Pharmacol.*, **9**(1), 185 (1966).
200. ATSDR, Toxicological Profile for Toluene, U.S. Dept. of Health and Human Services, Atlanta, GA, 1992.
201. A. A. Fisher, *Contact Dermatitis*, 2nd ed., Lea & Febiger, Philadelphia, 1973.
202. H. F. Smyth and H. F. Smyth, Jr., *J. Ind. Hyg.*, **10**, 261 (1928).
203. T. Dalhamn, M. C. Edfors, and R. Rylander, *Arch. Environ. Health*, **17**, 746 (1968).
204. E. Reisin, A. Teicher, R. Jaffe, and H. E. Eliahous, *Brit. J. Ind. Med.*, **32**, 163 (1975).
205. M. G. Casarett, "Social Poisons," Chapter 25 in *Toxicology, the Basic Science of Poisons*, L. J. Casarett and J. Doull, Eds., Macmillan, New York, 1975.
206. J. Soderlund, *Int. J. Occup. Health Safety*, **3**, 42–55 (1975).
207. I. Astrand, H. Ehrner-Samuel, A. Kilbom, and P. Ovrum, *Work Environ. Health*, **9**(3), 119 (1972).

208. F. Gamberale and M. Hultengren, *Work Environ. Health*, **9**(3), 131 (1972).
209. I. Elster, *Dtsch. Med. Wochenschr.*, **97**, 1887 (1972).
210. S. M. Taher, R. J. Anderson, R. McCartney, M. M. Popovtzer, and R. W. Schreier, *N. Engl. J. Med.*, **290**, 765 (1974).
211. W. J. O'Brien, W. B. Yeoman, and J. A. E. Hobby, *Brit. Med. Jr.*, **2**, 29 (1971).
212. T. W. Kelly, *Pediatrics*, **56**, 605 (1975).
213. A. Friborska, *Folia Haematol.*, **99**(2-3), 233 (1973).
214. A. Vlastiborova and A. Friborska, *Folia Haematol.*, **99**(2-3), 230 (1973).
215. D. Pey, *Arch. Mal. Prof. Med. Trav. Secur. Soc.*, **33**(10-11), 584 (1972).
216. I. Juzwiak and W. Fedorowicz, *Czas Stomatol.*, **27**(8), 855 (1974).
217. S. R. Cohen and A. A. Maier, *J. Occup. Med.*, **16**(3), 201 (1974).
218. L. Greenberg, M. R. Mayers, H. Heimann, and S. Moskowitz, *J. Am. Med. Assoc.*, **118**, 573 (1942).
219. W. F. Von Oettingen, P. A. Neal, and D. D. Donahue, *J. Am. Med. Assoc*, **118**(8), 579 (1942).
220. M. Sato, *Sangyo Igaku*, **15**(3), 261 (1973).
221. D. Szadkowski, R. Pett, J. Angerer, A. Manz, and G. Lehnert, *Int. Arch. Arbeitsmed.*, **31**(4), 265 (1973).
222. M. Ogata, K. Tomokuni, and Y. Takatsuka, *Brit. J. Ind. Med.*, **27**(1), 43 (1970).
223. M. Ogata, Y. Takatsuka, and K. Tomokuni, *Brit. J. Ind. Med.*, **28**, 382 (1971).
224. H. B. Elkins, *The Chemistry of Industrial Toxicology*, 2nd ed., J. Wiley, New York, 1959.
225. M. Ikeda and H. Ohtsuji, *Brit. J. Ind. Med.*, **26**, 244 (1969).
226. A. Capellini and L. Alessio, *Med. Lav.*, **62**(4), 196 (1971).
227. E. DeRosa, M. Mazzotta, F. Forin, and M. A. Corradina, *Lav. Um.*, **27**(1), 18 (1975).
228. D. M. Jerina, N. Kaubisch, and J. W. Daly, *Proc. Natl. Acad. Sci. U.S.*, **68**(10), 2545 (1971).
229. M. A. Q. Khan and J. P. Bederka, *Survival in Toxic Environments*, Academic, New York, 1974.
230. T. Hasegawa, S. Kira, and M. Ogata, *Igaku to Seibutsuzaku*, **89**(5), 291 (1974).
231. T. Kojima and H. Kobayashi, *Nippon Hoigaku Zasshi*, **27**(4), 258 (1973).
232. T. Kojima and H. Kobayashi, *Nippon Hoigaku Zasshi*, **29**(2), 82 (1975).
233. M. Ogata, T. Salki, S. Kira, T. Hasegewa, and S. Watanabe, *Sangyo Izoku*, **16**(1), 23 (1974).
234. O. H. Bakke and R. R. Scheline, *Toxicol. Appl. Pharmacol.*, **16**, 691 (1970).
235. G. Brenniman, R. Hartung, and W. J. Weber, Jr., *Water Res.*, **10**(2), 165 (1976).
236. J. E. Morrow, R. L. Gritz, and M. P. Kirton, *Copeia*, **2**, 326 (1975).
237. R. F. Lee, R. Sauerheber, and A. A. Benson, *Science*, **177**(4046), 344 (1972).
238. L. Braier, *Haematologica*, **58**(78), 491 (1973).
239. C. J. Jankins, R. A. Jones, and J. Seigel, *Toxicol. Appl. Pharmacol.*, **16**, 818 (1970).
240. C. P. Carpenter, D. L. Geary, Jr., R. C. Myers, D. J. Nachreiner, L. J. Sullivan, and J. M. King, *Toxicol. Appl. Pharmacol.*, **26**, 473 (1976).
241. T. T. Ishikawa and H. Schmidt, Jr., *Pharmacol. Biochim. Behav.*, **1**(5), 593 (1973).
242. T. Ikeda and H. Miyake, *Toxicol. Lett.*, **1**(4), 235 (1978).

243. S. Moeschlin and B. Speck, *Acta. Haematol.*, **38**, 104 (1967).
244. A. A. Lyapkalo, *Gig. Tr. Prof. Zabol.*, **3**, 14 (1973).
245. S. V. Durmishidze, D. S. Ugrekhelidze, and A. N. Dzhikya, *Prik. Biokhim. Mikrobiol.*, **10**(5), 673 (1974).
246. N. R. Vishwanath, R. B. Patil, and G. Rangaswami, *Zentralbl. Bakteriol. Parasitenkd. Infektionskir. Hyg., Abt. 2*, **130**(4), 348 (1975).
247. T. Miller, *Am. J. Vet. Res.*, **27**(121), 1755 (1966).
248. K. Yamada, S. Horiguchi, and J. Takahashi, *Agric. Biol. Chem.*, **29**(10), 943 (1965).
249. R. L. Raymond, V. W. Jamison, and J. O. Hudson, *Appl. Microbiol.*, **15**(4), 857 (1967).
250. E. T. McKenna and M. J. Coon, *J. Biol. Chem.*, **245**(15), 3882 (1970).
251. J. Nozaka and M. Kusunose, *Agric. Biol. Chem.*, **32**(12), 1484 (1968).
252. J. Nozaka and M. Kusunose, *Agric. Biol. Chem.*, **32**(8), 1033 (1968).
253. L. Meites, *Handbook of Analytical Chemistry*, 1st ed., McGraw-Hill, New York, 1963.
254. I. Viden, V. Kubelka, and J. Mostecky, *Z. Anal. Chem.*, **280**(5), 369 (1976).
255. D. L. Guertin and and H. W. Gerarde, *Am. Med. Assoc. Arch. Ind. Health*, **20**, 262 (1959).
256. A. Sato, T. Nakajima, and Y. Fugiwara, *Br. J. Ind. Med.*, **32**(3), 210 (1975).
257. T. Kojima and H. Kobayashi, *Nippon Hoigaku Zasshi*, **27**(4), 255 (1973).
258. *Criteria for a Recommended Standard—Occupational Exposure to Toluene*, U.S. Dept. Health, Education and Welfare, Cincinnati, OH, 1973.
259. I. Gontea, E. Bistriceanu, M. Draghicesu, and M. Manea, *Arch. Sci. Physiol.*, **22**(3), 397 (1968).
260. M. Guillemin, J. C. Murset, M. Lob, and J. Riquez, *Brit. J. Ind. Med.*, **31**, 310 (1974).
261. M. Sittig, *Handbook of Toxic and Hazardous Chemicals and Carcinogens*, 2nd ed., Noyes Data Corporation, Park Ridge, NJ, 1985.
262. W. J. Hayes, *Clinical Handbook on Economic Poisons*, U.S. Government Printing Office, Washington, DC, 1971.
263. Hygienic Guide Series, *Am. Ind. Hyg. Assoc. J.*, **29**, 702 (1971).
264. L. T. Fairhall, *Industrial Toxicology*, 2nd ed., Hafner, New York, 1969.
265. D. Hogger, *Schweiz. Med. Wochenschr.*, **97**, 368 (1967).
266. E. Ghislandi and A. Fabiani, *Med. Lav.*, **48**, 577 (1957).
267. C. P. Carpenter, E. R. Kinkead, D. L. Geary, Jr., L. J. Sullivan, and J. M. King, *Toxicol. Appl. Pharmacol.*, **33**(3), 543 (1975).
268. S. Gitelson, L. Aladjemoff, S. Ben-Hador, and R. Katznelson, *J. Am. Med. Assoc.*, **197**(10), 165 (1966).
269. W. B. Deichmann and H. W. Gerarde, *Toxicology of Drugs and Chemicals*, Academic, New York, 1969.
270. C. E. Searle, Ed., *Chemical Carcinogens*, American Chemical Society, Washington, DC, 1976.
271. S. T. Crooke, *Texas Med.*, **68**, 67 (1972).
272. J. L. Svirbely, R. C. Dunn, and W. F. Von Oettingen, *J. Ind. Hyg. Toxicol.*, **25**, 366 (1943).
273. R. E. Joyner and W. L. Pegues, *J. Occup. Med.*, **3**, 211 (1961).

274. R. Morley, D. W. Eccleston, C. P. Douglas, W. E. Greville, D. J. Scott, and J. Anderson, *Brit. Med. J.*, **3**, 442 (1970).
275. E. Lederer, *Muench. Med. Wochenschr.*, **144**(29/30), 1302 (1972).
276. V. Mathies, *Med. Klin.*, **63**, 463 (1970).
277. *API Toxicological Review of Xylene*, American Petroleum Institute, New York, 1960.
278. W. E. Clark, *Am. J. Clin. Pathol.*, **68**, 425 (1977).
279. P. I. Mikullski, R. Wiglusz, A. Bublewska, J. Uselis, *Brit. J. Ind. Med.*, **29**, 450 (1972).
280. L. Nelkin, *Zentralbl. Gewerbehyg.*, **18**, 182 (1931).
281. E. Rosenthal-Deussen, *Arch. Gewerbepathol. Gewerbehyg.*, **2**, 92 (1931).
282. J. Kucera, *J. Pediatr.*, **72**, 857 (1968).
283. K. Engstrom, K. Husman, and V. Riihimaki, *Int. Arch. Occup. Environ. Health*, **39**(3), 181 (1977).
284. R. R. Lauwerys, T. Dath, J. M. LaChapelle, J. P. Buchet, and H. Roels, *J. Occup. Med.*, **20**(1), 17 (1978).
285. V. Riihimaki and P. Pfaffli, *Scand. J. Work Environ. Health*, **4**(1), 73 (1978).
286. V. Sedivec and J. Flek, *Int. Arch. Occup. Environ. Health*, **37**(3), 205 (1976).
287. V. Sedivec and J. Flek, *Prac. Lek.*, **27**(3), 68 (1975).
288. V. Sedivec and J. Flek, *Prac. Lek.*, **26**(7), 243 (1974).
289. W. Senczuk and J. Orlowski, *Brit. J. Ind. Med.*, **35**(1), 50 (1978).
290. A. Winnicka, J. Chmielewski, and T. Mardkowicz, *Pol. Tyg. Lek.*, **32**(31), 1149 (1977).
291. G. Aschan, I. Bunnfors, D. Hyden, B. Larsby, L. M. Odkvist, and R. Tham, *Acta Otolaryngol.*, **84**(5-6), 370 (1977).
292. R. H. Rigdon, *Arch. Surg.*, **41**, 101 (1940).
293. W. Hann and P. Jensen, *Water Quality Characteristics of Hazardous Materials*, Vols. 1-4, Environmental Engineering Division, Civil Engineering Dept., Texas A&M University, 1974.
294. R. Fabre, R. Truhaut, and S. Laham, *Arch. Mal. Prof.*, **21**, 301 (1960).
295. W. E. Engelhardt, *Arch. Hyg. Bakteriol.*, **114**, 219 (1935).
296. L. M. Kashin, I. L. Kulinskaya, and L. F. Mikhailovskaya, *Vrach. Delo.*, **8**, 109 (1968).
297. Y. A. Krotov and N. A. Chebotar, *Gig. Tr. Prof. Zabol.*, **16**(6), 40 (1972).
298. *Criteria for a Recommended Standard—Occupational Exposure to Xylene* U.S. Dept. Health, Education, and Welfare, National Institute for Occupational Safety and Health, Washington, DC, 1975.
299. R. Fabre, R. Truhaut, and S. Laham, *C.R. Acad. Sci.*, **250**, 2655 (1960).
300. I. Fridlyand, *Farmakol. Toksikol. (Moscow)*, **33**(4), 499 (1970).
301. W. Senczuk, S. Litewka, J. Orlowski, and H. Pogorzelska, *Ann. Pharm. (Poznan)*, **9**, 3 (1971).
302. C. Harper, *Pharmacol. Fed. Proc.*, **34**(1), 785 (1975).
303. M. F. Carlone and J. R. Fouts, *Xenobiotica*, **4**(11), 705 (1974).
304. R. T. Drew and J. R. Fouts, *Toxicol. Appl. Pharmacol.*, **29**(1), 111 (1974).
305. T. Omori, S. Horiguchi, and K. Yamada, *Agric. Biol. Chem.*, **31**(11), 1337 (1967).
306. T. Omori and K. Yamada, *Agric. Biol. Chem.*, **33**(7), 979 (1969).
307. G. K. Skriabin, Kh. G. Ganbarov, L. A. Golovleva, I. I. Chervin, and V. M. Adanin, *Mikrobiologiia*, **45**(6), 951 (1976).

308. T. Omori and K. Yamada, *Agric. Biol. Chem.*, **34**(5), 659 (1970).
309. R. S. Davis, F. E. Hossler, and R. W. Stone, *Can. J. Microbiol.*, **14**(2), 1005 (1968).
310. D. T. Gibson, V. Mahadevan, and J. F. Davey, *J. Bacteriol.*, **119**(3), 930 (1974).
311. J. F. Davey and D. T. Gibson, *J. Bacteriol.*, **119**(3), 923 (1974).
312. V. W. Jamison, R. L. Raymond, and J. O. Hudson, *Appl. Microbiol.*, **17**(6), 853 (1969).
313. W. H. Stahl, *Compilation of Odor and Taste Threshold Values*, Data, American Society for Testing and Materials, Philadelphia, 1973.
314. ATSDR, *Toxicological Profile for Total Xylenes*, U.S. Dept. of Health and Human Services, Atlanta, GA, 1990.
315. B. Sova, *Cesk. Hyg.*, **20**(4), 214 (1975).
316. M. Ogata, K. Tomokuni, and Y. Takatsuka, *J. Ind. Med.*, **26**(4), 330 (1969).
317. J. P. Buchet and R. R. Lauwerys, *Brit. J. Ind. Med.*, **30**, 125 (1973).
318. J. Orlowski, *Bromatol. Chem. Toksykol*, **7**, 87 (1974).
319. H. Matsui, M. Kasao, and S. Imamura, *J. Chromatogr.*, **145**(2), 231 (1978).
320. J. Angerer, *Int. Arch. Occup. Environ. Health*, **36**(4), 287 (1976).
321. S. Kira, *Sangyo Igaku*, **19**(3), 126 (1977).
322. T. Arimma, T. Fukuda, and N. Tani, Jap. Pat. 75136282, October 29, 1975.
323. V. M. Bagnyuk, T. L. Oleinik, A. G. Brekhunets, and V. G. Koval, *Mater. Vses. Nauchn. Simp. Sovrem. Probl. Samoochishcheniya Regul. Kach. Vody*, 5th, **6**, 3 (1975).
324. D. L. Opdyke, *Food Cosmet. Toxicol.*, **13**, 681 (1975).
325. R. J. Sperber and S. L. Rose, *Soc. Plast. Eng. Tech.*, **21**, 521 (1975).
326. N. W. Lazarew, *Arch. Exp. Pathol. Pharmacol.*, **143**, 223 (1929).
327. P. I. Mikulski and R. Wiglusz, *Toxicol. Appl. Pharmacol.*, **31**(1), 21 (1975).
328. R. Wiglusz, M. Kienitz, G. Delag, E. Galuszko, and P. Mikulski, *Bull. Inst. Marit. Trop. Med. (Gdynia)*, **26**(3–4), 315 (1975).
329. R. Wiglusz, G. Delag, and P. Mikulski, *Bull. Inst. Marit. Trop. Med. (Gdynia)*, **26**(3–4), 303 (1975).
330. S. Laham and E. O. Matutina, *Arch. Toxikol.*, **30**(3), 199 (1973).
331. E. Bateman and C. Henningsen, *Proc. Am. Wood Preservers' Assoc.*, 136 (1923).
332. R. E. McDonald and W. R. Buford, *Plant Dis. Rep.*, **58**(12), 1143 (1974).
333. S. H. Safe, H. Plugge, B. Chittim, and J. F. S. Crocker, *Publ. Environ. Secr. Natl. Res. Counc. Can., Iss. NRCC/CNRC, 16073*, Department of Chemistry, Guelph University, Guelph, Ontario, 1977.
334. L. B. Brattsten and C. F. Wilkinson, *Pestic. Biochem. Physiol.*, **3**(4), 393 (1973).
335. L. B. Brattsten, C. F. Wilkinson, and M. M. Root, *Insect. Biochem.*, **6**(6), 615 (1976).
336. H. Dannenberg, I. Brachmann, and C. Thomas, *Z. Krebsforsch.*, **74**(1), 100 (1970).
337. ATSDR, *Toxicological Profile for Ethylbenzene* U.S. Dept. of Health and Human Services, Atlanta, GA, 1990.
338. H. F. Uhlig and W. C. Pfefferle, Chapter 12 in *Refining Petroleum for Chemicals*, R. F. Gould, Ed., American Chemical Society, Washington, DC, 1970.
339. M. S. Wolff, S. M. Daum, W. V. Lorimer, I. J. Selikoff, and B. B. Aubrey, *J. Toxicol. Environ. Health*, **2**(5), 997 (1977).
340. A. M. El Masri, J. N. Smith, and R. T. Williams, *Biochem. J.*, **64**, 50 (1956).

341. T. Dutkiewicz and H. Tyras, *Brit. J. Ind. Med.*, **24**, 330 (1967).
342. G. A. Maylin, M. J. Cooper, and M. H. Anders, *J. Med. Chem.*, **16**, 6 (1973).
343. D. T. Gibson, B. Gschwendt, W. K. Yeh, and M. Kobal, *Biochemistry*, **12**(8), 1520 (1973).
344. M. Tanabe, R. L. Dehn, and M. H. Kuo, *Biochemistry*, **10**(6), 1087 (1971).
345. K. F. Meleschenko, *Gig. Sanit.*, **6**, 90 (1975).
346. National Research Council, *Drinking Water and Health*, Vol. 1, National Academy Press, Washington, DC, 1977.
347. *Fed. Reg.*, **40**(196), 47262 (1976).
348. K. A. Nikogosyan, *Zh. Eksp. Klin. Med.*, **12**(6), 76 (1972).
349. A. G. Ruhrchemie, Neth. Appl. Pat. 75 02553, August 17, 1976.
350. H. W. Werner, R. C. Dunn, and W. F. Von Oettingen, *J. Ind. Health Toxicol.*, **26**, 264 (1974).
351. W. Senczuk and B. Litewka, *Bromatol. Chem. Toksykol.*, **7**(1), 93 (1974).
352. D. Robinson, J. N. Smith, and R. F. Williams, *Biochem. J.*, **56**, XI (1954).
353. T. Omori, Y. Jigami, and Y. Minoda, *Agric. Biol. Chem.*, **39**(9), 1775 (1975).
354. Y. Jigami, T. Omori, and Y. Minoda, *Agric. Biol. Chem.*, **39**(9), 1781 (1975).
355. S. M. Fridman, S. M. Safonnikova, S. A. Sakaeva, and R. F. Daukaeva, *Gig. Sanit.*, **12**, 78 (1977).
356. C. R. Noller, *Chemistry of Organic Compounds*, 3rd ed., W. B. Saunders, Philadelphia, 1966.
357. J. Wepierre, Y. Cohen, and G. Valetti, *Eur. J. Pharmacol.*, **3**(1), 47 (1968).
358. R. I. Levitt, *J. Gen. Microbiol.*, **49**(1), 411 (1967).
359. J. L. Palotay, K. Adachi, R. L. Dobson, and J. S. Pinto, *J. Natl. Cancer Inst.*, **57**(6), 1269 (1976).
360. L. Ahlstrom, *Kem. Tidkir.*, **89**(1-2), 28 (1977).
361. *API Toxicological Review of Styrene*, American Petroleum Institute, New York, 1962.
362. *Chemical Safety Data SD-37, Properties and Essential Information for Safe Handling and Use of Styrene Polymer*, Manufacturing Chemists' Association, Inc., Washington, DC, 1971.
363. R. Aries, Fr. Pat. 2,108,829, November, 1972.
364. H. H. Cornish, K. J. Hahn, and M. L. Barth, *Environ. Health Perspect.*, **11**, 191 (1975).
365. M. Chaigneau and G. LeMona, *Ann. Pharm. Fr.*, **32**, 485 (1974).
366. I. Astrand, A. Kilbom, P. Ovrum, I. Wahlberg, and O. Vesterberg, *Work Environ. Health*, **11**(2), 69 (1974).
367. F. Gamberale and M. Hultengren, *Work Environ. Health*, **11**(2), 86 (1974).
368. R. D. Stewart, H. C. Dodd, E. D. Baretta, and A. W. Shaffer, *Arch. Environ. Health*, **16**, 656-662 (1968).
369. J. M. Schwarzmann and N. P. Kutscha, *Res. Life Sci.*, **19**(3), 1 (1971).
370. H. C. Spencer, D. D. Irish, E. M. Adams, and V. K. Rowe, *J. Ind. Hyg. Toxicol.*, **24**, 295 (1942).
371. Y. Alarie, *Toxicol. Appl. Pharmacol.*, **24**(2), 279 (1973).
372. J. C. Rogers and C. C. Hooper, *Ind. Med. Surg.*, **26**, 32 (1957).

373. A. N. Kohn, *Am. J. Ophthalmol.*, **85**(4), 569 (1978).
374. N. I. Ponomareva and N. S. Zlobina, *Gig. Tr. Prof. Zabol.*, **15**(6), 22 (1971).
375. P. Gotell, O. Axelson, and B. Lindelof, *Work Environ. Health*, **9**(2), 76 (1973).
376. W. V. Lorimer, R. Lilis, W. J. Nicholson, H. Anderson, A. Fischbein, S. Daum, W. Rom, C. Rice, and I. J. Selikoff, *Environ. Health Perspect.*, **17**, 171 (1976).
377. A. A. Bashirov, *Ter. Arkh.*, **42**(12), 41 (1970).
378. Z. A. Volkova, L. E. Milkov, K. A. Lopukhova, L. M. Malyar, Y. L. Marenko, and T. K. Shakhova, *Gig. Tr. Prof. Zabol.*, **14**(1), 31 (1970).
379. I. I. Alekperov, *Gig. Tr. Prof. Zabol.*, **12**, 191 (1970).
380. A. A. Bashirov, *Gig. Tr. Prof. Zabol.*, **15**(5), 57 (1971).
381. L. P. Lukoshkina and I. I. Alikperov, *Gig. Tr. Prof. Zabol.*, **8**, 42 (1973).
382. A. A. Askalonov, *Farmakol. Toksikol.*, **36**(5), 611 (1973).
383. K. Linstrom, H. Harkonen, and S. Hernberg, *Scand. J. Work Environ. Health*, **2**(3), 129 (1976).
384. H. Harkonen, K. Lindstrom, A. M. Seppalainen, A. Asp, and S. Hernberg, *Scand. J. Work Environ. Health*, **4**(1), 53 (1978).
385. A. M. Seppalainen and H. Harkonen, *Scand. J. Work Environ. Health*, **2**(3), 140 (1976).
386. P. C. Holmberg, *Scand. J. Work Environ. Health*, **3**(4), 212 (1977).
387. V. G. Lappo and I. I. Krasnikova, *Gig. Vop. Proizvod. Primen. Polim. Mater.*, 222 (1969).
388. A. S. Izyumova, *Gig. Sanit.*, **37**(4), 29 (1972).
389. N. Ragule, *Gig. Sanit.*, **11**, 85 (1974).
390. H. Vainio, K. Hemminki, and E. Elovaara, *Toxicology*, **8**(3), 319 (1977).
391. P. Melvy and A. J. Garro, *Mutat. Res.*, **40**, 15 (1976).
392. H. Vainio, R. Paakonen, K. Ronholm, V. Raunio, and O. Pelkonen, *Scand. J. Work Environ. Health*, **2**(3), 147 (1976).
393. T. Meretoja, H. Vainio, and H. Jarventaus, *Toxicol. Letters*, **1**(5-6), 315 (1976).
394. *Report NCI Bioassay of a Solution of Beta-Nitro-Styrene and Styrene for Possible Carcinogenicity-TR170*, Office of Cancer Communications, National Cancer Institute, Bethesda, MD (U.S. Government Printing Office, 1979).
395. H. Van Rees, *Int. Arch. Arbeitsmed.*, **33**(1), 39 (1974).
396. K. C. Leibman, *Environ. Health Perspect.*, **11**, 115 (1975).
397. C. Burkewicz, J. Rybkowska, and H. Zielinska, *Med. Prac.*, **25**(3), 305 (1974).
398. M. Ikeda, T. Imamura, M. Bayashi, T. Tabuchi, and I. Hara, *Int. Arch. Arbeitsmed.*, **32**(1-2), 93 (1974).
399. A. Wink, *Ann. Occup. Hyg.*, **15**(2-4), 211 (1972).
400. A. M. El Masri, J. N. Smith, and R. T. Williams, *Biochem. J.*, **68**, 199 (1958).
401. I. Danishefsky and M. Willhite, *J. Biol. Chem.*, **211**, 549 (1954).
402. S. P. James and D. A. White, *Biochem. J.*, **104**, 914 (1967).
403. S. E. Ruvinskaya, *Gig. Tr. Prof. Zabol.*, **9**(11), 29 (1965).
404. ATSDR, *Toxicological Profile for Styrene*, in U.S. Dept. of Health and Human Services, Atlanta, GA, 1991.
405. S. M. Rappaport and D. A. Fraser, *J. Am. Ind. Hyg. Assoc.*, **5**, 205 (1977).

406. V. K. Rowe, G. J. Atchison, E. N. Luce, and E. M. Adams, *J. Ind. Hyg. Toxicol.*, **25**(8), 348 (1943).
407. A. Slob, *Brit. J. Ind. Med.*, **30**, 390 (1973).
408. H. H. Harkonen, P. Kalliokoski, S. Hietala, and S. Hernberg, *Work, Environ. Health*, **11**(3), 162 (1974).
409. J. Sollenberg and A. Baldeston, *J. Chromatogr.*, **132**(3), 469 (1977).
410. F. M. Gaszhiev and T. V. Aliev, *Ser. Biol. Nauk.*, **4**, 112 (1974).
411. W. Boehr, S. Drobnik, W. Hild, R. Kroebel, A. Meyer, and G. Naumann, Ger. Pat. 2363474, June 26, 1975.
412. R. Aries, Fr. Pat. 2104717, May 26, 1972.
413. R. E. Fearon, *Am. Heart J.*, **75**(5), 634 (1968).
414. A. Alpande de Morais, C. M. Andrade da Mata Rezende, M. V. Von Buelow, J. C. Mourao, O. R. Gottlieb, M. C. Marx, A. I. Da Rocha, and M. T. Magalhaes, *An. Acad. Brs. Ciene*, **44**, 303 (1972).
415. J. D. Peele, Jr. and E. O. Oswald, *Biochem. Biophys. Acta*, **497**(2), 598 (1977).
416. L. M. Sergeta, M. K. Byalko, N. I. Alberton, and G. M. Balan, *Gig. Tr. Prof. Zabol.*, **12**, 51 (1975).
417. V. M. Sergeta, M. I. Alberton, and V. P. Fomenko, *Zdravookhr Kaz.*, **1**, 50 (1977).
418. V. N. Bravre, *Narusheniya Metab., Tr. Nauchn. Konf. Med. Inst. Zapadn. Sib., 1st*, 259 (1974).
419. Z. M. Fadeeva and Y. N. Iekhler, *Nauch. Tr. Omsk. Med. Inst.*, **107**, 166 (1971).
420. L. M. Makareva, *Farmakol. Toksikol.*, **35**(4), 491 (1972).
421. G. M. Klimina, *Narusheniya Metab., Tr. Nauchn. Konf. Med. Inst. Zapadn. Sib., 1st*, 46 (1974).
422. G. M. Klimina, *Narusheniya Metab., Tr. Nauchn. Konf. Med. Inst. Zapadn. Sib., 1st*, 255 (1974).
423. I. I. Solovev, *Narusheniya Metab., Tr. Nauchn. Konf. Med. Inst. Zapadn. Sib., 1st*, 255 (1974).
424. A. A. Nikiforova, *Mater. Nauch. Sess. Posvyashd. 50-Letuju Obrazsov, USSR Omsk. Gos. Med. Inst.*, 871 (1972).
425. I. M. Mirzoyan and R. K. Zhakenova, *Vopr. Gig. Tr. Prof. Zabol.*, 247 (1972).
426. J. L. Weeks, M. B. Lentle, and B. C. Lentle, *J. Occup. Med.*, **12**(7), 246 (1970).
427. J. Nordal, *Nord Vet. Med.*, **82**(9), 469 (1970).
428. R. Mestres, *Proc. Int. Citrus Cymp., 1st Riv.*, **2**, 1035 (1968).
429. I. Hakkinen, E. Siltanen, S. Hernberg, A. M. Seppalainen, P. Karli, and E. Vikkula, *Arch. Environ. Health*, **26**, 70 (1973).
430. A. M. Seppalainen and I. Hakkinen, *J. Neurol. Neurosurg. Psychiatry*, **38**, 248 (1975).
431. W. B. Deichmann, K. V. Kitzmiller, M. Kierker, and S. Witherup, *J. Ind. Hyg. Toxicol.*, **29**(1), 1 (1947).
432. W. G. Levine, P. Milburn, R. L. Smith, and R. Williams, *Biochem. Pharmacol.*, **19**(1), 235 (1970).
433. P. Von Riag and R. Ammon, *Arzneim. Forsch.*, **22**(8), 1399 (1972).
434. H. Von Berninger, P. Ammon, and I. Berninger, *Arzneim. Forsch.*, **18**(2), 880 (1968).
435. F. McPherson, J. W. Bridges, and D. V. Parke, *Nature*, **252**(5483), 488 (1974).

436. P. J. Creaven, D. W. Parke, and R. T. Williams, *Biochem. J.*, **96**, 879 (1965).
437. M. D. Burke and J. W. Bridges, *Zenobiotica*, **5**(6), 357 (1975).
438. D. E. Willis and R. F. Addison, *Comp. Gen. Pharmacol.*, **5**(1), 77 (1974).
439. S. A. Atlas and D. W. Nebert, *Arch. Biochem. Biophys.*, **175**(2), 495 (1976).
440. P. Von Raig and R. Ammon, *Arzneim. Forsch.*, **9**, 1266 (1970).
441. F. J. McPherson, J. W. Bridges, and D. V. Parke, *Biochem. J.*, **154**, 773 (1976).
442. D. Catelani, G. Mosselmans, J. Nienhaus, C. Sorlini, and V. Trecanni, *Experientia*, **26**(8), 922 (1970).
443. T. Ohmori, T. Ikai, Y. Minoda, and K. Yamada, *Agric. Biol. Chem.*, **37**(7), 1599 (1973).
444. D. Lunt and W. C. Evans, *Biochem. J.*, **118**, 54 (1976).
445. D. Catelani, C. Sorlini, and V. Treccani, *Experientia*, **27**(10), 1173 (1971).
446. D. Catelani, A. Colombi, C. Sorlini, and V. Treccani, *Biochem. J.*, **134**, 1063 (1973).
447. M. D. Burke, D. J. Benford, J. W. Bridges, and D. V. Parke, *Biochem. Soc. Trans.*, **5**(5), 1370 (1977).
448. H. Kameoka, K. Nishikawa, and H. Wada, *Nippon Nogei Kagaku Kaishi*, **49**(10), 557 (1975).
449. M. M. Ranney, *Synthetic Lubricants*, Noyes Data Corp., Park Ridge, NJ, 1972.
450. D. L. Opdyke, *Food Cosmet. Toxicol.*, **12**(5-6), 705 (1974).
451. R. V. Subba-Rao and M. Alexander, *Appl. Environ. Microbiol.*, **33**(1), 101 (1977).
452. D. D. Focht and M. Alexander, *Appl. Microbiol.*, **20**(4), 608 (1970).
453. S. W. Stroud, *J. Endocrinol.*, **2**, 55 (1940).
454. R. R. Scheline, *Experientia*, **30**(8), 880 (1974).
455. J. E. Sinscheimer and R. V. Smith, *Biochem. J.*, **111**(1), 35 (1969).
456. L. K. Tay and J. E. Sinsheimer, *Drug Metab. Dispos.*, **4**(2), 154 (1976).
457. E. L. Docks and G. Krishan, *Biochem. Pharmacol.*, **24**(21), 1965 (1975).
458. T. Watabe and K. Akamatus, *Biochem. Pharmacol.*, **24**(3), 442 (1975).
459. J. S. Henderson and J. L. Weeks, *Ind. Med.*, **42**(2), 10 (1973).
460. H. Bradner, U.S. Pat. 3988437, October 26, 1976.
461. M. O. Amdur and D. A. Creasia, *Am. Ind. Hyg. Assoc. J.*, **27**(4), 349 (1966).
462. I. Y. R. Adamson, *Arch. Environ. Health*, **26**(4), 192 (1973).
463. P. Scoppa and K. Gerbaulet, *Boll. Soc. Ital. Biol. Sper.*, **47**(7), 194 (1971).
464. I. Y. R. Adamson and J. M. Furlong, *Arch. Environ. Health*, **28**(3), 155 (1974).
465. S. Kiriyama, M. Banjo, and H. Matsushima, *Nutr. Rep. Int.*, **19**(2), 79 (1974).
466. Z. F. Khromenko, V. D. Gostinskii, and N. G. Ivanov, *Nauch. Tr. Irkutsk. Med. Inst.*, **115**, 122 (1972).
467. *Criteria for a Recommended Standard—Occupational Exposure to Terphenyl*, U.S. Dept. of Health, Education, and Welfare, Public Health Service, NIOSH, Washington, DC, 1977.
468. C. Hesse, K. Hilp, H. Kating, and G. Schaden, *Arch. Pharm.*, **310**(10), 792 (1977).
469. R. A. Johnstone, *Nature*, **200**, 1184 (1963).
470. I. Schmeltz, J. Tosk, and D. Hoffmann, *Anal. Chem.*, **48**(4), 645 (1976).
471. J. B. Addison, P. J. Silk, and I. Unger, *Int. J. Environ. Anal. Chem.*, **4**(2), 135 (1975).

472. *API Toxicological Review of Naphthalene*, American Petroleum Institute, New York, 1959.
473. *Chemical Safety Data Sheet SD-58, Properties and Essential Information for Safe Handling and Use of Naphthalene*, Manufacturing Chemists' Association, Inc., Washington, DC, 1956.
474. M. Sherer, *J. Am. Osteopath. Assoc.*, **65**(1), 60 (1965).
475. N. I. Pavlivoda, *Zdravookhr. Beloruss.*, **18**(7), 81 (1971).
476. T. Sollman, *A Manual of Pharmacology*, 8th ed., Saunders, Phildelphia, 1957.
477. N. L. Sharma, R. N. Singh, and N. K. Natu, *J. Indian Med. Assoc.*, **48**(1), 20 (1967).
478. T. Valaes, S. A. Doxiadis, and P. Fessas, *J. Pediatr.*, **63**, 904 (1963).
479. W. M. Grant, *Toxicology of the Eye*, 3rd ed., Charles C Thomas, Springfield, IL, 1986.
480. J. Doull, Chapter 5 in *Toxicology, the Basic Science of Poisons*, L. J. Casarett and J. Doull, Eds., Macmillan, New York, 1975.
481. J. H. Sanderson, *Practitioner*, **2**, 216 (1976).
482. R. E. Gosselin, H. C. Hodge, R. P. Smith, and M. N. Gleason, *Clinical Toxicology of Commercial Products*, 4th ed., Williams and Wilkins, Baltimore, 1976.
483. A. Hamilton and H. L. Hardy, *Industrial Toxicology*, PSG Publishing Co., Littleton, MA, 1974.
484. *Industrial Hygiene Field Operational Manual*, U.S. Dept. Labor, OSHA Instruction CPL 2-2.20, Office of Field Coordination, Washington, DC, April 2, 1979.
485. W. M. Grant, *Toxicology of the Eye*, 2nd ed., Charles C Thomas, Springfield, IL, 1974.
486. J. P. Dawson, W. W. Thayer, and J. F. Desforges, *Blood*, **13**, 113 (1958).
487. W. B. Schafer, *Pediatrics*, **7**, 172 (1951).
488. W. G. Grigor, H. Robin, and J. D. Harley, *Med. J. Aust.*, **2**(2), 1229 (1966).
489. R. H. Gadsden, R. R. Mellette, and W. C. Miller, Jr., *J. Am. Med. Assoc.*, **168**, 1220 (1958).
490. D. R. Adams, *Brit. J. Ophthalmol.*, **14**, 544 (1930).
491. R. J. Meyer, *N. Engl. J. Med.*, **252**(15), 622 (1955).
492. G. Ghetti and L. Mariani, *Med. Lav.*, **57**, 533 (1956).
493. H. Hanssler, *Dtsch. Med. Wochenschr.*, **89**, 1794 (1964).
494. U. Irle, *Dtsch. Med. Wochenschr.*, **89**, 1798 (1964).
495. T. L. Naiman and M. H. Kosoy, *Can. Med. Assoc. J.*, **91**, 1243 (1964).
496. J. A. Anzuilewicz, H. J. Dick, and E. E. Chiarulli, *Am. J. Obstet. Gynecol.*, **78**, 519 (1959).
497. A. K. Brown, *Am. J. Dis. Child*, **94**, 510 (1957).
498. J. V. Mackell, F. Rieders, H. Brieger, and E. L. Bauer, *Pediatrics*, **7**, 722 (1951).
499. ATSDR, *Toxicological Profile for Naphthalene and 2-Methylnaphthalene*, U.S. Dept. of Health and Human Services, Atlanta, GA, 1990.
500. M. R. Juchau and M. J. Namkung, *Drug Metab. Dispos.*, **2**(4), 380 (1974).
501. H. R. Koch, K. Doldi, and O. Hockwin, *Doc. Ophthalmol. Proc. Ser.*, **8**, 293 (1976).
502. R. van Heyningen and A. Pirie, *Biochem. J.*, **102**, 842 (1967).
503. H. Ishizaka, *Showa Igakkai Zasshi*, **31**(9), 471 (1971).
504. K. Ikemoto, *Osaka City Med. J*, **17**(1), 1 (1971).
505. S. K. Srivastava and E. Beutler, *Biochem. J.*, **112**, 421 (1969).

506. A. M. Potts and L. M. Gonasum, Chapter 13 in *Toxicology, the Basic Science of Poisons*, L. J. Casarett and J. Doull, Eds., Macmillan, New York, 1975.
507. A. Pirie, *Exp. Eye Res.*, **7**(3), 354 (1968).
508. M. Shimotori, *Acta Soc. Ophthalmol. Jap.*, **76**(11), 1545 (1972).
509. K. Alexandrov and C. Frayssinet, *J. Natl. Cancer Inst.*, **51**(3), 1067 (1973).
510. J. Yoon and K. Kim, *Hanguk Sikmul Poho Hakkoe Chi*, **16**(2), 127 (1977).
511. R. B. Boose and L. Terriere, *J. Econ. Entomol.*, **60**(2), 580 (1967).
512. J. A. Schafer and L. Terriere, *J. Econ. Entomol.*, **63**(3), 787 (1970).
513. R. F. Lee, R. Sauerheber, and G. W. Dobbs, *Mar. Biol. (Berl.)*, **17**(3), 201 (1972).
514. M. J. Melancon, Jr. and J. J. Lech, *Arch. Environ. Contam. Toxicol.*, **7**(2), 207 (1978).
515. U. Varanasi, M. Uhler, and S. I. Stranahan, *Toxicol. Appl. Pharmacol.*, **44**, 277 (1978).
516. H. R. Sanborn and D. C. Malins, *Proc. Soc. Exp. Biol. Med.*, **154**, 151 (1977).
517. J. M. Neff, *Prepr. Div. Pet. Chem. Am. Chem. Soc.*, **20**(4), 839 (1975).
518. S. S. Rossi and J. W. Anderson, *Mar. Biol.*, **39**(1), 51 (1977).
519. M. G. Horning, C. D. Kary, P. A. Gregory, and W. G. Stillwell, *Toxicol. Appl. Pharmacol.*, **37**(1), 18 (1976).
520. O. Skjaeggestad, *Acta Pathol. Microbiol. Scand.*, **169**, 1 (1964).
521. P. Brookes and P. D. Lawley, *Brit. Empire Cancer Campaign Res.*, *41st Ann. Rep., Part II*, 77 (1963).
522. P. Brookes and P. D. Lawley, *Nature*, **202**, 781 (1964).
523. P. O. Ts's and P. Lu, *Proc. Natl. Acad. Sci. U.S.*, **51**(1), 17 (1964).
524. P. Daudel, M. Croisy-Delcey, P. Jacquignon, and P. Vigny, *C.R. Acad. Sci. [D]*, **277**, 2437 (1973).
525. M. Miko and L. Drobnica, *Folia Fac. Med. Univ. Comenianae Bratisl.*, **7**(1), 217 (1969).
526. I. Schmiltz, J. Tosk, J. Hilfrich, N. Hiroto, D. Hoffman, and E. L. Wynder, *Carcinog. Compr. Surv.*, **3**, 47 (1978).
527. E. D. S. Corner and L. Young, *Biochem. J.*, **61**, 132 (1955).
528. T. R. Norton, Chapter 4 in *Toxicology, the Basic Science of Poisons*, L. J. Casarett and J. Doull, Eds., Macmillan, New York, 1975.
529. J. W. Daly, D. M. Jerina, and B. Witkop, *Experientia*, **28**(10), 1129 (1972).
530. D. M. Jerina, J. W. Daly, B. Witkop, P. Zaltzman-Nirenberg, and S. Udenfriend, and S. Udenfriend, *Biochemistry*, **9**(1), 147 (1970).
531. E. Boyland, M. Kimura, and P. Sims, *Biochem. J.*, **92**, 631 (1964).
532. A. Pirie and R. van-Heyninger, *Biochem. J.*, **100**(3), 70 (1966).
533. R. van Heyningen, *Exp. Eye Res.*, **9**, 38 (1970).
534. K. W. Bock, G. V. Ackeren, F. Lorch, and F. W. Birke, *Biochem. Pharmacol.*, **25**(1), 2351 (1976).
535. W. G. Stillwell, G. W. Griffin, and M. G. Horning, *Biochem. Pharmacol.*, **4**, 1341 (1978).
536. D. M. Jerina, J. W. Daly, B. Witkop, P. Zaltzman-Nirenberg, and S. Udenfriend, *J. Am. Chem. Soc.*, **90**(23), 6535 (1968).
537. K. Netter, *Naunyn-Schmiedebergs Arch. Pharmakol. Exp. Pathol.*, **262**(3), 375 (1969).
538. H. D. Colby, R. E. Kramer, J. W. Greiner, D. A. Robinson, R. F. Krause, and W. J. Canady, *Biochem. Pharmacol.*, **24**(17), 1644 (1975).
539. R. D. Schonerod, M. A. Khan, L. C. Terriere, and F. W. Plapp, Jr., *Life Sci.*, **7**(13), 681 (1968).
540. B. Griffiths and W. C. Evans, *Biochem. Jr.*, **95**(3), 51 (1965).

541. E. I. Kvasnikov and N. Z. Tin'yanova, *Mikrobiol. Zh. (Keiv)*, **32**(4), 416 (1970).
542. M. Malesset-Bras and E. Azoulay, *Ann. Inst. Pasteur*, **109**(6), 894 (1965).
543. M. S. Twefix and Y. Hamdi, *Acta Microbiol. Pol., Ser. B.*, **19**(2), 133 (1970).
544. A. M. Cundell and R. W. Traxler, *Mater. Org. (Berl.)*, **11**(1), 1 (1976).
545. M. Martonova, B. Skarka, and Z. Radij, *Folia Microbiol.*, **17**(1), 63 (1972).
546. F. A. Catterall, K. Murray, and P. A. Williams, *Biochim. Biophys. Acta*, **237**(2), 361 (1971).
547. V. V. Modi and R. N. Patel, *Appl. Microbiol.*, **16**(1), 172 (1968).
548. O. Atlavinyte, A. Lugauskas, and J. Daciulyte, *Biosfera Chel., Mater. Vses. Simp., 1st*, **217**, 13–17 (1975).
549. M. N. Schroth and D.C. Hildebrand, *Phytopathology*, **58**, 848 (1968).
550. K. O. Kusk, *Physiol. Plant.*, **43**(1), 1 (1978).
551. C. Soto, J. A. Hellebust, and T. C. Hutchinson, *Verh.-Int. Ver. Theor. Ange. Limnol.*, **19**(3), 2145 (1975).
552. I. Schmeltz, D. Hoffman, and E. L. Wynder, *Trace Subst. Environ. Health*, **8**, 281 (1974).
553. K. Adachi, *Hyogo-ken Eisei Kenkyusho Kenkyu Hokoku*, **10**, 22 (1975).
554. C. A. Nau, J. Neal, and M. Thornton, *Arch. Environ. Health*, **12**, 382 (1966).
555. R. S. Caldwell, E. M. Calderone, and M. H. Mallon, *Fate Eff. Pet. Hydrocarbons Mar. Ecosyst. Org., Proc. Symp., 1976* (Mar. Sci. Cent., Oregon State Univ., Newport, OR) p. 210 (publ. 1977).
556. L. G. Andreikova and L. A. Kogau, *Koks Khim.*, **8**, 47 (1977).
557. P. E. Gaffney, *J. Water Pollut. Control Fed.*, **48**(11), 2590 (1976).
558. D. G. Crosby and C-S Tang, *J. Agric. Food Chem.*, **17**(6), 1291 (1969).
559. B. Rabussier and J. P. Mandon, Ger. Pat. 2412368, September 19, 1974.
560. L. Ashino, Jap. Pat. 77 25021, February 14, 1977.
561. P. Scoppa, *Z. Naturforsch.*, **21**(11), 1054 (1966).
562. L. N. Bolonova, *Farmakol. Toksikol. (Moscow)*, **30**(4), 484 (1967).
563. C. N. Statham, C. R. Elcombe, S. P. Szyjka, and J. J. Lech, *Xenobiotica*, **8**(2), 65 (1978).
564. K. Konya, T. Kitagaki, and Y. Konogai, Jap. Pat. 77 01022, January 6, 1977.
565. H. E. Tatem, *Fate Eff. Pet. Hydrocarbon. Mar. Ecosyst. Org. Proc. Symp., 1976* (Mar. Sci. Cen., Oregon State Univ., Newport, OR) p. 210 (publ. 1977).
566. S. Itoh, K. Endo, Y. Nakajima, K. Shimizu, and H. Narita, Jap. Pat. No. 76 11020, September 29, 1976.
567. P. Grasso and C. O'Hare, Chapter 14 in *Chemical Carcinogens*, C. E. Searle, Ed., American Chemical Society, Washington, DC, 1976.
568. I. Schmeltz and D. Hoffman, "Formation of Polynuclear Hydrocarbons," in *Carcinogenesis, A Comprehensive Survey*, Vol. 1, Polynuclear Aromatic Hydrocarbons, R. Freudenthal and P. W. Jones, Eds., Raven Press, New York, 1976.
569. G. Grimmer, H. Boehnke, and H. P. Harke, *Int. Arch. Occup. Environ. Health*, **40**(2), 93 (1977).
570. M. Argirova, *Khig. Zdraeopaz.*, **18**(5), 18 (1975).
571. W. Ziechmann, *Therapiewoche*, **28**(7), 1199 (1978).
572. J. K. Selkirk, *J. Toxicol. Environ. Health*, **2**, 1245 (1977).
573. A. Dipple, Chapter 5 in *Chemical Carcinogens*, C. E. Searle, Ed., American Chemical Society, Washington, DC, 1976.

574. S. Sung, *C.R. Acad. Sci., Ser. D.*, **274**(10), 1597 (1972).
575. D. L. Sanioto and S. Schreier, *Biochem. Biophys. Res. Commun.*, **67**(2), 530 (1975).
576. Z. *Krebsforsch.*, **69**, 103 (1976).
577. J. C. Arcos, *Am. Lab.*, July, 29 (1978).
578. P. O. Grover, "Polycyclic Hydrocarbon Epoxides: Formation and Further Metabolism by Animal and Human Tissues," in *Chemical Carcinogenesis Essays*, R. Montesano, L. Tomatis, and W. Davis, Eds., International Agency for Research on Cancer, Lyon, 1974.
579. D. R. Thakker, M. Nordqvist, H. Yagi, W. Levin, D. Ryan, P. Thomas, A. H. Conney, and D. M. Jerina, "Comparative Metabolism of PAH," in *Polynuclear Aromatic Hydrocarbons*, P. W. Jones and P. Leber, Eds., Third International Symposium on Chemistry and Biology—Carcinogenesis and Mutagenesis, Ann Arbor Science Publishers, Ann Arbor, MI, 1979.
580. S. K. Yang, D. W. McCourt, J. C. Lentz, and H. V. Gelboin, *Science*, **196**(4295), 1199 (1977).
581. H. V. Gelboin, F. J. Wiebel, and N. Kinoshita, "Microsomal Aryl Hydrocarbon Hydroxylases," in *Biological Hydroxylation Mechanisms*, G. S. Boyd and R. M. S. Smellie, Eds., Academic, London, 1972.
582. N. H. Sloane and T. K. Davis, *Arch. Biochem. Biophys.*, **163**(1), 46 (1974).
583. E. T. Cantrell, G. A. Warr, D. L. Busbee, and R. R. Martin, *J. Clin. Invest.*, **52**(8), 1181 (1973).
584. D. W. Nebert, *Clin. Pharmacol. Ther.*, **14**(4), 693 (1973).
585. P. H. Jellinck and G. Smith, *Biochem. Biophys. Acta*, **304**(2), 520 (1973).
586. A. Poland and A. Kende, *Cold Spring Harbor Conf. Cell Prolif.*, **4**(B), 847 (1977).
587. D. W. Nebert and H. V. Gelboin, *Arch. Biochem. Biophys.*, **134**(1), 76 (1969).
588. F. J. Wiebel and H. V. Gelboin, "Enzyme Induction and Metabolism," in *Chemical Carcinogenesis Essays, International Agency for Research on Cancer*, P. Montesano and L. Tomatis, Eds., Lyon, 1974.
589. P. E. Thomas, D. Ryan, and W. Levin, "Cytochrome P-450 and Epoxide Hydrase," in *Polynuclear Aromatic Hydrocarbons, Third International Symposium*, P. W. Jones and P. Leber, Eds., Ann Arbor Science Publishers, Ann Arbor, MI, 1979.
590. R. Bass, G. Bochert, H. J. Merker, and D. Neubert, *J. Toxicol. Environ. Health*, **2**, 1353 (1977).
591. ATSDR, *Toxicological Profile for Polycyclic Aromatic Hydrocarbons*, U.S. Dept. of Health and Human Services, Atlanta, GA, 1990.
592. P. E. Strup, R. D. Giammar, T. B. Stanford, and P. W. Jones, "PAH in Combustion Effluents," in *Carcinogenesis, A Comprehensive Survey*, Vol. 1, R. Freudenthal and P. W. Jones, Eds., Raven Press, New York, 1976.
593. A. Bjorseth and G. Lunde, *Am. Ind. Hyg. Assoc. J.*, **38**, 224 (1977).
594. A. Hase, P. H. Lin, and R. A. Hites, "Analysis of Complex Polycyclic Aromatic Hydrocarbon Mixtures by Computerized GC-MS," *Carcinogenesis, A Comprehensive Survey*, Vol. 1, R. Freudenthal and P. W. Jones, Eds., Raven Press, New York, 1976.
595. L. Weil, H. Berger, and K. E. Quntin, *Chem. Ing. Tech.*, **49**(5), 429 (1977).
596. J. F. Mesquita, *C.R. Acad. Sci. Paris, Ser. D.*, **265**(4), 322 (1967).
597. J. E. Thornton and R. R. Bell, *Southwest. Vet.*, **26**(3), 227 (1973).
598. D. C. Hittle and J. J. Stukel, *J. Am. Ind. Hyg. Assoc.*, **37**(4), 199 (1976).
599. J. Neiser and V. Masek, *Zentralbl. Arbeitsmed. Arbeits. Prophyl.*, **26**(7), 127 (1976).

600. B. N. Tarusov, Yu. M. Petrusevich, N. G. Kozhanov, D. A. Makeev, and V. E. Novikov, *Tr. Mosk. O-va. Ispyt. Prir.*, **52**, 183 (1975).

601. I. Purchase, E. Longstaff, J. Ashby, J. Styles, D. Anderson, P. Lafevre, and F. R. Westwood, *Nature*, **264**, 624 (1976).

602. J. A. Styles, *Brit. J. Cancer*, **37**(6), 931 (1978).

603. S. E. Herbes and G. F. Risi, *Bull. Environ. Contam. Toxicol.*, **19**(2), 147 (1978).

604. G. J. Mulkins-Phillips and J. E. Stewart, *Can. J. Microbiol.*, **20**, 955 (1974).

605. S. E. Herbes and L. R. Schwall, *Appl. Environ. Microbiol.*, **35**(2), 306 (1978).

606. A. S. Russell, N. Jarrett, M. J. Bruno, J. A. Remper, and L. K. King, U.S. Pat. 3977846, August 31, 1976.

607. J. S. Robertson and P. J. Dunstan, *Biochem. J.*, **124**(3), 543 (1971).

608. M. D. Kipling, Chapter 6 in *Chemical Carcinogens*, C. E. Searle, Ed., American Chemical Society, Washington, DC, 1974.

609. J. D. Scribner, *J. Natl. Cancer Inst.*, **50**, 1717 (1973).

610. N. C. Popescu, D. Turnbull, and J. A. Dipaolo, *J. Natl. Cancer Inst.*, **59**(1), 289 (1977).

611. U. Bayer and T. Bauknecht, *Experientia*, **33**(1), 25 (1977).

612. H. Buchenauer, *Phytopathol. Z.*, **72**(4), 291 (1971).

613. D. Hoffman and E. L. Wynder, Chapter 7 in *Chemical Carcinogens*, C. E. Searle, Ed., American Chemical Society, Washington, DC, 1974.

614. H. Kubota, W. H. Griest, and M. R. Guerin, *Trace Subst. Environ. Health*, **9**, 281 (1975).

615. L. J. Casarett and J. Doull, Eds., *Toxicology, the Basic Science of Poisons*, Macmillan, New York, 1975.

616. D. M. Jerina, *Fed. Proc.*, **37**(6), 1383 (1978).

617. E. C. Miller and J. A. Miller, Chapter 16 in *Chemical Carcinogens*, C. E. Searle, Ed., American Chemical Society, Washington, DC, 1974.

618. E. Bingham and H. L. Falk, *Arch. Environ. Health*, **19**, 779 (1969).

619. J. Pataki and C. B. Huggins, *Cancer Res.*, **29**(3), 506 (1969).

620. D. W. Jones and R. S. Matthews, Chapter 4 in *Progress in Medicinal Chemistry 10*, G. P. Ellis and G. B. West, Eds., Elsevier, New York, 1974.

621. B. Tierney, A. Hewer, C. Walsh, P. L. Grover, and P. Sims, *Chem. Biol. Interact.*, **18**(2), 179 (1977).

622. I. Chouroulinkov, A. Gentil, B. Tierney, P. Grover, and P. Sims, *Cancer Lett.*, **3**(5-6), 247 (1977).

623. W. Baird, P. L. Grover, P. Sims, and P. Brookes, *Cancer Res.*, **36**, 2306 (1976).

624. W. Baird, A. Dipple, P. L. Grover, P. Sims, and P. Brookes, *Cancer Res.*, **33**(10), 2386 (1973).

625. S. Levy, D. Papadopoulo, S. Nocentini, L. Chamaillard, O. Beesau, M. Hubert-Habart, and P. Makovits, *Eur. J. Cancer*, **12**(11), 871 (1976).

626. H. Marquardt, S. Baker, B. Tierney, P. L. Grover, and P. Sims, *Int. J. Cancer*, **19**(6), 828 (1977).

627. V. M. Bobr and Yu. P. Kozlov, *Tr. Mosk. Obshchest. Ispyt. Pir.*, **32**, 33 (1970).

628. G. Prodi, A. M. Ferreri, P. Rocchi, and S. Grilli, *Z. Krebsforsch. Klin. Onkol.*, **81**(2), 161 (1974).

629. B. Solymoss, A. Somogyi, and K. Kovacs, *Haematologia*, **5**(1-2), 87 (1971).

630. A. G. Schwartz and A. Perantoni, *Cancer Res.*, **35**(9), 2482 (1975).

631. F. Serri, G. Pisanu, and G. Cantu, *G. Ital. Dermatol./Minerva Dermatol.*, **108**(1), 73 (1973).
632. E. P. Shuba, *Ukr. Biokhim. Zh.*, **41**(3), 249 (1969).
633. C. W. Heizmann and H. J. Wyss, *Arch. Gynekol.*, **216**(1), 51 (1974).
634. R. W. Newburg, Chapter 12 in *Toxicology, the Basic Science of Poisons*, L. J. Casarett and J. Doull, Eds., Macmillan, New York, 1975.
635. J. W. Flesher, *Biochem. Pharmacol.*, **16**(9), 1821 (1969).
636. C. J. Grubbs and R. C. Moon, *Cancer Res.*, **33**(7), 1785 (1973).
637. J. A. Schmutz, A. C. Brownie, and A. P. Chaudhry, *Cancer Res.*, **34**(3), 578 (1974).
638. F. B. Daniel, L. K. Wong, C. T. Oravec, F. D. Cazer, C'L. Wang, S. M. D'Ambrosio, R. W. Hart, and D. T. Witiak, "Metabolism and DNA Binding of PAH," *Polynuclear Aromatic Hydrocarbons, Third International Symposium*, R. W. Jones and P. Leber, Eds., Ann Arbor Science Publishers, Ann Arbor, MI, 1979.
639. J. Booth, G. R. Keysell, and P. Sims, *Biochem. Pharmacol.*, **22**(14), 1781 (1973).
640. J. H. Weisburger, Chapter 15 in *Toxicology the Basic Science of Poisons*, L. J. Casarett and J. Doull, Eds., Macmillan, New York, 1975.
641. R. L. Carter, *Brit. J. Cancer*, **28**(1), 91 (1973).
642. V. DeCosta and R. F. Aguirre, *Rev. Cent. Cienc. Biomed., Univ. Fed. St. Maria*, **4**(3-4), 49 (1976).
643. K. Adachi, S. Yamasawa, and W. Montagna, *J. Natl. Cancer Inst.*, **42**(1), 61 (1969).
644. B. Toth, *Tumori*, **57**(3), 169 (1971).
645. E. Huberman and L. Sachs, *Int. J. Cancer*, **13**(3), 326 (1974).
646. N. Veda, H. Venaka, T. Akematsu, and T. Sugiyama, *Nature*, **262**(5569), 581 (1976).
647. T. V. Nikonova, *Buyll. Eksp. Biol. Med.*, **84**(7), 88 (1977).
648. D. W. Lindasy, J. R. Jones, W. H. Higgins, and P. W. Brown, *Exp. Lung Cancer: Carcinog. Bioassays Int. Sympt.* (Res. Div. Carreras Rothmans Ltd., Basildon/Essex, Engl.), 521 (1974).
649. B. L. Van Duuren and B. M. Goldshmidt, unpublished data in B. L. Van Duuren, Chapter 2 in *Chemical Carcinogens*, C. E. Searle, Ed., American Chemical Society, Washington, DC, 1974.
650. R. M. Coomes, *Colo. Sch. Mines*, **71**(4), 101 (1976).
651. A. P. Ilnitsky, V. S. Mischenko, and L. M. Shabad, *Cancer Lett.*, **3**(5-6), 227 (1977).
652. G. M. Badger and J. Novotny, *Nature*, **198**, 1086 (1963).
653. J. N. Neiser and V. Masek, *Sb. Pr. Pedagog. Fak. Ostrone Rada E*, **5**, 75 (1975).
654. J. Howard and T. Fazio, *Ind. Med. Surg.*, **39**(10), 46 (1970).
655. N. G. Turkiya, G. L. Chechelashvili, P. N. Krasnyanskaya, D. S. Beniashvili, and L. I. Dzagnidze, *Vopr. Pitan.*, **30**(1), 31 (1971).
656. M. Rohrlich and P. Suckow, *Getreide Mehl. Brot.*, **26**(4), 114 (1972).
657. A. Audere, Z. Lindbergs, and G. A. Smirnov, *Gig. Sanit.*, **4**, 98 (1975).
658. L. M. Shabad, Y. L. Cohan, A. P. Ilnitsky, A. Y. Khesina, N. P. Shcherbak, and G. A. Sonirnov, *J. Natl. Cancer, Inst.*, **47**(6), 1179 (1971).
659. F. A. Schmid, M. S. Demetriades, F. M. Schabel, and G. S. Tarnowski, *Cancer Res.*, **27**(3), 563 (1967).
660. L. M. Shabad, *Arch. Geschwulstforsch.*, **38**(3-4), 185, (1971).
661. T. D. Sterling and S. V. Pollack, *Am. J. Publ. Health*, **62**(2), 152 (1972).

662. R. C. Levitt, J. S. Felton, J. R. Robinson, and D. W. Nebert, *Pharmacologist*, **17**(2), 213 (1975).
663. H. Reznik-Schueller and U. Mohr, *Zentralbl. Bakteriol. Hyg.*, **159**(5–6), 493 (1974).
664. H. Reznik-Schueller and U. Mohr, *Zentralbl. Bakteriol. Hyg.*, **159**(5–6), 503 (1974).
665. T. G. Samsonidze, M. A. Tsartsidze, and G. G. Samsonidze, *Soobsch. Akad. Nauk. Gruz. USSR*, **73**(1), 217 (1974).
666. I. A. Shendrikova, M. N. Ivanov-Golitsyn, and A. Y. Likhachev, *Vopr. Onkol.*, **20**(7), 53 (1974).
667. T. Tanaka, *Teratology*, **16**(86), (1977).
668. M. M. Andiranova, *Bull. Exp. Biol. Med.*, **71**(6), 677 (1971).
669. M. R. Juchau, *Toxicol. Appl. Pharmacol.*, **18**(3), 655 (1971).
670. M. R. Juchau, D. L. Berry, P. K. Zachariah, M. J. Namkung, and T. J. Slaga, "Prenatal Biotransformation of Carcinogens," in *Carcinogenesis, A Comprehensive Survey*, Vol. 1, R. Freudenthal and P. W. Jones, Eds., Raven Press, New York, 1976.
671. P. Leber, G. Kerchner, and R. I. Freudenthal, "Species Comparison of BP Metabolism," in *Carcinogenesis, A Comprehensive Survey*, Vol. 1, R. Freudenthal and P. W. Jones, Eds., Raven Press, New York, 1976.
672. J. K. Selkirk and M. C. McLeod, "Metabolism of B(a)P and B(e)P," in *Polynuclear Aromatic Hydrocarbons, Third International Symposium*, P. W. Jones and P. Leber, Eds., Ann Arbor Science Publishers, Ann Arbor, MI, 1979.
673. C. A. Jones, B. P. Moore, G. M. Cohen, and J. W. Bridges, "Metabolism of PAH," in *Polynuclear Aromatic Hydrocarbons, Third International Symposium*, P. W. Jones and P. Leber, Eds., Ann Arbor Science Publishers, Ann Arbor, MI, 1979.
674. D. M. Jerina, H. Yagi, O. Hernandez, P. M. Dansette, A. W. Wood, W. Lefin, R. L. Chang, P. G. Wislocki, and A. H. Conney, "Biologic Activity of BP Metabolites," in *Carcinogenesis, A Comprehensive Survey*, Vol. 1, R. Freudenthal and P. W. Jones, Eds., Raven Press, New York, 1976.
675. G. A. Belitskii, T. P. Raybykh, and V. A. Kodlyakov, *Tsitologiya*, **19**(10), 1193 (1977).
676. J. W. Flesher and K. L. Syndor, *Proc. Am. Assoc. Cancer Res.*, **13**, 55 (1972).
677. Y. Ioki, M. Kodama, Y. Tagashiar, and C. Nagata, *Gann*, **65**(4), 379 (1974).
678. N. H. Sloane, H. Chen, B. Divan, R. Bedigan, and H. Meier, "6-Hydroxymethylbenzo(a)pyrene Synthetase," in *Carcinogenesis, A Comprehensive Survey*, Vol. 1, R. Freudenthal and P. W. Jones, Eds., Raven Press, New York, 1976.
679. I. Y. Wang, R. E. Rasmussen, R. Creasy, and T. T. Crocker, *Life Sci.*, **20**(7), 1265 (1977).
680. M. E. McManus, K. F. Ilett, *Drug Metab. Dispos.*, **5**(6), 503 (1977).
681. H. Vadi, Bengt. Jerstromm, and S. Orrenius, "BP Metabolism in Rat Liver and Lung," in *Carcinogenesis, A Comprehensive Survey*, Vol. 1, R. Freudenthal and P. W. Jones, Eds., Raven Press, New York, 1976.
682. D. I. Katz, R. J. Stenger, E. A. Johnson, R. K. Datta, and J. Rice, *Arch. Int. Pharmacodyn. Ther.*, **229**(2), 180 (1977).
683. N. Nemoto, S. Takayama, and V. H. Gelboin, *Biochem. Pharmakol.*, **26**(19), 1825 (1977).
684. W. A. Bornstein, B. Hassuck, H. A. Chuang, and E. Bresnick, *Fed. Proc.*, **37**(6), 1383 (1978).
685. S. K. Yang, D. W. McCourt, P. P. Noller, and V. H. Gelvoin, *Proc. Natl. Acad. Sci., U.S.*, **73**(8), 2594 (1976).

686. R. E. Kouri, P. A. Lubet, and D. A. Brown, *J. Natl. Cancer Inst.*, **49**(4), 993 (1972).
687. R. E. Rasmussen and I. Y. Wang, *Cancer Res.*, **34**(9), 2290 (1974).
688. T. Tang and M. A. Friedman, *Mutat. Res.*, **46**(6), 387 (1977).
689. Yu L. Kogan, *Gig. Sanit.*, **39**(7), 110 (1974).
690. M. N. Pozlazova, G. E. Fedoseeva, A. Y. Khesina, M. N. Meisel, and L. M. Shabad, *Life Sci.*, **6**(10), 1053 (1967).
691. M. N. Pozlazova, G. E. Fedoseeva, A. Y. Khesina, M. N. Meisel, and L. M. Shabad, *Dokl. Acad. Nauk. USSR Ser. Biol.*, **198**(5), 1211 (1971).
692. A. Y. Khesina, N. P. Scherbak, L. M. Shabad, and I. S. Vostrov, *Bull. Exp. Biol. Med.*, **68**(10), 70 (1969).
693. H. Lorbacher, H. D. Puels, and H. W. Schlipkoeter, *Zentralbl. Bakteriol Parasitenk. Infektionskr.*, **155**(2), 168 (1971).
694. D. B. Clayson and R. C. Garner, Chapter 8 in *Chemical Carcinogens*, C. E. Searle, Ed., American Chemical Society, Washington, DC, 1974.
695. C. D. Haagensen and O. F. Krehbiel, *Am. J. Cancer*, **27**, 474 (1936).
696. E. L. Wynder and D. Hoffman, *Cancer*, **12**, 1079 (1959).
697. J. Jacob and G. Grimmer, *Zentralbl. Bakteriol. Parasitenkd. Infektionskr.*, **165**(3–4), 305 (1977).
698. J. Neal and N. M. Trieff, *Health Lab. Sci.*, **9**(1), 32 (1972).
699. A. Gold, *Environ. Aspects Chem. Use Rubber Process, Oper., Conf. Proc.*, 137 (1975).
700. A. Eisenstadt and A. Gold, *Proc. Natl. Acad. Sci., U.S.*, **75**(4), 1667 (1978).
701. J. H. Weisburger, Chapter 1 in *Chemical Carcinogens*, C. E. Searle, Ed., American Chemical Society, Washington, DC, 1976.
702. K. Buerki, R. A. Seibert, and E. Bresnick, *Biochim. Biophys. Acta*, **260**(1), 98 (1972).
703. F. Oesch, *Xenobiotica*, **3**(5), 305 (1973).
704. J. C. Arcos, M. F. Argus, and G. Wolf, *Chemical Induction of Cancer*, Vol. 1, Academic Press, New York, 1968. Washington, DC, 1974.
705. M. A. Mehlman, R. E. Shapiro, and H. Blumenthal, *New Concepts in Safety Evaluation*, Hemisphere, Washington, DC, 1976.
706. W. F. Dunning, M. R. Curtis, and M. J. Eisin, *Am. J. Cancer*, **40**, 85 (1940).
707. T. L. Thomas, P. Decoufle, and R. Moure-Eraso, *J. Occup. Med.*, **22**(2), 97 (1980).
708. P. K. Gupta, T. S. Dikshith, and K. K. Datta, *Toxicology*, **7**(1), 57 (1977).
709. H. M. Smith, *Bur. Mines Rep. Invest. 6542*, U.S. Dept. Interior, Washington, DC, 1964.
710. D. I. Zul'Fugarly and N. S. Umakhanova, *Azerbaidzhan. Khim. Zhur.*, **1**, 65 (1960).
711. S. M. Katchenov and E. I. Flegontova, *Vestsi Akad. Navuk. Belarus. USSR Ser. Khim. Navuk.*, **16**(6), 95 (1970).
712. R. L. Erickson, A. T. Myers, and C. A. Horr, *Bull. Am. Assoc. Petrol. Geol.*, **38**, 2200 (1954).
713. L. D. Rowe, J. W. Dollahite, and B. J. Camp, *J. Am. Vet. Med. Assoc.*, **162**(1), 61 (1973).
714. S. S. Rossi, J. W. Anderson, and G. S. Ward, *Environ. Pollut.*, **10**(1), 9 (1976).
715. J. M. Neff, J. W. Anderson, B. A. Cox, R. B. Laughlin, Jr., S. S. Rossi, and H. E. Tatem, *Sources, Eff. Sinks. Hydrocarbons Aquat. Environ. Proc. Symp.*, 515 (1977).

716. R. Knowles and C. Wishart, *Environ. Pollut.*, **13**(2), 133 (1977).
717. R. F. Lee and M. Takashashi, *Cons. Int. Explor. Mer.*, **171**, 150 (1977).
718. W. Pulich, Jr., K. Winters, and C. Van Baalen, *Mar. Biol.*, **28**(2), 87 (1974).
719. *API Toxicological Review of Gasoline*, American Petroleum Institute, New York, 1967.
720. J. I. Tonge, R. N. Hurley, and J. Ferguson, *Lancet*, **1**, 1059 (1969).
721. A. Rothe, *Z. Aerztl. Fortbild.* (Jena), **66**(15), 758 (1972).
722. C. F. Phillips and R. K. Jones, *J. Am. Ind. Hyg. Assoc.*, **39**(2), 118 (1978).
723. K. P. Pandya, G. S. Rao, A. Dhasmana, and S. H. Saidi, *Ann. Occup. Hyg.*, **18**(4), 363 (1975).
724. J. Przybylowski, J. Wysocki, Z. Szczepanski, A. Sychlowy, and A. Podolecki, *Bromatol. Chem. Toksykol.*, **9**(1), 33 (1976).
725. L. Brandt, P. G. Nilsson, and F. Mitelman, *Brit. Med. J.*, **1**, 553 (1978).
726. A. Poklis and C. Burkett, *Clin. Toxicol.*, **11**(1), 35 (1977).
727. R. L. Boeckx, B. Postl, and F. J. Coodin, *Pediatrics*, 60(2), 140 (1977).
728. M. B. Chenoweth, *J. Ind. Hyg. Toxicol.*, **28**, 151 (1946).
729. M. Bass, *J. Am. Med. Assoc.*, **212**, 2075 (1970).
730. L. S. Goodman and A. Gilman, *The Pharmacological Basis of Therapeutics*, 4th ed., Macmillan, New Haven, CT, 1971.
731. J. Przybylowski, W. Kowalski, and A. Podalecki, *Patol. Pol.*, **27**(2), 149 (1976).
732. E. E. Gasanova and S. F. Fatalieva, *Azerb. Med. Zh.*, **48**(6), 29 (1971).
733. R. S. Sunargulov, A. K. Giniyatullina, and T. S. Ivanova, *Oftalmol. Zh.*, **31**(1), 20 (1976).
734. J. Karkos and J. Sikora, *Neuropatol. Pol.*, **11**(1), 99 (1973).
735. N. A. Minkina, E. G. Berliner, and S. A. Chernova, *Probl. Adapt. Gig. Tr.*, 50 (1973).
736. S. Urishibara, *Tokyo Jikeikai Ika Daigaku Zasshi*, **91**(2), 198 (1976).
737. I. Feeler, *Gig. Tr. Prof. Zabol.*, **16**(8), 25 (1972).
738. G. Soderman, *Hereditas*, **71**(2), 335 (1972).
739. C. A. Halder, C. E. Holdsworth, B. Y. Cockrell, and V. J. Piccirillo, *Toxicol. Indust. Health.*, **1**, 67–68 (1985).
740. D. R. Webb, G. M. Ridder, and C. L. Alden *Food Chem. Toxicol.*, **27**, 639–649 (1989).
741. L. C. Stone, M. S. McCracken, R. L. Kanerva, and C. L. Alden, *Food Chem. Toxicol.*, **25**, 35–41 (1987).
742. G. M. Ridder, E. C. von Bargen, C. L. Alden, and R. D. Parker, *Fundam Appl. Toxicol.*, **15**, 732–743 (1990).
743. C. L. Alden, R. L. Kanerva, G. M. Ridder, and L. C. Stone, *Renal Effects of Petroleum Hydrocarbons*, Princeton Scientific, Princeton, 1984, pp. 107–120.
744. C. L. Gaworski, C. C. Haun, J. D. MacEwen, E. H. Vernot, and R. H. Bruner, *Fundam. Appl. Toxicol.* **5**, 785–793 (1985).
745. J. A. Swenberg, B. Short, S. Borghoff, J. Strasser, and M. Charbonneau, *Toxicol. Appl. Pharmacol.* **97**, 35–47 (1989).
746. T. Nagata, M. Kagleura, K. Hara, and K. Totoki, *Nippon Hoegaku Zasshi*, **31**(3), 136 (1977).

747. R. Takahashi, T. Sone, and T. Hirata, Jap. Pat 77 19190, February 14, 1977.
748. H. Hanninen, L. Eskelinen, K. Husman, and M. Nurminen, *Scand. J. Work Environ. Health*, **2**(4), 240 (1976).
749. R. Rawson, F. Parker, and H. Jackson, *Science*, **93**, 2433 (1941).
750. C. H. Hine and H. H. Zuidema, *Ind. Med. Surg.*, **39**, 215 (1970).
751. L. D. White, D. G. Taylor, P. A. Mauer, and R. E. Kupel. *Am. Ind. Hyg. Assoc. J.*, **31**, 225 (1970).
752. V. J. Harris and R. Brown, *Am. J. Roentgenol. Radium Ther. Nucl. Med.*, **125**(3), 531 (1975).
753. *Criteria for a Recommended Standard—Occupational Exposure to Refined Petroleum Solvents*, U.S. Dept. Health, Education, and Welfare, Public Health Service, NIOSH, Washington, DC, 1977.
754. A. J. McMichael, R. Spiritas, L. L. Kupper, and J. F. Gamble, *J. Occup. Med.*, **17**(4), 234 (1975).
755. C. P. Carpenter, E. R. Kinkead, D. L. Geary, Jr., L. J. Sullivan, and J. M. King, *Toxicol. Appl. Pharmacol.*, **33**, 526 (1975).
756. F. W. Wilson, *J. Occup. Med.*, **18**, 821 (1976).
757. C. P. Carpenter, E. R. Kinkead, D. L. Geary, L. J. Sullivan, and J. M. King, *Toxicol. Appl. Pharmacol.*, **32**(2), 263 (1975).
758. G. D. DiVincenzo and W. J. Krasavege, *Am. Ind. Hyg. Assoc. J.*, **35**(1), 21 (1974).
759. C. P. Carpenter, E. R. Kinkead, D. L. Geary, Jr., L. J. Sullivan, and J. M. King, *Toxicol. Appl. Pharmacol.*, **34**, 413 (1975).
760. *API Toxicological Review of Petroleum Naphthas*, American Petroleum Institute, New York, 1969.
761. H. Altenkirch, J. Mager, G. Stoltenburg, and J. Helmbrecht, *J. Neurol.*, **214**(2), 137 (1977).
762. G. L. Choules and W. C. Russell, *Vet. Hum. Toxicol.*, **19**(4), 253 (1977).
763. J. A. Richardson and H. R. Pratt-Thomas, *Am. J. Med. Sci.*, **221**(5), 531 (1951).
764. R. H. Daffner and J. P. Jiminez, *Radiology*, **106**(2), 383 (1973).
765. L. Chin, A. Picchioni, and B. Duplisse, *J. Pharm. Sci.*, **58**, 1353 (1969).
766. Anon., *Brit. Med. J.*, **3**, 487 (1972).
767. R. C. Ng. H. Darwish, and D. A. Stewart, *Can. Med. Assoc. J.*, **111**, 537 (1974).
768. *API Toxicological Review of Kerosene*, American Petroleum Institute, New York, 1967.
769. H. Tagami and A. Ogino, *Dermatologica*, **146**, 123 (1973).
770. A. P. Luplescu, H. Pinkus, and D. J. Birmingham, *Proc. Electron Microsc. Soc. Am.*, **30**, 92 (1972).
771. A. P. Luplescu, D. J. Birmingham, and H. Pinkus, *J. Invest. Dermatol.*, **60**(1), 32 (1973).
772. A. P. Luplescu and D. J. Birmingham, *J. Invest. Dermatol.*, **65**(5), 419 (1975).
773. D. E. Johnston, *J. Am. Med. Women's Assoc.*, **10**, 421 (1955).
774. J. Wolfsdorf and H. Kundig, *S. Afr. Med. J.*, **46**, 619 (1972).
775. A. Volkova, V. Tsetlin, E. Zhuk, and E. Izotova, *Gig. Sanit.*, **34**, 24 (1969).
776. H. W. Gerarde, *Occup. Health Rev.*, **16**(3), 17 (1964).

777. P. Gross, J. M. McNerney, and M. A. Baleyak, *Am. Rev. Respir. Dis.*, **88**(5), 656 (1963).
778. M. D. Mann, D. J. Pirie, and J. Wolfsdorf, *J. Pediatr.*, **91**(3), 495 (1977).
779. R. W. Lewis, *Int. J. Biochem.*, **2**(11), 609 (1971).
780. K. Morihara, *Appl. Microbiol.*, **13**(5), 793 (1965).
781. K. Katsuri and D. V. Tamhane, *Ind. J. Exp. Biol.*, **9**(2), 235 (1971).
782. K. I. Markov and T. Kobarska, *Dokl. Akad. Sel-skokhoz. Nauk. Bolg.*, **4**(4), 413 (1971).
783. N. Gerasimova and M. Bekhtereva, *Mikrobiologiya*, **39**(4), 616 (1970).
784. C. P. Carpenter, D. L. Geary, Jr., R. C. Myers, D. J. Nachreiner, L. J. Sullivan, and J. M. King, *Toxicol. Appl. Pharmacol.*, **36**, 443 (1976).
785. N. E. Davies, *Aerosp. Med.*, **35**, 481 (1964).
786. C. L. Gaworski and H. F. Leahy, *Proc. 9th Ann. Conf. Environ. Toxicol.*, Govt. Rept. AMRL-TR-79-68, Aerospace Medical Research Laboratory, Wright-Patterson Air Force Base, OH, 1979.
787. C. C. Haun, *Aerosp. Med. Res. Lab. Rep. AMRL-TR (U.S.)*, **125**, 287 (1975).
788. S. A. Klein, D. Jenkins, and R. C. Cooper, *Aerosp. Med. Res. Lab. Rep., AMRL-TR (U.S.)*, **125**, 429 (1975).
789. J. Sula and V. Krol, *Prot. Vitae*, **16**(6), 266 (1971).
790. J. A. Berlin and D. W. Micks, *Ann. Entomol. Soc. Am.*, **66**(4), 775 (1973).
791. H. E. Tatem, B. A. Cox, and J. W. Anderson, *Estuarine Coastal Mar. Sci.*, **6**(4), 365 (1978).
792. R. C. Clark, Jr., and J. S. Finley, *Fish Bull.*, **73**(3), 508 (1975).
793. D. M. Stainken, *J. Fish Res. Board Can.*, **35**(5), 637 (1978).
794. K. Yamada and M. Yogo, *Agric. Biol. Chem.*, **34**(2), 296 (1970).
795. M. D. Kipling and H. A. Waldron, *Prev. Med.*, **5**, 262 (1976).
796. H. E. Burmeister, *Berufsdermatosen*, **21**(2), 69 (1973).
797. H. Luther and G. Bergmann, *Erdoel Kohle*, **8**, 298 (1955).
798. E. G. Ivanyuk and V. V. Vasilenko, *Gig. Sanit.*, (7), 82 (1976).
799. G. A. Gellin, *Ind. Med.*, **39**(2), 38 (1970).
800. Anonymous, *Occup. Health Safety*, Sept./Oct., 16 (1976).
801. G. A. Gellin, *J. Occup. Med.*, **11**(3), 128 (1969).
802. M. H. Samitz and S. A. Katz, *Contact Dermatitis*, **1**, 158 (1975).
803. W. Catchpole, E. MacMillan, and H. Powel, *Ann. Occup. Hyg.*, **14**(2), 171 (1971).
804. J. A. Waterhouse, *Ann. Occup. Hyg.*, **14**(2), 161 (1971).
805. T. H. F. Smith, *Ind. Med.*, **39**(2), 29 (1970).
806. J. R. Jepsen, S. Stoyanov, M. Unger, J. Clausen, and H. Christensen, *Acta Pathol. Microbiol. Scand.* **85**(5), 731 (1977).
807. R. Cabridenc, *Microb. Mater.*, 123 (1974).
808. J. Markind, J. Neri, and R. Stana, *AIChE Symp. Ser.*, **71**(151), 70 (1975).
809. H. Ramos, *J. Occup. Med.*, **16**(4), 273 (1974).
810. A. B. Balzan, *Fachh. Chemigr. Lithogr. Tiefdruck*, **5**(3), 173 (1974).
811. W. Lijinsky, U. Saffiotti, and P. Shubik, *Toxicol. Appl. Pharmacol.*, **8**(1), 113 (1966).

CHAPTER TWENTY-TWO

# Halogenated Benzenes

A. Philip Leber, Ph.D., D.A.B.T., and Theodore J. Benya, Ph.D., D.A.B.T.

## 1  INTRODUCTION

This group of chemicals is derived through substitutions of chlorine for hydrogen atoms on the benzene ring. The toxicologic properties of chlorinated benzenes differ substantially from benzene in that they lack the hematologic toxicity and leukemogenic activity that have been associated with benzene. The compounds have an aromatic odor with lower volatilities, higher densities, and lower flammabilities than the parent compound benzene. Some of the settings in which exposures may be encountered are in manufacturing sites, home and public areas where deodorants/disinfectants are present, and in the applications of the materials as fumigants, insecticides, lacquers, paints, and seed disinfection products.

Because a substantial number of studies have focused upon comparative evaluations of compounds in this chemical group, the results of these investigations are presented below prior to detailed discussions of individual compounds. In addition, a summary of various regulatory classifications of these compounds is provided.

### 1.1  Comparative Toxicological Properties of Chlorobenzenes

According to a Chemical Manufacturers Association (CMA) Technical Report (1), consistently observed signs of acute intoxication vary with the route of administration and the degree of exposure. Varshavskaya (2) reported the following effects for chlorobenzene and *o*- and *p*-dichlorobenzenes (DCB) in mice, rats, and guinea

---

*Patty's Industrial Hygiene and Toxicology, Fourth Edition, Volume 2, Part B*, Edited by George D. Clayton and Florence E. Clayton.
ISBN 0-471-54725-5   © 1994 John Wiley & Sons, Inc.

**Table 22.1.** Oral $LD_{50}$s for Chlorobenzenes (3)

| Species | $LD_{50}$ (mg/kg body weight) | | |
| --- | --- | --- | --- |
| | Chlorobenzene | $o$-Dichlorobenzene | $p$-Dichlorobenzene |
| White mice | 1145 | 2000 | 3220 |
| White rats | 2390 | 2138 | 2512 |
| Rabbits | 2250 | 1875 | 2812 |
| Guinea pigs | 5060 | 3375 | 7595 |

pigs: hyperemia of the visible mucous membranes, increased salivation and lacrimation, initial excitation followed by drowsiness, adynamia, ataxia, paraparesis, paraplegia, and dyspnea. Induced mortality, which usually occurs within 3 days, results from respiratory paralysis. Changes revealed by gross postmortem examination may include hypertrophy and necrosis of the liver and submucosal hemorrhages in the stomach. Histologically disclosed changes are edema of the brain and necrosis of the centrobular region of the liver, the proximal convoluted tubules of the kidneys, the bronchial and bronchiolar epithelium of the lungs, and the stomach mucosa. The liver generally showed the most severe damage.

Table 22.1 presents data on the acute oral toxicity of mono- and dichlorobenzenes in rodents. This information suggests greater toxicity for $o$-DCB relative to the other compounds in three of four animal species tested. The basis for this observation may lie within the differences seen for the metabolism of these compounds, and the formation of a more reactive metabolite for the ortho-substituted chemical (discussed further below).

The acute effects of intraperitoneal dosing of a series of chlorinated benzenes (mono, di, tri, tetra, and penta substituted) were assessed in rats (4). Following single doses at levels in excess of 100 mg/kg, animals were monitored for evidence of liver, kidney, and thyroid effects. The 1,2-di, 1,2,4-tri- and 1,2,4,5-tetrachlorobenzenes were most toxic based on body weight losses over 3 days. Liver function enzyme levels (alanine aminotransferase or ALT) were significantly elevated, as were microscopic signs of cellular damage following treatments with 225 and 450 mg/kg of the monochloro compound, 150 mg/kg and above of 1,2-dichlorobenzene, and 185 mg/kg and above of 1,2,4-trichlorobenzene. Kidney changes included protein droplet nephropathy for the 1,4 dichloro and the more highly substituted compounds. Serum thyroxine (T4) levels were suppressed at the lowest doses administered (1 mmol/kg) for the 1,2-dichloro and the higher-substituted benzenes. The greatest activity was seen for the tri and penta compounds. This activity appears to depend on formation of phenolic metabolites, which were shown to have the capacity to displace T4 from serum binding proteins in the rat. Alternatively, enhanced metabolic deiodination of T4 may result from induction of hepatic enzymes by the chlorinated benzene compounds. The precise mechanism for this phenomenon has not yet been established.

Hepatotoxicity evaluations of the three dichlorobenzene isomers (ortho, meta, para) were made following intraperitoneal injection into Sprague–Dawley and F344

Table 22.2. Results of Carcinogenesis Bioassays in Rodents Following Administration of Chlorinated Benzenes

| Compound | Tumors In | | Dose (mg/kg) | Ref. |
| --- | --- | --- | --- | --- |
| | Mice | Rats | | |
| MCB | None | Liver nodules (males) | 60, 120 oral 30, 60 (male mice) | 6 |
| o-DCB | None | None | 60, 120 (gavage) | 7 |
| p-DCB | Liver | Males—kidney | 150, 300—male rats 300, 600—others oral | 8 |
| | None | None | 75,500 ppm inhaln. 5 hr/day, 5 days/ week, 57 weeks— mice; 78 weeks—rat | 9 |
| HCB | Liver | Liver, males— kidney | 6–24 mg/kg daily (diet) | 10, 11 |
| | Liver, thyroid, angiosarcomas | **(Hamsters)** | 4–16 mg/kg daily | 12 |

male rats (5). Liver effects were monitored using plasma ALT levels, microscopic observations, liver nonprotein sulfhydryl (NPSH) concentration, and $^{14}$C-DCB binding to liver proteins. One day following a dose of 132 mg/kg of each isomer, ALT levels were increased for o-DCB only. As dose levels were increased up to five times, p-DCB exhibited no activity in this assay, whereas m-DCB induced a weak response beginning at 264 mg/kg. Test chemical binding to proteins paralleled this liver response. Results for phenobarbital-pretreated rats indicated that liver enzyme induction was associated with the hepatotoxicity of the ortho and meta but not the para isomer. Further evidence of enzyme mediation in the hepatotoxic mechanism was found when SKF-525A (a P450 enzyme inhibitor) blocked the response of the ortho and meta isomers. NPSH levels were depressed relative to controls following dosing of ortho and meta isomer dosing, but the para isomer did not cause any change in this parameter. The authors concluded that the relative hepatotoxicity of dichlorobenzene isomers is o- > m- $\gg$ p-, and that oxidative biotransformation plays an important role in this activities of these chemicals.

## 1.2 Carcinogenicity of Chlorinated Benzenes

The carcinogenic potentials of many of the chlorobenzenes have been assessed in animal studies. Table 22.2 summarizes these findings.

The findings for monochlorobenzene (MCB) were negative in mice whereas benign liver tumors were induced in rats. o-DCB did not induce rodent tumors in this bioassay. p-DCB was shown to be carcinogenic in both rodent species in oral gavage studies but the results were negative in inhalation tests. Owing to the relatively short periods of exposure in the latter study compared to the conventional 100+ weeks, the results may be considered equivocal. Hexachlorobenzene (HCB)

**Table 22.3.** Cancer Classifications for Chlorobenzenes

| Chemical | U.S. EPA (25) | IARC (26) | NTP (27) | ACGIH (28) |
|---|---|---|---|---|
| Chlorobenzene | D[a] | — | — | — |
| o-Dichlorobenzene | — | 3[a] | — | — |
| p-Dichlorobenzene | C[b] | 2B[d] | +[e] | — |
| Tri-, tetra-, pentachlorobenzenes | — | — | — | — |
| Hexachlorobenzene | B2[c] | 2B[d] | +[e] | — |

[a]Not classifiable as to carcinogenicity in humans.
[b]Possible human carcinogen.
[c]Probable human carcinogen (no human evidence).
[d]Possibly carcinogenic to humans.
[e]Reasonably anticipated to be carcinogenic to humans.

has been demonstrated to induce tumors in multiple species in multiple organs. These results have contributed to the higher suspicion of carcinogenic potential for HCB compared to other chlorinated benzenes (see Table 22.3 below).

The chlorobenzene class of compounds is devoid of significant evidence of genetic toxicity (13). MCB and o-DCB are the two members of the class most likely to form epoxides, which renders suspect genotoxic activities. However, MCB was negative in bacterial and *Aspergillus* gene mutation tests and in rat hepatic DNA repair assays but weakly positive in a mouse micronucleus assessment (14). The mutagenicity literature for o-DCB indicates some activity in higher plants but no effects in *Salmonella*, *E. coli*, and yeast (15). p-DCB literature (16) indicates that the compound lacks genotoxic activity in bacteria, hepatic DNA repair assays, mouse lymphoma cells, tests for chromosomal aberrations in lymphocytes, Chinese hamster ovary cells, and rat and mouse bone marrow cells. This compound did induce chromosomal changes in various higher plant assays although human health implications have not been established (further discussion below for p-DCB). HCB has been shown to be devoid of genotoxic activities in *Salmonella* gene mutation and rat dominant lethal testing (17). This lack of genetic activity suggests that alternative mechanisms of action (nongenotoxic) for the tumorigenic members of the class need to be considered.

Because of the induction of kidney tumors by p-DCB and HCB, these chemicals as well as other chlorobenzenes were investigated for evidence of a protein-mediated mechanism that is known to operate for numerous chemicals including unleaded gasoline and d-limonene. It was discovered that both p-DCB (18) and HCB (19) induced a male rat-specific protein that causes increased kidney cell turnover and necrosis (sometimes referred to as "hyalin droplet nephropathy") with kidney tumor formation a long-term outcome of this chronic injury. The implication of these findings is that kidney tumors formed in this manner have no relevance to human risk assessments because the protein ($\alpha$-2-microglobulin) responsible for the adverse effects in the kidney is unique to male rats. This concept is now recognized in certain regulatory agencies (20). Evaluations of other agents have been performed for this protein-induction activity, and results indicate that the 1,3,5 (21) and 1,2,4 (22) trichlorobenzenes, 1,2,4,5-tetrachlorobenzene (23), and pentachlorobenzene (24) are active. However, the actual carcinogenic potentials of these chemicals have not been determined in long-term bioassays.

The U.S. Environmental Protection Agency (EPA) and Occupational Safety and Health Administration (OSHA) regulate various chlorobenzenes as pesticides, as potential pollutants of air and water, or as workplace contaminants. These agencies as well as other nonregulatory bodies [International Agency for Research in Cancer (IARC), National Toxicology Program (NTP), American Conference of Governmental Industrial Hygienists (ACGIH)] issue classifications that are intended to indicate the level of evidence for carcinogenic potentials in humans (based on animal data) for each of these chemicals. These classifications are presented below in Table 22.3.

The precise relevance to human health of tumor findings in animal studies is a topic that continues to be actively debated. Although there is no substantive evidence that any of the chlorobenzenes have induced cancer in humans, positive results in animals indicate a need for awareness that exposure to these materials should be kept to a minimum.

## 2 CHLOROBENZENE [CAS # 108-90-7]

Chlorobenzene, also known as monochlorobenzene and MCB, has the formula $C_6H_5Cl$.

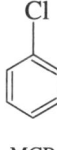

MCB

### 2.1 Uses and Industrial Exposure

Chlorobenzene is used extensively as a solvent for pesticide formulations and auto parts degreasing, and as a chemical intermediate in the production of diphenyl oxide, diisocyanates and nitrochlorobenzene. There is an occupational exposure potential to personnel engaged in the manufacture and use of this compound.

### 2.2 Physical and Chemical Properties

| | |
|---|---|
| Physical state | Colorless liquid |
| Molecular weight | 112.56 |
| Specific gravity | 1.1066 (20°C) |
| Melting point | −44.9°C |
| Boiling point | 132.0°C |

| | |
|---|---|
| Vapor density | 3.88 (air = 1) |
| Vapor pressure | 11.8 mm Hg (25°C) |
| Refractive index | 1.5216 (25°C) |
| Percent in "saturated" air | 1.55 (25°C) |
| Density of "saturated" air | 1.05 (air = 1) |
| Solubility | 49 mg/100 ml water at 20°C; soluble in alcohol, benzene, diethyl ether |
| Flash point | −90°F (closed cup) |
| Odor threshold | 1–8 mg/m$^3$ |
| Partition coefficient (log octanol/water) | 2.84 |

Conversion factors: 1 mg/l = 217 ppm (air); 1 ppm = 4.60 mg/m$^3$ at 25°C, 760 mm Hg

## 2.3 Effects in Animals

As is the case for many volatile organic chemicals, chlorobenzene is a nervous system depressant. Lesions of the liver and kidneys have also been observed following administration of toxic doses. Descriptions of these toxicities are presented in the following sections.

### 2.3.1 Acute Toxicities

Hollingsworth et al. (29) reported that rats survived an oral dose of MCB of 1.0 g/kg, but 4.0 g/kg was lethal to all animals dosed. Guinea pigs survived an oral dose of 1.6 g/kg, but administration of 2.8 g/kg was fatal. Additional oral LD$_{50}$ information is presented for MCB in Table 22.1.

The intraperitoneal LD$_{50}$ of chlorobenzene for rats was reported as 0.52 ml/kg (30), whereas subcutaneous injections of 4 to 5 or 7 to 8 g/kg to rats caused death in a few days to within a few hours, respectively. Necroses of the liver and kidneys were observed in terminal animals (31).

Limited information is available for acute toxicities resulting from inhalation exposures. Cat experiments conducted in 1904, and reported by Flury and Zernik (32), showed that exposures to 37 mg/l (8000 ppm) of MCB resulted in severe narcosis in 30 min and death in 2 hr. Exposure to 17 mg/l (3700 ppm) resulted in death in 7 hr, and 7 hr of exposure to 11 to 13 mg/l (2400 and 2900 ppm) caused restlessness, tremor, and muscular spasms, but no serious injury or fatalities. Definite narcotic effects were produced by exposure to 5.5 mg/l (1200 ppm) whereas concentrations of 1 to 3 mg/l (200 to 660 ppm) were tolerated for hours without significant effects (33). Rozenbaum et al. (34) reported that mice exhibited an approximate LC$_{50}$ (2 hr) of 4300 ppm.

Irish (33) found that topical application of MCB to the skin of rabbits caused slight reddening. Continuous contact for a week results in moderate erythema and slight superficial necrosis. In these studies, there was no indication of absorption of toxic levels, although prolonged skin contact is irritating.

Eye contact with liquid MCB results in transient conjunctival irritation to rabbits that resolves within 48 hr (35). No corneal injury was observed, although pain is likely to accompany contact with this chemical.

### 2.3.2 Subchronic Oral Toxicity

MCB was administered orally to rats 5 days/week for a total of 137 doses over a period of 192 days. A dose of 14.4 mg/kg body weight/day was survived without any observable effect. At 144 mg/kg, there was a slight decrease in growth from which test rats recovered. At both 144 mg/kg and 288 mg/kg, there were significant increases in liver and kidney weights with slight liver pathology. Blood and bone marrow were normal, as reported by Flury and Zernik (32, 33).

Similar observations were made by Hollingsworth et al. (29). No adverse side effects were observed in rats following the repeated administration of 18.8 mg/kg/day, 5 days/week for 192 days. A dosage of 188 mg/kg/day caused a slight increase in liver and kidney weights, and 376 mg/kg/day caused slight cirrhosis and focal necrosis of the liver and a slight decrease in average spleen weights. No other effects were observed.

Knapp et al. (36) administered MCB (by capsule) to dogs in doses of 27.2, 54.5, and 272.5 mg/kg/day, 5 days/week for 93 days. At the low and intermediate levels, the treatment was without effect. The highest dosage produced a reduction of blood sugar, but an increase in immature leukocytes, elevated serum glutamic–pyruvic transaminase and alkaline phosphatase, and increases in total bilirubin and total cholesterol. Four of eight dogs treated with the highest dose died after 14 to 21 doses. Gross and/or microscopic pathology was evident in the liver, kidney, gastroenteric mucosa, and hematopoietic tissue (1).

### 2.3.3 Subchronic Vapor Exposure

In a multi-species study reported by Irish (33), rats, rabbits, and guinea pigs were exposed 7 hr/day, 5 days/week for a total of 32 exposures over a period of 44 days. At a concentration of 1000 ppm in the air, there were histopathological changes of the lungs, liver, and kidneys. There was a slight depression of growth. Although there was no mortality in rats or rabbits, the guinea pigs did show a higher-than-normal mortality. These same species of animals survived exposures of 475 ppm, which induced a slight increase in liver weight and minimal liver pathology. The blood was essentially normal in all test animals. At a concentration of 200 ppm, all the animals appeared to be normal.

Khanin (37), after exposing rats to 24-hr inhalation of MCB in a concentration of 1.0 mg/m$^3$ for 70 to 82 days, noted encephalopathy and inflammation of the internal organs, protein dystrophy, and a scattering of regenerated cells on the liver lobes. Foci of giant-cell hyperplasia were also found in the kidneys (1). Tarkhova (38) exposed groups of 15 male rats continuously for 60 days to MCB. The air concentration of 1.0 mg/m$^3$ produced a lowering and distortion of the ratio of the chronaxie of the antagonistic muscles, increased blood cholineserase activity, and lowered the α-globulin blood serum content. Exposure to 0.1 mg/m$^3$ for only 60 days did not produce these changes (1).

A steady reduction of blood catalase and leukocytic indophenol oxidase activities were noted in guinea pigs following exposure to concentrations of 0.1 to 1.5 mg/l for 3 hr every second day for 62 weeks (39). In another investigation, male rats and rabbits were exposed to 75 or 250 ppm of MCB vapors for 7 hr/day for up to 120 exposures over 24 weeks (40). Groups of animals from both species were sacrificed after 5 and 11 weeks of exposure and examined for hematologic and other changes. Statistical analysis of the data suggested some treatment-related effects of red cell parameters including an increase in reticulocyte count, which was more apparent in rats than in rabbits. Clinical chemical changes in both species were nonspecific. Most tissues examined showed no remarkable changes; only the congestion of the liver and kidneys of the rabbits sacrificed after 5 weeks suggested a relationship to exposures.

### 2.3.4 Chronic Oral Exposure

The NTP evaluated toxic responses in rats following 13- and 102-week dietary exposures to MCB (6). Hepatic changes as evidenced by increases in liver weights and degenerative effects were observed, as were renal necrosis and degeneration of proximal tubules. Damage to these organs was postulated to be associated with covalent binding of the activated chemical (perhaps an epoxide) with proteins in the target tissues (41, 42).

### 2.3.5 Teratology and Reproductive Effects

The teratogenic potential of MCB was evaluated following inhalation exposures to rats and rabbits (43). Pregnant animals were exposed to 0, 75, 210, and 590 ppm MCB for 6 hr/day during the species' gestational periods of organogenesis. Evidence of maternal toxicities seen at the highest exposure level included increased liver-to-body weight ratios (both species) and body weight and feed consumption decreases in rats. Minor skeletal alterations in the 590-ppm rats (delayed ossification and bilobulation of centra of cervical vertebrae) were the only compound-related fetal changes. The fetal results in the first of two trials in rabbits were various major malformations (spina bifida, acephaly, omphalocele, heart anomalies) in MCB-exposure groups. However, because of a lack of dose-responsiveness for these findings, as well as a lack of evidence supportive of chemically induced teratogenicity in a follow-up rabbit study by the same investigators, MCB was considered to lack developmental (embryonic and fetal) toxicity in both rats and rabbits.

A multi-generation rat reproduction study was performed using vapor inhalation exposures to MCB (44). $F_0$ male and female rats were exposed to 0, 50, 150, or 450 ppm for 10 weeks prior to mating. Exposures continued during mating, gestation, and lactation phases of the study. $F_1$ rats were exposed 11 weeks premating through the lactation period of $F_2$ animals. Fertility and mating indexes, survival, and gross and microscopic observations were included in the study. No compound-related body weight or feed consumption deficits nor mortality were observed. No adverse changes were seen in any reproductive or offspring parameters such as survival or litter sizes. Hepatocellular hypertrophy and certain renal changes (tubular dilation, interstitial nephritis, and focal regenerative epithelium) were seen in mid- and high-dose males in the $F_0$ and $F_1$ generations. Germinal degeneration of the testes was observed in high-dose $F_0$ males and in mid- and high-dose $F_1$ animals. It was concluded that the no observable effect level (NOEL) for all effects in the study was 50 ppm MCB, and 450 ppm had no impact on reproductive performance or fertility of either sex of rats.

## 2.4 Absorption, Metabolism, and Excretion

MCB is readily absorbed from air in the lungs of rats (45). Their studies indicated that following single 8-hr exposures to the $^{14}$C-labeled compound, radioactivity was highest in epididymal and adipose tissues. In general, tissue levels of $^{14}$C-MCB equivalents increased in proportion to inhalation exposures from 100 to 700 ppm. Fatty tissue levels increased above linear rates, consistent with the octanol–water partition coefficient (2.84) for this chemical. Elimination of the test chemical via exhaled air occurred in a biphasic manner. The first phase of this route of elimination was lengthened at higher exposure concentrations as an apparent result of metabolic pathway saturation. Following multiple daily exposures, this initial phase was shortened, perhaps due to an induction of metabolic conversion of MCB in exposed animals.

Absorption from the gastrointestinal tract was also evaluated (46) with 31 and 18 percent of orally administered MCB doses being retained in humans and rats, respectively. Absorption through the skin appears to be negligible based on the low systemic toxicity of MCB following dermal application.

The metabolism and excretion of this compound have been studied and reported by Spencer and Williams (47) and by Smith et al. (48), who administered MCB to rabbits in an oral dose of 0.5 g/kg of body weight. They found that 27 percent of the administered dose was excreted unchanged in the expired air; 25 percent appeared in the urine as a glucuronide, 27 percent as an ethereal sulfate, and 20 percent as mercapturic acid. The recovery of administered radioactivity (99 percent) was nearly complete in this study.

Smith et al. (49) fed radioactive $^{14}$C-labeled MCB to rabbits in gavage doses of 0.5 g twice daily for 4 consecutive days. The urine and feces were collected from animals during and for 3 days after treatment. One rabbit was sacrificed, and its organs removed for $^{14}$C analysis. In this experiment, loss of the compound by expiration appeared to account for the overall low recovery (Tables 22.4 and 22.5).

The majority of metabolites from $^{14}$C-MCB-dosed rabbits is found in urine, and only small amounts of radioactivity are present in the tissues and feces; furthermore, the overall low recovery of radioactivity is almost certainly a consequence of loss of MCB by respiration. Similar observations were made by Parke and Williams (50), who studied the metabolism of $^{14}$C-labeled benzene. The relative amounts of conjugates obtained in this study (mercapturic acids 25 percent, ethereal sulfates

**Table 22.4.** Excretion and Distribution of the Radioactive Metabolites from Orally Administered [$^{14}$C]-Chlorobenzene in Rabbits[a] (49)

|  | % of Dose |
|---|---|
| Urine | 19.6 |
| Feces |  |
|   Methanol extract | 1.05 |
|   Dry burn | 1.55 |
| Tissues (one animal) | 0.053 |

[a]Combined total urinary, carcass, and fecal $^{14}$C collected during 4 days dosing and 3 subsequent days. Total dose = 8.28 g.

37 percent, and glucuronides 37 percent) are in excellent agreement with those of Spencer and Williams (47). Furthermore, the proportion of diphenolic conjugates (ethereal sulfates, 53:1, and glucuronides, 42:1) to their monophenolic counterparts is virtually identical to values reported by Smith et al. (48).

A small amount of various phenolic metabolites is excreted in the free state. Smith et al. (49) obtained proportionately more of these materials as monophenols in their later study than in the earlier study (48). It is possible that the analytic procedure of the earlier study, which estimates monophenols as toluene-*p*-sulfonate derivatives, leaves a proportion of these isomers undetected.

Ogata and Shimada (46) summarized information available on the metabolism of MCB. The compound is thought to be first oxidized to the 3,4-epoxide, which can then follow one of several pathways. One route leads to the formation of the *p*-mercapturic acid conjugate following glutathione conjugation. A second pathway results in formation of 4-chlorocatechol, and the third ends with formation of 4-

**Table 22.5.** Distribution of Radioactive Metabolites in the Urine of Rabbits Dosed Orally with $^{14}$C-Chlorobenzene (49)

| Metabolite | % of Total Urinary Radioactivity |
|---|---|
| 3,4-Dihydro-3,4-dihydroxychlorobenzene | 0.57 |
| Monophenols | 2.84 |
| Diphenols | 4.17 |
| Mercapturic acids | 23.80 |
| Ethereal sulfates | 33.88 |
| Glucuronides | 33.57 |
|   Total | 98.83 |

chlorophenol and its conjugates. The first two appear to be the primary pathways for humans as evidenced from results in volunteers and workers.

## 2.5 Human Experience

### 2.5.1 Intoxications

Reich (51) reported the case of a 2-year-old boy who swallowed 5 to 10 ml of Puran, a cleaning agent containing MCB. He showed no ill effects for 2.5 hr, but after eating lunch he quickly lost consciousness and suffered vascular paralysis and heart failure, but recovered and survived. The odor of MCB in breath and urine persisted for 5 to 6 days.

Severe anemia and medullary aplasia in a 70-year-old woman was related to her employment in hat making, which required the use of glue containing 70 percent MCB. Early complaints included headache and irritation of the upper respiratory tract and mucosa of the eyes (52).

### 2.5.2 Industrial Exposure

According to the CMA report (1), Rozenbaum et al. (34) reported in 1947 that they examined 52 people who were occupationally exposed to MCB; 28 were employed in a factory where the only chemical vapors they were exposed to were those of MCB. Many of the individuals who had worked there for 1 to 2 years suffered from headache, dizziness, somnolence, and dyspeptic disorders. Examination revealed acroparesthesia in eight people, spastic contractions of finger muscles in nine, hyperesthesia of the hands in four, spastic contraction of the gastrocnemius muscle in two, and a vasovegetative instability in eight. The remaining 24 individuals displayed no characteristic abnormalities.

### 2.5.3 Experimental Subjects

To establish the threshold of action of MCB on the electrical activity of the brain, Tarkhova (53) exposed four healthy persons, 17 to 24 years old, to the compound at air concentrations of 0.1, 0.2, and 0.3 mg/m$^3$ for 13 min. The electrical potentials of the brain in response to light excitations were recorded on an electroencephalogram, and the threshold of action was found to be 0.2 mg/m$^3$. From this study, the author suggested that the "maximum single-time permissible concentration" of MCB in atmospheric air be set at 0.1 mg/m$^3$ or 0.02 ppm.

## 2.6 Odor and Taste

The odor of MCB is often described as like that of mothballs. The odor threshold is given as 0.21 ppm in air in an Arthur D. Little report (54). Varshavskaya (2) gave the threshold concentration for odor and taste in water as 0.02 mg/l.

## 2.7 Hygienic Standards of Permissible Occupational Exposure

The threshold limit value (TLV) of MCB adopted by the ACGIH is a TLV–TWA (time-weighted average) of 10 ppm (26 mg/m$^3$) (28).

The permissible exposure limit, according to OSHA standards, is 75 ppm; the immediately dangerous to life or health (IDLH) concentration is 2400 ppm (55).

## 3  o-DICHLOROBENZENE [CAS # 95-50-1]

The formula for o-dichlorobenzene is $C_6H_4Cl_2$; it is also called 1,2-dichlorobenzene and o-DCB.

o—PCB

### 3.1  Uses and Industrial Exposures

o-Dichlorobenzene has been used as a solvent, fumigant, insecticide, and chemical intermediate. Industrial exposure has occurred in personnel engaged in the manufacturing and handling of this compound.

### 3.2  Physical and Chemical Properties

| | |
|---|---|
| Physical state | Colorless liquid |
| Molecular weight | 147.01 |
| Specific gravity | 1.2973 (25°C) |
| Melting point | −17.6°C |
| Boiling point | 180.48°C |
| Vapor density | 5.07 (air = 1) |
| Vapor pressure | 1.15 mm Hg(20°C) |
| | 1.56 mm Hg(25°C) |
| Refractive index | 1.5476 (25°C) |
| Percent in "saturated" air | 0.2 (25°C) |
| Density of "saturated" air | 1.01 (air = 1) |
| Solubility | Insoluble in water; soluble in ethanol, benzene, diethyl ether |
| Flash point | 155°F (open cup); 151°F (closed cup) |
| Partition coefficient | 3.38 (octanol–water) |

Explosive limits    2 to 9% by volume in air
Odor threshold     50 ppm

Conversion factors: 1 mg/l = 166.3 ppm; 1 ppm = 6.01 mg/m³ at 25°C, 760 mm Hg

### 3.3 Effects in Animals

The toxicologic effect of *o*-DCB is injury primarily to the liver and kidneys. This material is weakly anesthetic; a brief exposure to a high concentration results in depression of the central nervous system.

#### 3.3.1 Acute Toxicity

Varshavskaya (56) of the Sechenov First Moscow Medical Institute reported that the toxicity of MCB and dichlorobenzene (isomer not stated) is on the same level; increasing the number of chlorine atoms in a benzene molecule does not affect the toxic action, but only its degree of "expressivity." In large doses the toxicity of dichlorobenzenes depends more on spatial distribution of chlorine atoms than their number, and the ortho isomer is more toxic than the para isomer. The effects of the dichloro compounds in rats were found to be essentially the same. Conditioned reflex activity was depressed, indicating a cerebral cortical effect; erythropoiesis was significantly decreased, MCB produced eosinophilia, and *o*-DCB, neutropenia. *o*-DCB, more than MCB, led to a sharp rise in urinary steroids. Although both benzenes increased tissue acid phosphatase and sharply decreased tissue alkali phosphatase, no sign of carcinogenic action was noted macroscopically, histologically, or histochemically.

Intraperitoneal injection of 0.03 ml of *o*-DCB in male 180-g Sprague–Dawley rats resulted in loss of liver glycogen and minimal necrosis. The same effects were noted after an intraperitoneal injection of 0.04 ml of MCB, but little or no effect was noted following the intraperitoneal injection of 0.1 g of *p*-dichlorobenzene (57).

Extensive renal necrosis (coagulation necrosis of the proximal convoluted tubules) was produced in 48 hr in C57 Black 6J mice after an intraperitoneal dose of the ortho isomer of 1.47 g/kg, but not by an equimolar (10 mmol) dose of the para isomer. Similar histopathological changes were observed after an intraperitoneal dose of 6.75 mmol of MCB/kg. Sprague–Dawley rats were less sensitive. The pronounced toxicity to the liver of the ortho isomer has been associated with a higher level of adduct formation to liver proteins (41).

Hollingsworth et al. (58) placed undiluted *o*-DCB in the eyes of rabbits. The response was moderate pain and slight conjunctival irritation. There was no serious injury and the irritation cleared in a few days (33).

#### 3.3.2 Subchronic Oral Toxicity

Hollingsworth et al. (58) fed guinea pigs *o*-DCB as a solution in olive oil. All guinea pigs survived 0.8 g/kg body weight but all succumbed to 2.0 g/kg body weight. These authors also administered o-DCB to rats in aqueous gum arabic at a dose of 376 mg/kg body weight by stomach tube 5 days/week for a total of 138 doses in 192 days. Growth and mortality were not affected. There was a moderate increase in average liver weight and a slight increase in average kidney weight; slight histopathological changes were observed in the liver. At a dose of 188 mg/kg/day, there was a slight increase in the weights of the liver and kidneys but no apparent histopathology (33).

According to the 1974 IARC report (59), the maximum tolerated dose for rats of o-DCB administered by gavage 5 days/week for about 28 weeks ranged from 19 to 190 mg/kg/day. Minimal liver and kidney damage occurred at higher dosage levels. No adverse effect could be observed at 18.8 mg/kg/day.

o-DCB was administered to rats for periods of 10 to 90 days by oral gavage to assess subchronic toxic responses according to standard protocols (60). In the 10-day study, exposures to 300 mg/kg/day resulted in decreases in body and organ weights (heart, kidneys, spleen, testes, and thymus) and an increase in liver weights and hepatocellular necrosis in male animals. Lower-dose groups (150 mg/kg and below) were virtually free of toxic effects. In 90-day testing, 400 mg/kg/day induced body and spleen weight deficits whereas kidney, liver, heart, lung, brain, and testis weights increased significantly above control levels. Liver pathology indicated hepatocellular necrosis, centrolobular degeneration, and hypertrophy. Various clinical chemistry tests were elevated in the 400-mg/kg group (blood urea nitrogen, total bilirubin, ALT) with the latter enzyme also increased for the 100 mg/kg group. The NOEL for the 90-day study was considered to be 25 mg/kg.

### 3.3.3 Vapor Inhalation Exposures

Hollingsworth et al. (58) reported that rats exposed to a concentration of 977 ppm in the air survived for 2 hr but succumbed to a 7-hr exposure period. Rats survived a single 7-hr exposure to 539 ppm. These animals showed drowsiness, unsteadiness, and eye irritation. Definite organic injury was observed. The weight of liver and kidneys showed an increase. Microscopic examination of the liver revealed marked central lobular necrosis; cloudy swelling of the tubular epithelium was noted in the kidneys (33).

The CMA technical report (1) referred to studies conducted by Czajkowska et al. (61), who exposed two groups of rats to o-DCB vapors in concentrations of 20 and 100 mg/m$^3$ for 4 hr daily, 5 days/week. This treatment decreased body weight gain, cholinesterase activity, and the number of thrombocytes, but increased the number of eosinophiles observed. The effects were not marked in rats exposed to 100 mg/m$^3$. With the lower dosage, there was complete recovery within 1 month after cessation of exposure.

Dogs tolerated 1 hr exposure to o-DCB in a concentration of 2 ml/m$^3$; exposure of one dog to this concentration for 2 hr/day for 14 days caused no obvious signs of intoxication; however, exposure of dogs to 4 ml/m$^3$ resulted in somnolence during exposure [Reidel, in 1941, reported by von Oettingen (31)].

The only chronic vapor exposure studies reported are those of Hollingsworth et al. (58). They exposed animals for 7 hr/day, 5 days/week. At a concentration of 93 ppm of o-chlorobenzene in air, rats, guinea pigs, and rabbits survived for periods of 6 to 7 months. The animals showed no deleterious effects on growth, mortality, organ weights, hematology, and histopathology (8).

### 3.3.4 Chronic Oral Exposures

Long-term bioassays were performed on o-DCB in rats and mice using oral gavage administration (7). Dose levels of 0, 60 or 120 mg/kg/day for 103 weeks had no effects on survival, body weights, or microscopic appearances of tissues except for lower survival in high-dose males and increased renal tubular regeneration in high-dose male mice. Tumor results for this study are described above in the section on carcinogenicity.

### 3.3.5 Teratogenicity and Reproductive Effects

The teratogenicity of airborne o-DCB vapor was evaluated in rats and rabbits (62). Pregnant animals were exposed 6 hr/day to 0, 100, 200, or 400 ppm during periods of organogenesis. In rats, body weight deficits were observed in all test groups, as was an increase in liver weights in the high exposure group. Maternal effects in rabbits were limited to a slight decrease in body weights. No evidence of o-DCB-induced malformations or other fetotoxic effects were observed in either species in this study.

The reproductive toxicity potential of o-DCB was evaluated in a rat two-generation inhalation study (63). $F_0$ animals were exposed to concentrations of 0, 50, 150, and 400 ppm for 10 weeks prior to mating as well as during mating, gestation, and lactation. $F_1$ animals selected for breeding were similarly exposed. $F_2$ pups were observed through weaning. No compound-related mortality was noted although high-dose parental animals experienced depressed body weights. Liver weights were elevated in mid- and high-dose males whereas kidney weights only were increased in males at these exposure concentrations. These changes were accompanied by microscopic alterations in these tissues. A small increase in liver weight was seen in the 50-ppm males. No o-DCB induced effects were seen for reproductive performance or fertility indices in this study.

## 3.4 Metabolism

In 1923, Hele and Callow (64) reported that MCB and o-DCB are excreted in the dog as ethereal sulfates and as mercapturic metabolites. They suggested that these metabolites contained an aromatic nucleus, to which the chlorine was still attached, and that in no case was the chlorine liberated before the metabolites were formed.

Following the oral administration of the ortho isomer (1.47 to 2.94 g) and MCB (1.12 to 11.1 g) to dogs. Collow and Hele (65) observed a close correlation between the "organic" chlorine and the "extra" sulfur excreted in the urine. This observation was taken as evidence that mercapturic acids are formed from these compounds.

After its oral administration to rabbits, o-DCB is metabolized mainly to 3,4-

dichlorophenol; but 2,3-dichlorophenol, 3,4-dichlorophenylmercapturic acid, and 3,4- and 4,5-dichlorocatechol are also formed. $p$-DCB was found to be converted to 2,5-dichlorophenol and 2,5-dichloroquinol conjugated with glucuronic and/or sulfuric acid (66).

Oral administration of $o$-DCB of 2 or 4 mg/kg/day for 2 weeks to rats resulted in the accumulation of 80 to 100 ppm of the compound in fatty tissues (67).

New techniques were described that demonstrated the utility of dynamic in vitro assays for measuring enzymatic activities in human liver slices (68, 69). The three dichlorobenzenes were shown to be biotransformed to soluble metabolites; this transformation would involve both oxidative and conjugative changes in the test compounds. Such an approach would have significant utility for assessing metabolic pathways for exogenous chemicals in human tissues.

### 3.5 Human Experience

Hollingsworth et al. (58) recorded concentrations ranging from 1 to 44 ppm with an average of 15 ppm $o$-DCB in a workroom atmosphere where large quantities of this compound were handled. The odor of $o$-DCB was not detectable at these levels. Thorough physiological examination of all who worked in these areas failed to show any indication of injury (organic injury or hematologic effects) from the exposure (33).

Six months of industrial inhalation exposure of a young man to Orthosol, a product containing 95 percent of the ortho and 5 percent of the para isomer, resulted in severe pallor, exhaustion, and vomiting, with intense gastric pain and headache. Blood tests revealed a rapidly developing hemolytic anemia. Rapid and complete recovery followed cessation of exposure. Thirteen co-workers similarly exposed suffered no injury (70).

Eczematoid dermatitis of the hands, arms, and face of a 47-year-old glazier who worked with $o$-DCB was described by Downing (71). Intense erythema and edema appeared promptly when the compound was applied locally to the skin of one arm. A large bullous lesion developed somewhat later.

### 3.6 Odor and Warning Properties

The odor of $o$-DCB is detectable by the average person at 50 ppm in air. Eye and nose irritation are not noted at this level. The odor becomes strong and the irritation noticeable at concentrations around 100 ppm. The compound has fair warning properties at this level, but the possibility of adaptation must be recognized (33). According to Varshavskaya (56), the threshold concentrations for odor and taste in water for $o$-DCB are 0.002 and 0.0001 mg/l, respectively.

Information on individual case histories provides scant support for cause-and-effect relationships between chemical exposures and certain adverse health effects such as cancer. Such is the situation for the alleged association that has been made in the literature between a number of individual cases of leukemia and exposures to $o$-DCB. One reference (72) cites the case in which a 55-year-old woman developed acute myeloblastic leukemia after use for an unspecified time of a cleaning solvent that contained 80 percent *o*-DCB, 15 percent *p*-DCB, and 2 percent *m*-DCB. Tolot et al. (73) presented the case of a 40-year-old worker who had been exposed to *o*-DCB for 22 years in the preparation of dyestuffs. The man exhibited purpura, intense anemia, marked hepatomegaly, and discrete splenic enlargement. The disorder proved fatal in 4 months. The case was diagnosed as proliferating myelosis.

The overall evidence has not been considered sufficient to classify *o*-DCB a human carcinogen.

### 3.7 Hygienic Standards of Permissible Exposure

The recommended threshold limit value of *o*-DCB for ACGIH is 25 ppm (150 mg/m$^3$). The recommended short-term exposure limit (STEL) is 50 ppm (28).

OSHA has set the permissible exposure for *o*-DCB as 50 ppm (300 mg/m$^3$); the concentration considered immediately dangerous to life or health (IDLH) is 1000 ppm (55).

## 4  *p*-DICHLOROBENZENE [CAS # 106-46-7]

The formula for *p*-dichlorobenzene is $C_6H_5Cl_2$; it is also called 1,4-dichlorobenzene and *p*-DCB.

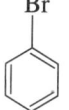

Bromobenzene

### 4.1 Uses and Industrial Exposures

*p*-Dichlorobenzene has been widely used as an insect (moth) repellent, fumigant, disinfectant, space deodorant for urinals, toilet bowls, diaper pails, and chemical intermediate (polyphenylene sulfide resins) and in waxes, abrasives, and finishes. The potential for exposure exists for those involved in the manufacture, handling, and use of this compound.

### 4.2 Physical and Chemical Properties

| | |
|---|---|
| Physical state | Colorless or white crystals |
| Molecular weight | 147.01 |

| | |
|---|---|
| Specific gravity | 1.248 (55°C) |
| Melting point | 53°C |
| Boiling point | 174°C |
| Vapor density | 5.07 (air = 1) |
| Vapor pressure | 1.18 mm Hg 25°C |
| Solubility | 80 mg/l water; soluble in benzene, alcohol, diethyl ether |
| Flash point | 53.8 C (open cup) |
| Flammability limits | 6.2 to 16% |
| Odor threshold | 0.18 to 30 ppm |
| Partition coefficient | 3.60 (octanol/water) |

Conversion factors: 1 mg/l = 166.3 ppm (air); 1 ppm = 6.01 mg/m$^3$ at 25°C, 760 mm Hg

## 4.3 Toxicity in Animals

### 4.3.1 Acute Effects

Oral administration of 1 g of p-DCB/kg body weight to dogs for the treatment of intestinal worms produced no deleterious effect (74); however, Dikmans (75) reported that single doses of 0.5 or 1.0 g/kg produced severe signs of intoxication, and in one dog, death.

Hollingsworth et al. (29) fed p-DCB to rats as a 20 percent solution in olive oil. The rats survived single doses of 1 g/kg of body weight, but all animals succumbed to a dose of 4 g/kg of body weight. Guinea pigs were fed a 50 percent solution, and survived a single dose of 1.6 g p-DCB/kg body weight, but succumbed to a dose of 2.8 g/kg (33).

The subcutaneous lethal dose in 22- to 26-day-old mice was reported as 5145 mg/kg. The animals developed tremors (within 2 to 3 hr of dosing), which continued for 3 or more days, and ultimately died of respiratory failure (76).

Solid p-DCB causes very little irritation to the skin. It does produce a burning sensation when held in close contact with the skin for a prolonged period. Fumes from the surface of hot p-DCB may irritate the skin slightly when the contact is repeated or prolonged. There is no evidence of significant absorption through the skin (33).

Pike (77) exposed rabbits repeatedly for 8 hr to p-DCB vapor concentrations of 4.6 to 4.8 mg/l (770 to 800 ppm). He also administered the compound orally to rabbits in repeated doses of 0.5 or 1.0 g/kg. His purpose for administering highly toxic doses was to observe possible harmful effects in the eyes of these animals. He noted "toxic eye-ground changes, but no lens changes" in the rabbits. He concluded, based on the results in rabbits and his clinical experience, that these observations suggest that cataracts are not produced by p-DCB. He also administered naphthalene orally to rabbits and produced cataracts in these animals; hence it is very likely that earlier reports of cataract formation may have had their origin in other compounds used in the manufacture of "mothproofing agents."

Solid particles, vapors, or fumes of p-DCB are very painful to the eyes and nose of humans and animals. But in order for vapors to be painful, it is usually necessary for the material to be heated, or to be dispersed in such a way that there is a very large surface area for evaporation in a poorly ventilated area. p-DCB is painful to most people in concentrations between 50 and 80 ppm; the discomfort is quite severe at 160 ppm (33).

### 4.3.2 Subchronic Effects

The intramuscular injection of 20 daily doses of 125 mg of p-DCB in guinea pigs produced a state of coagulative deficiency as a result of reduced activity of the prothrombin complex and thrombokinase. The concurrent administration of lipotropic agents (e.g., betaine, choline, or vitamin $B_{12}$) resulted in a protective action (78).

Totaro and Licari (79) also injected 125-mg daily doses for 20 days in guinea pigs and found this treatment caused loss of body weight and an increase in blood serum transaminase activity. The concurrent administration of lipotropic agents provided a protective action.

According to the CMA report (1), Rimington and Ziegler (80) reported that feeding 1140 mg MCB/kg, 455 mg o-DCB/kg, or 770 mg p-DCB/kg to male white rats by gastric intubation for 5, 15, and 5 days, respectively, induced experimental hepatic porphyria. The treatment caused increased urinary coproporphyrin excretion, followed by increases in porphobilinogen and α-aminolevulinic acid excretion. The mono and ortho compounds caused severe liver damage with intense necrosis and fatty changes. Rats with livers thus damaged displayed significantly decreased activities of liver catalase.

### 4.3.3 Chronic Toxicity

An extensive inhalation study in animals was reported by Hollingsworth et al. (29) in 1956. They exposed the animals 5 days/week, 7 hr/day for extended periods of time. Rats, guinea pigs, and rabbits exposed to 798 ppm in air showed definite reactions of toxicity. Exposures ran from a few to as many as 69 days; an occasional animal died. Signs of intoxication included tremors, weakness, weight loss, eye irritation, and unkempt appearance; some became comatose. Histopathologically, the liver showed cloudy swelling and central necrosis. There was also a slight cloudy swelling of the tubular epithelium of the kidneys in some animals. Rabbits showed some lung changes.

At a vapor concentration of 341 ppm, rats and guinea pigs survived for a period of 6 months. There was slight growth depression in male guinea pigs, a slight increase in average liver weight in male rats, and slight pathological changes in the liver of male guinea pigs. At a concentration of 158 ppm, animals survived exposures lasting from 137 to 219 days. No adverse effects on growth or mortality were observed in rats, mice, rabbits, or monkeys. There was slight growth depression seen in guinea pigs. Liver weights were slightly increased in male and female rats and in female guinea pigs; there were some questionable histopathological changes in the liver. At a concentration of 96 ppm, rats, guinea pigs, rabbits, mice, and a monkey were exposed for periods up to 6 or 7 months. No adverse effects were observed in other animal species (33).

Zupko and Edwards (81) exposed groups of rabbits, rats, and guinea pigs 20 to 30 min/day for up to 34 days to a $p$-DCB vapor concentration of approximately 100 mg/l of air. Each exposure induced intense irritation of eyes and nose, tremors and twitches of the extremities, loss of righting reflex, definite nystagmus, and rapid but labored respiration. The animals, as a rule, recovered from these signs of intoxication in 30 min to 2 hr; the hind leg function was the last to return to normal. Some animals died before the end of the exposure period. The compound was found to have a selective action on the granulocytes of the blood, producing a granulocytopenia in a majority of the animals. The total leukocyte (showing some increase in lymphocytes) and erythrocyte counts were not generally affected. Histological studies of 12 rabbits, 13 rats, and eight guinea pigs revealed lung damage and a definite selective action on the kidneys. Every animal showed marked and extensive kidney damage. There was comparatively little liver damage or evidence of hepatitis.

Hollingsworth et al. (29) studied the effect of repeated oral administration of $p$-DCB. Rats were fed 5 days/week for a total of 138 doses in 192 days. At a daily dose of 376 mg/kg, an increase in liver weights plus a slight increase in kidney weight were observed. Microscopic examination of the liver revealed slight cirrhosis and focal necrosis. At 188 mg/kg, a slight increase in the average weight of the liver and kidneys occurred. No effects could be observed in rats at a daily dose of 18.8 mg/kg. Similar liver and kidney changes were noted in rabbits following doses of 1000 mg/kg of body weight/dose; the effects were less intense, but there was definite injury at 500 mg/kg (33).

In 1958, Hollingsworth et al. (82) published their results on the effects of oral dosing (by stomach tube) of rats five times a week for a total of 138 doses of 18.8, 188, and 376 mg $p$-DCB/kg body weight. The lowest dose produced no deleterious effects. The 188-mg/kg dose caused a slight increase in liver and kidney weights. The repeated oral administration of the highest dose level resulted in a moderate increase in liver weight, together with slight cirrhosis and focal necrosis of the liver, a slight increase in kidney weight, and a slight decrease in the weight of the spleen.

Groups of rabbits were treated similarly with a dose of 500 mg/kg for as many as 263 doses over 367 days, and with 92 doses of 100 mg/kg over a period of 219 days. Both dose levels produced loss of body weight, tremors, weakness, cloudy swelling, and focal necrosis of the liver. The 1000-mg/kg dose caused some deaths (82).

### 4.3.4 Teratology and Reproductive Effects

To assess the teratogenic potential of $p$-DCB, pregnant rabbits were exposed on days 6 through 18 of gestation by inhalation to vapor concentrations of 0, 100, 300, or 800 ppm (62). Maternal animals exhibited depressed body weight gains at the high exposure level only. No compound-related adverse developmental (teratogenic or fetotoxic) effects were seen in fetuses in any of the test groups.

Assessment of the reproductive toxicity of $p$-DCB was performed in rats via inhalation exposures (83). Weanling rats were exposed 6 hr/day to 0, 66, 211, or 538 ppm vapor for 10 weeks prior to mating and during the 3-week mating period. Selected F1 weanling animals were exposed to the same air concentrations for 11 weeks prior to mating. All high-exposure parental groups exhibited depressed body weights and feed consumptions as well as other clinical signs of toxicity. There was no adverse impact noted on reproductive capacities for any $F_0$ or $F_1$ groups, although 538 ppm $F_1$ and $F_2$ offspring exhibited decreased body weights and elevated perinatal mortality. Evidence of increased kidney weights and hyaline droplet nephropathy was seen in $F_0$ and $F_1$ males in all exposure groups, and renal tubular cell hyperplasia was seen in the high-dose males only. Kidney weights were increased in females only in the high-dose $F_0$ group. Liver weights were increased in mid- and high-dose $F_0$ and $F_1$ rats of both sexes with hepatocellular hypertrophy observed at the high dose only. It was concluded that $p$-DCB vapor exposures had no adverse impact on reproductive function in rats. Postnatal toxicity was observed in offspring at inhalation exposures of 538 ppm. A NOEL for male rats was not observed owing to kidney changes seen at all exposure levels.

## 4.4 Mutagenicity

Because of the carcinogenicity of $p$-DCB in rodents, this compound has received considerable attention in terms of its potential mutagenicity. Further discussion of this compound is therefore warranted. Srivastava (84) studied the effect of $p$-DCB on somatic chromosomes of *Vicia faba*, *V. narbonensis*, *V. arvense*, and *Lathyrus sativus*. Various mitotic anomalies were encountered. These included shortening and thickening of chromosomes, precocious separation of chromatids, tetraploid cells, binucleate cells, chromosome bridges, and chromosome breakage. Chromosome breaks generally took place at the heterochromatic regions.

DNA binding of $p$-DCB was evaluated in multiple tissues of mice and rats following intraperitoneal injection of the compound (85). Covalent binding was observed in liver, kidney, lung, and stomach of mice but not rats. Mouse and rat microsomes from liver and lungs induced in vitro binding of $p$-DCB to calf thymus DNA. SKF525A inhibited this binding while glutathione enhanced adduct formation. The authors suggested that microsomal glutathione transferases may play an important role in the activation of $p$-DCB in terms of binding to genetic macromolecules.

In vivo assessments of $p$-DCB-induced unscheduled DNA synthesis and S-phase synthesis were made in tumor target tissues in rodents (86). Following oral administration of $p$-DCB at tumorigenic levels as determined in NTP studies (8), rat kidney and mouse liver cells were exposed to radio-labeled thymidine and evaluated by autoradiographic techniques. Dose levels of 300 to 1000 mg $p$-DCB/kg in mice increased hepatic cell turnover four or nine times in male and female mice, respectively, but the same doses either had no effect (females) or doubled (males) this cellular index in kidneys. There was no increase in unscheduled DNA synthesis in livers or kidneys in either species. The lack of evidence for genotoxicity as well as the increased cellular division for male rats make these findings consistent with tumorigenicity mechanistic evidence for male rat kidneys discussed above (section on carcinogenicity).

### 4.5 Absorption, Metabolism, and Excretion

The compound is not absorbed through intact skin in acutely hazardous amounts (31). Vapors of the compound are readily absorbed from the lungs. All indications are that absorption of the solid compound occurs readily from the gastroenteric tract. Fats and oils or organic solvents enhance absorption of the compound.

According to Azouz et al. (87), following single administration in the rabbit of 0.5 mg $p$-DCB/kg, the compound is oxidized mainly (approximately 30 percent) to 2,5-dichlorophenol, which is subsequently conjugated and excreted. 2,5-Dichloroquinol is also formed (approximately 6 percent of the dose), but in contrast to the ortho derivative, no mercapturic acid or dichlorocatechol is formed.

Repeated-exposure studies (daily for 10 days) in rats exposed to doses of $^{14}$C $p$-DCB of 1000 ppm (3-hr inhalation) or 250 mg/kg (oral, subcutaneous) indicated that tissue levels of $^{14}$C increased up to 6 days at which time they began to decline (88). Of the radioactivity eliminated from exposed animals during the 5 days following the final exposure, over 90 percent was found in the urine, and 1 to 6 percent exited the body via lung and fecal clearances. There was evidence for enterohepatic circulation as approximately 50 percent of administered radioactivity passed through the cannulated bile ducts. Sulfate and glucuronide conjugates of 2,5-dichlorophenol represented approximately 50 and 30 percent of the urinary $^{14}$C, and lesser amounts of dihydroxy and mercapturic acid metabolites of $p$-DCB were found. These investigators also showed that $^{14}$C accumulated in fat at levels 10 to 40 times higher than those in organs (kidneys, liver, lungs).

Two metabolites appeared in the blood of male Wistar rats fasted for more than 16 hr, then given oral doses of $p$-DCB of 200 or 800 mg/kg (5.0 ml/kg dose in a corn oil solution). The metabolites that remained in the blood for many hours after the parent compound had virtually disappeared were the methyl sulfoxide and methyl sulfone metabolites of $p$-DCB. The authors (89) suggest that the storage and slow release of $p$-DCB from the fatty body tissues is responsible for the prolonged presence of the metabolites in blood.

Carlson and Tardiff (90) studied the effect of chlorinated benzenes on the metabolism of foreign organic compounds. They found that $p$-DCB, 1,2,4-trichlorobenzene, 1-bromo-4-chlorobenzene, and HCB (but not MCB) decreased hexobarbital sleeping time immediately and/or up to 14 days following treatment. In addition, cytochrome $c$ reductase, cytochrome P450, EPN detoxification, glucuronyl transferase, benzopyrene hydroxylase, and azoreductase were increased to varying degrees by the administration for 14 days of these four chlorinated benzenes, at doses from 10 to 40 mg/kg/day. Administration of $p$-DCB or 1,2,4-trichlorobenzene for 90 days at these doses resulted in increased in EPN detoxification, benzopyrene hydroxylase, and azoreductase. Levels of the former two were still elevated 30 days after cessation of the administration of the compounds. 1,2,4-Trichlorobenzene was found to be a more potent inducer of cytochrome $c$ reductase. Cytochrome P450 was also elevated and remained so through the 30-day recovery period. The authors concluded that even simple chlorinated benzenes can enhance the metabolism of foreign organic compounds (62).

Pagnotto and Walkley (91) investigated the value of measuring the urinary excretion of $p$-DCB as an index of industrial exposure to $p$-DCB. Their study involved workers in a chemical manufacturing plant, in a household products packaging factory, and in a plant where $p$-DCB was employed in the manufacture of abrasive wheels. Exposure was almost exclusively due to vapor inhalation. The concentration of $p$-DCB in air averaged 8 to 34 ppm. Spot air samples were collected during the work shift (usually 3 shifts/day).

Excretion of dichlorophenol in urine occurred as expected, starting very shortly after the exposure began and then rising to a maximum at the end of the exposure period. This was followed by a rapid decrease, then a more gradual reduction over a period of several days. In one instance, small but detectable amounts of dichlorophenol were found several weeks after cessation of exposure (91).

The presence of dichlorophenol in the urine is revealed also by its distinctive odor, which is noticeable at a level of 100 mg/l and still detectable at about 20 mg/l. The concentration of dichlorophenol in the urine ranged from 10 to 233 mg/l (average 14 to 103 mg/l). No painful irritation of the eyes or nose was reported by the workers, except when there was direct contact with $p$-dichlorobenzene dust or crystals. These studies showed a fairly good correlation between the average air concentration of $p$-dichlorobenzene and the excretion of $p$-dichlorophenol at the end of an exposure period (91).

Because of sex differences for rats in toxic responses to $p$-DCB, an assessment of compound distributions following a 24-hr inhalation exposure was conducted (92). During and after exposure to 500 ppm vapor, chemical concentrations in serum, kidney, liver, and body fat were measured. Although levels in the serum did not appear to be different for the two sexes, $p$-DCB values for female liver samples were higher than for males. This is consistent with the observed higher hepatotoxicity in female rats. Similarly, higher levels of $p$-DCB in male kidneys corresponded to the greater toxicity in this organ in male rats. The renal results conform to evidence for involvement of $\alpha$-2-microglobulin in male rat responses to this chemical.

## 4.6 Human Experience

### 4.6.1 Nonoccupational Exposure

The CMA report (1) summarized early $p$-DCB intoxications, many of which were reported in the foreign literature.

A 62-year-old-man, presumably because of home use of $p$-DCB, began to suffer from dizziness, asthenia, anemia, and hypogranulocytosis. He recovered gradually after discontinuation of exposure (93).

A 19-year-old girl who worked for 18 months with an agent containing 90 percent of the para isomer and 10 percent of hexachlorethane suffered marked asthenia, dizziness, significant loss of body weight, a mild anemia, and hyperleukocytosis. She recovered rapidly after withdrawal from exposure. Two female workers who performed the same work suffered no abnormalities (94).

A 53-year-old woman who used *p*-DCB crystals extensively in her home for 12 to 15 years suffered pulmonary granulomatosis. A lung biopsy revealed numerous small lesions which contained crystals physically similar to those of *p*-DCB (95). An association with *p*-DCB exposure and the lung condition cannot be supported in the absence of chemical analyses of the crystals to confirm the chemical's identity.

A number of similar intoxications were presented in some detail in the CMA report (1). Briefly, the signs and symptoms these individuals experienced included periorbital swelling, intense headache, and profuse rhinitis; or acute illness with nausea, headache, vomiting, weight loss, numbness, clumsiness, and a burning sensation in the legs. One subject lost 22.5 kg in 3 months, developed ascites, and died. One man, a 52-year-old trapper who used the compound on animal hides, suffered weakness, nausea, subacute yellow atrophy of the liver, jaundice with elevated serum bilirubin, and elevated alkaline phosphatase (96).

Campbell and Davidson (97) reported the unusual case history of a pregnant 21-year-old woman who developed pica for the para isomer. She consumed, throughout pregnancy, one or two blocks of toilet air freshener per week, which were composed primarily of *p*-DCB. She developed a severe hypochromic, microcytic anemia, with excessive polychromasia, and marginal nuclear hypersegmentation of the neutrophils. She recovered completely after withdrawal of the chemicals. Neonatal examination of the child revealed no abnormalities.

Statistics of poisonings in children disclosed *p*-DCB is the least toxic of the active ingredients in moth repellent products. Other ingredients are naphthalene and camphor; the latter was used particularly in the past. Naphthalene products look dry and are slowly soluble in cold ethanol whereas *p*-DCB products appear wet and oily and are rapidly soluble in unheated ethanol. According to this report "ingestion of 29 g of *p*-DCB has been well tolerated in man."

### 4.6.2 Occupational Exposure

The many incidents of intoxication that resulted from the household use of products containing *p*-DCB contrast sharply with the scarcity of reports of industrial exposure of this compound.

According to the 1964 Hygienic Guide Series (98) on *p*-DCB voluntary industrial overexposure is unlikely. The most serious effects of overexposure are those of lung, liver, and kidney injury.

Hollingsworth et al. (29) reported on their extensive industrial experience in handling *p*-DCB. They reported on 58 men who had worked continuously or intermittently on operations involving the handling of this material for periods of 8 months to 25 years, and an average of 4.75 years. Early investigations showed air concentrations ranging from 10 to 550 ppm, with an average of 85 ppm. A later study separated the job area into two ranges: one with concentrations from 100 to 725 ppm, with an average of 380 ppm; and another with concentrations from 5 to 275 ppm, with an average of 90 ppm.

A third survey was made after there had been some major changes in operations. The concentrations varied from 50 to 170 ppm, with an average of 105 ppm. Under these conditions, there were still some complaints of eye and nose irritation by the workmen. There were no complaints when air concentrations ranged from 15 to 85 ppm with an average of 45 ppm. The men in these areas were examined thoroughly at various times throughout the period of study. There was no evidence of organic injury, hematologic effects, or eye changes (33).

Von Oettingen in 1955 (31), and also Hollingsworth et al. in 1956 (29), reviewed the literature. According to Berliner (99), cataract formation occurred following exposure to a "mothproofing agent" containing $p$-DCB. Considering also the publications of several other authors and the report by Pike (77) in particular, which states that $p$-DCB does not cause cataracts, and the report by Hollingsworth et al. (29) on industrial experience with this compound, it appears that Berliner's findings may have been due to chemicals other than $p$-DCB that were present in the "mothproofing agent" used (33).

Severe systemic effects were reported by Walgren (100) in eight workers employed in the production of moth deterrents with $p$-DCB as the primary ingredient. The individuals developed irritation of the mucous membranes of eyes and throat, methemoglobinemia, and loss of appetite and body weight. Only one of the workers experienced an exposure that exceeded 1 year, and one worker only 1 month. Methemoglobin concentrations ranged between 0.12 and 0.29 g/100 ml blood. All cases showed increased muscle reflexes, ankle clonus, and fine finger tremors. Blood pressures were within the normal range. In the blood, slight diminution of erythrocytes and hemoglobin was observed along with relative lymphocytosis up to 63 percent. Four workers showed thrombocytopenia, and one workman had granulocytopenia. Sternal marrow was essentially normal. The interpretation of these data is difficult in the absence of the specific compounds involved other than $p$-DCB in the working atmosphere.

### 4.7 Odor, Taste, and Warning Properties

$p$-DCB has a very distinctive aromatic odor. The threshold of detection varies from 15 to 30 ppm in air. The odor becomes very strong at concentrations between 30 and 60 ppm. The vapors are painful to the eyes and nose at concentrations of 80 to 160 ppm. Above 160 ppm, they are intolerable to any person who has not been exposed to the compound long enough to have developed tolerance.

The odor and irritating effects are good warnings to prevent overexposure to $p$-DCB. It should be recognized, however, that a person may become sufficiently accustomed to the odor to tolerate high concentrations (33).

Exposure of hens for 3 days to air containing 3.4 to 6.4 ppm of $p$-DCB imparted an unpleasant, sweetish taste to the egg yolks. Ill effects in the hens were not noted and there was no reduction in egg production (101).

The threshold concentrations for odor and taste for p-DCB in water were reported as 0.002 and 0.006 mg/l, respectively (3).

### 4.8 Hygienic Standards of Permissible Exposure

The ACGIH threshold limits for p-DCB are 75 ppm (450 mg/m) TLV–TWA, and 100 ppm (675 mg/m$^3$) TLV–STEL (28).

The current OSHA permissible exposure limit is 75 ppm, the short-term limit is 110 ppm, and the IDLH concentration immediately dangerous to life and health is 1000 ppm (55).

## 5 TRICHLOROBENZENES

Trichlorobenzene, $C_6H_3Cl_3$, exists as three isomers: 1,2,4-trichlorobenzene [CAS # 120-82-1], 1,2,3-trichlorobenzene [CAS # 87-61-6], and 1,3,5-trichlorobenzene [CAS # 108-70-3].

### 5.1 Uses

The 1,2,4 isomer is used as an intermediate in the manufacture of herbicides, as a dye carrier, dielectric fluid, and heat transfer medium, and as a solvent for high-temperature-melting products. The 1,2,3 isomer has similar uses except as chemical intermediate. 1,3,5 trichlorobenzene is used for some of these applications plus as a termite preparation and insecticide. These compounds are found as unintended by-products of the manufacture of the mono- and dichlorobenzenes.

### 5.2 Physical and Chemical Properties

The following properties are for the 1,2,4 isomer. The other two isomers of trichlorobenzene have similar properties.

| | |
|---|---|
| Physical state | Colorless liquid or white crystals |
| Molecular weight | 181.46 |
| Specific gravity | 1.45 |
| Melting point | 16.9°C |
| Boiling point | 213°C |
| Vapor density | 6.26 (air = 1) |
| Vapor pressure | 0.29 mm Hg 25°C |
| Solubility | 34.6 mg/l water; soluble in benzene, $CS_2$, diethyl ether; slightly soluble in alcohol |
| Flash point | 110°C (open cup) |

Partition coefficient   4.2 (octanol/water)

Conversion factors: 1 mg/l = 136.1 ppm (air); 1 ppm = 7.35 mg/m$^3$ at 25°C, 760 mm Hg

## 5.3  Toxicity Findings

### 5.3.1  Acute Toxicity

A review of sparse information (13) indicates that the trichlorobenzenes exhibit a relatively low degree of toxicity following oral exposures. Rats and mice exhibited an oral LD$_{50}$ value of 760 mg/kg for the 1,2,4 isomer. Repeated dosing of the 1,2,4, and 1,2,3 isomers at 500 mg/kg and above caused elevated excretion of urinary porphyrins in rats. Evidence of liver effects in other animal studies included fatty infiltration and necrotic changes, induction of microsomal enzymes, and increases in serum glutamate–pyruvate transaminase levels.

A percutaneous toxicity study of 1,2,4 isomer in rats showed that the single dose LD$_{50}$ for this compound was 6100 mg/kg.

Air levels of 3 to 5 ppm are associated with eye and respiratory irritation.

### 5.3.2  Subchronic Toxicity

A number or oral and inhalation subchronic studies were performed on members of this chemical class in rats, rabbits, dogs, mice, guinea pigs, and monkeys; the results are reviewed by the U.S. EPA (13). The majority of studies noted liver effects, although respiratory epithelial lesions were seen in an inhalation study.

In 90-day toxicity studies, rats of both sexes were fed 1,2,4-trichlorobenzene in the diet for 3 months at concentrations of 0, 200, 600, and 1800 ppm (102). Animals were monitored for clinical changes as well as subjected to eye and blood chemistry–hematology exams. Postmortem assessments of gross and microscopic tissue changes were made. No compound-related clinical findings were found. Slightly increased blood urea nitrogen was seen in high-dose animals, although values were within the normal range. Platelet counts were higher in high-dose males compared to controls. Liver weights were increased in a dose-related manner in both sexes with centrolobular cell hypertrophy being evident in the 600- and 1800-ppm groups. Kidney weights increased in a dose-responsive manner for all test groups in male rats. Microscopically, kidneys exhibited dilated tubules, hyaline droplets, mineral deposition, and regenerative tubular epithelium, characteristic of compounds inducing α-2-microglobulin (see section on carcinogenicity above). Testes weights for the high-dose group were above control levels, although no microscopic findings were reported for this tissue. The NOEL for this chemical in this study was 200 ppm for females but less than 200 ppm for males.

A similar 3-month dietary study was performed in B6C3F1 mice exposed to 0, 200, 3850, or 7700 ppm 1,2,4,-trichlorobenzene (103). Body weights and feed consumptions were suppressed in all test groups at various times during the study including terminal weights for 3850-ppm males and 7700-ppm males and females. Various liver function tests (elevated serum protein, albumin, globulin, alanine aminotransferase, and sorbitol dehydrogenase) indicated chemically induced hepatic effects that corresponded to higher liver weights in the mid- and high-dose groups. Livers exhibited hepatocellular cytomegaly along with hepatocellular compression, vacuolar degeneration, and necrosis. The NOEL for these liver changes in both sexes was 200 ppm, although the overall NOEL in males was below 200 ppm owing to body weight reductions at this exposure level late in the study.

### 5.3.3 Chronic Studies

Only one report of chronic toxicity evaluation was found. Mice were subjected to dermal applications of 1,2,4 isomer (in acetone solution) twice weekly for 2 years (13). Effects noticed in test animals included excitability, panting, dermal thickening, inflammation, and keratinization.

### 5.3.4 Teratology and Reproductive Effects

Teratogenicity of the three trichlorobenzene isomers was evaluated in rats following oral administration [75 to 600 mg/(kg)(day)] on days 6 to 15 of gestation (104). Maternal effects included thyroid and liver lesions as well as decreased hematocrits and hemoglobin levels. No major malformations were noted in fetuses although mild osteogenic (skeletal) changes were observed.

A second teratology study assessed the activity of the 1,2,4, isomer following oral gavage administration of 0, 36, 120, 360, or 1200 mg/kg to rats (105). Dosing was conducted on days 9 to 13 of gestation and sacrifices were performed on day 14. All high-dose dams died by day 11. Body weight gain deficits and 22 percent mortality were present in the 360 mg/kg dams. Hepatocellular hypertrophy was absent, slight, and moderate in the 0, 120, and 360 mg/kg dams, respectively. The uteri from controls and the 360-mg/kg groups were examined. Embryo lethality was seen in 25 percent of the litters of the 360 mg/kg group but none was observed in controls. No significant differences were reported for resorptions or anatomic abnormalities. Treated-group embryos exhibited altered head length, crown–rump length, and reduced (23%) total embryo protein contents.

A two-generation reproduction study was conducted in rats to evaluate the effects of 1,2,4-trichlorobenzene incorporated into drinking water (106). Test compound levels were 0, 25, 100, or 400 ppm with exposures being continuous from birth of the $F_0$ animals until the $F_2$ animals were 32 days old. Feed and water consumptions were depressed in $F_0$ rats but were sporadic and not seen in other generations. Enlarged adrenals were seen in 400-mg/kg adult animals that were later confirmed to be compound-related in parenteral dosing. No parental effects were observed for blood parameters, locomotor activity, and liver and kidney microscopy. There were also no effects in reproductive end points including fertility, neonatal weights, litter sizes, preweaning viability, and postweaning growth.

### 5.4 Metabolism and Distribution

$^{14}$C-Labeled 1,2,4-trichlorobenzene was orally administered to rats at a dose level of 50 mg/kg (107); 66 and 17 percent were excreted in the urine and feces, respectively, within 7 days. Even distribution of radioactivity was observed throughout tissues examined, with slightly higher levels in fat. Free 2,4,5- and 2,3,5-trichlorophenols and their conjugates were the main metabolites in the urine. 5- or 6-Sulfhydryl, methylthio, methylsulfoxide, and methylsulfone derivatives of 1,2,4-trichlorobenzene were minor metabolites. There was also evidence of reductive chlorination catalyzed by intestinal microflora.

Other studies reviewed by the EPA (13) indicate that species differences occur for the metabolism of the 1,2,4 isomer. Studies in rhesus monkeys (108) indicate that a reductive transformation takes place, for 3,4,6-trichloro-3,5-cyclohexadiene-1,2-diol glucuronides account for 50 to 60 percent of the 24-hr urinary metabolites.

It is concluded from available information that the three isomers have a common first step in metabolism, and that is production of the arene oxide. Subsequent metabolic steps (phenol formation, conjugation, reductions) are species dependent.

## 5.5 Effects in Humans

Two case reports suggest that one woman developed aplastic anemia following use of trichlorobenzene to clean her husband's work clothes, and a 60-year-old male who was exposed to a number of chlorinated chemicals, including trichlorobenzene, developed anemia (13).

## 5.6 Hygienic Standards of Permissible Exposure

ACGIH has established a TWA–TLV of 5 ppm (37 mg/m$^3$) for trichlorobenzene mixtures (28). No standards have been established for these compounds by the National Institute for Occupational Safety and Health (NIOSH/OSHA).

## 6 TETRACHLOROBENZENES

Tetrachlorobenzene, $C_6H_2Cl_4$, exists as the 1,2,4,5 isomer [CAS # 95-94-3], the 1,2,3,4 isomer [CAS # 634-66-2], the 1,2,3,5 isomer [CAS # 634-90-2], and as mixed isomers [CAS # 12408-10-5].

## 6.1 Uses and Occurrences

Much of the production of these materials is incidental to the manufacture of the lower-molecular-weight chlorobenzenes. The 1,2,3,4 isomer is used as a component of dielectric fluids and an intermediate in fungicide manufacturing. The 1,2,4,5 isomer is an intermediate in the syntheses of herbicides and defoliants, as an insecticide and moisture-resistant impregnant, in electrical insulation, and in packing protection. Commercial uses for the 1,2,3,5 isomer are not listed by the EPA (13).

## 6.2 Physical and Chemical Properties

The following properties are for the 1,2,4,5 isomer, which is estimated to be produced in the largest volume of the various tetrachlorobenzenes. The other two isomers of tetrachlorobenzene have similar but not identical properties.

| | |
|---|---|
| Physical state | White crystals |
| Molecular weight | 215.90 |
| Specific gravity | 1.86 |
| Melting point | 139.5°C |
| Boiling point | 246°C |
| Vapor density | 7.4 (air = 1) |
| Vapor pressure | 0.05 mm Hg 25 C |
| Solubility | 0.6 mg/l water; soluble in benzene, $CS_2$, chloroform; slightly soluble in alcohol |
| Flash point | 311°F (open cup) |
| Partition coefficient | 4.9 (octanol/water) |

Conversion factors: 1 mg/l = 114.4 ppm (air); 1 ppm = 8.74 mg/m³ at 25°C, 760 mm Hg

## 6.3 Toxicity Findings

### 6.3.1 Acute Effects

The oral $LD_{50}$ values for the three isomers in rats were reported to be in the range of 1200 to 3000 mg/kg, with the 1,2,3,4 isomer being most toxic and the 1,2,3,5 least toxic (109).

### 6.3.2 Subchronic and Chronic Effects

Rats of both sexes were exposed to diets containing 0, 0.5, 5.0, 50, or 500 ppm of the three isomers for 1 and 3 months. Results for the 1,2,4,5, isomer indicated increased liver and kidney weights, elevated microsomal enzyme activities, and marked microscopic changes in the liver (both sexes) and kidneys (males only). There were no effects on body weight or survival. The other two isomers were shown to be less toxic than the 1,2,4,5 isomer (109).

Other subchronic studies reviewed by the EPA (13) indicated that exposures to these compounds may induce alterations in hepatic glycogen formation, adrenal function, blood cholinesterase activities (increased), blood hemoglobin levels, and other hematologic parameters.

Dogs (two per sex) were exposed to 5 mg/kg of 1,2,4,5-tetrachlorobenzene via the diet for 2 years (110). The dogs were observed for 20 months post-exposure. Toxicity findings were limited to elevations in serum alkaline phosphatase and total bilirubin at the end of the exposure period with total recovery by 3 months post-exposure. Postmortem tissue findings revealed no compound-related effects.

### 6.3.3 Teratology and Reproductive Effects

Rats and rabbits were subjected to teratology assessments following oral gavage dosing of the 1,2,4,5-tetrachloro isomer (in corn oil) to pregnant animals (111). Rats received doses of 0, 25, 75, or 125 mg/kg/day on days 6 to 15 of gestation, and rabbits were dosed with 0, 5, 15, or 25 mg/kg/day on days 6 to 18 of gestation. Maternal animals were sacrificed on days 21 and 29, respectively, to assess fetal development. The highest dose in rats induced decreased body weight gains and feed consumption, urinary staining, nasal/ocular discharges, and respiratory sounds. Maternal liver weights were increased following 75 and 125 mg/kg treatments. At these two dose levels, increased incidences of reduced ossification were noted in fetuses. All rabbit findings lacked dose responsiveness but suggested compound-related toxicities including maternal effects at all doses including deaths (at 5 and 25 mg/kg), abortions (5 mg/kg), and transient reductions in body weight gains (5 and 15 mg/kg). Fetal effects reported in rabbits were one minor visceral and one skeletal variation at 15 and 5 mg/kg, respectively. It was concluded that the NOEL for developmental toxicity in rats was 25 mg/kg; a NOEL in rabbits was not obtained.

The three isomers were administered to pregnant rats on days 6 to 15 of gestation at dose levels of 0, 50, 100, or 200 mg/kg (112). Maternal toxicity for the 1,2,4,5 isomer included organ weight changes with elevations in microsomal enzyme activities. Fetal changes were limited to increased lethality of pups at the highest dose level for the 1,2,3,4 and 1,2,3,5 isomers, plus evidence of minor osteogenic delays for the latter compound.

1,2,4,5-Tetrachlorobenzene was subjected to reproductive toxicity testing in rats using dietary exposures (113). Weanling $F_0$ males and females were fed 0, 30, 300, or 1000 ppm of test compound for 10 weeks before and 3 weeks during mating. This exposure schedule was repeated for $F_1$ weanlings. 1000-ppm groups displayed significantly reduced body weight gains and feed consumption in exposed parents in both $F_0$ and $F_1$ generations. In addition, this exposure level increased the number of stillborn pups ($F_1$) and caused a high incidence of postnatal death. 300-ppm $F_1$ and $F_2$ groups experienced decreased pup weights and litter viability. Liver (300 and 1000 ppm) and kidney (30, 300, and 1000 ppm) effects (increased weights and microscopic lesions) were observed in $F_0$ and $F_1$ parental animals. The authors concluded that although parental and suckling rats experienced compound-related effects in the 300- and 1000-ppm groups, no reproductive parameters were affected.

### 6.4 Metabolism and Distribution

Evidence from rabbit studies indicate that the three isomers of tetrachlorobenzene are efficiently absorbed from the gastrointestinal tract as evidenced by the small percentages recovered in fecal excreta. The highest concentrations (6 days after exposure) of the unchanged compounds were found in fat with the lowest levels present in brain and liver (114).

In a 1,2,4,5 isomer dietary 2-year dog study, concentrations of the compound were monitored in fat and plasma (110). The fat/plasma ratio was 650 after 1 month of treatment, although this ratio decreased thereafter, reaching 280 by the end of the exposure phase. During the 20-month post-exposure period, the ratio again rose, consistent with a high affinity of the chemical for adipose tissue. The elimination half-lives for plasma and fat were approximately 110 days.

Metabolic transformations of the three isomers of tetrachlorobenzene were assessed in rabbits following oral administration of the compounds (115). Exposure to the 1,2,3,5 isomer yielded 2,3,4,5-, 2,3,5,6-, and 2,3,4,6-tetrachlorophenols. The 1,2,3,4 isomer was metabolized to 2,3,4,5- and 2,3,4,6-tetrachlorophenols and the 1,2,4,5-tetrachlorobenzene yielded 2,3,5,6-tetrachlorophenol. It is apparent that chlorine shifts on the benzene ring are needed to produce many of these metabolites. The 1,2,4,5-tetrachlorobenzene is the least, and the 1,2,3,4 compound the most metabolized of the three isomers. In addition to urinary excretion of phenolic forms of these compounds in rabbits, glucuronide and ethereal sulfate conjugates of the phenols were present in urines (114). Unchanged isomers were detected in feces, tissues, and expired air. Given the substantial metabolic differences between species seen for the trichlorobenzenes, it is likely that similar variations would be seen for the tetrachlorobenzene isomers.

### 6.5 Human Effects

A study of Hungarian factory workers engaged in 1,2,4,5-tetrachlorobenzene production was reviewed by the EPA (13). Chromosomal abnormalities in blood samples collected from workers who had at least 6 months in the production job were compared to those of controls who worked at the same location but were not exposed to the tetrachlorobenzene area of the plant. Face masks were said to be worn by the workers. The authors found an increased frequency of cells with less than 46 chromosomes, increases in polyploidy chromatid-type chromosome aberrations, labile chromosome-type aberrations, and stable chromosome-type aberrations. The authors considered the 1,2,4,5 isomer to be causative in producing these clastogenic changes. The strength of these conclusions is weakened by the lack of any other reported genotoxic findings for the tetrachlorobenzenes (13).

### 6.6 Hygienic Standards of Permissible Occupational Exposure

None have been established for these compounds by OSHA or other guideline-setting organizations.

## 7 PENTACHLOROBENZENE [CAS # 608-93-5]

Pentachlorobenzene, $C_6HCl_5$, is also known as quintochlorobenzene and QCB.

### 7.1 Uses and Occurrences

This chemical is no longer produced in the United States but is a by-product of the manufacture of the lower chlorobenzenes. Past uses included employment as a chemical intermediate, as a flame retardant, and as part of a pesticide to treat oyster drill. The chemical has been found in various environmental media including soils, ground and surface waters, and sediments, primarily as a result of waste emissions from manufacturing locations.

### 7.2 Physical and Chemical Properties

| | |
|---|---|
| Physical state | White needles |
| Molecular weight | 250.34 |
| Specific gravity | 1.85 at 16.5°C |
| Melting point | 86°C |
| Boiling point | 277°C |
| Vapor pressure | 1 mm at 98.6°C |
| Vapor density | 8.63 |
| Solubility | 0.56 mg/l water, 25°C; slightly soluble in benzene, chloroform, ether; soluble in hot alcohol and $CS_2$ |
| Partition coefficient | 5.63 |

Conversion factors: 1 mg/l = 98.8 ppm; 1 ppm = 11.25 mg/m³ at 25°C, 760 mm Hg

### 7.3 Effects in Animals

#### 7.3.1 Acute Effects

Oral gavage studies performed in rodents indicate that $LD_{50}$ values for this compound range from 940 mg/kg in weanling female rats to 1370 mg/kg in adult female mice (116). Clinical signs included tremors, hypoactivity, and hypersensitivity to touch.

Dermal application of 2500 mg QCB/kg to the shaved skin of rats resulted in no mortality or clinical indications of toxicity (116).

#### 7.3.2 Subchronic and Chronic Effects

In a combination general toxicity–reproduction study, weanling female rats were fed diets containing 0, 125, 250, 500, or 1000 ppm QCB for 6 months; males received 0, 125, or 1000 ppm for 100 days (116). There was no evidence of QCB-induced body weight losses, mortality, clinical abnormalities, or changes in feed consumption. Hematologic examinations indicated slightly lower values for hematocrit and erythrocyte counts in high-dose males, and hemoglobin reductions and leukocyte count increases in both sexes at the high dose. Visual examination (viscera with ultraviolet light) and biochemical analyses (livers of females) to assess potential porphyria indicated no significant evidence for this metabolic lesion. There were increases observed for liver weights in rats fed 500 and 1000 ppm with microscopic evidence for hepatocellular hypertrophy and vacuolization. Male rats had evidence in kidneys of hyaline droplet formation, atrophic tubules, and lymphocytic infiltration.

A study reviewed by the EPA (13) to assess liver effects in female rats was conducted by exposing animals to 500 ppm QCB in the diet for 2 months (117). Although no increases in excretion of porphyrins in the urine were observed, levels of microsomal cytochrome P450 were approximately doubled by this treatment, suggesting the likelihood that liver enzyme induction has occurred under these exposure conditions.

No long-term studies for QCB were found in the literature.

### 7.3.3 Teratology and Reproductive Effects

As an adjunct to the subchronic study by Linder (116) described above, treated males and females were mated to untreated rats. There were no adverse effects observed in litters of treated (125 and 1000 ppm) males. Offspring of 250-, 500-, and 1000-ppm treated females were adversely affected. Survival of $F_1$ pups and weanling body weights were depressed in the 500- and 1000-ppm groups. Liver to body weight increases as well as hepatocellular hypertrophy were seen in pups in the three highest dose groups, suggesting a transferral of compound from dams to offspring during gestation and/or lactation. 125 ppm was considered a NOEL in this study.

A rat teratology study was conducted in which pregnant females were gavaged with 0, 50, 100, or 200 mg/kg QCB in corn oil on days 6 to 15 of gestation (118). Maternal sacrifices occurred on day 22. No clinical signs of toxicity were observed in the dams. Numbers of live fetuses per litter and fetal body weights were somewhat depressed in the high-dose group. Anatomically, increases in extra or fused ribs (all test chemical levels) as well as sternal defects (200 mg/kg) were seen in dosed groups. Soft tissue anomalies were not associated with QCB exposures. The authors also found dose–response QCB residues in tissues of these fetuses. The study has demonstrated a lowest observed effect level (LOEL) for adverse effects on skeletal development of 50 mg/kg.

A teratogenicity evaluation in mice was reported following administration of 50 or 100 mg QCB/kg in corn oil on days 6 to 15 of gestation (119). Maternal effects were increased liver weights only. There was no adverse impact on fetal development observed in this study.

### 7.4 Metabolism and Distribution

Because of the high lipophilicity of QCB, the compound has a high potential to cross biologic membranes. Following oral gavage administration of radio-labeled QCB to rhesus monkeys, approximately 12 percent was recovered in the urine and 25 percent in feces within 40 days (120). Assessment of tissue levels 40 days after treatment showed that the highest concentrations were located in body fat and bone marrow. The estimated elimination half-life of QCB and metabolites in the monkey is 2 to 3 months. Measurements of residues in these animals showed that 54 percent of radioactive label in the blood was pentachlorophenol and 45 percent was QCB. Urinary metabolites were pentachlorophenol (58 percent), 2,3,4,5-tetrachlorophenol (32 percent), and 2,3,5,6-tetrachlorophenol (10 percent). No significant differences were observed for results in the male and female monkeys.

In other studies reviewed by EPA (13), results from rabbits and rats provided similar profiles of the pharmacokinetics and metabolism of QCB.

### 7.5 Human Effects

No information was located that discusses the impact of QCB exposures on human health.

### 7.6 Hygienic Standards of Permissible Occupational Exposure

None have been established for this compound by OSHA or other guideline-setting organizations.

## 8 HEXACHLOROBENZENE [CAS # 118-74-1]

Hexachlorobenzene, $C_6Cl_6$, is also called perchlorobenzene and HCB.

p—DCB

### 8.1 Uses and Industrial Exposure

HCB is found as a by-product in the manufacture of chlorinated $C_2$ solvents, chlorobenzenes, and certain chlorinated pesticides such as pentachloronitrobenzene. Although it is no longer registered in the United States for pesticidal purposes, HCB was used as a fungicide to control wheat bunt and smut fungi of cereal grains. The technical grade employed in agriculture contains 98 percent HCB, 1.8 percent pentachlorobenzene, and 0.2 percent 1,2,4,5-tetrachlorobenzene. Commercial formulations applied as dusts contain 10 to 40 percent HCB. Other applications include use as an additive for pyrotechnic compositions for the military, porosity controller in the manufacture of electrodes, intermediate in dye manufacture and organic synthesis, and wood preservative. Individuals engaged in the manufacture of chlorinated benzenes or the application of HCB pesticidally or for other uses, and workers involved in contaminated waste site remediation are populations potentially exposed to this chemical. Aerial dispersion appears to be the major pathway for this compound entering the marine environment (67). Trace levels have been found in lake sediments, fish, waste sites, and soils (67a).

## 8.2 Physical and Chemical Properties

| | |
|---|---|
| Physical state | White needles (monoclinic prisms) |
| Molecular weight | 284.80 |
| Specific gravity | 1.5691 at 23.6°C |
| Melting point | 230°C |
| Boiling point | 322°C (sublimes) |
| Vapor pressure | 1 mm at 114.4°C |
| Vapor density | 9.83 |
| Solubility | 0.005 mg/l water, 25°C; slightly soluble in alcohol; soluble in benzene, chloroform, ether |
| Flash point | 242°C |
| Fire hazard | Slight when exposed to heat or flame |
| Stability | Very stable, unreactive compound. There is no evidence that hexachlorobenzene is broken down by physical or chemical process in the environment. It is "dangerous when heated to decomposition it emits highly toxic fumes of chlorides" (EPA) (67) |
| Partition coefficient | 6.18 |

Conversion factors: 1 mg/l = 85.8 ppm; 1 ppm = 11.65 mg/m$^3$ at 25°C, 760 mm Hg

## 8.3 Effects In Animals

### 8.3.1 Acute Toxicity

HCB displays low acute toxicity in test animals. It is absorbed slowly from the gastroenteric tract with the compound primarily entering into the lymphatic system. Oral $LD_{50}$s for laboratory animals are cats, 1700 mg/kg; rabbits, 2600 mg/kg; rats, 3500 mg/kg; and mice, 4000 mg/kg. According to an EPA report (121), the average lethal oral dose for guinea pigs is 3000 mg/kg. For bluegill fish, rainbow trout, and channel catfish, the average lethal dose in water is given as more than 100 ppm.

### 8.3.2 Subchronic and Chronic Toxicities

Table 22.6 summarizes the subchronic toxicity and signs of intoxication produced in rats, rabbits, mice, guinea pigs, chickens, and Japanese quail. This table was prepared by the EPA and is based on information from a 1975 report by the National Academy of Sciences (122).

**Table 22.6.** Subchronic Toxicities of Hexachlorobenzene[a]

| Species | Dose | Test Duration | Effects Observed |
|---|---|---|---|
| Rats | 2 mg/kg/day | 13 days | No toxic effects |
| | 6 mg/kg/day | 13 days | Skin twitching and nervousness |
| | 20 mg/kg/day | 13 days | Neurotoxic symptoms; increase in liver weight |
| | 60 mg/kg/day and 200 mg/kg/day | 13 days | Neurotoxic symptoms; increases in liver and kidney weights |
| Rats | 10 mg/kg/day | 30 days | No toxic effects |
| | 30 mg/kg/day and 65 mg/kg/day | 30 days | Increases in feed consumptions/body weight gains; increased urinary coproporphyrin; increase in liver/body weight ratio |
| | 100 mg/kg/day | 30 days | Same as at 30 mg/kd/day plus elevated excretion of uroporphyrin |
| Rats | 100 mg/kg/day | 51 days | 13/33 deaths in 1 month; neurotoxic symptoms; increased liver weight; porphyria |
| Rats | 300 mg/kg/day | 10 days | 30% mortality |
| | 150 mg/kg/day | 30 days | 60% mortality |
| | 50 mg/kg/day | 30 days | 30% mortality |
| Male rats | 0.2% | 12 weeks | Retardation in weight gain; porphyria; degenerative changes in liver |
| Rats | 0.025 mg/kg/day[b] | 4–8 months | No toxic symptoms; possible effect on conditioned reflexes |
| Guinea pigs | 0.5% | 8–10 days | Marked neurological symptoms |
| Mice | 0.5% | 8–10 days | Marked neurological symptoms |
| Rabbits | 0.5% | 6 weeks | Increase in urinary porphyrins |
| | 0.5% | 8–12 weeks | Mortalities |
| Japanese quail | 1 ppm | 90 days | No toxic effects |
| | 5 ppm | 90 days | Slight increase in liver weight; slight porphyria |
| | 20 ppm | 90 days | Increased liver weight, decreased egg production; porphyria; liver/kidney pathological changes |
| | 80 ppm | 90 days | 5/15 deaths (18- to 62-day period); neurotoxic symptoms; porphyria; increased liver weight; decreased egg production and hatchability; liver/kidney pathological changes |

Table 22.6. (Continued)

| Species | Dose | Test Duration | Effects Observed |
|---|---|---|---|
| Japanese quail | 500 and 2500 ppm | 30 days | All 12 died in 30 days |
| | 100 ppm | 3 months | Mortality (1/12 on 20th day, 10 within 7 weeks, 1 in 10 weeks). One surviving cock showed marked loss of weight. Necrosis of liver cells; porphyria |
| Chickens | 120–480 ppm | 3 months | No toxic effects |

[a]This table was prepared by C. E. Mumma and E. W. Lawless for the EPA (121) and is based on information from a 1975 report by the National Academy of Sciences (122). The route is oral (in diet) unless otherwise indicated.
[b]Administered in water.

Owing to a serious human episode of HCB poisoning in Turkey (described in Section 8.8), the toxicity literature on HCB is voluminous, and the following discussions are organized according to the animal species tested.

*8.3.2.1 Rats.* Grant et al. (123) fed dietary concentrations of 10, 20, 40, 60, 80, and 160 ppm HCB for 9 or 10 months to weanling male and female Sprague–Dawley rats; the 200-ppm dosage was equal to approximately 1 mg/kg body weight. The response in the sexes differed to some extent, with the exception being chemical residues in the liver and the liver-to-body weight ratios in rats fed 80 and 160 ppm; these were dose-related but not sex-related. Also, the pharmacological action (narcosis) of pentobarbital and zoxazolamine was shortened in both males and females fed 20 ppm and higher levels. Consistent with these findings, males fed 40 ppm and higher concentrations displayed increased hepatic aniline hydroxylase and *N*-demethylase activities and cytochrome P450 levels; in females, these activities remained unaltered at all dietary levels. In females fed 80 or 160 ppm, weight gains were reduced and females, but not males, readily acquired chemical porphyria.

More recent reports include the investigation by Kuiper-Goodman et al. (124), who studied the subacute toxicity of HCB in Charles River strain rats fed diets providing dosages of 0.5, 2, 8, and 32 mg of the compound in corn oil per kilogram of body weight per day. Subgroups of rats were killed at 3, 6, 9, 12, and 15 days of feeding. At 15 weeks the remaining rats were then fed a compound-free diet, and sacrificed after 1, 2, 4, 16, and 33 weeks. Signs of intoxication were dose related and included excessive irritability, tremors, alopecia, and nonspecific dermal changes with slight scabbing at sites displaying alopecia. Ataxia with hind leg paralysis and loss of pain response in the legs were seen in a few females at the 32-mg/kg/day dose level after 6 and 9 weeks of treatment. In males and females, tremors disappeared after 3 to 4 weeks of feeding; all other signs of intoxication disappeared by 9 weeks.

Histological changes in these animals were confined to the liver and spleen.

Hepatomegaly and an increase in the size of centrolobular hepatocytes were noted. In males, serum sorbitol dehydrogenase activity, an indicator of liver damage, was first increased (maximally at 6 weeks) at the highest dose level, then subsequently dropped to near control levels. Histochemically, there was depletion of this enzyme in the liver. Liver succinic dehydrogenase and glucose-6-phosphatase were also equally depleted. Females particularly developed porphyria (together with porphyrin excretion) with high porphyrin values persisting in some rats, but not in others, after the animals were placed on a HCB-free diet. These studies support earlier investigations that female rats are more susceptible than males to the toxic effects of HCB. According to the investigators, a dose of 0.5 mg/kg appears to be the "no-effect" level in rats.

Boger et al. (125) administered similar dosages for a more prolonged period. They focused their investigation on liver changes in female Wistar rats after oral administration (gavage) of 0.5, 2.8, or 32 mg of HCB/kg body weight twice a week for 29 weeks, They reported a dose-related increased retention of the compound in the liver, which was enlarged. There was also a dose-dependent enlargement of hepatocytes, along with porphyrin deposition, increases of glycogen, and alterations of endoplasmic reticulum. At the 32-mg/kg dose level, the liver contained 273 µg of HCB per gram of tissue.

Dose-related changes in the hepatic ultrastructure were reported by Mollenhauer et al. (126). They fed outbred Sprague-Dawley rats dietary concentrations of 5 to 25 ppm of HCB for 3 to 12 months. In some animals, the smooth endoplasmic reticulum proliferated to become the predominant feature in the cell, whereas in others, the reticulum was apparently replaced by quantities of a storage product, presumably glycogen. The proliferation of the smooth endoplasmic reticulum occurred most often in animals fed the smaller doses, whereas storage product accumulations were most often observed in rats given larger doses. The most significant changes were noted in mitochondria, which became elongated and swollen. Storage bodies of 1 to 4 mm in diameter appeared in some cells. These bodies were surrounded by a double membrane derived from the endoplasmic reticulum, and may have been the partly digested remains of degenerated mitochondria.

Lissner et al. (127) fed Wistar rats a diet containing 0.2 percent of HCB. During the first week, the levels of microsomal cytochrome P450 and the activity of aniline hydroxylase in the liver were elevated. After the initial threefold rise in cytochrome P450, there was a second steep rise after the thirtieth day of treatment. Aniline hydroxylase activity showed a similar pattern. After 40 days of feeding, the urinary excretion of 5-aminolevulinic acid and porphyrin were markedly elevated. According to the authors, during an early period of HCB exposure, microsomal enzyme induction could lead to the oxidation of the compound, with the resulting metabolite possibly being the actual porphyrogenic agent and a more potent inducer of the microsomal enzymes than unmetabolized HCB.

Timme et al. (128) studied the effect of siderosis in rats (following the intraperitoneal injection of the iron-containing compound imferon, five doses of 10 mg each) after an interval of 2 months between imferon injections and HCB feeding of a dietary concentration of 0.3 percent. Tremors and urinary porphyrin were noted soon after the feeding was started, but porphyrins appeared earlier in the siderotic than in the control, nonsiderotic HCB-fed rats. In the nonsiderotic group, the liver contained only traces of stainable iron whereas in the siderotic rats, iron was initially diffuse in the lobule, but later the centrolobular areas contained only minimal quantities. In both groups, red porphyrin fluorescence appeared first in the centrolobular zone, but in the siderotic group it became widespread throughout the lobule. The authors concluded that the administration of iron prior to HCB feeding induced a more extensive development of liver lesions, and they suggest that this could account for the higher levels of excreted porphyrins in the iron-laden animals.

Food deprivation and its effect on HCB toxicity was studied by Villeneuve et al. (129) in female (30 to 90 g) rats. They fed dietary concentrations of 4, 20, 100, and 500 ppm ad libitum to four groups of 10 rats each. Similarly, second groups of rats were fed concentrations of 8, 40, 200, and 1000 ppm, but the food intake of these animals was reduced to 50 percent of that of the first groups. The investigators found that the reduced food intake, combined with the higher dietary concentrations of HCB, increased the following: the degree of liver hypertrophy, microsomal enzyme activity and porphyrin accumulation in the liver, excretion of porphyrin in the urine, and accumulation of HCB in brain and liver, as well as its rate of excretion in the feces.

Zawirska and Dzit (130) reported that spectrophotometric and chromatographic investigations of porphyrins isolated from rat livers indicated that the most efficient effect of synporphyrinogenesis, which was manifested through increased levels of uroporphyrin, had occurred in the groups treated simultaneously with HCB and testosterone, and with HCB, testosterone, and stilbestrol.

*8.3.2.2 Beagles.* Studies in 6.3- to 10.3-kg male and female beagles were performed by Gralla et al. (131), who administered HCB daily for up to 12 months in gelatin capsules at dose levels of 1, 10, 100, and 1000 mg per dog. Signs of intoxication included anorexia, loss of body weight, and mortality at the 1000-mg dose. These signs were observed to a lesser degree in the dogs treated with the 100-mg dose. Clinical laboratory changes in the animals at the two highest dose levels included anemia, hypoglycemia, and hypocalcemia. Testicular degeneration was considered by the investigators to be related to malnutrition.

Pathological lesions were confined to the abdomen and included serositis, necrosis, fibrosis, and steatitis of the omentum. Nodular hyperplasia of gastric lymphoid tissue was found in all dogs including those at the 1 mg/day dose. A dose-related neutrophilia appeared in the animals receiving the two highest dosages. Four severely affected dogs receiving the highest dose showed generalized vasculitis and one dog showed amyloidosis. No hepatic fluorescence was found at necropsy, indicating that these dogs were free of porphyria and suggesting that the dog, unlike the rat, is insensitive to this well-known toxic effect of HCB.

Luthra et al. (132) studied the effect of HCB on the serum lipoprotein pattern in beagles. They administered the compound in gelatin capsules in doses of 1, 10, and 100 mg/day for 13 weeks. Sera were collected at the termination of the treatment period following a 12-hr fast. They found that the percentage distribution in untreated control dogs was 37 ± 6 for β-lipoprotein and 47 ± 11 for α-lipoprotein. Utilizing statistical analysis by the Dunnett's multiple comparison method, a non-dose-related decrease of 13 to 40 percent in β-lipoprotein and an increase of 28 to 46 percent in the α-lipoprotein fractions were found. These results support the suggestion that HCB interferes with the lipid metabolism and/or transport.

*8.3.2.3 Monkeys.* Mueller et al. (133) studied the kinetic, metabolic, and histopathological effects of HCB and related compounds in rhesus monkeys. One male monkey was fed in the diet a daily dose of 10 ppm of $^{14}$C-labeled HCB for 540 days. The compound was absorbed more slowly from the gastroenteric tract than dieldrin or pentachloronitrobenzene (PCNB), which were studied concurrently. Absorption occurred with minor involvement of the portal venous system; major absorption was by the lymphatic system. Deposition of chlorobenzene took place in the fat, thymus, and bone marrow. The only affected organs in the dieldrin monkeys were the liver and kidneys, which showed in the liver the advanced parenchymatous degeneration associated with nodular hyperplasia. In the PCNB animals, the kidneys were the only organs involved. In the HCB monkey, degenerative changes were evident in the kidneys and cerebellum, and there was thymic cortical atrophy.

In another investigation with rhesus monkeys, Rozman et al. (134) concluded that none of the monitored parameters indicated harmful effects from a dose of 110 μg/day of HCB for 550 days.

*8.3.2.4 Pigs.* A 90-day pig toxicity study was conducted by Tonkelaar et al. (135) in which the animals were fed 0.05, 0.5, 5, and 50 mg/kg/day. Only the highest dose caused signs of intoxication, including porphyria, increased liver weights, and death during the experiment. All pigs excreted coproporphyrin and showed histopathological liver changes. Retention of HCB was highest in body fat (approximately 500 times the concentration in blood), followed by the liver, kidneys, and brain, in the order given. Under the conditions of this experiment with pigs, the no-effect level was judged to be 0.05 mg/kg/day.

Fassbender et al. (136) found that dietary ingestion of HCB (1 ppm) by pigs did not adversely affect their growth or health. Residues of the compound were found in significantly higher concentrations in liver and kidneys than in fat or muscle.

HCB distribution and retention in the tissues of swine were studied by Hansen et al. (137). They administered HCB to third-litter sows in dietary concentrations of 1 or 20 ppm throughout gestation and nursing. Swine receiving 1 ppm were not adversely affected and residue concentrations in tissues other than fat and bone marrow remained at or below the dietary concentration. Residues of HCB in the dissectable fat of these pigs accumulated to concentrations five- to sevenfold higher than dietary concentrations. Piglets accumulated fat residues that were higher that those of the sows through both placental transfer and nursing. A similar proportional accumulation of HCB in fat occurred in sows receiving 20 ppm in the diet.

Signs of toxicity included neutrophilia, gastric irritation, fatty replacement of Brunner's gland, and pancreatic periductal fibrosis. Hepatotoxicity was not apparent. There was a tendency toward neutrophilia in the sows fed 20 ppm, and that effect was also observed in beagle dogs, but only at higher HCB doses (131). The propensity toward gastric irritation and ulceration had also been previously observed in swine by Hansen et al. (138).

The authors indicated that the dietary concentration of 1 ppm is equivalent to an exposure of about 0.025 mg HCB/kg/day, which induces no discernible toxic response. This exposure level is still well above the conditional upper human intake of 0.6 µg/kg/day suggested by the Food and Agricultural Organization and World Health Organization (FAO/WHO) (139). Hansen et al. cautioned that the accumulation of HCB in fat could occur when exposing food-producing animals to levels above recommended dietary concentrations. At a dietary concentration of 20 ppm (approximately 0.5 mg/kg/day), fat residues accumulated to values considered hazardous. The effects noted in this investigation were very similar to those observed when 20 ppm Arochlor 1242 (PCB) was fed to swine (138), but HCB accumulated to concentrations that were severalfold higher than those of PCB. Kuiper-Goodman et al. (124) found that 0.5 mg/kg/day for 15 weeks is a "no-effect" level for HCB in rats, and Gralla et al. (131) found that this dose level was not toxic in dogs, but swine appear to be somewhat more susceptible in spite of their large fat reserve for storage of this compound.

### 8.4 Reproduction and Teratogenicity

Placental transfer of HCB has been reported in mice and rats.

Courtney and Andrews (140) performed studies with CD-1 mice that were treated with HCB by gastric intubation at levels of 10, 50, or 100 mg/kg/day. Two groups were treated before transplantation from day 0 through 5 of gestation and sacrificed at day 12 or 17. Two other groups were treated after implantation from day 6 through 11 or from day 6 through 16, and were sacrificed 24 hr after final treatment. The pesticide was mobilized from maternal depots and transferred across the placenta and deposited in the fetus. Concentration in whole fetuses ranged from 0.76 ppm to as high as 19.09 ppm and appeared to be related to the level of dose received by the mother. Fetal HCB levels were higher in the mice treated before implantation than in those treated on days 6 to 11, indicating the effect of maternal body burden.

Villeneuve et al. (129) studied placental transfer of HCB in pregnant nulliparous Wistar rats fed doses of HCB in the range of 5 to 120 mg/kg/day in corn oil from day 6 through day 16 of pregnancy. They found that this compound crosses the placenta and accumulated in the fetuses on day 22 of gestation in a dose-related manner. At all dose levels, the highest concentrations were found in the maternal liver, followed by the fetal liver, whole fetus, and fetal brain. They noted no fetopathological effects, suggesting that HCB, in the amounts fed to pregnant rats, caused no apparent adverse effect on fetal development.

In a four-generation test, groups of 10 male and 20 female Sprague–Dawley rats were fed diets containing 10, 20, 40, 80, 160, 320, or 640 ppm of HCB from weaning (141). The two highest dose levels were highly toxic to dams, leading to decreased fertility indexes in these groups. Litter sizes in these groups were decreased in the $F_{1b}$, $F_{2a}$, and $F_{2b}$ generations. Suckling pups in the F1 generation were particularly sensitive, and many died prior to weaning. There was an increased number of stillbirths but no gross abnormalities were seen in these pups. The 40-ppm diet induced increases in liver weights of 21-day-old pups and 20 ppm was shown to be a NOEL in this study.

HCB decreased survival of chicks of Japanese quail treated with a 20-ppm diet for 90 days (142). These results confirmed those of another study in which administration of 80 ppm for 90 days reduced egg production and hatchability (143, 144).

Iatropoulos et al. (145) studied the response of nursing rhesus monkeys to HCB given to their lactating mothers. The compound was given daily by gavage to three lactating monkeys in a dose of 64 mg/kg for 22, 38, or 60 days. Their infants were 140, 99, and 81 days old, respectively, when necropsied. A 99-day-old infant was listless for 24 hr before it died. Its mother was asymptomatic. The infant necropsy revealed lung edema and extensive engorgement of all CNS vessels with multiple extravasations. A 140-day-old infant showed lethargy, depression, and ataxia shortly before it died. Its mother was also asymptomatic. The major infant necropsy findings were lung edema with bronchopneumonia. The youngest of the infants was asymptomatic although its body weight was below the normal range. The body weight of the 99-day-old infant was likewise below normal, whereas the body weight of the 140-day-old infant was within normal range.

The microscopic changes in all monkey infants were minor or negligible. They consisted mainly of a mild centrolobular hepatocellular hypertrophy, a diffuse fatty metamorphosis, a coarse vacuolation of slight degree within the proximal renal tubular cells, and a mild gliosis in the cerebrum. The data indicate that infant rhesus monkeys are very susceptible to the toxic effects of HCB administered to their lactating mothers.

## 8.5 Immunotoxicity

The immunotoxic potential of HCB has been documented in experimental animals. Alteration of host resistance to infectious agents has been demonstrated in mice fed a diet containing 167 ppm HCB for 3 and 6 weeks, and then infected with the malarial parasite *Plasmodium berghei*, resulting in reductions in mean survival times of 24 percent and 31 percent, respectively (146).

The pre- and postnatal exposure of rats to a diet containing 50 and 150 mg/kg until 5 weeks of age resulted in a twofold and threefold increase, respectively, in the susceptibility to a lethal challenge with *Listeria monocytogenes* (143). Alteration of humoral immunity has also been reported. Dietary feeding of 167 ppm to mice for 6 weeks resulted in a twofold reduction in the number of spleen cells that produce specific antibodies during the primary antibody response to sheep erythrocyte antigen (147).

Alteration of cell-mediated immune responses has also been investigated. Vos et al. (143) reported that rats exposed pre- and postnatally to up to 150 mg/kg until

5 weeks of age did not show any alteration in skin transplant rejection times. However, in a chronic toxicity study, Silkworth (148) reported that dietary administration of 167 ppm HCB to mice for 37 weeks resulted in a 20 percent reduction in the graft-versus-host activity of isolated spleen cells. In the same study, an 80 percent reduction in the specific cell-mediated lymphotoxicity directed against cell surface alloantigen by sensitized spleen cells from mice exposed to HCB for 6 weeks was observed; this was associated with the time when the highest concentration of this compound was detected in the spleen. Lymphocytic blastogenesis induced by the mitogens lipopolysaccharide, phytohemagglutinin, and concanavalin A is not consistently influenced by HCB (143, 148); however, an increased rate of background DNA synthesis which increased with continued exposure was reported by Silkworth (148). Inhibition of the rosette-forming ability of alveolar macrophages has been reported in rats fed 250 ppm HCB for 10 weeks (149).

The immunotoxic potential of HCB may be related to the ability of the thymus and spleen to concentrate this compound to levels above the serum concentration, and it has been suggested that the mechanism of action exists within the effector phase of the immune response (143, 148).

### 8.6 Mutagenicity

In a dominant lethal test conducted by Khera (150), male rats that received 20, 40, or 60 mg HCB/kg body weight orally for 10 days were mated sequentially with untreated females (one male × two females; 14 mating periods, each of 5 days duration). There were no significant differences between the test and control groups with regard to the incidence of pregnancies, corpora lutea, live implants, or deciduomas, at any dose level or in any of the mating periods.

Also using the dominant lethal ray assay procedure, Simon et al. (151) treated groups of 10 male rats with oral doses of 70 or 221 mg of HCB/kg/day for 5 consecutive days. Ten additional rats received a single oral dose of the positive control, 0.5 mg triethylenemelamine/kg. The test compounds were dissolved in olive oil. Each male was mated with two naive nulliparous females each week for 14 weeks. The females were then sacrificed on the fourteenth day of gestation, and their uteri and ovaries examined for the number of corpora lutea and total viable and nonviable implantations.

HCB failed to induce any compound-related effects. The authors concluded that HCB does not display mutagenic activity.

### 8.7 Distribution and Metabolism Studies

In the 15-week feeding study described above (124) for the subchronic toxicity evaluation in rats, tissue HCB concentrations rose rapidly, and by 3 weeks of treatment were equilibrated. Concentrations in adipose tissue and liver were about 200- to 500-fold and 10- to 20-fold higher, respectively, than serum levels. At the 32-mg/kg/day level, the serum level, by week 13, reached 19.4 µg/g tissue in males and 30.9 µg/g in females. Concentrations particularly in adipose tissue and liver, but also in brain, kidney, and spleen, were significantly higher in females than in males.

Koss et al. (152) studied the toxicity of HCB in rats over a 53-week period. From the first through fifteenth week, the compound was given orally every other day at a dose of 178 mol/kg (50 mg/kg). Nine weeks after the start of treatment, an equilibrium was reached between intake and elimination of the compound and its metabolites. At this time, 1 g of liver contained approximately 1 μmol HCB, 50 nmol pentachlorophenol, 5 nmol tetrachlorohydroquinone, and approximately 0.1 nmol pentachlorothiophenol. Thirty-eight weeks after cessation of administration, the biologic half-life was found to be 4 to 5 months. Liver porphyrin, urinary porphyrin, δ-aminolevulenic acid, and porphobilinogen increased up to week 15. During the subsequent 38-week period, the porphyrin content of the liver continued to increase, but the porphyrin content of the urine and porphyrin precursors had decreased to almost normal levels.

The metabolism of HCB also was studied by Engst et al. (153) in male Wistar rats, fed by stomach tube 8 mg/kg (in sunflower oil) for 19 days. Tissue concentrations at the termination of the treatment period were for body fat, 82 ppm; muscle, 17 ppm; total liver, 125 μg; total kidney, 21 μg; total spleen, 9 μg; total heart, 1.5 μg; and adrenal, 0.5 μg each. The urine contained HCB and its primary metabolite, pentachlorophenol, which was present alone or sometimes associated with 2,3,4,6-tetrachlorophenol and/or 2,3,5,6-tetrachlorophenol, or 2,4,5-trichlorophenol. Trace amounts of related compounds were also detected. The feces contained high amounts of HCB and smaller amounts of pentachlorobenzene.

Schuster and Renner (154), who dosed rats with a total of 300 mg/kg of HCB for 10 months, found pentachlorophenol and 2,4,5-trichlorophenol to be the primary urinary metabolites. 1.2.4.5-Tetrachlorobenzene and certain other chlorobenzenes were detected, but not completely identified.

Lui and Sweeney (155) concurred that pentachlorophenol is the primary metabolite in rats. They also detected more polar compounds, presumably derivatives of HCB. Formation of pentachlorophenol from HCB may proceed either through a free-radical mechanism or by initial formation of an arene oxide. In either event, reactive intermediates may form covalent bonds with cellular constituents leading to possible irreversible damage.

Koszo et al. (156) reported that when rats are given an oral dose (by stomach tube) of 0.2 g HCB, the compound can be found in urine and feces 1 day later in 32 and 43,260 μg amounts, respectively. The urinary and fecal excretion of pentachlorophenol was 30 μg and less than 1 μg, respectively.

In experiments by Koss et al. (157) with labeled HCB, they obtained evidence for an unexpectedly high degree of conversion to metabolites. Four weeks after administration, 7 percent of the radioactivity was excreted in rats via the kidneys and 27 percent in feces. Nearly the total in urine was contained in metabolites of HCB, and 69 percent of the radioactivity in the feces was represented by unchanged compound. It was calculated that the rat eliminates almost half of HCB in the form of metabolites including pentachlorophenol, which accounts for about one-fifth of the total excreted. The other major metabolites identified are tetrachlorohydroquinone and pentachlorothiophenol. These substances represent 3 and 16 percent, respectively, of the label excreted.

More recently, Koss et al. (158) described identification of pentachlorothiophenol and pentachlorothioanisole in the livers of rats treated with HCB. In order to clarify the fate of these metabolites, they were administered to rats, and the conversion products excreted in urine and feces were isolated. The metabolites of pentachlorothiophenol and pentachlorothioanisole were found to be excreted in both conjugated and free forms. From extracts of excreta, tetra- and trichlorobenzene with two or three sulfur-containing substituents on the ring were isolated; these were analogous compounds in which thiol groups were converted into sulfoxide and sulfone groups, as well as analogous compounds in which chlorine was replaced by hydrogen. Following administration of the sulfoxide and sulfone of pentachlorothioanisole under analogous conditions, pentachlorothiophenol and pentachlorothioanisole and their metabolites were detected in the excreta of the animals. No evidence was obtained that the parent compounds were excreted in the unchanged form.

Goerz et al. (159) studied the possible role of the two HCB metabolites, pentachlorophenol and pentachlorobenzene, in the production of porphyria in Wistar rats. They fed HCB and QCB, each in a dietary concentration of 0.05 percent, for 80 days. A third group of rats, prior to HCB feeding, ingested a diet containing 0.5 percent of pentachlorophenol for 40 days and subsequently HCB for 54 days. Because the urinary porphyrin excretion remained the same throughout the entire experimental period in rats fed QCB and pentachlorophenol, and because a distinct porphyria was established after 60 days by HCB, the authors considered it quite unlikely that the two HCB metabolites are the porphyrogenic agents of HCB. In spite of the pentachlorophenol pretreatment, HCB led to porphyria after only 50 days, the same as for the HCB controls. The investigators consider it unlikely that HCB-induced porphyria has a distinct pathogenic connection to the induction of the cytochrome P450 system.

To elucidate the relationship between chemical structure and their biologic activities, the contents of cytochromes and hepatic constituents in addition to the activities of drug-metabolizing enzymes and $\delta$-aminolevulinic acid (ALA) synthetase were examined in rats treated with various chlorinated benzenes, that is, MCB, $p$-DCB, 1,3,5-trichlorobenzene, 1,2,4,5-tetrachlorobenzene, pentachlorobenzene, and HCB (160). The content of cytochrome P450 and activities of aminopyrine demethylase and aniline hydroxylase were increased by oral administration of all chlorobenzenes except MCB as a daily dose of 250 mg/kg, once daily, for 3 days. The contents of microsomal protein and phospholipids also showed a similar tendency to those described above. The activity of ALA synthetase was increased by treatment with all compounds used. The content of cytochrome P450 and activity of aminopyrine demethylase were decreased in 24 hr after a single administration of MCB in doses of 125, 250, 500, and 1000 mg/kg, whereas the activity of ALA synthetase was increased markedly by all doses used. The activity of aniline hydroxylase was increased by a dosing of 1000 mg/kg MCB.

In the time course after a single administration of MCB in a dose of 250 mg/kg, the activity of ALA synthetase decreased in 6 hr after administration, was subsequently restored to normal levels in 12 hr, and then increased markedly in 24 hr. The opposite changes were noted in the content of cytochrome P450.

After 5 to 12 months of feeding beagles dose levels of 1, 10, 100, and 1000 mg HCB/dog (131), concentrations of the compound in adipose tissue of test animals were 11 ppm at a dose of 1 mg/kg, 67 ppm at 10 mg/kg, 714 ppm at 100 mg/kg, and 1216 at 1000 mg/kg. The concentrations in bile were 0.52, 4.87, and 50 ppm, respectively.

Mull et al. (161) studied the distribution and excretion of HCB in castrated growing lambs fed dietary concentrations of 0.01, 0.1, or 1 ppm for 90-day periods. Biopsies were taken at days 7, 15, 30, 45, and 60, and every 30 days thereafter through day 300. The treatment did not affect the growth rate of the animals.

By the end of the 90-day exposure period adipose tissue concentrations of HCB had reached a level 10 times that in the diet. The omental fat was found to contain higher concentrations of HCB residues than did perirenal fat at 90 days, but not at 300 days. The apparent decrease of the compound in the fat following cessation of treatment was primarily due to dilution brought about by increases in the quantities of fat in the carcass. The studies indicated a non-dose-related HCB half-life of approximately 90 days.

In an investigation with laying hens given seven consecutive daily doses ranging from 1 to 100 mg/kg, Hansen et al. (137) determined the half-time of HCB in fat to be 24 to 27 days. They found that more than half of the compound is excreted in the yolks of eggs. Tissue concentrations of the compound paralleled the concentrations in body fat. The skin provided a significant reservoir for HCB.

Yang et al. (162), who administered $^{14}$C-hexachlorobenzene intravenously to three rhesus monkeys, found that this treatment led to the highest level of radioactivity in body fat and bone marrow. One year after an intravenous injection, the cumulative fecal and urinary metabolites were estimated to be only 2.8 and 1.6 percent, respectively, of the dose administered. They explained the long-term storage of HCB in fat as the basic reason for the slow elimination of this compound from the body.

In monkeys, pentachlorophenol was the major metabolite, but traces of pentachlorobenzene were also found in the feces. All urinary radioactivity was in the form of unidentified polar metabolites of HCB; they were neither pentachlorobenzene or pentachlorophenol. Parenthetically, rat findings indicate that polar metabolites formed were estimated to be less than 0.2 percent of the injected dose.

In summary, HCB is absorbed from the gastrointestinal tract more rapidly from a lipophilic vehicle (80 percent in olive oil) versus a low rate from aqueous formulations (<20 percent in 1 percent methyl cellulose) with absorption taking place primarily through lymphatic channels. Distribution occurs toward tissues with higher fat content. HCB is metabolized to tetra- and pentachlorobenzene and their phenols/thiophenols and corresponding conjugates (13).

## 8.8 Human Experience

### 8.8.1 Nonoccupational Exposure

An epidemic of HCB intoxication occurred during the years 1955 to 1959 in Turkey. Wheat that was intended to be used as seed was treated (2 kg/1000 kg grain) with this fungicide. Accidentally, the wheat was ground and used as flour for bread. Some 5000 people suffered a vesicular and bullous disease resembling porphyria cutanea tarda, which was frequently associated with hepatomegaly, porphyria, and death. The annual mortality in the exposed population ranged from 3 to 11 percent. According to Cabral et al. (12, 163), young children were predominantly affected. It was reported that all children born to porphyric mothers died (164).

In mild cases, lesions developed early, with the vesicles and bullae appearing particularly in skin areas exposed to sunlight. Peak instances occurred during the summer months. In children, the initial lesions resembled comedones and milia, whereas in adults, bullous lesions developed promptly. Frequently, the lesions became crusted and at times ulcerated. There were many clinical variations, which included hyperpigmentation, hypertrichosis, alopecia (in some instances permanent) or excessive growth of hair (occasionally the entire body showed excess hair growth), corneal opacities, deformation of the exposed parts, notably the digits, loss of body weight, wasting of skeletal muscles, and hepatomegaly. The urine frequently showed the characteristic port wine color associated with porphyria. Weakness and convulsions were also reported in infants (165).

Following cessation of ingestion of contaminated bread, recovery usually occurred. The daily intake of HCB was estimated as having ranged 0.05 to 0.2 g/day per person for months [See Cam and Nigogosyan (166), Cetingil and Ozen (167), and DeMatteis et al. (168).] Twenty years after the incident in Turkey, some of the individuals were still suffering from the effects of HCB. Cripps et al. (169) examined 32 porphyric Turks with an average age of 29.5 years. These people still suffered the following: hyperpigmentation (53 percent of the patients), hirsutism (41 percent), scarring of hands and face (50 percent), pinched facies and rhagades (22 percent), fragile skin on hands and face (12.5 percent), enlarged liver (9 percent), ascites and jaundice (9.4 percent), recent red urine (9.4 percent), weakness and parathesias (43.7 percent), small hands, sclerodermoid thickening, shortening of distal phalanx and painless arthritis (44 percent), and enlarged thyroid (38 percent), compared to 5 percent in the general Turkish population of this area. In 10 patients the excretion of uroporphyrin in urine and feces was elevated. A porphyric's maternal milk was found to contain more than 0.7 ppm HCB. The cutaneous effects were frequently precipitated by sunlight; differences in susceptibility could be expected and have been reported.

These studies and a report by Peters (170) demonstrate that symptoms of subchronic or chronic HCB intoxication may involve cutaneous, hepatic, arthritic, visceral, urinary, and neurological effects that may persist for 20 and more years. Unfortunately, precise information is not available on the severity of the original exposures (determined by blood concentrations) of the individuals who showed symptoms after 20 years.

A therapeutic trial of ethylenediaminetetraacetic (EDTA) administered orally (with initial intravenous administration in some patients) was performed in seven patients who, following treatment, remained symptomatic with porphyria. Four of the EDTA-treated patients became asymptomatic within 3 months, and all patients were completely asymptomatic within 1 year. After cessation of treatment, one patient underwent relapse 10 months later, followed by abrupt recovery on resumption of EDTA therapy (170).

Pederson and Rimington (171), in seeking an explanation of the mode of action of HCB, suggested that this compound forms a complex with one or several porphyrin compounds in the heme biosynthetic pathway of the liver; the sizes of the porphyrin ring and the HCB molecule are very similar, and the flat nature and high polarizability of each could lead to a considerable van der Waals interaction.

For concentrations of HCB in human tissue in various countries of the world, and in food and drink, refer to the 1979 IARC monograph (144).

### 8.8.2 Occupational Exposure

The "toxic hazard" rating of HCB in the industrial environment is "1" as it applies to acute and chronic local and acute systemic effects, which means that the compound causes readily reversible changes that disappear after the end of exposure (121). As a result of industrial exposure the chronic systemic effects are very limited. Ehrlicher (172) reported no serious illnesses or changes of liver function on the workers of a manufacturing plant who had been exposed to airborne HCB over a 40-year period.

According to Sairtskii (173), a "hexachlorobenzene concentration of 0.1 mg/liter could be assumed to represent the threshold toxicity value, and that 1/100 of that value may then represent the limit of permissible concentration of hexachlorobenzene in air for workers" (121).

Porphyria is one of the longest lasting effects of hexachlorobenzene intoxication in humans. According to Hayes (174), porphyrins are probably the cause of more frequent and more serious photosensitization in humans than all other materials combined. But photosensitization does not occur in all cases of increased porphyrin concentration in blood, urine, or feces. The authors also indicated that for some people who suffer from latent acute intermittent porphyria, an acute attack may be brought on by alcohol or barbiturates. [Porphyria was apparently first produced in rabbits 1895 by Stokvis (175) by administration of sulfomethylmethane; in 1941, Deichmann (176) produced it in rabbits, cats, rats, and mice by treatment with methyl, ethyl, and *n*-butyl methacrylate. Exposure of porphyric rats and mice to sunlight produced marked edema in areas of the body unprotected by hair.]

## 8.9 Cancer

Further details are presented on the carcinogenesis results described above in the section on cancer. In 1977 Cabral et al. (12) reported on the life-long feeding of this compound to Syrian golden hamsters. A dietary concentration of 50 ppm was fed to 36 males and 30 females, 100 ppm to 30 males and 30 females, and 200 ppm to 60 males and 60 females. The control group consisted of 40 males and 40 females.

The observations for animals at 80 weeks of age demonstrated a significant production of hepatomas and liver hemangioendotheliomas. The capacity of HCB to induce hepatomas, and their incidence, are as follows: low dose, males 27.2 and females 37.5 percent; medium dose, males 75 and females 42 percent; high dose, males 87 and females 81 percent. The latency period appeared to be dose related. No metastases were found, and no hepatomas were noted in the controls.

The incidence of hemangioendotheliomas in the hamsters fed 200 ppm was 34 percent in males and 9 percent in females. Three of the hemangioendotheliomas occurring in the 200-ppm males caused metastases. The investigators accepted these observations as evidence that HCB in the dosages fed is carcinogenic in the Syrian golden hamster.

Cabral et al. (12) also reported on groups of Swiss mice fed for life a diet containing 50 ppm, 100 ppm, and 200 ppm HCB. Hepatoma incidences in mice fed 200 ppm were 4.0 percent in the males and 29.7 percent in the females. In both males and females fed 100 ppm, the incidence of hepatomas was 10 percent. None of the hepatomas metastasized, and none was noted in the control mice.

In groups receiving 100, 200, and 300 mg/kg diets, the incidences of liver-cell tumors in survivors (males and females) at the time the first liver-cell tumor was observed were 3/12 and 3/12, 7/29 and 14/26, and 1/3 and 1/10, respectively. The effective intake of HCB that induced liver-cell tumors was 12 to 24 mg/kg/day (12, 177). The results of these studies support the observations in the Syrian Golden hamster.

A second long-term study was conducted in Sprague–Dawley rats using dietary exposures of 0, 75, and 150 ppm (178, 179). Liver tumor (benign and malignant) incidences were increased, as were bile duct tumors in both low- and high-dose groups, with females exhibiting higher rates for these tumors. Males, on the other hand, exhibited higher rates of renal cell adenomas than females and kidney carcinoma rates were indistinguishable from control rates.

Follow-up studies have been performed on the population exposed to high levels of HCB from bread contamination in Turkey. In the period of at least 20 years since the accidental exposures, there was no evidence of increased rates of cancer development in 161 (180) or 204 (181) individuals.

### 8.10 Hygienic Standards of Permissible Exposure

According to the 1978 FAO/WHO Food Standards Programme Codex Committee on Pesticide Residues (139), the conditional acceptable daily intake of HCB is 0.0006 mg/kg body weight. The maximum residue limits of HCB in milled cereal products is 0.01 mg/kg, in raw cereals 0.50 mg/kg, in milk and milk products 0.5 mg/kg, and in eggs and in the carcass fat of cattle, goats, pigs, sheep, and poultry, 1 mg/kg (95).

OSHA and ACGIH have not published a standard or recommended a TLV for this compound.

## 9 BROMOBENZENE [CAS # 108-86-1]

Bromobenzene, $C_6H_5Br$, is also called phenyl bromide.

HCB

### 9.1 Uses

The compound is employed as a starting material in organic syntheses in which a Grignard intermediate (phenyl magnesium bromide) is used. The material is a chemical precursor for certain agricultural products and has been used as an additive to motor oils. Bromobenzene has also been used as a high-density solvent for chemical recrystallization processes.

### 9.2 Physical and Chemical Properties

| | |
|---|---|
| Appearance | Clear liquid |
| Molecular weight | 157.02 |
| Density (20°C) | 1.49 |
| Melting point | −30.6°C |
| Boiling point | 156.2°C |
| Odor | Aromatic |
| DOT hazard | Combustible liquid |
| Vapor density | 5.41 |
| Solubility | 450 ppm in water; miscible in chloroform, benzene, petroleum hydrocarbons; soluble in alcohol, ether |

### 9.3 Toxic Effects

#### 9.3.1 Acute Toxicity

There is little published information on bromobenzene relating to its short-term toxic effects. NIOSH (182) has cited data primarily from foreign sources which indicate that the 2-hr vapor inhalation $LC_{50}$ in mice was 21 $g/m^3$ with clinical evidence of somnolence and muscle spasticity. The same $LC_{50}$ was given for rats. The $LD_{50}$ following intraperitoneal injection of bromobenzene in immature rats was 206 mg/kg, whereas for adult rats the $LD_{50}$ was 3.87 g/kg. Oral $LD_{50}$ values for both mice and rats are reported as 2.7 g/kg. The chemical is also listed as being irritating to the skin (183), and is therefore likely to be irritating to the eyes and respiratory tract.

### 9.3.2 Subchronic and Chronic Toxicity

Early work on the chemical focused upon examination of bromobenezene's liver toxicity, and later included assessment of kidney and lung effects. Pioneering work showed that bromobenzene is relatively inert until it is biologically transformed to reactive chemical species by liver enzymes. Bioactivation occurs via transformation by oxidative or Phase I hepatic pathways involving cytochrome P450. The induction of microsomal enzymes was shown to enhance the liver toxicity of bromobenzene, as did depletion of cellular glutathione. Conversely, inhibitors of P450 (e.g., SKF-525A) decreased hepatic necrosis induced by the chemical (184).

Metabolites that were found in rat urine following administration of bromobenzene included the 3,4-dihydrodiol, 3- and 4-S-glutathionyl derivatives of the 3,4-dihydrodiol compound, and 4-bromophenol. This led to the proposal that the active toxin in the liver was the 3,4-epoxide of bromobenzene (185). It was hypothesized that the epoxide reacted with tissue proteins, forming covalent adducts that rendered vital enzymes and membranes biologically inactive.

Subsequent work demonstrated that both the 2,3- and 3,4-epoxides were reactive intermediates, for 2-, 3-, and 4-S-(bromophenyl)cysteines were found following hydrolysis of liver proteins from $^{14}$C-bromobenzene-treated rats (186). Because only 2 percent of the $^{14}$C-binding to liver protein was accounted for by these cysteine metabolites, other pathways that are involved in the hepatotoxic mechanism are being examined. Evidence exists for protein adducts being formed via quinone metabolites of bromobenzene (187) based on results indicating formation of 2,5-diydroxyphenol mercapturic acid. Other metabolites have been identified that indicate multiple pathways of bromobenzene metabolism (188).

Kidney effects may be associated with metabolites different from those that are active in the liver. o-Bromophenol administration to rats decreased renal glutathione levels and increased blood urea nitrogen, suggestive of adverse effects to the kidney (184).

### 9.3.3 Mutagenicity and Carcinogenicity

NIOSH (182) cited mutagenicity results from four assays. All were considered negative with the exception of one which was considered inconclusive. No citations were located for chronic bioassays in laboratory animals to assess tumorigenic potential.

## 10 OTHER HALOGENATED BENZENE COMPOUNDS

### 10.1 Uses

There are many halogenated benzene chemicals that are manufactured and sold but only in small amounts (<1 million pounds) relative to those for the lower chlorinated benzenes. These materials are sold for the manufacture of specialty products (e.g., pharmaceuticals, agricultural chemicals, high performance polymers). A partial listing of available chemicals in this class is provided below:

o-, m-, and p-dibromobenzene
tri- and tetrabromobenzenes
1,2-, 1,3-, and 1,4-bromochlorobenzene
2,4- and 3,5-dibromochlorobenzene
o-, m-, and p-bromofluorobenzene
2,4,- and 3,4-dibromofluorobenzene
o-, m-, and p-bromotoluene
o-, m-, and p-bromophenol

### 10.2  Physical and Chemical Properties

A potential source of information is the Material Safety Data Sheets from the suppliers of these chemicals. Further data may be available from the chemical literature.

### 10.3  Toxicity Information

Virtually no data have been published on the health effects of these chemicals. Producers may have information which is unpublished.

## REFERENCES

1. Chemical Manufacturers Association, Technical Report Worldwide Literature Search on Chlorobenzenes, submitted to the Manufacturing Chemists Association, Washington, DC, March 4, 1976.
2. S. P. Varshavskaya, *Hyg. Sanit.*, **33**(10), 17 (1967).
3. S. P. Varshavskaya, *Nauchn. Tr. Aspir. Ordinatorov. 1-i Mosk. Med. Inst.*, 175 (1967); through Ref. 1 and *Chem. Abstr.*, **51**, 4816 (1970).
4. C. den Besten, J. J. R. M. Vet, H. T. Besselink, G. S. Kiel, B. J. M. van Berkel, R. Beems, and P. J. van Bladeren, *Toxicol. Appl. Pharmacol.*, **111**, 69 (1991).
5. E. R. Stine, L. Gunawardhana, and I. G. Sipes, *Toxicol. Appl. Pharmacol.*, **109**, 472 (1991).
6. National Toxicology Program, "Toxicology and Carcinogenesis Study on Monochlorobenzene," Technical Report 261, 1985.
7. National Toxicology Program, "Toxicology and Carcinogenesis Study on 1,2-Dichlorobenzene," Technical Report 255, 1985.
8. National Toxicology Program, "Toxicology and Carcinogenesis Study on 1,4-Dichlorobenzene," Technical Report 319, 1987.
9. E. Loeser and M. H. Litchfield, *Food Chem. Toxicol.*, **21**, 825 (1983).
10. R. W. Lambrecht, E. Erturk, E. E. Grunden, H. A. Peters, C. R. Morris, and G. T. Bryan, *Fed. Proc.*, **42**, 782 (1983).
11. R. W. Lambrecht, E. Erturk, E. E. Grunden, H. A. Peters, C. R. Morris, and G. T. Bryan, *Proc. Am. Assoc. Cancer Res.*, **24**, 59 (1983).

12. J. R. P. Cabral, *Toxicol. Appl. Pharmacol.*, *Abstr.*, **41**, 155 (1977); J. G. Voss, M. J. Van Logten, J. G. Kreeftenberg, and W. Kruizinga, *N.Y. Acad. Sci.* (June 21, 1987).
13. U.S. EPA, "Health Assessment Document for Chlorinated Benzenes," NTIS PB85-150332, 1985.
14. Agency for Toxic Substances and Disease Registry, U.S. Public Health Service, "Toxicological Profile for Chlorobenzene, 1990.
15. U.S. EPA, "Drinking Water Health Document on *o*-Dichlorobenzene," NTIS PB89-192231, 1988.
16. Agency for Toxic Substances and Disease Registry, U.S. Public Health Service, "Toxicological Profile for 1,4 Dichlorobenzene," 1989.
17. Agency for Toxic Substances and Disease Registry, U.S. Public Health Service, "Toxicological Profile for Hexachlorobenzene, 1990.
18. M. Charbonneau, J. Strasser, E. A. Lock, M. J. Turner, and J. A. Swenberg, *Toxicol. Appl. Pharmacol.*, **99**, 122 (1989).
19. L. Bouthillier, E. Greselin, J. Brodeur, C. Viau, and M. Charbonneau, *Toxicol. Appl. Pharmacol.*, **110**, 315 (1991).
20. U.S. Environmental Protection Agency, "Alpha 2μ-Globulin: Association with Chemically Induced Renal Toxicity and Neoplasia in the Male Rat," EPA/625/3-91/019F, 1991.
21. M. Cote, I. Chu, D. C. Villeneuve, S. E. Secours, and V. E. Valli, *Drug Chem. Toxicol.*, **11**, 11 (1988).
22. W. B. Coate, W. H. Schoenfish, R. I. Trent, and W. M. Busey, *Arch. Environ. Health*, **35**, 249 (1977).
23. I. Chu, D. C. Villeneuve, C. R. Barnes, R. M. Benoit, and Y. H. Qin, *J. Toxicol. Environ. Health*, **13**, 777 (1983).
24. R. Linder, J. Scotti, K. Goldstein, and K. McElroy, *J. Environ. Pathol. Toxicol.*, **4**, 1983 (1980).
25. U.S. EPA, "Guidelines for Carcinogen Risk Assessment," *Fed. Reg.*, **51**(185), 33992 (Sept. 1986).
26. *IARC Monographs on the Evaluation of Carcinogenic Risk to Humans*, Supplement 7, International Agency for Research on Cancer, World Health Association, (1987).
27. National Toxicology Program, *6th Annual Report on Carcinogens*, U.S. Public Health Service, 1991.
28. *TLVs, Threshold Limit Values and Biological Exposure Indices for 1992–1993*, American Conference of Governmental Industrial Hygienists, Cincinnati, OH.
29. R. L. Hollingsworth, V. K. Rowe, F. Oyen, H. R. Hoyle, and H. C. Spencer, *A.M.A. Arch. Ind. Health*, **14**, 138 (1956); also through *Hygienic Guide Series*, American Industrial Hygiene Association, May–June, 1964.
30. J. J. Kocsis, S. Harkaway, and R. Snyder, *Ann. N.Y. Acad. Sci.*, **243**, 104 (1975).
31. W. F. von Oettingen, *The Halogenated Hydrocarbons, Toxicity and Potential Dangers*, U.S. Publ. Health Serv. Publ. No. 414, 1955.
32. F. Flury and F. Zernik, *Schadliche Gase*, Springer, Berlin, 1931.
33. D. D. Irish, "Halogenated Hydrocarbons: II. Cyclic," *Industrial Hygiene and Toxicology*, 2nd rev. ed., Wiley-Interscience, New York, 1963, pp. 1333–1345.
34. N. D. Rozenbaum, R. S. Block, S. N. Kremneva, S. L. Ginzburg, and I. V. Pozhariskii, *Gig. Sanit.*, **12**(1), 21, (1947); through Ref 1.

35. Hygienic Guide Series, American Industrial Hygiene Association, 1964; American Conference of Governmental Industrial Hygienist, "Threshold Limit Values for 1963," *AMA Arch. Environ. Health.*, **7**, 592 (1963).
36. W. K. Knapp, Jr., W. M. Busey, and W. Kundzins, *Toxicol. Appl. Pharmacol.*, **19**, 393 (1971).
37. A. G. Khanin, Tr. *Tsent. Inst. Usoversh. Vrachei.*, **135**, 97 (1969); through Ref. 1.
38. L. P. Tarkova, *Hyg. Sanit.*, **30**(3), 327 (1965); through Ref 1.
39. M. Lecca-Radu, *Igiena*, **8**, 231 (1959); through Ref. 1.
40. W. A. Skinner, G. W. Newell, and J. V. Dilley, *Toxic Evaluation of Inhaled Chlorobenzene*, Final Report prepared for the Division of Biomedical & Behavioral Sciences, National Institute of Occupational Safety and Health, Cincinnati, Ohio, June 15, 1977.
41. W. D. Reid, G. Krishna, J. R. Gillette, and B. Bernard, *Pharmacology* **10**(4), 193 (1973).
42. W. D. Reid, *Exp. Mol. Pathol.*, **19**, 197 (1973).
43. J. A. John, W. C. Hayes, T. R. Hanley, Jr., K. A. Johnson, T. S. Gushow, and K. S. Rao, *Toxicol. Appl. Pharmacol.*, **76**, 365 (1984).
44. R. S. Nair, J. A. Barter, R. E. Schroeder, A. Knezevich, and C. R. Stack, *Fundam. Appl. Toxicol.*, **9**, 678 (1987).
45. T. M. Sullivan, G. S. Born, and G. P. Carlson, *Toxicol. Appl. Pharmacol.*, **71**, 194 (1983).
46. M. Ogata and Y. Shimada, *Int. Arch. Occup. Environ. Health*, **53**, 51 (1983).
47. B. Spencer and R. T. Williams, *Biochem. J.*, **46**(4), XV (1950); **47**, 279 (1950).
48. J. R. Lindsay Smith, B. Spencer, and R. T. Williams, *Biochem J.*, **47**(3), 284 (1950).
49. J. R. Lindsay Smith, A. J. Shaw, and D. M Foulkes, *Xenobiotica*, **2**(3), 215 (1972).
50. D. V. Parke and R. T. Williams, *Biochem. J.*, **54**, 231 (1953).
51. H. Reich, *Samml. Vergiftungsfallen*, **5**, 193 (1934)
52. R. Girard, T. Tolot, P. Martin, and P. Bourret, *J. Med. Lyon*, **50**(1164), 771 (1969); through Ref. 1.
53. L. P. Tarkhova, *Hyg. Sanit.*, **30**(3), 327 (1965); through Ref. 1.
54. Arthur D. Little, Inc., Research on Chemical Odors, Part 1, Odor Threshold for 53 Commercial Chemicals, Research report for the Manufacturing Chemists Association, Washington, DC, October, 1968.
55. NIOSH/OSHA Pocket Guide to Chemical Hazards, U.S. Department of Health, Education and Welfare, Public Health Service, Center for Disease Control, National Institute for Occupational Safety and Health, September, 1985.
56. S. P. Varshavskaya, *Gig. Sanit.*, **33**(10), 15 (1968); through abstract in *Toxicol. Appl. Pharm.*, 69-0069 (1971).
57. B. B. Brodie, W. D. Reid, A. K. Cho, G. Sipes, G. Krishna, and J. R. Gillette, *Proc. Natl. Acad. Sci.*, **68**(1), 160 (1971).
58. R. L. Hollingsworth, V. K. Rowe, F. Oyen, T. R. Torkelson, and E. M. Adams, *A.M.A. Arch. Ind. Health*, **17**, 180 (1958).
59. IARC Monographs on the Evaluation of Carcinogenic Risk of Chemicals to Man, Vol. 7, International Agency for Research on Cancer, World Health Association, Lyon, 1974.

60. K. S. Robinson, R. J. Kavlock, N. Chernoff and L. E. Gray, *J. Toxicol. Environ. Health*, **8**, 489 (1981).
61. T. Czajkowska, U. Ruta, S. Szendzikowski, and Z. Swierchowski, *Med. Pracy*, **21**(5), 450 (1970), through Ref. 1.
62. W. C. Hayes, T. R. Hanley, Jr., T. S. Gushow, K. A. Johnson, and J. A. John, *Fundam. Appl. Toxicol.*, **5**, 190 (1985).
63. R. Nair, J. Barter, H. Bolte, R. Schroeder, and C. Stack, Fifth International Congress of Toxicology, Abstract, 1989.
64. T. S. Hele and E. H. Callow, *Proc. Physiol. Soc. J. Physiol.*, **57**, xlii (1923).
65. E. H. Callow and T. S. Hele, *Biochem. J.*, **20**, 598 (1926).
66. W. M. Azouz, D. V. Parke, and R. T. Williams, *Biochem. J.*, **59**(3), 410 (1955).
67. A. Jacobs, M. Blangetti, and E. Hellmund, *Vom Wasser*, **43**, 259 (1974); through G. R. Pielmeier.
68. J. Barr, A. J. Weir, K. Brendel, and I. G. Sipes, *Xenobiotica*, **21**, 331 (1991).
69. J. Barr, A. J. Weir, K. Brendel, and I. G. Sipes, *Xenobiotica* **21**, 341 (1991).
70. J. Gadrat, J. Monnier, A. Ribet, and R. Bourse, *Arch. Mal. Prof. Med. Trav. Secur. Soc.*, **23**(10/11), 710 (1962); through Ref. 1.
71. J. G. Downing, *J. Am. Med. Assoc.*, **112**, 1457 (1939).
72. R. Girard, F. Tolot, P. Martin, and P. Bourret, *J. Med. Lyon*, **50**(1164), 771 (1969); through Ref. 1.
73. F. Tolot, B. Soubrier, J. R. Bresson, and P. Martin, *J. Med. Lyon* **50**(1164), 761 (1969); through Ref. 1.
74. T. Sollmann, *A Manual of Pharmacology and Its Applications to Therapeutics and Toxicology*, 8th ed., W. B. Saunders, Philadelphia, 1957.
75. G. Dikmans, *J. Agric. Res.*, **35**, 645 (1927).
76. D. Irie, T. Sasaki, and R. Ito, *Tohoku Igaku Zasshi*, **20**(5/6), 772 (1973); through Ref. 1.
77. M. H. Pike, *J. Mich. Med. Soc.*, **33**, 581 (1944).
78. L. Salamone and A. Coppola, *Folia Med.*, **43**, 259 (1960); through Ref. 1.
79. S. Totaro and G. Licari, *Folia Med.*, **67**(5), 507 (1964); through Ref. 1.
80. C. Rimington and G. Ziegler, *Biochem. Pharmacol.*, **12**(12), 1387 (1963).
81. A. G. Zupko and L. D. Edwards, *J. Am. Pharmacol. Assoc. Sci. Ed.*, **38**(3), 124 (1949).
82. R. L. Hollingsworth, V. K. Rowe, F. Oyen, T. R. Torkelson, and E. A. Adams, *A.M.A. Arch. Ind. Health*, **17**(1), 180 (1958).
83. T. L. Neeper-Bradley, R. W. Tyl, L. C. Fisher, D. L. Fait, D. E. Dodd, I. M. Pritts, R. H. Garman, and J. A. Barter, *Teratology*, **39**, 470, Abstr. 59 (1989).
84. L. M. Srivastava, *Cytologia*, **31**, 166 (1966).
85. G. Lattanzi, S. Bartolli, B. Bonora, A. Colacci, S. Grilli, A. Niero, and M. Mazzullo, *Tumori*, **75**, 305 (1989).
86. K. L. Steinmetz, J. P. Bakke, C. M. Hamilton, K. C. Pardo, M. Ramsey, and J. C. Mirsalis, *The Toxicologist*, Abst. 636 (1988).
87. W. M. Azouz, D. V. Parke, and R. T. Williams, *Biochem. J.*, **59**, 410 (1955).
88. D. R. Hawkins, L. F. Chasseaud, R. N. Woodhouse, and D. G. Cresswell, *Xenobiotica*, **10**, 81 (1980).

89. R. Kimura, T. Kayashi, M. Sato, A. T. Aimoto, and T. Murata, *J. Pharmacobiodyn.*, **2**(4), 237 (1979).
90. G. P. Carlson and R. G. Tardiff, *Toxicol. Appl. Pharmacol.*, **36**, 383 (1976).
91. L. D. Pagnotto and J. E. Walkley, *Am. Ind. Hyg. Assoc. J.*, **26**, 137 (1965).
92. T. Umemura, K. Takada, Y. Ogawa, E. Kamata, M. Saito, and Y. Kurokawa, *Toxicol. Lett.*, **52**, 209 (1990).
93. M. Perrin, *Bull. Acad. Med.*, **125**(302) (1941); through Ref. 1.
94. G. Petit and J. Champaix, *Arch. Mal. Prof.*, **9**, 311
95. R. W. Weller and A. J. Crellin, *Arch. Intern. Med.*, **91**, 408 (1953).
96. L. H. Cotter, *N.Y. State J. Med.*, **53**, 1690 (1953).
97. D. M. Campbell and R. J. L. Davidson, *J. Obstet. Gynaecol. Br. Commonw.*, **77**, 657 (1970).
98. American Industrial Hygiene Association, *Hygienic Guide Series*, *p-Dichlorobenzene*, May–June, (1964).
99. M. L. Berliner, *A.M.A. Arch. Ophthalmol.*, **22**, 1023 (1939).
100. K. Walgren, *Zent. Arbeitmed. Arbeitsschutz*, **3**, 14 (1953).
101. H. J. Langner and H. G. Hillinger, *Berl. Munench. Tieraerztl. Wochenschr.*, **84**(18), 351 (1971).
102. Biodynamics, Inc., A Three Month Dietary Range-Finding Study of 1,2,4 Trichlorobenzene in Rats, Project Report #86-3122 to Chemical Manufacturers Association, 1989.
103. Hazelton Laboratories America, Inc., 13-Week Toxicity Study with 1,2,4 Trichlorobenzene in Mice, Project #HLA 6221-104 to Chemical Manufacturers Association, 1989.
104. W. D. Black, V. E. O. Valli, J. A. Ruddick, and D. C. Villeneuve, *The Toxicologist*, **3**, 30 (1983), Abstract.
105. K. T. Kitchen and M. T. Ebron, *Environ. Res.*, **31**, 362 (1983).
106. M. Robinson, J. P. Bercz, H. P. Ringhand, L. W. Condie, and M. J. Parnell, *Drug Chem. Toxicol.*, **14**, 83 (1991).
107. A. Tanaka, M. Sata, T. Tsuchiya, T. Adachi, T. Niimura, and T. Yamaha, *Arch. Toxicol.*, **59**, 82 (1986).
108. R. D. Lingg, W. H. Kaylor, and S. M. Pyle, *Drug Metabol. Dispos.*, **10**, 134 (1982).
109. I. Chu, D. Villeneuve, and V. Secours, *J. Toxicol. Envir. Health*, **11**, 663 (1983).
110. W. H. Braun, L. Y. Sung, D. G. Keyes, and R. J. Kociba. *J. Toxicol. Environ. Health*, **4**, 727 (1978).
111. L. C. Fisher, R. W. Tyl, B. L. Butler, M. A. Vrbanic, J. P. Van Miller, and C. R. Stack, *Teratology*, **41**, 556, Abstr. 153 (1990).
112. J. A. Ruddick, D. C. Villeneuve, I. Chu, S. Kacew, and V. E. Valli, *Teratology*, **23**, Abstr. 59A (1981).
113. T. L. Neeper-Bradley, R. W. Tyl, L. C. Fisher, D. L. Fait, M. A. Vrbanic, J. P. Van Miller, P. E. Losco, and C. R. Stack, *Teratology*, **41**, 580, Abstr. 52 (1990).
114. W. R. Jondorf, D. V. Parke, and R. T. Williams, *Biochem. J.*, **69**, 181 (1958).
115. J. Kohli, D. Jones, and S. Safe, *Can. J. Biochem.*, **54**, 203 (1976).
116. R. Linder, T. Scotti, J. Goldstein, K. McElroy, and D. Walsh, *J. Environ. Pathol. Toxicol.*, **4**, 183 (1980).

117. G. Goerz, W. Vizethum, K. Bolsen, Th. Krieg, and R. Lissner, *Arch. Dermatol. Res.*, **263**, 189 (1978).
118. K. S. Khera and D. C. Villeneuve, *Toxicology*, **5**, 117 (1975).
119. K. D. Courtney, J. E. Andrews, and M. T. Ebron, U.S. EPA Report EPA 600-J-77-123, NTIS PB284-762, 1975.
120. K. Rozman, J. Williams, W. F. Mueller, F. Coulston, and F. Korte, *Bull. Environ. Contam. Toxicol.*, **22**, 190 (1979).
121. "EPA Survey of Industrial Processing Data, Hexachlorobenzene and Hexachlorobutadiene Pollution from Chlorocarbon Processes," 560/3-76-003, Environmental Protection Agency, Washington, DC, prepared by C. E. Mumma and E. W. Lawless of the Midwest Research Institute, Kansas City, MO, 1975.
122. National Academy of Sciences, "Assessing Potential Ocean Pollutants: A Report of the Study Panel on Assessing Potential Ocean Pollutants," Ocean Affairs Board Commission on National Resources, National Research Council, Washington, DC, 1975.
123. D. L. Grant, F. Iverson, G. V. Hatina, and D. C. Villeneuve, *Environ. Physiol. Biochem.*, **4**(4), 159 (1974).
124. T. Kuiper-Goodman, D. L. Grant, C. A. Moodie, G. O. Korsrud, and I. C. Munro, *Toxicol. Appl. Pharmacol.*, **40**, 529 (1977).
125. A. Boger, G. Koss, W. Koransky, R. Naumann, and H. Frenzel, *Arch. Pathol. Anat. Physiol.*, **382**(2), 127 (1979).
126. H. H. Mollenhauer, J. H. Johnson, R. L. Younger, and R. L. Clark, *Am. J. Vet. Res.*, **36**(12), 1777 (1975).
127. R. Lissner, G. Goerz, M. G. Eichnauer, and H. Ippen, *Biochem. Pharmacol.*, **24**(18), 1729 (1975).
128. A. H. Timme, J. J. F. Taljaard, B. C. Shanley, and S. M. Joubert, *S. Afr. Med. J.*, **48**(43), 1833 (1974).
129. D. C. Villeneuve, G. J. A. Speijers, E. M. den Tonkelaar, M. J. von Logten, J. G. Vos, P. A. Greve, and J. G. van Esch, *Toxicol. Appl. Pharmacol.*, **45**(1), 341 (1978).
130. B. Zawirska and D. Dzik, *Patol. Pol.*, **28**(3), 349 (1977); through Pestic. Abstr., U.S. Environ. Prot. Agency, 78-0386, 1978.
131. E. J. Gralla, R. W. Fleischman, Y. K, Luthra, M. Hagopian, J. R. Baker, E. Esber, and W. Marcus, *Toxicol. Appl. Pharmacol.* **40**, 227 (1977).
132. Y. K. Luthra, J. J. Esber, E. J. Gralla, M. Hagopian, and W. Marcus, *Fed. Proc. Fed. Am. Soc. Exp. Bio.*, **36**(3), 356 (1977).
133. W. F. Mueller, M. J. Iatropoulos, K. Rozman, F. Korte, and F. Coulston, *Toxicol. Appl. Pharmacol.*, **45**(1), 283 (1978).
134. K. Rozman, W. W. F. Mueller, F. Coulston, and F. Korte, *Chemosphere*, **7**(2), 177 (1978).
135. E. M. den Tonkelaar, H. G. Verschuuren, J. Bankovska, T. DeVries, R. Kroes, and G. J. Van Esch, *Toxicol. Appl. Pharmacol.*, **43**(1), 137 (1978).
136. C. P. Fassbender, A. R. Alvarez, and S. Wenzel, *Arch. Lebensmittelhyg.*, **28**(6), 201 (1977).
137. L. G. Hansen, S. B. Dorn, S. M. Sundlof, and R. S. Vogel, *J. Agric. Food Chem.*, **26**(6), 1369 (1978).
138. L. G. Hansen, C. S. Byerly, R. L. Metcalf, and R. F. Bevill, *Am. J. Vet. Res.*, **36**, 23 (1975).

139. Codex Alimentarius Commission, Joint FAO/WHO Food Standards Programme Codex Committee on Pesticides Residues, Tenth Session, The Hague, 29 May–5 June, 1978.
140. K. D. Courtney and J. E. Andrews, *Toxicol. Lett.*, **3**(6), 357 (1979).
141. D. L. Grant, W. E. J. Phillips, and G. V. Hatina, *Arch. Environ. Contam. Toxicol.*, **5**, 207 (1977).
142. B. A. Schwetz, J. M. Norris, R. J. Kociba, P. A. Keeler, R. F. Cornier, and P. J. Gehring, *Toxicol. Appl. Pharmacol.*, **30**, 255 (1974).
143. J. G. Vos, M. J. van Logten, J. G. Kreeftenberg, P. A. Steerenberg, and W. Kruizinga, *Drug. Chem.*, **2**, 61 (1979).
144. IARC Monographs on the Evaluation of the Carcinogenic Risk of Chemicals to Humans, Vol. 20, Hexachlorobenzene, International Agency for Research and Cancer, World Health Organization, 1979.
145. M. J. Iatropoulos, J. Bailey, H. P. Adams, F. Coulston, and W. Hobson, *Environ. Res.*, **16**(1–3), 38 (1987).
146. L. D. Loose, J. B. Silkworth, K. A. Pittman, K. F. Benitz, and W. Mueller, *Infect. Immunol.*, **20**, 30 (1978).
147. L. D. Loose, K. A. Pittman, K. F. Benitz, and J. B. Silkworth, *J. Reticuloendothel. Soc.*, **22**, 253 (1977).
148. J. B. Silkworth, "Modification of Cell-mediated Immunity by Polychlorinated Biphenyl (Arochlor 1016) and Hexachlorobenzene," Thesis Dissertation, Center for Experimental Pathology and Toxicology, Albany, New York. Condensed version to be published by the Environmental Protection Agency Office of Pesticides and Toxic Substances Series 560/Research and Development Series 600, 1979.
149. R. L. Ziprin and S. R. Fowler, *Toxicol. Appl. Pharmacol.*, **39**, 105 (1977).
150. K. S. Khera, Food Cosmet. *Toxicol.*, **12**, 471 (1974).
151. G. S. Simon, B. R. Kipps, R. G. Tardiff, and J. F. Borzelleca, *Toxicol. Appl. Pharmacol.*, **45**(1), 330 (1978).
152. G. Koss, S. Seubert, W. Koransky, and H. Ippen, *Arch. Toxicol.*, **40**(4), 285 (1978).
153. R. Engst, R. M. Macholz, and M. Kujawa, *Bull. Environ. Contam. Toxicol.*, **16**(2), 248 (1976).
154. K. P. Schuster and G. Renner, *N. Arch. Exp. Pathol. Pharmakol.*, **297**(2), R5 (1977).
155. H. Lui and G. D. Sweeney, *FEBS (Fed. Eur. Biochem. Soc.) Lett.*, **51**(1), 225 (1975).
156. F. Koszo, C. Ziklosi, and N. Simon, *Biochem. Biophys. Res. Commun.*, **80**(4), 781 (1978).
157. G. W. Koss, W. Koransky, and Steinbach, *Arch. Toxicol.*, **35**, 107 (1976).
158. G. Koss, W. Koransky, and K. Steinbach, *Arch. Toxicol.*, **42**(1), 19 (1979); through Pestic. Abstr. U.S. Environ. Prot. Agency 79-2108, 1979.
159. G. Goerz, W. Vizethum, K. Bolsen, and Th. Krieg, *Arch. Dermatol. Res.*, **263**(2), 189 (1978).
160. T. Ariyoshie, K. Idegichi, Y. Ishizika, K. Iwasaki, and M. Akakan, *Chem. Pharm. Bull.*, **23**(4), 817 (1975).
161. R. L. Mull, W. L. Winterlin, S. A. Peoples, and L. Ocampo, *J. Environ. Pathol. Toxicol.*, **1**(6), 865 (1978).
162. R. S. Yang, K. A. Pittman, D. R. Rourke, and V. B. Stein, *J. Agric. Food Chem.*, **26**(5), 1076 (1978).

163. J. R. P. Cabral, T. Mollner, F. Raitano, and P. Shubik, *Toxicol. Appl. Pharmacol. Abstr.*, **45**(1), 323 (1978).
164. H. A. Peters, A. Gocmen, D. J. Cripps, G. T. Bryan, and I. Dogramaci, *Arch. Neurol.*, **39**, 744 (1982).
165. D. J. Cripps, H. A. Peters, and A. Gocmen, *Br. J. Dermatol.*, 111, 413 (1984).
166. C. Cam and G. Nigogosyan, *J. Am. Med. Assoc.*, **183**, 88 (1963).
167. A. L. Cetingil and M. A. Ozen, *Blood*, **16**, 1002 (1960).
168. F. De Matteis, B. E. Prior, and C. Rimington, *Nature*, **191**, 363 (1961).
169. D. J. Cripps, H. A. Peters, and A. Gocmen, *J. Invest. Dermatol.*, **71**(4), 277 (1978); *Clin. Res.*, **26**(3), 489A (1978).
170. H. A. Peters, *Fed. Proc. Fed. Am. Soc. Exp. Biol.*, **35**(12), 2400 (1976).
171. L. G. Pederson and G. L. Rimington, *Nature*, **191**, 363 (1961).
172. H. Ehrlicher, *Zentrabl. Arbeitsmed. Arbeitsschutz*, **18**(7), 204 (1968).
173. I. V. Sairtskii, *Vopr. Prom. Sel'skokhoz. Toksikol. Kievsk. Med. Inst.*, **158**, 173 (1964); through *Chem. Abstr.*, 63, 8952d (1965) and Ref. 121.
174. W. J. Hayes, Jr., *Toxicology of Pesticides*, Williams and Wilkins, Baltimore, 1975.
175. B. J. Stokvis, *Z. Klin. Med.*, **28**, 1 (1895); through Ref. 174.
176. W. B. Deichmann, *J. Ind. Hyg. Toxicol.*, **23**(7), 343 (1941), and unpublished studies conducted in 1935.
177. J. G. Vos, H. L. van der Maas, A. Musch, and E. Ram, *Toxicol. Appl. Pharmacol.*, **18**, 944 (1971).
178. R. W. Lambrecht, E. Erturk, E. E. Grunden, H. A. Peters, C. R. Morris, and G. T. Bryan, *Fed. Proc.*, **42**, 782 (1983).
179. R. W. Lambrecht, E. Erturk, E. E. Grunden, H. A. Peters, C. R. Morris, and G. T. Bryan, *Proc. Am. Assoc. Cancer Res.*, **24**, 59 (1983).
180. H. A. Peters, A. Gocmen, and D. J. Cripps, *Arch. Neurol.*, **39**, 744 (1982).
181. D. J. Cripps, H. A. Peters, and A. Gocmen, *Brit. J. Dermatol.*, 111, 413 (1984).
182. NIOSH, *Registry of Toxic Effects of Chemical Substances 1985–1986*, National Institute for Occupational Safety and Health.
183. M. Windholz, Ed., *The Merck Index*, 9th ed. Merck & Co., Rahway, NJ, 1976.
184. T. J. Monks and S. S. Lau, *Toxicology*, **52**, 1 (1988).
185. B. B. Brodie, W. D. Reid, A. K. Cho, G. Sipes, G. Krishna, and J. R. Gillette, *Proc. Natl. Acad. Sci. U.S.A.*, **68**, 160 (1971).
186. P. E. Weller and R. P. Hanzlik, *Chem. Res. Toxicol.*, **4**, 17 (1991).
187. J. Zheng and R. P. Hanzlik, *Chem. Res. Toxicol.*, **5**, 561 (1992).
188. J. Zheng and R. P. Hanzlik, *Drug. Metab. Dispos.*, **20**, 688 (1992).

CHAPTER TWENTY-THREE

# Chlorinated Hydrocarbon Insecticides

A. Philip Leber, Ph.D., D.A.B.T., and Theodore J. Benya, Ph.D., D.A.B.T.

## 1 INTRODUCTION

The organochlorine insecticides represent the first group of synthetic compounds to have a significant impact on the control of infectious diseases transmitted via insect vectors. These insecticides were used extensively in the United States and other western countries, and are still used in third-world regions both as agricultural insecticides and as agents to combat such vector-borne diseases as malaria, typhus, plague, Chagas' disease, yellow fever, dengue, encephalitis, filariasis, and African trypanosomiasis (sleeping sickness) (1). Of these insecticides, DDT (dichlorodiphenyltrichloroethane) is credited as the primary compound which, for the first time in history, brought epidemics of malaria, typhus, and plague to a complete stop. DDT was introduced in 1943, with related insecticides following shortly thereafter. This chemical is still used extensively in tropical regions to combat malarial mosquitoes, and substitution of this pesticide with others such as malathion would be most expensive.

From 1950 to 1972, the U.S. production of DDT in thousand kilogram units equaled 1,204,700. The production of the aldrin–toxaphene group (aldrin, chlordane, dieldrin, endrin, heptachlor, toxaphene) over the years 1952 to 1972 totaled $865,600 \times 10^3$ kg units; a total of $41,500 \times 10^3$ kg units of benzene hexachloride (lindane) was manufactured in the United States from 1950 to 1963. Of these

---

*Patty's Industrial Hygiene and Toxicology, Fourth Edition, Volume 2, Part B*, Edited by George D. Clayton and Florence E. Clayton.
ISBN 0-471-54725-5 © 1994 John Wiley & Sons, Inc.

**Table 23.1.** Regulatory Standings of Chlorinated Hydrocarbon Insecticides (3, 4)

| Chemical | U.S. Registration | Foreign Use |
|---|---|---|
| Aldrin/dieldrin | EPA canceled all products | Yes |
| Chlordane/ heptachlor | Voluntarily canceled. Use for fire ants in power transformers, existing stocks for termites in homeowners' possession allowed. May reinstate for termite use pending findings of air monitoring tests | Use in termite control, some agricultural applications |
| DDT | Canceled all products | Yes |
| Endrin | Voluntary cancellation, all products | Unknown |
| Kepone | Canceled all products | Unknown |
| Lindane | Restricted uses (plant nurseries, pet shampoos, livestock sprays, seed treatment, household sprays, flea collars, hardwood logs, etc.). Avocados, pecans are only food crop use allowed | Yes |
| Mirex | Canceled all products. Existing stocks used for ants on pineapples in Hawaii | Unknown |
| Toxaphene | All products canceled. Existing stocks use allowed for cattle dip, pineapples in Puerto Rico, bananas in Virgin Islands, emergency use on corn, cotton, small grains | Unknown |

quantities, approximately 50 percent of the DDT and 80 percent of the aldrin–toxaphene group were used in the United States (2).

## 1.1 Comparative Properties of Chlorohydrocarbon Insecticides

One attribute that contributes to the effectiveness of this chemical class is the persistence in the environment, providing not only an immediate impact on insect populations but a prolonged insecticidal presence extending well beyond the time of application. This persistence is now generally considered an undesirable feature owing to findings suggesting delayed adverse impacts on nontarget populations of insects as well as birds. In addition, increased cancer risks for humans are alleged to result from exposures to these chemicals such as those resulting from pesticide applications and ingestion of contaminated fish and other food species. Although the actual balance of risks versus benefits associated with the use of these insecticides is debated, regulatory action has virtually eliminated their use in the United States and other western countries. Table 23.1 summarizes the regulatory status of these products.

The persistence of these insecticides in the environment and their prolonged activity against pests following application can be attributed to a combination of their insolubility in water and high solubility in fats, absorption and adsorption onto particulate matter, and resistance to chemical, physical, and microbiological

# CHLORINATED HYDROCARBON INSECTICIDES

**Table 23.2.** Acute Toxicity of Organochlorine Insecticides[a]

| Acute Toxicity | Pesticide | CASRN | Rat Oral $LD_{50}$ (per kg Body Weight) Males and Females Combined (6) |
|---|---|---|---|
| Highly toxic | Endrin | 72-20-8 | 13 mg |
| | Dieldrin | 60-57-1 | 46 mg |
| | Aldrin | 309-00-2 | 50 mg |
| Moderately toxic | Toxaphene | 8001-35-2 | 85 mg |
| | Lindane | 58-89-9 | 90 mg |
| | Heptachlor | 76-44-8 | 131 mg |
| | Kepone | 143-50-0 | 95 mg |
| | p,p'-DDT | 50-29-3 | 116 mg |
| | Chlordane | 57-74-9 | 283 mg |
| Slightly toxic | Kelthane | 115-32-2 | 1.05 g |
| | p,p'-DDE | 72-55-9 | 1.16 g |
| | Perthane | 77-47-4 | 4.0 g |
| | Hexachloropentadiene | | (corrosive) |

[a]Reproduced with permission by American Medical Association, August 29, 1979.

degradation. From target crops and surrounding soil and water, these compounds have entered the food chains of mammals, birds, fishes, and other animal species. DDT in particular was implicated in inducing acute and perhaps chronic insecticide intoxications in fish and birds as a result of bioaccumulation. In 1970, at the height of insecticidal use in the United States, Canada, and European countries, significant decreases in eggshell thickness were found in 15 of 22 species of aquatic birds, particularly in those feeding in fresh and brackish waters near agricultural areas. According to King et al. (5), population declines of the following species were observed, and continued study was proposed: brown pelican, reddish egret, white-faced ibis, laughing gull, and Forster's tern.

The acute lethal potencies of these compounds in laboratory rats vary considerably, as shown in Table 23.2.

Smith (7) discusses the fact that the primary acute toxicity noted in animals and humans following excessive exposures to chlorinated insecticides is neurological hyperactivity. With DDT and related compounds, the effects progress gradually from mild tremors to convulsions, whereas convulsions are the first sign of intoxication for compounds such as lindane, aldrin, dieldrin, endrin, toxaphene, and related materials. The latter can produce incoordination, weakness, and an ataxic state that is not associated with tremor, discriminating the intoxication induced by these substances from that of DDT. In general, the acute effects have not been shown to pose significant hazards to exposed populations, and current concern over the toxicities of these compounds is linked primarily to chronic low-level exposures discussed below.

Table 23.3. Carcinogenicity Classifications[a] of Chlorinated Insecticides

| Chemical | U.S. EPA (11, 12) | NTP (13) | IARC (14) | Basis |
|---|---|---|---|---|
| Aldrin | B2 | — | 3 | Mouse liver tumors |
| Chlordane/ heptachlor | B2 | — | 3 | Mouse liver tumors |
| Chlordecone (Kepone) | — | e | 2B | Rat, mouse liver tumors |
| DDT | B2 | e | 2B | Mouse liver, lung tumors, lymphomas; rat liver tumors; no tumors in three hamster studies |
| Dieldrin | B2 | — | 3 | Mouse liver tumors |
| Endrin | — | — | 3 | No evidence |
| Lindane | B2/C | e | — | Mouse liver tumors |
| Mirex | B2 | e | 2B | Mouse, rat liver, and thyroid tumors |
| Toxaphene | B2 | e | 2B | Mouse, rat liver tumors |

[a]B2 = Probable human carcinogen (no human evidence).
2B = Possibly carcinogenic to humans.
C = Possible human carcinogen.
3 = Not classifiable as to carcinogenicity in humans.
e = Reasonably anticipated to be carcinogenic to humans.

## 1.2 Carcinogenicity of Chlorohydrocarbon Insecticides

As mentioned above, because of reported environmental effects of these pesticides plus their classifications in the early 1970s as "potential human carcinogens," the use of most organochlorine pesticides was discontinued or markedly curtailed in the United States, Canada, and most European countries (8–10). The alleged human hazard, cancer, was based on observations of tumor induction in laboratory animals, primarily in mice, in which these compounds produced benign and malignant liver cell tumors. Table 23.3 lists carcinogenicity classifications by the U.S. Environmental Protection Agency (EPA), the National Toxicology Program (NTP), and the International Agency for Cancer Research (IARC); these classifications are intended to reflect human cancer risks related to chemical exposures to the chlorinated pesticides listed.

One reason for this debate is that many chemicals that induce rodent tumors do not appear to have human activity. In 1975, Kraybill (15) wrote "None of the pesticide chemicals thus far have been shown to be carcinogenic to man." Workers (mostly men) who have been engaged in the manufacture, handling, and spraying of DDT, aldrin, dieldrin, toxaphene, chlordane, and heptachlor have been exposed to considerably higher concentrations and quantities of these insecticides than the general population of the United States. Among the exposed groups, only acute effects such as eye, skin, or respiratory irritation were reported, particularly following exposures to dusty formulations of the compound. High worker exposure occurring during the early years of production of some of these compounds was associated with induction of liver microsomal enzymes and the ability of some highly exposed workers to increase their drug-metabolizing capacity.

However, frank and undisputed injury to the liver or other human organs has not been reported in the United States, Canadian, and western European literature. To the best of our knowledge the organochlorine insecticides (individually and in combination) that have been ingested with home- and restaurant-prepared food and drink by the U.S. population for more than 35 years (DDT was introduced in 1943), followed by a period of greatly reduced intake, have caused no recognized or clearly defined harmful effects. As of 1992 no reports relating exposure to the chlorinated pesticides to cancer in humans have appeared. The significance of rodent liver tumors as indicators of human cancer risks continues to be debated in scientific circles (16–18).

Parenthetically, although the organochlorine insecticides are considered potential human carcinogens, it is noted that there has been a significant, almost constant decrease in the incidence of liver cancer deaths (males and females, classified since 1949 as primary, secondary, and not stated whether primary or secondary) in the continental United States, namely, from 8.8 (per 100,000 population) in 1930 to 8.4 in 1944 (when DDT was introduced for use) and to 5.6 in 1972. In the period 1985 to 1987, the liver cancer death rate in the United States was 4.1. Based on U.S. Vital Statistics for the general U.S. population, this almost steady decline in total liver cancer deaths for a 55-year period is even more significant in light of the constantly increasing life-span of the people of the United States, which in turn has resulted in a constant increasing percentage of the population "at risk" for liver cancer (2).

## 2 DDT, DICHLORODIPHENYL TRICHLOROETHANE [CAS # 50-29-3]

The formula for DDT is $C_{14}H_9Cl_5$; it is also known as 1,1,1-trichloro-2,2'-bis(*p*-chlorophenyl)ethane and *p,p'*-DDT.

*p,p'*-DDT

### 2.1 Uses and Occupational Exposures

DDT was released for commercial use in the United States on August 31, 1945; 1946 was the first full year of use. At that time, DDT was introduced in the malaria-stricken regions of the world and remains as one of the primary insecticides for the control of malaria through the eradication of vectors. From 1946 to 1972, DDT was one of the most widely used agriculture insecticides in the United States and other countries.

The toxicologic properties of DDT have not contributed to unusual human health problems during the manufacture, handling, and use of this insecticide. The only confirmed serious incidents of injury are those resulting from massive accidental or suicidal ingestion of DDT. The many years of manufacture, agricultural use, and malaria spraying operations involving this insecticide have not provided hazardous to the health of highly diversified occupational groups thus engaged.

Because of its removal (in the United States) from the market, there are virtually no risks related to occupational exposures to DDT. The toxicology of this chemical does remain important in the consideration of the presence of DDT in chemical waste sites, its migration to environmental media, and the resulting potentials for public exposures.

## 2.2 Physical and Chemical Properties

| | |
|---|---|
| Physical state | Colorless crystals or white to slight off-white powder with odorless or slight aromatic odor |
| Molecular weight | 354.5 |
| Specific gravity | 0.98 to 0.99 |
| Melting point | 108.5 to 109°C |
| Boiling point | 260°C |
| Vapor pressure | $1.5 \times 10^{-7}$ mm Hg (at 20°C) |
| Solubility | Practically insoluble in water, diluted acids, alkalies; 78 g/100 ml benzene; 58 g/100 ml acetone; 45 g/100 ml carbon tetrachloride; 116 g/100 ml cyclohexanone |
| Partition coefficient (log octanol/water) | 6.19 |

Conversion factors: 1 mg/l (air) = 69 ppm; 1 ppm (air) = 14.5 mg/m$^3$ at 25 C, 760 mm Hg

## 2.3 Effect in Animals

### 2.3.1 Acute Inhalation Toxicities

From an occupational perspective, dry formulations of DDT display particle sizes sufficiently large to result in upper respiratory tract deposition only, leading to oral ingestion of the compound. Other formulations were examined by Neal et al. (19), who reported on the effects of animal exposures to aerosols of DDT. Dogs, rats, and guinea pigs were exposed to an initial concentration of DDT of 54.4 mg/l air for a period of 45 min. No indications of toxicity were observed. They also reported that the oil used as a carrier had some effects on the response. Mice tolerated 6.22 mg of DDT/l air without causing signs of toxicity when the solution contained 6 percent sesame oil; when the concentration of oil was 9.5 percent, toxic effects were observed.

Cameron and Burgess (20) exposed rats, guinea pigs, and rabbits to a concentration approximating 1000 mg/m$^3$ for a period of 2 hr daily. The animals showed signs of intoxication (muscular weakness and tremor), and deaths occurred after four to 10 exposures.

A rhesus monkey showed no signs of intoxication during or after two 7-hr periods of exposure to 0.13 mg DDT/l air on the first day, and to 0.4 mg/l on the second day. Six rats concurrently exposed on the first day showed mild tremors; six other rats exposed on the second day demonstrated tremors that lasted 3 days. All rats survived. A rabbit, a cat, and a guinea pig exposed 7 hr/day for 3 days to 0.2, 0.3, and 0.45 mg DDT/l air, respectively, demonstrated no ill effects (21).

### 2.3.2 Acute Oral Exposure

Woodard et al. (22) investigated the acute oral toxicity of DDT. They reported some deaths in adult rats at a dosage of 140 mg/kg. Nonlethal and lethal (partial survival) doses were as follows: rabbits—260 mg/kg, 400 mg/kg; mice—399 mg/kg, 448 mg/kg; guinea pigs—178 mg/kg, 224 mg/kg. Lehman (23) reported the approximate $LD_{50}$ for rats as 250 mg/kg. DDT has one-tenth the lethal potency in newborn rats as it does for adult animals (7).

Deichmann et al. (21) reported $LD_{50}$ values (in mg/kg) for DDT in rats (Wistar) to vary with the solvent as follows: olive oil, 240; cyclohexanone, 280; corn oil, 420; propylene glycol, mineral oil, or cream (18 percent fat), 940; and DDT in 40 percent ethanol, aqueous methylcellulose, or tri-o-cresyl phosphate, 1400 mg/kg. They further reported that a room temperature of 30°C retarded, whereas a temperature of 5°C shortened, the onset of tremors. The lethal dose was not altered by environmental temperatures.

### 2.3.3 Chronic Oral Exposure

Fitzhugh and Nelson (24) fed rats a diet containing DDT for 2 years. At concentrations of 600 and 800 ppm in the diet, the animals showed moderately severe tremors, particularly during the early months of exposure. An occasional animal had tremors at 400, but rarely at 200 ppm. Concentrations of 400 ppm and above produced a higher than normal rate of mortality, and an increase in weight of the liver. At concentrations of 600 or 800 ppm, there were indications of increased kidney weights. Histopathological studies showed moderate liver damage at concentrations of 200 ppm and above. There were indications of slight liver changes at 100 ppm.

Treon and Cleveland (25) indicated that rats maintained for 18 months to 2 years on a diet containing 25 ppm DDT showed an increase in liver weight. At 12.5 ppm in the diet for a similar period, they reported no effects. Dogs on a diet containing 30 ppm of DDT for a period of 15.7 months showed no effect. Laug et al. (26) indicated that hepatic cell alterations were seen in rats at 5 ppm in the diet (and higher) but not at 1 ppm. These authors observed an accumulation of DDT in animal body fat following ingestion at concentrations as low as 1 ppm in the diet.

Long-term studies have been conducted in rats, dogs, and monkeys in which no observed effect levels were demonstrated, and reported in Reference 7 to be 0.05, 8, and 2.2 mg/kg/day, respectively. For additional information on early studies on the acute or chronic toxicity and the mechanism of action of DDT, refer to Reference 21. More recent reviews also address the toxicity of this compound (2, 7, 27).

### 2.3.4 Skin Absorption

Experiments by Draize et al. (28) indicated that dusts and solutions of DDT may cause a slight to moderate erythema on the skin of rabbits. Using a dry dust of 5 percent DDT in talc, they were unable to find any indication of a systemic toxic effect due to absorption through the skin; neither was there evidence of systemic effects following application of a 10 percent solution of DDT in corn oil in doses up to 940 mg of DDT/kg of body weight. Skin application of solutions of 25 and 30 percent of DDT in dimethyl phthalate or in dibutyl phthalate in doses up to 9.4 ml/kg for a 24-hr period induced signs of toxicity but no fatalities.

### 2.3.5 Absorption, Metabolism, and Excretion

Uptake of DDT from the gastrointestinal tract occurs slowly as evidenced by a 2 to 6 hr delay in the onset of convulsions following single high-dose oral administration to rats. Intravenous dosing of DDT can lead to convulsions within 20 min in this species. DDT absorption from the gastrointestinal tract is significantly enhanced when incorporated into vegetable oil. Lymph cannulation experiments demonstrated that the pesticide is taken up from the gastrointestinal tract primarily via the lymph system as opposed to the bloodstream, for recovery of orally administered compound radioactivity by this technique was 65 percent (29). Less than 0.1 percent of administered material appeared in excreted urine, up to 37 percent in the feces/gastrointestinal tract, and 67 percent in the carcass.

The metabolism of $p,p'$-DDT has been extensively examined in the rat, and approximately nine metabolites have been identified (7). In humans, the degradation of DDT proceeds by dehydrochlorination to the unsaturated $p,p'$-DDE [1,1-dichloro-2,2-bis($p$-chlorophenyl)ethylene] and/or by substitution of a hydrogen for one ethyl-binding chlorine atom yielding $p,p'$-DDD [1,1-dichloro-2,2-bis ($p$-chlorophenyl)ethane]. DDD is further metabolized through a series of intermediates yielding the carboxy acid form, $p,p'$-DDA [2,2-bis($p$-chlorophenyl)acetic acid]. DDA is relatively water soluble and is excreted primarily in the urine (30).

Roan et al. (31) studied urinary excretion following ingestion of DDT and DDT metabolites in six volunteers given technical DDT (5, 10, 20 mg/day) or $p,p'$-DDE, $p,p'$-DDD, or $p,p'$-DDA (5 mg/day for 21 to 183 days), demonstrating that within 24 hr of ingestion of DDT, urinary DDA excretion increased detectably. According to these investigators, excretion of DDT as DDA appeared to depend totally on the preferential reductive dechlorination of DDT to DDD (rather than DDE), and thence to DDA.

DDT and DDE are fat soluble, and may be retained in the body fat of humans for years. It is generally recognized that of the DDT isomers, metabolites, and degradation products, DDE is the compound most widely distributed in nature.

Human hair has been found to contain chlorinated hydrocarbon insecticides and other halogenated compounds in concentrations less than 1 ppm. In experiments with rats, chlorinated hydrocarbon insecticides and polychlorinated biphenyls (PCBs) were excreted in hair in all instances, suggesting that this route of excretion may be more important in eliminating certain chlorinated hydrocarbons than was formerly recognized (32).

Review of the literature indicates that DDT accumulates in all mammalian tissues, with the highest levels found in body fat. The compound can be transferred to fetuses via the placental route and is readily excreted by mothers into the milk.

## 2.4 Human Experience

### 2.4.1 Malaria

According to Khambata (33), by the end of 1974 more than one billion people had been "freed" from malaria due to chlorinated pesticide use. It is because of this enviable record that DDT continues to be used as an antimalarial agent in countries where this and related diseases are endemic, despite the fact that some vectors have become resistant to DDT. The millions of people in tropical areas (e.g., India) for whose benefit DDT is being used will continue to be the primary nonoccupational exposure group. In these individuals the compound is absorbed by inhalation of airborne pesticide in their dwellings and crop fields, as well as by ingestion with their food and water. Monitoring of DDT residues near Delhi, India (where DDT is manufactured) indicates moderate levels still exist in fauna including fish, birds, and buffalo, whereas human fat and blood samples show evidence of excessive and/ or recent exposures (34). In general, there is evidence of a gradual decline in these residues. Because of the extensive use of DDT for the control of malaria, typhus, and related diseases, it will continue to be found in trophic food chains. Planktons, bivalves, fishes, birds, and mammals, as well as the populace in parts of the world distant from the areas sprayed will experience some degree of exposure (7).

### 2.4.2 Occupational Exposure

Ortelee (35) reported on a well-designed study of 40 men exposed to DDT during its manufacture and formulation. Their exposure was followed by the analysis of DDT concentrations in urine, which was compared with the excretion of DDT by men whose oral intake of DDT was known. The author concluded that it is unlikely that any illness will occur from DDT at the prevailing dietary levels, for the men studied showed no effects from occupational exposure for up to 6.5 years. During this time, they absorbed an average of 200 times the quantity absorbed by the 1950 general population in the daily diet.

A study conducted by Laws et al. (36) of 35 men exposed for 11 to 19 years in a plant that produced DDT continuously and exclusively (2.7 million kg/month) from 1947 to 1967 did not reveal any ill effects in workers attributable to DDT. The overall range of storage of all isomers and metabolites of DDT in the body fat of these men ranged from 38 to 647 ppm, as compared to an average of 8 ppm in the general population. Based on the storage of total DDT in the body fat and excretion of DDA in the urine, it was estimated that the average daily intake of DDT by each of the 20 workers with "high" exposure was 17.5 to 18.0 mg/day, as compared to an average of 0.05 mg/day for the general population. The concentrations of DDT in body fat averaged 344 times the concentrations in serum. The workers stored less DDE than DDT, which is related primarily to the intensity rather than to the duration of the exposure. DDA is much more important as an excretory product in those occupationally exposed to DDT than it is in members of the general population who absorb DDT primarily with their diet.

### 2.4.3 Studies in Volunteers

Hayes (37, 38) has reviewed the literature on DDT and other pesticides, and found that acute poisoning in humans has occurred. However, chronic intoxication in humans has not been confirmed. An oral dose of 10 mg/kg produced illness in some men but not all; 285 mg/kg has been ingested without fatal results, although with toxic response. The tolerated chronic dose in humans is not known, but from animal experiments, 2.5 to 5 mg/kg body weight/day may approximate a lowest effect level.

In 1956, Hayes et al. (39) reported on the ingestion of repeated oral doses of DDT. Three men completed 1 year at 35 mg/day (0.5 mg/kg body weight). This dose was about 200 times the daily rate at which the average person in the United States ingested DDT in the diet at that time. In this limited study, no complaints of illness related to DDT were reported by the subjects, nor were adverse clinical changes found during careful medical examination.

Hayes et al. (40) repeated and extended this investigation. Twenty-four male volunteers, aged 24 to 29, were given either technical DDT or $p,p'$-DDT at the rate of 0.05 or 0.5 mg/(kg)(day) for 21.5 months. They were then observed for an additional 25.5 months; 16 were followed for 5 years. The men on the high dosage of DDT for 21.5 months ingested the equivalent of approximately 1680 times the 1965 to 1970 U.S. dietary intake of $p,p'$-DDT. DDT concentrations in biopsied body fat of volunteers ingesting 0.5 mg/kg/day of technical or $p,p'$-DDT reached a mean of 325 and 281 ppm, respectively, in 18.8 months. Loss of tissue DDT progressed slowly after ingestion of the high dose of DDT was discontinued. At a point 22.5 months following the final doses of DDT, the concentrations in body fat in the two respective groups were reduced to 32 and 35 percent of the original values, respectively. DDE fat concentrations increased throughout the dosing period and continued to rise slightly even after discontinuation of the 0.5-mg/kg/day dose. DDA excretion in the volunteers was marked. All men were subjected to extensive physical, neurological, biochemical, hematologic, and organ function tests, which showed "no definite chemical or laboratory evidence of injury by DDT," indicating "a high degree of safety of DDT for the general population."

# CHLORINATED HYDROCARBON INSECTICIDES 1513

In a similar investigation, Morgan and Roan (30) administered DDT and certain closely related compounds to nonoccupationally exposed men. One volunteer ingested 10 mg of DDT per day for 183 days. Another received a daily dose equivalent to approximately 400 times the average (1965 to 1970) daily dietary intake of total DDT, namely, 20 mg/day for 183 days. A third received 5 mg $p,p'$-DDE/day for 92 days, and a fourth ingested 5 mg $p,p'$-DDD for 81 days. Extensive clinical tests were conducted before, during, and after the periods of pesticide intake, but abnormalities or harmful effects were not detected in any of the men.

Ensberg et al. (41) found "no shortening of antipyrine life" in volunteers who ingested 100 or 200 µg DDT for 21 days. Enzyme stimulation to a degree that reduced the half-life of one dose of 400 mg phenylbutazone was reported by Poland et al. (42). This was noted in 18 men of the Montrose Chemical Corporation whose serum and fat total DDT concentrations reached levels 20 and 30 times those of matched controls. Evidence of liver microsomal enzyme induction was observed as the serum half-life of phenylbutazone dropped 19 percent, whereas the urinary excretion of 6-hydroxycortisol rose 57 percent above normal. There were considerable variations in individual susceptibilities of these men to enzyme induction.

## 2.4.4 *Findings Related to Public Exposures*

Kraybill (43) estimated in 1966 that 90 percent of total persistent pesticides exposure to the general U.S. and European populations originate with the diet. Of all foods consumed, those of animal origin, meat (particularly fatty meat), seafood, poultry (eggs), and dairy products, provided the major portion of these pesticides, including DDT. This appeared to be significant because these food sources were subject to only minimal direct application of insecticides; therefore, the presence of DDT, DDE, and related organochlorine compounds in foods of animal origin were related primarily to environmental sources. U.S. market basket surveys performed in the period October 1980 to March 1982 (44) indicated that DDT was detected (<0.6 ppb) only in leafy and root vegetables whereas DDE was more commonly found, with highest levels also found in leafy and root vegetables (2.4 and 4.6 ppb). Potential sources of DDT are imported foods from countries in which DDT use is not prohibited.

Concentrations of DDT and its metabolites in fatty tissues of the general U.S. population have shown a steady decline, from approximately 5.3 to 20.0 ppm for total DDT at the height of the DDT use in the 1960s to approximately 8.0 ppm (total DDT) in the early 1970s. Indications are that these concentrations are still declining. Body fat concentrations of the organochlorine insecticides and of DDT in particular have varied widely among the U.S. population. In general, they have been considerably lower in infants and the very young than in adults, higher in adult males than adult females, higher in Negroes than Caucasians, and higher in population groups who live in or near agricultural areas and in those who used the insecticide in home or garden. Generally, the more body fat, the greater the total body burden of DDT and other chlorinated pesticides is likely to be, but the lower the concentrations (45).

Human breast milk contains about 3 percent fat, which contributes to the fact that infants are likely to ingest, over a short period of time, higher concentrations of DDT and related compounds than do their mothers. However, as Hayes (46) recently reported, "Infants are in no danger from DDT in their mother's milk unless the dosage to the mother is one that approaches or perhaps reaches a level toxic to her." A similar conclusion was reached by Smith (7) 15 years later in his review of DDT toxicology.

Organochlorine pesticides are absorbed during cigarette smoking. In 1969, the total chlorinated insecticide residues in cigarettes ranged from 12.5 to 63.4 ppm. DDT and DDD residues accounted for more than 97 percent; the remainder consisted of dieldrin and endrin. In 1974 the residues ranged from 3.6 to 7.1 ppm. Smoke condensate from one brand contained an average of 8.3 ng of DDT residues per cigarette (47). Commercial filters and increased butt length reduce the mainstream smoke residues considerably (48).

An epidemiologic study was reported by Fowler (49) in which a sizable population group in Mississippi was studied before and after a number of years of widespread application of DDT, both on crops and for mosquito control. It was concluded that in general, the health of the population had improved over this period of time. This could be attributed to improved sanitation, but it should also be noted that a number of disease-carrying insects were effectively controlled by the use of DDT. There was no indication that DDT produced deleterious effects on the population. Misuse of the compound did cause a few cases of acute intoxication.

Numerous studies were reviewed in which associations between biologic markers of DDT exposures and adverse health effects were noted (27). These correlations appeared for DDE and cancer, uterine levels of DDT and leiomyomas, serum levels of organochlorine insecticides including DDT and DDE with hypertension and vascular disease, blood and placental levels of DDT and metabolites and premature births and spontaneous abortions, and maternal milk levels of DDE and hyporeflexia in nursing infants. Many other studies have been cited (7, 27) that do not confirm these results, and thus any causal relationships between these adverse findings with DDT exposures are considered tenuous.

## 2.5 Cancer

At this time there is no documented evidence that the dietary absorption of DDT, alone or in combination with insecticides of the aldrin–toxaphene group, has caused cancer in the general population (45, 50, 51). No evidence has as yet been presented that DDT has caused cancer among millions of individuals (almost entirely men) who have been occupationally engaged for as long as 35 years in the manufacture and handling or spraying of this insecticide (as dust, solution, or suspension) in all parts of the world and under all possible climatic conditions. Therefore, at this time, there is no evidence to indicate that DDT is a human carcinogen.

It is recognized that general population studies are not likely to uncover a low-potency carcinogen, particularly if it targets an organ with an historically high background incidence of tumors, but if we accept the data reported by Vital Statistics of the United States, which provides cancer incident statistics, then it will be of interest and some degree of comfort to know that the death rates from cancer of the liver and its biliary passages have shown an almost constant decline in the United States over the past 42 years, namely, from 8.8 per 100,000 population in 1930, to 8.4 in 1944 when DDT was introduced, to about 4.1 in 1987 (52), indicating an apparent lack of an adverse negative impact on human liver cancer rates from use of chlorinated pesticides.

The extensive literature on the possible carcinogenicity of DDT in animal species reveals that DDT, when ingested in high concentrations with the diet (as compared to human intake), causes hepatocellular carcinomas in several strains of mice (53–59). Non-metastasizing liver tumors, according to the 1979 *IARC Monographs on the Evaluation of the Carcinogenic Risk of Chemicals to Humans*, occur in rats fed DDT (60, 61).

A report entitled "Bioassays of DDT, TDE and $p,p'$-DDE for Possible Carcinogenicity" was published in 1978 by the Carcinogenesis Testing Program of the National Cancer Institute (62). The summary reads as follows:

There was no evidence for the carcinogenicity of DDT in Osborne-Mendel rats or B6C3F1 mice, of TDE in female Osborne-Mendel rats, although $p,p'$-DDE was hepatotoxic in Osborne-Mendel rats. The finding suggests a possible carcinogenic effect of TDE in male Osborne–Mendel rats based on the induction of combined follicular cell carcinomas and follicular cell adenomas of the thyroid. Because of the variation of these tumors in control male rats in this study, the evidence does not permit a more conclusive interpretation of these lesions; $p,p'$-DDE was carcinogenic in B6C3F1 mice, causing hepatocellular carcinomas in both sexes.

According to the 1979 World Health Organization (WHO) Environmental Health Criteria Report,

"The evidence for the carcinogenicity of DDT in rats is not convincing and is negative in hamsters, . . . negative results in dogs and monkeys are inconclusive because of the small groups studied and short duration of treatments; . . .the carcinogenicity of $p,p'$-DDE is similar to that of DDT, but DDT produced a significant incidence of lung tumors."

A number of studies have addressed the mechanism involved in the development of rodent tumors induced by DDT. Cancer promotional activity has been demonstrated by DDT, which enhanced tumor yields in aflatoxin-pretreated rats (63). This observation, along with the compound's lack of significant genotoxicity, is consistent with DDT's inhibition of intracellular communication activity induced in cultured rodents cells (64).

All available evidence indicates that humans do not appear to be susceptible to the tumorigenic action of the organochlorine insecticides and phenobarbital. No increase in the occurrence of tumors has been found in heavily exposed populations.

This includes groups of workers who manufacture and formulate DDT and dieldrin and who have been examined carefully for tumors (24, 67, 68).

## 2.6 Mutagenicity

As pointed out in the NCI carcinogenesis report (62) DDT and its metabolites, DDE and DDA, have been tested for mutagenicity in a variety of systems. DDT and DDE failed to induce mutations in histidine-requiring strains of *S. typhimurium* to bacteria devoid of histidine requirements, and along with DDA, proved nonmutagenic in host-mediated bioassays in mice (69). DDT proved positive for mutagenicity in *D. melanogaster*. DDT itself may be a very weak mutagen in *Drosophila* (70).

Yoder et al. (71) examined lymphocyte cultures from agricultural workers handling various insecticides, including DDT, for chromosomal aberrations during the peak spraying season and again during the winter. In the nonexposed controls, no apparent differences in the number of chromatid breaks in 25 cells from each person examined were noted, but in the pesticide workers, a fivefold increase in these lesions was noted during the peak spraying season. The health implications of chromosomal findings such as these are not clear.

In reviews of the genotoxic potential of DDT and metabolites (7, 27), it was concluded that these materials were without genetic activity in bacteria and *Neurospora*. In other in vitro tests, unscheduled DNA synthesis and gene mutation assays in mammalian cells were generally negative, whereas chromosomal aberrations were reported for human and hamster cells. The positive findings for dominant lethality in rodents were questioned given the dubious relevance to human health of the exposure conditions employed.

## 2.7 Teratology and Reproductive Effects

The results of teratogenicity evaluations for DDT indicate that the compound does not produce anatomic anomalies in offspring of treated dams. Decreases in fetal body and organ weights were observed in rabbits following dosing of 1 mg/kg/day to maternal animals (72). Other studies were reported that indicated decreased lengths of gestation, increased resorption rates, preweaning mortality, learning impairment in mice, and an inhibition of a number of female rodent developmental effects (27).

Teratogenic effects of DDT were also absent in studies of reproduction, including those for two generations of rats, six generations of mice, and three generations of dogs (65). Teratogenic effects of DDT were not found in a six-generation reproduction study in Swiss mice (73). In this investigation, feeding of the parent generation (four males and 14 females) was continued through weaning of the second litter. Individual records were kept of each mouse of the first and second litters of each generation, and from the data collected, indexes for fertility, gestation, viability, lactation, and survival were calculated. At a feeding level of 25 ppm, the effects on lactation were questionable in the second, fourth, and sixth generations. At 100 ppm, the lactation index was lowered significantly in the first, third, and sixth generations. The survival index was lowered significantly at this feeding level in the first, third, fifth, and sixth generations, and the viability index was lowered in the third and fifth generations. The reproduction study with 250 ppm DDT was discontinued after the second generation because of a high rate of fatalities among the pups. In general, effects were more severe when DDT was fed in combination with related organochlorine pesticides.

The estrogenic activity of DDT causes reproductive impairment in male rats as evidenced by decreased testicular weights and spermatogenesis in addition to decreased fertility following DDT administration during the first 23 days following birth. In adult rats, females experienced decreases in fertility and ovarian weights, and in a multigeneration study, no litters were produced by the second-generation animals with effects occurring at dose levels as low as 0.35 mg/kg/day (74).

## 2.8 Hygienic Standards of Permissible Occupational Exposure

The threshold limit value (TLV) for DDT for the American Conference of Governmental Industrial Hygienists (ACGIH) is a TLV–TWA (time-weighted average) of 1.0 mg/m$^3$ (75). No limit was set in 1977 for the concentration immediately dangerous to life and health (76).

According to the 1978 Joint Food and Agricultural Organization–World Health Organization (FAO/WHO) Food Standards Programme Codex Committee on Pesticides Residues, the conditional acceptable daily intake (ADI) of DDT is 0.005 mg/kg body weight (77).

## 3 ALDRIN [CAS # 309-00-2]

The principal component of aldrin, $C_{12}H_8Cl_6$, is hexachlorohexahydro-*endo,exo*-dimethanonaphthalene, known as HHDN.

Aldrin

Technical aldrin contains not less than 95 percent of HHDN and not more than 5 percent of insecticidally active related compounds (7).

## 3.1 Uses and Industrial Exposures

Aldrin has been used as an insecticide for soil insects and the control of termites around buildings. Industrial exposures occur among groups that have been involved in the manufacture of aldrin, and in the handling and spraying of suspensions and emulsions of this compound (see also dieldrin).

### 3.2 Physical and Chemical Properties

| | |
|---|---|
| Molecular weight HHDN (principal constituent) | 364.93 |
| Physical state at 25°C | Tan to dark brown, solid |
| Odor | Mild chemical |
| Melting point, °C | 104 |
| Setting range, °C | 54.4 to 65.6 |
| Density | 1.7 |
| Vapor pressure, mm of Hg at 25°C | $6 \times 10^{-6}$ |
| Flammability | Nonflammable |
| Chlorine content | 57 to 59 percent |
| Solubility | Moderately soluble in paraffins, aromatic, halogenated solvents, esters, ketones; sparingly soluble in alcohols; 11 ppb in water at 20°C |
| Corrosive action | Noncorrosive to steel, brass, monel, copper, nickel, aluminum |
| Stability | Stable in presence of ordinary organic bases, inorganic bases, alkaline oxidizing agents; stable with dilute acids but reacts with concentrated mineral acids, acid catalysts, acid oxidizing agents, phenols, active metals |
| Partition coefficient (log octanol/water) | 3.01 |

### 3.3 Effect in Animals

Aldrin has broad-spectrum central nervous system (CNS) stimulating activity. Toxic or lethal doses produce nausea, vomiting, hyperexcitability, convulsions, and/or coma, followed by death initiated by respiratory failure. Responses from chronic administration may include anorexia, loss of body weight, and degenerative changes in the liver (7).

#### 3.3.1 Acute Oral Toxicity

Lehman (23) gave the approximate $LD_{50}$ for aldrin in rats as 67 mg/kg body weight. Tremors and convulsions were the characteristic responses to a toxic dose. Death may be delayed for several days.

Treon and Cleveland (25) reported the $LD_{50}$ of a solution of aldrin in peanut oil in female rats to be 45.9 mg/kg, and in male rats to be 49 mg/kg. They reported the $LD_{50}$ in dogs to range from 65 to 95 mg/kg. The toxic dose varies with the solvent in which it is given. Ball et al. (78) indicated that the toxicity of aldrin varies with the quality of the product and the nature of the formulation. Oral $LD_{50}$ values for rodents varied (from 10.6 to 59.6 mg/kg) following aldrin administration using various dosing formulations.

Clinical observations (CNS effects) in laboratory animals following short-term exposures are described in detail in Section 1.1 above.

### 3.3.2 Chronic Oral Toxicity

Smith (7) summarized findings for subchronic and chronic studies performed in rodents and dogs. Typically dietary exposures to aldrin in rats caused liver weight increases and liver tumors in mice. Treon and Cleveland (79) fed aldrin to rats for a period of 2 years at concentrations of 2.5, 12.5, and 25 ppm by weight in the diet. There was no increase in mortality or decrease in growth at any of the levels fed. At 12.5 ppm and above, some animals showed increased liver weights and some degenerative hepatic cell changes. Dogs treated by gavage were less tolerant to the effects of aldrin as dose levels down to 1 mg/kg/day but not 0.2 mg/kg/day were fatal within 1 year.

Ball et al. (78) fed rats concentrations of 5, 10, and 20 ppm in the diet. They saw no response over a 6-week period in which they fed the pure compound. The diet was then changed to comparable doses of aldrin as a commercial wettable powder. At 20 ppm, the growth rate showed an initial drop, then a gain to a level higher than normal. They observed no signs of neuromuscular involvement during the first 6 weeks, but thereafter nervous system responses at levels of 10 and 20 ppm. Ball et al. also reported some disturbance of the estrous cycle of rats at concentrations of 10 and 20 ppm. Kitselman (80) reported parenchymatous degeneration of the liver and kidneys of dogs fed 0.02 and 0.06 mg/kg/day for periods up to 1 year.

### 3.3.3 Skin Exposure

Lehman (23) reported that a 4 percent aldrin solution in dimethyl phthalate had an approximate $LD_{50}$ of less than 150 mg/kg (single dose) in rabbits. However, when applied in 5 mg/kg doses for 10 consecutive days, no animals survived. The chemical produced no skin irritation, but the animals developed severe convulsions and died.

A review of toxicokinetic information (81) suggests that dermal absorption occurs rapidly enough to cause toxic effects, although quantitative estimates for uptake rates are not available.

### 3.3.4 Metabolism

Bann et al. (82) reported that aldrin is readily converted to dieldrin in the body and that it is stored as such, primarily in the fatty tissues. This conversion takes place in the liver as well as the lungs and other extrahepatic tissues (83). Aldrin is also converted to dieldrin in plants, microorganisms, and lower animal species.

Aldrin and dieldrin are absorbed from the gastrointestinal tract, lungs, and skin. Because of the rapid conversion to dieldrin, aldrin is rarely detected in tissues. Following aldrin exposure, dieldrin is soon detected in blood, and concentrates in fatty tissue, the brain, and liver. Dieldrin is further oxidized to its 9-hydroxy metabolite, which has been found in fecal (via biliary excretion) samples from production workers. The biological half-life in blood of dieldrin in humans has been estimated to be in the range of 0.7 to 1 year (81).

### 3.3.5 Reproductive Effects and Teratogenicity

Deleterious effects on fertility, gestation, viability, lactation, and survival indexes in mice were noted in the first- and second-generation parental animals, and in their offspring (first and second litters) of test groups fed aldrin and dieldrin at 25 ppm each, and aldrin (100 ppm) plus 100 ppm chlordane. Less marked, but still significant, effects were found in the first and second generations and their offspring after feeding aldrin at 3, 5, and 10 ppm, dieldrin at 3 and 10 ppm, and DDT at 100 and 250 ppm (84).

Quantification of germ cells at various developmental stages in the testicular seminiferous epithelium cycle along with androgen and gonadotrophin assays [(follicle stimulating hormone) FSH, and (lutenizing hormone) LH] were performed on rats treated with aldrin for 13 or 26 days (one and two cycles of seminiferous epithelium) (85). Degeneration of a broad spectrum of germ cells as well as reductions of sperm counts and LH/testosterone were observed in treated rats. These effects were maximal after aldrin treatment for two cycles. FSH was also reduced at the end of two cycles. Administration of human chorionic gonadotrophin along with aldrin treatment suppressed the degeneration of germ cells and enhanced testicular testosterone levels. The authors proposed that aldrin may inhibit gonadotrophin release from the pituitary, or alternatively, may have a direct toxic effect on the testes.

A study was reported to identify birth defects following single oral doses of aldrin on days 7, 8, and 9 (hamsters) or day 9 (mice) of pregnancy (86). These anomalies included cleft palate, webbed feet, and eye opening abnormalities. Owing to protocol inadequacies (use of one-dose level, treatment on one day of gestation, inadequate number of test animals), this study was considered inappropriate for teratogenicity assessment of aldrin (81).

### 3.3.6 Starvation

Weanling, adult, and old Osborne–Mendel rats were fed for 4 weeks diets supplemented with 50 ppm DDT, 7.5 ppm aldrin, or a combination of 50 ppm DDT plus 7.5 ppm aldrin. They were subsequently starved for 6 days, with free access to water. This period of starvation resulted in marked loss of total body lipids, decreased liver to body weight ratio, and a decreased total body lipid to body weight ratio.

As a result of severe starvation, "total DDT" (simple summation of DDT, DDE, and DDD) and dieldrin concentrations decreased in the blood of male and female rats of all ages, regardless of the pesticide supplement fed before starvation, but only in weanling male and female rats were reductions marked and statistically significant.

There was no distinct pattern to the effects of starvation on the concentration and retention of pesticides in the brain and kidney of male and female rats of all ages. In general, in all rats, except for adult and old females, starvation induced a decrease in both the concentration and the total quantity of "total DDT" and dieldrin in the liver. In these groups of old females, the opposite occurred. In male rats particularly, there was a marked conversion of DDT to DDD in the liver as a result of starvation.

The total quantity of pesticides in the total body decreased during the period of severe starvation, regardless of sex, age, or the pesticide supplement fed before starvation. On the whole, the effects were most marked in weanlings and least marked in old rats.

In females of all ages, starvation induced a moderate increase in the concentrations of DDT, its metabolites, and dieldrin in the abdominal fat. In male rats, "total DDT" increased, but dieldrin decreased in the abdominal fat. With the exception of weanling male rats, starvation increased hepatic microsomal enzyme activity for the substrates tested: EPN, $p$-nitroanisole, and methyl orange.

Feeding of the DDT and/or aldrin supplement for 4 weeks to male and female weanling rats resulted in a significant increase in growth rate above that of weanling rats fed a control diet (87).

### 3.3.7 Enzyme Effects from Aldrin plus Parathion Exposures

In female Osborne–Mendel rats, a single oral dose of 20 mg/kg of aldrin (approximately 25 percent of an $LD_{50}$) increased liver enzyme activities as follows: $O$-demethylase, 320 percent; $O$-dearylase, 10 percent; $N$-demethylase, 350 percent; azoreductase, 190 percent, and nitroreductase, 300 percent. Oral administration of 1.5 mg/kg of parathion 1 hr after administration of 20 mg/kg aldrin decreased the above enzyme activities significantly. The highest levels reached were 110, 100, 105, 95, and 81 percent, respectively. Similar studies with DDT, chlordane, methoxychlor, and toxaphene indicated that the net results of hepatic microsomal enzyme activities of a combination dose of an organochlorine compound plus parathion is not predictable (88). Pretreatment of male Osborne–Mendel rats with an oral dose of 30 mg/kg of aldrin or dieldrin provided a significant degree of protection against the toxic effects of an $LD_{50}$ of parathion (89).

### 2.3.8 Effects of Aldrin and DDT

Deichmann et al. (90) fed aldrin (by capsule) and DDT alone and in combination (five times a week) to pure-bred beagles for 10 months. The administration of aldrin (0.6 mg/kg/day) resulted in a constantly increasing concentration of dieldrin in blood and in body fat which, after 10 months, reached a body fat concentration of 75 ppm. Discontinuation of aldrin administration resulted in a gradual decline in dieldrin fat concentration to 25 ppm after 12 additional months. The administration of aldrin at half this dosage (0.3 mg/kg per day) but in combination with DDT (12 mg/kg/day) resulted in the retention of roughly the same concentration of dieldrin in body fat (70 ppm). At 10 months, DDT alone (24 mg/kg/day) had produced a retention of 550 ppm of $p,p'$-DDT in body fat. Feeding of only half this dose of DDT (12 mg/kg/day), but in combination with aldrin (0.3 mg/kg/day), resulted in a body fat concentration of 1290 ppm of DDT. These beagle studies demonstrated that in the dog, the fate of aldrin (dieldrin) is significantly influenced by the presence of DDT, and vice versa.

## 3.4 Human Experience

The effects of aldrin and dieldrin are similar, both quantitatively and qualitatively, in animals, and this, according to Hayes (6), appears to be true for humans also:

> Persons exposed to oral dosages which exceed 10 mg/kg (of aldrin) frequently become acutely ill. A dosage of about 44 mg/kg led to convulsions in a child. Symptoms may appear within 20 minutes, and in no instance has a latent period of more than 12 hours been confirmed in connection with a single exposure. The most thoroughly described related case involved an attempted suicide by ingesting aldrin at an estimated dosage of 25.6 mg/kg. There have been at least two deaths caused by the ingestion of undissolved dieldrin and several caused by drinking emulsions or solutions. The dosage in these cases is unknown.

No quantitative investigative work has been done with air dispersions of aldrin. Some clinical studies have been conducted on employees engaged in packaging and/or handling this material. Princi and Spurbeck (91) studied workers who were exposed through packaging the material, which undoubtedly led to some degree of absorption of aldrin through the skin and possibly by way of the respiratory tract. Actually, the exposure was to a mixture of chlordane, aldrin, and dieldrin. Analysis of the air showed concentrations of 5 to 57 mg/m$^3$, analyzed as organochlorine and calculated as aldrin. By a special absorption technique, the actual aldrin concentration was determined to be between 1 and 2.6 mg/m$^3$. No evidence of a harmful response from the exposure was noted.

Nelson (92) studied a group of workers exposed to dusts of aldrin. He recorded complaints of headache, dizziness, nausea, and vomiting, but found no evidence of liver injury in these individuals.

According to Van Raalte (93), "It is certain that in industrial and other occupational (agricultural and public health use) situations, the principal route of intake is percutaneous. The amount deposited on the skin is much greater and much more important than the amount inhaled. This has been measured for endrin, and, considering the physicochemical properties, the same must be true for aldrin and dieldrin. The respiratory exposure (intake) is only a few percent of the total intake."

Only minor erythema is observed from skin contact with aldrin. It should be recognized that commercial preparations for agricultural use may contain other more irritating ingredients.

### 3.5 Carcinogenicity

Aldrin induces rodent liver tumors typically observed in chlorinated pesticide bioassays. See Section 4.6 under dieldrin for further information.

### 3.6 Hygienic Standards of Permissible Exposure

The threshold limit of aldrin (skin) adopted by the ACGIH for a TLV–TWA is 0.25 mg/m$^3$ (75). The IDLH (immediately dangerous to life or health), according to the 1978 National Institute for Occupational Safety and Health–Occupational Safety and Health Administration (NIOSH–OSHA) Pocket Guide to Chemical Hazards, is 100 mg/m$^3$ (76).

According to the 1978 FAO/WHO Standards Programme Codex Committee on Pesticide Residues, the ADI of aldrin and dieldrin (singly and in combination) is 0.0001 mg/kg body weight (77).

## 4 DIELDRIN [CAS # 60-57-1]

The principal component of dieldrin, $C_{12}H_8Cl_6$, is hexachloroepoxyoctahydro-*endo,exo*-dimethanonaphthalene, known as HEOD.

Dieldrin

Dieldrin contains not less than 85 percent by weight HEOD and not more than 15 percent by weight of insecticidally related compounds.

Some of the discussions in the literature include comments on related chlorinated cyclopentadiene pesticides (e.g., aldrin and endrin) owing to their frequent testing concomitant with other members of the chemical class as well as the ubiquitous bioconversion of aldrin to dieldrin.

### 4.1 Uses and Industrial Exposures

Dieldrin was first used by cotton growers in the 1950s; it has subsequently been used on other crops for the control of vector-borne diseases and for mothproofing woolen goods. Dieldrin, as well as other cyclodiene insecticides, is uniquely suited for the control of termites. In 1974 the registration of products containing aldrin and dieldrin was canceled.

Occupational exposures have occurred among all groups that have been involved in the manufacture or handling of the compound, and in the spraying of dieldrin suspensions and emulsions. Overexposure, resulting in acute intoxication, occurred primarily in the early days of dieldrin, aldrin, and endrin manufacture and in spraying operations with these compounds in Kenya, India, Iran, and other malaria-ridden countries (6, 66).

## 4.2 Physical and Chemical Properties

| | |
|---|---|
| Molecular weight HEOD (principal component) | 380.93 |
| Physical state at 25°C | Buff to light brown, solid dry flakes |
| Odor | Mild chemical |
| Melting point | 176°C (pure) |
| Density | 1.75 |
| Vapor pressure, mm Hg at 25°C | $3.1 \times 10^{-6}$ |
| Flammability | Nonflammable |
| Solubility | Moderately soluble in aromatics, halogenated solvents, esters, ketones, sparingly soluble in aliphatic hydrocarbons and alcohol; 110 ppb in water at 20°C |
| Corrosive action | Noncorrosive to steel, brass, monel, copper, nickel, aluminum |
| Stability | Stable in presence of ordinary organic bases, inorganic bases, alkaline oxidizing agents; stable with dilute acids, but reacts with concentrated mineral acids, acid catalysts, acid-oxidizing agents, phenols, active metals |
| Partition coefficient (log oil/water) | 4.08 |

## 4.3 Effects in Animals

Dieldrin is a CNS stimulant, as are other chlorinated cyclodiene insecticides. It is stored unchanged, primarily in the fatty tissues. It is excreted as such, and in the form of several metabolites. Traces are secreted in the urine.

### 4.3.1 Acute Effects

Review of the toxicities of single oral exposures indicates that $LD_{50}$s range from 25 to 170 mg/kg for rats, mice, and hamsters (7, 81). Dermal application of the chemical has comparable toxic potency ($LD_{50}$ 60 to 90 mg/kg), and these studies show that the finely divided powder is readily absorbed, as is dieldrin in solution. Certain powder formulations (wettable or dusts) decrease the acute dermal toxicity in rabbits. In animals, dieldrin in xylene is roughly 40 times more toxic than DDT following dermal application. Tests with other solvents indicate a smaller relative factor of toxicity. The basis for the differences in dermal toxicities is that undissolved DDT is not absorbed from the skin surface whereas undissolved dieldrin is readily absorbed (6).

The human $LD_{50}$ is estimated to be 5 mg/kg (81) with signs of intoxication including headache, dizziness, malaise, hyperexcitability, muscle twitching, loss of consciousness, convulsions, and depression. Heart rates are also slowed, with returns to normal following a recovery period.

### 4.3.2 Subchronic/Chronic Effects

Central nervous system and cardiac effects following chlorinated cyclodiene insecticide exposures to rats appear to result from alterations of normal calcium transport in CNS neural synaptosomes and sarcoplasmic reticulum of the heart (94). The pesticides (aldrin, dieldrin, endrin) depressed calcium-pump activity while causing decreases in calmodulin levels in these tissues, suggesting that the compounds' activities result from interferences with calmodulin's regulation of calcium-dependent events in neurons and cardiac tissue.

Histopathological organ and tissue changes have been studied primarily in rodents. Many investigators have reported the production of characteristic hepatic "chlorinated insecticide" lesions (6, 21, 24, 95). These include increased organ weight, enlarged liver cells, and proliferation of the smooth endoplasmic reticulum with no or minimal evidence of necrotic lesions (increased serum enzymes).

Virgo and Bellward (96), who fed various dietary concentrations of dieldrin to female Swiss–Vancouver mice, noted dose-related hepatomegaly, increases in cytochrome P450 and microsomal protein, and a decrease in pentobarbital sleeping time. Hurkat (97) reported a decrease in liver glycogen and an increase in liver cholesterol in rabbits treated with dieldrin. The decrease in glycogen appeared to be due to the destruction of glucose phosphatase in the membranes of the endoplasmic reticulum. These and related studies emphasize the specific toxicity of dieldrin for the liver in some species.

Wright and co-workers published two informative reports that dealt with the effects of dieldrin on the subcellular structure and function of mammalian liver cells, and on the effects of prolonged ingestion of dieldrin on the liver of male rhesus monkeys (98, 99). In 1964, the first report discussed the significance of observations in rats, mice, beagles, and rhesus monkeys following administration of dieldrin and the carcinogen 4-amino-2,3-dimethylazobenzene (ADAB). Ingestion of dieldrin resulted in proliferation of the smooth endoplasmic reticulum of liver parenchymal cells, which was associated with an enhanced activity of the liver microsomal mixed function oxidative system. The dieldrin-induced alterations in liver subcellular structure and function were reversible in the rat, mouse, and dog. The effects of ADAB on mouse liver contrasted with the effects of dieldrin and phenobarbital, the former inducing a depression of liver glucose phosphatase activity with no increase in the activity of the liver microsomal mixed function oxidative system.

The second report deals with the biochemical findings of an extensive study on the effects of an approximately 6-year exposure of rhesus monkeys to dieldrin at dietary concentrations of 0.01 to 5 ppm. Increases in the activity of liver microsomal monooxygenase system was seen at an exposure of 25 to 30 µg/kg/day, which is a dose approximately 300 times greater than the daily intake of dieldrin by the general 1966–1967 population. The results obtained in these monkeys and the absence of detectable changes in human liver, particularly in the industrial population, point not only to a slow rate of metabolic clearance of dieldrin in these primate species, but also to a low degree of sensitivity of the liver to this compound (98, 99).

### 4.3.3 Reproductive Toxicity and Teratogenicity

In a six-generation study, white Swiss mice were fed various concentrations of aldrin, dieldrin, DDT, chlordane, or toxaphene. Few or no adverse effects were noted through five or six generations fed 25 ppm toxaphene, chlordane, or DDT. Marked effects in fertility, gestation, viability, lactation, or survival indexes were noted in the parent (first) and second generations and in their offspring (first and second litters) fed 25 ppm aldrin, 25 ppm dieldrin, or 10 ppm aldrin plus 100 ppm chlordane.

Less marked, but still significant, effects were found in the first and second generations and their offspring after feeding 3, 5, or 10 ppm aldrin, 3 or 10 ppm dieldrin, 50 or 100 ppm chlordane, and 100 or 250 ppm DDT. Histological examination of the organs and tissues of mice revealed changes in the livers of all groups and in the kidneys, lungs, and brains of most groups. Typical liver changes included fatty metamorphosis with an increased amount of basophilic substances, hepatic cell necrosis throughout the parenchyma but particularly near the central vein, and also moderate congestion. The kidneys demonstrated moderate vascular congestion, focal glomerulonephritis, and slight to moderate nodular lymphocytic infiltration. Also there was frequently slight dilatation of the convoluted tubules, cloudy swelling, desquamation, and pale basophilic masses in the tubular lumina. The lungs showed moderate congestion and mild alveolar emphysema with minute hemorrhages. Some brain sections showed slight vascular congestion, edema in the parenchyma, and swollen upper motor neurons (84).

There was no evidence of birth defects in studies in mice and rats administered dieldrin orally at the maternal and fetotoxic (supernumerary ribs) dose level of 6 mg/kg/day on days 7 to 16 of gestation (100). 3 mg/kg/day was a no observable effect level (NOEL) for all toxic effects in this study.

Dix and Wilson (101) gave pregnant rabbits 0.2 to 0.6 HEOD/kg by oral gavage dosing daily from day 6 through day 18 of their gestation period. No indication of a teratogenic effect was noted. Ottolenghi et al. (86) administered aldrin 50 mg/kg or dieldrin 30 mg/kg each in a single dose to pregnant hamsters, and aldrin or dieldrin, 25 and 15 mg/kg, respectively, to pregnant CDI mice. They noted an increase in fetal deaths, congenital anomalies, and retardation of growth. It was also shown that the vehicle (corn oil) produced fetotoxicity. In a later teratology study, dieldrin and photodieldrin were orally administered in doses of 1.5, 3.0, and 6.0 mg/kg/day on days 7 to 16 of gestation to CD1 mice and rats. In mice, the highest dose produced an increased percentage of supernumerary ribs and a decrease in the number of caudal ossification centers. No such changes were observed in the rats. Both dieldrin and photodieldrin induced significant liver/body weight increases in the mothers.

In another teratology study by Dix et al. cited by Jager (66), CFI mice were given oral doses of 0.25, 0.5, and 1.0 mg HEOD. Some maternal and fetal toxicity was noted in the mice dosed with dimethyl sulfoxide (DMSO) and with HEOD in DMSO. No teratogenic effect occurred in either group.

### 4.3.4 Forced Elimination from Tissues

Starvation proved to be the only practical method for augmenting dieldrin elimination in chickens (102). In another study, Sell et al. (103) found that when turkeys were subjected to three successive periods of fasting (7, 7, and 4 days), interrupted by periods of feeding (7, 12, and 24 days, respectively), there was acceleration in the decline of both the concentration of dieldrin on body fat and the total amount of dieldrin in the carcass. To be effective, starvation must be severe enough to reduce body lipids to approximately 10 percent or less of the carcass dry matter. Sodium barbital, charcoal, two anion-exchange resins, Colestipol and cholestyramine, a high-fiber diet, and a high-energy protein diet were ineffective in augmenting dieldrin excretion in these birds.

### 4.4 Human Experience

Dieldrin is readily absorbed through the skin and the gastroenteric tract, and by the respiratory tract following inhalation exposure. In humans, overdoses have produced headache, vertigo, nausea, vomiting, and fatigue, followed somewhat later (depending on the dose) by muscle twitchings, myoclonic jerks, and convulsions (66). Hayes reported acute intoxications and some fatalities following excessive skin and inhalation exposures, plus some ingestion by spray personnel. Symptoms included "sudden falls and convulsions with loss of consciousness." Hayes also reported on men who had one or more "fits" 15 to 120 days following their last dieldrin exposure (104).

The potential hazards associated with the agricultural use of dieldrin and endrin in the Pacific Northwest were investigated by Wolfe et al. (105); in these studies these compounds were used as dust for the control of pests on potatoes, apples, and orchard cover sprays. Application was by hand nozzles attached to portable sprayers, by spray machines, and by air-blast machines. Few workers wore gloves during dusting operations. In all operations studied, calculations indicated that the potential dermal exposure was greater than the potential respiratory exposure. Respiratory exposure from dusting potatoes with 1 percent endrin dust was calculated to be 2.2 percent of dermal exposure. Although spraying orchard cover crops with a liquid endrin formulation for mouse control, the spraymen were subject to only 0.4 percent as much respiratory exposure as dermal exposure. When they sprayed pears with dieldrin, respiratory exposure was found to be 1.8 percent of the dermal contamination. The data presented indicate that the hazard from agricultural use of dieldrin and endrin, as practiced in the Pacific Northwest, is not particularly great when compared with the hazard associated with the use of the more toxic organic phosphorus compounds. The greatest hazard probably occurred when dieldrin and endrin, as emulsifiable concentrates, were measured and poured.

Jegier (106) arrived at similar conclusions, based on studies conducted with the agricultural use of endrin in the Canadian province of Quebec. He found the mean dermal exposure in six subjects to be 0.66 mg endrin/hr, whereas the respiratory exposure measured only 0.04 mg/hr.

Jager (66) summarized the data related to the medical supervision of workers exposed during the manufacture of dieldrin from 1954 to 1968; by 1968, 233 men had sustained exposure to aldrin or dieldrin for more than 4 years and 35 men for 10 to 13 years, totaling 1768 man-years of exposure. Geometric mean blood dieldrin concentrations (by gas–liquid chromatography) in these men for the 6 years 1964 through 1969 were 69, 59, 49, 31, 32, and 24 ppb, respectively. This, according to Hunter and Robinson (107), is equivalent to an approximate average daily dieldrin intake of 407 μg/day, or about 58 times the 1968 daily intake by the general U.K. population. Extensive clinical and laboratory tests conducted on these employees "revealed no abnormalities other than those that would be expected in any group of 233 workers. Clinical findings, body weights, and blood pressure monitoring results compared before and after 10 years of exposure to those of a control group, did not show any effects from long-term intensive exposure (to the combination) of . . . insecticides aldrin, dieldrin, endrin, and telodrin" (66).

Jager established the half-time of dieldrin in blood at approximately 8.5 months, and a no-effect level for aldrin and dieldrin at a blood level of 0.105 mg/ml (105 ppb). This level was later modified by Van Raalte (108) to 200 ppb dieldrin in blood. This level corresponds to a total equivalent oral intake of 33 μg dieldrin/day, or a total intake of more than 2000 μg/day (107).

### 4.4.1 Nonoccupational Exposure

General population groups have absorbed dieldrin primarily by way of food and drink. Of all foods consumed, those of animal origin, that is, meat (particularly fatty meat), seafood, poultry (eggs), and dairy products, provided the major portion of this compound. The home and garden use of this insecticide also contributed to the absorption and retention of dieldrin. For the years 1965 to 1970, the average daily dietary intake by a 70-kg individual in the United States was 0.7 μg aldrin,

4.9 μg dieldrin, and 0.3 μg endrin (109). There is no significant evidence that dietary intake of dieldrin by these population groups delivers via mothers' milk quantities of dieldrin that considerably exceed the ADI.

During the early 1970s the dieldrin blood level in the general U.S. population was approximately 0.3 ppb. The body fat concentration of the U.S. adult population (1966 to 1972) was approximately 0.18 ppm for dieldrin and less than 0.02 ppm for endrin (45).

### 4.5 Environmental Levels

The local application and volatilization of dieldrin and related organochlorine pesticides contribute to their partial retention in soils and in surface and ground waters, thus contaminating the local environment as well as distant locations.

During its period of maximum use (1950 to 1975), dieldrin had a deleterious effect in susceptible wildlife, birds, marine life, and insects. Some species of insects developed resistance. In countries where dieldrin is used for the control of malaria and related vector-borne diseases, the question of risk versus benefits needs to receive continued consideration (7).

Summaries of environmental levels of dieldrin (air, water, soil) have been prepared (81) along with identification of populations with greatest potentials for significant exposures.

#### 4.5.1 Photodieldrin

Photodieldrin is an "environmental metabolite" of dieldrin that has been found to be partially formed by microorganisms. Following the spraying of 5.6 kg/ha of dieldrin on pasture land, photodieldrin residues were detected in the grasses the day following the spraying. Five days after application, these residues had accumulated to a maximum concentration of 51 ppm, then declined to 9 ppm after 107 days. Photodieldrin accounted for one-third to one-half the total dieldrin residue after the first 23 days. About 26 g/ha of photodieldrin volatilized during the first 3 weeks after spraying, demonstrating that photodieldrin residues are less volatile than the parent compound dieldrin (110). Fifteen months after spraying the soil with 5 ppm of dieldrin (per soil dry weight), Weisberger et al. (111) identified three conversion products of photodieldrin in a sample of soil.

Photodieldrin appears to be metabolized by the leaves of plants. One metabolic product was found in the leaves of kidney beans that had been exposed for 2 to 4 days to sunlight, but not to artificial light. At 20°C, only small losses occurred due to volatilization of photodieldrin; metabolites were not produced at this temperature. In water, with and without algae, $^{14}$C-photodieldrin was persistent, but in the presence of algae, 40 percent of the added photodieldrin was adsorbed or absorbed by the algae (112).

### 4.6 Carcinogenicity of Aldrin and Dieldrin

Aldrin and dieldrin are discussed together in this section because (1) the latter is a major metabolite of aldrin, (2) the ultimate metabolic products of both compounds are similar, and (3) the two compounds were frequently tested together and reported in the same bioassay reports.

In an investigation by Deichmann et al. (113), a total of 1100 Osborne–Mendel rats were fed a Purina diet supplemented with aldrin or dieldrin (20, 30, and 50 ppm per compound) or endrin (2, 6, and 12 ppm). The doses fed were intended to be excessive, short of producing signs of severe chronic intoxication. The overall tumor incidence (benign and malignant) in male and female rats fed aldrin or dieldrin was significantly lower than the tumor incidence of the control rats. However, there was no significant difference in the tumor incidence of the control rats and the experimental rats fed endrin. In the 956 rats examined histologically, no primary malignant hepatic tumor was found and only two benign hepatic tumors (hemangiomas), one in a male control rat and the other in a female rat fed endrin at 6 ppm.

The diet control male and female rats survived for a mean of 19.7 and 19.5 months, respectively. Because of chronic toxicity, six experimental groups showed a reduced life-span. The remaining 12 experimental groups of 50 animals each provided significant information on the noncarcinogenicity of these compounds to the rat. The mean survivals of groups fed aldrin, dieldrin, or endrin are as follows. Male and female rats fed 20 ppm aldrin survived 19.4 and 18.7 months, respectively; those fed 30 ppm, 19.7 and 18.5 months, respectively; and male rats fed 50 ppm, 20.2 months. Male and female rats fed 20 ppm dieldrin survived 19.5 and 20.5 months, respectively; and male rats fed 30 ppm, 19.8 months. Finally, male and female rats fed 2 ppm endrin, survived 19.6 and 19.3 months.

These negative carcinogenic effects support earlier reports on aldrin and dieldrin rat feeding studies by Cleveland (79), who fed 2.5 to 25 ppm, and by Song and Harville (114), who fed levels up to 285 ppm. Negative carcinogenic effects in rats were also reported by Walker et al. (115), who fed dieldrin for a period of 2 years in concentrations of 0.1, 1.0, and 10 ppm. They observed "no tumorigenic activity which could be related to the feeding of dieldrin."

A second carcinogenic dietary study was undertaken by Deichmann et al. (116) with two strains of female rats (Osborne–Mendel and Sprague–Dawley). It was shown that chronic feeding of female Osborne–Mendel and Sprague–Dawley rats with aldrin at 20 ppm and 50 ppm in the diet supported their previous findings; that is, aldrin does not produce benign or malignant liver tumors in these two strains of rats. Also, as noted before, aldrin at a feeding level of 50 ppm causes systemic toxicity as evidence by a reduced survival rate.

In the Osborne–Mendel rats, there was an elevated incidence of malignant lymphoreticular tumors in the 20-ppm test group. The significance of this finding is questionable, because these tumors were not noted in the female Osborne–Mendel rats fed 50 ppm, nor in the female Sprague–Dawley rats fed 20 ppm or 50 ppm of aldrin.

The most recent bioassay of aldrin and dieldrin for possible carcinogenicity was reported by the National Cancer Institute (NCI) (117). Aldrin was fed to male and female Osborne–Mendel strain rats at dietary concentrations of 30 and 60 ppm, to B6C3F1 hybrid male mice at 4 and 8 ppm concentrations, and to female B6C3F1 mice at 3 and 6 ppm. Dieldrin was fed to male and Osborne–Mendel rats at dietary concentrations of 29 and 65 ppm, and to male and female B6C3F1 mice at 2.5 and 5 ppm. All concentrations were TWA doses.

### 4.6.1 Aldrin

Mouse results from the NCI study (117) showed that there was a significant dose-related increase in the incidence of hepatocellular carcinomas in males. In the female mice, there was a trend in dose-related mortality, primarily with early deaths in the high-dose groups (117). NCI's chronic bioassay results indicated an increase in the incidences of hepatocellular carcinomas in male B6C3F1 mice (but not females) following 80-week exposures to 3 or 6 ppm aldrin (118). Positive results were reported for two other mouse studies in which test animals received 10 ppm aldrin in the diet for up to 2 years (119). It was determined that malignant liver tumors were induced in these bioassays.

Rat results from the NCI study suggested a lack of liver tumorigenicity following 30- and 60-ppm dietary exposures. In both male and female rats there was an increased combined incidence of follicular cell adenoma and carcinoma of the thyroid. The incidence was significant in low-dose, but not in high-dose groups when compared with the pooled controls; however, when compared with matched controls, the incidence was not significant. In addition, cortical adenoma of the adrenal gland was observed in the aldrin-treated rats in significant proportions in the low-dose, but not in the high-dose female when compared with pooled controls. Because this increased incidence was not consistently significant, it is questionable whether the incidence of any of these adrenal tumors was related to the feeding of aldrin. The non-dose-responsive increases in thyroid tumors in males may be considered suggestive by current NTP criteria.

### 4.6.2 Dieldrin

Dietary exposures to rats in the NCI bioassay (117) showed that there was a significant difference in the combined incidence of adrenal cortical adenoma or carcinoma in the low-dose females and the pooled controls: "Although this tumor was also found in animals treated with aldrin, it is not clearly associated with treatment. . . ."

In male mice, there was a significant increase in the incidence of hepatocellular carcinomas in the high-dose group, which may be associated with treatment. The incidence of neoplasm in the experimental female mice was much lower and "probably not biologically significant." In the female mice there was a significant dose-related rate of mortality (149).

### 4.6.3 Overall Conclusions from Cancer Studies

The NCI report (117) concluded:

> Under the conditions of these bioassays none of the tumors occurring in Osborne–Mendel rats treated with aldrin or dieldrin could clearly be associated with treatment.

Aldrin was carcinogenic for the liver of male B6C3F1 mice, producing hepatocellular carcinomas. With dieldrin, there was a significant increase in the incidence of hepatocellular carcinomas in the high-dose males which may be associated with treatment.

Based on a review of the earlier aldrin–dieldrin rat or mouse cancer studies by Fitzhugh et al. (95), Deichmann et al. (113), Walker et al. (115), Stevenson et al. (120), Cleveland (79), Davis and Fitzhugh (121), and Thorpe and Walker (122), the NCI (117) report concluded that "there was no convincing evidence that aldrin or dieldrin was carcinogenic. However, several of the studies in mice showed an increase in liver lesions, usually termed 'hepatoma' in the species. Evaluation was not always possible because detailed data were lacking."

Aldrin and dieldrin have not been found to produce liver tumors in dogs (80) or Syrian hamsters (123).

In a study to examine potential mechanisms of tumorigenesis for rodent liver, gap junction communication was assessed for aldrin, dieldrin, and toxaphene (124). All three compounds inhibited intercellular communication, consistent with the chemicals' tumor-promoting and neurotoxic effects.

Mortality data in a group of 232 high-exposure workers engaged in the manufacture and formulation of aldrin, dieldrin, and endrin for a mean of 11 years and a mean of 24 years of observation were assessed (125). There were 25 deaths in the workers versus 38 expected, based on the Dutch male population. Although exposures were considered high and observation periods long to allow meaningful evaluations, there was no evidence that these workers were subjected to excess risk of cancer.

### 4.7 Mutagenicity

The mutagenic potential of dieldrin was investigated through direct bacterial tests with and without microsomal activation, host-mediated assay, blood, and urine analysis for active metabolites, micronuclei test, metaphase analysis, and dominant lethal test. The doses of dieldrin investigated were 0.08, 0.8, and 8 mg/kg per mouse. Overall evaluation of the data indicated that dieldrin was negative in all four animal tests. No increase in the number of mutants was found in any of the bacterial tests (126).

Dean et al. (127) conducted similar studies with HEOD, the major constituent of dieldrin, in mice and Chinese hamsters. Three test systems showed no evidence of induction of dominant lethality, chromosome breakage, or gene conversion in the animals. Additional studies were conducted by these investigators on short-term lymphocyte cultures from 21 workers currently or previously employed in the manufacture of dieldrin. The degree of chromosome damage in these workers did not differ significantly from that found in a control group of workers. These findings suggest that HEOD does not present a mutagenic hazard in mammals.

The results of these investigators were substantiated by Bidwell et al. (126) in comprehensive analyses designed to test the mutagenic potential of dieldrin.

## 4.8 Hygienic Standards of Permissible Exposure for Dieldrin

The threshold limit for dieldrin (skin) adopted by the ACGIH is 0.25 mg/m$^3$ (75). The IDLH concentration, according to the NIOSH–OSHA Pocket Guide to Chemical Hazards, is 450 mg/m$^3$ (76).

According to the 1978 FAO/WHO Standards Programme Codex Committee on Pesticides Residues, the ADI of dieldrin and aldrin (singly and in combination) is 0.0001 mg/kg body weight. The maximum recommended limits (MRLs) for agricultural commodities, set by the Codex Committee in 1978, range from 0.02 to 0.15 mg/kg (77).

## 5 CHLORDANE [CAS # 12789-03-6, technical]

The formula for chlordane, or 1,2,4,5,6,7,8,8-octochloro-3a,4,7,7a-tetrahydro-4, 7-methanoindane, is $C_{10}H_6Cl_8$.

Chlordane

Technical chlordane contains chlordane isomers (approximately 60 percent) together with heptachlor (4 to 10 percent) and a variety of side-reaction products containing from six to nine chlorines (128, 129). The major chlordane isomers are designed as *cis*- and *trans*-chlordane (alpha or gamma), respectively; they occur in the ratio of approximately 1:1 (130).

### 5.1 Use and Industrial Exposure

Chlordane has found use as an agricultural insecticide, mainly for corn, and for the control of cutworms, ants, root weevils, rose beetles, grasshoppers, and grubs. It also has use as a home insecticide. It is most effective in killing termites as a single application provides termite protection for more than 26 years. Related agents offer protection as follows: dieldrin, more than 25 years; heptachlor, more than 22 years; 8 percent DDT, maximum of 13 years; lindane, maximum of 12 years; and pentachlorophenol (in heavy oil), not more than 4 years (131). Beginning in 1987, chlordane could be used only on the exterior of buildings to control termites; registrations for all other uses were canceled in 1983 (132).

## 5.2 Physical and Chemical Properties

| | |
|---|---|
| Physical state | Colorless to amber, viscous liquid; the commercial product is a mixture containing 60 to 75 percent of the pure compound and 25 to 40 percent of related compounds. Chlorine content is 64 to 67 percent |
| Molecular weight | 409.80 |
| Specific gravity | 1.59 to 1.63 (at 25°C) |
| Boiling point | 175°C |
| Vapor pressure | $2.2 \times 10^{-5}$ mm Hg (supercooled liquid *cis*-chlordane) |
| Refractive index | 1.56 to 1.57 (at 25°C) |
| Solubility | Insoluble in water; miscible with aliphatic and aromatic hydrocarbon solvents, including deodorized kerosine; decomposes in weak alkalies |
| Partition coefficient (log octanol/water) | 5.54 (estd. for pure compound) |

Conversion factors: 1 mg/l = 59.7 ppm; 1 ppm = 16.76 mg/m at 25°C; 760 mm Hg

## 5.3 Effects in Animals

Response to the absorption of chlordane is not unlike that to other members of the chlorinated cyclodiene group of insecticides. The primary acute response is in the central nervous system. The signs of intoxication following absorption of a toxic or lethal oral dose include loss of appetite, irritability, hyperexcitability, vomiting, and tremors, leading to convulsions and death. Anorexia and loss of body weight may be marked if death is delayed. Poisoning from chronic exposure also produced effects on the central nervous system. According to Hyde and Falkenburg (133), electrocerebral disturbances serve as an early sensitive indicator of chlordane intoxication. Cellular changes in the liver may occur. Edema of the lungs and irritation of the gastroenteric tract have also been reported.

### 5.3.1 Acute Oral Toxicity

In 1952, Ingle (134) reported the $LD_{50}$ for the rat to be 250 mg/kg body weight when dissolved in corn oil. Lehman (135) in 1952 reported the $LD_{50}$ for the rat to be 457 mg/kg body weight.

Chlordane manufactured before 1951 was more toxic than that manufactured during and after 1951. The greater toxicity of the early technical chlordane was partly due to the presence of hexachlorocyclopentadiene, an intermediate in the manufacture of this insecticide. The rat oral $LD_{50}$ value reported for purified chlordane in 1953 was 500 mg/kg (136), and for Velsicol technical chlordane (reported in 1954) was 570 mg/kg (137).

Chlordane, as well as heptachlor, is very toxic to many species of invertebrates. Both compounds are detrimental to populations of the pollinating species (130).

### 5.3.2 Vapor Exposure

According to Ingle's monograph (138), Nickerson and Radeleff published reports on the effects of prolonged inhalation of vapors when chlordane was used as a residual premise insecticide. Their conclusions were that "no ill effect was observed in pigeons continuously exposed for 60 days to the vapors arising from surfaces treated with chlordane at the rate of 1,000 mg per square foot. . . . Histopathological studies revealed no lesions attributable to chlordane." Their conclusions were the same for chicks exposed to chlordane vapors for 30 days. Fog applications of 7 percent chlordane into a room housing rabbits, guinea pigs, white rats, mice, and poultry produced no obvious deleterious effects (139). Toxicity observed from vapors was not associated with chlordane but with volatile contaminants, particularly hexachlorocyclopentadiene.

### 5.3.3 Skin Exposure

Eight cattle sprayed 12 times at 2-week intervals with 2 percent of the compound manufactured in 1953 showed no signs of local irritation or systemic intoxication. Spraying cattle with chlordane of earlier manufacture caused death in 3 of 10 cattle after three applications (140). Injury to the skin and mucous membranes by the early chlordane was significant. Later chlordane has not been shown to cause such irritation and is also more slowly and less completely absorbed (138).

### 5.3.4 Chronic Oral Toxicity

Ingle (138), in a 2-year chronic rat-feeding study, noted retardation of growth at concentrations of 150 and 300 ppm in the diet, but not at concentrations of 5, 10, and 30 ppm. Liver damage was marked at 150 and 300 ppm, but slight at 30, minimal at 10, and absent at 5 ppm. There was no injury to the kidneys at 5, 10, or 30 ppm, but there was marked injury at 150 and 300 ppm. The lung showed marked damage at 300 ppm, mild injury at 150, and no injury at lower concentrations.

In 1952, Lehman (135), after feeding the pre-1951 chlordane, reported that the minimal effective dose was 2.5 ppm and the maximum tolerated dose for rats over a 2-year period was 0.125 mg/(kg)(day). In 1953, Ambrose (136), in similar chronic rat feeding studies, reported growth depression at 320 ppm, but normal growth at 160 ppm and less. He observed enlargement of the liver at 80 ppm, but no effect at 10 ppm.

A reinvestigation of technical chlordane was conducted by Ingle in 1955 (138), when the material was fed to rats for 2 years at dosage levels of 2.5, 5, 10, 25, 50, 75, 150, and 300 ppm. The study revealed the following:

1. Hepatic cells (the most sensitive target tissue) showed alterations at 50 ppm; there were no alterations and no variation from control at 2.5, 5, 10, and 25 ppm.
2. Body growth retardation was apparent at 300 ppm, but not at 150 and lower concentrations.
3. Mortality increased at 300 ppm, but not at 150 or lower concentration. The "no-effect level" for clinical effects was above 150 ppm.
4. No tissue changes were noted that would suggest a carcinogenic effect.

Ninety-day inhalation studies were performed in rats and cynomolgus monkeys during which animals were exposed to 0 to 10 mg/m$^3$ (141). Rats exhibited minor alteration of the liver but similar effects were absent in the primates.

### 5.3.5 Absorption, Metabolism, and Excretion

Chlordane is absorbed through the skin, more readily via the lungs, and from the gastroenteric tract. It is retained primarily in body fat. Both isomers of chlordane are oxidatively degraded to a series of mono- and dihydroxy derivatives which are excreted in the feces and, to a lesser extent, in the urine (142). Nonachlor forms a small portion of the residue in the fat of rats (130).

Balba and Saha (143) fed 1700 mg of $^{14}$C-labeled alpha and gamma isomers of chlordane to four male rabbits, in four doses each. Of the alpha isomer, 77 percent was excreted in the feces and urine, and 84 percent of the gamma isomer. The alpha isomer was retained primarily in body fat, and successively less in kidney, muscle, liver, and brain; the highest concentrations of the gamma isomer were found in the kidney, followed by fat, liver, muscle, and brain. The concentrations of oxychlordane in the tissues (primarily in fat) were higher than those of the parent compound.

## 5.4 Human Experience

Chlordane is considered moderately toxic. The very few reported incidents of acute chlordane intoxication in humans or animals have resulted from gross negligence or misuse (6, 37, 45). Use of early chlordane often resulted in irritation of the eyes, mucous membranes, and/or skin of industrial and agricultural workers. This does not appear to have been a problem with the product manufactured since 1951.

Smith (7) surveyed a number of episodes involving accidental exposures and suicidal intent. Deaths have occurred following dermal exposure to a chlordane–DDT solution as well as following ingestion of a chlordane–talc formulation. Serious poisonings are characterized by onset of violent convulsions within 0.5 to 3 h with death or recovery following shortly thereafter.

Human exposure to vapors of 7 percent chlordane, for 15 min at 3-day intervals for periods of 12 weeks and repeated a year later, did not result in symptoms of toxicity (139).

According to the Council for Agricultural Science and Technology (CAST)

(130), the long-term effect on health due to chlordane in foods is "very small." Neither is there any evidence to indicate that the home and garden use of chlordane has constituted a significant hazard. Levels of chlordane in human body fat have been in the parts per billion range.

### 5.5 Volatility

Chlordane is translocated from plants and soils to which it is applied, and enters the surrounding atmosphere and waters. Stauffer (129) found the vapor concentration over a 72 percent water emulsion of pooled liquid technical-grade chlordane to be 213 ng/l.

### 5.6 Persistence in Soil and Water

Chlordane is relatively persistent in the environment. It is absorbed on soil solids although it is readily released for uptake into plants. Dorough and Pass (144) treated soils (at corn planting) with 455 and 910 of "active-ingredient" chlordane, or high-purity chlordane. After 1 year, 50 to 70 percent of the residues had "dissipated" from the top 10 cm of the soil. The whole corn plant, harvested for silage 102 days after planting, contained 0.03 to 0.04 ppm of chlordane (alpha and gamma isomers combined). The mature corn grain and cobs were free of detectable residues (less than 0.008 ppm). According to CAST (130), soil concentrations ranging from traces to mean values of 1.5 ppm have been reported. Residues rarely persist in detectable concentrations for more than 5 years.

Technical chlordane was added to a freshwater lake (initially free of detectable pesticide residues) to create a level of 10 ppb of the pesticide. After 7 days, the lake water contained 4610 part per trillion (ppt) total technical chlordane; by 421 days the level had dropped to 9.5 ppt or 0.095 percent of the initial concentration. Mean chlordane residues in sediments of this lake were 35.3, 19.4, 33.9, 31.8, and 10.3 ppb 7, 24, 52, 279, and 421 days, respectively, after treatment. Neither heptachlor nor heptachlor epoxide was detected in sediments 279 days after treatment (145). Bioaccumulation of chlordane is similar to that of other organochlorine insecticides. In a terrestrial aquatic model system, algae, snails, mosquitoes, and fish accumulated $^{14}$C-chlordane 98, 286, 613, 6132, and 8258 times the $^{14}$C-chlordane concentration in the water (146).

There has been little evidence that fish absorb chlordane directly from water, but rather chlordane residues accumulate as a result of food chain contamination. Average concentrations of chlordane residues in freshwater fish ranged from 0.16 to 1.01 ppm (147).

### 5.7 Cancer

In 1977, the National Cancer Institute (148) reported the results of a bioassay of chlordane, which was examined for possible carcinogenicity. Groups of 50 male and female Osborne–Mendel rats were fed TWA doses of chlordane: 203.5 and

407.0 ppm for males and 120.8 and 241.5 ppm for females for 80 weeks. They were then observed for 29 weeks. Groups of 50 B6C3F1 male and female mice were fed TWA diets containing 29.9 and 56.2 ppm (males) and 30.1 and 63.8 ppm (females) chlordane.

The effect on survival rates indicated that mortality was dose-related for female rats and for male mice. Male control rats, for reasons unknown, showed an abnormally low survival rate.

In the experimental rats, there was statistical evidence for the induction of proliferative lesions of follicular cells of the thyroid, and of malignant fibrous histiocytoma, but these findings were discounted because the rates of incidence were comparatively low and/or are known to be variable in control rat populations.

Hepatocellular carcinoma failed to appear at a significant rate in the rats fed chlordane. Further, the number of lesions of the liver did not become significant with the inclusion of nodular hyperplasia, or with the application of life-table adjustments to the data.

In mice, hepatocellular carcinoma showed a highly significant dose-related trend. These high levels of significance were maintained when hepatocellular carcinoma was combined with nodular hyperplasia, or when the data were subjected to life-table adjustment. No other tumors were found in mice in sufficient numbers to justify analysis. The report concluded that "under conditions of this bioassay, chlordane is carcinogenic for the liver in mice."

On June 22, 1976, the administrator of the U.S. EPA initiated hearings in the continued use of chlordane and heptachlor, hearings that were terminated on February 21, 1978 (149). The testimony in the hearings dealt primarily with chronic laboratory studies involving chlordane and/or heptachlor with respect to whether one or both compounds cause cancer in laboratory animals, and what these results mean in terms of cancer risk in humans. Studies that were reviewed included those published by the NCI (148, 150), U.S. Food and Drug Administration (151), International Research and Development Corporation (IRDC) (152), and Witherup et al. (153).

Because there was an obvious dispute among pathologists in the interpretation of the experimental data and diagnosis of certain tissue slides, the Administrative Law Judge requested the National Academy of Sciences to review the data and render a report. This function was performed by the Pesticide Information Review and Evaluation Committee for the Advisory Center on Toxicology, Assembly of Life Sciences, National Research Council (154). This committee concluded that "chlordane and heptachlor epoxide, a metabolite of heptachlor, are carcinogens in the mouse." With respect to heptachlor, the evidence of hepatocellular carcinoma "is not so clear." But studies of other possible neoplastic and preneoplastic changes in the liver "suggest the probability that it too is carcinogenic in the mouse."

With respect to rats, the committee report stated that although a report of one previous bioassay "suggests that heptachlor epoxide is carcinogenic in the rat, examination of the slide made available to the committee did not confirm this" and that "there is no statistically significant evidence that any of the compounds are carcinogenic in rats" (154).

According to FAO/WHO (155), in considering the production of hepatomas in certain strains of mice, "These liver tumors have not been found to develop in any species other than mouse as the result of exposure to dieldrin or chlordane, and these chemicals have been shown to be nonmutagenic in a variety of studies." With regard to long-term studies, "Chlordane caused hepatocellular carcinoma in mice at a dose of 60 ppm in the diet, but not in rats at doses as high as 400 ppm for males and 240 ppm for females."

As to the question of risk to humans, the National Academy of Science Advisory Committee reported (154), in part:

> There are no adequate data to show that these compounds are carcinogenic in humans, but because of their carcinogenicity in certain mouse strains and the extensive similarity of the carcinogenic action of chemicals in animals and in humans, the Committee concluded that chlordane, heptachlor and/or their metabolites may be carcinogenic in humans. Although the magnitude of risk is greater than if no carcinogenicity had been found in certain mouse strains, in the opinion of the Committee, the magnitude of risk cannot be reliably estimated because of the uncertainties in the available data and in the extrapolation of carcinogenicity data from laboratory animals to humans.

The National Academy of Science advisory committee (154), commenting on two ongoing epidemiologic studies, stated that "the very limited data available do not indicate an increased risk of cancer in chlordane plant workers or in pest control operators," but emphasized that "the duration of these studies and the number of workers involved are not great enough to assess adequately the carcinogenic potential of chlordane and heptachlor."

More recent studies in rodents fed diets containing levels of 12.5 and 25 ppm chlordane were conducted (156, 157). There were no tumorigenic responses in either rats (dosed for 30 months) or mice (dosed for 24 months). Both chlordane and heptachlor have been shown to cause tumor promotion in mice following initiation with diethylnitrosamine (158).

Based on the scientific literature, which includes retrospective mortality studies of workers, there is no evidence that chlordane or heptachlor has caused cancer in humans (7).

## 5.8 Hygienic Standards of Permissible Exposure

The threshold limit value for chlordane (skin) adopted by ACGIH is 0.5 mg/m$^3$ TLV–TWA (75). The IDLH concentration of chlordane according to the NIOSH–OSHA Pocket Guide to Chemical Hazards is 500 mg/m$^3$ (76).

According to the 1978 Joint FAO/WHO Food Standards Programme Codex Committee on Pesticide Residues, the ADI of chlordane was recommended to be 0.001 mg/kg body weight. The MRLs of chlordane in various agricultural commodities range from 0.02 to 0.5 mg/kg (77).

On March 11, 1978, the EPA administrator signed a cancellation order for chlordane and heptachlor, but it provided for the continued use of chlordane and heptachlor for termite control and for the continued use of chlordane and/or heptachlor on a phase-out basis for a number of crops and insects (149).

## 6 HEPTACHLOR [CAS # 76-44-8]

Heptachlor, or 1,4,5,6,7,8,8-heptachloro-3a,4,7,7a-tetrahydro-4, 7-methanoindene, has the formula $C_{10}H_5Cl_7$.

Heptachlor

### 6.1 Uses and Potential Exposures

Heptachlor is an insecticide formerly used in the control of certain soil-inhabiting insects that attack corn and other field crops. Heptachlor was also used for seed treatment, cotton insects, and grasshoppers. This insecticide and other low-volatility cyclodiene insecticides are uniquely suited for the control of termites. Its only permitted use in the United States since 1983 is for subsurface soil treatments for termites and the dipping of nonfood plants.

The potential for occupational exposure exists for personnel engaged in the manufacture, handling, and application of heptachlor. Indoor air environments of treated dwellings have been shown to contain traces levels of this pesticide.

### 6.2 Physical and Chemical Properties

Heptachlor is the chlorination product of chlordane. The technical-grade product contains approximately 73 percent heptachlor, 22 percent *trans*-chlordane, and 5 percent nonachlor. Heptachlor epoxide is an oxidation product of heptachlor, and is found as a environmental residue and a mammalian metabolite following exposure to heptachlor.

| | |
|---|---|
| Physical state | White crystalline solid |
| Molecular weight | 373.32 |
| Specific gravity | 1.57 to 1.59 |
| Melting point | 95 to 96°C (pure) |
| Vapor pressure | $3 \times 10^{-4}$ mm at 25°C |
| Solubility | 56 ppm in water; soluble in alcohol; slightly soluble in xylene, carbon tetrachloride, cyclohexane |

Partition coefficient     5.05
(log hexane/water)

Conversion factors: 1 mg/l (air) = 65.1 ppm; 1 ppm = 15.35 mg/m$^3$ at 25°C, 760 mm Hg

### 6.3 Toxic Effects in Animals

Heptachlor is a chlorinated derivative of methanoindene with a structure similar to that of chlordane. However, there are noted differences in the biologic activity and in the degradation products of heptachlor and chlordane.

Extensive reviews of the toxicology literature on heptachlor have recently been published (7, 159).

#### 6.3.1 Acute Effects

Lehman (23) reported that heptachlor fed to rats in a single dose leads to an LD$_{50}$ of 90 mg/kg body weight. The principal toxic responses were tremors and convulsions. Acutely toxic oral doses reported by Buck et al. (160) are for the rat, 40 mg/kg (LD$_{50}$); mallard, 2000 mg/kg (LD$_{50}$); rabbit, 2000 mg/kg (LD$_{50}$); calf, 20 mg/kg; and sheep, 50 mg/kg.

The particle size of dry formulations and the solvents employed for solutions of the pesticide play a role in the dermal toxicity of heptachlor. Lehman (23) reported that when the dried powder of heptachlor was applied to the skin of rabbits, the approximate LD$_{50}$ was 2000 mg/kg body weight. When heptachlor was applied as a 20 percent solution in dimethyl phthalate, the approximate LD$_{50}$ was less than 780 mg/kg. However, when the chemical was applied in repeated smaller doses, the approximate LD$_{50}$ was less than 20 mg/kg/day. There were no survivors after 14 doses of 28 mg/kg. No skin irritation from the materials was observed. Gaines's study, as reviewed by the Agency of Toxic Substances and Disease Registry (ATSDR) (159), indicated that heptachlor in a xylene solution produced a single-dose dermal LD$_{50}$ of 195 mg/kg in rats.

Systemic effects observed in rats following acute exposures included hepatic changes including increased liver weights, necrosis, cellular vacuolization, steatosis, and elevated serum enzymes (serum glutamate–pyruvate transaminase, alkaline phosphatase), cholesterol, and bilirubin (159).

In earlier years, when heptachlor was being used on extensive areas in the southeastern United States for fire ant control, ingestion of residues killed a number of birds and resulted in the accumulation of sublethal doses in others. There were no reports of heptachlor residues in fish-eating birds even when these birds contained residues of other chlorinated hydrocarbon insecticides (130).

#### 6.3.2 Subchronic and Chronic Toxicity

Results from 6-week dietary studies in rats and mice indicated that the NOELs for lethality were 160 and 40 ppm, respectively (150).

The highest concentration that rats survived for 6 months was a dietary level of

30 ppm heptachlor. At this exposure, a significant amount of heptachlor was stored in the fat of exposed animals (161).

The ATSDR review (159) of the Enan study indicated that the investigators fed 10 ppm heptachlor in the diet to rats 5 days/week for 4 weeks. Liver changes included increases in microsomal enzymes, reduction in liver glycogen, and increased serum glucose and white blood cell counts. In another subchronic study by Akay and Alp described in this review, 10-week exposures to 100- and 200-ppm dietary levels of heptachlor induced tremors and ataxia in rats, and 50 ppm represented a NOEL for these effects. Granulomas were also induced in kidneys at the highest dose level.

Unpublished results from the Kettering Laboratory (162) and from the Food and Drug Administration (163) indicated that at least five chronic toxicity studies were conducted with rats, one with dogs, and one with mice. Liver damage similar to that produced by chlordane was produced in some test animals. The maximum demonstrated no-effect level was a dietary concentration of 5 ppm heptachlor, which was equivalent "to approximately 0.125 mg/kg per day for the dog, 0.25 mg/kg per day for the rat, and 0.75 mg/kg per day for the mouse" (130).

Results of long-term studies to assess chronic effects in rodents are described in Section 6.7.

### 6.3.3 Absorption, Metabolism, and Excretion

Heptachlor can be absorbed through the skin and via the lungs and gastroenteric tract. In the rat, heptachlor is metabolized to heptachlor epoxide (161, 164, 165). Heptachlor epoxide is also the oxidation product of heptachlor in other animals, as well as in plants and microorganisms, but not all heptachlor is converted to the epoxide, which is more toxic and more stable than the parent compound (166–168). The epoxide is partially stored in body fat, where it may remain for prolonged periods. Dehydrochlorination of heptachlor epoxide, followed by hydroxylation and double-bond rearrangement, leads to the formation of a metabolite that is the principal form in which heptachlor is excreted in the feces (130, 169).

Induction of the Phase II hepatic enzymes, epoxide hydrolase and glutathione S-transferase, by *trans*-stilbene oxide significantly enhanced the clearance of heptachlor from rats (170). This treatment also reduced the storage of heptachlor in rat tissues as an apparent result of enhanced formation of conjugated (and excretable) metabolites derived from the heptachlor epoxide precursor.

### 6.3.4 Reproductive Toxicity and Teratogenicity

Mice of both sexes were exposed to diets containing heptachlor up to 200 ppm for 10 weeks prior to mating (Akay and Alp study reviewed in Reference 159). Although no microscopic changes were observed for testes or ovaries, no offspring were produced in the testing. Other reproduction studies conducted in rats showed that heptachlor exposures increased resorption of fetuses and decreased fertility with each succeeding generation.

In a developmental toxicity assay, pregnant rats were treated orally on gestation days 7 to 17 with doses up to 20 mg/(kg/day) of heptachlor. This study revealed no increases in fetal mortality or anatomic anomalies at any dose level (171, as described in Reference 7).

### 6.4 Human Experience

Acute intoxication syndrome from heptachlor exposure includes abnormal behavior, hyperirritability, tremors, and convulsions. Acute heptachlor intoxications have been uncommon. Heptachlor has been handled and applied extensively in industry and agriculture (heptachlor since 1952, and chlordane since 1945) with a record of safety. Although chlordane was used extensively in homes and gardens, heptachlor was not registered for this purpose in the United States.

Ingestion of heptachlor from the human diet has not been shown to represent a significant source of exposure. According to a report by IARC (172), the average daily intake of heptachlor and heptachlor epoxide residues in the United States decreased from 2.3 µg/day in 1965 to 1.4 µg in 1970; at present, the dietary concentrations are further decreasing. Only five of 79 foods of animal origin collected and analyzed in Louisiana from 1968 to 1972 contained heptachlor (including heptachlor epoxide) residues in detectable concentrations. The range was from 0.01 to 0.15 ppm, and the average was 0.06 ppm (173). There is little evidence of heptachlor accumulation in the food chain. In the urine of the general population, the epoxide has been found in extremely low concentrations (174).

In the early 1980s, bovine milk on Oahu, Hawaii was shown to be contaminated with heptachlor as an apparent result of exposures of cow herds to pineapple foliage previously sprayed with the pesticide (175). Levels in the milk ranged from 0.12 to 5 ppm, and women were potentially exposed for up to 29 months. Studies of pregnancies and offspring indicated that there were no increases in fetal or neonatal deaths nor evidence of lower birth weights in offspring. Although a slight increase in the incidence of abdominal wall anomalies was seen relative to non-concurrent control rates, cause and effect could not be established because there was not sufficient analytic information to assess actual heptachlor exposures for the exposed and control populations.

### 6.5 Persistence in Soil and Water

Heptachlor and chlordane are relatively persistent in the environment. Their half-life in the soil is approximately 0.8 and 1 year, respectively. When these insecticides were applied at agricultural rates, residues rarely persisted in detectable quantities for more than 5 years after the last application (130).

Both heptachlor and chlordane are absorbed on soil solids and hence tend to remain near the site of application; only small amounts reported in plant tissues rarely exceeded 1 ppm. In surface waters, heptachlor and chlordane either have been undetectable or present traces up to mean concentrations of 6 ppt (130).

The volatility of heptachlor exceeds that of chlordane and is a major pathway of loss of these insecticides from soil (130).

## 6.6 Invertebrates

Heptachlor and chlordane are highly toxic to many species of invertebrates, and they are detrimental to populations of pollinating species. Both compounds are highly toxic to beneficial insects (130), and at high rates of application become highly detrimental to earthworms. However, Doane (176), Polinke (177), Schread (178), and Smith (179) found that field application at a rate of 1 kg/acre had no deleterious effect on earthworms.

For details on the hazard and toxicity of heptachlor to invertebrates and vertebrates, as well as on the behavior of heptachlor in the environment, refer to the report prepared by CAST (130).

## 6.7 Cancer

In 1977, NCI released a bioassay report on the possible carcinogenicity of heptachlor (150). This report included reviews of earlier feeding studies that were designed to investigate a possible carcinogenic action of this compound. In three of five unpublished long-term feeding studies, rats received diets containing 12.5 ppm of the test material for at least 2 years but showed no increase in tumor incidence attributable to treatment. In a fourth study, an increase in liver weights was the only reported effect in rats receiving 10 and 20 ppm heptachlor. In a fifth study with rats, the incidences of liver tumors in test animals given 0.5 to 10 ppm were elevated in males and females versus rates in controls. In one study with C3Heb/Fe/J mice, the feeding of heptachlor or heptachlor epoxide at a level of 10 ppm in the diet was associated with an increase in liver tumors including carcinomas (180).

In the 1977 NCI heptachlor study (150), groups of 50 male Osborne–Mendel rats were fed TWA doses of 38.9 and 77.9 ppm for 80 weeks, and groups of female rats were fed 25.7 and 51.3 ppm. All animals were observed for 30 weeks following exposure periods. Additionally, groups of 50 B6C3F1 mice of each sex were fed TWA doses of 6.1 and 13.8 ppm (males), and 9 and 18 ppm (females).

No hepatic tumors were observed in rats administered heptachlor. There was statistical evidence for induction of proliferative lesions of follicular cells of the thyroid in female experimental rats, but this finding was discounted because the rates were comparatively low and known to vary in control rat populations.

In mice fed heptachlor, hepatocellular carcinoma showed a highly significant dose-related trend in both males and females. No other tumors were found in these mice in sufficient numbers to justify statistical analysis.

ATSDR (159) reviewed human studies that included more than 16,000 pesticide (including chlordane/heptachlor) applicators and termite control operators. Results indicated that there was a 35 percent increased rate of lung cancer (as compared to controls) for the pesticide applicators only. Other large studies cited by ATSDR for workers in pesticide manufacturing were either negative or inconclusive. Because of methodological deficiencies, carcinogenicity could not be attributed to work exposures in these populations.

It was concluded that under the conditions of this bioassay, heptachlor is carcinogenic for the liver in mice. Based partially on this finding, EPA restricted the use of this pesticide (180).

### 6.8 Hygienic Standards for Permissible Exposures

The threshold limits of heptachlor (skin) adopted by the ACGIH were TLV–TWA 0.5 mg/m³ (75). The IDLH concentration of heptachlor according to the NIOSH–OSHA Pocket guide to Chemical Hazards is 700 mg/m³ (76).

According to the 1978 FAO/WHO Standards Programme Codex Committee on Pesticide Residues, the ADI of heptachlor is 0.00005 mg/kg body weight. According to this 1978 report, the MRLs of heptachlor and its epoxide, expressed as heptachlor for agricultural commodities, range from 0.01 to 0.2 mg/kg (77).

## 7 CHLORDECONE (KEPONE, [CAS # 143-50-0])

Kepone, or decachlorooctahydro-1,3,4-metheno-2$H$-cyclobuta($c$,$d$)pentalen-2-one, has the formula $C_{10}Cl_{10}O$.

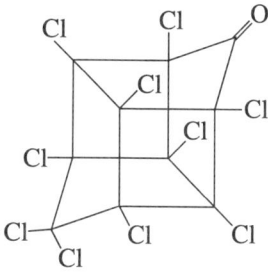

Chlordecone

### 7.1 Uses and Industrial Exposures

Chlordecone, a chlorinated dicyclodiene compound, was introduced under the trade name Kepone in 1958. In the United States, Kepone found use as an ant and cockroach poison. Outside the United States, it was used primarily as a pesticide against the banana root borer. The compound is the ketone analogue of mirex; it is a contaminant of mirex and is also a product of mirex degradation. Production of Kepone in the United States was discontinued on July 24, 1975 because of severe industrial exposure and intoxications in men employed in the Hopewell, Virginia plant, the only plant in the United States that manufactured Kepone.

### 7.2 Physical and Chemical Properties

        Physical state       Crystals
        Molecular weight   490.68

| | |
|---|---|
| Melting point | Decomposes at 350°C |
| Vapor density | 16.94 |
| Vapor pressure | $<3 \times 10^{-7}$ mm at 20 to 25°C |
| Solubility | 2 ppm in water; slightly soluble in hydrocarbon solvents; soluble in corn oil, alcohols, ketones, acetic acid |

Conversion factors: 1 mg/l = 49.8 ppm; 1 ppm = 20.07 mg/m³ at 25°C, 760 mm Hg

## 7.3 Effects in Animals

### 7.3.1 Acute Toxicity

Oral $LD_{50}$ values for Kepone are 132 mg/kg for male rats, 126 mg/kg for female rats, 71 mg/kg for male rabbits, and approximately 250 mg/kg for dogs. In these species, the outstanding sign of intoxication was the development of severe DDT-like tremors. These reached maximum intensity (in surviving animals) 2 to 3 days following treatment. The tremors gradually subsided over a period of a week or longer. Exacerbation of the tremors occurred whenever an animal became excited, for example, when handled (181). It was also noted that rats and pigs can develop a muscular weakness the second week after treatment which progresses through the seventh week with a final phase of recovery within 6 months (182).

Hyperexcitability and tremors were noted in mice the day following one oral dose of 50 mg Kepone/kg; mortality occurred on the fifth day (183).

### 7.3.2 Skin Exposure

The percutaneous $LD_{50}$ for male rabbits was 410 mg/kg (181). Application of 0.25 and 0.50 g of the active ingredient (in ant bait) to the skin of rabbits (five 24-hr exposures over a period of 3 weeks) produced no signs of local irritation, no effect on body weight, and no significant micropathological organ changes.

The potentiation of the hepatoxicity of carbon tetrachloride that was demonstrated by pretreating rats with Kepone (184) likely occurs as a result of enhancement of microsomal enzyme induction by Kepone. For an assessment of the neurotoxicity of Kepone, refer to the reports by End et al. (183).

### 7.3.3 Subchronic and Chronic Toxicity

Dogs exposed to Kepone for 17 months at dietary levels of 1, 5, or 25 ppm did not display significant in-life clinical changes except for a small suppression of body weight gains at 25 ppm. Examination of tissues indicated increased organ weights for the heart, kidneys, and liver at the highest dose level. No other changes were attributable to Kepone administration (185).

Chronic exposures to Kepone may include some or all of the following effects: tremors, loss of body weight, increase in oxygen consumption, neurological impairment, disturbance of the estrous cycle, and fetotoxicity. Inhibition of muscle lactate dehydrogenase has been observed in *in vitro* studies of the rabbit (186). Anatomic organ changes may include liver injury (congestion, hyperplasia, hypertrophy, focal necrosis, hepatomegaly), nephropathy, atrophy of the testes, and proteinuria (7). Results of chronic carcinogenicity bioassays are presented in Section 7.8.

## 7.4 Human Exposure

### 7.4.1 Occupational Exposure

The first recorded incidents of Kepone poisoning were recognized between March 1974 and July 1975, when 76 of 113 workers (57 percent) at the Hopewell, Virginia plant showed signs and symptoms of intoxication that included skin rash, nervousness, brain damage (loss of memory and behavioral changes), weight loss, pleuric and joint pains, oligospermia, tremors, ataxia, opsoclonus (rapid twitching of the eyes in horizontal, vertical, oblique, or rotary directions), and sterility (187). See also Cannon et al. (188), Cohn et al. (189), and Huff and Gerstner (119). The mean latency period between the start of employment and onset of symptoms was 6 weeks. When the effects were recognized, a detailed report was prepared in 1976 by a team of investigators of the U.S. Public Health Service (191). The symptoms persisted for as long as 6 months after termination of employment. The mean blood Kepone level for workers with illness was 2.53 ppm, and for those without disease, 0.60 ppm.

Smith (7) reviewed clinical and pathology findings from poisoned workers. From biopsied specimens it was determined that sural nerves displayed microscopic damage primarily to nonmyelinated and smaller myelinated fibers, sparing the larger myelinated fibers. In addition, skeletal muscle and liver samples revealed changes that were attributable to Kepone exposures.

Guzelian (192) has written an in-depth review of the comparison of toxic responses to Kepone in animals and humans.

### 7.4.2 Nonoccupational Exposure

The U.S. Department of Agriculture (1977) reported 56 incidents of nonoccupational human exposure to Kepone. Children under 5 years of age were exposed in 52 of the incidents, two incidents involved adults, and two cases were of unspecified age. All but nine of the children suffered Kepone exposure at home, primarily by devices used for the control of ants and cockroaches (193).

Blood Kepone levels for workers in nearby businesses and for residents of a community located within a distance of 1.6 km of the Kepone manufacturing plant ranged from undetectable to 32.5 ppb. There was no apparent association between the frequency of symptoms and proximity to the plant. Illnesses attributable to kepone were found in two wives of Kepone workers (188, 190).

It was estimated that consumption of seafood (finfish, crabs, oysters) caught in the James River and Chesapeake Bay led to an "average daily intake on the order of one microgram of Kepone involving 5 to 10 million people" (187). According to calculations (of Lu and Deichmann), the estimated daily intake of Kepone from finfish from the Chesapeake–James River waters during the years 1975 through 1976 were 0.3 µg for adults of the general population, and 2.1 µg/day for those whose diet consisted primarily of seafood (194).

Kepone residues in the tissues of the general U.S. population have apparently not been reported. Members of the EPA collected 298 samples of mothers' milk in nine southern states in 1976. Samples from three states (Alabama, Georgia, and North Carolina) showed Kepone levels ranging from less than 1 ppm to 5.8 ppb. Possible sources of exposure included the spraying of mirex for the control of fire ants and the subsequent degradation of mirex into Kepone (195) and the direct exposure to ant or cockroach traps.

### 7.5 Absorption, Metabolism, and Excretion

Kepone is absorbed from the gastrointestinal tract and by way of the lungs and skin. It is retained primarily in the liver but also in the brain, kidneys, and fat. Blanke et al (196) identified the conjugate of decachlorooctahydro1,3,4-metheno-2$H$-cyclobuta($c,d$)pentalen-2-ol (chlordecone alcohol) in fecal specimens from patients diagnosed as suffering from Kepone poisoning. This metabolite is a result of reduction of the ketonic group by an aldo–keto reductase enzyme that was identified in the gerbil, rabbit, and human but not in rodents (197).

Kepone crosses the placenta of mice and accumulated in the fetus. It is apparently not metabolized in the mouse (198). Kepone is rapidly excreted in the feces, with traces in the urine and milk of humans and cows (199).

According to Boylan et al. (200), oral administration of Cholestyramine (an anion-exchange resin) markedly increased the excretion of Kepone. Richter et al. (201) administered 8 percent light liquid paraffin for 24 days to rats fed previously a diet containing 3.3 ppm of $^{14}$C-Kepone for 3 days. This resulted in a significantly increased excretion of Kepone in feces compared to controls. The paraffin-tested rats had significantly lower concentrations of radioactivity in 14 of the 18 tissues analyzed. Excretion of radioactivity in the urine was of minor importance.

### 7.6 Biologic Magnification

Kepone is susceptible to transfer from particulate or food-web processes to higher trophic levels (199). Certain aquatic plants, animals, and fish readily accumulate the compound. The bioconcentration in finfish was found to be species specific.

### 7.7 Decontamination

Analysis of soil samples indicated that in 1978 approximately 450 kg of residual Kepone existed around Hopewell, Virginia (202). Biodegradation of Kepone in the James River has been insignificant.

A process is being developed for the destruction of Kepone in which the compound is vaporized at a low temperature and then passed for 1 sec through a high-temperature quartz tube at about 1000°C. Preliminary results are satisfactory. Other cleanup procedures under consideration range from dredging and destroying the compound by ozonation and ultraviolet light, adding activated carbon to the river to decrease the availability of Kepone to the food chain, and soil fixation with the sediment.

### 7.8 Cancer

In chronic feeding studies with Wistar strain rats conducted at the Medical College of Virginia, Kepone was found to be hepatotoxic and nephrotoxic. Liver tumors were noted in some of the animals. Representative liver sections submitted to four pathologists provided diagnoses for tumors that ranged from "hyperplasia," "suspicious carcinoma," "no carcinoma," to "carcinoma." Based on these studies, the carcinogenicity of Kepone in the rat remained debatable (181, 194).

In 1976, NCI (203) published a report on the carcinogenesis of technical-grade chlordecone (Kepone) with the following summary:

> A carcinogenesis bioassay of technical grade chlordecone (Kepone) was conducted using Osborne–Mendel rats and B6C3F1 mice. Chlordecone was administered in the diet for 80 weeks at two dose levels. The rats were sacrificed at study week 112 and the mice at week 90. The starting dose levels were 15 and 30 ppm for male rats, 30 and 60 for female rats, 40 ppm for male mice and 40 and 80 for female mice. As these dose levels were not well tolerated, the dose levels were reduced during the course of the experiment such that the average dose levels were as follows: 8 and 24 ppm for the male rats, 18 and 26 ppm for female rats, 20 and 23 ppm for male mice and 20 and 40 for female mice.
>
> Clinical signs of toxicity were observed in both species, including generalized tremors and dermatologic changes. A significant increase ($p < .05$) was found in the incidence of hepatocellular carcinomas of high dose level rats and of mice at both dose levels of chlordecone. The incidence in the high dose groups were 7 percent and 22 percent for male and female rats (compared with 0 in controls for both sexes) and 88 percent and 47 percent for male and female mice (compared with 16 percent for male room controls and 0 in females); for the low dose groups of mice the incidences were 81 percent for males and 52 percent for females. In addition, the time to detection of the first hepatocellular carcinoma observed at death was shorter for treated than control mice and, in both sexes and both species, it appeared inversely related to the dose. In chlordecone-treated mice and rats extensive hyperplasia of the liver was also found. The incidence of tumors other than in the liver for chlordecone-treated groups did not appear significantly different from controls.

The significance of the hepatocellular carcinomas noted in the experimental rats and mice is not clear. According to the report, the term "hepatocellular carcinoma" was used to diagnose proliferative lesions of the liver because, in the judgment of the pathologists, these lesions "had the potential or the capacity for progressive growth, invasion, metastasis, and for causing the death of the host." However, neither in the experimental rats nor mice were vascular invasion and/or metastases observed "in the material examined." It appears, therefore, that the diagnosis of hepatocellular carcinoma for proliferative lesions was of uncertain validity.

Additional comments regarding this NCI report by Newberne (204) include the following:

> It is unfortunate that dose levels were not determined more accurately before the chronic studies were initiated. One can, therefore, only speculate about the influence of early injury. . . . I do not, of course, agree with a diagnosis of "neoplastic nodule" since I don't know what this means. . . . No vascular invasion nor metastases were observed in the material examined. . . . I am thus at a loss to know whether any of the tumors were cancer or simply nodular hyperplasia.

As with many other chlorinated pesticides, Kepone appears to possess tumor promotional activity following nitrosamine initiation, as well as inhibitory action on intercellular communication (205, 206).

Based on the Medical College of Virginia, the NCI, and other reports of human and animal studies, it has been established that Kepone, in toxic doses, produces severe liver and kidney injury and damage to the CNS and reproductive organs. The risk of human cancer has not been established.

### 7.9 Hygienic Standards for Permissible Exposures

No TLVs or tolerances have been established for Kepone. The production of Kepone was discontinued in the United States in 1975 after closing of the plant in Hopewell, Virginia.

## 8 HEXACHLOROCYCLOHEXANE

The formula is $C_6H_6Cl_6$. 1,2,3,4,5,6-Hexachlorocyclohexane occurs as mixed isomers, BHC or benzene hexachoride [CAS #608-73-1]. The gamma isomer (CAS #58-89-9) is called lindane.

Lindane

Although BHC is a complex mixture of isomers, this review focuses upon the effects of technical lindane, which is composed of 99 percent gamma isomer.

## 8.1 Uses and Industrial Exposure

Lindane is used as an insecticide for the control of insects on cotton and other foliar plants, for soil and seed treatment of fruit and vegetable crops, and for control of termites. Lindane has been found effective for the control of insects on livestock and pets, and is important to public welfare in the control of lice, mosquitoes, and flies that have become resistant to DDT. In humans, lindane is used in lotions, creams, and shampoos to treat scabies and lice.

Serious health-related exposures have not been reported from the use, manufacturing, and application of technical lindane.

## 8.2 Physical and Chemical Properties

BHC occurs as eight isomers, with four of these found in sufficient quantity to be considered of insecticidal importance. A crude pesticidal mixture of six isomers is available under the common name benzene hexachloride (BHC). The most effective isomer is the gamma, and this, purified to 99 percent, is available under the common name lindane. Properties of technical lindane and pure gamma isomer are indicated below:

| | |
|---|---|
| Physical state | Off-white powder |
| Molecular weight | 290.80 |
| Specific gravity | 1.89 (pure isomer) |
| Melting point | 112.5 C (pure isomer) |
| Vapor pressure | $9.4 \times 10^{-6}$ mm at 20°C |
| Solubility | 17 ppm in water; 6.4% in alcohol; moderately soluble in acetone, benzene, chlorinated hydrocarbon solvents; poorly soluble in kerosene |
| Partition coefficient (log octanol/water) | 3.3 |

## 8.3 Effects in Animals

The primary response to the alpha and gamma isomers of BHC is stimulation of the CNS, resulting in hyperexcitability and convulsions, whereas the beta and delta isomers cause central depression. Detailed reviews have recently been published on the toxic effects of this chemical (7, 207).

The toxicity of BHC varies and is related to the amount of lindane present, which is the most acutely toxic isomer. Chronically, lindane is the least likely to bioaccumulate in exposed animals. The alpha, beta, and delta isomers have a low degree of acute toxicity, but are retained in body tissues for a longer period than lindane.

Micropathological changes are seen in the liver and kidneys and, to some degree, in the lungs following chronic exposure to toxic doses of these materials.

### 8.3.1 Acute Toxicity

Lehman (23) reported the oral $LD_{50}$ of lindane for rats as 125 mg/kg body weight, whereas Hayes (6) reported the oral $LD_{50}$ for male rats to be 88 and for female rats 91-mg/kg. The oral $LD_{50}$ for the rabbit ranges from 50 to 60 mg/kg; the oral $LD_{50}$ for the guinea pig has been given as 127 mg/kg.

Lehman (135) also found that the approximate $LD_{50}$ for rabbits, following application of lindane in dry form, was greater than 4000 mg/kg. Severe signs of intoxication and moderate skin irritation were observed at this level, but the animals survived. When lindane was applied as a 2 percent solution in dimethyl phthalate, the approximate $LD_{50}$ was greater than 188 mg/kg. As a 1 percent solution in a vanishing cream base, the approximate $LD_{50}$ was 50 mg/kg. Concentrations up to 4 ppm of BHC have been found in the milk of ewes after they were dipped with Entomoxan (0.5 percent BHC) (208).

Smith (7) reviewed acute toxicity results, and showed that dermal applications of lindane to rats were 10 to 50 percent less toxic based on lethality information. The review also indicates that young rabbits, calves, and lambs are particularly susceptible to dermally applied lindane.

### 8.3.2 Chronic Toxicity

The chronic toxicity for isomers of BHC decreases in the order beta > alpha > gamma > delta and is directly related to their tissue retention, and inversely to rates of metabolism (7). This contrasts with the order of acute toxicities, which are in the decreasing order of gamma > alpha > delta > beta.

Lehman (209) indicated that the highest tolerated daily dose of lindane in rats over a 2-year period without effect was 5 mg/kg or a dietary feeding level of 50 ppm (210). According to Fitzhugh et al. (211), rats can tolerate long-term feeding of diets containing 800 ppm of lindane, but showed some liver enlargement at this level. A comparable response was observed from BHC at a level of 100 ppm in the diet.

As with other chlorinated pesticides, lindane exhibits substantial neurotoxic effects following acute or chronic exposures. The symptoms may include violent epileptiform convulsions, rapid in onset and quickly leading to either death or recovery. Smith (7) has thoroughly reviewed mechanistic testing on this compound. There is evidence that chronic lindane exposure may lead to decreases in muscarinic receptors in the brain, increases in calcium uptake in rat brain synaptosomes, and possible inhibition of Na,K-ATPase-linked calcium extrusion in neural tissue. There is also an indication that lindane interferes with transmission of γ-aminobutyric and (GABA) transmission at neuronal junctions.

Two opposite effects of lindane on the threshold from pentylenetetrazol-induced convulsions have been reported (212). The threshold is lowered a few hours after the administration of a single small oral dose of lindane to mice (213). In earlier experiments, Herken and Klempau (214) and Coper at al. (215) could show that rats given large, near-lethal doses of lindane are protected from the effects of convulsive doses of pentylenetetrazol when examined a couple of days after the lindane administration.

In a more recent study Hulth et al. (212) studied the convulsive properties of lindane on the convulsive threshold of pentylenetetrazol and brain content of GABA in the mouse. They found that following the administration of a single dose of lindane, the threshold first decreases and then, after 2 days, increases above the normal level. None of the tested metabolites of lindane altered the convulsive threshold. When lindane elevated the convulsive threshold it also elevated the content of GABA in the brain. Lindane elevated both convulsive threshold and the GABA content. The smallest single oral dose of lindane lowering the convulsive threshold for pentylenetetrazol in the mouse, 3 hr after administration, corresponded to a lindane concentration of 56 ppb in whole blood.

### 8.3.3 Reproductive Toxicity and Teratology

Tests for teratogenicity were conducted by Palmer et al. (216) using groups of rabbits and rats. Lindane was administered to these animals during pregnancy at dosages of 5, 10, or 20 mg/kg daily for 12 days. Based on body weight changes, litter size, and mean fetal weights, and other observations, no evidence of teratology or toxicity to the embryos by lindane was noted.

Smith (7) reviewed a number of studies that indicated that lindane possesses estrogenic activity in rodents, with the effect of causing atrophy of seminiferous tubules and Leydig cells in males. In addition, female rats exhibited reduced sexual behavior at proestrus via an unidentified antiestrogenic mechanism. Citations for standard two-generation reproduction studies were not located.

### 8.3.4 Airborne Exposures

Queen (217) reported on the use of lindane by vaporization. Canaries were exposed to vapors of lindane (liberated as a cloud when a solid formulation was heated) in air concentration of 0.34 mg/l; the birds died after 6 to 16 days of exposure. Some home lindane-generating devices produced air concentrations that exceeded those in manufacturing plants. The possibility of absorption of lindane or BHC by way of the lungs exists also when these materials are used as dusts or liquid sprays.

### 8.3.5 Absorption, Metabolism, and Excretion

Lindane is stored in rodents (body fat) at a concentration similar to that found in the diet (218). Maximum storage is accomplished within 6 weeks and disappears from the fat depot within 3 weeks (219).

Van Asperen and Oppernoorth (220) reported that 1 mg of lindane injected subcutaneously in the mouse was eliminated in 4 days. After an intravenous injection of lindane, the compound was not noted in the urine or feces, but traces were found in several organs.

The metabolic pathway of the gamma isomer of BHC (lindane) in rats is highly complex. Metabolic routes include ring hydroxylation, dechlorination and/or reduction with double bond formation, and others that lead to chlorinated phenols, chlorinated mercaptobenzenes, and various cyclohexenes and cyclohexadienes. Phenolic and hydroxy metabolites typically form excretable glucuronide and sulfate conjugates (7).

Compared to several other chlorinated hydrocarbon pesticides, lindane is rapidly metabolized in the mammalian body (221). A degradation scheme for lindane, based on experiments in the rat, has been proposed by Engst et al. (222).

Seidler (223) studied the distribution, metabolism, and excretion of $^{14}$C-lindane in rats after they had become adapted to lindane administered orally. They found that body fat, kidneys, and the musculature were the primary sites of accumulation. Marked differences in radioactivity levels were found between the cortex, brainstem, and cerebellum.

The metabolite $\gamma$-2,3,4,5,6-pentachlorocyclohexane amounted to 1 to 3 percent of the hexane extract after 24 hr, and 5 to 8 percent after 72 hr, in certain organs. Feces, urine, and organs contained large amounts of conjugates and highly polar, hexane-soluble metabolites. The level of extractable (i.e., unbound) $^{14}$C activity in the urine was 8 percent, and in the feces, 43 percent. The level of extractable activity was 35 percent in both urine and feces after 72 hr. An additional 3 and 8 percent after 24 hr, and 18 and 14 percent, respectively, after 72 hr, were present as glucuronides, and the remainder as unidentified water-soluble conjugates. In the urine, 1 to 3 percent of lindane was present in free form; 3 to 6 percent of the total lindane was extracted unmetabolized. Half of the lindane administered was excreted within 3 to 4 days (223).

In 1975, Chadwick et al. (224) presented evidence for the hydrogenation of lindane by the hepatic mixed function oxidase system. A few years later, in 1978, Chadwick et al. (225) identified three lindane metabolites in the urine of Sprague–Dawley rats fed diets containing 400 ppm of lindane. Two of these, M-1 and M-2, were excreted primarily as glucuronide and sulfate conjugates. Mass spectral studies tentatively identified M-2 as a configurational isomer of 2,4,5,6-tetrachloro-2-cyclohexan-1-ol (2,4,5,6-TCCOL) and M-1 as 2,3,4,6-TCCOL.

The release of enzyme by rat liver lysosomes was studied by Okazaki and Kimura (226) because the lysosomes play a role in the elimination of foreign compounds. When damaged by chemicals or radiation, the lysosomal membrane breaks, releasing enzymes. It was found that BHC produced severe damage to lysosomes in vitro and in vivo and that the ratio of "activity of released enzyme" to "total enzyme" of $\beta$-glucuronidase, after BHC application in vitro, amounted to almost 100 percent after 90 min at 37°C. In vitamin E-deficient animals, the lysosomal membrane became more unstable, and the damage by chemicals increased.

## 8.4 Human Experience

### 8.4.1 Tissue Concentrations

From 1965 to 1970, the ADI of BHC (primarily lindane) averaged 5.6 μg for a 70-kg adult individual in the United States (109). Body fat concentrations in the U.S.

adult population groups generally ranged from 0.2 to 0.6 ppm (45). Fiserova-Bergerova et al. (227), who analyzed (by gas–liquid chromatography) the body fat of stillborns and fetuses in Dade County, Florida (where pesticides are used more vigorously than in most parts of the United States), recorded concentrations of BHC of 0.14 ppm, β-BHC of 0.26 ppm, and γ-BHC of 0.11 ppm. Little or no lindane or BHC was detected in the blood of individuals not occupationally exposed to lindane (45).

In certain parts of Argentina, crude BHC and BHC containing a high ratio of the beta isomer were used extensively by the general population. As a result, people of all ages were found to have approximately 16 times the blood concentration of β-BHC found in the U.S. population. Similar high blood levels were found in Asiatics, principally in Japanese and Formosans (228).

In general, human milk in the United States contained only traces of BHC, even during the height of lindane or BHC use. During the period 1972 to 1977, concentrations in the milk of lactating mothers in Okinawa increased because of the increased consumption of pork and American and Australian beef. In 1972 total BHC in 100-ml samples of human milk was 0.029 ppm; in 1977 it had risen to 0.075 ppm (229).

*8.4.2 Intoxications*

Danopoulos et al. (230) reported on clinical cases of exposure to technical BHC in Greece. A 40 percent dry powder of BHC, or the same powder mixed with either water or a petroleum solvent, was sprinkled on the ground, onto walls, and over clothing, bedding, and bodies of a local population. Seventy-nine persons were affected, 18 seriously. Five were treated in a clinic; all survived. Symptoms were related primarily to the gastroenteric tract and the central nervous system. There were also indications of electrocardiographic and hematologic changes.

The dermal application of 1 percent lotions of lindane for the treatment of scabies resulted in severe intoxication in some children and infants. Marked mental and motor retardation was noted 2 days after treatment in a 4-month-old child (231). Local application of 1 percent lindane proved particularly dangerous following a hot soapy bath, because this increased the percutaneous absorption of the compound (232–234).

Accidental ingestion by a 1-year-old boy of one teaspoonful of a 1 percent lindane solution, in addition to the local application of this solution for the treatment of scabies, resulted in irritability, hyperactivity, and vomiting, followed by recovery in the hospital (235).

Recently, eight incidents of nonfatal BHC poisoning (grand mal epilepsy) were reported by Nag et al. (236). The intoxications occurred in a family in a village in Uttar, India. These people had ingested wheat bread containing 0.005 percent BHC. It was found that the wheat flour had been contaminated with BHC during storage.

*8.4.3 Occupational Exposure*

According to Deichmann (237) in *Pesticides and the Environment* (reproduced with permission by Symposia Specialists, Inc. Miami, Florida,

Milby et al. (238) were apparently the first to establish that the presence of lindane in blood is a reflection of recent lindane absorption. Among groups occupationally exposed to this pesticide, blood concentrations did not appear to increase with increasing duration of exposure. However, blood lindane concentrations did increase as air and skin exposure became more intense. Air concentrations of lindane equal to 11 to 1170 micrograms/m$^3$, in operations which also provided ample skin contact, to which men were exposed during working hours for years, were related to blood lindane concentrations of 30.6 ppm (6.0 to 93.0 ppb).

Lindane has been implicated as an agent that may have been responsible for causing blood dyscrasias, primarily aplastic anemia (239–241). As Milby et al. (238) pointed out, in most cases the hematologic diagnosis is supported by convincing evidence, whereas the relationship to specific lindane exposure is less well documented.

Samuels and Milby (242) examined the possible adverse effects from working a few weeks to several years in lindane processing plants. Mean lindane blood concentrations of these workers were as follows: nonproduction workers, 0.93 ppb; production workers in two plants with little or no skin contact, 4.1 and 4.6 ppb, respectively; and workers in a plant where ample skin contact was recognized, 30.6 ppb. The study included seven children and one adult who lived in homes in which the mean lindane level was 2.2 ppb. The authors could not detect any clinical symptomatology or physical evidence of disease that was attributable to lindane exposure. However, because certain isolated abnormalities were noted in occasional instances, they decided to conduct an additional study in which an exposed population group was matched by age, sex, and race with individuals who suffered no exposure to lindane.

The Milby and Samuels study (238, 243) dealt with 80 individuals, 40 of whom were employed in a plant in which lindane was processed. All were Caucasians; 12 were female. Each of these people was matched with a control having no occupational exposure. The authors pointed out that only blood lindane concentrations were found to differ in the control and occupational groups; they were 0.1 ppb and 11.9 ppb, respectively. Furthermore, "Significant differences were not observed in regard to blood uric acid, alkaline phosphathase, platelet count, hemoglobin and lymphocyte, eosinophil and monocyte counts. There was no evidence of pancytopenia with reticulopenia, the hallmark of hypoplastic or aplastic anemia." They did not rule out the possibility of an idiosyncratic or a hypersensitive response in a few individuals. Although they questioned meaningful conclusions that are based on statistical analysis of relatively small groups of people, they conclude that "within the limitations of this study, lindane does not appear to produce hematologic disorders on a basis of toxic suppression of hematopoiesis" (243).

## 8.5 Cancer

A bioassay of lindane of the NCI was published in 1977 (244). The compound was administered for 80 weeks to groups of Osborne–Mendel rats and B6C3F1 mice. Time-weighted dietary levels for male rats were 236 or 472 ppm, for female rats

135 or 270 ppm, and for male and female mice 80 or 160 ppm. Results were as follows: (1) In rats, no tumors occurred at a statistically significant incidence in the treated groups of either sex. (2) In mice, the incidence of hepatocellular carcinoma in low-dose males was significant when compared with that in the pooled controls (controls 5/49, low dose 19/49, $p = .001$). This finding, by itself, is insufficient to establish the carcinogenicity of lindane. The incidence of hepatocellular carcinoma in high-dose male mice (9/46) was not significantly different from that in matched (2/10) or pooled controls. It is concluded that under the conditions of this bioassay lindane was not carcinogenic for Osborne–Mendel rats or B6C3F1 mice.

As with most other chlorinated insecticides, lindane has been shown to exhibit tumor promotional activity, possibly via inhibition of intercellular communication (7). Lindane exposures do inhibit formation of liver tumors following exposures to aflatoxin B1, suggesting an activity other than tumor promotion for lindane. Enhancement of enzymatic deactivation of aflatoxin via induction of microsomal enzymes may play a role in this phenomenon.

The combination of lindane's lack of significant mutagenic activity, its tumor promotional properties, and the lack of epidemiologic evidence suggest that lindane poses a minimal to nonexistent risk of cancer to humans under reasonable and prescribed use conditions.

### 8.6 Hygienic Standards for Permissible Exposures

The threshold limit of lindane (skin) adopted by the ACGIH was a TLV–TWA of 0.5 mg/m$^3$ (75). The IDLH concentration of lindane, according to the NIOSH–OHSA Pocket Guide to Chemical Hazards, is 1000 mg/m$^3$ (76). According to the 1978 FAO/WHO Standards Programme Codex Committee on Pesticides Residues, the ADI of lindane is 0.01 mg/kg body weight. The MRLs of lindane for agricultural commodities range from 0.05 to 3 mg/kg (77).

## REFERENCES

1. World Health Organization, Official Records No. 190, Geneva, April 1971.
2. W. B. Deichmann and W. E. MacDonald, *Ecotoxicol. Environ. Safety*, **1**, 89 (1977); *Am. Ind. Hyg. Assoc. J.*, **37**, 495 (1976).
3. U.S. EPA, "Suspended, Cancelled, and Restricted Pesticides," NTIS PB92-231240, 1990.
4. R. T. Meister, *1992 Farm Chemicals Handbook*, Meister Publishing Co., Willoughby, OH, 1992.
5. K. A. King, E. L. Flickinger, and H. Hildebrand, *Pestic. Monit. J.*, **12**(1), 16 (1978).
6. W. J. Hayes, Jr., *Clinical Handbook of Economic Poisons*, U.S. Dept. of Health, Education and Welfare, Communicable Disease Center—Toxicology Section, Atlanta, GA, 1963.
7. A. G. Smith, in *Handbook of Pesticide Toxicology*, Vol. 2, W. J. Hayes and E. R. Laws, Jr., Eds., Academic Press, New York, 1991.

8. "Brief for Respondent in Environmental Defense Fund," *Fed. Reg.*, **439**, 584 (1971).
9. Food and Drug Administration, *Fed. Reg.*, **37**, 13369 (July 7, 1972).
10. Environmental Protection Agency, Before the Administration, in re. Shell Chemical Company, Registrants FIFRA, Dockets No. 145, Recommended Decision, September 20, 1974.
11. U.S. EPA, "Health Effects Assessment Summary Tables," Annual FY-1991, NTIS PB91-921199, 1991.
12. U.S. EPA, "List of Food Use Pesticides Evaluated for Carcinogenicity, compiled as of June 1991," *Chem. Regul. Rep.*, 1000 (Aug. 28, 1992).
13. National Toxicology Program, *6th Annual Report on Carcinogens*, U.S. Public Health Service, 1991.
14. *IARC Monographs on the Evaluation of Carcinogenic Risks to Humans*, Suppl. 7, International Agency for Research on Cancer, World Health Organization, 1987.
15. H. F. Kraybill, U.S. Department of Health, Education and Welfare, National Institute of Health, Pest Control, Lecture, Houston, TX, Oct. 22, 1975.
16. "Mouse Liver Tumors-Relevance to Human Cancer Risk," Symposium of the European Society of Toxicology, *Arch. Toxikol.*, Suppl. 10 (1987).
17. "The Relevance of Mouse Liver Hepatoma to Human Carcinogenic Risk," The Nutrition Foundation, Washington, DC, 1983.
18. "Mouse Liver Neoplasia" symposium, proceedings published in *Mouse Liver Neoplasia*, J. A. Popp, Ed., Hemisphere, New York, 1984.
19. P. A. Neal, W. F. von Oettingen, W. W. Smith, R. B. Malmo, R. C. Dunn, H. E. Morann, T. R. Sweeney, D. W. Armstrong, and W. C. White, *Publ. Health Rep. Suppl.*, 177 (1944).
20. G. R. Cameron and F. Burgess, *Brit. Med. J.*, **1**, 865 (1945).
21. W. B. Deichmann, S. W. Witherup, K. V. Kitzmiller, and F. F. Heyroth, "The Toxicity of DDT," Kettering Laboratory, College of Medicine, University of Cincinnati, OH, 1950.
22. G. Woodward, A. A. Nelson, and H. O. Calvary, *J. Pharmacol. Exp. Ther.*, **82**, 152 (1944).
23. A. J. Lehman, *Assoc. Food Drug Offic. U.S. Q. Bull.*, **15**, 122 (1951).
24. O. G. Fitzhugh and A. A. Nelson, *J. Pharmacol. Exp. Ther.*, **89**, 18 (1947).
25. J. R. Treon and F. P. Cleveland, *J. Agric. Food Chem.*, **3**, 402 (1955).
26. E. P. Laug, A. A. Nelson, O. G. Fitzhugh, and F. M. Kunze, *J. Pharmacol. Exp. Ther.*, **98**, 268 (1950).
27. Toxicological Profile for DDT, DDE, and DDD, Agency for Toxic Substances and Disease Registry, U.S. Dept. of Commerce, NTIS PB90-182171, 1989.
28. J. H. Draize, A. A. Nelson, and H. O. Calvary, *J. Pharmacol. Exp. Ther.*, **82**, 159 (1944).
29. C. F. Rothe, A. M. Mattson, R. M. Nueslein, and W. J. Hayes, Jr., *Arch. Ind. Health*, **16**, 82 (1957).
30. D. P. Morgan and C. C. Roan, *J. Occup. Med.*, **15**(1), 26 (1973).
31. C. Roan, D. Morgan, and E. H. Paschal, *Arch. Environ. Health*, **22**, 309 (1971).
32. H. B. Matthews, J. J. Domanski, and F. E. Guthrie, *Xenobiotica*, **6**(7), 425 (1976).
33. S. R. Khambata, *Pesticides*, **8**(12), 59 (1974).

34. A. Nair and M. K. K. Pillai, *Sci. Total Environ.*, **121**, 145 (1992).
35. M. F. Ortelee, *A.M.A. Arch. Ind. Health*, **18**, 433 (1958).
36. E. R. Laws, A. Curley, and F. J. Biros, *Arch. Environ. Health*, **15**, 766 (1967).
37. W. J. Hayes, Jr., in *Pharmacology and Toxicology of DDT:DDT Insecticides*, P. Muller, Ed., Vol. 2, Birkhauser, Basel, 1955, p. 11.
38. W. J. Hayes, *Toxicology of Pesticides*, Williams and Wilkins, Baltimore, 1975.
39. W. J. Hayes, Jr., W. F. Durham, and C. Cueto, *J. Am. Med. Assoc.*, **162**(9), 890 (1956).
40. W. J. Hayes, Jr., W. E. Dale, and C. T. Pirkle, *Arch. Environ. Health*, **22**, 119 (1971).
41. I. F. G. Ensberg, A. deBruin, and R. L. Zielhuis, Coronel Laboratory, University of Amsterdam, personal communication, 1971.
42. A. Poland, D. Smith, R. Kuntzman, M. Jacobson, and A. H. Conney, *Clin. Pharmacol. Ther.*, **11**(5), 724 (1970).
43. H. F. Kraybill, Presentation at Am. Meat Sci. Assoc. Conf., June 22–24, 1966, Cornell University, Ithaca, NY.
44. M. Gartrell, J. Craun, and D. Podrebarac, *J. Assoc. Offic. Anal. Chem.*, **69**, 146 (1986).
45. W. B. Deichmann, *Environmental Problems in Medicine*, W. D. McKee, Ed., Charles C. Thomas, St. Louis, MO, 1974, pp. 347–420.
46. W. J. Hayes, Jr., *Toxicol. Appl. Pharmacol.*, **38**, 19 (1976).
47. J. H. Thorstenson and H. W. Dorough, *Tob. Sci.*, **20**(5), 25 (1976).
48. H. W. Dorough and Y. H. Atallah, *Bull. Environ. Contam. Toxicol.*, **13**(1), 101 (1975).
49. F. E. L. Fowler, *J. Agric. Food Chem.*, **1**, 469 (1953).
50. T. H. Jukes, *Arch. Intern. Med.*, **138**(5), 772 (1978).
51. H. F. Kraybill, "Pesticide Toxicity and Potential for Cancer: A Proper Perspective," presented at the National Pest Controllers Association Convention, Houston, Texas, October 1975.
52. "Cancer Facts & Figures—1992," American Cancer Society, Atlanta, GA, 1992.
53. L. Tomatis and V. Turusov, *GNN*, **17**, 219 (1975).
54. S. K. Kashyap, S. K. Nigam, A. B. Karnik, R. C. Gupta, and S. K. Chatterjee, *Int. J. Cancer*, **19**(5), 725 (1977).
55. M. D. Reuber, *Sci. Total Environ.*, **10**(2), 105 (1978).
56. IARC, *Some Organochlorine Pesticides*, Vol. 5, International Agency for Research on Cancer, World Health Organization, Lyon, France, 1974.
57. E. Thorpe and A. I. T. Walker, *Food Cosmet. Toxicol.*, **11**, 433 (1973).
58. J. R. M. Innes, B. M. Ulland, M. G. Valerio, L. Petrucelli, L. Fishbein, E. R. Hart, A. J. Pallatta, R. R. Bates, H. L. Falk, J. J. Gart, M. Klein, I. Mitchell, and J. Peters, *J. Natl. Cancer Inst.*, **42**, 1101 (1969).
59. L. Rossi, M. Ravera, G. Repetti, and L. Santi, *Int. J. Cancer*, **19**, 179 (1977).
60. IARC Monographs, Vols. 1–20, International Agency for Research Monographs Supplement, Lyon, September 1979.
61. J. R. P. Cabral, R. K. Hall, and P. Shubik, abstract presented at 20th Congress, European Society of Toxicologists, West Berlin, June 25–28, 1978.
62. NCI, "Bioassays of DDT, TDE and *p,p'*-DDE for Possible Carcinogenicity, CAS No.

50-29-3; 72-54-8, 72-55-9," NCI-CG-TR-131, U.S. Dept. of Health, Education and Welfare, Public Health Service, 1978.
63. C. Peraino, R. J. M. Fry, E. Staffeldt, and J. P. Christopher. *Cancer Res.*, **35**, 2884 (1975).
64. G. M. Williams, S. Telang, and C. Tong, *Cancer Lett.*, **11**, 339 (1981).
65. WHO, "DDT and Its Derivatives," published under joint sponsorship of United Nations Environmental Programme and World Health Organization, Geneva, 1979, pp. 106–107, 114–115.
66. K. W. Jager, *Aldrin, Dieldrin, Endrin, and Telodrin: An Epidemiological and Toxicological Study of Long Occupational Exposure*, Elsevier, New York, 1970.
67. A. Blair et al., *J. Natl. Cancer Inst.*, **71**, 31 (1983).
68. Wong et al., *Brit. J. Ind. Med.*, **41**, 15 (1984).
69. W. Buselmaier, G. Roehrborn, and P. Propping, *Mutat. Res.*, **21**, 25 (1973).
70. E. Vogel, *Mutat. Res.*, **16**, 157 (1972).
71. J. Yoder, M. Watson, and V. V. Benson, *Mutat. Res.*, **21**, 25 (1973).
72. S. Fabro et al., *Am. J. Obstet. Gynecol.*, **148**, 929 (1984).
73. M. L. Keplinger, W. B. Deichmann, and F. Sala, "Effects of Combinations of Pesticides on Reproduction in Mice," in *Pesticides Symposia*, Halos and Associates, Miami, 1970, pp. 125–138.
74. V. Green, "Trace Substances in Environmental Health," D. Hemphill, Ed., *Proc. Univ. Missouri 3rd Ann. Conf. Trace Substances Environ. Health*, **2**, 183 (1969).
75. "TLVs, Threshold Limit Values and Biological Exposure Indices for 1992–1993," American Conference of Governmental Industrial Hygienists, Cincinnati, OH, 1992.
76. *NIOSH/OSHA Pocket Guide to Chemical Hazards*, U.S. Dept. of Health, Education and Welfare, National Institute for Occupational Safety and Health and U.S. Dept. of Labor, Occupational Safety and Health Administration, June 1990.
77. Codex Alimentarius Commission, Joint Food and Agricultural Organization of the United Nations World Health Organization, Codex Committee on Pesticide Residues, Tenth Session, The Hague, May 29–June 5, 1978.
78. W. L. Ball, K. Kay, and J. W. Sinclair, *A.M.A. Arch. Ind. Hyg. Occup. Med.*, **7**, 292 (1953).
79. F. P. Cleveland, *Arch. Environ. Health*, **13**, 195 (1966).
80. C. H. Kitselman, *J. Am. Vet. Med. Assoc.*, **123**, 28 (1953).
82. J. M. Bann, T. J. DeCino, N. W. Earle, and Y. P. Sun, *J. Agric. Food Chem.*, **4**, 937 (1956).
81. Toxicological Profile for Aldrin/Dieldrin, Agency for Toxic Substances and Disease Registry, U.S. Dept. of Commerce, 1988.
83. B. Lang, K. Frei, and P. Maier, *Biochem. Pharmacol.*, **35**, 3643 (1986).
84. J. C. Street, *Science*, **146**, 1580 (1964).
85. S. Chatterjee, A. Ray, S. Ghosh, K. Bhattacharya, A. Pakrashi, and C. Deb, *J. Endocrinol.*, **119**, 75 (1988).
86. A. D. Ottolenghi, J. K. Haseman, and F. Suggs, *Teratology*, **9**, 11 (1974).
87. W. B. Deichmann, W. E. MacDonald, E. Blum, M. Bevilacqua, J. Radomski, M. Keplinger, and M. Balkus, *Ind. Med.*, **39**, 37 (1970).

88. W. E. MacDonald, J. MacQueen, W. B. Deichmann, T. Hamill, and R. Copsey, *Int. Arch. Arbeitsmed.*, **26**, 31 (1970).
89. W. B. Deichmann and M. L. Keplinger, *Pesticides Symposia*, Halos and Associates, Inc., Miami, 1970, pp. 121–123.
90. W. B. Deichmann, W. E. MacDonald, D. A. Cubit, *Science*, **172**, 275 (1971).
91. F. Princi and G. H. Spurbeck, *Arch. Ind. Hyg. Occup. Med.*, **3**, 64 (1951).
92. E. Nelson, *Rocky Mt. Med. J.*, **50**, 483 (1953).
93. H. G. S. Van Raalte, *Ecotoxicol. Environ. Safety*, **1**, 203 (1977) and personal communication, 1979.
94. B. D. Mehotra, K. S. Moorthy, S. R. Reddy, and D. Desiah, *Toxicology*, **54**, 17 (1989).
95. O. G. Fitzhugh, A. A. Nelson, and M. L. Quaife, *Food Cosmet. Toxicol.*, **2**, 551 (1964).
96. B. B. Virgo and G. D. Bellward, *Can. J. Physiol. Pharmacol.*, **53**(5), 903 (1975).
97. P. C. Hurkat, *Indian J. Anim. Sci.*, **47**(10), 671 (1977).
98. A. S. Wright, D. Potter, M. F. Wooder, and C. Donninger, *Food Cosmet. Toxicol.*, **2**, 551 (1964).
99. A. S. Wright, C. Donninger, R. D. Greenland, K. L. Stemmer, and M. R. Zavon, *Ecotox. Environ. Safety*, **1**(4), 477 (1978).
100. N. Chernoff, R. J. Kavlock, J. R. Katherin, J. M. Dunn, and J. K. Haseman, *Toxicol. Appl. Toxicol.*, **31**, 302 (1975).
101. K. M. Dix and A. B. Wilson, Shell, Tunstall Lab. Rep., TLGR 0051.71, 1971.
102. K. L. Davison and J. L. Sell, *Arch. Environ. Contam. Toxicol.*, **7**(2), (1978); **7**(3), 369 (1978).
103. J. L. Sell, K. L. Davison, and D. W. Bristol, *Poultry Sci.*, **56**(6), 2045 (1977).
104. W. J. Hayes, Jr., *Publ. Health Rep. U.S.*, **72**, 1087 (1957).
105. H. Wolfe, W. F. Durham, and J. F. Armstrong, *Arch. Environ. Health*, **6**, 458 (1963).
106. Z. Jegier, *Arch. Environ. Health*, **8**, 670 (1964).
107. C. G. Hunter and J. Robinson, *Food Cosmet. Toxicol.*, **6**, 253 (1968).
108. H. G. S. Van Raalte, *Ecotoxicol. Environ. Safety*, **1**(29), 203 (1977).
109. R. E. Duggan, H. C. Barry, and L. Y. Johnson, *Science*, **151**, 101 (1966); *Pest. Monit. J.*, **11**(2), 2 (1977).
110. B. C. Turner, D. E. Glotfelty, and A. W. Turner, *Agric. Food Chem.*, **25**(3), 548 (1977).
111. I. Weisberger, D. Bieniek, J. Kohli, and W. Klein, *J. Agric. Food Chem.*, **23**(5), 873 (1975).
112. G. Reddy and M. A. Q. Khan, *Bull. Environ. Contam. Toxicol.*, **13**(1), 64 (1975).
113. W. B. Deichmann, W. E. MacDonald, E. Blum, M. Bevilacqua, J. Radomski, M. Keplinger, and M. Balkus, *Ind. Med. Surg.*, **39**, 426 (1970).
114. J. Song and W. C. Harville, Abstr., *Fed. Proc.*, **23**, 336 (1964).
115. A. I. T. Walker, D. E. Stevenson, J. Robinson, E. Thorpe, and M. Roberts, *Toxicol. Appl. Pharmacol.*, **15**, 345 (1968).
116. W. B. Deichmann, W. E. MacDonald, and F. C. Lu, *Toxicology and Occupational*

*Medicine*, W. B. Deichmann, Ed., Elsevier-North Holland, New York, 1977, pp. 407–413.

117. National Cancer Institute, "Bioassay of Aldrin and Dieldrin for Possible Carcinogenicity," Carcinogenesis Tech. Rep. Ser. No. 21 (1978), NCI-CG-TR-21, 1978.
118. National Cancer Institute, "Bioassay of Dieldrin for Possible Carcinogenicity," DHEW Pub. No. (NIH) 78-822, National Cancer Institute Carcinogenesis Technical Report Series, No. 22 NCI-CG-TR-21, 1978.
119. S. S. Epstein, "The Carcinogenicity of Dieldrin," Part I, *Sci. Total Environ.*, **4**, 1 (1975).
120. D. E. Stevenson, E. Thorpe, P. F. Hunt, and A. I. T. Walker, *Toxicol. Appl. Pharmacol.*, **38**(2), 247 (1976).
121. K. J. Davis and O. G. Fitzhugh, *Toxicol. Appl. Pharmacol.*, **4**, 187 (1972).
122. E. Thorpe and A. I. T. Walker, *Food Cosmet. Toxicol.*, **11**, 443 (1973).
123. J. R. P. Cabral, P. Shubik, S. A. Bronczyk, and R. K. Hall, *Proc. Assoc. Cancer Res. 68th Ann. Meet. Am. Soc. Clin. Oncology (13th Ann. Meet.), Denver, Colorado, May 1977*, Abstr. No. 111, p. 28.
124. J. E. Trosko, C. Jone, and C. C. Chang, *Mol. Toxicol.*, **1**, 83 (1987).
125. P. H. Ribbens, *Int. Arch. Occup. Environ. Health*, **56**, 75 (1985).
126. K. Bidwell, E. Weber, I. Nienhold, T. Connor, and M. S. Legator, *Mutat. Res.*, **31**(5), 314 (1975).
127. B. J. Dean, S. M. A. Doak, and H. Somerville, *Food Cosmet. Toxicol.*, **13**(1), 317 (1975).
128. J. G. Saha and Y. W. Lee, *Bull. Environ. Contam. Toxicol.*, **4**(5), 283 (1969).
129. T. B. Stauffer, *Natl. Tech. Inform. Serv. AD-A049*, 627 (1977).
130. Council for Agricultural Science and Technology (CAST) (Iowa State University, Ames, Iowa), *Farm Chem.*, **138**(13), 36 (1975); Rep. No. 59, September 15, 1976.
131. V. K. Smith, personal communication to Dr. John V. Osmum, Purdue University; through Ref. 160.
132. Agency for Toxic Substances and Disease Registry, "Toxicological Profile for Chlordane," NTIS PB90-168709, 1989.
133. K. M. Hyde and R. L. Falkenburg, *Toxicol. Appl. Pharmacol.*, **37**(3), 499 (1976).
134. L. Ingle, *Arch. Ind. Hyg. Occup. Med.*, **6**, 354 (1952).
135. A. Lehman, *Assoc. Food Drug Offic. U.S. Q. Bull.*, **16**, 3 (1952).
136. A. M. Ambrose, *Fed. Proc.*, **12**, 298 (1953).
137. L. Ingle, Chem. Specialties Mfrs. Assn. Proc., December, 1954.
139. D. W. DeLong and P. J. Ludwig, *J. Econ. Entomol.*, **47**, 1056 (1954).
140. R. D. Radeleff, G. T. Woodward, W. J. Nickerson, and R. C. Bushland, *USDA Tech. Bull.*, **1122**, 7 (1955).
141. A. M. Khasawinah, C. J. Hardy, and G. C. Clark, *J. Toxicol. Environ. Health*, **28**, 327 (1989).
142. N. H. Poonawalla and F. Korte, *J. Agric. Food Chem.*, **19**(3), 467 (1971).
143. H. M. Balba and J. G. Saha, *J. Environ. Sci. Health*, **B13**, 211 (1978).
144. H. W. Dorough and B. C. Pass, *J. Econ. Entomol.*, **65**(4), 976 (1972).
145. P. C. Oloffs, L. J. Albright, and S. Y. Szeto, Simon Fraser University, Burnaby, Canada; through *Pestic. Abstr. No. 5*, 78–1020 (1978).

146. J. L. Sanborn, R. L. Metcalf, W. N. Bruce, and P. Y. Lu, *Environ. Entomol.*, **5**(3), 533 (1976).
147. C. Henderson, A. Inglis, and W. L. Johnson, *Pestic. Monit. J.*, **5**, 1 (1971).
148. National Cancer Institute, "Bioassay of Chlordane for Possible Carcinogenicity," Carcinogenesis Tech. Rep. Ser. No. 8, CAS No. 57-74-9, NCI-CG-TR-8, 1977.
149. Velsicol: Summary of the Toxicological Evidence of Heptachlor and Chlordane presented in Administrative Hearings called by the United States Environmental Protection Agency, November 18, 1974–March 6, 1878.
150. National Cancer Institute, "Bioassay of Heptachlor for Possible Carcinogenicity," Carcinogenesis Tech. Rep. Ser. No. 9, CAS No. 76-44-8, NCI-CG-TR-9, 1977.
151. U.S. Food and Drug Administration, Internal memorandum K. J. Davis to A. J. Lehman, "Pathology Report on Mice fed Aldrin, Dieldrin, Heptachlor and Heptachlor Epoxide for 2 years" (G. Fitzhugh and K. J. Davis), July 19, 1965; through Ref. 179.
152. International Research & Development Corporation, Mattowan, Michigan, Report No. 163-084, "75 Percent Heptachlor Epoxide–25 percent Heptachlor, 2-Acetoamidofluorine, Eighteen Month Oral Carcinogenic Study in Mice," September 26, 1973; through Ref. 179.
153. S. Witherup, F. P. Cleveland, and K. Stemmer, "The Psychological Effects of the Introduction of Heptachlor Epoxide in Varying Levels of Concentration into the Diet of CFN Rats," unpublished report, The Kettering Laboratory, University of Cincinnati, November 10, 1959.
154. Pesticide Information Review and Evaluation Committee for the Advisory Center on Toxicology, Assembly of Life Sciences, National Research Council, National Academy of Science, Washington, DC, "An Evaluation of the Carcinogenicity of Chlordane and Heptachlor," October 1977; through Ref. 179.
155. Food and Drug Organization of the United Nations, "FAO Plant Production and Protection Paper 10 Rev. Pesticides Residues in Food—1977," Report of the Joint Meeting of the FAO Panel of Experts on Pesticide Residues and Environment and the WHO Expert Committee on Pesticide Residues, Rome, 1978.
156. A. M. Khasawinah and J. F. Grutsch, *Reg. Toxicol. Pharmacol.*, **10**, 95 (1989a).
157. A. M. Khasawinah and J. F. Grutsch, *Reg. Toxicol. Pharmacol.*, **10**, 244 (1989b).
158. G. M. Williams and S. Numoto, *Carcinogenesis* (London), **5**, 1689 (1984).
159. Agency of Toxic Substances and Disease Registry, "Toxicological Profile for Heptachlor/Heptachlor Epoxide," NTIS PB89-194492, 1989.
160. W. B. Buck, G. D. Osweiler, and G. A. Van Gelder, *Clinical and Diagnostic Veterinary Toxicology*, Kendall Hunt, Dubuque, IA, 1973; through Ref. 160.
161. J. L. Radomski and B. Davidow, *J. Pharmacol. Exp. Ther.*, **107**, 3, 259 (1953).
162. H. E. Fairchild, "Heptachlor—A Review of its Uses, Chemistry, Environmental Hazards and Toxicology," Office of Pesticide Programs, Environmental Protection Agency, Washington, DC, 1972.
163. K. J. Davis, Unpublished internal memorandum, Food and Drug Administration; through Ref. 160.
164. B. Davidow and J. L. Radomski, *J. Pharmacol. Exp. Ther.*, **107**, 259 (1953).
165. R. D. O'Brien, *Insecticides, Action and Metabolism*, Academic Press, New York, 1967.
166. G. T. Brooks, *Residue Rev.*, **27**, 81 (1969).

167. K. P. Bovart, J. P. Fontenot, and B. M. Priode, *J. Anim. Sci.*, **33**(1), 127 (1971).
168. M. N. Melnikov, *Chemistry of Pesticides*, Springer, New York, 1971; through Ref. 160.
169. F. Matsamura and J. O. Nelson, *Bull. Environ. Contam. Toxicol.*, **5**(96), 489 (1970).
170. K. K. Rozman. *Toxicol. Lett.*, **20**, 5 (1984).
171. M. Yamaguchi, S. Tanaka, K. Kawashima, S. Nakaura, and A. Tananaka, *Eisei Shikensho Hokoku*, **105**, 33 (1987) (in Japanese).
172. International Agency for Research on Cancer, Report of Working Committees, Vol. 5, Lyon, France, 1974.
173. Louisiana Agric. Exp. Stn. Rep. No. 1, unpublished, 1972; through Ref. 160.
174. C. Cueto and F. J. Biros, *Toxicol. Appl. Pharmacol.*, **10**, 261 (1967).
175. P. A. Stehr-Green, J. C. Wohlieb, W. Royce, and S. L. Head, *J. Am. Med. Assoc.*, **259**, 374 (1988).
176. C. C. Doane, *J. Econ. Entomol.*, **55**, 416 (1962).
177. J. B. Polinke, *Ohio J. Sci.*, **51**, 195 (1951).
178. J. C. Schread, *Coun. Agric. Exp. Stn. Bull.*, 556 (1952).
179. R. D. Smith, M. S. Thesis, Louisiana State University; through Ref. 160.
180. Environmental Protection Agency, *Fed. Reg.*, **39**(229), 41298 (1974).
181. P. S. Larson, J. L. Egle, Jr., G. R. Hennigar, R. W. Lane, and J. F. Borzelleca, *Toxicol. Appl. Pharmacol.*, **48**, 29 (1979), and personal communication from P. S. Larson, 1976.
182. P. J. Soine, R. V. Blanke, and C. C. Schwartz, *Toxicol. Lett.*, **17**, 35 (1983).
183. D. End, R. A. Carchman, and W. L. Dewey, *Fed. Proc. Fed. Am. Soc. Exp. Biol.*, **38** (3, Pt 1), 845 (1979).
184. L. R. Curtis and H. M. Mehenhale, *Pharmacologist*, **20**(3), 187 (1978).
185. P. S. Larson, J. L. Egle, Jr., G. R. Henniger, R. W. Lane, and J. F. Borzelleca, *Toxicol. Appl. Pharmacol.*, **48**, 29 (1979).
186. B. M. Anderson, S. T. Kohler, and R. W. Young, *J. Agric. Food Chem.*, **26**(1), 130 (1978).
187. "Kepone/Mirex/Hexachlorocyclopentadiene: An Environmental Assessment," A Report prepared by the Panel on Kepone/Mirex/Hexachlorocyclopentadiene, Coordinating Committee for Scientific and Technical Assessments of Environmental Pollutants, National Academy of Sciences, Washington, DC, 1978.
188. S. B. Cannon, J. M. Veazey, Jr., R. S. Jackson, V. W. Burse, C. Hayes, W. E. Straub, P. J. Landrigan, and J. A. Liddle, *Am. J. Epidemiol.*, **107**, 529 (1978).
189. W. J. Cohn, R. V. Blanke, F. D. Griffith, Jr., and P. S. Guzelian, *Gastroenterology*, **71**, 901 (1976).
190. J. E. Huff, and H. B. Gerstner, *J. Environ. Pathol. Toxicol.*, **1**(4), 377 (1978).
191. Public Health Service, Cancer and Birth Defects Division, Bureau of Epidemiology, CDC, Atlanta, EPI-76-7-2 (1976).
192. P. S. Guzelian, *Ann. Rev. Pharmacol. Toxicol.*, **22**, 89 (1982).
193. U.S. Department of Agriculture (1977), Comments of the Secretary of Agriculture in Response to the Notice of Intent to Cancel Pesticide Products Containing Chlordecone, Trade Name, Kepone, January 11, 1977, Washington, DC; through Ref. 219.

194. W. B. Deichmann, Kepone, report prepared for Senator H. H. Bateman and the National Fisheries Institute, Inc., Washington, DC, November 10, 1976.
195. D. A. Carlson, K. D. Konyha, W. B. Wheller, G. P. Marshall, and R. G. Zaylskie, *Science*, **194**(4268), 939 (1976); through Ref. 219.
196. R. V. Blanke, M. W. Fariss, P. S. Guzelian, P. S. Patterson, and D. E. Smith, *Bull. Environ. Contam. Toxicol.*, **20**(6), 782 (1978).
197. D. T. Molowa, S. A. Wrighton, R. V. Blanke, and P. S. Guzelian, *J. Toxicol. Environ. Health*, **17**, 375 (1986).
198. J. J. Huber, *Toxicol. Appl. Pharmacol.*, **7**, 516 (1965).
199. Environmental Protection Agency, EPA News (February 27, 1976) and Review of the Environmental Effects of Mirex and Kepone (1978), prepared by the U.S. Environmental Protection Agency, Office of Research and Development, by Battelle Columbus Laboratories (M. A. Bell, R. A. Ewing, and G. A. Lutz), EPA 600/1-78-013, Washington, DC; through Ref. 219.
200. J. J. Boylan, J. L. Egle, and P. S. Guzelian, *Science*, **199**, 893 (1978).
201. E. Richter, J. P. Lay, W. Klein, and F. Korte, *Agric. Food Chem.*, **27**(1), 187 (1979).
202. *Sci. News*, **113**(25), 405 (1978).
203. National Cancer Institute, Report on Carcinogenesis Bioassay of Technical Grade Chlordecone (Kepone), U.S. Department of Commerce, National Technical Information Service, January 1976.
204. P. M. Newberne, personal communication, November 6, 1979.
205. G. Tsushimoto, J. E. Trosko, C. C. Chang, and F. Matsumura, *Toxicol. Appl. Pharmacol.*, **64**, 550 (1982).
206. A. E. Sirica, C. S. Wilkerson, L. L. Wu, R. Fitzgerald, R. V. Blanke, and P. S. Guzelian, *Carcinogenesis* (London), **10**, 1047 (1989).
207. Toxicological Profile for Alpha-, Gamma- and Delta-Hexachlorocyclohexane, Agency for Toxic Substances and Disease Registry, U.S. Dept. of Commerce, NTIS PB90-171406, 1989.
208. S. Floru, A. Polizu, M. Ripeanu, and S. Cusa, *An. Inst. Cercet. Prot. Plant.*, Acad. Stiinte Agr. Silvice., **9**, 587 (1973); through Pestic. Abstr. U.S. Environ. Prot. Agency, No. 75-2974 (1975).
209. A. J. Lehman, *Assoc. Food Drug Offic. U.S.Q. Bull.*, **183**, (1954).
210. A. J. Lehman, Summary of Pesticide Toxicity, Association of Food and Drug Officials of the United States, Topeka, Kansas, 1965.
211. O. J. Fitzhugh, A. A. Nelson, and J. P. Frawley, *J. Pharmacol. Exp. Ther.*, **100**, 59 (1950).
212. L. Hulth, L. Hoglund, A. Bergman, and L. Moller, *Toxicol. Appl. Pharmacol.*, **46**, 101 (1978).
213. L. Hulth, M. Larson, R. Carlsson, and J. E. Kihlstrom, *Bull Environ. Contam. Toxicol.*, **16**(2), 133 (1976).
214. H. Herken and I. Klempau, *Naturwissenchaften*, **37**, 493 (1950).
215. H. Coper, H. Herken, and I. Klempau, *Naturwissenchaften*, **38**, 69 (1951).
216. A. K. Palmer, A. M. Bottomley, A. Warden, H. Frohberg, and A. Bauer, *Toxicology*, **9**(3), 239 (1978).
217. W. A. Queen, *Assoc. Food Drug Offic. U.S.Q. Bull.*, **17**, 127 (1953).

218. A. J. Lehman, *Assoc. Food Drug Offic. Q. Bull.*, **20**, 95 (1956).
219. B. Davidow and J. P. Frawley, *Proc. Soc. Exp. Biol. Med.*, **76**, 780 (1951).
220. K. van Asperen and F. J. Oppernoorth, *Nature*, **173**, 1000 (1954).
221. W. Koransky and J. Portig, *Arch. Exp. Pathol. Pharmakol.*, 244, 564 (1963).
222. R. Engst, R. Machloz, M. Kujawa, H-J. Lewerenz, and R. Plass, *J. Environ. Sci. Health B.*, **11**(2), 95 (1976).
223. H. Seidler, R. M. Machloz, M. Haertig, M. Kujawa, and R. Engst, *Nahrung*, **19**(5/6), 473 (1975).
224. R. W. Chadwick, T. L. Chuang, K. Williams, *Pestic. Biochem. Physiol.*, **5**(6), 575 (1975).
225. R. W. Chadwick, C. C. Brydan, M. F. Copeland, and J. J. Freal, *Chemosphere*, **7**(8), 633 (1978).
226. H. Okazaki and S. Kimura, *Ecyo To Shokuryo* (J. Jap. Soc. Food Nutr.), **25**(3), 201 (1972); through *Pestic. Abstr.*, U.S. EPA. No. 72-2174, 1972.
227. V. Fiserova-Bergerova, J. L. Radomski, J. E. Davies, and J. H. Davis, *Ind. Med. Surg.*, **36**, 65 (1967).
228. J. L. Radomski, E. Astolfi, W. B. Deichmann, and A. A. Rey, *Toxicol. Appl. Pharmacol.*, **20**, 186 (1971).
229. M. Tagami and N. Oshiro, Okinawa-ken Kogai-Eisei Kenkyus hoho, *Ann. Rept. Okinawa Pref. Inst. Publ. Health*, **11**, 76 (1977).
230. E. Danopoulos, K. Melissinos, and G. Katas, *Arch. Ind. Hyg. Occup. Med.*, **8**, 582 (1953).
231. J. J. Hutter, *Clin. Res.*, **27**(1), 97A (1979).
232. L. M. Solomon, L. Fahrner, and D. P. West, *Arch. Dermatol.*, **113**(3), 353 (1977).
233. W. E. Pace and J. Purres, *Can. Med. Assoc. J.*, **104**, 719 (1971).
234. C. M. Ginsberg, W. Lowry, and J. S. Reisch, *J. Pediatr.*, **91**(6), 998 (1977).
235. M. Wheeler, *Mest. J. Med.*, **127**(6), 518 (1977).
236. D. Nag, G. C. Singh, and S. Senon, *Trop. Geogr. Med.*, **29**(3), 229 (1977).
237. W. B. Deichmann, *Pesticides and the Environment. A Continuing Controversy*, Symposia Specialists, Miami, FL, 1973, pp. 397–398.
238. T. H. Milby, A. J. Samuels, and F. Ottoboni, *J. Occup. Med.*, **10**, 584 (1968).
239. W. R. Best, *J. Am. Med. Assoc.*, **185**, 286 (1963).
240. L. Sanchez-Medal, J. P. Castenado, and F. Garcia-Rojas, *N. Engl. J. Med.*, **269**, 1365 (1963).
241. J. P. Loge, *J. Am. Med. Assoc.*, **193**, 110 (1965).
242. A. J. Samuels and T. H. Milby, *J. Occup. Med.*, **13**(3), 147 (1971).
243. T. H. Milby and A. J. Samuels, *J. Occup. Med.*, **13**(5), 256 (1971).
244. National Cancer Institute, "Bioassay of Lindane for Possible Carcinogenicity," Carcinogenesis Tech. Rep. Ser. No. 14, CAS No. 58-89-9, NIC-CG-TR-14, 1977.

CHAPTER TWENTY-FOUR

# Phenols and Phenolic Compounds

Ralph E. Allan, J.D., C.I.H.

## 1 PHENOL [CAS # 108-95-2]

### 1.1 Source, Uses, and Industrial Exposures

Of the top chemicals produced in the United States by volume, phenol ranked 33rd for both 1988 and 1989. Although there was a reduction in production of the top 50 organic chemicals for the years 1987–1988 and 1988–1989, the annual production of phenol increased by 9.9 percent and 9.3 percent, respectively (1). The production of phenol for the years 1989 and 1990 was 3806 million pounds and 3512 million pounds, respectively, and the estimated production for 1991 and 1992 was 2915 million pounds and 3031 million pounds (2).

Phenol (hydroxybenzene, carbolic acid), $C_6H_5OH$, is one of the many aromatic compounds present in coal tar. It is separated from other substances by fractional distillation (170 to 230°C) and by other methods of purification until "gray phenic acid" or a pure grade of phenol has been obtained. Synthetic processes developed for the production of phenol include fusion of sodium benzenesulfonate with sodium hydroxide and hydrolysis of chlorobenzene.

Phenol is used in the production or manufacture of a large variety of aromatic compounds, including explosives, fertilizers, coke, illuminating gas, lampblack, paints, paint removers, rubber, asbestos goods, wood preservatives, synthetic resins, textiles, drugs, pharmaceutical preparations, perfumes, bakelite, and other

---

*Patty's Industrial Hygiene and Toxicology*, Fourth Edition, *Volume 2, Part B*, Edited by George D. Clayton and Florence E. Clayton.
ISBN 0-471-54725-5  © 1994 John Wiley & Sons, Inc.

plastics (phenol–formaldehyde resins). Phenol also finds use in the petroleum, leather, paper, soap, toy, tanning, dye, and agricultural industries (3).

Phenol also finds uses as a general disinfectant. In solution or mixed with slaked lime, its cleansing properties are effective for maintaining floors, stables, cesspools, and the like (4). Phenol is bacteriostatic in a 0.2 percent concentration, bactericidal above 1 percent, and fungicidal above 1.3 percent (5). These qualities facilitate its addition to certain germicidal paints and slimicides (6).

With rare exceptions, human exposure in industry has been limited to accidental contact of phenol with the skin or to inhalation of phenol vapors. Following the introduction of phenol spray by Lister in 1867, the compound became very popular and was used extensively for a number of years. Its medicinal uses are now limited chiefly to its application as an agent for relieving itching, as a disinfectant for septic wounds, as a cauterizing agent, and for the treatment of severe disability (muscle spasms, paralysis, and related disorders) resulting from multiple sclerosis (7–17).

## 1.2 Physical and Chemical Properties

The properties are summarized as follows:

| | |
|---|---|
| Physical state | White crystalline mass of hygroscopic, translucent, needle-shaped crystals |
| Molecular weight | 94.11 |
| Specific gravity | 1.072 |
| Melting point | 41°C |
| Boiling point | 182°C |
| Vapor density | 3.24 (air = 1) |
| Vapor pressure | 0.3513 mm Hg (25°C) |
| Refractive index | 1.54 (45°C) |
| Concentration in "saturated" air | 0.046% by volume (25°C) |
| Density of "saturated" air | 1.00104 (air = 1) |
| Flash point | |
|   Closed cup | 80°C (175°F) |
|   Open cup | 85°C (185°F) |

Solubility characteristics of phenol are as follows: added to water (25°C) it forms a true solution when present in concentrations up to 8 percent, and also in concentrations ranging from about 71 to 97 percent, in terms of both weight and volume. The compound is soluble to more than 50 percent (18) in ethyl alcohol, chloroform, ethyl ether, ethyl acetate, toluene, glycerol, and olive oil. Its solubility in mineral oil is about 0.2 percent (25°C), in petroleum ether 5.5 percent (31°C), and in rabbit fat 40 percent (34°C) (19). According to Pilcher and Sollmann (20), one part of crystallized phenol dissolves in 8 to 9 parts of petrolatum, 20 to 21 parts of gasoline, 23 to 24 parts of solid petrolatum, and 45 to 50 parts of liquid petrolatum.

## 1.3 Determination in the Atmosphere and Biological Material

An early method for the evaluation of phenol in the air involved collection in an absorbing solution of dilute alkali such as sodium hydroxide or 0.5 percent sodium bicarbonate. The quantity of phenol is then determined with diazotized *p*-nitroaniline reagent (21), Folin–Ciocalteu reagent, 2,6-dibromoquinone chloroimide, or *p*-aminodimethylaniline sulfate (22). The last-named reagent used is when the concentration in air is very low (2 to 10 ppb). The phenol also may be absorbed in spectrograde alcohol with direct determination by ultraviolet spectrophotometry.

The current sampling and analytical method used by the Occupational Safety and Health Administration (OSHA) is described in method No. 32 and involves collection of the atmospheric sample using an XAD-7 sampling tube followed by desorbtion of the collected sample with methanol. The analysis is performed by high-performance liquid chromatography (HPLC) with an ultraviolet detection system at 218 nm (23).

The National Institute for Occupational Safety and Health (NIOSH) phenol method also uses an aqueous bubbler to collect vapors. The collected sample is analyzed using gas chromatography with flame ionizing detection (24).

It should be emphasized that in order to determine the phenol content of any material a specific analytic method should be employed by a competent chemist. Early qualitative tests and many of the early quantitative procedures used in the past were not specific for phenol.

Briefly, biologic material to be analyzed is extracted with ether and the amount of "free" and "conjugated" phenol is determined spectrophotometrically utilizing the color developed with various reagents such as those mentioned above. Simple, gross, qualitative tests can be made with various color-producing reagents, but separation by paper chromatography gives more specific results.

A critical review of analytic methods with their usefulness and limitations in estimating "free" and "conjugated" phenol in blood, organs, urine, saliva, sweat, and feces was published in 1942 (25). In addition to the analytic procedures recommended therein, the reader is advised to consider also those that have been published by Schmidt (26), Tucker (27), Chirkov (28), Baernstein (29), Lykken et al. (30), Glick (31), Armstrong et al. (32), Tompsett (33), Sherwood and Carter (34), and Van Haften and Sie (35).

Heistand and Todd (36) published a fully automated method for the determination of total phenol in 1- to 2-ml specimens of urine, a method that has certain advantages over earlier sulfate ratio procedures that are used for assessing exposure to benzene (37–39). This "total" phenol method involves hydrolysis of the conjugated phenols and steam distillation, followed by the colorimetric determination (with 4-aminoantipyrine) of the phenol in the distillate. Normal urinary excretion of free and conjugated phenol varies widely and is related to diet, gastroenteric activity in health and disease, and the absorption of certain medicines and chemicals. It is therefore important to understand that when benzene exposure is suspected, a high urinary value for phenol is not proof of benzene exposure. According to Fishbeck et al. (40), "The NIOSH proposed standard for benzene implies that only benzene exposure can elevate urinary phenol levels in excess of 75 mg/liter (ppm)." In the studies of Fishbeck et al., medications such as Chloraseptic lozenges and Pepto-Bismol raised the total urinary phenol level in their subjects to 270 and 472 ppm, respectively. Studies in dogs confirmed these observations.

### 1.4 Physiological Response

Regardless of the mode of administration, the signs of acute illness induced by phenol in experimental animals resemble those observed in humans. However, in humans, phenol usually exerts (directly and indirectly) a predominant action upon the higher centers resulting in sudden collapse.

The systemic effects of phenol poisoning in humans are coma, hypothermia, loss of vasoconstriction, cardiac depression, and respiratory arrest. Other respiratory complications, such as stertorous breathing are not generally recognized. In addition, renal complication may also be seen and may progress to acute failure (41).

In other mammals the predominant effects are exerted upon motor centers in the spinal cord, inducing marked twitchings and severe convulsions. Following absorption of a toxic dose the heart rate first increases, then becomes slow and irregular. The blood pressure increases slightly at first, then falls markedly. There may be salivation and marked dyspnea and the body temperature usually decreases. Prolonged oral or subcutaneous administration to animals can cause damage to the lungs, liver, kidneys, heart, and genitourinary tract (9, 17, 42–47).

After phenol ingestion, white or brownish stains and areas of cell death can be observed about the gastrointestinal tract. These are caused by the corrosive effect of phenol (48).

Prolonged inhalation of vapors (30 to 60 ppm) has induced respiratory difficulties, lung damage, loss of weight, and paralysis (49). Lethal doses for experimental animals are presented in Tables 24.1 and 24.2.

#### 1.4.1 Pathology

The pathological changes produced by phenol in animals vary with the route of absorption, vehicle employed, concentration, and duration of exposure. Evidence has now suggested that phenol in humans can be less toxic by ingestion than by absorption through wounds, body cavities, or intact skin (57).

Local damages to the skin include eczema, inflammation, discoloration, papillomas, necrosis, sloughing, and gangrene. Following oral ingestion the mucous membranes of the throat and esophagus may show swelling, corrosion, and necroses, with hemorrhage and serous infiltration of the surrounding areas (58).

In a severe intoxication the lungs may show hyperemia, infarcts, bronchopneumonia, purulent bronchitis, and hyperplasia of the peribronchial tissues. There can be myocardial degeneration and necrosis.

The hepatic cells may be enlarged, pale, and coarsely granular with swollen fragmented, and pyknotic nuclei (59). Prolonged administration of phenol may cause, in the kidney, parenchymatous nephritis (49), hyperemia of the glomerular

**Table 24.1.** Toxicity of 2 to 7 Percent Aqueous Solutions of Phenol for Adult Experimental Animals

| Species | Route[a] | Dose Killing 50% of Animals (g/kg) | Ref. |
|---|---|---|---|
| Mouse | SC | 0.3–0.35 | Tollens (50), Duplay, and Cazin (51) |
| Rat | SC | 0.45 | Deichmann and Witherup (52) |
| | Oral | 0.53[b] | Deichmann and Witherup (52) |
| | Cutaneous | 2.5 | Deichmann and Witherup (52) |
| Guinea pig | SC | 0.68 | Duplay and Cazin (51) |
| Rabbit | IV | 0.18 | Deichmann and Witherup (52) |
| | SC | 0.5–0.6 | Tauber and Tauber (53) and Tollens (50) |
| | Oral | 0.6 | Clarke and Brown (54) |
| | Oral | 0.4–0.6 | Deichmann and Witherup (52) |
| | IP | 0.5–0.6 | Deichmann and Witherup (52) |
| Cat | SC | 0.09 | Tollens (50) |
| | Oral | 0.1 | Macht (55) |
| Dog | Oral | 0.5 | Macht (55) |
| Monkey | Toxicity is of similar order to that for rabbit | | Smith, Elvove, and Frazier (56) |

[a] SC = subcutaneous; IV = intravenous; IP = intraperitoneal.
[b] $LD_{50}$.

and cortical region, cloudy swelling, edema of the convoluted tubules, and degenerative changes of the glomeruli. Blood cells become hyalinized, vacuolated, or filled with granules. Muscle fibers show marked striation.

### 1.4.2 Absorption, Excretion, and Metabolism

Phenol is readily absorbed through the intact and abraded skin and from the stomach, enteric tract, uterus, intraperitoneal cavity, and subcutaneous tissues of man and animals (52). Vapors of phenol are readily absorbed into the pulmonary

**Table 24.2.** Comparative Toxicity of Aqueous Preparations of Phenol in Various Concentrations[a] Applied to the Abdominal Skin of Rabbits (52)

| Administration of 2 g/kg Dose | Number of Rabbits | Deaths (%) |
|---|---|---|
| Emulsion: 10 g phenol and 90 g water | 10 | 100 |
| Emulsion: 25 g phenol and 75 g water | 10 | 90 |
| Emulsion: 50 g phenol and 50 g water | 10 | 90 |
| Solution: 75 g phenol and 25 g water | 10 | 80 |
| Solution: 90 g phenol and 10 g water | 10 | 50 |
| Solution: 95 g phenol and 5 g water | 17 | 53 |
| Melted phenol reagent heated to 40°C | 15 | 30 |

[a] Standard dose for all concentrations, 2 g phenol/kg of rabbit.

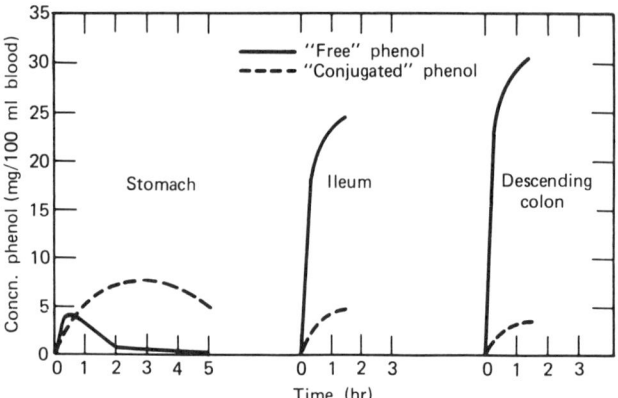

**Figure 24.1.** Comparative rate of absorption of phenol from the stomach, ileum, and descending colon of the rabbit. 0.3 g phenol/kg as a 5 percent aqueous solution was placed into ligatured sections of the gastroenteric tract.

circulation. Figure 24.1 presents the comparative rates of absorption from the gastroenteric tract of rabbits.

After absorption into the body, most of the phenol is oxidized and conjugated with sulfuric, glucuronic, and other acids. It is excreted in the urine as "free" and as "conjugated" phenol (61). Traces of "free" phenol are eliminated with the feces and expired air; Figures 24.2 and 24.3, respectively, show graphically the fate of a lethal and a sublethal dose of phenol (61, 62).

Additional information regarding the absorption (60, 63–66) and metabolism (62, 67–78) of phenol can be found in the works of other investigators.

### 1.4.3 Mode of Action

The primary site of stimulation in the central nervous system is the spinal cord. The twitchings are due to reflex stimulation (79), whereas the clonic convulsions are probably due to an increased excitation of the motor mechanism of the anterior horn cells (80). Local application of phenol solutions to the cerebellar cortex (81) did not produce the effects seen when phenol was applied to the spinal cord. When applied directly to the cerebral cortex, a dilute solution increased reflex excitability whereas more concentrated solutions destroyed the tissues (82). The heart apparently is slowed by a direct myocardial depression, not by stimulation of the vagus (83).

Cardiac toxicity first manifests itself by tachycardia and premature contractions. In severe cases, these symptoms progress to ventricular tachycardia or atrial fibrillation. And with gradual decrease in serum phenol level, the arrhythmias reverse back to isolated premature contractions before coming to normal rhythm (84).

The slight rise in blood pressure appears to be due to peripheral vasoconstriction. In humans, the more prominent fall of pressure is caused by its direct toxic action

# PHENOLS AND PHENOLIC COMPOUNDS

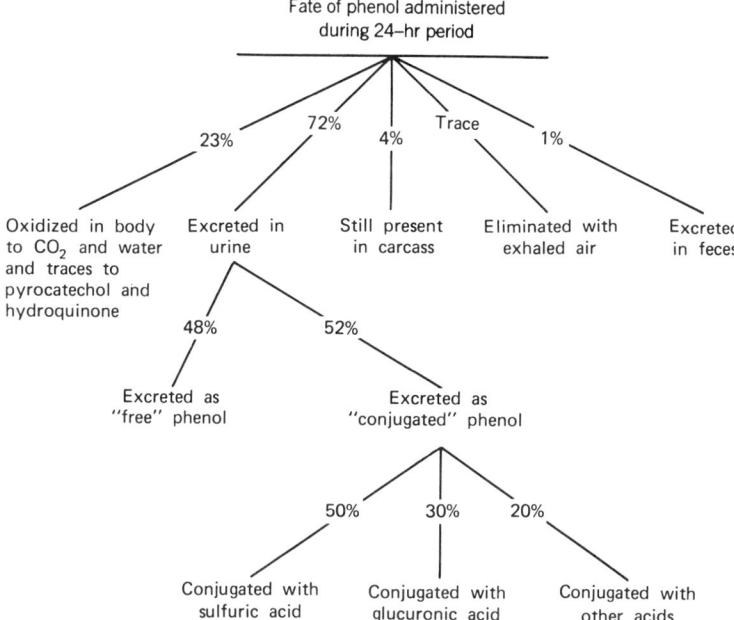

**Figure 24.2.** Metabolism of phenol in a rabbit given a sublethal oral dose (0.3 g/kg) (61).

on the myocardium, failure of the vasomotor centers, and local toxic action on the small blood vessels.

The toxic effects of phenol are related directly to the amount of "free" phenol in the blood.

### 1.4.4 Cause of Death

In an acute intoxication, death is usually due to respiratory failure.

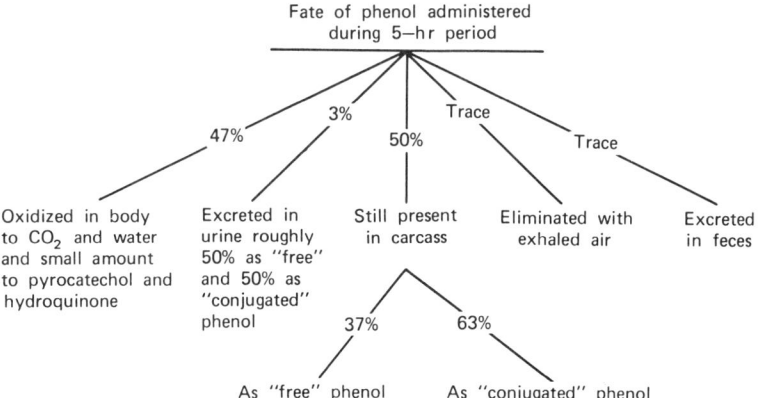

**Figure 24.3.** Metabolism of phenol in a rabbit given a lethal oral dose (0.5 g/kg) (62).

## 1.4.5 Acute Effects in Humans

Fatalities from poisoning by phenol occurred much more frequently the past decades than in recent years, although accidental poisonings still occur, particularly in the home.

An oral dose of 1 g of phenol may be lethal to humans; however, in exceptional cases, patients have survived the ingestion of 65 g of pure phenol or 120 g of the crude product. Roughly 50 percent of all reported cases have terminated fatally (11, 12).

The swallowing of phenol causes intense burning of the mouth and throat followed by marked abdominal pain. The breath has the odor of phenol, the face is pale and usually covered with cold sweat, the pupils may be contracted or dilated, and cyanosis is usually marked. Collapse, manifested by muscular weakness and unconsciousness, occurs in many cases a few minutes after the poison is swallowed. The pulse is usually weak and slow, but occasionally it is racing. Respiration may be increased in rate in the early stage of poisoning, but later it decreases in both rate and magnitude. In the early stage of an intoxication the temperature of the body may fluctuate above or below normal, but in the later stages there is more frequently antipyresis. Reflex activity is lost. General tremors, tonoclonic convulsions, or twitchings of isolated muscles of the face or limbs are occasionally observed, but they are never marked. Death usually results from respiratory failure (12, 17, 85–88). An example fatality points out the rapid rate of absorption of phenol through the skin. A technical assistant, in the absence of the plant maintenance staff and not wearing protective clothing, decided to check and adjust a faulty valve, which released a spray of liquid phenol over his thighs (approximately 25 percent of the body surface). A futile attempt was made to "splash water over his thighs from a nearby tap." He collapsed on the way to the ambulance room, and died 10 min after the exposure. His blood contained 7 mg percent of phenol (89).

B. W. Brown (90) reported on "Fatal Phenol Poisoning from Improperly Laundered Diapers." Actually, the offending compounds were sodium pentachlorophenate, sodium salts of other chlorophenols, and 3,4,4-trichlorocarbanilide, sold as Loxene. It cannot be overemphasized that the structure–activity relationships of phenol and phenol derivatives vary widely, and that to accept the properties of individual phenolic compounds as being those of phenol is a misconception and leads to error and confusion.

The phenol in medications, such as Castellani's paint, carbolated Vaseline, P&S Liquid, and Panscol Lotion, has been shown to produce irritant dermatitis in some persons (91).

The phenol use for facial peel develops the staphylococcal infection over the chemobrasion, with toxic effects resulting in toxic shock syndrome. This syndrome includes low blood pressure, rapid pulse, high temperature, erythroderma of the skin, and skin desquamation of palms and soles. The resulting toxic shock syndrome is fatal, with 8 percent of the cases ending in death (92).

## 1.4.6 Chronic Effects in Humans

In Lister's day (93), cases of chronic or subacute phenol poisoning were not uncommon among surgeons and their assistants. As Kobert (11) remarks, many of those doctors must have possessed great tolerance considering that they used, applied, and inhaled sprays of carbolic acid for years without illness or apparent discomfort. No doubt the intermittent character of their exposure was a factor in their favor. Data that gives us an insight into the quantities of phenol apparently tolerated by humans were reported in 1881 by Falkson (94), who stated that after inhaling these vapors for several hours his own urine contained phenol in amounts up to 2 g/day, and the urine of patients who absorbed the compound from the inspired air, as well as from the skin or from open wounds, contained up to 5 g/day.

More recently, chronic phenol poisoning has been infrequently reported. According to Zangger (17) severe chronic poisoning in humans is characterized by systemic disorders such as digestive disturbances, including vomiting, difficulty of swallowing, ptyalism, diarrhea, and anorexia; by nervous disorders, with headache, fainting, vertigo, and mental disturbances; and possibly by an eruption on the skin. The disease is usually fatal when there is extensive damage to the liver and kidneys. Prolonged cutaneous exposure to preparations of phenol may result in ochronosis.

In 1972 Merliss (95) encountered a case of classical phenol marasmus, reminiscent of the chronic carbolic acid poisonings that were common in physicians 100 or so years ago. The patient, age 44, had worked for years in a laboratory, distilling and handling phenol, cresol, and xylenol. From time to time he was exposed to the vapors of these compounds and, in addition, "often spilled phenol on his trousers and clothes," resulting in both inhalation and skin exposure. Signs and symptoms, which appeared slowly, included gradual deterioration without specific symptoms, loss of appetite and body weight, muscle pain in legs and arms, weakness, and excretion of dark urine. Physical examination, following 13.5 years of exposure, revealed an emaciated individual with an enlarged liver and elevated serum lactic dehydrogenase, glutamate–oxaloacetate transaminase, and glutamate–pyruvate transaminase. Although he was removed from the site of exposure, recovery was slow; 7 months later, he had gained 7.6 kg and his liver was no longer palpable.

Piotrowski (96) reported on an inhalation study with seven men and one woman, who were exposed for 8 hr to phenol vapors produced by the evaporation of phenol from a vessel kept at a constant temperature in an exposure chamber. To avoid skin exposure, these subjects remained outside the chamber and inhaled the phenol vapors, ranging from 6 to 20 mg/m$^3$, through a face mask connected to the exposure chamber (with two breaks of 30 min each).

It was found that the retention of phenol in the respiratory tract is related to the duration of exposure. Retention decreased from about 80 percent at initiation of the exposure to approximately 70 percent at the conclusion of the exposure period. The normal urinary levels and fluctuations were taken into consideration in determining the "experimental" quantities excreted. The excretion rate rose rapidly during the exposure, with maximum excretion "directly after the end of the exposure," and returned to the physiological level within 24 hr. Almost 100 percent of the phenol inhaled was excreted in the urine within 24 hr. Piotrowski also reported on skin exposure studies in which experimental subjects, wearing underwear and overalls, were exposed in the exposure chamber for 6 hr (with one break) to phenol vapors of 5, 10, and 25 mg/m$^3$. They inhaled uncontaminated air supplied from outside the exposure chamber. No essential differences were noted in the kinetics of phenol excretion following skin exposure when compared with exposure by inhalation. The skin absorption rate was roughly proportional to the concentration of phenol vapors, characterized by an absorption coefficient of approximately 0.35 m$^3$/hr, meaning that the subjects absorbed through the skin per hour the amount of phenol in 0.35 m$^3$ of air. It is of interest and significance that the clothing worn seemed to provide no protection. These studies support previous reports that phenol is rapidly excreted from the humans and animal body. Based on Piotrowski's studies, it appears that there is little, if any, retention of phenol as long as the "permissible" industrial daily exposure levels are not exceeded.

Little is known about the chronic respiratory effect of exposure to phenolic resin fumes. Schjenberg and Mitchell (97) studied 11 workers who had been exposed to these fumes, which when determined in the working atmosphere ranged in concentration from 7 to 10 mg phenol/m$^3$ and from 0.5 to 1.0 mg formaldehyde/m$^3$. On another occasion, the levels of formaldehyde ranged from 10.6 to 16.3 mg/m$^3$, but the authors believe that these latter concentrations were not typical. The limited measurements made do not permit an estimate of the cumulative exposure of the employees who worked with phenolic resins for periods of less than 1 year to more than 5 years. Pulmonary function tests performed on these subjects gave evidence that "relatively low concentrations of phenol–formaldehyde resin fumes might cause chronic airway disease." According to the authors, this study is the first to call attention to these effects.

The first reports of depigmentation of the skin of industrial workers exposed to certain phenolic compounds appeared in 1939 (98, 99). Recently, Kahn (100) reported patchy depigmentation in hospital employees who used two phenolic detergent germicides: O-Syl, containing *o*-benzylparachlorophenol and *p*(*tert*-butyl)phenol, and VesPhene, containing *o*-phenylphenol, *o*-benzyl-*p*-chlorophenol, and *p*(*tert*-amyl)phenol. In each group, initial signs appeared 6 months after the first use of these products.

The onset of depigmentation is insidious and may be preceded by mild erythema and pruritus until actual leukoderma appears several months after exposure. Depigmentation is related to the concentration of the offending material and the frequency of skin exposure. The color of the skin is apparently of no consequence. Repigmentation, that is, recovery—after cessation of exposure—varies and may take several months to a year or two (101). Bentley-Phillips (102) reports that O-Syl is safe and will not produce leukoderma if used according to the manufacturer's instructions. McGuire and Hendee (103) report on the biochemical basis for the depigmentation of the phenolic germicides.

## 1.5 Hygienic Standard of Permissible Exposure

The American Conference of Governmental Industrial Hygienists (ACGIH) threshold limit value (TLV–TWA, time-weighted average) (1992–1993) for phenol is 5 ppm (19 mg/m$^3$) (104). The OSHA permissible exposure limit (PEL) is also 5 ppm (105). The "S" skin notation in the listing refers to "the potential significant contribution to the over all exposure by the cutaneous route, including mucous membranes and the eyes, either by contact with vapors or, of probable greater significance, by direct skin contact with the substance" (104).

The ACGIH biologic exposure index (1992–1993) for phenol is 250 mg/g creatinine in urine collected at the end of a work shift. The determinant is usually present in biologic specimens collected from subjects who have not been occupationally exposed and the determinant is nonspecific, because it is observed after exposure to some other chemicals (104).

Beinhart (106) indicates that concentrations of about 4 ppm will impart a decided phenol taste to vegetables and fruit grown within a radius of more than a mile from a plant where such vapors escape.

Threshold limits (107) for the detection of odors in aqueous solutions are as follows: phenol, 25 mg/l; chlorophenols, 0.001 to 0.0005 mg/l; cresol, 0.0025 mg/l; chlorinated cresols, 0.001 to 0.0002 mg/l; thymol, 0.05 mg/l; resorcinol, 40.0 mg/l; creosote, 0.125 mg/l; chlorinated creosote, 0.01 mg/l; naphthol, 7.0 mg/l; and chlorinated naphthol, 0.5 mg/l. Hydroquinone imparted no odor to the water.

## 1.6 Flammability

Phenol, with a closed cup flash point of 79°C, presents a marked fire hazard. Phenol fires can be extinguished with water, carbon dioxide, or dry chemicals. Mixtures of air and 3 to 10 percent phenol are explosive (108).

## 1.7 Odor and Warning Properties

Phenol has a distinct, aromatic, somewhat sickening sweet and acrid odor discernable at 3.8 mg/m$^3$ (109). It has a sharp and burning taste.

## 1.8 Comments on the Use of Phenol in Medicine

The once favorite local application of dressings or compresses saturated with dilute (1 to 5 percent) aqueous solutions of phenol is no longer practiced. Many cases of gangrene have resulted from this type of therapy and as a result fingers and toes required amputation. An incident of gangrene with recovery was reported by Abraham (110). This involved an individual who unknowingly wore a rubber glove which contained, in one thumb, a few crystals of pure phenol. Gangrene has been observed less frequently upon the trunk.

An efficient antipruritic is obtained by incorporation of 1 to 2 percent of phenol into a salve, ointment, vegetable oil, or calamine lotion. However, preparations of this type should not be covered by a tight bandage after application to the skin.

Addition of camphor to phenol brings about a marked reduction of causticity and disturbs but little the analgesic action of phenol. Claims that the causticity is totally eliminated in mixtures containing more than 30 percent each of phenol and camphor are exaggerations. The causticity remaining can be traced in part to an aqueous phenol phase. This is formed because of the equilibration of the phenol between the phenol–camphor and the tissue fluids or perspiration of which no wound or skin remains free any length of time.

Phenol has been used for the treatment of the severe disability of painful muscle spasms or paralysis of the lower limbs (muscle spasticity in flexion) associated with multiple sclerosis. To relieve this condition, phenol (0.75 ml of 5 percent to 1.0 ml of 10 percent in glycerol) is injected intraspinally to reduce afferent sensory stimulation or to produce permanent destruction of the specific anterior nerve roots. In patients treated by Browne and Catton (111), excellent results were obtained in one case, but blocks had to be repeated after 5 months. In a second patient good results were obtained in the right leg; however, mild but painless spasms continued in the muscles of the left leg.

Fyfe and Quin (112), who treated 11 patients with intermittent claudication by injection of phenol (5 ml of 1:15 solution) into the lumbar sympathetic chain, concluded that this treatment produces no subjective or objective benefit, and that "such patients exhibit a marked placebo response (113) and account should be taken of this when assessing results of sympathectomy." Feldman and Yeung (114) treated 28 patients suffering from intermittent claudication by means of paravertebral lumbar somatic nerve block with 5 to 10 ml of 7.5 percent phenol in Myodil (iophendylate). Their results indicate that this treatment is less disturbing and more effective than other forms of conservative treatment.

Swerdlow (115), who employed subarachnoid injections of both phenol and chlorocresol in the treatment of intractable pain of 100 patients, found that the results "with chlorocresol were somewhat better than those produced by phenol." However, chlorocresol produced a higher incidence of complications (retention of urine, paralysis of anal sphincter). An 80 percent solution of phenol was injected locally by Shorey (116) for the treatment of pilonidal sinus in 25 patients. The solution was left in situ for 2 min, then the excess was expressed by pressure. In 21 patients, this treatment was a success.

Complications were reported by Galizia and Lahiri (117), such as paraplegia following block of the celiac plexus with 6 percent aqueous phenol for carcinoma of the pancreas. Superville-Sovak et al. (118), after stellate ganglion blockade with doses of 3 ml of 6 percent phenol in glycerol, caused triparesis in one patient and phenolic neurolysis of the cervical posterior roots, which led to respiratory arrest, in another. These complications illustrate the dangers of the spread of phenol beyond the intended site of neurolysis, with resultant local anesthesia, neural and vascular damage, and infarcts. Hughes (119) reported an incident of spinal cord infarction (with recovery), due to thrombosis of both posterior cerebral arteries secondary to an intraspinal injection of 1.0 ml of 5 percent phenol in glycerol; and

Ambasta (120) described a patient with Buerger's disease who, after left-sided lumbar sympathetic ganglion block with 8 ml of 5 percent phenol, suffered complications (presumably phenol-induced necrosis of the ureter) followed by recovery.

Little attention has been paid to the use of phenol in the treatment of massive intractable hematuria. Susan and Marsh (121) presented a case of intractable hemorrhagic cystitis secondary to cyclophosphamide therapy, which they treated for exactly 1 min by suprapubic bladder instillation of "30 cc of 100 percent phenol and 30 cc glycerin," with no untoward effects. The phenol–glycerol was removed by suction and replaced by alcohol which, after 1 min, was aspirated and replaced by saline. The authors feel that the favorable observations following local treatment with concentrated phenol are encouraging for patients who are extremely ill and poor surgical risks.

The reports by Wrench and Britten (122) suggests that phenolic fractions of creosote may be of value in the treatment of chronic psoriasis. Luntz (123) described a method for producing experimental primary glaucoma (in rabbits) by means of the subconjunctival injection of phenol.

### 1.9 Prevention of Poisoning by Phenol

The manufacture and use of phenol in industry need not present a hazard as long as management installs effective safety measures and employees are instructed and supervised in the handling of this compound. Employees should be questioned as to any known susceptibility to phenol or to closely related materials. Those affected with hepatic or kidney diseases should not be exposed to phenol for any length of time, because even intermittent exposure to vapors of phenol may become dangerous, particularly when the material is handled at elevated temperatures. Liquid phenol is combustible and mixtures of air containing 3 to 10 percent phenol are explosive.

Measures of safety include at least the following:

1. Publication and distribution of a manual outlining individual steps to be followed in the manufacture, handling, storage, and transport of phenol.
2. Effective ventilation.
3. Readily accessible showers that provide a flood of water for accidental skin exposures.
4. Proper disposal of phenol waste.
5. Proper confined space entry procedures for cleaning of tanks.
6. Continuous vigilance on the part of the hygienist or physician for signs and symptoms of chronic (local or systemic) intoxication.
7. A first aid station, fully equipped to handle acute intoxications 24 hr/day, 365/year.

## 1.10 Treatment of Phenol Poisoning

Phenol spilled on the skin is removed most efficiently by polyethylene glycol-300 (PEG-300) and by water, as follows.

To remove phenol from a finger, hand, or arm (or another part of the body, not exceeding approximately 10 percent of the body surface), swab the area promptly and repeatedly with cotton soaked in PEG-300, or if possible immerse the part directly into the PEG-300. If a larger body area has been contaminated with concentrated phenol, rush to remove all phenol-soaked clothing under a shower at full blast. After several minutes under the shower, continue decontamination with repeated swabbings or spraying with PEG-300 until the danger of collapse has passed. The recommendation to use PEG is based on the following studies. Schutz (124) discovered that polyethylene glycol-400 (PEG-400) is an effective agent for the treatment of rats exposed cutaneously for a period of 4 hr to a lethal dose of phenol (phenolum liquefactum). Ten animals were used per group. The fur was removed by shaving, and the animals were treated after the skin had healed ("geheilt"). After 4 hr of exposure, the animals were dipped into the decontaminating liquid. All rats, prior to dipping, showed convulsions. All rats treated with PEG-400 survived.

Conning and Hayes (125) studied the removal and decontamination of reagent-grade phenol, which they applied at 40°C in various solvents and concentrations to the shorn backs of Alderly Park (Wistar derived) rats. For first-aid treatment, based on these studies, they recommend—after removal of all clothing—swabbing all contaminated skin for at least 10 min with cotton wool soaked in glycerol, PEG-300, or a PEG/methylated spirit mixture (ethyl alcohol, denatured with methane). They found swabbing alone to be ineffective, but that a substantial therapeutic effect was achieved when the swabs were soaked in the above solvents. They believe that "the best emergency treatment will be to force an effective diluent into the epidermis before phenol, which is retained here, can be absorbed through the necrotic barrier it has created."

In 1972, Henschler (126) extended these investigations and used a sufficiently larger number of rats to permit statistical evaluation of the data. He applied liquid phenol (phenolum liquefactum 89 percent phenol) at 500°C "in hoher Schicht" to the belly area (12 $cm^2$) of Wistar rats under light anesthesia for exposure periods of 8 to 290 sec. Decontamination was initiated 1 min after the end of the exposure. This consisted of dipping the rats, with frequent agitation, for 5 min in approximately 300 ml of the decontaminating liquid. In some instances, phenol was also removed by swabbing the skin for 5 min with a decontaminating liquid. Periods (in seconds) in which application of liquid phenol to an abdominal skin area of 12 $cm^2$ caused death in 50 percent of rats were as follows:

| Experimental Conditions | Time of Lethal Exposure (sec) |
| --- | --- |
| Controls | 12.0 (10.16–14.16) |
| Immersed in water | 33.6 (27.5–41.0) |
| Immersed in glycerol | 52.8 (40.2–69.3) |

| | |
|---|---|
| Immersed in benzalkonium chloride | 76.2 (61.6–94.2) |
| Immersed in sodium bicarbonate 10% | 76.8 (61.6–94.2) |
| Immersed in THAM 4% | 78.0 (63.6–95.5) |
| Immersed in PEG-400 | 270.0 (216.5–336.0) |
| Swabbed with PEG-400 | 300.0 (242.0–373.0) |
| Immersed in water for 10 sec, followed by immersion in PEG-400 for 5 min | 300.0 (242.0–373.0) |
| Same as above, but in PEG-300 | 219.0 (176.5–272.0) |

As indicated, decontamination with water is of unquestionable value; it increased the period of lethal exposure of the treated rats nearly threefold. Glycerol, benzalkonium chloride, sodium bicarbonate, and THAM were even more effective. PEG-400 and PEG-300 were the most effective decontaminating agents, increasing the period of lethal exposure of the rats almost 25-fold; it made little difference whether the animals were immersed in PEG or swabbed with this compound.

Henschler's studies made an additional important contribution to the treatment of acute phenol intoxication. The intravenous injection of a 10 percent solution of sodium bicarbonate (0.9 g/kg) was effective in increasing the survival period of the phenol treated rats from 12 to 44 sec; it mattered little whether the bicarbonate was administered before or after the cutaneous application of phenol. Although the mechanism of action of the bicarbonate is not clear, these favorable results with rats (and also with dogs) suggest that sodium bicarbonate be considered in the treatment of severe human phenol intoxications.

Brown et al. (127) conducted similar studies, using Carworth Farm "E" strain rats. The hair was removed from the dorsolumbar region of these animals with a fine-tooth clipper, and phenol AR was applied at 45°C in a dose of 1 ml/kg body weight. A collar was placed around the neck of the animals to prevent licking; in some the exposed skin was covered with an occlusive dressing ($LD_{50}$ of phenol in these rats 0.68 ml/kg), and in others it remained uncovered ($LD_{50}$ of phenol in these rats 0.5 ml/kg). After 1 min of skin contact with phenol, a decontaminant was sprayed onto the treated area for 45 sec, or quickly wiped over, or swabbed for 10 to 120 sec. Each treatment group consisted of five males and five female rats.

Of 90 rats decontaminated with water, only one died. All animals developed convulsions and local "burns" ranged in severity from "faint" to "severe." Treatment with PEG-300/methylated spirits (MS) 2:1 mixture) provided by far the best protection. Only in two of the nine groups were "slight" convulsions noted. There were no deaths among the rats decontaminated with PEG-300/MS. Swabbing with PEG-300, PEG-400, or PEG-400/MS (2:1 by volume) was as effective as swabbing with PEG-300/MS (2:1 mixture). Decontamination with glycerol or methylated spirits was less effective. The use of cetrimide (cetylmethyl ammonium bromide), a synthetic detergent, was clearly contraindicated.

These results substantiate those of Conning and Hayes (125) and Henschler (126) that PEG is an effective decontaminating agent for the removal of phenol from the skin of rats and, presumably, also from the skin of humans. Brown et al. also tested the PEG-300/MS mixture for eye irritation in rabbits and found that very slight immediate irritation was caused by the MS in the PEG/MS mixture.

Dathe (128), one of Henschler's graduate students, recently repeated and extended somewhat Henschler's original phenol decontamination studies. His results added credence to what has already been reported for PEG.

It is well to cover phenol burns with wet dressings, such as compresses of saturated sodium sulfate. Healing of cutaneous phenol burns on animals was not expedited by treatment with various vegetable oils or glycerol. Application of a solution of 50 percent camphor in ethyl alcohol irritated the skin of rabbits and consequently retarded healing.

Speed is equally essential in the treatment of oral poisoning. If a patient is conscious and can be induced to vomit easily, 15 to 30 ml of castor oil or some other vegetable oil may well be administered. If vomiting cannot be induced readily and promptly, gastric lavage should be initiated without delay, preferably with an aqueous solution of 40 percent of Bacto-Peptone or milk; if these are not available, water may be used. Washing should be continued, employing 300 to 400 ml of liquid at a time, until the odor of phenol is no longer detectable.

After the stomach has been washed out, an oral dose of 15 to 50 ml of castor or another vegetable oil may be administered to advantage. As Jackson (64) and others (19, 129) have pointed out, oils retard the absorption of phenol and tend to reduce the local damage. [Animal experiments have shown that phenol is less toxic when administered in castor oil than in other vegetable oils (54)]. However, caution is in order in the face of serious alimentary damage, and it may be inadvisable to give a cathartic such as castor oil if much time has elapsed since the ingestion of phenol.

Symptomatic treatment consists primarily of the administration of circulatory stimulants. These may augment the systemic poisoning temporarily, but they counteract stasis and therefore tend to prevent thrombosis and necrosis. Transfusions also may be indicated. In collapse, heat should be applied. The treatment of chronic phenol poisoning is symptomatic after the patient has been removed from the site of exposures.

## 1.11 Factors Relating to the Legal Aspect of Poisoning by Phenol

The gross appearance of the lesions is suggestive, but not sufficiently specific to exclude other substances. Chemical analyses will provide positive or negative evidence in a case of suspected poisoning. Stomach contents (in oral poisoning) and blood and urine are best suited for analysis in the surviving patient. The relative concentrations of "free" and "conjugated" phenol will furnish useful information in relation to quantities ingested and absorbed, time interval between poisoning and drawing of blood, and the prognosis. In fatal cases any tissue or excretion may be employed for analysis.

Analysis of tissues should be carried out as soon as feasible after death because glycolysis will alter the respective concentrations of "free" and "conjugated" phenol.

Rising and Lynn (130) suggest ethyl alcohol as a preservative for samples of viscera containing phenol. These recommendations are supported by observations (19) that indicated that only 10 to 15 percent of phenol was lost during a period of 12 months when rat tissues (containing originally 25 mg of phenol) were kept in ethyl alcohol, at 7°C, in paraffin-stoppered containers. Fifty percent of the original content of phenol was lost from corresponding tissues preserved without alcohol under otherwise identical conditions.

Embalming destroys the evidence of poisoning by phenol because embalming fluids usually contain this or very closely related compounds.

## 1.12 Phenols in Water Supply

The presence of phenol and related compounds in a public water supply generally presents a serious problem to the engineer of a water plant. The treatment of a public water supply for the purpose of removing phenolic tastes and odors is not the answer to the problem. The phenols must be eliminated at the source of entry into the water (131).

Chase (131) recommends chlorine dioxide (1 to 10 lb/million gallons of water) to remove phenolic tastes. Caution is in order, however, because a high dosage of chlorine dioxide produces a breaking or sloughing off of the iron and manganese in the distribution system, and the resulting red water may cause as many, if not more, complaints as undesirable tastes and odors.

Not all tastes are due to phenol or phenols. According to Kinney (132), organic chemicals other than phenols are more responsible for the production of undesirable tastes and odors in some of our waters. Some of the nonphenols give phenolic tastes, and some of these chemicals apparently are not removed by treatment. Hence developing limits on phenols and requiring that phenols be treated to meet certain limits will not necessarily assure taste-free water. The publication by the American Public Health Association entitled *Standard Methods For Water Analysis* is suggested as a source of analytic methods (133).

Schneider (134) and Thornton and Wilhm (135) have tested phenol and phenolic compounds in waste waters (natural or artificial fresh or seawater) with regard to zoospore activity, sporangia development, dry weight of *Thraustochytrium striatum* Schneider, or the survival of the life stages of *Chironomus attenuatus*.

## 1.13 Action of Phenol and Related Compounds in Tumor Formation

Interesting and valuable contributions to this subject were provided by Boutwell, Rusch, Bosch, and others (136–140). Their reports deal with the action of phenol and about 50 phenol derivatives containing one to five halogen atoms, hydroxyl, carbonyl, carboxyl, and nitro radicals, as well as β-naphthol, phenyl acetate, anisole, certain cresols, and other related compounds.

Pretreatment with a single application of about 75 μg of 9,10-dimethyl-1,2-benzanthracene (DMBA) was not essential for the induction of tumors (140). Phenol and certain other materials alone induced both papillomas and carcinomas, but the time required for the production of tumors was extended when pretreatment with DMBA was excluded. The maximum response was produced with a 10 percent solution of phenol, while a less marked, but positive, effect was caused by a 5 percent solution. A more concentrated (20 percent) solution caused death from systemic toxicity before tumors were produced. By using reagent-grade, USP, and purified phenol, it was shown that this action was due to phenol itself, not to a contaminant.

Certain inferences concerning the chemical structure required for tumor formation certainly may be made from the results obtained with the derivatives of phenol. Monohalogenated phenols, methylphenols, and dimethylphenols were as potent in promoting papillomas as phenol itself (with the exception of 2,6-dimethylphenol, which was inactive). The addition of a nitro, carbonyl, carboxyl, or second phenolic group essentially eliminated the carcinogenicity, although resorcinol showed some activity. It appears that there must be at least one unsubstituted position ortho to the phenolic group for papilloma-promoting activity.

There seems to be no correlation between the known biologic effects and the carcinogenicity of phenol or its derivatives.

Several reports have appeared on the tumor-promoting activity of certain phenolic compounds found in tars, soots, oils, cigarette smoke, and tea. Certain strains of rodents and the seeds of tumor-prone interspecific hybrids of the diploid *Nicotiana glauca* x *N. langdorffii* and the tetraploid *N. Suaveolens* x *N. langdorffii* were used in these studies. Evaluation of the observations in terms of humans is difficult (141, 142).

There is no specific evidence of human cancer attributable to phenol or related compounds; however, the carcinogenicity to mice certainly emphasizes the precautions necessary in handling these materials.

## 2 PYROCATECHOL [CAS # 120-80-9]

### 2.1 Source, Uses, and Industrial Exposures

Pyrocatechol (catechol, *o*-dihydroxylbenezene, 1,2-benzenediol, pyrocatechin), $C_6H_4(OH)_2$, may be obtained by the fusion of *o*-phenolsulfonic acid with alkali, by heating chlorophenol with a solution of sodium hydroxide at 200°C in an autoclave, or by cleavage of the methyl ether group of guaiacol (obtained from beechwood tar) with hydroidic acid (143).

Pyrocatechol is used for various purposes, but particularly as an antioxidant in the rubber, chemical, photographic, dye, fat, and oil industries. It is also employed in cosmetics and in some pharmaceuticals.

Cases of industrial or accidental poisoning have been rare.

### 2.2 Physical and Chemical Properties

The properties are listed below:

| | |
|---|---|
| Physical state | Colorless, crystalline solid; sublimes readily; volatile in steam |

| | |
|---|---|
| Molecular weight | 110.11 |
| Specific gravity | 1.344 (4°C) |
| Melting point | 105°C |
| Boiling point | 245.6°C (decomposes at 240 to 245°C( (144) |
| Vapor density | 3.79 (air = 1) |
| Solubility | Soluble in water, alcohol, ether |
| Flash point (closed cup) | 127.2°C (261°F) |

## 2.3 Determination in the Atmosphere

Methods for the determination of catechol in air have not been developed. The methods of Baernstein (29) or Tompsett (145) can be used to determine pyrocatechol in urine and in other biologic materials.

## 2.4 Physiological Response

Phenol-like signs of illness are induced in experimental animals given toxic or lethal doses. Unlike phenol, large doses of pyrocatechol can cause a predominant depression of the central nervous system (CNS) and a prolonged rise of blood pressure (146, 147). In anesthetized rats it can cause CNS stimulation (148).

The repeated absorption of sublethal doses by animals may also induce methemoglobinemia, leukopenia, and anemia.

Pyrocatechol is more toxic than phenol except by inhalation (149). The approximate lethal oral dosages (in g/kg) for various experimental animals are as follows: dog, 0.3 (150); rabbit, 0.2 (151); cat, 0.1; and guinea pig, 0.16 (152). Most rats and guinea pigs die when given a single subcutaneous injection of about 0.22 g/kg, and the lethal intravenous dose for dogs is about 0.04 g/kg (153).

The oral $LD_{50}$ in rats is 0.3 g/kg. The dermal $LD_{50}$ in rabbits is 0.8 g/kg. It is an irritant to eyes and skin, but less irritating to the skin than phenol. After 8 hr of inhalation at concentrations of 2 or 2.8 $g/m^3$ rats showed signs of intoxication (irritation and tremors) for about 24 hr after exposure. At 1.5 $g/m^3$ no signs were observed (149).

### 2.4.1 Pathology

Dietering (154) reported degenerative changes in the tubuli of the kidney, with red blood cells and fibrin clots in the lumina. Flickinger (149) reported hyperemia of the stomach and intestines after lethal oral doses in rats and loss of toes and tips of tails of rats after exposure to high concentrations (2 or 2.8 $g/m^3$) in a chamber.

### 2.4.2 Absorption, Excretion, and Metabolism

Catechol is readily absorbed from the gastroenteric tract and through the intact skin of mice (155). It seems quite probable that this compound can be absorbed from the lungs, but substantiating data are not available.

Inhalation results in a burning sensation in the throat and lungs and subsequently, a pronounced increase in the rate of breathing (156).

After absorption, part of the catechol is oxidized with polyphenol oxidase to benzoquinone (157). Another fraction conjugates in the body with hexuronic, sulfuric, and other acids. A small amount is excreted in the urine as "free" pyrocatechol. The "conjugated" fraction hydrolyzes easily in the urine (29, 158) with the liberation of the "free" compound; this is oxidized with the formation of dark-colored substances that impart to the urine a "smoky" appearance.

### 2.4.3 Mode of Action

Apparently pyrocatechol acts by mechanisms similar to those reported for phenol. The rise of blood pressure appears to be due to peripheral vasoconstriction (147).

### 2.4.4 Cause of Death

Death apparently is initiated by respiratory failure.

### 2.4.5 Effects Observed in Humans

Cases of industrial or accidental poisoning have been rare. Contact with the skin has been known to cause an eczematous dermatitis, and absorption through the skin in a few instances has resulted in symptoms of illness resembling closely those induced by phenol, except for certain central effects (convulsions) which were more marked (159).

## 2.5  Hygienic Standard of Permissible Exposure

The ACGIH TLV-TWA for 1992–1993 for pyrocatechol is 5 ppm (23 mg/m$^3$) (104). The OSHA PEL is also 5 ppm (23 mg/m$^3$) (105). The "S" skin notation in the listing refers to the potential significant contribution to the overall exposure by the cutaneous route, including mucous membrane and the eyes, either by contact with vapors or, of probable greater significance, by direct skin contact with the substance (104).

## 2.6  Flammability

Pyrocatechol fires should be extinguished with water, carbon dioxide, or dry chemicals.

## 2.7  Odor and Warning Properties

There is a faint, characteristic odor.

## 3  RESORCINOL [CAS # 108-46-3]

### 3.1  Source, Uses, and Industrial Exposures

Resorcinol (*m*-dihydroxybenzene, resorcin), $C_6H_4(OH)_2$, is usually prepared by fusing sodium *m*-benzenedisulfonate with sodium hydroxide. It can be made in several other ways, by destructive distillation of brazilin or by the fusion of galbanum, ammoniac, sagapenum, asafetida, or acroides with caustic potash (160). It is used in tanning, in photography, and in the manufacture of explosives, dyes, cosmetics, organic chemicals, antiseptics, resins, and adhesives (128). It has also been suggested as an aerial bactericide.

In addition, resorcinol is a cross-linking agent commonly used in the production of neoprene. It is also an effective rubber tackifier and can be used as a chemical intermediate in the synthesis of resorcinol–formaldehyde resins, resin progenitors, and wood adhesive resins (161).

Industrial exposures are rather rare, but could occur in any industry if the compound is heated beyond 300°F.

## 3.2 Physical and Chemical Properties

The properties are summarized as follows:

| | |
|---|---|
| Physical state | White, needle-shaped crystals or rhombic tablets and pyramids, which turn pink on exposure to light and air |
| Molecular weight | 110.11 |
| Specific gravity | 1.272 (15°C) |
| Melting point | 110.7°C |
| Boiling point | 276°C |
| Vapor density | 3.79 (air = 1) |
| Percent in "saturated" air | 2.64% by volume (25.1°C) |
| Density of "saturated" air | 1.0739 (air = 1) |
| Solubility | Soluble in water, alcohol, glycerol, ether (162) |
| Flash point (closed cup) | 127.2°C (261°F) |

## 3.3 Determination in the Atmosphere

Air samples can be collected with impingers containing distilled water. If millipore filters are used, about 50 percent may pass through the filters (149). Analysis can be performed with ultraviolet spectroscopy at 273.5 nm using a 10-cm cell.

### 3.3.1 Determination within the body

Detection of free resorcinol in plasma and urine requires the use of HPLC and a simple ethanol extraction. This method is useful to concentrations as low as 0.5 percent, at which it gives recoveries of greater than 90 percent with good reproducibility (163).

## 3.4 Physiological Response

The signs of intoxication resemble those induced by phenol, but the antipyretic action of resorcinol is more marked. Resorcinol is less toxic than pyrocatechol.

The approximate lethal oral dose of resorcinol, in aqueous solution, for rabbits is 0.75 g/kg (151) and for rats and guinea pigs 0.37 g/kg (164). Dogs were killed by intravenous injections of 0.7 to 1.0 g/kg (153).

The oral $LD_{50}$ in rats is 0.98 g/kg and the dermal $LD_{50}$ in rabbits is 3.36 g/kg (149). It is irritating to the eyes and skin. At high dermal doses (1 to 7.95 g/kg or 2.5 to 19.3 g total dose) it causes irritation and necrosis in a dose-related response (149). Inhalation by rats for 1 hr at 7.8 $g/m^3$ or 8 hr at 2.8 $g/m^3$ caused no signs of intoxication. In subacute inhalation studies rats, rabbits, and guinea pigs were exposed to 34 $mg/m^3$ 6 hr daily for 2 weeks without toxic effects (149).

### 3.4.1 Pathology

Lethal oral doses in rats produced hyperemia and distention of the stomach and intestines (149).

Pathology reported for humans includes marked siderosis of the spleen (165) and marked tubular injury in the kidney. Becker (166) reported fatty changes and anemia of the liver, degenerative changes in the kidney, fatty changes of the heart muscle, moderate enlargement and pigmentations of the spleen, and edema and emphysema of the lungs.

Upon cutaneous contact with resorcinol, the erythematous areas turn a reddish color. When exposure persists for 2 to 3 days, areas red-violet in color emerge. Prolonged exposure of greater than 3 days results in a rusty red lesion. Upon discontinuation of contact, the skin will have assumed normal pigmentation after about a week (169).

### 3.4.2 Absorption, Excretion, and Metabolism

Resorcinol is readily absorbed from the gastroenteric tract and, in a suitable solvent, is readily absorbed through the human skin. The compound is excreted in the urine, as are other phenols, in a free state and conjugated with hexuronic, sulfuric, or other acids.

### 3.4.3 Mode of Action

Resorcinol causes local damage by direct irritation of the tissues. It causes systemic effects in much the same manner as pyrocatechol.

### 3.4.4 Cause of Death

Death apparently is initiated by respiratory failure.

### 3.4.5 Effects Observed in Humans

Resorcinol is antifungal, antibacterial, antipruritic, antiseptic, surfactant, mildly keratolytic, and a local irritant. It is commonly employed in the treatment of acne, psoriasis, eczema, seborrheic dermatitis, other cutaneous lesions, and ringworms. Resorcinol is used to remove warts, corns, and calluses (167).

# PHENOLS AND PHENOLIC COMPOUNDS

A 1 to 5 percent concentration is common in various lotions, creams, and ointments. A higher concentration of 10 percent is found in products used to treat athlete's foot. Resorcinol is most effective when delivered as an aerosol spray germicide (167).

The cutaneous application of solutions or salves (168) containing from 3 to 25 percent of this compound may result in local hyperemia, itching, dermatitis, edema, corrosion, and the loss of the superficial layers of the skin. The allergic/sensitization reactions also include eczematous reaction of erythema, edema, and the formation of vesicles. Burning sensations may also be noted (169).

These changes, if they are severe, may be associated with some or all of the following effects: enlargement of regional lymph glands, restlessness, methemoglobinemia, cyanosis, convulsions, tachycardia, dyspnea, and death (12, 159, 170, 171). Ingestion of resorcinol induces similar signs and symptoms. Thus a child, after accidentally swallowing 4 g, complained of dizziness and somnolence. The ingestion of 8 g, in another case, induced an almost immediate hypothermia, fall in blood pressure, and decrease in the rate of respiration, with tremors, icterus, and hemoglobinuria. Recovery was noted 2 hr after the poisoning (12). Other cases are on record in which similar doses apparently had no ill effects (172).

Resorcinol has been reported to cause sensitization and cross-sensitization with other phenolic materials (99, 173) and to cause goiter (174).

Resorcinol in certain resins was reported to cause respiratory problems in the rubber industry (175, 176).

An epidemiologic study of rubber workers exposed to a hexamethylenetetramine–resorcinol rubber system revealed no specific symptoms caused by resorcinol. The concentrations of resorcinol in air were less than 0.3 mg/m$^3$ (177). In another study there were no reports or irritation or discomfort by workers when concentrations were 10 ppm or less for periods of at least 30 min (149).

## 3.5 Hygienic Standards of Permissible Exposure

The TLV–TWA is 10 ppm (45 mg/m$^3$). The STEL (short-term exposure limit) is 20 ppm (90 mg/m$^3$) (104).

## 3.6 Flammability

Resorcinol fires may be extinguished with water, carbon dioxide, or dry chemicals.

## 3.7 Odor and Warning Properties

Resorcinol has a faint, characteristic odor and a sweetish, followed by a bitter taste.

## 3.8 Treatment of Poisoning

Resorcinol should be thoroughly washed from the skin with ample volumes of water. If it is swallowed, the patient should be forced to vomit or gastric lavage should be performed. Saline cathartics will remove unabsorbed portions from the enteric tract. Additional treatment is symptomatic. If a patient has malfunction of the kidneys, he should be observed with extra care.

## 4 HYDROQUINONE [CAS # 123-31-9]

### 4.1 Source, Uses, and Industrial Exposures

Hydroquinone ($p$-dihydroxylbenzene, $p$-hydroxyphenyl), $C_6H_4(OH)_2$, is usually prepared by the reduction of quinone with sulfurous acid.

Hydroquinone occurs naturally in several arthropods and assorted species of plants. It is present in the African plant, Noogoora burr (*Xanthuim pungens*), at concentrations high enough to poison the pigs and cattle whose inexperience allows them to venture a taste (178). Quinone is additionally present in traceable levels in cigarette smoke (179).

Its industrial uses are similar to those of pyrocatechol and resorcinol. It is a reducing agent used extensively as a photographic developer and as an antioxidant or stabilizer of certain materials that polymerize in the presence of oxidizing agents (143). Many derivatives of hydroquinone have been used as bacteriostatic agents. Certain hydroquinone derivatives, specifically 2,5-bis(ethylenimino)hydroquinone, have been reported to be good antimitotic and tumor-inhibiting agents. Harmful effects are likely to occur to workers who inhale the dust or vapors or whose skin or eyes come in contact with the dust or vapors for prolonged periods (180).

### 4.2 Physical and Chemical Properties

The properties are summarized below:

| | |
|---|---|
| Physical state | Crystallizes from water in hexagonal prisms |
| Molecular weight | 110.11 |
| Specific gravity | 1.332 (15°C) |
| Melting point | 172°C |
| Boiling point | 285°C (730 mm Hg) |
| Vapor density | 3.81 (air = 1) |
| Vapor pressure | 4.0 mm Hg (150°C) |
| Density of "saturated" air | 1.011 (150°C) (air = 1) |
| Solubility | 7% in water at 25°C; soluble in hot water, alcohol, ether |
| Flash point (closed cup) | 165°C (329°F) |

### 4.3 Determination in the Atmosphere

Hydroquinone aerosols can be collected by drawing contaminated air through a 37-mm cassette containing a 0.8 μm cellulose ester membrane filter and cellulose backup pad. Owing to the unstable nature of hydroquinone on the collection media, immediately following sampling, transfer the filter to an ointment jar and stabilize by dissolution in 1 percent acetic acid. These samples can then be analyzed using either HPLC or ultraviolet spectrophotometry (181).

Hydroquinone in biologic materials can be determined by the methods reported by Baernstein (29) or by others (182–185).

## 4.4 Physiological Response

A toxic concentration of hydroquinone in the tissues of experimental animals induces signs of illness that resemble, in many respects, those induced by phenol. In acute poisoning, there is increased motor activity, hypersensitivity to external stimuli, hyperactive reflexes, dyspnea, and cyanosis. These signs are followed by marked clonic convulsions and later by complete exhaustion, hypothermia, paralysis, loss of reflexes, coma, and death. Marked formation of methemoglobin occurs. Subacute poisoning may be characterized by hemolytic icterus, anemia, leukocytosis, reticulocytosis, increased cell fragility, hypoglycemia, depigmentation of fur, and marked cachexia (186).

Hydroquinone is more toxic than phenol. The approximate lethal oral dosages of this compound in aqueous solution are 0.2 g/kg for the rabbit (159) and 0.08 g/kg for the cat (187). The lethal intravenous dose for the dog is about 0.09 g/kg (117) and the lethal subcutaneous dose for mice is about 0.16 g/kg (188).

### 4.4.1 Pathology

Very little information concerning the pathology in animals is available. The changes reported for humans include hyperemia of abdominal organs that are rich in pigments, pathological changes in the liver and kidney, and bronchopneumonia in the lungs (188, 189).

### 4.4.2 Absorption, Excretion, and Metabolism

Hydroquinone is absorbed more rapidly than phenol from the gastroenteric tract (rats) (190). The compound may be absorbed through the skin, but this is questionable (155). Analysis of the blood or urine (29, 182–185) demonstrates the absorption of this compound.

Little is known concerning the metabolism of hydroquinone except that it is oxidized to the more toxic quinone (188, 191–195). (Hydroquinone also oxidizes to quinone when exposed to light or air.) Hydroquinone and quinone are partially excreted as such, and in conjugation with hexuronic, sulfuric, and other acids.

### 4.4.3 Mode of Action

Hydroquinone appears to act on the body by first being oxidized to quinone.

### 4.4.4 Cause of Death

Death apparently is initiated by a type of respiratory failure. The oxygen-carrying capacity of the blood is so markedly reduced that anoxia results.

### 4.4.5 Effects Observed in Humans

Hydroquinone causes signs and symptoms of an illness resembling that induced by pyrocatechol or resorcinol. Ingestion of 1 g by an adult (smaller quantity by a child) may cause tinnitus, nausea, dizziness, a sense of suffocation, an increased rate of respiration, vomiting, pallor, muscular twitchings, headache, dyspnea, cyanosis, delirium, and collapse. The urine is usually green or brownish green in color and continues to darken on standing (12). Fatal cases have been reported after ingestion of 5 to 12 g (196). Certain "health teas" have been prepared from leaves of blueberry, red whortleberry, cranberry, or bearberry. Their ingestion should be avoided because the leaves may contain hydroquinone in a concentration (sometimes exceeding 1 percent) capable of producing irritation of the intestinal mucosa and systemic poisoning (12). Cases of dermatitis have resulted from skin contact with hydroquinone. Lapin (197) reports such findings after application of an "antiseptic oil" that apparently contained traces of hydroquinone as an antioxidant.

Hypomelanosis–hyperpigmentation of the skin (198) and depigmentation of the skin of black persons have been reported (199).

Anderson (200), Velhagen (201), Sterner (202), and others (203) have reported cases of keratitis and discoloration of the conjunctiva among men exposed to concentrations ranging from 10 to 30 mg of vapor or dust of hydroquinone per cubic meter of air. Anderson and Oglesby (204) reported corneal changes, particularly alteration of the curvature, that were manifested long after exposure and after the stain and pigment had disappeared.

### 4.5 Hygienic Standards of Permissible Exposure

The TLV–TWA (1992–1993) is 2 mg/m$^3$. The STEL is 4 mg/m$^3$ (104).

### 4.6 Odor and Warning Properties

Hydroquinone has a sweet taste.

### 4.7 Safe Handling Procedures

In view of its potential toxicity, hydroquinone should be handled with caution. Contact with the eyes or inhalation of its dust or vapors, liberated particularly at elevated temperatures, must be avoided. Although quinone vapors are considerably more toxic than hydroquinone dust, one circumstance that makes a dust such as hydroquinone a greater hazard is that the extent of eye injury depends largely upon the concentration of the compound in the fluids bathing the eye. The irritating vapors of quinone become diluted and washed away with the tears, whereas dust particles of hydroquinone are apt to remain for a longer period of time and, in dissolving, produce localized areas of high concentration. As with many substances, it is advisable not to wear contact lenses when working with this chemical.

## 5 QUINONE [CAS # 106-51-4]

### 5.1 Source, Uses, and Industrial Exposures

Quinone (*p*-benzoquinone, 1,4-benzoquinone), $OC_6H_4O$, can be prepared by oxidation starting with aniline or by the reduction of hydroquinone with bromic acid (143). The compound has found wide application in the dye, textile, chemical, tanning, and cosmetic industries primarily because of its ability to transform certain nitrogen-containing compounds into a variety of colored substances. Severe local damage to the skin and mucous membranes may occur following contact with solid quinone, solutions of quinone, and quinone vapors condensing upon exposed parts of the body (particularly moist surfaces) (205).

### 5.2 Physical and Chemical Properties

The properties are listed below:

| | |
|---|---|
| Physical state | Large, yellow, monoclinic prisms, with an acrid, chlorine-like odor |
| Molecular weight | 108.09 |
| Specific gravity | 1.318 (20°C) |
| Melting point | 115.7°C |
| Boiling point | Sublimes |
| Vapor pressure | Considerable; sublimes readily upon gentle heating |
| Solubility | Soluble in hot water, alcohol, ether, alkalies |

### 5.3 Determination in the Atmosphere

Air samples can be collected in a midget impinger containing isopropranol and then a color, developed with phloroglucinol, read at 520 nm (206).

### 5.4 Physiological Response

Absorption of large doses of quinone from the gastroenteric tract or from subcutaneous tissues of animals induced local changes, crying, clonic convulsions, respiratory difficulties, drop of blood pressure, and death by paralysis of the medullary centers. Asphyxia appears to play an important role in the terminal picture, both because of pulmonary damage resulting from excretion of quinone into the alveoli and because of certain not too well-defined effects of quinone upon hemoglobin (207). The urine of severely poisoned animals may contain protein, blood, casts, and "free" and "conjugated" hydroquinone.

#### *5.4.1 Pathology*

See Section 4.4.1.

### 5.4.2 Absorption, Excretion, and Metabolism

Quinone is readily absorbed from the gastroenteric tract and subcutaneous tissues. It is partially excreted unchanged, but the bulk is eliminated in conjugation with hexuronic, sulfuric, and other acids.

### 5.4.3 Mode of Action

Quinone apparently has a direct effect on the medulla and the oxygen-carrying capacity of the blood.

### 5.4.4 Cause of Death

Paralysis of the medullary centers is caused by lethal dose of quinone.

### 5.4.5 Effects Observed in Humans

The local changes may include discoloration, severe irritation, erythema, swelling, and the formation of papules and vesicles. Prolonged contact may lead to necrosis. Vapors condensing upon the eyes are capable of inducing serious disturbances of vision (200, 208–212, 214). According to Sterner et al. (213), the injury usually extends through the entire layer of the conjuctiva and is characterized by a deposit of pigment. The staining, varying from diffuse brown to globules of brownish black, is located primarily in the zones extending from the canthi medially to the edges of the cornea. All layers of the cornea are involved in the injury with a resultant discoloration that may be white and opaque or brownish green and translucent. Alteration of the cornea can occur after the pigment has disappeared (204). Ulceration of the cornea has resulted from one brief exposure to a high concentration of the vapor of quinone, as well as from repeated exposures to moderately high concentrations. Following discontinuation of exposure, recovery occurred promptly and spontaneously and appeared to be nearly complete. There were no systemic effects or changes in the composition of the blood or urine (202). Cases of oral poisoning apparently have not been reported (9).

## 5.5 Hygienic Standards of Permissible Exposure

The TLV–TWA (1992–1993) is 0.1 ppm (0.44 mg/m$^3$). The STEL is 0.3 ppm (1 mg/m$^3$) (104). The OSHA PEL is also 0.1 ppm (0.44 mg/m$^3$) (105).

## 5.6 Odor and Warning Properties

Quinone has a pungent odor. The vapors are irritating enough to cause sneezing.

## 5.7 Prevention of Poisoning

Control of the exposure is largely a matter of adequate ventilation. Avoid skin contact and upon contact with clothing, remove contaminated apparel immediately.

## PHENOLS AND PHENOLIC COMPOUNDS

Personal protection (full face, air supplied respirator) may be necessary for short exposures in operations where other controls may not be feasible (205).

## 6 PYROGALLOL [CAS # 87-66-1]

### 6.1 Source, Uses, and Industrial Exposures

Pyrogallol (pyrogallic acid, pyro, 1,2,3-trihydroxybenzene), $C_6H_3(OH)_3$, is prepared by heating dried gallic acid at about 200°C. Carbon dioxide splits off, leaving pyrogallol (160).

Pyrogallol's usefulness in various industries is based primarily on its property of being easily oxidized in alkaline solutions (even by atmospheric oxygen) so that such solutions become potent reducing agents. It is used specifically as a developer in photography and for maintaining anaerobic conditions for bacterial growth. It is additionally used in dying operations, process engraving, and as a topical antibacterial agent (215).

### 6.2 Physical and Chemical Properties

The properties are summarized as follows:

| | |
|---|---|
| Physical state | White, or nearly white needle or leaf-shaped crystals or crystalline powder |
| Molecular weight | 126.11 |
| Specific gravity | 1.453 (4°C) |
| Melting point | 133°C |
| Boiling point | 309°C (decomposes at 293°C) |
| Solubility | Soluble in water (1:2), alcohol (1:1.5), and ether (1:2) at 25°C (145) |

### 6.3 Determination in the Atmosphere

Specific methods for the determination of pyrogallol in air are not available. The content of pyrogallol in the urine can be determined by the method described by Tompsett (33).

### 6.4 Physiological Response

Signs of intoxication include vomiting, hypothermia, fine tremors, weakness, muscular incoordination, diarrhea, loss of reflexes, coma, and asphyxia (12).

Because of its marked reducing action, pyrogallol has a tremendous affinity for the oxygen of the blood. Heyroth and Largent (151) observed that the intravenous injection of 0.3 g/kg into a rabbit provided a sufficient quantity of pyrogallol to unite with all of the oxygen of the blood, thereby causing the death of the animal. There was extensive destruction and fragmentation of the erythrocytes.

The urine of poisoned animals may contain casts, glucose, hemoglobin, methemoglobin, urobilin, and other compounds that cause discoloration.

Repeated absorptions of toxic but sublethal concentrations into the tissues of animals have been found to cause severe anemia, icterus, nephritis, and uremia. The approximate lethal dosages of pyrogallol in aqueous solution for various animal species, under varying conditions of administration, are as follows: rabbit, 1.1 g/kg (orally) (151); rabbit or guinea pig, 1.0 g/kg (subcutaneously) (9); dog or cat, 0.35 g/kg (subcutaneously); and dog, 0.09 g/kg (intravenously) (153).

### 6.4.1 Pathology

Pathological changes in animals caused by pyrogallol include edema and hyperedema of the lungs, and moderate fatty degeneration, round cell infiltration, and necrosis of the liver. The kidneys may show hyperemia, necrosis of the epithelium, granular pigmentation, and glomerular nephritis (216–217). The nuclei of striated muscle disappear, the striation of the muscle is lost, and the muscle plasma swells, coagulates, and decomposes. In the heart, there is separation of fibers of the myocardium, interfibrillary hemorrhages, infiltration of the endocardium, lesions of the endocardium, and fibrous deposits in the valves (218). Changes of the bone marrow and myeloid changes in the spleen were noted after chronic administration of this compound (219).

### 6.4.2 Absorption, Excretion, and Metabolism

Pyrogallol is readily absorbed from the gastroenteric tract and from parenteral sites of injection. Little is absorbed through the intact skin. The bulk of absorbed pyrogallol is readily conjugated with hexuronic, sulfuric, or other acids and excreted within 24 hr via the kidneys (220). A fraction is excreted unchanged.

### 6.4.3 Mode of Action

Most of its pharmacological effects can be traced directly to its potent reducing action.

### 6.4.4 Cause of Death

Death is initiated by respiratory failure. Pyrogallol ties up the oxygen in the blood.

### 6.4.5 Effects Observed in Humans

Cases of human poisoning have not been frequent. Cases reported in the older literature (9) include one man who ingested an aqueous solution containing 8 g of pyrogallol and who recovered after suffering an acute intoxication; another, who ingested 15 g of this compound, died despite prompt vomiting. When applied to the human skin in the form of a salve, it can cause local discoloration, irritation, eczema, and even death (221). Repeated contact with the skin can cause sensitization (222). The symptoms observed in acute intoxications in humans resemble closely the signs of illness displayed by experimental animals.

# PHENOLS AND PHENOLIC COMPOUNDS

## 6.5 Hygienic Standard of Permissible Exposure

No standard has been established for pyrogallol.

## 6.6 Odor and Warning Properties

Pyrogallol is practically odorless.

## 6.7 Prevention and Treatment of Poisoning

Avoid skin contact. Wash affected areas thoroughly with water. If pyrogallol has been swallowed, induce vomiting or administer gastric lavage. Further treatment is symptomatic. Phloroglucinol (1,3,5-trihydroxybenzene) and benzenetriol (1,2,4-trihydroxybenzene) are of little industrial or hygienic importance.

# 7 CRESOL [CAS # 1319-77-3]

Pure cresol (cresylic acid, tricresol, methylphenol), $CH_3C_6H_4OH$, is a mixture of ortho, meta, and para isomers (o-cresol, m-cresol, and p-cresol). Crude cresol (commercial cresol) is a mixture of aromatic compounds containing about 20 percent of o-cresol, 40 percent of m-cresol, and 30 percent of p-cresol with small amounts of phenol and xylenols.

## 7.1 Source, Uses, and Industrial Exposures

Crude cresol is obtained by distilling "gray phenic acid" at a temperature within the ranges of about 180 to 205°C. o-Cresol may be separated from the crude or purified mixture by repeated fractional distillation in vacuo. The m- and p-cresols are separated by treating the mixture with sulfuric acid. This yields the crystallized disulfonic acid of p-cresol, and p-cresol upon hydrolysis. The liquid fraction, when hydrolyzed with steam at 130°C, yields m-cresol, which may be further purified by double distillation in vacuo. Each of the cresols can be prepared synthetically by diazotization of the specific toluidine or by fusion of the corresponding toluenesulfonic acid with sodium hydroxide.

The cresols have found wide application in synthetic resin, explosives, petroleum, photographic, paint, and agricultural industries. They have been used for years as antiseptics, disinfectants, and insecticides and additives to lubricating oils. Cresol is used in the manufacturing of tricresyl phosphate and additionally finds use as an organic intermediate for the manufacture of salicylaldehyde, coumarin, herbicides, and surfactants (223).

The o-, m-, and p-cresols are marketed individually ("practical grade," purity, 98 or 99 percent), in a mixture as cresylic acid, or as crude cresol.

Industrial exposure has occurred in industries utilizing cresol. Fatalities can occur from prolonged and/or extensive skin contact. Because of its low vapor pressure

Table 24.3. Physicochemical Properties of Cresols

| Property | Commercial Cresol | o-Cresol | m-Cresol | p-Cresol |
|---|---|---|---|---|
| Molecular weight | 108.13 | 108.13 | 108.13 | 108.13 |
| Specific gravity | 1.030–1.045 | 1.048 | 1.034 | 1.035 |
| Melting point (°C) | Liquid | 30.9 | 12.0 | 34.8 |
| Boiling point (°C) | 185–205 | 191 | 202.7 | 201.9 |
| Vapor density | 3.72 | 3.72 | 3.72 | 3.72 |
| Vapor pressure (mm Hg at 25°C) | — | 0.2453 | 0.1528 | 0.1080 |
| Refractive index | 1.5353 (31) | 1.5372 (55) | 1.5425 (25) | 1.5 (50) |
| Percent in "saturated" air at 25°C | — | 0.0323 | 0.0201 | 0.0142 |
| Density of "saturated" air at 25°C (air = 1) | — | 1.00089 | 1.00054 | 1.00039 |
| Solubility in water (%) | 1.0 | 2.5 | 0.5 | 1.8 |
| Solubility in mineral oil (%) | — | 2.0 | 2.5 | 0.7 |
| Flash point (closed cup) (°C) (°F) | 43.5 (110.3)[a] | 81 (178) | 86 (187) | 86 (187) |

[a]Closed or open cup not indicated.

and disagreeable odor, cresol usually does not present an acute inhalation hazard. Data are not available concerning vapor concentrations that are toxic to humans.

## 7.2 Physical and Chemical Properties

Crude cresol is a colorless liquid that turns brown upon exposure to air or light; cresol, a colorless crystalline compound; $m$-cresol, a yellowish liquid; and $p$-cresol, a white crystalline compound. Cresol is readily soluble or miscible with organic solvents, vegetable oils, or ether and alcohol ($o$-cresol at 30°C, $p$-cresol at 36°C). Other pertinent physicochemical properties are presented in Table 24.3.

## 7.3 Determination in the Atmosphere

The amount of cresol in contaminated air can be determined in essentially the same manner as that described for phenol. The cresol is absorbed in dilute alkali and determined colorimetrically with diazotized $p$-nitroaniline reagent (21) or absorbed in spectrograde alcohol with direct determination by ultraviolet spectrophotometry (224). OSHA recommends gas–liquid chromatography (NTIS P.B. 250159).

Sampling procedures consist of drawing a known volume of air through a silica gel tube consisting of two 20/40 mesh silica gel sections, 150 and 75 mg, separated by a 2-mm portion of urethane foam. The collected sample is desorbed with acetone and analyzed by gas chromatography (225).

Acetone desorbed samples are analyzed using a gas chromatograph equipped with a flame ionization detector. The column is packed with 10 percent free fatty acid polymer in 80/100 mesh, acid washed DMCS chromosorb W. The useful range of this method is 5 to 60 mg/m³ (226).

Ultraviolet spectrophotometry has been used to detect cresol in air samples. Cresol is measured by detecting particular absorption bands, but interference from other air contaminants with absorption bands in the same range reduced the sensitivity and precision of the ultraviolet spectrophotometric method. Paper and thin-layer chromatography have been suggested as methods for the separation and analysis of cresol and structurally similar compounds (227).

Infrared spectrophotometry is shown to be effective in isolating cresol from other structurally similar substances. Identification and quantitative analyses of the cresol isomers, phenol, and xylenols are possible when either cyclohexane or carbon disulfide is used as the solvent. In the direct-reading infrared analyzers, cresol reportedly absorbs at 8.6 μm, with a sensitivity of 0.3 ppm (228).

The concentration of phenol and cresol in alkaline solutions are determined quantitatively using a colorimetric procedure. When reacted with Folin–Denis reagent, the compound yields distinct colors whose intensities can be measured. The concentration of the cresol isomers is determined as total cresol. This method has been applied to analysis of cresol and phenol in biologic fluids, such as blood and urine (225, 229).

## 7.4 Physiological Response

The predominant signs of local and systemic intoxication produced by these compounds are very similar to those produced by phenol (170, 230, 231). Acute exposures by all routes of absorption may cause muscular weakness, gastroenteric disturbances, severe depression, collapse, and death. Although the effects are primarily on the central nervous system, edema of the lungs and injury of the kidneys, liver, pancreas, and spleen also may occur. Repeated exposures may result in copious salivation, diarrhea, loss of appetite, digestive disturbances, damage to the liver and kidneys, and skin eruptions. Cresol has a marked corrosive action on tissues, producing burns and dermatitis.

Campbell (232) exposed mice to air saturated with vapors of cresylic acid. Single brief exposures did not seem to be harmful, whereas repeated exposures caused fatalities. Rats were reported to survive 8 hr of inhalation of air substantially saturated with vapors at room temperature (233).

*m*-Cresol is generally considered the least toxic; however, reports differ as to whether *o*- or *p*-cresol is the more toxic of the two (Table 24.4). For all practical purposes the three isomers can be considered as having essentially the same degree of toxicity.

### 7.4.1 Pathology

The pathological changes induced by these compounds are similar to those caused by phenol. These include irritation, corrosion, hemorrhages, and destruction of cellular protoplasm of the gastroenteric tract (following oral administration), kidney

Table 24.4. Lethal Dosages of Cresols for Experimental Animals

| Species | Route | Dose (g/kg) | | | Ref. |
|---|---|---|---|---|---|
| | | o-Cresol | m-Cresol | p-Cresol | |
| Mouse | SC (dil. aq. suspensions) | 0.35 | 0.45 | 0.15 | Tollens (50) |
| Rat | Oral (10% soln. in olive oil) | 1.35 | 2.02 | 1.8 | Deichmann and Witherup (52) |
| Rabbit | IV (0.5% aq. soln.) | 0.2 | 0.28 | 0.16 | Deichmann and Witherup (52) |
| | SC (dil. sq. suspensions) | 0.45 | 0.5 | 0.3 | Meili (233), Tollens (50) |
| | Oral (10% soln. in olive oil) | 0.8 | 1.1 | 1.1 | Deichmann and Witherup (52) |
| Cat | SC (20% aq. suspensions) | 0.6 | 0.15 | 0.08 | Deichmann and Witherup (52) |

tubule damage, nodular pneumonia, and congestion of the liver with pallor and necrosis of hepatic cells (46).

In humans (234), Kathe reported that the lungs showed hyperemia, emphysema, edema, and bronchopneumonia with petechial hemorrhages in the pleura. The liver showed turbidity, inflammatory reactions, and fatty degeneration; the kidney showed parenchymatous and hemorrhagic nephritis; the myocardium was degenerated, and there were small hemorrhages in the epicardium and endocardium.

### 7.4.2 Absorption, Excretion, and Metabolism

Cresol is absorbed through the skin, open wounds, and mucous membranes of the gastroenteric and respiratory tracts. The rate of absorption through the skin depends more upon the size of the area exposed than on the concentration of the material applied (43).

The major route of excretion of the cresols is in the urine, where o- and m-cresol can be found as sulfo-conjugated derivatives. The p-cresol is oxidized to p-hydroxybenzoic acid (236). Amounts may be excreted in bile (38) and traces in the exhaled air.

The information available as to the normal metabolism and the rate of absorption, detoxication, and excretion of o-, m-, and p-cresols indicates that these compounds behave very much like phenol (43, 237–239). They are oxidized and conjugated with sulfuric and glucuronic acids. According to Bray et al. (240), cresols are excreted by the rabbit primarily as oxygen conjugates, 60 to 72 percent as ether glucuronides and 10 to 15 percent as ethereal sulfates. The authors could show by paper chromatography that o- and m-cresols are hydroxylated to a small extent and that p-cresol gives rise to the formation of some p-hydroxybenzoic acid. From the urine of animals fed o- and m-cresols they isolated 2,5-dihydroxytoluene, and from those fed p-cresol, p-cresylglucuronide.

Cresols may be determined in biologic material, particularly urine, by extracting the urine with ether and then distilling at 100°C. Phenol and all three isomers of cresol are recovered. The *p*-cresol can be separated by treatment of the mixture with 1-nitroso-2-naphthol. The ortho and meta isomers can be separated from phenol and *p*-cresol by a reaction with 4-aminoantipyrine (26). *o*-Cresol may be separated from *m*-cresol by paper chromatography (241).

When evaluating analyses it should be remembered that the cresols, particularly *p*-cresol, are normally present in human urine. It has been reported that the normal human excretes in the urine from 16 to 39 mg of *p*-cresol/day (26).

### 7.4.3 Mode of Action

Very meager data are available concerning the exact mechanism of action of cresols; however, they appear to act in much the same manner as phenol.

### 7.4.4 Cause of Death

In most cases, death appears to have been caused by respiratory failure.

### 7.4.5 Effects Observed in Humans

Cutaneous application can cause local tissue corrosion, burns, and dermatitis. Upon initial contact, the skin becomes red and gradually fades to white and develops blisters. Upon contact with the eyes, cresol causes irritation, conjunctivitis, corneal burns, and keratitis (236). Certain individuals are hypersensitive to cresol.

The predominant signs of intoxication closely resemble those induced by phenol. The treatment of poisoning and the precautions required for the safe handling and use of cresol and its various preparations are the same as those recommended for phenol.

The degree of systemic illness that may be induced by one of the commercial preparations depends primarily on its content of cresol.

### 7.5 Hygienic Standard of Permissible Exposure

The TLV–TWA (1992–1993) is 5 ppm (22 mg/m$^3$) (104). The OSHA permissible exposure limit is also 5 ppm (22 mg/m$^3$) (105). The "S" skin notation in the listing refers to the potential significant contribution to the overall exposure by the cutaneous route, including mucous membrane and the eyes, either by contact with vapors or, of probable greater significance, by direct skin contact with the substance (104).

### 7.6 Odor and Warning Properties

Cresols have an odor much like that of phenol or creosote. With concentrations in the air as low as 5 ppm the odors are easily recognized.

### 7.7 Safe Handling Procedures

The prevention and treatment of poisoning by cresol should be considered the same as that described for phenol.

## 8 CREOSOTE [CAS # 37331-87-6]

### 8.1 Source, Uses, and Industrial Exposures

Creosote (creosotum, creosote oil, brick oil) is obtained by distillation (205 to 220°C) of the tar obtained from beech and other woods, as well as from coal (200 to 250°C). It has also been prepared from the residue of olives (242). Creosote obtained from beechwood is composed almost entirely of guaiacol ($C_6H_4OHOCH_3$) and cresol ($C_6H_3OHCH_3OCH_3$), whereas creosote obtained from coal tar contains, in addition to these, phenol, cresols, pyrol, pyridine, and other aromatic compounds. Purification of the crude preparation is accomplished by distillation and extraction from suitable oils.

Creosote has found application as an antiseptic, disinfectant, antipyretic, astringent, styptic, germicide, constituent of fuel oil, and therapeutic agent. It is also used as a lubricant for die molds, as an agent for waterproofing, as a wood preservative, as an animal dip, and in the manufacture of chemicals. Creosote is commonly applied to railroad ties, poles, fence posts, marine pilings, and similar wood products exposed to the elements (243).

Workers most likely to be exposed are carpenters, railroad workers, farmers, tar distillers, glass- and steel-furnace attendants, and engineers.

Injuries to the skin or eyes have occurred mainly among men engaged in dipping or in "pickling" and handling "sleepers," mine timbers, and woods for floors and other purposes. Jonas calls attention to burns induced by fine particles of sawdust from creosote-treated lumber. He observed that men with fair skin were very sensitive, whereas dark-skinned workers demonstrated a remarkable resistance. The burns were reduced to a minimum on rainy days, probably because of the decreased dispersion of both the wood particles and creosote. Engels (244) considered the use of creosote-treated timber in mines inadvisable because of the fire hazard and because of the contamination of the air.

### 8.2 Physical and Chemical Properties

The properties are as follows:

| | |
|---|---|
| Physical state | Yellowish or colorless, flammable, oily liquid (often brownish because of impurities or oxidation) |
| Molecular weight | Varies with purity |
| Specific gravity | 1.07 to 1.08 |
| Boiling point | 195 to 400°C |
| Solubility | Slightly soluble in water; miscible with alcohol, ether, fixed or volatile oils; soluble in glycerin and in solutions of fixed alkali hydroxides |

Flash point
   Closed cup    73.9°C (165°F)
   Open cup      85°C (185°F)

### 8.3 Determination in the Atmosphere

The American Wood-Preservers' Association reported the use of AWPA Standard A6-83 (a method for the determination of oil-type preservatives and water in wood) for the determination of creosote levels in preserved wood. This method involves the refluxing and extraction of creosote containing wood borings or shavings using toluene, xylene, or a mixed toluene–xylene solvent. Creosote content is calculated as the difference between the sample weight before and after extraction and is expressed as pounds per cubic foot (lb/ft$^3$) of treated wood (245).

### 8.4 Physiological Response

The signs of intoxication produced in animals resemble those for humans (see below). When administered orally to various animal species the approximate lethal dosages of creosote are 0.1 g/kg for the pigeon and 0.6 to 0.8 g/kg for the rabbit, cat, and dog (12). Cattle have been fatally poisoned by licking creosote from treated telephone poles (244, 246–247).

    The carcinogenicity of creosote oils has been studied quite thoroughly using mice (248–250). It was found that the high carcinogenic potency could not be correlated with the content of benzopyrene. This led to the conclusion that something other than (or in conjunction with) benzopyrene caused the skin cancers.

#### 8.4.1 Pathology

Following oral ingestion the resultant lesions are those of intense irritation and congestion of the entire gastroenteric tract.

#### 8.4.2 Absorption, Excretion, and Metabolism

Creosote is rapidly absorbed from the gastroenteric tract and through the skin. It appears to be excreted in the urine mainly in conjugation with sulfuric, hexuronic, and other acids (208, 251, 252). Oxidation also occurs with the formation of compounds that impart a "smoky" appearance to the urine. Traces are excreted by way of the lungs. Analytic methods for the estimation of creosote apparently have not been applied to analysis of biologic material.

#### 8.4.3 Mode of Action

The mechanism of action has not been well defined.

#### 8.4.4 Cause of Death

Death from large doses of creosote appears to be due largely to cardiovascular collapse.

### 8.4.5 Effects Observed in Humans

Fatalities have occurred 14 to 36 hr after the ingestion of about 7 g by adults or 1 to 2 g by children (12). The symptoms of systemic illness included salivation, vomiting, respiratory difficulties, thready pulse, vertigo, headache, loss of pupillary reflexes, hypothermia, cyanosis, and mild convulsions. The repeated absorption of therapeutic doses from the gastroenteric tract may induce signs of chronic intoxication, characterized by disturbances of vision and digestion (increased peristalsis and excretion of bloody feces). In isolated cases of "self-medication," hypertension (253) and also general cardiovascular collapse (12) have been described.

Contact of creosote with the skin or condensation of vapors of creosote on the skin or mucous membranes may induce an intense burning and itching with local erythema, grayish yellow to bronze pigmentation (254, 255), papular and vesicular eruptions, gangrene (256, 257), and in isolated instances cancer (212, 257–261). A review of this phase of the problem is given by Hueper (209). Heinz bodies have been noted in the blood of a patient 1 year after his exposure to creosote (262). Jonas (255) and Goldenberg (263) made similar observations following percutaneous absorption of this preparation. Upon contact with the eyes, creosote caused protracted keratoconjunctivitis. This involves loss of corneal epithelium, clouding of the cornea, miosis, irritability, and photophobia. Subsequently, both blurring of vision and superficial keratitis can occur (264).

According to Jonas permanent corneal scars result in about one third of such cases. Photosensitization has been reported by Schwartz and Tulipan (171) and severe systemic illness by Lewin (12).

### 8.5  Hygienic Standard of Permissible Exposure

A TLV for creosote has not been established.

### 8.6  Flammability

Creosote is flammable and burns with a luminous, smoky flame.

### 8.7  Odor and Warning Properties

Creosote has a penetrating smoky odor and a burning, caustic taste.

### 8.8  Prevention of Poisoning by Creosote

Protective measures include adequate ventilation, use and frequent change of protective garments, wearing of goggles, application of a heavy layer of petroleum jelly (255) or lanolin–castor oil ointment upon the skin of the face, thorough cleansing of all parts of the body that have become exposed accidentally, and thorough washing of the face and hands after every working period. Jonas recommended a preparation containing calcium salts, benzocaine, sulfur, and a vegetable oil base for the treatment of creosote burns. For inflamed eyes he suggested washing with aqueous boric acid or, if pain is a factor, applying Metaphen ophthalmic ointment. [Coal tar ointments will intensify the local damage (261).] Caution is in order when old creosote-treated lumber is handled or sawed because it retains a considerable portion of the oil for periods up to 25 or 30 years (265, 266).

## 9 PENTACHLOROPHENOL [CAS # 87-86-5] AND SODIUM PENTACHLOROPHENATE [CAS # 131-52-2]

The solubilities of pentachlorophenol and its sodium salt differ, but their toxic effects are the same. Therefore, for practical purposes, the following information pertains to both compounds.

### 9.1 Source, Uses, and Industrial Exposures

Pentachlorophenol (Santophen 20, Penta, Dowicide 7, Penchloral, PCP, Cuprinol, Evrisan, Santobrite), $C_6Cl_5OH$, is prepared by the chlorination of phenol. Pentachlorophenol and its sodium salt, $C_6Cl_5ONa$, are used to prevent the growth of fungi, mold, algae, lichens, mosses, and other microorganisms. Materials treated include wood and other cellulosic products, textiles, paints, adhesives, leather, pulp, paper, and industrial waste systems (267).

Pentachlorophenol is commonly utilized for its qualities as a general disinfectant for trays on which mushrooms are grown. It might also be used as a preemergence herbicide (268).

Technical-grade pentachlorophenol contains 4 to 12 percent tetrachlorophenols, which are pesticides (fungicides) registered with the Environmental Protection Agency. The technical material finds extensive use in cooling towers of electric plants as additives to adhesives based on starch and vegetable and animal protein, in shingles, roof tiles, brick walls, concrete blocks, insulation, pipe sealant compounds, photographic solutions, textiles, and drilling muds in the petroleum industry, among other applications (269).

After prolonged or repeated contact with the skin, the dust or solutions of pentachlorophenol can cause dermatitis and systemic intoxication (270, 271). Low concentrations of the dust (as little as $0.3$ $mg/m^3$) irritate the mucous membranes of the nose, throat, and eyes. Fatalities have occurred from absorption of solutions through the skin.

### 9.2 Physical and Chemical Properties

The properties of pentachlorophenol are listed below:

| | |
|---|---|
| Physical state | White, monoclinic, crystalline solid (technical-grade, dark gray to brown) |
| Molecular weight | 266.3 |

Table 24.5. Solubility of Pentachlorophenol and Its Sodium Salt

| Solvent | Solubility at 20°C (% Wt./Vol.) |
|---|---|
| *Pentachlorophenol* | |
| Organic solvents | |
|   Methanol | 57.0 |
|   Ethanol | 53.0 |
|   Ethanol (95%) | 47.5 |
|   Diethylene glycol | 27.5 |
|   Acetone | 21.5 |
|   Xylene | 14.0 |
|   Benzene | 11.0 |
|   Ethylene glycol | 6.0 |
|   Carbon tetrachloride | 2.0 |
| Oils of vegetable origin | |
|   Pine oil | 32.0 |
|   Corn oil | 14.8 |
|   Linseed oil (raw) | 13.9 |
|   Tung oil | 11.2 |
|   Turpentine | 3.0 |
| Water | 14[a] |
| *Sodium Pentachlorophenate* | |
| Water | 21 at 5°C |
|  | 29 at 40°C |

[a] Parts per million.

| | |
|---|---|
| Specific gravity | 1.85 (22°C) |
| Melting point | 188 to 189°C |
| Boiling point | 310°C (decomposes) |
| Vapor pressure | 0.00011 mm Hg (20°C) |
| Percent in "saturated" air | 0.0000145% by volume (20°C) |
| Density of "saturated" air | 1.0000011 (air = 1) |

The solubilities of these materials are of particular interest, because they are generally used in solution (Table 24.5).

Irradiation of a dilute aqueous solution (100 ppm) with sunlight or ultraviolet light results in the formation of photodegradation products such as chlorinated phenols, tetrachlorodihydroxyl benzenes, and nonaromatic fragments such as dichloromaleic acid. The prolonged irradiation of pentachlorophenol or its photodegradation products will yield colorless solutions containing no ether-extractable volatile materials (272).

## 9.3 Determination in the Atmosphere and in Biologic Material

An electrostatic precipitator or glass fiber filter paper can be used to collect dust or fumes of pentachlorophenol from contaminated air (273). The compound can then be determined by the spectrophotometric method of Deichmann and Schaefer (274), which was developed for the estimation of pentachlorophenol in tissue and water, or by chromatography.

Determination of sodium pentachlorophenol in the atmosphere is created out by thin-layer chromatography. This method is based on the reaction between sodium pentachlorophenolate and silver on a thin layer of silica gel (275).

Pentachlorophenol can be measured quantitatively by other methods after first being isolated by solvent partitioning or by steam distillation. The quantitative methods include the copper-pyridine, the lime-ignition, the methylene blue, spectroscopic, and direct titration methods (267). In blood, urine, adipose tissue, or clothing it can be measured at levels of detection in the low parts per billion range using small amounts of sample (276–282).

### 9.4 Physiological Response

When absorbed in sufficient quantity into the tissues of dogs, rabbits, rats, and guinea pigs, pentachlorophenol and sodium pentachlorophenate (283, 284) produce an acute toxic state characterized by accelerated respiration, moderately elevated blood pressure, hyperpyrexia, hyperglycemia, glycosuria, and hyperperistalsis (vomiting was observed in dogs after a subcutaneous dose). The urinary output is increased at first, but diminished later. There is a rapidly developing motor weakness which, in fatal cases, terminates in asphyxial convulsions with cardiac and muscular collapse. Rigor mortis is immediate and most profound.

A gradual loss of weight, but no significant changes in the hemoglobin content of the blood, is noted in rabbits given repeated, slightly sublethal doses of the salt. The number of erythrocytes and various types of leukocytes remains within normal limits.

However, studies also show that triglyceride levels are higher in the exposed group, and total cholesterol levels tend to be lower in that group. In addition, lower levels of bilirubin are present in the exposed subjects (285).

Rectal temperature and blood sugar values rise markedly after the administration of a single slightly sublethal dose, but they fail to show such elevations when the doses are repeated.

The general and local effects produced by the application of solutions of pentachlorophenol to the skin of animals vary with the period of exposure and the vehicle employed. Irritation of the skin and marked local damage followed by complete recovery is the usual result of cutaneous application of single or repeated doses of pentachlorophenol in fuel oils. Single and repeated applications of aqueous solutions of the salt may cause mild or no local damage. Table 24.6 presents acutely lethal doses in experimental animals.

In a chronic (2-year) oral study with rats there were decreased body weights, increased serum glutamic–pyruvate transaminase activity, and increased urinary specific gravity at 30 mg/kg/day. There was pigment in the liver and kidneys of males and females at 30 mg/kg/day and of females only at 10 mg/kg/day. Pentachlorophenol was not carcinogenic. There were no effects in males at 10 mg/kg/day or less or in females at 3 mg/kg/day or less. In a reproduction study with rats

**Table 24.6.** Acute Toxicity of Pentachlorophenol and Its Sodium Salt in Experimental Animals (288–290)

| Species | Route | Solvent | Dose Killing ~50% of Animals (mg/kg) | Ref. |
|---|---|---|---|---|
| | | *Pentachlorophenol* | | |
| Rabbit | Cut. | 11% in olive oil | (450 was not fatal) | 288 |
| | | 5% in Stanolex fuel oil | 60 | 289 |
| | | 5% in Shell fuel oil | 130 | 289 |
| | | 1.8% in pine oil | 40 | 289 |
| | SC | 5% in olive oil | 70 | 288 |
| | Oral | 5% in Stanolex fuel oil | 70 | 288 |
| | | 11% in olive oil | 100 | 288 |
| Rat | Oral | 0.5% in Stanolex fuel oil | 27 | 288 |
| | | 1% in olive oil | 78 | 288 |
| | | *Sodium Pentachlorophenate (in terms of Pentachlorophenol)* | | |
| Rabbit | Cut. | 10% aqueous | 250 | 288 |
| | SC | 10% aqueous | 100 | 288 |
| | Oral | 2% aqueous | 275 | 290 |
| | IV | 1% aqueous | 22 | 288 |
| Rat | SC | 2% aqueous | 66 | 288 |
| | Oral | 2% aqueous | 210 | 288 |
| | | 2, 25% aqueous | 125–200 | |
| | | *Copper Pentachlorophenate* | | |
| Rat | Oral | 2–20% suspension in Tween 20 | 600 | |

there were decreased body weights of parents and progeny, and decreased pup survival at 30 mg/kg/day, but there were no effects at 3 mg/kg/day (286).

Pine shavings, used as litter material for cats, of lumber treated with pentachlorophenol caused acute PCP intoxication and deaths (287).

### 9.4.1 Pathology

In animals, the postmortem evidences of injury consist largely of extensive damage to the vascular system, with heart failure, and with involvement of the parenchymatous tissues as a result of the heart failure (congestion and edema) and the injury to small blood vessels (edema and proliferation of endothelium, dilatation of arterioles and capillaries, and hemorrhage).

### 9.4.2 Absorption, Excretion, and Metabolism

Pentachlorophenol and its sodium salt can be absorbed through the skin and from the gastroenteric tract.

Table 24.7. Distribution of Pentachlorophenol in the Tissues of a Rabbit 24 Hr after Oral Administration of One Dose of Sodium Pentachlorophenate Equivalent to 94 mg of Pentachlorophenol (288)

| Tissue or Fluid | Total Weight of Tissue of Fluid (g) | Pentachlorophenol | | |
|---|---|---|---|---|
| | | In 10-g Sample Analyzed (mg) | In Total Tissue or Fluid (mg) | Recovered (% of Amount Ingested) |
| Urine and feces | 199.1 | 3.40 | 66.21 | 70.6 |
| Stomach and intestine (wall and contents) | 575.7 | 0.11 | 6.33 | 6.76 |
| Muscle | 931.0 | 0.065 | 6.05 | 6.42 |
| Bones | 382.6 | 0.075 | 2.86 | 3.06 |
| Skin | 260.9 | 0.08 | 2.08 | 2.22 |
| Blood | 222.6 | 0.08 | 1.78 | 1.90 |
| Liver and gallbladder | 100.0 | 0.14 | 0.70 | 0.743 |
| Kidneys | 17.7 | 0.185 | 0.25 | 0.267 |
| Heart, lungs, and testes | 25.9 | 0.06 | 0.16 | 0.171 |
| Central nervous system | 12.8 | 0.075$^a$ | 0.09 | 0.096 |
| Total recovery | | | 86.51 | 92.23 |

$^a$Entire sample of 12.8 g analyzed.

Complete data on the distribution of pentachlorophenol in the tissues were obtained from two rabbits. Each was given one oral dose of the sodium salt (0.25 percent aqueous solution) equivalent to 37 mg pentachlorophenol/kg; it was then placed in a metabolism cage for 24 hr, after which it was sacrificed for analysis of the tissues. Of the administered quantities of pentachlorophenol, 92 and 91 percent, respectively, were recovered. The bulk of the material was found in the urine, and about 4 and 7 percent, respectively, were present in the gastroenteric tract. The remainder was well distributed throughout the tissues (Table 24.7). Only traces of pentachlorophenol are excreted in conjugated form in the urine of the rat and rabbit.

Because the principal route of elimination of pentachlorophenol from the body of the rabbit is by the urine, urinary concentration might offer a means of determining the severity of human exposure. After the development of very sensitive analytical methods, considerable work has been done to measure the urinary concentration of man after occupational exposure (291–293). Urinary concentrations vary considerably because of type of exposure and time from exposure to sampling, for example; but concentrations of the order of 36 ppm do not appear to be associated with adverse effects. It is important that background levels of the order of 10 to 50 ppb urine are in the general population (291, 294) as well as in untreated control rats (279).

Some attempts to correlate levels in blood with those in urine have been made in rats (279) and in humans (293). The ratio of blood to urine appears to be about 10 to 2.5. Cattle fed 0.05, 0.5, or 5.0 mg/kg pentachlorophenol for 14 days had blood PCP levels of 2, 432 to 454, and 494 ppb. There were no signs of toxicity. Blood PCP had a half-life of less than 2 days. For calves, doses of 4 mg/kg for 14 days were toxic (295).

### 9.4.3 Mode of Action

Apparently, pentachlorophenol causes a radical uncoupling of oxidation and phosphorylation cycles in the tissues. This produces a markedly increased basal metabolic rate and a marked temperature increase (273, 274).

In vitro tests have shown that $10^6$ to $10^3$ (or greater) concentrations of pentachlorophenol uncouple oxidative phosphorylation, inhibit mitochondrial and myosin adenosine triphosphatase, inhibit glycolytic phosphorylation, inactivate respiratory enzymes, and cause gross damage to mitochondria (296).

### 9.4.4 Cause of Death

Hyperpyrexia and cardiac failure apparently are the cause of death in pentachlorophenol poisoning.

### 9.4.5 Effects Observed in Humans

If not handled with proper precautions, pentachlorophenol and its sodium salt are capable of inducing discomfort and local or systemic effects. Manufacturing, handling, and use have shown that irritation of the skin is likely to occur as a result of relatively brief, single exposures to solutions containing more than approximately 10 percent of the material.

Low-level exposure caused urticaria or irritation of nose, throat, and lungs as well as skin (285).

Solutions as dilute as about 1 percent may cause irritation if contacts are prolonged or repeated frequently. Fever, perspiration, headache, malaise, abdominal pains, and a feeling of thirst are common symptoms. In severe exposures, these symptoms escalate to states of coma, convulsions, and vascular collapse (285).

However, solutions containing 0.1 percent or less are not likely to cause any adverse local effects.

Dusts and sprays of pentachlorophenol or sodium pentachlorophenate will cause painful irritation to the eyes and upper respiratory tract. Atmospheric concentrations appreciably greater than 1.0 mg/m$^3$ of air will cause pain in the uninitiated person. Concentrations as high as 2.4 mg/m$^3$ can be tolerated by those who are accustomed to the material. When concentrations are appreciably below 1.0 mg/m$^3$, there are no noticeable effects, even among persons unaccustomed to the material. Severe effects, including fatalities, have been reported in several countries (297–307).

Most of these have resulted from the practically uncontrolled use of pentachlorophenol as an oil solution for the destruction of weeds or termites. In some of the cases, the men dipped planks of wood into 1.5 to 3 percent solutions of a mixture of 80 percent pentachlorophenol and 20 percent sodium pentachlorophenate. Examination of these reports reveals that systemic effects were noted particularly in uninstructed workers who used knapsack sprayers and who drenched their backs as well as walked with bare feet through puddles of the material or who, in dipping operations, disregarded nearly every safe handling precaution. With proper ventilation and protective equipment, pentachlorophenol can be used safely (307).

The signs and symptoms of intoxication include loss of appetite, respiratory difficulties, sweating, hyperpyrexia, gastrointestinal upset, and in severe cases, a rapidly progressive coma.

Excessive pentachlorophenol use causes acute pancreatitis, intravascular hemolysis, and aplastic anemia. In addition, the chronic effects of pentachlorophenols include the reduction of humoral and cell-mediated immunity (296).

Autopsy revealed inflamed gastric mucosa, congestion of the lungs, edema in the brain, cardiac dilatation, centrilobular degeneration in the liver, and mild degeneration of renal tubules. When analyzed, pentachlorophenol was found in the liver, blood, and urine.

Although not directly related to industrial exposures, pentachlorophenol in drinking water (12.5 ppm) and in bath water has caused intoxication (308–310). It has also been shown that babies are more susceptible than adults to pentachlorophenols (311).

Fatal poisoning of infants was traced to improper laundering of diapers and bedding with a material containing sodium pentachlorophenate (and other chlorinated phenols) (90). The tissue levels of one infant that died were about 20 to 34 ppm, with blood levels of the order of 120 ppm or more. Asymptomatic infants had levels in the serum up to 26 ppm.

With most industrial chemicals, the toxicity of impurities is not considered, at least in much detail. Commercial pentachlorophenol has impurities including chlorodibenzodioxins and chlorodibenzofurans, some of which are very toxic materials (312, 313). A purified pentachlorophenol can be made with little or not impurities, but at a higher cost.

Owing to recent publicity it is deemed appropriate to mention that there are highly toxic impurities, but their overall contribution to the toxic effects from occupational exposures to technical pentachlorophenol are not well known. Pentachlorophenol must be handled with appropriate precautions to prevent overexposure.

## 9.5 Hygienic Standard for Permissible Exposure

The TLV–TWA (1992–1993) for pentachlorophenol is 0.5 mg/m$^3$. The STEL is 1.5 mg/m$^3$ (104). The OSHA PEL is also 0.5 mg/m$^3$ (105). The skin notation refers to "the potential significant contribution to the overall exposure by the cutaneous route, including mucous membranes and the eyes, either by contact with vapors or, of probable greater significance by direct skin contact with the substance" (104).

The ACGIH biological exposure index (1992–1993) for pentachlorophenol is 2 mg/g of creatinine in urine collected prior to the last shift of the work week and 5 mg/l of free pentachlorophenol in plasma collected at the end of the work shift (104).

Based upon the epidemiological studies, there is no convincing evidence to indicate that the inhalation of pentachlorophenol in any form results in human carcinogenesis. Similarly there are no studies relating cancer to humans (314) after oral exposure to pentachlorophenol. There is some evidence, however, that pentachlorophenol is carcinogenesic to mice following ingestion (315).

The *Threshold Limit Values for 1992–1993* does not list pentachlorophenol in its carcinogenesis classification but does indicate that it is identified by other sources as a suspected or confirmed human carcinogenesis.

### 9.6 Odor and Warning Properties

There is a rather marked, characteristic odor.

### 9.7 Instructions for Handling Pentachlorophenol

Pentachlorophenol and its sodium salt must be handled with caution. The dust of these materials is very irritating to the eyes, skin, and mucous membranes, and if inhaled, will induce violent coughing and sneezing.

Skin contact with solutions of pentachlorophenol must be avoided in order to prevent absorption and systemic intoxication. Inhalation and ingestion should equally be avoided. Individuals suffering from kidney and liver diseases have a lowered resistance and should not be permitted to suffer occupational exposure.

If exposure to the dust cannot be eliminated completely, then the nose and mouth must be protected by a respirator or folded gauze and the eyes by goggles. Protective clothing should be worn, including a cap or hat and gloves to be replaced immediately if they become soaked with a solution of pentachlorophenol. All clothing worn during one spraying operation should be left at the workplace and laundered before reuse. Washing with soap and water is a must before eating, drinking, or smoking. At the end of each day, a workman should shower and change into clean clothing.

### 9.8 Treatment of Local Effects and Systemic Intoxication

Solutions spilled on the skin are best removed by prompt and thorough washing with soap and water. If the material gets in the eyes, they should be washed in flowing water for 5 to 30 min, and then brought to the attention of an ophthalmologist who may apply ophthalmic drops or medication, if indicated.

If a solution of pentachlorophenol or sodium pentachlorophenate is accidentally swallowed, vomiting should be induced immediately. This first-aid treatment should be followed by thorough gastric lavage with water or milk and the administration of a saline cathartic.

Treatment of a systemic intoxication should be directed toward promoting loss of heat, allaying anxiety, and replacing lost fluids and electrolytes.

# PHENOLS AND PHENOLIC COMPOUNDS

Because death is partly caused by the fever, every effort must be made to keep the body temperature from rising dangerously. Sponging with alcohol is of very limited usefulness. Cooling the entire body with ice-cold towels is more effective.

However, if the temperature continues to rise above 40°C (104°F), then nothing but heroic treatment will save the victim. If possible, place the individual in a tub filled with cool water, then add crushed ice (DO NOT USE DRY ICE). This is an emergency; call a physician! Antipyretic drugs are of no value; they can even be dangerous. Atropine, the drug of choice is an organophosphate intoxication, is decidedly dangerous and contraindicated. Two recent reports suggest forced diuresis to increase the urinary excretion of pentachlorophenol (316).

## 10 OTHER CHLOROPHENOLS

There are many mono-, di-, tri-, and tetrachlorophenols. Obviously, each compound cannot be discussed separately in a report of this type. Therefore, some of the physical and chemical properties are given for at least one representative in each group (di, trichlorophenols, etc.).

The toxicity is discussed for the chlorophenols as a whole. Individual differences are emphasized.

### 10.1 Source, Uses, and Industrial Exposures

Most of the chlorophenols are prepared by the chlorination of phenol.

The tetrachlorophenols (and their sodium salts) have been used as fungicides and wood preservatives. The trichlorophenols are used as fungicides and bactericides. 2,4-Dichlorophenol [CAS # 120-83-2] is used in organic synthesis, 2,4-Dichlorophenoxyacetic acid [CAS # 94-75-7] is used as a weed killer.

### 10.2 Physical and Chemical Properties

Some of the pertinent chemical and physical properties of the chlorinated phenols are presented in Table 24.8.

### 10.3 Determination in the Atmosphere

Methods for the determination of chlorophenols in air apparently are not available.

### 10.4 Physiological Response

In rats, oral, subcutaneous, and intraperitoneal lethal doses of the chlorophenols produce similar signs of poisoning. Oral administration, however, results in fatal poisoning in smaller dosage and in a shorter period of time than subcutaneous administration.

Restlessness and an increased rate of respiration appear a few minutes after

**Table 24.8. Physicochemical Properties of Some Chlorophenols (18)**

| Compound | Empirical Formula | Physical State | Molecular Weight | Specific Gravity | M.P. (°C) | B.P. (°C) | Solubility (Parts per 100 Parts) | | |
|---|---|---|---|---|---|---|---|---|---|
| | | | | | | | Water | Alcohol | Ether |
| o-Chlorophenol[a] | ClC$_6$H$_4$OH | Colorless liquid | 128.56 | 1.241 (18.2/15°C) | α, 7; β, 0; γ, 4.1 | 175–176 | 2.85[b] | Sol. | Sol. alk |
| m-Chlorophenol | ClC$_6$H$_4$OH | Needles | 128.56 | 1.268 (25°C) | 32–33 | 214 | 2.6[b] | Sol. | Sol. |
| p-Chlorophenol | ClC$_6$H$_4$OH | Needles | 128.56 | 1.306 (20/4°C) | 41–43 | 217 | 2.71[b] | V. sol. | V. sol. |
| 2,4-Dichlorophenol | Cl$_2$C$_6$H$_3$OH | Needles/benzene | 163.01 | 1.383 (60/25°C) | 45 | 209–210 | 0.45[b] | V. sol. | V. sol. |
| 2,3,5-Trichlorophenol | Cl$_3$C$_6$H$_2$OH | Needles/aq. alcohol | 197.46 | | 55 | 249–250 | Sl. sol. hot Sol. | | Sol. |
| 2,4,5-Trichlorophenol | Cl$_3$C$_6$H$_2$OH | Colorless/petroleum ether | 197.46 | | 61–63 | 252 | Insol. | Sol. | Sol. |
| 2,4,6 Trichlorophenol | Cl$_3$C$_6$H$_2$OH | Needles | 197.46 | 1.490 (75/4°C) | 68–69 | 246 | 0.09[c] | V. sol. | V. sol. |
| 2,3,4,6-Tetrachlorophenol | Cl$_4$C$_6$HOH | Needles | 231.90 | 1.6 (60/4°C) | 69–70 | 164 (23 mm) | V. sl. sol. | V. sol. | V. sol. |

[a] Vapor pressure 1.393 (18–80°C).
[b] 20°C.
[c] 25°C.

Table 24.9. Lethal Dosages of Some Chlorophenols to Rats

| Compound | IP MLD (312) | Oral $LD_{50}$ (61) | SC $LD_{50}$ (61) |
|---|---|---|---|
| o-Chlorophenol | 0.23 | 0.67 | 0.95 |
| m-Chlorophenol | 0.36 | 0.57 | 1.39 |
| p-Chlorophenol | 0.28 | 0.67 | 1.03 |
| 2,4-Dichlorophenol | 0.43 | $0.58^a$ | $1.73^b$ |
| 2,4,5-Trichlorophenol | 0.36 | $0.82^a$ | $2.26^b$ |
| Tetrachlorophenol | 0.13 | $0.14^a$ | $0.21^b$ |

[a]Doses expressed as g/kg; compounds administered in olive oil.
[b]Fuel oil was the solvent.

administration of o- and m-chlorophenols and are followed a few minutes later by a rapidly developing motor weakness. Tremors, clonic convulsions (which can be induced by noise or touch), dyspnea, and coma set in promptly and continue until death. Similar signs are produced by p-chlorophenol, but the convulsions are more severe. 2,4- and 2,6-dichlorophenols and 2,4,6-, and 2,4,5-trichlorophenols [CAS # 95-95-4] produce these signs also, but decreased activity and motor weakness do not appear quite so promptly. The tremors are much less severe, but in this case, also, they continue until a few minutes before death. Tetrachlorophenols take an intermediate place between the lower homologues and pentachlorophenol. The signs they produce are similar to those caused by the mono-, di-, and trichlorophenols, except that tremors and convulsions probably are due to asphyxia or hypoglycemia and result from a different mechanism than those noted with the lower chlorinated phenols. Deichmann (61) reported that no hyperpyrexia was produced by these chlorophenols in rats and rabbits, but Farquharson et al. (317) reported hyperpyrexia from injections of tri- and tetrachlorophenols.

The lethal dosages ($LD_{50}$) of some of these compounds are presented in Table 24.9.

The higher chlorinated phenols produce contraction of the isolated rat phrenic nerve diaphragm preparation and stimulate oxygen uptake of rat brain homogenate (317).

From the work of several investigators (61, 317, 318), it can be concluded that the general effect of increasing the chlorination of phenol is a reduction of the convulsant action but an increase in the inhibition of oxidative phosphorylation. (Pentachlorophenol does not produce convulsions.)

### 10.4.1 Pathology

In the rat, monochlorophenols caused marked injury to the kidneys with red blood cell casts in the tubules, fatty infiltration of the liver, and hemorrhages in the intestines (59).

### 10.4.2 Absorption, Excretion, and Metabolism

The compounds are readily absorbed from the gastroenteric tract and from parenteral sites of injection.

The monochlorophenols are excreted as conjugates of sulfuric and glucuronic acids. The urine darkens after standing (319).

### 10.4.3 Mode of Action

It has been postulated that the convulsant action of the lower chlorinated phenols is due to the undissociated molecule (320). The interference with oxidative phosphorylation evidently caused by tri-, tetra-, and pentachlorophenols may be due to the dissociated chlorophenate ion.

### 10.4.4 Effects Observed in Humans

Dermatoses, including photoallergic contact dermatitis, have been reported in humans after exposure to 2,4,5-trichlorophenol, chloro-2-phenylphenol, and tetrachlorophenols; these included papulofollicular lesions, comedones, sebaceous cysts, and marked hyperkeratosis (320–323).

## 10.5 Hygienic Standards for Permissible Exposure

No standards for permissible exposure have been set.

## 10.6 Odor and Warning Properties

The chlorophenols have a distinct odor.

## 10.7 Related Compounds

There are many phenol derivatives that are of some importance in industrial toxicology. Because of the large number of such compounds and their varied uses a full treatment of each compound does not seem to be merited. Therefore, the discussion of these compounds is limited to a brief presentation of toxicity data.

# 11 BROMO- AND IODOPHENOLS

## 11.1 Physiological Response in Animals

These compounds are rapidly absorbed from the gastroenteric tract. The approximate (oral) $LD_{50}$ in the rat in mg/kg of pentabromophenol is slightly more than 200; of 2,4,6-tribromophenol, somewhat less than 2000; and of 2,4,6-triiodophenol, from 2000 to 2500. The compounds were administered as sodium salts in 15 to 40 percent aqueous solutions. In rats and guinea pigs the subcutaneous $LD_{50}$s of $o$-bromophenol are 1.5 and 1.8 g/kg, respectively.

In general, the symptoms produced by all the compounds except pentabromophenol were the same as those produced by pentachlorophenol and consisted of an increased respiratory rate and amplitude followed by loss of muscle tone, collapse, and death. The signs and symptoms from pentabromophenol included increased respiratory rate and amplitude with general body tremors, occasional convulsions, and death.

Evidence of some cumulative toxicity was obtained with 2,4,6-tribromophenol and 2,4,6-triiodophenol. An increase of one- to threefold in the $LD_{50}$ was obtained from five daily doses of these compounds (324). Herdt et al. (325) studied the effects of three potential molluscicides, pentabromophenol and sodium and copper pentachlorophenate. These were given in drinking water to three young bulls at a dosage of 7.6 mg/(kg)(day) for 5 weeks. No significant signs of intoxication and no micropathological changes were noted. The authors believe that these halogenated phenols can be used with safety as molluscicides in the field, provided reasonable precautions are taken in their application.

## 11.2 Pathology in Animals

The pathological changes were most marked in the lungs. All compounds produced mild to severe congestion and petechial hemorrhages. In addition, large doses of 2,4,6-triiodophenol produced severe inflammation of the mucous membrane of the pylorus and fundus of the stomach with corrosion and hemorrhages. These were also produced, but in a considerably less severe form, by the larger doses of 2,4,6-tribromophenol. Smaller doses of these two compounds or larger doses of the others produced only moderate inflammation or congestion of the mucosa of the stomach and intestines.

## 12  *O*-PHENYLPHENOL [CAS # 90-43-7]

### 12.1  Use

This compound, $C_6H_5C_6H_5OH$, also known as 2-hydroxydiphenyl, Orthoxenol, and Dowicide I, is being used as a fungicide, germicide, household disinfectant, intermediate for dyes, and preservative in water–oil emulsions such as are used in the rubber, textile, and metalworking industries. The water–soluble sodium salt of *o*-phenylphenol is used for protecting water-extendible paints against decomposition prior to use and is employed as a preservative for proteins and other types of decomposable adhesives. It is used in the treatment of citrus fruit to prevent mold, in dishwashing formulations, in a fungistatic wax for coating vegetables, and for dipping paper and similar containers employed for the storage of foods (326).

### 12.2  Physiological Response in Animals

Hodge et al. (326) have found that *o*-phenylphenol has a low acute oral toxicity for male rats; the $LD_{50}$ was found to be 2.7 g/kg. When tested on 200 human subjects as a 5.0 percent solution in sesame oil and as a 0.1 percent aqueous solution of the sodium salt, *o*-phenylphenol failed to cause either primary skin irritation or skin sensitization. Male and female rats (25 of each sex per group) that were maintained for 2 years on diets containing 0.02 or 0.2 percent o-phenylphenol showed no adverse effects as judged by gross appearance, growth, hematology, rate of mortality, organ weights, and histopathological changes in various tissues. Similar groups of rats maintained for 2 years on a diet containing 2 percent of the material deviated from the controls by exhibiting slight retardation of growth, histopathological kidney changes (marked tubular dilation), and the presence of small amounts of o-phenylphenol in tissues of the kidney.

Additional studies on cats have shown that ingestion results in hemorrhagic gastroenteritis and hemorrhages in the liver, lungs, and myocardium. Similar studies have also shown phenylphenol to be a weak mutagen (327).

Dogs that received oral doses of 0.02, 0.2, and 0.5 g/kg/day of o-phenylphenol for a period of 1 year showed no adverse effects as judged by body weights, hematologic values, organ weights, and histopathological changes in various tissues.

### 12.3 Effects Observed in Humans

It is recognized that phenylphenol isn't absorbed in toxic amounts through the skin. In fact, oil solutions of up to 5.0 percent result in no ill responses when applied to human skin. However, aqueous solutions of the sodium salt are irritating in concentrations greater than 0.5 percent (328).

Concentrated solutions (20 percent or more) can cause skin irritation; therefore protective gloves and clothing should be worn.

Eye protection is especially recommended since contact can cause corneal injury. This holds particularly true for the material in sodium salt form (329).

## 13 DI-*tert*-BUTYLMETHYLPHENOL [CAS # 29759-28-2]

This compound is also known as DBMP, 4-methyl-2,6-di-*tert*-butylphenol, 2,6-di-*tert*-butyl-*p*-cresol, di-*tert*-butylhydroxytoluene, Deenax, Paranox, DBPC Antioxidant, and Ionil.

### 13.1 Uses

Di-*tert*-butylmethylphenol (DBMP) is an antioxidant that in small amounts prevents the deterioration of a variety of materials, including fats, oils, waxes, resins, and plastic films. It is incorporated into many edible vegetable or animal fats and oils, into baked and fried foods, and into waxes or plastic films used for coating food wrappers or containers.

It acts as an antiskidding agent when added to paints and inks, and its oxidation retardive qualities allow it to be employed in various cosmetics and pharmaceuticals (330).

## 13.2 Physical Properties

The properties are as follows:

Physical state    Slightly yellow, crystalline solid
Melting point    70°C
Boiling point    265°C

## 13.3 Physiological Response in Animals

When absorbed in toxic concentrations into the tissues of unanesthetized animals, DBMP induced signs of intoxication resembling those seen after absorption of a toxic dose of a parasympathetic drug (salivation, a mild degree of miosis, unsteadiness, restlessness, hyperexcitability, diarrhea, and tremors) (331). When given intravenously to a dog under pentobarbital anesthesia, DBMP (25 mg/kg) induced a prompt reduction of blood pressure. Atropine sulfate partially antagonized this depressor effect. Large doses of DBMP produced a gross disturbance of sodium, potassium, and water balance in the rabbit (332). It was concluded that the increase in sodium and aldosterone excretion was due to pyelonephritis, and that death was due to potassium depletion.

From the results of chronic toxicity studies using dogs and rats (331), it was concluded that DBMP is a relatively innocuous compound for occupational handling.

Metabolism studies in rats, dogs, and humans showed some differences in excretion patterns (333–341). In the rat there was slower excretion than in humans, and evidence of enterohepatic circulation was not apparent in humans. In humans more metabolites are excreted in the urine and overall excretion is more rapid than in rats (342). The oral feeding of DBMP increased the detoxification and inhibited cancer induction by known carcinogens (343, 344).

## 13.4 DETERMINATION IN THE ATMOSPHERE

The compound is extracted from a product, a ready-to-eat breakfast cereal for instance, using C52 and determined utilizing gas–liquid chromatography with flame ionization detection methods. Detection and determination of the compound in edible oil are also carried out by gas chromatography and thin-layer chromatography (345).

## 14 p-tert-BUTYLPHENOL [CAS # 27178-34-3]

### 14.1 Uses

p-tert-Butylphenol and other phenolics, such as p-tert-amylphenol, o-benzyl-p-chlorophenol, and p-tert-butylcatechol, are used as germicides. They also are used as plasticizers, emulsion breakers, pour point depressants, in organic synthesis, and/or resins for shoe adhesives.

### 14.2 Physiological Response in Animals

After repeated, prolonged contact with the skin these compounds can cause erythema, ulceration, crusting, and depigmentation.

### 14.3 Effects Observed in Humans

Irritation to the skin and depigmentation have been reported with most of these compounds. In addition, *p-tert*-butylphenol has been reported to cause contact dermatitis (100, 346–350). The irritation usually subsides rapidly after the initial effect; however, the depigmentation with a slower onset likewise has much more long-lasting effects.

## 15 DODECYLTHIOPHENOL [CAS # 36612-94-9]

The acute toxicity of this material, $C_6H_4(C_{12}H_{26})SH$, is of a low order. An intramuscular dose of 20 g/kg is not fatal in the rat; lethal oral doses range from 20 to 30 g/kg. When applied to the skin of the rat, rabbit, and guinea pig, the material induces loss of hair in 6 to 12 days. When applied to the human skin, the material may induce local eczema but apparently does not cause loss of hair (346).

## ACKNOWLEDGMENT

The author wishes to acknowledge and thank Vickie Mercer, Andrew Tan, and Robert F. Rukstalis, who assisted in the preparation of this chapter.

## REFERENCES

1. *Chem. Eng. News*, **12**, April 9, 1990.
2. *Chem. Eng. News*, **29**, December 9, 1991.
3. T. F. Mancuso, *Ohio State Med. J.*, **51**, 672 (1955).
4. *The Merck Index*, 1983, p. 1043.
5. A. G. Gilman, L. S. Goodman, and A. Gilman, Eds, *Goodman and Gilman's The Pharmacological Basis of Therapeutics*, Macmillan Publishing Co., New York, 1985, p. 969.
6. American Conference of Governmental Industrial Hygienists, *Documentation of the Threshold Limit Values and Biological Exposure Indices*, 1986, p. 469.
7. Bacelli, cited by H. C. Wood, Jr., *J. Am. Med. Assoc.*, **32**, 1249 (1899).
8. U. Conforti, *Policlin. Rome*, **44** 1381 (1909).
9. A. Ellinger in A. Heffter, *Handbuch der experimentellen Pharmakologie*, Springer, Berlin, Vol. 1, 1923, p. 893.
10. H. Herding, *Bull. Med.*, **53**, 281 (1939).

11. R. Kobert, *Lehrbuch der Intoxikationen*, Ende, Stuttgart, 1906.
12. L. Lewin, *Cifte und Vergiftungen*, Stilke, Berlin, 1929.
13. B. Mistretta, *Policlin. Rome, Sez. Prat.*, **44**, 2287 (1937).
14. A. Pisani, *Gazz. Med. Lomb.*, **87**, 99 (1928).
15. M. Sein, *Indian Med. Gas.*, **74**, 270 (1939).
16. C. Sironi, *Gazz. Osp. Clin.*, **51**, 395 (1930).
17. H. Zangger, in *Lehrbuch der Toxikologie*, F. Flury and H. Zangger, Eds., Springer, Berlin, 1928; *Occupation and Health*, International Labor Office, Geneva, 1934.
18. N. A. Lange and G. M. Forker, Eds., *Handbook of Chemistry*, Handbook Publishers, Sandusky, OH, 1956.
19. W. B. Deichmann, S. Witherup, and M. Christian, unpublished observations.
20. J. D. Pilcher and T. Sollmann, *J. Pharmacol. Exp. Ther.*, **6**, 377 (1914).
21. H. B. Elkins, *The Chemistry of Industrial Toxicology*, Wiley, New York, 1950.
22. M. M. Braverman, S. Hockheiser, and M. B. Jacobs, *Am. Ind. Hyg. Assoc. Q.*, **18**, 132 (1957).
23. *OSHA Analytical Methods Manual*, 2nd ed., Part 1, 1990.
24. *NIOSH Manual of Analytical Methods*, 3rd ed., Vol. II, Method 3502, 1984.
25. W. Deichmann and L. Schafer, *Am. J. Clin. Pathol.*, **12**, 129 (1942).
26. E. G. Schmidt, *J. Biol. Chem.*, **145**, 533 (1952); **150**, 69 (1943); **179**, 211 (1949).
27. I. W. Tucker, *J. Assoc. Off. Agric. Chem.*, **25**, 779 (1942).
28. S. K. Chirkov, *J. Appl. Chem. U.S.S.R.*, **17**, 31 (1944).
29. H. D. Baernstein, *J. Biol. Chem.*, **161**, 685 (1945).
30. L. Lykken, R. S. Treseder, and V. Zahn, *Ind. Eng. Chem.*, **18**, 103 (1946).
31. D. Glick, Ed., *Methods of Biochemical Analysis*, Vol. 1, Interscience Publishers, York, 1954, p. 27–52.
32. M. D. Armstrong, K. N. F. Shaw, and P. E. Wall, *J. Biol. Chem.*, **218**, 293 (1956).
33. S. L. Tompsett, *Clin. Chem.*, **4**, 237 (1958).
34. R. J. Sherwood and F. W. G. Carter, *Ann. Occup. Hyg.*, **13**, 125 (1970).
35. A. A. Van Haften and S. T. Sie, *Am. Ind. Hyg. Assoc. J.*, **26**, 52 (1965).
36. R. N. Heistand and A. S. Todd, *Am. Ind. Hyg. Assoc. J.*, **33**, 378 (1972).
37. J. E. Walkley, L. D. Pagnatto, and H. B. Elkins, *Am. Ind. Hyg. Assoc. J.*, **22** 362 (1961).
38. J. Teisinger and V. Fiserova-Bergerova, *Arch. Mol. Prof. Med. Trav.*, **16**, 221 (1955).
39. H. J. Docter and R. L. Zielhuis, *Ann. Occup. Hyg.*, **10**, 317 (1967).
40. W. A. Fishbeck, R. R. Langner, and R. J. Kociba, *Am. Ind. Hyg. Assoc. J.*, **36**, 820 (1975).
41. R. E. Grosselin, H. C. Hodge, and R. P. Smith, *Clinical Toxicology of Commercial Products*, Williams & Wilkins, Baltimore, Vol. III, 1984, p. 271.
42. M. Biebl, *Beitr. Pathol. Anat. Allg. Pathol.*, **84**, 257 (1930); *Z. Ges. Exp. Med.*, **87**, 436 (1933); **93**, 515 (1934).
43. V. G. Heller and L. Pursell, *J. Pharmacol. Exp. Ther.*, **63**, 99 (1938).
44. W. B. Deichmann and P. Oesper, *Ind. Med.*, **9**, 296 (1940).
45. L. Wachholz, *Deut. Med. Wochenschr.*, **21**, 146 (1895).

46. O. Wandel, *Arch. Exp. Pathol. Pharmakol.*, **56**, 161 (1907).
47. W. Hesselbach, Inaugural Dissertation, Halle, 1890.
48. R. E. Grosselin, H. C. Hodge, and R. P. Smith, *Clinical Toxicology of Commercial Products*, Williams & Wilkins, Baltimore, Vol III, 1984, p. 271.
49. W. B. Deichmann, K. V. Kitzmiller, and S. Witherup, *Am. J. Clin. Pathol.*, **14**, 273 (1944).
50. K. Tollens, *Arch. Exp. Pathol. Pharmakol.*, **52**, 220 (1905).
51. S. Duplay and M. Cazin, *Compt. Rend.*, **112**, 672 (1891).
52. W. B. Deichmann and S. Witherup, *J. Pharmacol. Exp. Ther.*, **80**, 233 (1944).
53. Tauber and S. Tauber, *Z. Physiol. Chem.*, **2**, 366 (1878–1879); *Arch. Exp. Pathol. Pharmakol.*, **36**, 197, 211 (1895).
54. T. W. Clarke and E. D. Brown, *J. Am. Med. Assoc.*, **46**, 782 (1906).
55. D. I. Macht, *Bull. Johns Hopkins Hosp.*, **26**, 98 (1915).
56. M. I. Smith, E. Elvove, and W. H. Frazier, *Publ. Health Rep. U.S.*, **45**, 2509 (1930).
57. R. E. Grosselin, H. C. Hodge, and R. P. Smith, *Clinical Toxicology of Commercial Products*, Vol III, 1984, p. 271.
58. A. Lesser, *Arch. Pathol. Anat. Physiol.*, **83**, 230 (1881).
59. P. Binet, *Rev. Med. Suisse. Romande*, **15**, 561 (1895); **16**, 449 (1896); cited by W. F. von Oettingen, *Natl. Inst. Health Bull.*, **190** (1949).
60. W. B. Deichmann, S. Witherup, and M. Dierker, *J. Pharmacol. Exp. Ther.*, **105**, 265 (1952).
61. W. Deichmann, *Fed. Proc.*, **2**, 77 (1943); *Arch. Biochem.*, **3**, 345 (1944).
62. E. Baumann, *Arch. Ges. Physiol.*, **13**, 285 (1876); *Z. Physiol. Chem.*, **2**, 335 (1878–1879), **3**, 149, 250 (1879), **6**, 183 (1882), **10**, 123 (1886); *Arch. Anat. Physiol., Physiol. Abt.*, 1879, p. 245.
63. W. B. Deichmann, T. Miller, and J. B. Roberts, *Arch. Ind. Hyg. Occup. Med.*, **2**, 254 (1950).
64. D. E. Jackson, *Experimental Pharmacology and Materia Media*, Mosby, St. Louis, 1939.
65. H. Nicolai, *Klin. Wochenschr.*, **18**, 123 (1939); **20**, 80 (1941).
66. T. Sollmann, P. J. Hanzlik, and J. D. Pilcher, *J. Pharmacol. Exp. Ther.*, **1**, 409 (1910).
67. L. Brieger, *Z. Physiol. Chem.*, **2**, 241 (1878–1879), **3**, 134 (1879).
68. F. Muller, *Berlin. Klin. Wochenschr.*, **24**, 405, 433, 436 (1887).
69. W. F. Rogers, Jr., M. P. Burdick, and C. R. Burnett, *J. Lab. Clin. Med.*, **45**, 87 (1955).
70. W. Deichmann and L. Schafer, *Am. J. Clin. Pathol.*, **12** 129 (1942).
71. H. C. Bray et al., *Biochem. J.*, **52**, 422 (1955).
72. S. Suzaki, N. Takahaski, and F. Egami, *Biochim. Biophys. Acta*, **24**, 444 (1957).
73. R. H. de Meio and M. Wizerkaniuk, *Biochim. Biophys. Acta*, **20**, 428 (1956).
74. J. D. Gregory and F. Lipmann, *J. Biol. Chem.*, **229**, 1081 (1957).
75. E. Becher, S. Litzner, W. Taglich, and F. Doenecke, *Munch. Med. Wochenschr.*, **1925**, 1676; **1927**, 1656; *Klin. Wochenschr.*, **1926**, 147; *Z. Klin. Med.*, **104**, 182, 195 (1926); *Z. Physiol. Chem.*, **47**, 173 (1906).

76. W. J. Darby, R. H. de Meio, M. L. C. Bernheim, and F. Bernheim, *J. Biol. Chem.*, **158**, 67 (1945).
77. R. H. de Meio and R. I. Arnold, *J. Biol. Chem.*, **156**, 577 (1944); *J. Pharmacol. Exp. Ther.*, **84**, 64 (1945); *Rev. Soc. Arg. Biol.*, **17**, 570 (1941); **18**, 158 (1942).
78. L. I. Ipschitz and E. Bueding, *J. Biol. Chem.*, **129**, 333 (1939).
79. S. Balioni and M. Magnini, *Arch. Fisiol.*, **6**, 240 (1909), abstr., *Biochem. Z.*, **10**, 97 (1910).
80. M. Magnini et al. *Arch. Fisiol.*, **8**, 111, 157, 166 (1910); abstr., *Biochem. Z.*, **10**, 831 (1910).
81. K. Lobker, *Deut. Med. Wochenschr.*, **15**, 219 (1889).
82. J. W. C. Gunn, *J. Pharmacol. Exp. Ther.*, **29**, 297 (1926).
83. G. Haas and E. F. Schlesinger, *Arch. Exp. Pathol. Pharmakol.*, **104**, 56 (1924).
84. S. A. Botta et al., *Cardiac Arhythemias in Phenol Face Peeling: A Suggested Protocol for Prevention-Aesth. Plastic Surg.*, **12**, 115 (1988).
85. E. Salkowski, *Arch. Pathol. Anat. Physiol.*, **73**, 409 (1878); *Arch. Ges. Physiol.*, **5**, 33 (1872); *Z. Physiol. Chem.*, **7**, 161 (1882–1883), **13**, 264 (1889).
86. T. Sollman, *A Manual of Pharmacology*, Saunders, Philadelphia, 1927.
87. F. P. Underhill, *Toxicology or the Effects of Poisons*, Blakiston, Philadelphia, 1928.
88. A. R. Cushny, C. W. Edmunds, and J. A. Gunn, *Pharmacology and Therapeutics*, Lea and Febiger, Philadelphia, 1940.
89. C. J. Griffiths, *Med. Sci. Law*, **13**, 46 (1973).
90. B. W. Brown, *Am. J. Publ. Health*, **60**(5), 901 (1970).
91. A. Fisher, *Cutis*, **26**(4), 363 (1980).
92. I. Dymtryshyn, Jr., *J. Am. Acad. Dermatol.*, **9**(1), 163 (1983).
93. J. Lister, *Lancet*, **2**, 95, 353, 668 (1867).
94. R. Falkson, *Arch. Klin. Chir.*, **26**, 204 (1881).
95. R. R. Merliss, *J. Occup. Med.*, **14**(1), 55 (1972).
96. J. K. Piotrowski, *Brit. J. Ind. Med.*, **28**, 172 (1971).
97. J. B. Schjoenberg and C. A. Mitchell, *Arch. Environ. Health*, **30**, 574 (1975).
98. E. A. Oliver, L. Schwartz, and L. H. Warren, *J. Am. Med. Assoc.*, **113**, 927 (1939).
99. R. Adams, *Occupational Contact Dermatitis*, Lippincott, Philadelphia, 1969.
100. G. Kahn, *Arch. Dermatol.*, **102**, 177 (1970).
101. F. D. Arundell, *Arch. Dermatol.*, **108**, 848 (1973).
102. B. Bentley-Phillips, *Arch. Dermatol.*, **110**, 296 (1974).
103. J. McGuire and J. Hendee, *J. Invest. Dermatol.*, **57**, 256 (1971).
104. American Conference of Governmental Industrial Hygienists, *TLVs, Threshold Limit Values for Chemical Substances in Workroom Air Adopted by ACGIH for 1992–1993*.
105. 29 CFR 1910.
106. E. G. Beinhart, *Science*, **103**, 207 (1946).
107. M. S. Nesmeyanova, *Gig. Sanit.*, **7**, 11 (1953); through *Chem. Abstr.*, **48**, 913d (1954).
108. *The Merck Index*, 11th ed., Merck, Rahway, NJ, p. 1150, 1983.
109. *Criteria for Recommended Standard for Phenol*, Dept. HHS, Public Health Services, CDC, NIOSH Pub. No. 761976, July 1976.

110. A. J. Abraham, *Brit. J. Plastic Surg.*, **25**, 282 (1972).
111. R. A. Browne and D. V. Catton, *Can. Anaesth. Soc. J.*, **22**, 208 (1975).
112. T. Fyfe and R. O. Quinn, *Brit. J. Surg.*, **62**, 68 (1975).
113. M. Hamilton and G. M. Wilson, *Q. J. Med.*, **21**, 169 (1952).
114. S. A. Feldman and M. L. Yeung, *Anaesthesia*, **30**, 174 (1975).
115. M. Swerdlow, *Anaesthesia*, **28**, 297 (1973).
116. B. A. Shorey, *Brit. J. Surg.*, **62**, 407 (1975).
117. E. J. Galizia and S. K. Lahiri, *Brit. J. Anaesth.*, **46**, 539 (1974).
118. B. Superville-Sovak, M. Rasminsky, and M. H. Finlayson, *Arch. Neurol.*, **32**, 226 (1975).
119. J. T. Hughes, *Neurology*, **20**, 659 (1970).
120. S. S. Ambasta, *Brit. J. Urology*, **47**, 276 (1975).
121. L. P. Susan and R. J. Marsh, *Urology*, **5**, 119 (1975).
122. R. Wrench and A. Z. Britten, *Brit. J. Dermatol.*, **92**, 575 (1975); **93**, 67 (1975).
123. M. H. Luntz, *Am. J. Ophthalmol.*, **61**, 665 (1966).
124. E. Schutz, *Arch. Exp. Pathol. Pharmacol.*, **232**, 237 (1957).
125. D. M. Conning and M. J. Hayes, *Brit. J. Ind. Med.*, **27**, 155 (1970).
126. D. Henschler, *Medichem*, **27**, 211 (1972).
127. V. K. H. Brown, V. L. Box, and B. J. Simpson, *Arch. Environ. Health*, **30**, 1 (1975).
128. K. Dathe, Inaugural Dissertation, University of Wurzburg, 1976.
129. R. Kobert, *Lehrbuch der Intoxikationen*, Enke, Stuttgart, 1906.
130. L. W. Rising and E. V. Lynn, *J. Am. Pharm. Assoc., Sci. Ed.*, **21**, 138 (1932).
131. D. E. Chase, *J. Am. Water Works Assoc.*, **45**, 493 (1953).
132. J. E. Kinney, *J. Am. Water Works Assoc.*, **45**, 499 (1953).
133 *Standard Methods for Water Analysis*, APHA, Washington, DC, 1978.
134. J. Schneider, *Mar. Biol.*, **16**, 214 (1972).
135. K. Thornton and J. Wilhm, *Hydrobiologia*, **45**, 261 (1974).
136. R. K. Boutwell, H. P. Rusch, and B. Bosch, *Proc. Am. Assoc. Cancer Res.*, **2**, 6 (1955).
137. H. P. Rusch, D. Bosch, and R. K. Boutwell, *Acta Unio Int Contra Cancrum*, **11**, 699 (1955).
138. R. K. Boutwell, K. P. Rusch, and B. Booth, *Proc. Am. Assoc. Cancer Res.*, **2**, 96 (1956).
139. R. K. Boutwell and D. K. Bosch, *Cancer Res.*, **19**, 413 (1959).
140. M. H. Salaman and O. M. Glendenning, *Brit. J. Cancer*, **11**, 433 (1957).
141. H. E. Kaiser, *Cancer*, **20**(1), 615 (1967).
142. R. A. Anderson, *Cancer Res.*, **33**, 2450 (1973).
143. R. Q. Brewster, *Organic Chemistry*, Prentice-Hall, New York, 1948.
144. *International Critical Tables of Numerical Data, Physics, Chemistry and Technology*, Vol. 7, McGraw-Hill, New York, 1930.
145. S. L. Tompsett, *J. Pharm. Pharmacol.*, **10**, 20 (1958).
146. G. Barger and H. H. Dale, *J. Physiol.*, **41**, 19 (1910).

147. C. H. H. Harald, M. Nierenstein, and H. E. Roar, *J. Physiol.*, **41**, 308 (1910).
148. A. Angel and R. N. Lemon, *J. Physiol.*, **225**, 56 (1972).
149. C. W. Flickinger, *Am. Ind. Hyg. Assoc. J.*, **37**, 596 (1976).
150. G. Colasanti and L. Moscatelli, *Boll. Ed Atti Accad. Med. Roma* (1887–1888).
151. F. Heyroth and E. Largent, personal communication.
152. A. Masing, Inaugural Dissertation, Dorpat, 1882.
153. W. Gibbs and H. A. Hare, *Dubois Arch. Physiol.*, **1890**, 352.
154. H. Dietering, *Arch. Exp. Pathol. Pharmakol.*, **188**, 493 (1938).
155. F. Sander, Inaugural Dissertation, Koln, 1933; cited by W. F. von Oettingen, *Natl. Inst. Health Bull.*, **190** (1949).
156. R. E. Grosselin, H. C. Hodge, and R. P. Smith, *Clinical Toxicology of Commercial Products*, Vol. III, 1984, p. 271.
157. W. G. C. Forsyth and V. C. Quainel, *Biochim. Biophys. Acta*, **26**, 155 (1957).
158. F. Vorsatz, *Collegium*, **1942**, 424; *Chem. Abstr.*, **37**, 6927 (1943).
159. A. R. Cushny, C. W. Edmunds, and J. A. Gunn, *Pharmacology and Therapeutics*, Lea and Febiger, Philadelphia, 1940.
160. E. F. Cook and E. W. Martin, *Remington's Practice of Pharmacy*, Mack, Easton, PA, 1948.
161. G. G. Hawley, *The Condensed Chemical Dictionary*, 1981, p. 892.
162. *Merck Index*, Merck, Rahway, NJ, 1940.
163. D. Yeung et al., *J. Chromatogr.*, **224**(3), 513 (1981).
164. L. Brieger, *Z. Physiol. Chem.*, **2** 241 (1878–1879); **8**, 134 (1879).
165. J. Brudzinski. *Ther. Monatsh.*, **13**, 517 (1899); cited by W. F. von Oettingen, *Natl. Inst. Health Bull.*, **190** (1949).
166. I. Becker, *Samml. Vergiftungsfallen*, **4**, 7 (1933); cited by W. F. von Oettingen, *Natl. Inst. Health Bull.*, **190** (1949).
167. A. Osol, Ed., *Remington's Pharmaceutical Sciences*, Mack, Easton, PA, 1980, p. 1107.
168. E. A. Strakosch, *Arch. Dermatol. Syph.*, **48** 384 (1943).
169. L. R. Braathen, *Clinical Toxicology*, **16**, 126 (1985).
170. M. Haenelt, *Munch. Med. Wochenschr.*, **72**, 386 (1925).
171. L. Schwartz and L. Tulipan, *Occupational Disease of The Skin*, Lea and Febiger, Philadelphia, 1939.
172. V. Surbeck, *Deut. Arch. Klin. Med.*, **32**, 515 (1883).
173. A. A. Fisher, *Contact Dermatitis*, 2nd ed., Lea and Febiger, Philadelphia, 1973.
174. P. Quintet, *Ann. Endocrinol.*, **28**, 199 (1967).
175. E. Mastromatteo, *J. Occup. Med.*, **1**, 502 (1965).
176. J. Rankin, *A.M.A. Air Pollut. Med. Res. Conf.*, October 6, 1970.
177. J. F. Gamble, A. J. McMichael, T. Williams, and M. Battigelli, *Am. Ind. Hyg. Assoc. J.*, **37**, 499 (1976).
178. M. L. Clark, D. G. Harvey, D. J. Harvey, D. J. Humphreys, *Veterinary Toxicology*, 2nd Ed., C.R.C. Press, Inc., Boca Raton, Fla., 1981, p. 206.
179. K. G. Harbison and R. T. Kelly, *Environmental Toxicology*, 3rd ed., Vol. 13, p. 39, 1987.

180. Hygienic Guide Series, *Am. Ind. Hyg. Assoc. J.*, **24**, 194 (1963).
181. *NIOSH Manual of Analytical Methods*, 3rd ed., Method 5004, 1984.
182. W. Deichmann, *J. Lab. Clin. Med.*, **28**, 770 (1943).
183. E. Ergriwe, *Z. Anal. Chem.*, **125**, 241 (1943).
184. J. G. Scott, *J. Soc. Motion Picture Engrs.*, **39**, 37 (1942).
185. J. F. Treon and W. Crutchfleld, Jr., *Ind. Eng. Chem.*, **14**, 119 (1942).
186. D. W. Fassett, *Fed. Proc.*, **8**, 290 (1949).
187. H. Oettel, *Arch. Exp. Pathol. Pharmakol.*, **183**, 319 (1936).
188. S. Busatto, *Deut. Z. Ges. Gerichtl. Med.*, **31**, 285 (1939); *Arch. Antropol. Crim. Psichiatr. Med. Leg.*, **60**, 620 (1940).
189. I. Zeidman and R. Deutl, *Am. J. Med. Sci.*, **210**, 328 (1945).
190. M. L. Mareque and A. D. Marenzi, *Compt. Rend. Soc. Biol.*, **127**, 153 (1938).
191. R. Honorato and R. E. Ortuzar, *Reu. Med. Climen. Santiago, Chile*, **5**, 223 (1943).
192. R. Labes, *Arch. Exp. Pathol. Pharmakol.*, **146**, 44 (1929); **152**, 111 (1930).
193. P. Marquardt, *Arch. Exp. Pathol. Pharmakol.*, **201**, 234 (1943).
194. S. Sato and M. Ugai, *Okayama-Igakkai-Zasshi*, **51**, 829 (1939).
195. E. N. Speranskaya-Stepanova, *J. Physiol. U.S.S.R.*, **29**, 334 (1940).
196. L. Zedman and R. Deute, *Am. J. Med. Sci.*, **210**, 328 (1945).
197. J. H. Lapin, *Am. J. Dis. Children*, **63**, 89 (1942).
198. B. Bentley-Phillips and M. A. Bayler, *Brit. J. Dermatol.*, **90**, 232 (1974).
199. A. M. Kligman and I. Willis, *Arch. Dermatol.*, **111**, 49 (1975).
200. B. Anderson, *A.M.A. Arch. Ophthalmol.*, **38**, 812 (1947).
201. T. C. Velhagen, cited in L. Schwartz and L. Tulipan, *Occupational Diseases of The Skin*, Lea and Febiger, Philadelphia, 1939.
202. J. H. Sterner, personal communication.
203. A. L. Dashevskiy and F. F. Marmorshteyn, *Vestnik Oftalmol.*, **19**(1–2), 50 (1941).
204. B. Anderson and F. Olgesby, *A.M.A. Arch. Ophthalmol.*, **59**, 495 (1958).
205. Hygienic Guide Series, *Am. Ind. Hyg. J.*, **24**, 192 (1963).
206. J. H. Sterner, F. L. Oslesby, and B. Anderson, *J. Ind. Hyg. Toxicol.*, **29**, 60 (1947).
207. S. Liu, *Biochem. Z.*, **195**, 248 (1928).
208. A. Hamilton, *Industrial Poisons in The United States*, Macmillan, New York, 1925.
209. W. C. Hueper, *Occupational Tumors and Allied Disease*, Thomas, Springfield, IL, 1942.
210. R. L. Mayer, *Klin. Wochenschr.*, **7**, 1958 (1928); *Arch. Dermatol. Syph.*, **158**, 266 (1929).
211. N. Takizawa, *Proc. Imp. Acad. Tokyo*, **16**, 309 (1940).
212. R. P. White, *The Dermatergoses or Occupational Affections of The Skin*, 4th ed., Lewis, London, 1934.
213. J. H. Sterner, F. L. Oglesby, and B. Anderson, *J. Ind. Hyg. Toxicol.*, **29**, 74 (1947).
214. E. R. Plunkett, *Handbook of Industrial Toxicology*, Chemical Publishing Co., New York, 1966.
215. *The Merck Index*, Vol. 9, Merck, Rahway, NJ, 1976, p. 1038.
216. E. Bauman and E. Herter, *Z. Physiol. Chem.*, **1**, 244 (1877).

217. M. A. Aranassiew, *Arch. Pathol. Anat. Physiol.*, **98**, 472 (1884).
218. A. Natason, Inaugural Dissertation, Dorpat, 1888; cited by A. Ellinger, Herrter's *Handbuch der experimentellen Pharmakologie*, Vol. 1, Springer, Berlin, 1923, p. 891.
219. A. von Domarus, *Arch. Exp. Pathol. Pharmakol.*, **58**, 335 (1908).
220. D. Vitali, *Jahresber. Tierchem.*, **29**, 827 (1804); cited by W. F. von Oettingen, *Natl. Inst. Health Bull.*, **190** (1949).
221. A. Neisser, *Z. Klin. Med.*, **1**, 88 (1880).
222. R. J. Lewis Sr., and N. I. Sax, *Hawley's Condensed Chem. Dictionary*, Van Nostrand Reinhold Co., New York, 1993.
223. E. Zurhelle and S. K. de Boer, *Arch. Dermatol. Syph.*, **183**, 130 (1942).
224. D. W. Fassett, personal communication.
225. Criteria Document: Cresol, NIOSH 92-103, 1978.
226. Criteria Document: Cresol, NIOSH 63, 1978.
227. Criteria Document: Cresol, NIOSH 64, 1978.
228. Criteria Document: Cresol, NIOSH 92-103, 1978.
229. R. M. Chapin, *J. Biol. Chem.*, **47**, 309 (1921).
230. J. Hasenbach, *Zentr. Chir.*, **68**, 67 (1941).
231. F. Beran, *Nachrbl. Dtsch. Pflanzenschutzdienst Berlin*, **20**, 33 (1940).
232. I. Campbell, *Soap Sanit. Chem.*, **17**(4), 103 (1941).
233. H. F. Smyth, Jr., *Am. Ind. Hyg. Assoc. Q.*, **17**, 2 (1956).
234. W. Meili, Dissertation, Bern, 1891.
235. E. Kathe, *Arch. Pathol. Anat. Physiol.*, **185**, 132 (1906).
236. R. LeFaux, *Practical Toxicology of Plastics*, 123 (1968).
237. R. D. Embody et al., *Trans. Am. Fish. Soc.*, **70**, 304 (1940).
238. M. Hunaki, *Mitt. Med. Akad. Kioto*, **29**, 99 (1940).
239. M. E. Klinger and J. F. Norton, *Ind. Hyg. Dig.*, **9**, 355 (1945).
240. H. C. Bray, W. V. Thorpe, and K. White, *Biochem. J.*, **46**, 275 (1950).
241. R. L. Hossfeld, *J. Am. Chem. Soc.*, **73**, 852 (1951).
242. G. de B. Camps, *An. Real. Acad. Farm.*, **2**, 353 (1941).
243. Considine, *Chemical And Process Technology Encyclopedia*, McGraw-Hill, New York, 1974, p. 300.
244. W. Engels, *Chem. Ztg.*, **55**, 285 (1931).
245. World Health Organization, International Agency for Research on Cancer, Monographs on the Evaluations of the Carcinographic Risk of Chemicals to Man, Vol. 35, WHO, Geneva, 1985.
246. G. Hanlon, *Aust. Vet. J.*, **14**, 73 (1938).
247. K. Kasai, *Chem. Abstr.*, **2**, 2583 (1908).
248. W. E. Poel, A. G. Kammer, L. J. Sullivan, and C. B. Willingham, *Proc. Am. Assoc. Cancer Res.*, **2**, 30 (1955).
249. W. E. Poel and A. C. Kammer, *J. Natl. Cancer Inst.*, **18**, 41 (1957).
250. W. Lijinsky, U. Saffiotti, and P. Shubik, *J. Natl. Cancer Inst.*, **18**, 687 (1957).
251. E. J. Fellows, *Proc. Soc. Exp. Biol. Med.*, **42**, 103 (1939).
252. E. J. Fellows, *Am. J. Med. Sci.*, **197**, 683 (1939).

253. S. X. Robinson, *Ill. Med. J.*, **74**, 278 (1938).
254. Hudelo et al., *Bull. Soc. Fr. Dermatol. Syph.*, **34**, 144 (1927).
255. A. D. Jonas, *J. Ind. Hyg. Toxicol.*, **25**, 418 (1943).
256. G. Michel et al., *Rev. med. de l'est*, **63**, 775 (1935).
257. D. Schapiro, *Vrach. Delo*, **11**, 631 (1928).
258. S. Cabot, N. Shear, and M. J. Shear, *Am. J. Pathol.*, **16**, 301 (1940).
259. M. Knallinsky, *Rev. Arg. Dermatoslf*, **23**, 313 (1939).
260. R. D. Sall et al., *J. Natl. Cancer Inst.*, **45** (1940).
261. L. Schwartz, *Ind. Med.*, **11**, 387 (1942).
262. N. Lenson, *N. Engl. J. Med.*, **254**, 520 (1956).
263. E. Y. Goldenberg, *Vrach. Delo*, **10–11**, 663 (1939).
264. W. Grant, *Toxicology of the Eye*, 283 (1986).
265. N. A. Richardson, *Chem. Ind.*, **1934**, 710.
266. H. von Shrenk, A. L. Kammerer, and H. Schmitz, *Proc. Am. Wood. Preservers Assoc.*, 1936, 167.
267. *Santobrite and Monsanto Penta*, Monsanto Application Manual, 09-2(E)M-E-1, 1973.
268. S. B. Walker, C. R. Worthing, Eds. *The Pesticide Manual—A World Compendium*, Thornton Health, UK: British Crop Protection Council, 1987, p. 641.
269. D. P. Cirelli, *Pentachlorophenol*, K. R. Rao, Ed., Plenum, New York, 1977, pp. 13–18.
270. *Pentachlorophenol—Technical*, Monsanto Tech. Bull., IC/PS-8, 1971.
271. *Handling Precautions for Penta and Santobrite*, Monsanto Tech. Bull., IC/PS-3, 1971.
272. A. S. Wong and D. G. Crosby, *Pentachlorophenol*, K. R. Rao, Ed., Plenum, New York, 1977, pp. 19–25.
273. Hygienic Guide Series, *Am. Ind. Hyg. Assoc. Q.*, **18**, 274 (1958).
274. W. B. Deichmann and L. J. Schaefer, *Ind. Eng. Chem.*, **14**, 310 (1942).
275. E. F. Malygina et al., *Gig Sanit.*, **3**, 71 (1972).
276. K. Erne, *Acta Pharmacol. Toxicol.*, **14**, 158 (1958).
277. A. Benevue, J. R. Wilson, E. F. Potter, M. K. Song, H. Beckman, and C. Mallett, *Bull. Environ. Contam. Toxicol.*, **1**, 257 (1966).
278. A. Bevenue, M. L. Emerson, L. J. Casarett, and W. L. Yauger, Jr., *J. Chromatogr.*, **38**, 467 (1968).
279. W. F. Barthel, A. Curley, C. L. Thrasher, V. A. Sedlak, and R. Armstrong, *J. AOAC*, **52**, 294 (1969).
280. M. Cranmer and J. Freal, *Life Sci.*, **9**, 121 (1970).
281. J. R. Rivers, *Bull. Environ. Contam. Toxicol.*, **8**, 294 (1972).
282. T. M. Shafik, *Bull. Environ. Contam. Toxicol.*, **10**, 57 (1973).
283. F. Vallier, L. Roche, and A. Brune, *Compt. Rend. Soc. Biol.*, **148**, 374 (1954).
284. F. Vallier, L. Roche, and A. Brune, *Compt. Rend. Soc. Biol.*, **148**, 690 (1954).
285. W. J. Hayes, Jr., *Pesticides Studied in Man*, Williams & Wilkins, Baltimore, p. 1098.
286. B. A. Schwartz, J. G. Quast, P. A. Keeler, C. C. Humiston, and R. J. Kociba, *Symposium on Pentachlorophenol*, K. R. Rao and L. N. Richards, Eds. Plenum Press, Pensacola, FL, 1978.
287. R. L. Peet, C. MacDonald, and A. Keefe, *Aust. Vet. J.*, **53**(12), (1977).

288. W. B. Deichmann, W. Machle, K. V. Kitzmiller, and C. Thomas, *J. Pharmacol. Exp. Ther.*, **76**, 104 (1942).
289. R. A. Kehoe, W. B. Deichmann, and K. V. Kitzmiller, *J. Ind. Hyg. Toxicol.*, **21**, 160 (1939).
290. W. Machle, W. B. Deichmann, and C. Thomas, *J. Ind. Hyg. Toxicol.*, **25**, 192 (1943).
291. A. Benevue, T. J. Haley, and H. W. Klemmer, *Bull. Environ. Contam. Toxicol.*, **2**, 293 (1967).
292. A. Benevue, J. Wilson, J. Casarett, and H. W. Klemmer, *Bull. Environ. Contam. Toxicol.*, **2**, 319 (1967).
293. L. J. Casarett, A. Benevue, W. L. Yauger, Jr., and S. A. Whalen, *Am. Ind. Hyg. Assoc. J.*, **30**, 360 (1969).
294. T. Akisada, *Jap. Anal.*, **14**, 101 (1965).
295. G. A. Van Celder, *J. Am. Vet. Med. Assoc.*, **173**(7), 885 (1978).
296. E. C. Weinbach, *Proc. Natl. Acad. Sci. U.S.A.*, **43**, 393 (1957); through *Chem. Abstr.*, **51**, 132326 (1957).
297. R. Truhart, P. L. Epee, and E. Poussemart, *Arch. Mal. Prof. Med. Trav. Sec. Soc.*, **13**, 567 (1952).
298. R. Truhart, G. Vitte, and E. Poussemart, *Arch. Mal. Prof. Med. Trav. Sec. Soc.*, **13**, 570 (1952).
299. S. Nomura, *J. Sci. Labour Tokyo*, **29**, 274 (1953).
300. D. Gordon, *Med. J. Aust.*, **2**, 485 (1956).
301. J. A. Menon, *Brit. J. Med.*, **1**, 1156 (1958).
302. A. Benevue and H. Beckman, *Residue Rev.*, **19**, 83 (1967).
303. D. M. Blair, *Bull. WHO*, **25**, 597 (1961).
304. L. Epee, P. Blanc, M. Lavignolle, A. Lazarini, and M. Fave, *Arch. Mal. Prof.*, **24**, 310 (1963).
305. H. Bergner, P. Constantinidis, and J. H. Martin, *Can. Med. Assoc. J.*, **92**, 448 (1965).
306. M. F. Mason, S. M. Wallace, E. Foerster, and W. Drummond, *J. Forensic Sci.*, **10**, 136 (1965).
307. J. Levin, C. Rappe, and C. Nilsson, *Scand. J. Work Environ. Health*, **2**, 71 (1976).
308. K. Uede, M. Nagai, and M. Osafune, *Osaka Sheriti Eisei Kenkyo-Sho, Kenleyu, Hokoku*, **7**, 19 (1962).
309. J. E. Smith, L. E. Loveless, and E. A. Belden, *Morb. Mortal. Wk. Rep.* (IUSDHEW, Atlanta, GA), **16**, 334 (1967).
310. J. B. Chapman and P. Robson, *Lancet*, 1266 (1965).
311. R. Baxter, *Occup. Hyg.*, **28**, 429 (1984).
312. E. C. Villanueva, V. W. Burse, and R. W. Jennings, *J. Agric. Food Chem.*, **21**, 739 (1973).
313. K. R. Rao and N. L. Richards, Ed., *Symposium on Pentachlorophenol*, Pensacola, Florida, June 27–29, 1977.
314. *Toxicological Profile Pentachlorophenol*, Agency for Toxic Substances and Disease Registry, U.S. Public Health Service, 1989.
315. National Toxicology Program, "Technical Report on the Toxicology and Carcinogenesis Studies of Pentachlorophenol (CAS # 07-5) in B6C3F1 Mice (Feed Studies), NIH Publication No. 89-2804.

316. T. J. Haley, *Ecotoxicol Environ. Saf.*, **1**(2), 343 (1977); J. F. Young and T. J. Haley, *Clin. Toxicol.*, **12**(1), 41 (1978).
317. M. E. Farquharson, J. C. Gage, and J. Northover, *Brit. J. Pharmacol.*, **13**, 20 (1958).
318. H. Bechold and P. Ehrlich, *Z. Physiol. Chem.*, **47**, 173 (1906).
319. W. F. von Oettingen, *Natl. Inst. Health Bull.*, **190** (1949).
320. M. G. Butler, *Arch. Dermatol. Syphilol.*, **35**, 251 (1937).
321. K. O. Stingily, *South Med. J.*, **33**, 1268 (1941).
322. A. K. Zelikov and L. N. Danilov, *Sov. Med.*, **7**, 145 (1974).
323. R. M. Adams, *Arch. Dermatol.*, **106**, 711 (1972).
324. E. F. Stohlman, *Publ. Health Rep. U.S.*, **66**, 1303 (1951).
325. J. R. Herdt, L. N. Loomis, and M. O. Nolan, *Publ. Health Rep.*, **66**, 1313 (1951).
326. H. C. Hodge, E. A. Maynard, H. J. Blanchet, Jr., H. C. Spencer and V. K. Rowe, *J. Pharmacol. Exp. and Ther.*, **104**, 202 (1952).
327. R. Gosselin, R. Smith, and H. Hodge, *Clinical Toxicology of Commercial Products*, Williams & Wilkins, Baltimore, Vol. II, 1984, p. 189.
328. *Ibid*.
329. *Ibid*.
330. *The Merck Index*, Merck, Rahway, NJ, 1976, p. 198.
331. W. B. Deichmann, J. J. Clemmer, R. Rakoczy, and J. Bianchine, *A.M.A. Arch. Ind. Health*, **11**, 93 (1955).
332. F. A. Denz and J. G. Llaurado, *Brit. J. Exp. Pathol.*, **38**, 515 (1957).
333. J. C. Dacre, *Biochem. J.*, **78**, 758 (1961).
334. M. Akagi and I. Aoki, *Chem. Pharm. Bull. Tokyo*, **10**, 101 (1962).
335. I. Aoki, *Chem. Pharm. Bull. Tokyo*, **10**, 105 (1962).
336. J. W. Daniel and J. C. Cage, *Food Cosmet. Toxicol.*, **3**, 405 (1965).
337. L. C. Landomery, A. J. Ryan, and S. E. Wright, *J. Pharmacol. London*, **19**, 383 (1967).
338. L. C. Landomery, A. J. Ryan, and S. E. Wright, *J. Pharmacol. London*, **19**, 388 (1967).
339. J. W. Daniel, J. C. Cage, D. I. Jonas, and M. A. Stevens, *Food Cosmet. Toxicol.*, **5**, 475 (1967).
340. Anonymous, *Food Cosmet. Toxicol.*, **6**, 79 (1968).
341. Anonymous, *Food Cosmet. Toxicol.*, **6**, 533 (1968).
342. J. W. Daniel, J. C. Cage, and D. I. Jones, *Biochem. J.*, **106**, 783 (1968).
343. B. M. Ulland, J. H. Weisberger, R. S. Yamamoto, and E. K. Weisburger, *Food Cosmet. Toxicol.*, **11**, 199 (1973).
344. P. H. Grantham, J. H. Weisburger, and E. IC. Weisburger, *Food Cosmet. Toxicol.*, **11**, 209 (1973).
345. Association of Official Analytical Chemists, *Official Methods of Analysis*, **13**, 324 20.009, 1982.
346. A. Enders and A. Moench, *Arzneim. Forsch.*, **2**, 587 (1952).
347. K. E. Malten, *Dermatologica*, **117**, 103 (1958).
348. K. E. Malten, *Dermatologica*, **135**, 54 (1967).
349. D. Beetz, *Dermatol. Monatsschr.*, **157**, 42 (1971).
350. K. Ito, K. Nishitani, and I. Hara, *Bull. Pharm. Res. Inst.*, **76**, 6 (1968).

# Subject Index

Acenaphthene:
 physical properties, 1379
 toxicity, 1385
Acetaminophen:
 toxicity, 972
 uses, 972
4-Acetaminostilbene, 999
Acetanilide:
 methemoglobin formation, 950
  capacity for, 952
 sensitivity, 951
 toxicity, 972
 uses, 972
Acetanilide hydroxylase, benzene and
  production of, 1323
Acetoin, butadiene synthesis, 1251
Acetophenetidin:
 methemoglobin formation, capacity for, 952
 sensitivity, 951
2-Acetylaminofluorene:
 hygienic standards, 970
 physical and chemical properties, 969
 toxicity, 969
 uses, 968–970
2-Acetylaminofluorine, Zymbal gland tumors
  and, 999
N-Acetyl-p-aminophenol, 972
Acetylene:
 physiochemical properties, 1254–1255
 toxicity, 1254–1255
N-Acetylethanolamine, 1164–1165
Acrolein, allylamine and production of, 1134
Aerosol propellants:

 abuse of, 1202–1204
 blood levels in exposed persons, 1201–1202
 bronchodilator, asthmatic deaths from,
  1204–1207
 cardiopulmonary effects, 1200–1202
 respiratory uptake, 1201
 use of, 1199–1200
Alanine aminotransferase, xylene and
  production of, 1337
Aldrin, 1517–1523, 1529–1532
 acute toxicity, 1505, 1518–1519
 carcinogenicity, 1523, 1529–1532
 classification, 1506
 chronic toxicity, 1519
 dieldrin interactions, 1521–1522
 concentrations after starvation, 1520–1521
 human exposures, 1522–1523
 hygienic standards, 1523
 liver enzyme effects, 1521
 metabolism, 1519–1520
 physical and chemical properties, 1518
 regulatory standings, 1504
 reproductive effects, 1520
 skin exposure, 1519
 teratogenicity, 1520
 uses and industrial exposures, 1517–1518
Alicyclic amines:
 absorption, 1093
 biotransformation routes, 1094–1096
 in human environment, 1089
 manufacture, 1092
 metabolism, distribution, and elimination,
  1093–1104

Alicyclic amines (Continued)
  nitrosation by nitrite ions, 1096
  physical and chemical properties, 1087–1092
    monamines, 1090
    polyamines, 1091
  physiological and pathological effects:
    in animals, 1104, 1108–1110
    in humans, 1110–1111
    local irritation, 1108
    Smyth-Carpenter range finding tests, 1105–1107
    sympathomimetic activity, 1108–1109
  uses, 1092–1093
Alicyclic hydrocarbons, see Cycloalkanes; Cycloalkenes; Cycloolefins; Dicyclic alkenes
Aliphatic amines, see also Aliphatic monoamines; Aliphatic polyamines
  absorption, 1093
  N-acetylation, 1097
  biotransformation routes, 1094–1096
  histamine release, 1109–1110
  in human environment, 1089
  identified as carcinogens, 960
  manufacture, 1092
  metabolism, distribution, and elimination, 1093–1104
    long- and branched chains, 1101
  microsomal N-oxidation, 1095–1096
  nitrosation by nitrite ions, 1096
  physical and chemical properties, 1087–1092
  physiological and pathological effects:
    in animals, 1104, 1108–1110
    in humans, 1110–1111
    local irritation, 1108
    Smyth-Carpenter range finding tests, 1104, 1105–1107
    sympathomimetic activity, 1108–1109
  uses, 1092–1093
Aliphatic hydrocarbons, saturated:
  industrial hygiene, 1225
  microbiological characteristics, 1225
  occurrence and use, 1221
  physical characteristics:
    flammability, 1222
    lipid solubility, 1222
    volatility, 1222
  threshold limit values, 1226
  toxicity, 1223–1225
Aliphatic monoamines, physical and chemical properties, 1090
Aliphatic polyamines, 1140–1149
  as acyl acceptors, 1140
  metabolism, distribution and elimination, 1101–1103

physical and chemical properties, 1091
toxicity, 1140
uses, 1093
Alkanes:
  boiling range, 1222
  hazard properties, 1223
  physicochemical properties, 1226–1227
Alkanolamines, 1149–1157
  basicity, 1088–1089
  physical characteristics and toxicologic levels, 1112
  toxicity, 1150
  uses, 1150
Alkatrienes, toxicity, 1253–1254
Alkenes:
  hygienic standards, 1244
  occurrence and use, 1240
  physical and chemical properties, 1241, 1242–1244
  toxicity, 1241
Alkylalkanolamines, 1157–1165
Alkylamines:
  basicity, 1088–1089
  commercial production, 1092
Alkylamines, higher, 1129–1132
  toxicity, 1130
  uses, 1129–1130
Alkyldiphenyls, toxicity, 1365
Alkylnaphthalenes, toxicity, 1377
Alkynes:
  physiochemical properties, 1254
  toxicity, 1254–1256
Allene, see Propadiene
Allylamine:
  pathological changes, 1110
  physical and chemical properties, 1090
  toxicity, 1106, 1111, 1132–1133, 1133–1134
  uses, 1133
Allylbenzene:
  physical properties, 1304–1305
  toxicity, 1360
American Conference of Governmental Industrial Hygienists (ACGIH), carcinogen standards, 959–960
Amine oxidation enzyme system, 1095–1096
Amino compounds, acute toxic response, 947
3-Aminoacetanilide, methemoglobin formation, 1021
3-Aminoaniline, see m-Phenylenediamine
4-Aminoaniline, see m-Aminoazotoluene
1,2-Aminoaniline, see o-Phenylenediamine
p-Aminoaniline see p-Phenylenediamine
o-Aminoazotoluene:
  physical and chemical properties, 1038
  toxicity, 1038

uses, 1038
Aminobenzene, see Aniline
4-Aminobiphenyl:
  N-oxidation, 1040
  physical and chemical properties, 1036
  toxicity, 1036–1037
  uses, 1036
Aminobiphenyls, 1035–1037
γ-Aminobutyric acid, ethylenediamine and, 1141, 1142
4-Aminodiphenyl, carcinogenicity, 959
p-Aminodiphenylamine:
  physical and chemical properties, 1038
  toxicity, 1038
  uses, 1037
2-Aminoethanol, toxicity, 1106
Aminoethylethanolamine, toxicity, 1165
1-Aminoheptane:
  toxicity, 1128
  uses, 1128
2-Aminoheptane:
  toxicity, 1128
  uses, 1128
3-Aminoheptane:
  toxicity, 1129
  uses, 1129
4-Aminoheptane:
  toxicity, 1129
  uses, 1129
o-Aminolevulinic acid synthetase, in HCB metabolism, 1488
o-Aminolevulinic acid:
  benzene and, 1324
  in HCB metabolism, 1487
5-Aminolevulinic acid, elevation with HCB, 1481
m-Aminonitrobenzene, methemoglobin formation, capacity for, 952
2-Amino-5-nitrophenol:
  physical and chemical properties, 1004
  toxicity, 1004
4-Amino-2-nitrophenol:
  physical and chemical properties, 1004
  toxicity, 1004
m-Aminophenol:
  physical and chemical properties, 971
  toxicity, 971
  uses, 971
o-Aminophenol:
  methemoglobin formation, 950
    capacity for, 952
  physical and chemical properties, 970
  toxicity, 970
  uses, 970

p-Aminophenol:
  methemoglobin formation, 950
    capacity for, 952
  physical and chemical properties, 973
  reproductive effects, 964
  teratogenicity, 972
  toxicity, 973
  uses, 971–972
Aminophenols, biogradation, 967
4-Aminophenyl ether:
  physical and chemical properties, 1014
  toxicity, 1014–1015
  uses, 1014
1-Amino-2-propanol, see Isopropanolamine
3-Amino-1-propanol:
  physical characteristics, 1112
  toxicity, 1107, 1112
Aminopyrine N-demethylase, benzene and production of, 1323
Aminopyrine demethylase, HCB and increases in, 1488
4-Aminostilbene, 999
11-Aminoundecanoic acid, 1131
Ammonia:
  basicity, 1089
  in manufacture of aliphatic amines, 1092
n-Amylamine:
  metabolism, distribution, and elimination, 1101
  physical and chemical properties, 1090
  toxicity, 1125
  uses, 1125
Amylamines, 1125–1126
  toxicity, 1106
Amylbenzene, toxicity, 1351
α-n-Amylene, see 1-Pentene
Aniline:
  analgesic and antipyretic effects, 972
  biogradation, 966
  bladder cancer risk, 1044
  carcinogenicity, 984
  cross-sensitization with p-phenylenediamine, 1008
  in diaper marking, 953
  hygienic standards, 984
  metabolism, 955
  methemoglobin formation, 950
  physical and chemical properties, 983
  toxicity, 983–984
  uses, 983
Aniline hydrochloride, teratogenicity, 965
Aniline hydroxylase:
  elevation with HCB, 1481
  HCB and increases in, 1488

Aniline tumors, 983-984
Anosmia, from aliphatic amines, 1088
Anthracene:
  derivatives, 1386
  hygienic standards, 1385-1386
  microbial studies, 1385
  physical properties, 1379, 1385
  toxicity, 1380, 1385
Antiobesity agents, toxic effects, 977
Antioxidants, isobutene in production of, 1247
Apnea, isobutane-related, 1231
Aromatic amines:
  acetylation, 955
  demethylation, 955
  identified as carcinogens, 960
  known to be carcinogens, 959
Aromatic amino compounds:
  aquatic toxicity, 967-968
  biogradation, 966-967
  carcinogenicity, 956-963
  environmental concerns, 965-968
  $N$-hydroxylation, 956-957
  metabolism, 953-956
  methemoglobin formation, 953-954
    alternate pathways, 951
    capacity for, 952
  production and use, 948, 949
  reproductive effects, 963-965
  toxicity, 948
  toxicology, 949-968
    methemoglobinemia, 949-953
Aromatic hydrocarbons:
  monocyclic:
    accumulation in marine species, 1306
    characteristics, 1301-1302
    chemical structure, 1301-1302
    industrial hygiene, 1306
    medical surveillance programs, 1306
    metabolism, 1303
    microbial effects, 1306
    origin and sources, 1302
    photochemical degradation, 1302
    physical properties, 1302
    toxicity, 1302-1303
    utilization, 1302
  physical properties, 1304-1305
  polycyclic, 1378-1393
    physical properties, 1379
Aromatic nitro compounds:
  acute toxic response, 947
  aquatic toxicity, 967-968
  biogradation, 966-967
  carcinogenicity, 956-963
  environmental concerns, 965-968

  enzymatic reduction, 957
  $N$-hydroxylation, 956-957
  metabolism, 953-956
  methemoglobin formation, 953-954
    alternate pathways, 951
    capacity for, 952
  physical properties, 949
  production and use, 948
  reproductive effects, 963-965
  toxicity, 948
  toxicology, 949-968
    methemoglobinemia, 949-953
Aromatic petroleum naphtha:
  hygienic standards, 1410
  physical properties, 1397
  toxicity, 1399-1400, 1409-1410
Arsine, in acetylene production, 1254
$N$-Arylhydroxylamines, reactivity, 957
$N$-Arylnitrosamines, formation from nitrite reactions, 957
Aspiration pneumonitis, from aliphatic hydrocarbons, 1223
Asthmatic deaths, from bronchodilator aerosols, 1204-1207
Axonopathy, aliphatic hydrocarbon-induced, 1224-1225

Benzanthracene:
  physical properties, 1379
  toxicity, 1380-1381, 1387
  uses, 1386
Benzene, 1309. *See also* Halogenated benzene
  acute toxicity:
    cellular effects, 1309, 1318
    dermal, 1308, 1318
    eye, 1318
    inhalation, 1308-1309, 1318-1319
    oral, 1307-1308, 1318
  biogradation, 1325
  biologic monitoring, 1325-1326
  carcinogenicity, 1325
  chemical properties, 1306-1307
  chronic toxicity, 1319-1325
    clinical determination, 1322-1323
    inhalation, 1319-1320
    prophylaxis, 1323
    sex differences, 1325
    subcellular effects, 1322
  cytotoxic effects, 1324
  fetotoxicity, 1321
  hematologic and subcellular effects, 1324-1325
  hygienic standards, 1325
  immunologic response, 1322, 1325

# SUBJECT INDEX 1635

metabolism, 1323–1324
  acute exposures, 1319
  chronic exposures, 1320
mutagenicity, 1325
nitration, 948
occurrence and preparation, 1307
origin and sources, 1302
physical properties, 1304–1305, 1306–1307
precautionary measures, 1326
toxicity:
  acute, 1309, 1310
  animal studies, 1310, 1314–1317
  chronic, 1312–1313, 1314–1317
  human exposures, 1309, 1312–1313
uses, 1307
Benzidine:
  biotransformation, 955
  carcinogenicity, 959
    epidemiologic studies, 961–962
  physical and chemical properties, 1033
  toxicity, 1033
  uses, 1033
  Zymbal gland tumors, 999
Benzo(a)anthracene:
  metabolism, 1384
  occurrence, 1378
Benzoic acid, in toluene metabolism, 1330
1,2-Benzophenanthrene, physical properties, 1379
3,4-Benzophenanthrene:
  physical properties, 1379
  toxicity, 1381, 1389
Benzophenanthrenes, toxicity, 1389
Benzo(a)pyrene:
  hygienic standards, 1391
  metabolism, 1391
  microbial studies, 1391
  mutagenicity, 1391
  occurrence, 1378
  production, sources, and uses, 1390
  teratogenicity and mutagenicity, 1384
  toxicity, 1378, 1380, 1382–1383, 1384, 1390–1391, 1392
Benzo(e)pyrene:
  physical properties, 1379
  toxicity, 1383
3,4-Benzopyrene, tetralin and, 1285
p-Benzoquinone, see Quinone
1,4-Benzoquinone, see Quinone
Bicyclo[4.4.0]decane, see Decalin
Bladder cancer:
  2-Naphthylamine-related, 1041–1042
  cyclohexylamine-induced, 1135, 1138
  toluidine-related, 1044

Bottoms, 1394
Brick oil, see Creosote
Bromobenzene:
  acute toxicity, 1493
  carcinogenicity, 1494
  mutagenicity, 1494
  physical and chemical properties, 1493
  subchronic and chronic toxicity, 1494
  uses, 1493
Bromochlorodifluoromethane, fire extinguishing application, 1198
2-Bromo-2-chloro-1,1,1-trifluoroethane, see Halothane
Bromophenols, physiological response, 1616–1617
Bromotrifluoromethane, fire extinguishing application, 1198
Bronchopulmonary toxicity, from chlorofluorocarbons, 1183
Butadiene:
  carcinogenicity, 1251
  hygienic standards, 1252
  physicochemical properties, 1242–1243
  toxicity, 1250–1252
Butadiyne:
  physiochemical properties, 1254–1255
  toxicity, 1256
Butane:
  hygienic standards, 1230
  physicochemical properties, 1226–1227
  toxicity, 1229–1230
n-Butane, hazard properties, 1223
1,3-Butanediamine:
  physical and chemical properties, 1091
  toxicity, 1107
cis-Butene-2, physicochemical properties, 1242–1243
trans-Butene-2, physicochemical properties, 1242–1243
1-Butene:
  hygienic standards, 1246
  physicochemical properties, 1242–1243
  toxicity, 1246
2-Butene:
  hygienic standards, 1247
  toxicity, 1247
2-Butylamine:
  toxicity, 1122
  uses, 1122
n-Butylamine:
  basicity, 1089
  metabolism, distribution, and elimination, 1100
  physical and chemical properties, 1090

n-Butylamine (*Continued*)
  toxicity, 1121–1122
    amine bases and hydrochloride salts, 1109
  uses, 1121
*tert*-Butylamine:
  toxicity, 1124
  uses, 1124
Butylamines, 1121–1125
  toxicity, 1105
Butylbenzene, polysubstituted, 1350–1351
*sec*-Butylbenzene, toxicity, 1349
*tert*-Butylbenzene, toxicity, 1349
Butylbenzene, toxicity, 1350–1351
*n*-Butylbenzene, physical properties, 1304–1305
*sec*-Butylbenzene, physical properties, 1304–1305
*tert*-Butylbenzene, physical properties, 1304–1305
*n*-Butylbenzene, toxicity, 1349
β-Butylene, *see* 2-Butene
Butylene, *see* 2-Butene
α-Butylene, *see* Isobutene
*p-tert*-Butylphenol:
  human exposures, 1620
  physiological response, 1620
  uses, 1619
*tert*-Butyltoluene:
  physical properties, 1304–1305
  toxicity, 1349
1-Butyne:
  physiochemical properties, 1254–1255
  toxicity, 1256
2-Butyne:
  physiochemical properties, 1254–1255
  toxicity, 1256

Calcium carbide, in acetylene production, 1254
Calcium oxalate, naphthalene and production of, 1375
Camphene:
  physical and chemical properties, 1276–1277
  toxicity, 1291–1292
Camphor, with phenol, 1578
Carbon black, in adhesives, 953
Carbon dioxide, toxicity, 1226
Carboxylic acids, as products of aromatic hydrocarbon metabolism, 1303
Carcinogens:
  aromatic nitro and amino compounds, 956–963
  definition, 958
  epidemiologic studies, 961
  standards, 958–960
Carcinomas:
  from ozone depletion, 1178
  hereditary factor, 958
Cardiac arrhythmia, chlorofluorocarbon-related, 1180–1181, 1190
Cardiotoxicity:
  allylamine-related, 1133–1134
  chlorofluorocarbons, 1180–1183
  cyclopropane, 1269
  fluorine compounds, 1180–1183, 1191
  propane-related, 1229
β-Carotene, physicochemical properties, 1242–1243
Caryophyllene:
  physical and chemical properties, 1276–1277
  toxicity, 1292
Cataracts:
  from dichlorobenzene exposure, 1460, 1467
  from dinitrophenols, 977
  from ozone depletion, 1178
  TNT-related, 1062
Chemical pneumonitis, from aliphatic hydrocarbons, 1224
Chlordane:
  acute toxicity, 1505, 1534–1535
  carcinogenicity, 1537–1539
  classification, 1506
  chronic toxicity, 1535–1536
  environmental levels, 1537
  human exposures, 1536–1537
  hygienic standards, 1539–1540
  liver enzyme effects, 1521
  metabolism, 1536
  physical and chemical properties, 1534
  regulatory standings, 1504
  skin exposure, 1535
  use and industrial exposure, 1533
Chlordecone, *see* Kepone
Chlorinated hydrocarbon insecticides:
  acute toxicity, 1505
  carcinogenicity, 1506–1507
  classifications, 1506
  comparative properties, 1504–1505
  in infectious disease control, 1503
  persistence in environment, 1504
  regulatory standings, 1504
*m*-Chloroaniline:
  carcinogenicity, 993–994
  metabolism, 993
  methemoglobin formation, 993
  physical and chemical properties, 992–993
  toxicity, 993–994
  uses, 991
*o*-Chloroaniline:
  physical and chemical properties, 991
  toxicity, 991

# SUBJECT INDEX

uses, 990
p-Chloroaniline, toxicity, 993
Chlorobenzenes:
  cancer classifications, 1446–1447
  carcinogenicity, 1445, 1445–1447
  comparative toxic properties, 1443–1445
  protein-induction activity, 1446–1447
  toxicity, 1443, 1444
Chlorocresol, phenol and, 1578
Chlorodifluoroethane:
  refrigerant poisoning, 1196
  toxicity, 1190
1-Chloro-1,1-difluoroethane:
  carcinogenicity, 1192
  toxicity, 1188
Chlorodifluoromethane:
  carcinogenicity, 1193
  mutagenicity, 1192–1193
  refrigerant poisoning, 1196
  threshold limit values, 1195
  toxicity, 1188, 1190
Chlorodinitrobenzene isomers, 1027–1031
2-Chloro-1,3-dinitrobenzene, physical and chemical properties, 1027
2-Chloro-1,4-dinitrobenzene:
  physical and chemical properties, 1028
  uses, 1028
3-Chloro-1,2-dinitrobenzene:
  physical and chemical properties, 1028
  toxicity, 1028
  uses, 1028
4-Chloro-1,2-dinitrobenzene:
  physical and chemical properties, 1029
  toxicity, 1029
4-Chloro-1,3-dinitrobenzene:
  metabolism, 1030
  physical and chemical properties, 1029–1030
  toxicity, 1030
  uses, 1029
5-Chloro-1,3-dinitrobenzene:
  physical and chemical properties, 1030–1031
  uses, 1030
Chlorofluorocarbon, see also Hydrogenated chlorofluorocarbon
  chemical structure, completely halogenated compounds, 1187–1188
  classification, 1178
  dermal contact hazards, 1198
  in pharmaceuticals, synthetic oxygen transport fluids, 1207–1208
  photodissociation and ozone destruction, 1178
  threshold limit values, 1195
Chlorofluorocarbon, prototype, 1179–1188

bronchopulmonary and respiratory toxicity, 1183–1185
carcinogenicity, 1186
cardiac arrhythmia, 1180–1181
cardiac rate changes, 1182
cardiotoxicity, 1180–1183
  mechanism of, 1183
changing airway resistance, 1184
decreased pulmonary compliance, 1184
extrathoracic toxicity, 1185–1186
mutagenicity, 1186
myocardial contractility and, 1182
respiratory minute volume changes, 1184–1185
tetratogenicity, 1186
4-Chloro-2-hydroxyacetanilide, 1026
4-Chloro-2-nitroaniline:
  physical and chemical properties, 997
  toxicity, 997
Chloronitrobenzene isomers, 1024–1025
m-Chloronitrobenzene:
  methemoglobin formation, capacity for, 952
  physical and chemical properties, 1025
  toxicity, 1025
  uses, 1025
o-Chloronitrobenzene:
  physical and chemical properties, 1024
  toxicity, 1024–1025
  uses, 1024
p-Chloronitrobenzene:
  hygienic standards, 1027
  physical and chemical properties, 1026
  teratogenicity, 1026–1027
  toxicity, 1026–1027
  uses, 1026
1-Chloro-1,1,2,2,2-pentafluoroethane:
  refrigerant poisoning, 1196
  threshold limit values, 1195
  toxicity, 1187
Chloropentafluoroethane, toxicity, 1190
m-Chlorophenol:
  lethal dosages, 1615
  physical and chemical properties, 1614
o-Chlorophenol:
  lethal dosages, 1615
  physical and chemical properties, 1614
p-Chlorophenol:
  lethal dosages, 1615
  physical and chemical properties, 1614
  physiological response, 1615
Chlorophenols:
  human exposures, 1616
  lethal dosages, 1615
  physical and chemical properties, 1614

Chlorophenols (*Continued*)
  physiological response, 1613, 1615
    absorption, excretion, and metabolism,
      1615–1616
    mode of action, 1616
    pathology, 1615
  source, uses, and industrial exposures, 1613
2-Chloro-1,4-phenylenediamine, toxicity, 1012
4-Chloro-1,2-phenylenediamine:
  hygienic standards, 1011
  physical and chemical properties, 1011
  toxicity, 1011
  uses, 1011
4-Chloro-1,3-phenylenediamine:
  physical and chemical properties, 1011–1012
  toxicity, 1012
  uses, 1011
1-Chloro-1,2,2,2-tetrafluoroethane, toxicity, 1188
4-Chloro-*o*-toluidine:
  physical and chemical properties, 1044–1045
  toxicity, 1045
  uses, 1044
5-Chloro-*o*-toluidine:
  physical and chemical properties, 1045
  toxicity, 1045
  uses, 1045
6-Chloro-*o*-toluidine, 1045
2-Chloro-1,1,1-trifluoroethane:
  carcinogenicity, 1193
  reproductive effects, 1194
1-Chloro-2,2,2-trifluoroethyl difluoromethyl
  ether, *see* Isoflurane
2-Chloro-1,1,2-trifluoroethyldifluoromethyl
  ether, *see* Enflurane
Chlorotrifluoromethane, 1187
Chlorpromazine, in xylene metabolism, 1338
Choline, alkylalkanolamines and metabolism of,
  1160, 1161
Cholinesterase:
  ethene inhibition of, 1244
  methylaminoethanol inhibition of, 1158
  xylene and production of, 1337
Chrysene, toxicity, 1381, 1389
Cigarette smoke, benzene in, 1307
Clonic-tonic convulsions, dibutylethanolamine-
  related, 1163
CNS depression:
  aliphatic hydrocarbon-related, 1224
  aromatic hydrocarbon-related, 1303
  by cycloalkanes, 1267
  cyclohexane-induced, 1273
  pyrocatechol-induced, 1585
Coal oil, *see* Kerosine
Coal tar, benzene in, 1307

Convulsions, cyclopentane-induced, 1273
Creosote:
  flammability, 1604
  human exposures, 1604
  odor and warning properties, 1604
  physical and chemical properties, 1602–1603
  physiological response:
    absorption, excretion, and metabolism, 1603
    cause of death, 1603
    pathology, 1603
  prevention of poisoning, 1604–1605
  source, uses, and industrial exposures, 1602
Creosotum, *see* Creosote
Cresol:
  atmospheric determination, 1598–1599
  human exposures, 1601
  hygienic standards, 1601
  lethal dosages, 1600
  microbial effects, 1306
  odor and warning properties, 1601
  physical and chemical properties, 1598
  physiological response:
    absorption, excretion, and metabolism,
      1600–1601
    cause of death, 1601
    mode of action, 1601
    pathology, 1599–1600
  safe handling procedures, 1601
  source, uses, and industrial exposures,
    1597–1598
Crown gall neoplasms, tetralin-related, 1288
Crude oil:
  characteristics and uses, 1393–1394
  hygienic standards, 1395
  physical properties, 1396
  toxicity, 1394–1395, 1398
Cumene:
  physical properties, 1304–1305
  toxicity, 1348–1349
Cumulenes, *see* Alkatrienes
Cutting oils:
  hygienic standards, 1416
  toxicity, 1415–1416
*N*-(Cyanoethyl)diethylenetriamine, toxicity, 1147
Cyclamate:
  carcinogenicity, 1138–1139
  metabolized to cyclohexylamine, 1135
Cyclic guanosine monophosphate, 1065, 1066
Cyclic olefins, toxicity, 1291–1292
Cycloalkanes:
  carcinogenicity, 1268
  inhalation toxicity, 1270–1271
  physical and chemical properties, 1268–1269
  toxicity, 1267–1268, 1272

SUBJECT INDEX 1639

uses and production, 1267
Cycloalkenes:
  alkenyl, 1281–1283
  inhalation toxicity, 1280
  physical and chemical properties, 1276–1277
  toxicity, 1275–1276, 1278–1279
Cycloamines, metabolism, distribution and elimination, 1101
Cyclobutane:
  physical and chemical properties, 1268–1269
  toxicity, 1269, 1272
Cyclododecane, toxicity, 1272, 1275
Cyclododecatriene, toxicity, 1278–1279, 1284
Cycloheptane:
  physical and chemical properties, 1268–1269
  toxicity, 1275
Cycloheptatriene:
  physical and chemical properties, 1276–1277
  toxicity, 1278–1279, 1284
Cycloheptene:
  physical and chemical properties, 1276–1277
  toxicity, 1281
Cyclohexane:
  inhalation toxicity, 1270–1271
  nephropathy, 1268
  photodecomposition, 1274
  physical and chemical properties, 1268–1269
  toxicity, 1272, 1273–1274
Cyclohexene:
  hygienic standards, 1281
  inhalation toxicity, 1280
  physical and chemical properties, 1276–1277
  toxicity, 1277, 1281
Cyclohexenylene, see 1-Vinylcyclohex-1-ene
Cyclohexylamine, 1135–1140
  basicity, 1089
  carcinogenicity, 1138–1139
  cardiac and blood pressure effects, 1137
  hygienic standards, 1137
  metabolism, 1136
  physical and chemical properties, 1090
  reproductive effects, 1136–1138
  toxicity, 1106, 1135–1139
  uses, 1135
Cyclononane:
  physical and chemical properties, 1268–1269
  toxicity, 1275
Cyclooctadiene:
  physical and chemical properties, 1276–1277
  toxicity, 1278–1279, 1283–1284
Cyclooctane:
  physical and chemical properties, 1268–1269
  toxicity, 1275

Cyclooctatetraene, physical and chemical properties, 1276–1277
Cyclooctene, toxicity, 1281
Cycloolefins, see also Cycloalkenes
  toxicity, 1283–1285
Cycloparaffins, see Cycloalkanes
Cyclopentadiene:
  physical and chemical properties, 1276–1277
  toxicity, 1278–1279, 1283
Cyclopentane:
  hygienic standards, 1272–1273
  inhalation toxicity, 1270–1271
  physical and chemical properties, 1268–1269, 1276–1277
  toxicity, 1272–1273
Cyclopentapyrene, 1392
  physical properties, 1379
Cyclopentene:
  hygienic standards, 1276
  inhalation toxicity, 1280
  toxicity, 1276, 1278–1279
Cyclopropane:
  inhalation toxicity, 1270–1271
  physical and chemical properties, 1268–1269
  toxicity, 1269
Cymene:
  physical properties, 1304–1305
  toxicity, 1349
Cytochrome $b_5$ reductase, see NADH-methemoglobin reductase

$p,p'$-DDE:
  acute toxicity, 1505
  metabolism, 1510
Deaner's psychostimulant effects, 1159
Deanol, choline stimulation, 1159
Decahydronaphthalene, see Decalin
Decalin:
  napthalane and, 1284
  inhalation toxicity, 1286–1287
  nephropathy, 1268
  toxicity, 1278–1279, 1284–1285
$iso$-Decalin, physical and chemical properties, 1276–1277
$trans$-Decalin, physical and chemical properties, 1276–1277
Decamethylenediamine, toxicity, 1146
Decane:
  hazard properties, 1223
  physicochemical properties, 1226–1227
  toxicity, 1238
Decene, physicochemical properties, 1242–1243
1-Decyne:
  physicochemical properties, 1254–1255

1-Decyne (*Continued*)
  toxicity, 1256
Depigmentation, phenol-related, 1576
Dermal irritation:
  from aliphatic hydrocarbons, 1224, 1236–1237, 1237, 1251
  limonene-related, 1282
Diallylamine:
  physical and chemical properties, 1090
  toxicity, 1106, 1111, 1134
  uses, 1134
4,4'-Diaminobiphenyl, *N*-oxidation, 1040
1,5-Diaminonaphthalene:
  physical and chemical properties, 1042–1043
  uses, 1042
Diaminopropane, pathological changes, 1110
Di-*n*-amylamine:
  toxicity, 1125
  uses, 1125
Diaphorase, *see* NADH-methemoglobin reductase
Dibenz(*a,h*)anthracene, toxicity, 1383, 1393
Dibenzoanthracene, occurrence, 1378
Dibenzopyrene, occurrence, 1378
Di-*sec*-butanolamine, 1157
Dibutylamine:
  physical and chemical properties, 1090
  toxicity, 1105, 1123
Di-*n*-butylamine, uses, 1123
2-Dibutylaminoethanol:
  physical characteristics, 1112
  toxicity, 1107, 1112
Dibutylethanolamine, *see also* 2-Dibutylaminoethanol
  clonic-tonic convulsions, 1163
  hygienic standards, 1163
  toxicity, 1163
  uses, 1162
Di-*tert*-butylmethylphenol:
  atmospheric determination, 1619
  physical properties, 1619
  physiological response, 1619
  uses, 1618
2,3-Dichloroaniline, physical and chemical properties, 995–996
2,4-Dichloroaniline, physical and chemical properties, 995
2,5-Dichloroaniline, physical and chemical properties, 995
2,6-Dichloroaniline, physical and chemical properties, 996
3,4-Dichloroaniline:
  physical and chemical properties, 994
  toxicity, 994

3,5-Dichloroaniline, physical and chemical properties, 996
1,2-Dichlorobenzene, toxicity, 1444
*o*-Dichlorobenzene:
  acute toxicity, 1455
  carcinogenicity, 1445, 1445–1446
  chronic toxicity, 1457
  human exposures, 1458
  hygienic standards, 1459
  metabolism, 1457–1458
  odor and warning properties, 1458–1459
  physical and chemical properties, 1454–1455
  reproductive effects, 1457
  subchronic toxicity, 1455–1457
  teratogenicity, 1457
  toxicity, 1443, 1444
*p*-Dichlorobenzene:
  acute toxicity, 1460–1461
  biologic activation, 1488
  carcinogenicity, 1445, 1445–1446
  chronic toxicity, 1461–1462
  human exposures:
    nonoccupational, 1465–1466
    occupational, 1466–1467
  hygienic standards, 1468
  metabolism, 1458, 1464–1465
  mutagenicity, 1463–1464
  odor, taste, and warning properties, 1467–1468
  physical and chemical properties, 1459–1460
  reproductive effects, 1462–1463
  subchronic toxicity, 1461
  teratogenicity, 1462–1463
  toxicity, 1443, 1444
3,3'-Dichlorobenzidine:
  hygienic standards, 1034
  physical and chemical properties, 1034
  toxicity, 1034
  uses, 1034
Dichlorocatechol, in chlorobenzene metabolism, 1458–1459
Dichlorodifluoroethane, in aerosols, cardiopulmonary effects, 1200
1,2-Dichloro-1,1-difluoroethane:
  reproductive effects, 1194
  toxicity, 1188
2,2-Dichloro-1,1-difluoroethyl methyl ether, *see* Methoxyflurane
Dichlorodifluoromethane:
  in aerosols, 1199
    blood levels, 1202
    tissue levels, 1202–1203
  carcinogenicity, 1193
  threshold limit values, 1195
  toxicity, 1187, 1190

# SUBJECT INDEX 1641

Dichlorodiphenyltrichloroethane
  acute toxicity, 1505, 1508–1509
  as an antimalarial agent, 1511
  carcinogenicity, 1514–1516
    classification, 1506
  chronic toxicity, 1509–1510
  concentrations after starvation, 1522
  excretion through air, 1511
  history, 1503–1504
  human exposures:
    occupational, 1511–1512
    public, 1513–1514
    volunteer studies, 1512–1513
  hygienic standards, 1517
  liver enzyme effects, 1521
  metabolism, 1510–1511
  mutagenicity, 1516
  physical and chemical properties, 1508
  regulatory standings, 1504
  reproductive effects, 1516–1517
  teratogenicity, 1516–1517
  uses and occupational exposures, 1507–1508
1,1-Dichloro-1-fluoroethane, toxicity, 1188
Dichlorofluoromethane:
  carcinogenicity, 1193
  cardiopulmonary effects, 1200
  refrigerant poisoning, 1196
  threshold limit values, 1195, 1196
  toxicity, 1188, 1189
2,6-Dichloro-4-nitroaniline:
  physical and chemical properties, 997
  toxicity, 997
  uses, 996
2,3-Dichlorophenol, in chlorobenzene metabolism, 1458–1459
2,4-Dichlorophenol:
  lethal dosages, 1615
  physical and chemical properties, 1614
  physiological response, 1615
2,5-Dichlorophenol:
  in chlorobenzene metabolism, 1458–1459
  in dichlorobenzene metabolism, 1464
2,6-Dichlorophenol, physiological response, 1615
3,4-Dichlorophenol, in chlorobenzene metabolism, 1458–1459
2,6-Dichloro-1,4-phenylenediamine, toxicity, 1012–1013
2,5-Dichloroquinol, in chlorobenzene metabolism, 1458–1459, 1464
1,2-Dichloro-1,1,2,2-tetrafluoroethane:
  refrigerant poisoning, 1196
  threshold limit values, 1195
  toxicity, 1187

1,1-Dichloro-1,2,2,2-tetrafluoroethane, toxicity, 1187
Dichlorotetrafluoroethane, toxicity, 1190
2,2-Dichloro-1,1,1-trifluoroethane, toxicity, 1188
3,4-Dichlorphenylmercapturic acid, in chlorobenzene metabolism, 1458–1459
Dicyclic alkanes, toxicity, 1284–1285
Dicyclic alkenes, toxicity, 1285, 1288–1291
Dicyclic hydrocarbons:
  inhalation toxicity, 1286–1287
  physical and chemical properties, 1276–1277
  toxicity, 1278–1279
Dicyclohexylamine:
  physical and chemical properties, 1090
  toxicity, 1106, 1139
  uses, 1139
Dicyclopentadiene:
  hygienic standards, 1288–1289
  inhalation toxicity, 1286–1287
  physical and chemical properties, 1276–1277
  toxicity, 1278–1279, 1288–1289
Dieldrin, 1523–1533
  acute toxicity, 1505, 1525
  carcinogenicity, 1529–1532
    classification, 1506
  chronic toxicity, aldrin interactions, 1521–1522
  environmental levels, 1529
  human exposures, 1522–1523, 1527–1528
    nonoccupational, 1528–1529
  hygienic standards, 1533
  liver enzyme effects, 1521
  metabolism, 1519–1520
  mutagenicity, 1532
  physical and chemical properties, 1524
  regulatory standings, 1504
  reproductive toxicity, 1526–1527
  subchronic and chronic toxicity, 1483, 1525–1526
  teratogenicity, 1526–1527
  uses and industrial exposures, 1523–1524
Diels-Alder reactions, cycloolefins in, 1283
Diesel fuel:
  physical properties, 1413
  toxicity, 1412, 1414
Diethanolamine:
  hygienic standards, 1154
  physical characteristics, 1112
  toxicity, 1112, 1153
  uses, 1153
Diethylamine:
  basicity, 1089
  hygienic standards, 1118

Diethylamine (*Continued*)
  metabolism, distribution, and elimination, 1100
  nitrosation, 1097
  physical and chemical properties, 1090
  toxicity, 1105, 1117–1118
  uses, 1117
2-Diethylaminoethanol:
  physical characteristics, 1112
  toxicity, 1107, 1112
N,N'-Diethylaniline:
  physical and chemical properties, 987
  toxicity, 987
  uses, 987
Diethylbenzene:
  physical properties, 1304–1305
  toxicity, 1345, 1346
2,2-Diethyldihexylamine, toxicity, 1106
N,N'-Diethylenediamine, physical and chemical properties, 1091
Diethylenediamine piperazine, toxicity, 1107
Diethylenetriamine:
  hygienic standards, 1147
  physical and chemical properties, 1091
  toxicity, 1107, 1110–1111, 1146–1147
  uses, 1146
Diethylenetriaminepentaacetic acid:
  toxicity, 1148
  uses, 1148
Diethylethanolamine, *see also* 2-Diethylaminoethanol
  choline metabolism and, 1160, 1161
  hygienic standards, 1161
  toxicity, 1161
Di(-2-ethylhexyl)amine, physical and chemical properties, 1090
Difluoroethane, refrigerant poisoning, 1196
1,1-Difluoroethane:
  mutagenicity, 1192
  toxicity, 1188, 1189
Difluoromethane, toxicity, 1189
Di-*n*-heptylamine:
  physical and chemical properties, 1090
  toxicity, 1129
  uses, 1129
Dihexylamine:
  toxicity, 1127
  uses, 1127
Diisobutylamine, 1123
  toxicity, 1105
Diisopropanolamine, uses, 1156
Diisopropylamine:
  hygienic standards, 1120
  physical and chemical properties, 1090
  toxicity, 1105, 1120
  uses, 1120
Diisopropylbenzenes, toxicity, 1350
Diisopropylethanolamine, 1162
Dimethylamine:
  absorption, 1096
  basicity, 1089
  hygienic standards, 1115
  metabolism, distribution, and elimination, 1098
  nitrosation, 1097
  physical and chemical properties, 1090
  toxicity, 1105, 1114–1115
  uses, 1113–1114
4-Dimethylaminoazobenzene:
  physical and chemical properties, 1039
  toxicity, 1039
  use, 1039
2-Dimethylaminoethanol:
  physical characteristics, 1112
  toxicity, 1107, 1112
1-Dimethylamino-2-propanol:
  physical characteristics, 1112
  toxicity, 1112
4-Dimethylaminostilbene, 999
Dimethylaniline:
  demethylation, 955
  methemoglobin formation, 950
N,N'-Dimethylaniline:
  carcinogenicity, 985
  hygienic standards, 986
  metabolism, 985–986
  *N*-oxide formation, 985
  physical and chemical properties, 985
  toxicity, 985–986
  uses, 985
Dimethylbenzanthracene:
  hygienic standards, 1388
  mutagenicity, 1388
  toxicity, 1387–1388
  tumorigenesis, 1388, 1583–1584
3,3'-Dimethylbenzidine, 999
Dimethylbutane, 1235
  physicochemical properties, 1226–1227
Dimethylbutylamine, 1125
Dimethylcyclohexane:
  inhalation toxicity, 1270–1271
  physical and chemical properties, 1268–1269
  toxicity, 1275
Dimethylcyclohexylamine:
  physical and chemical properties, 1090
  toxicity, 1139
  uses, 1139
Dimethylenemethane, toxicity, 1250

# SUBJECT INDEX 1643

Dimethylethanolamine, *see also*
  2-Dimethylaminoethanol
  metabolism, distribution and elimination, 1103–1104
  toxicity, 1158–1159
  uses, 1158
2,5-Dimethylhexane, toxicity, 1237
Dimethylisopropanolamine, 1164
3,3-Dimethyl-2-methylene norcamphone, *see* Camphene
1,4-Dimethylnaphthalene, physical properties, 1304–1305
Dimethylnaphthalene, toxicity, 1378
2,7-Dimethyloctane, 1239
  physicochemical properties, 1226–1227
Dimethyl-*p*-phenylenediamine, myotoxicity, 1008
2,2-Dimethylpropane, toxicity, 1233
Dinitrobenzene:
  aquatic toxicity, 968
  isomers, 1019–1023
  metabolic pathway reduction, 967
  methemoglobinemia from, 953
1,2-Dinitrobenzene, metabolism, 954
1,3-Dinitrobenzene:
  metabolism, 954, 964
  methemoglobin formation, species differences, 964
  reproductive effects, 963–964, 965
1,4-Dinitrobenzene:
  in adhesives, 953
  metabolism, 954
*m*-Dinitrobenzene:
  hygienic standards, 1023
  metabolism, 1020–1021
  methemoglobin formation, 1021
    capacity for, 952
  physical and chemical properties, 1020
  reproductive effects, 1021–1022
    age- and species-related differences, 1022–1023
  toxicity, 1020–1023
  uses, 1019–1020
*o*-Dinitrobenzene:
  hygienic standards, 1019
  methemoglobin formation, capacity for, 952
  physical and chemical properties, 1019
  toxicity, 1019
  uses, 1019
*p*-Dinitrobenzene:
  in adhesives, 953
  methemoglobin formation, 951
    capacity for, 952

2,4-Dinitrochlorobenzene, methemoglobin formation, 952
4,6-Dinitro-*o*-cresol:
  hygienic standards, 982
  physical and chemical properties, 981
  toxicity, 981–982
  uses, 981
Dinitrophenol isomers, uses, 976–979
2,3-Dinitrophenol:
  physical and chemical properties, 976
  toxicity, 976
2,4-Dinitrophenol:
  metabolism, 978
  physical and chemical properties, 977
  toxicity, 977–978
  uses, 976
2,5-Dinitrophenol:
  physical and chemical properties, 978
  toxicity, 978
2,6-Dinitrophenol:
  physical and chemical properties, 978–979
  toxicity, 979
3,4-Dinitrophenol:
  physical and chemical properties, 979
  toxicity, 979
3,5-Dinitrophenol:
  physical and chemical properties, 979
  toxicity, 979
Dinitrotoluene:
  carcinogenicity, 960–961
  initiating vs. promoting activity, 1060
  metabolites, 954–955
Dinitrotoluene technical grade:
  carcinogenicity, 1055–1056
  physical and chemical characteristics, 1055
  reproductive effects, 1055
  toxicity, 1055–1056
  uses, 1054
2,3-Dinitrotoluene, 1061
2,4-Dinitrotoluene:
  carcinogenicity, 1058–1059
  hygienic standards, 1059
  metabolism, 955
  physical and chemical properties, 1057
  reproductive effects, 964–965, 1058
  toxicity, 1057–1059
  uses, 1056
2,5-Dinitrotoluene, 1061
2,6-Dinitrotoluene:
  hygienic standards, 1060
  metabolism, 1059–1060
  physical and chemical properties, 1059
  reproductive effects, 964–965
  toxicity, 1059–1060

2,6-Dinitrotoluene (*Continued*)
  uses, 1059
3,4-Dinitrotoluene, 1061
3,5-Dinitrotoluene, 1061
Dinitrotoluol, methemoglobin formation, 952
Dioctylamine, 1131
Diphenyl:
  hygienic standards, 1365
  metabolism, 1364–1365
  microbial studies, 1365
  physical properties, 1304–1305
  preparation and uses, 1364
  toxicity, 1366–1367
Diphenylalkanes, toxicity, 1365, 1370
Diphenylalkenes, toxicity, 1370
Diphenylamine:
  carcinogenicity, 957
  nephrotoxicity, 1013–1014
  physical and chemical properties, 1013
  toxicity, 1013–1014
  uses, 1013
Diphenylethylene, toxicity, 1370
Diphenylmethane:
  physical properties, 1304–1305
  toxicity, 1365, 1370
Dipropylamine:
  basicity, 1089
  physical and chemical properties, 1090
  toxicity, 1105, 1120
Divinylbenzene:
  hygienic standards, 1362
  physical properties, 1304–1305
  toxicity, 1360, 1362
Divinylstyrene, production and uses, 1362
Dodecane:
  hygienic standards, 1239
  physicochemical properties, 1226–1227
  toxicity, 1239
Dodecene, physicochemical properties, 1242–1243
Dodecylamine:
  toxicity, 1131
  uses, 1131
Dodecylbenzene:
  physical properties, 1304–1305
  toxicity, 1349, 1351
Dodecyldimethylamine, 1132
Dodecylthiophenol, toxicity, 1620

Eicosane:
  physicochemical properties, 1226–1227
  toxicity, 1239
Endrin:
  acute toxicity, 1505

  carcinogenicity, classification, 1506
  regulatory standings, 1504
Enflurane, in general anesthetics, 1207
Environmental contamination, 966
Environmental Protection Agency (EPA),
  carcinogen regulation, 961
Epinephrine:
  chlorofluorocarbon sensitization and, 1181,
    1183, 1190
  pentane and, 1231
  relationship of aliphatic amines to, 1108–1109
Epoxy resins, acute toxicity, 1110–1111
Epping jaundice, 1002
Erythroblastosis, pinene-related, 1289
Erythromyelosis, benzene-induced, 1318
Ethane:
  hazard properties, 1223
  physicochemical properties, 1226–1227
  toxicity, 1228
Ethanolamine:
  metabolism, distribution and elimination, 1103
  physical characteristics, 1112
  toxicity, 1112
Ethene:
  physicochemical properties, 1242–1243
  toxicity, 1241–1244
1-Ethenylcyclohexene, *see* 1-Vinylcyclohex-1-ene
Ethereal sulfate, from MCB metabolism, 1451–1452
Ethylamine, 1116–1117
  basicity, 1089
  hygienic standards, 1116–1117
  metabolism, distribution, and elimination,
    1099–1100
  pathological changes, 1110
  physical and chemical properties, 1090
  toxicity, 1105, 1116
    amine bases and hydrochloride salts, 1109
  uses, 1116
*N*-Ethylaminoethanol, antimetabolite activity,
  1159–1160
2-Ethylaminoethanol:
  physical characteristics, 1112
  toxicity, 1107, 1112
*N*-Ethylaniline:
  physical and chemical properties, 986–987
  toxicity, 987
  uses, 986
Ethylbenzene:
  hygienic standards, 1343, 1346
  metabolism, 1343
  physical properties, 1304–1305
  reproductive effects, 1343
  toxicity, 1343, 1344–1345
Ethyl-*n*-butylamine, 1124–1125

# SUBJECT INDEX 1645

2-Ethylbutylamine:
  physical and chemical properties, 1090
  toxicity, 1106
Ethylcyclohexane:
  inhalation toxicity, 1270-1271
  physical and chemical properties, 1268-1269
Ethylcyclopentane:
  physical and chemical properties, 1268-1269
  toxicity, 1273
Ethylene, see Ethene
Ethylene glycol dinitrate:
  hygienic standards, 1064
  physical and chemical properties, 1063
  toxicity, 1063-1064
  uses, 1063
Ethylenediamine:
  acute toxicity, 1110-1111
  GABA-mimetic effect, 1142
  hygienic standards, 1143
  neurotoxicity, 1142
  pathological changes, 1110
  physical and chemical properties, 1091
  reproductive effects, 1142-1143
  toxicity, 1107, 1141-1143
    amine bases and hydrochloride salts, 1109
  uses, 1141
Ethylenediaminetetraacetic acid:
  metabolism, distribution and elimination, 1103
  toxicity, 1144
  treatment for HCB intoxication, 1491
  uses, 1144
Ethylenimine, pathological changes, 1110
Ethylethanolamine, see 2-Ethylaminoethanol
2-Ethyl-1-hexene, toxicity, 1250
2-Ethylhexylamine:
  physical and chemical properties, 1090
  toxicity, 1127
  uses, 1127
Ethylidenenorbornene, toxicity, 1291
2,2'-(Ethylimino)diethanol, toxicity, 1107
N-Ethyl-2,2'-iminodiethanol:
  physical characteristics, 1112
  toxicity, 1112
Ethylparathion, 948

Ferrihemoglobin, secondary reduction pathway, 951
Ferrihemoglobin reductase, see NADH-methemoglobin reductase
Fish odor syndrome, 1098
140 Flash naphtha, 1402
  physical properties, 1397
  toxicity, 1402, 1409
Fluorine-containing organic compounds:
  in aerosol propellants, 1199-1207
  animal toxicologic studies, 1186-1194
    dog, 1191, 1192
    mouse, 1190-1191
    rat, 1191
  blood levels in exposed persons, 1201-1202
  in bronchodilator aerosols, 1204-1207
  carcinogenicity, animal studies, 1192-1193
  classification of, 1178-1179
  dermal contact hazards, 1198
  evaluating developmental alternatives, 1189-1190
  fire extinguishing application, 1198-1199
  inhalation hazards, 1195
  mutagenicity, animal studies, 1192-1193
  occupational exposure, 1194-1199
    foam blowing agents, 1197
    solvents, 1197
  ozone depletion, 1177-1178
  in pharmaceuticals, general anesthetics, 1207-1208
  pyrolysis of polymers, 1199
  refrigerant poisoning, 1196-1197
  reproductive effects, 1193-1194
  respiratory uptake, 1201
  threshold limit values, 1195-1196
Fluorocarbon, see also Hydrogenated fluorocarbon
  bromine-containing, fire extinguishing application, 1198
  classification, 1179
Fluoroform, mutagenicity, 1192
Follicle-stimulating hormone (FSH), 1058
  reduction with aldrin treatment, 1520
Food and Drug Administration (FDA), carcinogen regulation, 961
Friedel-Crafts reaction, diphenylalkanes, 1370
Fuel oil no. 2, see Diesel fuel

Gangrene, phenol in treatment of, 1577
Gas oil, fractionation, 1394
Gasoline:
  chemical properties, 1395
  fractionation, 1394
  hygienic standards, 1406
  physical properties, 1395, 1396
  toxicity, 1395, 1398, 1405-1406
Glucose-6-phosphatase, depletion with HCB, 1481
Glucose-6-phosphate dehydrogenase
  glutathione reduction, 951
  hemolytic anemia with deficiency of, 993
Glue sniffing:
  aliphatic hydrocarbons, 1224, 1234

Glue sniffing (*Continued*)
  aromatic hydrocarbons, 1301
  rubber cement, 1309
  toluene, 1329
Glutathione:
  ferrihemoglobin reduction, 951
  naphthalene and, 1374
Guanylate cyclase, 1065, 1066

Hair dyes, PPDA-containing, 1008
Hair spray, toxicity of, 1200
Halogenated benzene:
  physical and chemical properties, 1495
  toxicity, 1495
  uses, 1494–1495
Halons, toxicity, 1199
Halothane, in general anesthetics, 1207
Health teas, hydroquinone in, 1592
Heating oils:
  physical properties, 1413
  toxicity, 1414
Hematopoiesis, benzene-induced, 1318
Hematuria, intractable, phenol in treatment of, 1579
Hemoglobin M disease, 951
Hemolytic anemia, G6PD deficiency and, 993
Hepatitis, methylenedianiline-related, 1002
Hepatocarcinogenicity:
  dinitrotoluenes, 1056, 1058–1059
  toluenediamine-related, 1049–1050
Hepatoxicity:
  chlorofluorocarbon-induced, 1185–1186
  dichlorobenzenes, 1444–1445
  HCB-related, 1481
Heptachlor:
  acute toxicity, 1505, 1541
  carcinogenicity, 1544–1545
    classification, 1506
  environmental levels, 1543
  human exposures, 1543
  hygienic standards, 1545
  metabolism, 1542
  physical and chemical properties, 1540–1541
  regulatory standings, 1504
  reproductive effects, 1542–1543
  subchronic and chronic toxicity, 1541–1542
  teratogenicity, 1542–1543
  uses and potential exposures, 1540
Heptadecane, physicochemical properties, 1226–1227
*n*-Heptadecane, toxicity, 1239
1,4-Heptadiene, physicochemical properties, 1242–1243
1,6-Heptadiyne:
  physiochemical properties, 1254–1255
  toxicity, 1256
Heptafluoropropane, toxicity, 1189
1-Heptane, physicochemical properties, 1242–1243
*n*-Heptane:
  hazard properties, 1223
  hygienic standards, 1236
  in toluene production, 1326
  toxicity, 1235–1236
*cis*-Heptene-2, physicochemical properties, 1242–1243
*trans*-Heptene-3, physicochemical properties, 1242–1243
Heptenes, toxicity, 1249
*n*-Heptylamine, physical and chemical properties, 1090
Heptylamines, 1128–1129
Hexachlorobenzene:
  acute toxicity, 1478
    sex differences, 1481
  carcinogenicity, 1445, 1445–1446, 1491–1492
  human exposures:
    nonoccupational, 1490–1491
    occupational, 1491
    Turkey episode, 1490
  hygienic standards, 1492
  immunotoxic potential, 1485–1486
  metabolism, 1486–1489
  mutagenicity, 1486
  photosensitization, 1491
  physical and chemical properties, 1478
  reproductive effects, 1484–1485
  subchronic and chronic toxicities, 1478–1484, 1479–1480
    beagles, 1482–1483
    monkeys, 1483
    pigs, 1483–1484
    rats, 1480–1482
  technical grade, 1477
  teratogenicity, 1484–1485
  uses and industrial exposure, 1477–1478
Hexachlorocyclohexane:
  acute toxicity, 1552
  carcinogenicity, 1556–1557
  chronic toxicity, 1552–1553
  human exposures:
    intoxications, 1555
    occupational, 1555–1556
    tissue concentrations, 1554–1555
  hygienic standards, 1557
  metabolism, 1553–1554
  physical and chemical properties, 1551
  reproductive toxicity, 1553
  teratogenicity, 1553

## SUBJECT INDEX

uses and industrial exposure, 1551
Hexachloropentadiene, acute toxicity, 1505
Hexadecane, physicochemical properties, 1226–1227
$n$-Hexadecane, toxicity, 1239
Hexadecene, physicochemical properties, 1242–1243
Hexadecylmethylenediamine, 1146
Hexadiene:
 physicochemical properties, 1242–1243
 toxicity, 1253
Hexafluoropropane, toxicity, 1189
Hexahydrobenzene, see Cyclohexane
Hexahydrotoluene, see Methylcyclohexane
Hexamethylbenzene, 1342
Hexamethylene, see Cyclohexane
Hexamethylenediamine:
 basicity, 1089
 physical and chemical properties, 1091
 toxicity, 1146
Hexane, physicochemical properties, 1226–1227
$n$-Hexane:
 hazard properties, 1223
 hygienic standards, 1234
 toxicity, 1224, 1233–1234
2,5-Hexanedione, axonopathy from, 1224, 1224–1225
1-Hexene:
 physicochemical properties, 1242–1243
 toxicity, 1249
2-Hexene, 1249
cis-Hexene-2, physicochemical properties, 1242–1243
trans-Hexene-2, physicochemical properties, 1242–1243
Hexylamine, 1126–1128
$n$-Hexylamine:
 physical and chemical properties, 1090
 toxicity, 1127
 uses, 1127
Hexylbenzene, toxicity, 1351
2-Hexylene, see 2-Hexene
High flash naphtha, physical properties, 1397
High petroleum naphtha, physical properties, 1397
Hyalin droplet nephropathy, chlorobenzene-related, 1446
Hydrogen cyanide, in manufacture of aliphatic amines, 1092
Hydrogen peroxide, PPDA and, 1008–1009
Hydrogenated chlorofluorocarbon:
 carcinogenicity, 1192–1193
 chemical structure, 1188
 classification, 1178

dermal contact hazards, 1198
threshold limit values, 1195
Hydrogenated fluorocarbon:
 classification, 1179
 nonchlorinated, 1190
 chemical structure, 1188–1189
Hydroquinone:
 atmospheric determination, 1590–1591
 in benzene metabolism, 1308
 cross-sensitization, 1008
 human exposures, 1592
 hygienic standards, 1592
 physical and chemical properties, 1590
 physiological response:
  absorption, excretion, and metabolism, 1591
  cause of death, 1591
  mode of action, 1591
  pathology, 1591
 safe handling procedures, 1592
 source, uses, and industrial exposures, 1590
$N$(Hydroxyethyl)diethylenetriamine, toxicity, 1147
Hydroxylamine, methemoglobin formation, 950
Hypertension, ethene-related, 1244
Hypoglycemia, ethene-related, 1244

2-Iminodiethanol, toxicity, 1106
Immunologic diseases, from ozone depletion, 1178
Intermittent claudication, phenol in treatment of, 1578
International Association of Research Chemists (IARC), carcinogen standards, 959
Iodophenols, physiological response, 1616–1617
Iomex, toxicity, 1393
Isoamylamine:
 physical and chemical properties, 1090
 toxicity, 1126
 uses, 1126
$n$-Isoamylamine, metabolism, distribution, and elimination, 1101
Isobutane:
 hygienic standards, 1231
 physicochemical properties, 1226–1227
 toxicity, 1230–1231
Isobutene:
 hygienic standards, 1248
 physicochemical properties, 1242–1243
 toxicity, 1247–1248
Isobutylamine:
 physical and chemical properties, 1090
 toxicity, 1123
Isobutylbenzene, physical properties, 1304–1305
Isobutylene, see Isobutene

Isoflurane, in general anesthetics, 1207
Isoheptane, toxicity, 1236
Isohexane:
  hygienic standards, 1235
  physicochemical properties, 1226-1227
  toxicity, 1235
Isohexane-2, physicochemical properties, 1242-1243
Isohexene, toxicity, 1249
Isohexylamine:
  toxicity, 1127
  uses, 1127
β-Isohexylene, see Isohexene
Isoniazid, acetylation, 1097
Isooctanes, toxicity, 1237
Isopentane:
  physicochemical properties, 1226-1227
  toxicity, 1232-1233
Isopentene:
  hygienic standards, 1248
  toxicity, 1248
Isopentyne:
  physiochemical properties, 1254-1255
  toxicity, 1256
Isoprene, 1252-1253
  physicochemical properties, 1242-1243
  toxicity, 1252-1253
Isopropanolamine:
  physical characteristics, 1112
  toxicity, 1107, 1112
Isopropylamine:
  basicity, 1089
  physical and chemical properties, 1090
  toxicity, 1105
4-Isopropylaminodiphenylamine, cross-sensitization, 1008
Isopropylbenzene:
  hygienic standards, 1347
  metabolism, 1347
  production and uses, 1346-1347
  toxicity, 1347
Isopropyltoluene, toxicity, 1350

Jet fuel:
  physical properties, 1413
  toxicity, 1412

Kelthane, acute toxicity, 1505
Kepone:
  acute toxicity, 1505, 1546
  carcinogenicity, 1549-1550
    classification, 1506
  decontamination, 1548-1549
  environmental levels, 1548
  human exposures:
    nonoccupational, 1547-1548
    occupational, 1547
  hygienic standards, 1550
  metabolism, 1548
  physical and chemical properties, 1545-1546
  regulatory standings, 1504
  subchronic and chronic toxicity, 1546-1547
  uses and industrial exposures, 1545
Kerosene, see Kerosine
Kerosine:
  fractionation, 1394
  hygienic standards, 1412
  physical properties, 1397, 1413
  production and uses, 1410-1411
  toxicity, 1404, 1411-1412

Lactation, heptadecane concentrations, 1240
Lactic dehydrogenase, reduction with dinitrobenzene, 965
Leukemia, benzene-related, 1321
Leukocytosis, dicyclopentadiene-related, 1288
Limonene:
  physical and chemical properties, 1276-1277
  teratogenicity, 1282
  toxicity, 1278-1279, 1282-1283
  uses and production, 1282
d-Limonene, nephropathy, 1268
Lindane:
  acute toxicity, 1505
  carcinogenicity, classification, 1506
  production, 1503-1504
  regulatory standings, 1504
Lipid peroxidation, ethane production, 1228
Lipid synthesis, butane-inhanced, 1230
Liver mixed-function oxidase system, methemoglobin formation, 950
Locked-lung syndrome, 1206
Lubricating oils:
  fractionation, 1394
  hygienic standards, 1415
  toxicity, 1415
Luteinizing hormone (LH), 1058
  reduction with aldrin treatment, 1520
Lycopene, physicochemical properties, 1242-1243

Malaria, DDT and eradication of, 1511
Marasmus, phenol, 1575
Medicinal oils, see White oils
Mercapturic acid:
  in benzene metabolism, 1319
  in chlorobenzene metabolism, 1451-1452, 1458-1459

# SUBJECT INDEX 1649

Mesitylene, microbial effects, 1306
Methane:
   hazard properties, 1223
   physicochemical properties, 1226–1227
   toxicity, 1226–1228
Methanol, toxicity, 1226
Methemoglobin formation:
   chloroanilines, 993
   compounds, 950
   nitroanilines, 989–990
Methemoglobinemia:
   from aromatic nitro and amino compounds, 949–953
   with G6PD deficiency, 951
   in newborn infants, 953
   symptoms, 949–953
Methoxychlor, liver enzyme effects, 1521
Methoxyflurane, in general anesthetics, 1207
Methyl $n$-butyl ketone, axonopathy from, 1224
Methylamine:
   basicity, 1089
   hygienic standards, 1113
   metabolism, distribution, and elimination, 1098
   physical and chemical properties, 1090
   toxicity, 1105, 1113
     amine bases and hydrochloride salts, 1109
   uses, 1113
Methylamines, biosynthesis, 1093
2-Methylaminoethanol:
   physical characteristics, 1112
   toxicity, 1107, 1112
$N$-Methylaniline:
   hygienic standards, 985
   physical and chemical properties, 984–985
   toxicity, 985
Methylbenzanthracene:
   physical properties, 1379
   toxicity, 1387
2-Methylbutane, toxicity, 1232–1233
3-Methylbutyne, toxicity, 1256
Methylcholanthrene:
   physical properties, 1379
   toxicity, 1383
3-Methylcholanthrene:
   metabolism, 1392
   mutagenicity and tumorigenesis, 1392–1393
   toxicity, 1392
   in xylene metabolism, 1338
Methylchrysene, toxicity, 1389
3-Methylchrysene, physical properties, 1379
5-Methylcyclohexadiene-1,3, physical and chemical properties, 1276–1277
Methylcyclohexadiene, toxicity, 1283

Methylcyclohexane:
   hygienic standards, 1275
   inhalation toxicity, 1270–1271
   physical and chemical properties, 1268–1269
   in toluene production, 1326
   toxicity, 1272, 1274–1275
Methylcyclopentadiene, toxicity, 1283
Methylcyclopentane:
   physical and chemical properties, 1268–1269
   toxicity, 1273
4,4'-Methylenebis(2-chloroaniline):
   biogradation, 967
   carcinogenicity, 999–1000
1,1'-Methylenebisbenzene, *see* Diphenylmethane
4,4'-Methylenebis(2-chloroaniline):
   carcinogenicity, 957–958, 961–962
   mechanism, 999–1000
   hygienic standards, 1000
   physical and chemical properties, 999
   toxicity, 999–1000
   uses, 999
4,4'-Methylenedianiline:
   carcinogenicity, 1002–1003
   hygienic standards, 1003
   physical and chemical properties, 1001
   thyroid effects, 1002, 1003
   toxicity, 1001–1003
   uses, 1000–1001
Methylethanolamine, choline metabolism and, 1160
Methylethylbenzene, 1346
   physical properties, 1304–1305
   toxicity, 1345
Methylethylene, *see* 2-Butene; Propene
3-Methylhexane, toxicity, 1236
2-Methylhexane, toxicity, 1236
$N$-Methyl-2,2'-iminodiethanol, physical characteristics, 1112
2,2'-(Methylimino)diethanol, toxicity, 1107
$N$-Methyl-2,2'-iminodiethanol, toxicity, 1112
Methylnaphthalene:
   physical properties, 1304–1305
   production and uses, 1377
   toxicity, 1377
Methylparathion, 948
2-Methylpentane:
   hygienic standards, 1235
   toxicity, 1235
3-Methylpentane, 1235
   physicochemical properties, 1226–1227
2-Methylpent-2-ene, *see* Isohexene
2-Methylpropane, *see also* Isobutene
   hygienic standards, 1231
   toxicity, 1230–1231

α-Methylstyrene, 1360–1361
  hygienic standards, 1363
  toxicity, 1360–1361, 1363
∞-Methylstyrene, physical properties, 1304–1305
p-Methylstyrene, physical properties, 1304–1305
Mineral seal oil, see Kerosine
Mineral spirits, 1400–1401
  physical properties, 1396
  toxicity, 1400–1401
Mirex:
  carcinogenicity, classification, 1506
  regulatory standings, 1504
Monoallylamine, toxicity, 1111
Monoamine oxidase:
  in aliphatic amine metabolism, 1094, 1095
  gasoline and inhibition of, 1406
Mono-sec-butanolamine, 1157
Monobutylbenzenes, toxicity, 1350
Monobutylethanolamine, toxicity, 1162
Monochlorobenzene:
  acute toxicities, 1448–1449
  carcinogenicity, 1445, 1445–1446
  chronic toxicity, 1450
  distribution of metabolites, 1452
  human exposures:
    experimental, 1453
    industrial, 1453
    intoxications, 1453
  hygienic standards, 1454
  metabolism, 1451–1453
  odor and taste, 1453
  physical and chemical properties, 1447–1448
  reproductive effects, 1450–1451
  subchronic toxicity, 1449–1450
  teratogenicity, 1450–1451
Monoethanolamine:
  hygienic standards, 1153
  toxicity, 1151–1152
  uses, 1151
Monoethyldiethanolamine, 1164
Monoethylethanolamine, toxicity, 1159–1160
Monoisopropanolamine, uses, 1156
Monoisopropylethanolamine, 1161–1162
Monomethyldiethanolamine, 1164
Monomethylethanolamine, toxicity, 1158
Muconic acid, in benzene metabolism, 1319
Myelotoxicity, aromatic hydrocarbon-related, 1303
Myocardial contractility, chlorofluorcarbons and, 1182

NADH-methemoglobin reductase, 950–951
NADPH oxidase, benzene and, 1324
Naphtha:
  butadiene synthesis, 1251
  in toluene production, 1326
Naphthalane, see Decalin
Naphthalene:
  hygienic standards, 1376–1377
  metabolism, 1374, 1376
  microbial studies, 1376
  mutagenicity and carcinogenicity, 1375–1376
  physical properties, 1304–1305
  production and uses, 1371, 1373
  toxicity, 1372
    acute, 1373–1374, 1374–1375
    animal studies, 1374–1376
    chronic, 1374, 1375
    eye and dermal, 1373, 1374
    human exposures, 1373–1374
    inhalation, 1373–1374, 1374
    oral effects, 1373, 1374
    to marine species, 1375
Naphthane, see Decalin
Naphthenic aromatic solvents, toxicity, 1410
Naphthylamine, 1039–1042
2-Naphthylamine:
  carcinogenicity, 961–962
  $N$-oxidation, 1040
α-Naphthylamine:
  methemoglobin formation, 950
  physical and chemical properties, 1039–1040
  toxicity, 1040
  uses, 1039
β-Naphthylamine:
  bladder cancer risk, 1041–1042
  carcinogenicity, 959
  $N$-hydroxylation, 956
  metabolism, 956
  methemoglobin formation, 950
  physical and chemical properties, 1041
  toxicity, 1041–1042
  uses, 1041
National Institute for Occupational Safety and Health (NIOSH):
  carcinogen standards, 959–960
  Ohio rubber plant investigation, 953
  phenol sampling method, 1569
Natural gas, fractionation, 1394
Neoheptane, 1236
Neohexane, 1235
  physicochemical properties, 1226–1227
Neononane, toxicity, 1238
Neopentane:
  physicochemical properties, 1226–1227
  toxicity, 1233
Nephrotoxicity:
  of aromatic nitro and amino compounds, 973
  of cycloalkanes, 1268

diphenylamine-related, 1013-1014
triethanolamine-induced, 1155
Neurotoxicity:
heptane-related, 1236
pentane-related, 1232
2,2',2"-Nitrilotriethanol, toxicity, 1107
m-Nitroacetanilide, testicular toxicity, 964
m-Nitroaniline:
metabolism, 955
physical and chemical properties, 988-989
synthesis, 1019-1021
toxicity, 989
o-Nitroaniline:
aquatic toxicity, 968
biogradation, 966-967
metabolism, 955
physical and chemical properties, 988
toxicity, 988
p-Nitroaniline:
hygienic standards, 990
metabolism, 955
methemoglobin formation, 950, 989-990
physical and chemical properties, 989
teratogenicity, 965
toxicity, 989-990
uses, 989
Nitrobenzene:
biogradation, 966
hygienic standards, 1018
metabolism, 953-954, 1016-1017
methemoglobin formation, 950
capacity for, 952
photochemical formation from benzene, 1307
physical and chemical properties, 1016
reduction, 1017
reproductive effects, 963-964, 1017-1018
toxicity, 1016-1018
systemic, 1016
uses, 1016
p-Nitrobenzene:
hygienic standards, 1023
physical and chemical properties, 1023
teratogenicity, 965
toxicity, 1023
Nitrobenzoic acid, biogradation, 966
2-Nitrobenzoic acid, 1052
2-Nitrobenzyl alcohol, 1052
Nitrobiphenyl, 1035-1037
2-Nitrobiphenyl:
physical and chemical properties, 1035
uses, 1035
4-Nitrobiphenyl:
physical and chemical properties, 1035-1036
toxicity, 1036

uses, 1035-1036
Nitroglycerin:
hygienic standards, 1067
methemoglobin formation, 950
pharmacokinetics, 1066
physical and chemical properties, 1065
toxicity, 1065-1066
uses, 1064
2-Nitronaphthalene, carcinogenicity, 960
Nitrophenol:
biogradation, 966-967
photochemical formation from benzene, 1307
2,4-Nitrophenol, metabolism, 954
m-Nitrophenol:
physical and chemical properties, 975
toxicity, 975
uses, 974
o-Nitrophenol:
physical and chemical properties, 974
toxicity, 974
p-Nitrophenol, 948
physical and chemical properties, 975
toxicity, 975-976
2-Nitro-1,4-phenylenediamine:
physical and chemical properties, 1009
toxicity, 1010
uses, 1009
3-Nitro-1,2-phenylenediamine, 1010
4-Nitro-1,2-phenylenediamine
physical and chemical properties, 1010
toxicity, 1010
uses, 1010
Nitroreductase activity, 954
Nitrosamines:
carcinogenic, in vivo biosynthesis, 1096
formation of, pH requirement, 1096-1097
Nitrosation:
nitrite concentration and, 1097
of primary amines, 1096-1097
of secondary amines, 1097
N-Nitroso compounds, formation from nitrite reactions, 957
Nitrosobenzene, methemoglobin formation, capacity for, 952
Nitrotoluene, 1051-1054
m-Nitrotoluene:
hygienic standards, 1053
physical and chemical properties, 1053
toxicity, 1053
uses, 1052
o-Nitrotoluene:
hygienic standards, 1052
physical and chemical properties, 1052
toxicity, 1052

*o*-Nitrotoluene (*Continued*)
  uses, 1051
*p*-Nitrotoluene:
  hygienic standards, 1054
  physical and chemical properties, 1053–1054
  toxicity, 1054
  uses, 1053
5-Nitro-*o*-toluidine:
  physical and chemical properties, 1046
  toxicity, 1046
  uses, 1046
Nitrotoluol, methemoglobin formation, capacity for, 952
No. 1 fuel oil, *see* Kerosine
Nonadecane, physicochemical properties, 1226–1227
*n*-Nonadecane, toxicity, 1239
Nonane, physicochemical properties, 1226–1227
*n*-Nonane:
  hazard properties, 1223
  toxicity, 1237–1238
Nonene, physicochemical properties, 1242–1243

Occupational Safety and Health Administration (OSHA):
  alkane classifications, 1226
  carcinogen standards, 959–960
  phenol sampling method, 1569
Octadecane, physicochemical properties, 1226–1227
*n*-Octadecane, toxicity, 1239
Octadecene, physicochemical properties, 1242–1243
Octadecylamine:
  physical and chemical properties, 1090
  toxicity, 1132
  uses, 1132
1,5-Octadiene, physicochemical properties, 1242–1243
Octafluorocyclobutane, mutagenicity, 1192
Octane, physicochemical properties, 1226–1227
*n*-Octane:
  hazard properties, 1223
  hygienic standards, 1237
  toxicity, 1236–1237
Octanol/water partition coefficient, 1306
Octene:
  hygienic standards, 1250
  physicochemical properties, 1242–1243
  toxicity, 1249–1125
1-Octylamine, toxicity, 1130
2-Octylamine, toxicity, 1130–1131
Olefins, *see* Alkenes
Olive knot neoplasms, tetralin-related, 1288

Organochlorine insecticides, *see* Chlorinated hydrocarbon insecticides
Ozone, reaction of alkenes with, 1241
Ozone depletion, from chlorofluorocarbons, 1177–1178

Pancytopenia:
  benzene-related, 1321
  TNT-related, 1062
Paraffins:
  fractionation, 1394
  hygienic standards, 1415
  physical properties, 1413
  toxicity, 1414–1415
Pelthane, acute toxicity, 1505
Pentabromophenol, physiological response, 1616–1617
Pentachlorobenzene:
  acute toxicity, 1475
  biologic activation, 1488
  carcinogenicity, 1447
  in HCB metabolism, 1489
  in hexachlorobenzene, 1477
  hygienic standards, 1477
  metabolism, 1476–1477
  physical and chemical properties, 1475
  in porphyria production, 1488
  reproductive effects, 1476
  subchronic and chronic effects, 1475–1476
  teratogenicity, 1476
  uses and occurrences, 1474–1475
Pentachloronitrobenzene:
  carcinogenicity, 1032–1033
  hygienic standards, 1033
  pharmacokinetics, 1032
  physical and chemical properties, 1031
  subchronic and chronic toxicity, 1483
  teratogenicity, 1032
  toxicity, 1031–1033
  uses, 1031
Pentachlorophenol:
  acute toxicity, 1608
  atmospheric and biologic determination, 1606–1607
  distribution in the tissues, 1609
  handling instructions, 1612
  in HCB metabolism, 1487, 1489
  human exposures, 1610–1611
  hygienic standards, 1611–1612
  in pentachlorobenzene metabolism, 1477
  physical and chemical properties, 1605–1606
  physiological response, 1607–1608
    absorption, excretion, and metabolism, 1608–1610

cause of death, 1610
  mode of action, 1610
  pathology, 1608
  in porphyria production, 1488
  solubility, 1606
  source, uses, and industrial exposures, 1605
  treatment of local and systemic intoxication, 1612
Pentachlorothiophenol, in HCB metabolism, 1487, 1488
Pentadecane, physicochemical properties, 1226-1227
$n$-Pentadecane, toxicity, 1239
1,1,1,2,2-Pentafluorodimethyl ether, toxicity, 1189
1,1,1,2,2-Pentafluoropropane, toxicity, 1189
1,1,2,2,2-Pentafluoroethane, toxicity, 1189
1,1,2,2,3-Pentafluoropropane, toxicity, 1188
1,1,3,3-Pentafluoropropane, toxicity, 1188
Pentamethylbenzene, 1342
Pentamethylenediamine:
  physical and chemical properties, 1091
  toxicity, 1145
Pentane, physicochemical properties, 1226-1227
$n$-Pentane:
  hazard properties, 1223
  odor intensity, 1232
  toxicity, 1231-1232
1-Pentene:
  hygienic standards, 1248
  physicochemical properties, 1242-1243
  toxicity, 1248
2-Pentene:
  toxicity, 1248
  hygienic standards, 1248
$iso$-Pentene-2, physicochemical properties, 1242-1243
$trans$-Pentene-2, physicochemical properties, 1242-1243
$p$-Pentine, see Cyclopentadiene
Pentole, see Cyclopentadiene
Pentyl-1-pentamine, toxicity, 1106
1-Pentyne:
  physiochemical properties, 1254-1255
  toxicity, 1256
Perfluorobutene, mutagenicity, 1192
Petrol, see Gasoline
Petrolatum, 1416
Petroleum, see also Gasoline; Natural gas
  fractionation, 1394
  as a hydrocarbon source, 1393
  liquefied, 1395
    fractionation, 1394
Petroleum cracking, alkene production, 1240

Petroleum distillates, uses, 1222
Petroleum ether:
  fractionation, 1394
  hygienic standards, 1407
  physical properties, 1396
  toxicity, 1398, 1407
Petroleum jelly, see Petrolatum
Petroleum naphthas:
  benzene in, 1307
  fractionation, 1394
  hygienic standards, 1407
  physical properties, 1396-1397
  production and uses, 1406
  toxicity, 1406-1407
Petroleum solvents, toxicity, 1398-1404
Petroleum spirits:
  hygienic standards, 1409
  toxicity, 1408-1409
Phenacetin:
  toxicity, 972
  uses, 972
Phenanthrene:
  derivatives, 1386
  physical properties, 1379
  tetralin production and, 1285
  toxicity, 1380, 1386
Phenergan, cross-sensitization, 1008
Phenol:
  acute toxicity, 1574
  atmospheric and biological determination, 1569-1570
  with camphor, 1578
  chronic respiratory effect, 1576
  chronic toxicity, 1574-1575
  comparative absorption, 1572
  flammability, 1577
  hygienic standards, 1577
  metabolism, 1573
  microbial effects, 1306
  odor and warning properties, 1577
  physical and chemical properties, 1568
  physiological response:
    absorption, excretion, and metabolism, 1571-1572
    cause of death, 1573
    mode of action, 1572-1573
    pathology, 1570-1571
  poisoning:
    legal aspects, 1582-1583
    prevention of, 1579-1580
    treatment, 1580-1582
  skin depigmentation, 1576
  source, uses, and industrial exposures, 1567-1568

Phenol (*Continued*)
  toxicity, 1571
    comparative, 1571
    in tumor formation, 1583–1584
    use in medicine, 1577–1579
    in water supply, 1583
Phenylacetylene, 1364
  physical properties, 1304–1305
Phenylbutene, 1363
1-Phenylbutene-2, 1361
  physical properties, 1304–1305
  toxicity, 1361
4-Phenylbutene-1, 1361
  toxicity, 1361
Phenylenediamine:
  derivatives, 1009–1013
  isomers, 1004–1009
1,3-Phenylenediamine, synthesis, 1019–1021
*m*-Phenylenediamine, 1005–1007
  carcinogenicity, 1006–1007
  hygienic standards, 1007
  methemoglobin formation, 952
  physical and chemical properties, 1005–1006
  toxicity, 1006–1007
  uses, 1005
*o*-Phenylenediamine:
  hygienic standards, 1005
  physical and chemical properties, 1005
  toxicity, 1005
  uses, 1005
*p*-Phenylenediamine:
  carcinogenicity, 1009
  cross-sensitization, 1001, 1007–1008
  hygienic standards, 1009
  myotoxicity, 1008
  physical and chemical properties, 1007
  toxicity, 1007–1009
  uses, 1007
Phenylglyoxylic acid, in styrene metabolism, 1357
Phenylhydroxylamine, methemoglobin
    formation, 950
  capacity for, 952
Phenyl-β-naphthylamine, carcinogenicity, 960
*o*-Phenylphenol:
  human exposures, 1618
  physiological response, 1617–1618
  use, 1617
Phosphine, in acetylene production, 1254
Photodieldrin, 1529
Picric acid:
  hygienic standards, 981
  physical and chemical properties, 980
  toxicity, 980–981
  uses, 980

α-Pinene, physical and chemical properties, 1276–1277
Pinenes:
  hygienic standards, 1289
  toxicity, 1289
Polyarylhydrocarbons:
  carcinogenicity, 1384
  hygienic standards, 1384–1385
  metabolism, 1384
  occurrence, 1378
  teratogenicity and mutagenicity, 1384
  toxicity, 1378, 1384
Polycyclic aromatics, origin and sources, 1302
Polyenes, toxicity, 1283–1285
Polymorphonuclear leukocytic reactions, octane-induced, 1237
Polyneuropathy, *n*-hexane-related, 1233
Polynitrotoluenes, 1054–1063
Polyphenols, origin and sources, 1302
Polytetrafluoroethylene, carcinogenicity, 1199
Porphyrin, HCB metabolism and, 1481, 1487, 1488, 1491
Pristane:
  physicochemical properties, 1226–1227
  toxicity, 1239
Propadiene:
  physicochemical properties, 1242–1243
  toxicity, 1250
Propane:
  hygienic standards, 1229
  physicochemical properties, 1226–1227
  toxicity, 1228–1229
  urban air concentrations, 1228
*n*-Propane, hazard properties, 1223
1,2-Propanediamine:
  physical and chemical properties, 1091
  toxicity, 1107, 1144
1,3-Propanediamine, toxicity, 1107, 1144
Propene:
  physicochemical properties, 1242–1243
  toxicity, 1244–1245
Propenylbenzenes, toxicity, 1362
Prophobilinogen, benzene and, 1324
Propylamine:
  physical and chemical properties, 1090
  toxicity, 1105
*n*-Propylamine:
  basicity, 1089
  metabolism, distribution, and elimination, 1100
  toxicity, 1119
    amine bases and hydrochloride salts, 1109
  uses, 1119
Propylbenzene, 1346

## SUBJECT INDEX

n-Propylbenzene:
  physical properties, 1304–1305
  toxicity, 1348
Propylene, see also Propene
  toxicity, 1229
Propylenediamine, see 1,2-Propanediamine
1-Propylethylene, see 1-Pentene
Propyne:
  physiochemical properties, 1254–1255
  toxicity, 1256
Protoporphyrin, benzene, and, 1324
Pulmonary pneumonitis, decane-induced, 1238
Pyrene, toxicity, 1381, 1389–1390
Pyrocatechol:
  atmospheric determination, 1585
  in benzene metabolism, 1308
  flammability, 1586
  human exposures, 1586
  hygienic standards, 1586
  odor and warning properties, 1586
  physical and chemical properties, 1584–1585
  physiological response:
    absorption, excretion, and metabolism, 1585–1586
    cause of death, 1586
    mode of action, 1586
    pathology, 1585
  source, uses, and industrial exposures, 1584
Pyrogallol:
  atmospheric determination, 1595
  human exposures, 1596
  physical and chemical properties, 1595
  physiological response, 1595–1596
    absorption, excretion, and metabolism, 1596
    mode of action, 1596
    pathology, 1596
  poisoning prevention and treatment, 1597
  source, uses, and industrial exposures, 1595
Pyrolysis of paraffin, in alkene production, 1240
Pyropentylene, see Cyclopentadiene
Pyrrole, formation from 2,5-hexanedione, 1225

Quinone:
  atmospheric determination, 1593
  human exposures, 1594
  hygienic standards, 1594
  odor and warning properties, 1594
  physical and chemical properties, 1593
  physiological response:
    absorption, excretion, and metabolism, 1594
    mode of action, 1594
  poisoning prevention, 1594–1595
  source, uses, and industrial exposures, 1593

Residual oils, 1416–1417
Resorcinol:
  atmospheric determination, 1587
  flammability, 1589
  human exposures, 1588–1589
  hygienic standards, 1589
  odor and warning properties, 1589
  physical and chemical properties, 1587
  physiological response:
    absorption, excretion, and metabolism, 1588
    cause of death, 1588
    pathology, 1588
  source, uses, and industrial exposures, 1586–1587
  treatment of poisoning, 1589–1590
Respiratory paralysis:
  aromatic hydrocarbon-related, 1303
  butadiene-induced, 1251
  dibutylethanolamine-induced, 1163
Respiratory toxicity, from chlorofluorocarbons, 1183
Retinopathy, from 4,4-Methylenedianiline, 1001
Retinyl acetate, 1253
Righting reflex, heptene and loss of, 1249
Rubber solvent:
  physical properties, 1396
  toxicity, 1400, 1408

Sodium bicarbonate, to treat phenol intoxication, 1581
Sodium dichromate, methemoglobin formation, 950
Sodium nitrite, methemoglobin formation, 950
Sodium pentachlorophenate:
  acute toxicity, 1608
  atmospheric and biologic determination, 1606–1607
  handling instructions, 1612
  human exposures, 1610–1611
  hygienic standards, 1611–1612
  physical and chemical properties, 1605–1606
  physiological response, 1607–1608
    absorption, excretion and metabolism, 1608–1610
    cause of death, 1610
    mode of action, 1610
    pathology, 1608
  solubility, 1606
  source, uses, and industrial exposures, 1605
  treatment of local and systemic intoxication, 1612
Sorbitol dehydrogenase, depletion with HCB, 1481

Squalene, physicochemical properties, 1242–1243
Stilbene, toxicity, 1370
cis-Stilbene, physical properties, 1304–1305
trans-Stilbene, physical properties, 1304–1305
Stoddard solvent:
    physical properties, 1396
    toxicity, 1401, 1409
Stove oil, 141
Styrene:
    absorption and metabolism, 1357
    hygienic standards, 1357–1358
    physical properties, 1304–1305
    production and uses, 1351–1352
    teratogenicity, 1356
    toxicity:
        acute, 1352, 1353
        subacute and chronic, 1352, 1354–1355, 1356–1357
Succinic dehydrogenase, depletion with HCB, 1481
Sudden sniffing death syndrome, 1203

Terpenes, toxicity, 1282–1283
Terphenyls:
    physical properties, 1304–1305
    toxicity, 1368–1369
Tetanic convulsions, $n$-heptane-induced, 1236
Tetrachlorobenzene:
    acute toxicity, 1472
    human exposures, 1474
    hygienic standards, 1474
    metabolism, 1473–1474
    physical and chemical properties, 1472
    reproductive effects, 1473
    subchronic and chronic toxicity, 1472
    teratogenicity, 1473
    uses and occurrences, 1471
1,2,4,5-Tetrachlorobenzene:
    biologic activation, 1488
    carcinogenicity, 1447
    in hexachlorobenzene, 1477
    toxicity, 1444
1,1,1,2-Tetrachloro-2,2-difluoroethane:
    threshold limit values, 1195
    toxicity, 1187
1,1,2,2-Tetrachloro-1,2-difluoroethane:
    threshold limit values, 1195
    toxicity, 1187
Tetrachloroethane, CFC and, toxicity studies, 1185–1186
Tetrachlorohydroquinone, in HCB metabolism, 1487–1488
Tetrachlorophenol, lethal dosages, 1615

1,2,4,5-Tetrachlorophenol, in HCB metabolism, 1487
2,3,4,5-Tetrachlorophenol, in pentachlorobenzene metabolism, 1477
2,3,4,6-Tetrachlorophenol:
    in HCB metabolism, 1487
    physical and chemical properties, 1614
2,3,5,6-Tetrachlorophenol:
    in HCB metabolism, 1487
    in pentachlorobenzene metabolism, 1477
Tetradecane, physicochemical properties, 1226–1227
$n$-Tetradecane, toxicity, 1239
Tetradecyldimethylamine, 1132
Tetraethylenepentamine:
    physical and chemical properties, 1091
    toxicity, 1149
1,2,2,2-Tetrafluoroethane, toxicity, 1189
Tetrafluoroethylene, carcinogenicity, 1199
1,1,2,2-Tetrafluoropropane, toxicity, 1189
1,2,3,4-Tetrahydrobenzene, see Cyclohexene
1,2,3,4-Tetrahydronaphthalene, see Tetralin
Tetrahydrostyrene, see 1-Vinylcyclohex-1-ene
Tetralin:
    chemical properties, 1276–1277
    nephropathy, 1268
    physical properties, 1276–1277, 1285
    toxicity, 1278–1279, 1285, 1288
Tetramethylbenzene, 1342
    physical properties, 1304–1305
    toxicity, 1341
Tetramethylene, see Cyclobutane
Tetramethylenediamine:
    metabolism, distribution and elimination, 1103
    physical and chemical properties, 1091
    toxicity, 1145
2,3,5,6-Tetramethyl-$p$-phenylenediamine, myotoxicity, 1008
Tetryl:
    hygienic standards, 998
    physical and chemical properties, 998
    toxicity, 998
    uses, 998
Thalidomide, teratogenic mechanism, 1140
Thinners:
    physical properties, 1397
    toxicity, 1402–1403, 1410
Thyroid tumors, butadiene-induced, 1251
Thyroxine, DNP and production of, 977–978
TIBA, see Triisobutylamine
TNPA, see Tripropylamine
TNT syndrome, 1062
Toluene:
    hygienic standards, 1331–1332

metabolism, 1330, 1331
microbial effects, 1331
oral and dermal, 1331
physical properties, 1304-1305
precautionary measures, 1332
properties, sources, and use, 1326
sensitization, 1330
toxicity, 1326
  acute, 1327
  animal studies, 1330-1331
  chronic, 1328, 1329
  human exposures, 1329-1330
  inhalation, 1329, 1330
  oral and dermal, 1329-1330
  to marine life, 1331
Toluene diisocyanate, 2,4-DNT in manufacture of, 1056
Toluenediamine, 1048-1051
2,3-Toluenediamine, physical and chemical properties, 1048
2,4-Toluenediamine:
  physical and chemical properties, 1049
  reproductive effects, 1049
  toxicity, 1049-1050
  uses, 1049
2,5-Toluenediamine:
  physical and chemical properties, 1050
  toxicity, 1050
2,6-Toluenediamine:
  physical and chemical properties, 1050-1051
  toxicity, 1051
  uses, 1050
3,4-Toluenediamine:
  physical and chemical properties, 1051
  toxicity, 1051
3,5-Toluenediamine, 1051
Toluic acid, in xylene metabolism, 1338
Toluidine, 1043-1048
$m$-Toluidine:
  hygienic standards, 1047
  physical and chemical properties, 1046
  toxicity, 1046-1047
  uses, 1046
$o$-Toluidine:
  carcinogenicity, 984, 1043-1044
  hygienic standards, 1044
  physical and chemical properties, 1043
  toxicity, 1043
  uses, 1043
  Zymbal gland tumors, 999
$p$-Toluidine:
  physical and chemical properties, 1047
  toxicity, 1047
  uses, 1047

Total phenol method, 1569
Toxaphene:
  acute toxicity, 1505
  carcinogenicity, classification, 1506
  liver enzyme effects, 1521
  regulatory standings, 1504
Toxic shock syndrome, phenol-induced, 1574
Transglutaminase, polyamines as acyl acceptors for, 1140
Triallylamine:
  physical and chemical properties, 1090
  toxicity, 1106, 1111, 1135
  uses, 1135
Triaminotoluene, 967
Tri-$n$-amylamine:
  toxicity, 1126
  uses, 1126
2,4,6-Tribromophenol, physiological response, 1616-1617
Tri-$sec$-butanolamine, 1157
Tributylamine, toxicity, 1105
Tri-$n$-butylamine:
  physical and chemical properties, 1090
  toxicity, 1124
Trichlorobenzene:
  acute toxicity, 1469
  chronic toxicity, 1470
  human exposures, 1471
  hygienic standards, 1471
  metabolism, 1470-1471
  physical and chemical properties, 1468-1469
  reproductive effects, 1470
  subchronic toxicity, 1469-1470
  teratogenicity, 1470
  uses, 1468
1,2,4-Trichlorobenzene:
  carcinogenicity, 1447
  effect on foreign organic compound metabolism, 1464
  toxicity, 1444
1,3,5-Trichlorobenzene:
  biologic activation, 1488
  carcinogenicity, 1447
Trichlorofluoroethane, toxicity, 1189
Trichlorofluoromethane:
  in aerosols, 1199
    blood levels, 1202
    cardiopulmonary effects, 1200
    respiratory uptake, 1201
    tissue levels, 1202-1203
  carcinogenicity, 1193
  cardiotoxicity, 1179
  foam blowing application, 1197
  mutagenicity, 1192

Trichlorofluoromethane (*Continued*)
  refrigerant poisoning, 1196
  threshold limit values, 1195, 1196
  toxicity, 1187, 1189
2,3,5-Trichlorophenol, physical and chemical properties, 1614
2,4,5-Trichlorophenol:
  in HCB metabolism, 1487
  physical and chemical properties, 1614
  physiological response, 1615
2,4,6-Trichlorophenol:
  physical and chemical properties, 1614
  physiological response, 1615
1,1,2-Trichloro-1,2,2-trifluoroethane:
  carcinogenicity, 1192
  refrigerant poisoning, 1196
  in solvents, 1197
  threshold limit values, 1195
  toxicity, 1187
Tri-*o*-cresyl phosphate, cyclohexane potentiation of, 1273
Tridecane:
  metabolism, 1240
  physicochemical properties, 1226–1227
  toxicity, 1239–1240
Triethanolamine:
  carcinogenicity, 1155
  physical characteristics, 1112
  toxicity, 1112, 1154–1156
  uses, 1154
Triethylamine:
  basicity, 1089
  hygienic standards, 1118
  metabolism, distribution, and elimination, 1100
  physical and chemical properties, 1090
  toxicity, 1105, 1118
  uses, 1118
*N*-Triethylaminoethanol, choline metabolism and, 1160
Triethylenetetramine:
  metabolism, distribution and elimination, 1103
  physical and chemical properties, 1091
  toxicity, 1107, 1148
  uses, 1148
Triethylenetriamine, acute toxicity, 1110–1111
1,1,1-Trifluorodimethyl ether, toxicity, 1189
2,4,6-Triiodophenol, physiological response, 1616–1617
Triisobutylamine, 1124
Triisopropanolamine, 1156–1157
  physical characteristics, 1112
  toxicity, 1107, 1112
Trimethylamine:
  basicity, 1089

  hygienic standards, 1116
  metabolism, distribution, and elimination, 1099
  odor, 1115
  *N*-oxidation capacity, 1097–1098
  physical and chemical properties, 1090
  toxicity, 1115
  uses, 1115
Trimethylbenzene:
  hygienic standards, 1340
  metabolism, 1340
  physical properties, 1304–1305
  production and uses, 1340
  toxicity, 1340, 1341
Trimethylenediamine, physicochemical properties, 1091
Trimethylethanolamine, metabolism, 1103–1104
2,2,5-Trimethylhexane, physicochemical properties, 1226–1227
Trimethylhexane, toxicity, 1238
Trimethylnaphthalene, toxicity, 1378
2,2,4-Trimethylpentane:
  nephropathy, 1268
  physicochemical properties, 1226–1227
  toxicity, 1237
2,3,4-Trimethylpentane:
  physicochemical properties, 1226–1227
  toxicity, 1237
Trimethylpentene, toxicity, 1250
Trimethylphenanthrene, toxicity, 1380
Trinitrobenzene:
  methemoglobin formation, capacity for, 952
  physical and chemical properties, 1024
  use, 1024
1,3,5-Trinitrobenzene, 968
*p*-Trinitrobenzene, methemoglobin formation, capacity for, 952
2,4,6-Trinitrotoluene:
  hygienic standards, 1063
  metabolism, 1063
  methemoglobin formation, 952
  physical and chemical properties, 1061
  toxicity, 1061–1063
  uses, 1061
Trinitrotoluene (TNT), metabolic pathway reduction, 967
Triolefins, *see* Alkatrienes
Triphenylamine:
  hygienic standards, 1015
  physical and chemical properties, 1015
  toxicity, 1015
  uses, 1015
Triphenyls:
  hygienic standards, 1371

metabolism, 1371
production and uses, 1370
toxicity, 1370–1371
Tripropylamine, 1120–1121
toxicity, 1105
Turpentine:
hygienic standards, 1291
inhalation toxicity, 1286–1287
physical and chemical properties, 1276–1277
toxicity, 1278–1279, 1289–1291

U.S. Department of Health and Human Services (USPHS), carcinogen standards, 958–959
Undecane, physicochemical properties, 1226–1227
$n$-Undecane, 1239

Vacuum distillation fractions, 1414
Varnish Makers' and Painters' Naphtha:
hygienic standards, 1408
physical properties, 1396
toxicity, 1400, 1408
Vasodilation:
endothelium-dependent, 1066
nitroglycerin-related, 1065–1066
Vertigo, hexene-related, 1249
Vinyl bromide, 999
Vinyl chloride, 999, 1351
Vinylcyclohexene, toxicity, 1281–1282
1-Vinylcyclohexene:
inhalation toxicity, 1280
physical and chemical properties, 1276–1277
4-Vinylcyclohexene:

inhalation toxicity, 1280
physical and chemical properties, 1276–1277
toxicity, 1278–1279
Vinyltoluene:
hygienic standards, 1358
toxicity, 1358, 1359–1360

White oils:
physical properties, 1413
toxicity, 1414
White spirits, see Stoddard solvent

Xylene:
absorption and metabolism, 1337
biologic monitoring, 1339
hygienic standards, 1339
isomeric forms, 1302
physical properties, 1304–1305
metabolism, 1338
microbial studies, 1338–1339
mixed, physical properties, 1304–1305
physical properties, 1332
production and use, 1332–1333
toxicity:
acute, 1333, 1334–1335, 1336, 1338
animal studies, 1334–1335, 1337–1338
chronic, 1333, 1336–1337, 1338
human exposures, 1333, 1336–1337

Yellow vision, 980

Zinc, elevations in, TNT-related, 1063
Zymbal gland tumors, 999

# Chemical Index

2-AAF, see 2-Acetylaminofluorene
Acenaphthene *[83-32-9]*, 1379, 1385
Acepyrene, see Cyclopentapyrene
2-Acetaminofluorene, see 2-Acetylaminofluorene
Acetaminophen *[103-90-2]*, 972
Acetanilide *[103-84-4]*, 950, 951, 952, 972
Acetanilide hydroxylase *[9012-80-0]*, 1323
Acetophenetidin *[62-44-2]*, 951, 952. See also Phenacetin
2-Acetylaminofluorene *[53-96-3]*, 968–970
N-Acetylaminophenathrene, see 2-Acetylaminofluorene
Acetylene *[74-86-2]*, 1254–1255, 1254–1255
N-Acetylethanolamine *[142-26-7]*, 1164–1165
Alanine aminotransferase *[9000-86-6]*, 1337
Aldrin *[309-00-2]*, 1504, 1505, 1506
Allene *[463-49-0]*, 1250
Allylamine *[107-11-9]*, 1090, 1106, 1133
Allylbenzene *[300-57-2]*, 1360
3-Aminoacetanilide *[102-28-3]*, 954, 1021
1,2-Aminoaniline, see o-Phenylenediamine
3-Aminoaniline, see m-Phenylenediamine
4-Aminoaniline, see p-Phenylenediamine
p-Aminoaniline, see p-Phenylenediamine
4'-Amino-3:2'-azotoluene, see o-Aminoazotoluene
o-Aminoazotoluene *[97-56-3]*, 1038
Aminobenzene, see Aniline
4-Aminobiphenyl *[92-67-1]*, 1036–1037, 1040
1-Aminobutane, see n-Butylamine
2-Aminobutane, see 2-Butylamine
1-Amino-2-butanol, see Mono-sec-butanolamine
γ-Aminobutyric acid *[56-12-2]*, 1141, 1142

Aminocyclohexane, see Cyclohexylamine
4'-Amino-2,3'-dimethylazobenzene, see o-Aminoazotoluene
4-Aminodiphenyl *[92-67-1]*, 959, 984
p-Aminodiphenyl, see 4-Aminobiphenyl
4-Aminodiphenylamine, see p-Aminodiphenylamine
p-Aminodiphenylamine *[101-54-2]*, 1037–1038
Aminoethane, see Ethylamine
Aminoethanol, see Monoethanolamine
2-Aminoethanol *[141-43-5]*, 1106. See also Monoethanolamine
2-(2-Aminoethylamino)ethanol, see Aminoethylethanolamine
1-Amino-2-ethylbutane, see Isohexylamine
Aminoethylethanolamine *[111-41-1]*, 1165
1-Aminoheptane *[111-68-21]*, 1128
2-Aminoheptane *[123-82-0]*, 1128
3-Aminoheptane *[2829-42-4]*, 1128–1129
4-Aminoheptane *[16751-59-0]*, 1129
1-Aminohexane, see n-Hexylamine
p-Aminohippurate, 991
3-Amino-1-Hydroxybenzene, see m-Aminophenol
4-Amino-1-hydroxybenzene, see p-Aminophenol
2-Aminoisobutane, see tert-Butylamine
o-Aminolevulenic acid *[106-60-5]*, 1324, 1487
5-Aminolevulinic acid *[106-60-5]*, 1481
o-Aminolevulinic acid synthetase *[9037-14-3]*, 1488
Aminomethane, see Methylamine
1-Amino-2-methylbenzene, see o-Toluidine
1-Amino-3-methylbenzene, see m-Toluidine

1-Amino-4-methylbenzene, see p-Toluidine
1-Amino-3-methylbutane, see Isoamylamine
1-Amino-2-methylpropane, see Isobutylamine
1-Aminonaphthalene, see α-Naphthylamine
2-Aminonaphthalene, see β-Naphthylamine
m-Aminonitrobenzene [99-09-2], 952
1-Amino-2-nitrobenzene, see o-Nitroaniline
1-Amino-3-nitrobenzene, see m-Nitroaniline
1-Amino-4-nitrobenzene, see p-Nitroaniline
4-Aminonitrobenzene, see p-Nitroaniline
2-Amino-4-nitrobenzoic acid, 954–955, 1057
2-Amino-6-nitrobenzoic acid, 1060
2-Amino-6-nitrobenzyl alcohol, 1060
2-Amino-5-nitrophenol [121-88-0], 1003–1004
4-Amino-2-nitrophenol [119-34-6], 1004
p-Amino-5-nitrophenol [121-88-0], 955
1-Aminononane [112-20-9], 1132
1-Aminooctane, see 1-Octylamine
2-Aminooctane, see 2-Octylamine
1-Aminopentane, see n-Amylamine
3-Aminopentene, see Allylamine
Aminophen, see Aniline
2-Aminophenol [95-55-6], 1017. See also
    o-Aminophenol
3-Aminophenol [591-27-5], 1017. See also
    m-Aminophenol
4-Aminophenol [123-30-8], 1017. See also
    p-Aminophenol
m-Aminophenol [591-27-5], 965, 967, 970–972
o-Aminophenol [95-55-6], 950, 952, 967, 970
p-Aminophenol [123-30-8], 950, 952, 954, 964, 967, 971–974
4-Aminophenyl ether [101-80-4], 1014
1-Aminopropane, see Propylamine
1-Amino-2-propanol, see Isopropanolamine; Monoisopropanolamine
3-Amino-1-propanol [156-87-6], 1107, 1112
3-Aminopropylene, see Allylamine
Aminopyrine N-demethylase [9037-69-8], 1323
Aminopyrine demethylase [9037-69-8], 1488
2-Aminotoluene, see o-Toluidine
3-Aminotoluene, see m-Toluidine
4-Aminotoluene, see p-Toluidine
o-Aminotoluene, see o-Toluidine
p-Aminotoluene, see p-Toluidine
11-Aminoundecanoic acid [2432-99-7], 1131
Ammonia [7664-41-7], 1089, 1092
Amylamine [110-58-7], 1106
n-Amylamine [110-58-7], 1090, 1125
Amylbenzene [538-68-1], 1351
2-Amylene, see 2-Pentene
β-n-Amylene, see 2-Pentene
p-tert-Amylphenol, see p-tert-Butylphenol
Anhydrous monomethylamine, see Methylamine

Aniline [62-53-3], 950, 952, 955, 966, 972, 982–984, 1008, 1044
Aniline hydroxylase [9012-80-0], 1481, 1488
Aniline oil, see Aniline
Anilinobenzene, see Diphenylamine
Anthracene [120-12-7], 1379, 1380, 1385–1386
Arsine [7784-42-1], 1254
Arylamine, see Aniline

Benzanthracene [56-55-3], 1380, 1386–1387
Benz(a)anthracene, see Benzanthracene
1,2-Benzanthracene, see Benzanthracene
1,2-Benzanthracene [56-55-3], 1379
Benzenamine, see Aniline
Benzene [71-43-2], 948, 1301–1302, 1304–1305, 1306–1326, 1310, 1312–1313, 1314–1317
Benzene hexachloride [608-73-1], 1550–1557
Benzeneazodimethylaniline, see
    4-Dimethylaminoazobenzene
1,2-Benzenediamine, see o-Phenylenediamine
p-Benzenediamine, see p-Phenylenediamine
1,4-Benzenediamine, see p-Phenylenediamine
1,2-Benzenediol, see Pyrocatechol
Benzidine [92-87-5], 955, 959, 961, 984, 1033, 1040
Benzo(a)anthracene [56-55-3], 1378, 1384
Benzoic acid [65-85-0], 1330
1,2-Benzophenanthrene [218-01-0], 1379
3,4-Benzophenanthrene [95-19-7], 1379, 1381, 1389
Benzo(a)pyrene [192-97-2], 1378, 1380, 1382–1383, 1384, 1390–1391
Benzo(e)pyrene [192-97-2], 1379, 1383, 1392
1,2-Benzopyrene, see Benzo(e)pyrene
3,4-Benzopyrene [50-12-8], 1285, 1379. See also
    Benzoa(a)pyrene
o-Benzyl-p-chlorophenol, see p-tert-Butylphenol
o-Benzyltoluene [713-36-0], 1365
Biethylene, see Butadiene
4-Biphenylamine, see 4-Aminobiphenyl
p-Biphenylamine, see 4-Aminobiphenyl
4,4'-Biphenyldiamine, see Benzidine
Biphenyline, see 4-Aminobiphenyl
Bis-2-aminoethylamine, see Diethylenetriamine
Bis(p-aminophenyl)methane, see 4,4'-Methylenedianiline
Bis(4-aminophenyl)methane, see 4,4'-Methylenedianiline
Bis(hydroxyethyl)amine, see Diethanolamine
Blasting gelatin, see Nitroglycerin
Blasting oil, see Nitroglycerin
Botrilex, see Pentachloronitrobenzene
Brassico, see Pentachloronitrobenzene
α-Butadiene, see Butadiene

Butadiene *[106-99-0]*, 1250–1252
1,3-Butadiene *[106-99-0]*, 1242–1243. See also Butadiene
1,2-Butadiene *[590-19-2]*, 1242–1243
1-Butadiyne, 1254–1255
Butadiyne *[460-12-8]*, 1256
Butafume, see 2-Butylamine
2-Butanamine, see 2-Butylamine
1-Butanamine, see n-Butylamine
Butane *[106-97-8]*, 1226–1227, 1229–1230
n-Butane *[106-97-8]*, 1223
1,4-Butanediamine, see Tetramethylenediamine
1,3-Butanediamine *[590-88-5]*, 1091, 1107
1-Butene *[106-98-9]*, 1242–1243, 1246
2-Butene *[107-01-7]*, 1247
cis-Butene-2 *[590-18-1]*, 1242–1243
trans-Butene-2 *[624-64-6]*, 1242–1243
sec-Butylamine, see 2-Butylamine
t-Butylamine, see tert-Butylamine
Butylamine *[109-73-9]*, 1105
n-Butylamine *[109-73-9]*, 1089, 1090, 1109, 1121–1122
2-Butylamine *[13952-84-6]*, 1122
tert-Butylamine *[75-64-9]*, 1124
2-Butylaminoethanol, see Monobutylethanolamine
Butylaminoethanol, see Monobutylethanolamine
n-Butylbenzene *[104-51-8]*, 1304–1305, 1349
sec-Butylbenzene *[135-98-8]*, 1304–1305, 1349, 1350
tert-Butylbenzene *[98-06-6]*, 1304–1305, 1349, 1350
p-tert-Butylcatechol, see p-tert-Butylphenol
α-Butylene, see Isobutene
p-tert-Butylphenol *[27178-34-3]*, 1619–1620
tert-Butyltoluene *[98-51-5]*, 1304–1305, 1349, 1350–1351
1-Butyne *[107-00-6]*, 1254–1255, 1256
2-Butyne *[503-17-3]*, 1254–1255, 1256

Cadaverine, see Pentamethylenediamine
Calcium carbide *[75-20-7]*, 1254
Calcium oxalate *[563-72-4]*, 1375
Camphene *[79-92-5]*, 1276–1277, 1291–1292
Camphor *[76-22-2]*, 1578
Carbazotic acid, see Picric acid
Carbon dioxide *[124-38-9]*, 1226
β-Carotene *[7488-99-5]*, 1242–1243
Carvene, see Limonene
α-Caryophyllene *[6753-98-6]*, 1276–1277
Caryophyllene *[87-44-5]*, 1292
Catechol, see Pyrocatechol
CHA, see Cyclohexylamine
Chlordane *[57-74-9]*, 1504, 1505, 1521

Chlordane, technical *[12789-03-6]*, 1533–1540
Chlordecone, see Kepone
Chlordimeform *[6164-98-3]*, 1044
p-Chloroacetanilide *[539-03-7]*, 993, 1026
3-Chloroaniline, see m-Chloroaniline
2-Chloroaniline, see o-Chloroaniline
p-Chloroaniline *[106-47-8]*, 1026
m-Chloroaniline *[108-42-9]*, 991–992
o-Chloroaniline *[95-52-2]*, 990–991, 1025
Chlorobenzene *[108-90-7]*, 1443, 1444. See also Monochlorobenzene
Chlorocresol *[59-50-7]*, 1578
Chlorodifluoroethane *[25497-29-4]*, 1190, 1196
1-Chloro-1,1-difluoroethane *[55949-44-5]*, 1188, 1193
Chlorodifluoromethane *[75-45-6]*, 1188, 1190, 1192, 1193, 1195, 1196
1-Chloro-2,3-dinitrobenzene, see 3-Chloro-1,2-dinitrobenzene
1-Chloro-2,4-dinitrobenzene, see 4-Chloro-1,3-dinitrobenzene
1-Chloro-2,5-dinitrobenzene, see 2-Chloro-1,4-dinitrobenzene
1-Chloro-2,6-dinitrobenzene, see 2-Chloro-1,3-dinitrobenzene
1-Chloro-3,4-dinitrobenzene, see 4-Chloro-1,2-nitrobenzene
1-Chloro-3,5-dinitrobenzene, see 5-Chloro-1,3-dinitrobenzene
2-Chloro-1,3-dinitrobenzene *[606-21-3]*, 1027–1028
2-Chloro-1,4-dinitrobenzene *[619-16-9]*, 1028
3-Chloro-1,2-dinitrobenzene *[602-02-8]*, 1028
4-Chloro-1,2-dinitrobenzene *[610-40-2]*, 1029
4-Chloro-1,3-dinitrobenzene *[97-00-7]*, 1029–1030
5-Chloro-1,3-dinitrobenzene *[618-86-0]*, 1030–1031
6-Chloro-1,3-dinitrobenzene, see 4-Chloro-1,3-dinitrobenzene
4-Chloro-2-methylaniline, see 4-Chloro-o-toluidine
5-Chloro-2-methylaniline, see 5-Chloro-o-toluidine
2-Chloro-4-nitroaniline, see 4-Chloro-2-nitroaniline
2-Chloro-5-nitroaniline, see 4-Chloro-2-nitroaniline
4-Chloro-2-nitroaniline *[89-63-4]*, 997–998
4-Chloro-3-nitroaniline, see 4-Chloro-2-nitroaniline
5-Chloro-2-nitroaniline, see 4-Chloro-2-nitroaniline

1-Chloro-2-nitrobenzene, see
    o-Chloronitrobenzene
1-Chloro-3-nitrobenzene, see
    m-Chloronitrobenzene
1-Chloro-4-nitrobenzene, see
    p-Chloronitrobenzene
2-Chloronitrobenzene, see o-Chloronitrobenzene
2-Chloro-1-nitrobenzene, see
    o-Chloronitrobenzene
3-Chloronitrobenzene, see
    m-Chloronitrobenzene
4-Chloronitrobenzene, see p-Chloronitrobenzene
m-Chloronitrobenzene [121-73-3], 952,
    1025–1026
o-Chloronitrobenzene [88-73-3], 1024–1026
p-Chloronitrobenzene [100-00-5], 1026–1027
1-Chloro-1,1,2,2,2-pentafluoroethane [76-15-3],
    1187
Chloropentafluoroethane [76-15-3], 1190
1-Chloro-1,1,2,2,2-pentafluoroethane [76-15-3],
    1195, 1196
p-Chlorophenol [106-48-9], 1614, 1615
m-Chlorophenol [108-43-0], 1614
o-Chlorophenol [95-57-8], 1614, 1615
3-Chlorophenylamine, see m-Chloroaniline
2-Chlorophenylamine, see o-Chloroaniline
2-Chloro-1,4-phenylenediamine [615-66-7], 1012
2-Chloro-p-phenylenediamine, see 2-Chloro-1,4-
    phenylenediamine
4-Chloro-1,2-phenylenediamine [95-83-0], 1011
4-Chloro-1,3-phenylenediamine [5131-60-2],
    1011–1012
4-Chloro-m-phenylenediamine, see 4-Chloro-1,3-
    phenylenediamine
4-Chloro-o-phenylenediamine, see 4-Chloro-1,2-
    phenylenediamine
Chlorotetrafluoroethane [63938-10-3], 1188
p-Chloro-o-toluidine, see 4-Chloro-o-toluidine
6-Chloro-o-toluidine [87-63-8], 1045
4-Chloro-o-toluidine [95-69-2], 1044–1045
5-Chloro-o-toluidine [95-79-4], 1045
Chlorotrifluoroethane [1330-45-6], 1193, 1194
Chlorotrifluoromethane [75-72-9], 1187
Chlorpromazine [50-53-3], 1338
Choline [62-49-7], 1158, 1160
Cholinesterase [9001-08-5], 1337
Chrysene [218-01-9], 1381, 1389
Cl-MDA, see 4,4'-Methylenebis(2-chloroaniline)
m-CNB, see m-Chloronitrobenzene
o-CNB, see o-Chloronitrobenzene
p-CNB, see p-Chloronitrobenzene
Creosote [37331-87-6], 1602–1605
Cresol [1319-77-3], 1597–1601
p-Cresol [106-44-5], 1339

Cresylic acid, see Cresol
Cumene [98-82-8], 1304–1305, 1348–1349. See
    also Isopropylbenzene
Cuprinol, see Pentachlorophenol
(Cyanoethyl)diethylenetriamine [78704-97-9],
    1147
Cyclobutane [287-23-0], 1268–1269, 1269, 1272
Cyclododecane [294-62-2], 1272, 1275
Cyclododecatriene [27070-59-3], 1278–1279,
    1284
Cycloheptane [291-64-5], 1268–1269, 1275
1,3,5-Cycloheptatriene [544-25-2], 1276–1277
Cycloheptatriene [544-25-2], 1278–1279, 1284
Cycloheptene [628-92-2], 1276–1277, 1281
Cyclohexanamine, see Cyclohexylamine
Cyclohexane [110-82-7], 1268–1269, 1270–1271,
    1272, 1274
Cyclohexene [110-83-8], 1276–1277, 1277, 1280,
    1281
Cyclohexylamine [108-91-8], 1089, 1090, 1106,
    1135
Cyclononane [293-55-0], 1268–1269, 1275
Cyclooctadiene [29965-97-7], 1278–1279,
    1283–1284
Cycloocta-1,5-diene [111-78-4], 1276–1277
Cyclooctane [292-64-8], 1275
Cyclooctatetraene [629-20-9], 1276–1277
Cyclooctene [931-88-4], 1281
Cyclopentadiene [542-92-7], 1278–1279, 1283
1,3-Cyclopentadiene [542-92-7], 1276–1277
Cyclopentane [287-92-3], 1268–1269, 1270–1271
Cyclopentapyrene [83381-96-8], 1379, 1392
Cyclopentene [142-29-0], 1276–1277, 1276,
    1278–1279, 1280
Cyclopropane [75-19-4], 1268–1269, 1269,
    1270–1271
Cymene, see Isopropyltoluene
o-Cymene [527-84-4], 1304–1305
m-Cymene [535-77-3], 1304–1305
p-Cymene [99-87-6], 1304–1305

DACPM, see 4,4'-Methylenebis(2-chloroaniline)
DADPM, see 4,4'-Methylenedianiline
DAPM, see 4,4'-Methylenedianiline
2,4-DAT, see 2,4-Toluenediamine
DBMP, see Di-tert-butylmethylphenol
DBPC Antioxidant, see Di-tert-
    butylmethylphenol
o-DCB, see o-Dichlorobenzene
p-DCB, see p-Dichlorobenzene
DCNA, see 2,6-Dichloro-4-nitroaniline
p,p'-DDE [72-55-9], 1505, 1510
DDT, see Dichlorodiphenyltrichloroethane
DEAE, see Diethylethanolamine

# CHEMICAL INDEX

Deanol, see Dimethylethanolamine
trans-Decalin [493-02-7], 1276–1277
cis-Decalin [91-17-8], 1276–1277
Decalin [91-17-8], 1268, 1278–1279, 1284–1285, 1286–1287
Decamethylenediamine [646-25-3], 1146
Decane [124-18-5], 1226–1227
n-Decane [124-18-5], 1223, 1238
1,10-Decanediamine, see Decamethylenediamine
Decanediamine, see Decamethylenediamine
Decene [25339-53-1], 1242–1243
Decylamine [2016-57-1], 1132
Decyldimethylamine [1120-24-7], 1132
1-Decyne [764-93-2], 1254–1255, 1256
Deenax, see Di-tert-butylmethylphenol
DETA, see Diethylenetriamine
2-N-Di-n-butylaminoethanol, see Dibutylethanolamine
1,3-Diacetamidobenzene [10268-78-7], 954
1,4-Diacetamidobenzene [140-50-1], 954
Diallylamine [124-02-7], 1090, 1106, 1111, 1134
1,2-Diaminobenzene, see o-Phenylenediamine
1,3-Diaminobenzene, see m-Phenylenediamine
1,4-Diaminobenzene, see p-Phenylenediamine
4-Diaminobenzene, see p-Phenylenediamine
m-Diaminobenzene, see m-Phenylenediamine
o-Diaminobenzene, see o-Phenylenediamine
p-Diaminobenzene, see p-Phenylenediamine
4,4'-Diaminobiphenyl, see Benzidine
1,4-Diaminobutane, see Tetramethylenediamine
4,4'-Diamino-3,3'-dichlorobiphenyl, see 3,3'-Dichlorobenzidine
4,4'-Diaminodiphenyl [92-87-5], 984
p-Diaminodiphenyl, see Benzidine
4,4'-Diaminodiphenyl oxide, see 4-Aminophenyl ether
4,4'-Diaminodiphenylmethane, see 4,4'-Methylenedianiline
1,2-Diaminoethane, see Ethylenediamine
1,6-Diaminohexane, see Hexamethylenediamine
1,5-Diaminonaphthalene [2243-62-1], 1042–1043
1,2-Diamino-4-nitrobenzene, see 4-Nitro-1,2-phenylenediamine
1,5-Diaminopentane [462-94-2], 1142. See also Pentamethylenediamine
1,2-Diaminopropane [78-90-0], 1110, 1142. See also 1,2-Propanediamine
1,3-Diaminopropane [109-76-2], 1110, 1142. See also 1,3-Propanediamine
2,3-Diaminotoluene, see 2,5-Toluenediamine
2,4-Diaminotoluene, see 2,4-Toluenediamine
2,5-Diaminotoluene, see 2,5-Toluenediamine
2,6-Diaminotoluene, see 2,6-Toluenediamine
3,4-Diaminotoluene, see 3,4-Toluenediamine

Di-n-amylamine [2050-92-2], 1125–1126
DIBA, see Diisobutylamine
Dibenz(a,h)anthracene [53-70-3], 1383, 1393
Dibenzoanthracene [53-70-3], 1378
Dibenzopyrene [58615-36-4], 1378
Di-sec-butanolamine [21838-75-5], 1157
Dibutylamine [111-92-2], 1105
Di-n-butylamine [111-92-2], 1090, 1123
2-(Dibutylamino)ethanol, see Dibutylethanolamine
2-N-Dibutylaminoethanol, see Dibutylethanolamine
2-Dibutylaminoethanol [102-81-8], 1107, 1112
2,6-Di-tert-butyl-p-cresol, see Di-tert-butylmethylphenol
Dibutylethanolamine [102-81-8], 1162–1164
Di-tert-butylhydroxytoluene, see Di-tert-butylmethylphenol
Di-tert-butylmethylphenol [29759-28-2], 1618–1619
2,6-Di-tert-butylphenol, see Di-tert-butylmethylphenol
Dichloran, see 2,6-Dichloronitroaniline
2,4-Dichloroaniline [554-00-7], 995, 1026
2,3-Dichloroaniline [608-27-5], 994–995
2,6-Dichloroaniline [608-31-1], 996
3,5-Dichloroaniline [626-43-7], 996
3,4-Dichloroaniline [95-76-1], 994
2,5-Dichloroaniline [95-82-9], 995–996
3,4-Dichlorobenzenamine, see 3, 4-Dichloroaniline
1,2-Dichlorobenzene, see o-Dichlorobenzene
1,4-Dichlorobenzene, see p-Dichlorobenzene
p-Dichlorobenzene [106-46-7], 1443, 1444, 1445–1446, 1458, 1459–1468, 1488
o-Dichlorobenzene [95-50-1], 1443, 1444, 1444, 1445–1446, 1454–1459
3,3'-Dichlorobenzidine [91-94-1], 999, 1034
3,3'-Dichlorobiphenyl-4,4'-diamine, see 3,3'-Dichlorobenzidine
4,5-Dichlorocatechol [3428-24-8], 1458
3,4-Dichlorocatechol [3978-67-4], 1458
Dichlorodifluoroethane [25915-78-0], 1200
1,2-Dichloro-1,1-difluoroethane [55949-45-6], 1188, 1194
Dichlorodifluoromethane [75-71-8], 1187, 1190, 1193, 1195, 1199, 1200, 1202
p,p'-Dichlorodiphenyltrichloroethane [50-29-3], 1505
Dichlorodiphenyltrichloroethane [50-29-3], 1503, 1504 1506, 1507–1517, 1521, 1522
1,1-Dichloro-1-fluoroethane, 1188
Dichlorofluoromethane [75-43-4], 1188, 1189, 1193, 1195, 1196, 1200

2,6-Dichloro-4-nitroaniline *[99-30-9]*, 997
2,3-Dichlorophenol *[576-24-9]*, 1458
2,4-Dichlorophenol *[120-83-2]*, 1613, 1614, 1615
2,5-Dichlorophenol *[583-78-8]*, 1458, 1464
2,6-Dichlorophenol *[87-65-0]*, 1615
3,4-Dichlorophenol *[95-77-2]*, 1458
2,4-Dichlorophenoxyacetic acid *[94-75-7]*, 1613
2,5-Dichloro-*p*-phenylenediamine, see 2,5-Dichloro-1,4-phenylenediamine
2,6-Dichloro-*o*-phenylenediamine, see 2,6-Dichloro-1,2-phenylenediamine
2,6-Dichloro-*p*-phenylenediamine, see 2,6-Dichloro-1,4-phenylenediamine
2,6-Dichloro-1,2-phenylenediamine *[122-39-4]*, 1013
2,5-Dichloro-1,4-phenylenediamine *[20103-09-7]*, 1013
2,6-Dichloro-1,4-phenylenediamine *[609-20-1]*, 1012–1013
3′,4′-Dichloropropionanilide *[709-98-8]*, 994
2,5-Dichloroquinoline *[59412-12-3]*, 1458, 1464
1,1-Dichloro-1,2,2,2-tetrafluoroethane, 1187
Dichlorotetrafluoroethane *[1320-37-2]*, 1190
1,2-Dichlorotetrafluoroethane *[76-14-2]*, 1187, 1195, 1196
Dichlorotrifluoroethane *[34077-87-7]*, 1188
Dicyclohexylamine *[101-83-7]*, 1090, 1106, 1139
Dicyclopentadiene *[77-73-6]*, 1276–1277, 1278–1279, 1286–1287, 1288–1289
Didecylamine *[1120-49-6]*, 1132
Dieldrin *[60-57-1]*, 1483, 1504, 1505, 1506, 1519–1520, 1521–1522
Diesel fuel *[68334-30-5]*, 1412, 1413, 1414
Diethamine, see Diethylamine
Diethanolamine, see 2,2′-Iminodiethanol
*N,N*′-Diethylamine, see Diethylamine
Diethylamine *[109-89-7]*, 1089, 1090, 1097, 1105, 1117–1118
Diethylaminoethanol, see Diethylethanolamine
2-Diethylaminoethanol *[100-37-8]*, 1107, 1112, 1160
*N,N*′-Diethylaniline *[91-66-7]*, 987
1,4-Diethylbenzene *[105-05-5]*, 1304–1305
1,2-Diethylbenzene *[135-01-3]*, 1304–1305
1,3-Diethylbenzene *[141-93-5]*, 1304–1305
Diethylbenzene *[25340-17-4]*, 1304–1305, 1345, 1346
2,2-Diethyldihexylamine *[106-20-7]*, 1106
*N,N*′-Diethylenediamine *[110-85-0]*, 1091
Diethylenediamine piperazine *[110-85-0]*, 1107
Diethylenetriamine *[111-40-0]*, 1091, 1107, 1110–1111, 1146–1147
Diethylenetriaminepentaacetic acid *[67-43-6]*, 1147–1148

*N,N*′-Diethylethanamine, see Triethylamine
*N,N*′-Diethylethanolamine, see Diethylethanolamine
*N*-Diethylethanolamine, see Diethylethanolamine
Diethylethanolamine *[100-37-8]*, 1160–1161
Di(-2-ethylhexyl)amine *[106-20-7]*, 1090. See also Dioctylamine
Difluoroethane *[25497-28-3]*, 1196
1,1-Difluoroethane *[75-37-6]*, 1188, 1189, 1192
Difluoromethane *[75-10-5]*, 1189
Diheptylamine, see Di-*n*-heptylamine
Di-*n*-heptylamine *[2470-68-0]*, 1090, 1129
Di-*n*-hexylamine, see Dihexylamine
*N,N*′-Dihexylamine, see Dihexylamine
Dihexylamine *[143-16-8]*, 1127
*m*-Dihydroxybenzene, see Resorcinol
3,4-Dihydroxybenzoic acid *[99-50-3]*, 1339
2,2′-Dihydroxydiethylamine, see Diethanolamine
*o*-Dihydroxybenzene, see Pyrocatechol
*p*-Dihydroxybenzene, see Hydroquinone
Dihydroxy-*p*-toluic acid *[34344-80-4]*, 1339
Diisobutylamine *[110-96-3]*, 1105, 1123
Diisopropanolamine *[110-97-4]*, 1156
*N,N*-Diisopropylamine, see Diisopropylamine
Diisopropylamine *[108-18-9]*, 1090, 1105, 1120
2-(Diisopropylamino)ethanol, see Diisopropylethanolamine
Diisopropylbenzene *[25321-09-9]*, 1350
Diisopropylethanolamine *[96-80-0]*, 1162
Dimethylamine *[124-40-3]*, 1089, 1090, 1096, 1097, 1105, 1113–1115
4-Dimethylaminoazobenzene *[60-11-7]*, 1038–1039
*N,N*-Dimethyl-4-aminobenzene, see 4-Dimethylaminoazobenzene
2-Dimethylaminoethanol *[109-01-0]*, 1107, 1112. See also Dimethylethanolamine
1-Dimethylamino-2-propanol, see Dimethylisopropanolamine
*N,N*-Dimethylaniline *[121-69-7]*, 985–986
Dimethylaniline *[121-69-7]*, 950, 955
3,9-Dimethylbenzanthracene, 1380
4,9-Dimethylbenzanthracene, 1380
7,12-Dimethylbenzanthracene, 1380
Dimethyl-1,2-benzanthracene *[[43178-07-0]*, 1387–1388
9,10-Dimethyl(*a*)benzanthracene *[58429-99-5]*, 1583–1584
Dimethylbenzene, see Xylene
2,9-Dimethyl-3,4-benzophenanthrene, 1381
2,2-Dimethylbutane *[75-83-2]*, 1235
2,3-Dimethylbutane *[79-29-8]*, 1226–1227

N,N-Dimethylbutylamine, see Dimethylbutylamine
Dimethylbutylamine [927-62-8], 1125
Dimethylcyclohexane [98-94-2], 1268–1269, 1270–1271, 1275
N,N-Dimethylcyclohexylamine, see Dimethylcyclohexylamine
Dimethylcyclohexylamine [98-94-2], 1090, 1139
Dimethylenediamine, see Ethylenediamine
Dimethylenemethane [463-49-0], 1250
N,N-Dimethylethanolamine, see Dimethylethanolamine
Dimethylethanolamine [108-01-0], 1158–1159, 1159
1,1-Dimethylethylamine, see tert-Butylamine
2,5-Dimethylhexane [592-13-2], 1237
Dimethylisopropanolamine [108-16-7], 1164
Dimethylisopropanolamine [108-16-7], 1112
Dimethylnaphthalene [28804-88-8], 1378
1,6-Dimethylnaphthalene [575-43-9], 1372
2,7-Dimethyloctane [1072-16-8], 1226–1227, 1239
2,3-Dimethylpentane [565-59-3], 1226–1227
2,4-Dimethylpentane, 1226–1227
N,N-Dimethyl-4-phenylazoaniline, see 4-Dimethylaminoazobenzene
N,N-Dimethyl-4-(phenylazo)benzenamine, see 4-Dimethylaminoazobenzene
Dimethyl-p-phenylenediamine [99-98-0], 1008
2,2-Dimethylpropane [463-82-1], 1233
Dimethylpropanolamine [3179-63-3], 1112. See also 1-Dimethylamino-2-propanol
3,6-Dimethylpyrocatechol [2785-78-6], 1339
Dinitrobenzene [25154-54-5], 953
1,2-Dinitrobenzene [528-29-0], 954, 968. See also o-initrobenzene
1,3-Dinitrobenzene [99-65-0], 954, 963–964, 965, 968. See also m-Dinitrobenzene
1,4-Dinitrobenzene [100-25-4], 953, 954, 968. See also p-Nitrobenzene
m-Dinitrobenzene [99-65-0], 952, 1019–1023
o-Dinitrobenzene [528-29-0], 952, 1019
p-Dinitrobenzene [100-25-4], 951, 952, 953
2,4-Dinitrobenzoic acid [610-30-0], 954–955, 1057
2,6-Dinitrobenzoic acid [603-12-3], 1060
2,6-Dinitrobenzyl alcohol [96839-34-8], 1059
4,4'-Dinitrobiphenyl [1528-74-1], 957
1,3-Dinitro-4-chlorobenzene, see 4-Chloro-1,3-dinitrobenzene
2,4-Dinitrochlorobenzene, see 4-Chloro-1,3-dinitrobenzene
2,4-Dinitro-1-chlorobenzene, see 4-Chloro-1,3-dinitrobenzene

2,4-Dinitrochlorobenzene [97-00-7], 952
2,6-Dinitro-o-cresol [609-93-8], 982
1,2-Dinitroethane, see Ethylene glycol dinitrate
2,3-Dinitrophenol [66-56-8], 976
2,4-Dinitrophenol [51-28-5], 976–978
2,5-Dinitrophenol [329-71-5], 978
2,6-Dinitrophenol [573-56-8], 978–979
3,4-Dinitrophenol [577-71-9], 979
3,5-Dinitrophenol [586-11-8], 979
2,4-Dinitro-1-toluene, see 2,4-Dinitrotoluene
Dinitrotoluene [25321-14-6], 954–955, 960–961, 1054–1056
2,3-Dinitrotoluene [602-01-7], 1061
2,4-Dinitrotoluene [121-14-2], 952, 955, 964–965, 1056–1059
2,5-Dinitrotoluene [619-15-8], 1056, 1061
2,6-Dinitrotoluene [606-20-2], 952, 964–965, 1056, 1059–1060
3,4-Dinitrotoluene [610-39-9], 1056, 1061
3,5-Dinitrotoluene [618-85-9], 1061
Dinitrotoluol, see 2,4-Dinitrotoluene; 2,6-Dinitrotoluene
Dioctylamine [1120-48-5], 1131
DIPA, see Diisopropylamine
Di-2-pentenamine, see Diallylamine
Dipentene, see Limonene
Dipentylamine, see Di-n-amylamine; Pentyl-1-pentamine
Diphenyl [95-52-4], 1304–1305, 1364–1365, 1366–1367
Diphenylamine [122-39-4], 957, 1013–1014
N-Diphenylaniline, see Diphenylamine
4,4'-Diphenylenediamine, see Benzidine
Diphenylethylene, see Stilbene
Diphenylmethane [101-81-5], 1304–1305, 1365, 1370
Di-2-propenylamine, see Diallylamine
Di-n-propylamine, see n-Dipropylamine
Dipropylamine [142-84-7], 1105
n-Dipropylamine [142-84-7], 1089, 1090, 1119–1120
Divinyl erythrene, see Butadiene
Divinylbenzene [1321-74-0], 1360, 1362
m-Divinylbenzene [108-57-6], 1304–1305
o-Divinylbenzene [98-83-9], 1304–1305
p-Divinylbenzene [105-06-6], 1304–1305
DMBA, see Dimethylbutylamine
DMEA, see Dimethylethanolamine
1,2-DNB, see o-Dinitrobenzene
1,4-DNB, see p-Nitrobenzene
p-DNB, see p-Nitrobenzene
DNBA, see Di-n-butylamine
DNCB, see 4-Chloro-1,3-dinitrobenzene
DNPA, see n-Dipropylamine

Dodecahydrodiphenylamine, see Dicyclohexylamine
Dodecanamine, see Dodecylamine
Dodecane [112-40-3], 1226–1227, 1239
Dodecene [25378-22-7], 1242–1243
Dodecylamine [124-22-1], 1130, 1131–1132
Dodecylbenzene [123-01-3], 1304–1305, 1349, 1351
Dodecylthiophenol [36612-94-9], 1620
Dowicide 7, see Pentachlorophenol
Dowicide I, see o-Phenylphenol
DPA, see Diphenylamine
DPTA, see Diethylenetriaminepentaacetic acid

EA, see Monoethanolamine
EBA, see Ethyl-n-butylamine
EDTA, see Ethylenediaminetetraacetic acid
EGDN, see Ethylene glycol dinitrate
Eicosane [112-95-8], 1226–1227
n-Eicosane [112-95-8], 1239
Endrin [72-20-8], 1504, 1505, 1506
Enflurane [13838-16-9], 1207
Epinephrine [51-43-4], 1108–1109, 1181, 1183
Essence of mirbane, see Nitrobenzene
Ethane [74-84-0], 1223 1226–1227, 1228
1,2-Ethanediamine, see Ethylenediamine
1,2-Ethanediol dinitrate, see Ethylene glycol dinitrate
Ethanolamine [141-43-5], 1112. See also 2-Aminoethanol
Ethene [74-85-1], 1242–1243
4-Ethenylcyclohexene, see 4-Vinylcyclohex-1-ene
Ethereal sulfate, 1451–1452
Ethine, see Acetylene
Ethyl parathion [56-38-2], 948
Ethylamine [75-04-7], 1089, 1090, 1105, 1109, 1116–1117
2-Ethylaminoethanol [110-73-6], 1107, 1112. See also Monoethylethanolamine
N-Ethylaniline [103-69-5], 986–987
Ethylbenzene [100-41-4], 1342–1346, 1344–1345
Ethyl-n-butylamine [13360-63-9], 1124–1125
2-Ethylbutylamine [617-79-8], 1090, 1106. See also Isohexylamine
N-Ethylbutylamine, see Ethyl-n-butylamine
Ethylcyclohexane [1678-91-7], 1268–1269, 1270–1271
Ethylcyclopentane [2146-38-5], 1268–1269, 1273
N-Ethyldiethanolamine, see Monoethyldiethanolamine
Ethyldiethanolamine [139-87-7], 1112. See also N-Ethyl-2,2'-iminodiethanol
Ethylene dinitrate, see Ethylene glycol dinitrate
Ethylene glycol dinitrate [628-96-6], 1063–1064

Ethylenediamine [107-15-3], 1091, 1107, 1109, 1110, 1110–1111
Ethylenediamine [98-94-2], 1140–1143
Ethylenediaminetetraacetic acid [60-00-4], 1143–1144, 1491
(Ethylenedinitrilo)tetraacetic acid, see Ethylenediaminetetraacetic acid
Ethylenimine [151-56-4], 1110
Ethylenzene [100-41-4], 1304–1305
N-Ethylethanamine, see Diethylamine
2-Ethylhexene [1632-16-2], 1250
2-Ethylhexylamine [104-75-6], 1090, 1127–1128
Ethylidenenorbornene [16219-75-3], 1291
2,2'-Ethyliminodiethanol, see Monoethyldiethanolamine
N-Ethyl-2,2'-iminodiethanol, 1112
2,2'-(Ethylimino)diethanol [139-87-7], 1107
Ethylmethyl ethylene, see 2-Pentene
1-Ethylpentylamine, see 3-Aminoheptane
Ethylphenylamine, see N-Ethylaniline
Ethyne, see Acetylene
Evrisan, see Pentachlorophenol

N-2-Fluorenylacetamide, see 2-Acetylaminofluorene
Fluoroform [75-46-7], 1192
Follicle-stimulating hormone [9002-68-0], 1520
Folosan, see Pentachloronitrobenzene

Glonoin, see Nitroglycerin
Glucose-6-phosphatase [9001-39-2], 1481
Glucose-6-phosphate dehydrogenase [9001-40-5], 951
Glutathione [70-18-8], 1374
Glyceryl trinitrate, see Nitroglycerin
Glycol dinitrate, see Ethylene glycol dinitrate
GTN, see Nitroglycerin

Halothane [151-67-7], 1207
HCB, see Hexachlorobenzene
Hemimellitine, see 1,2,3-Trimethylbenzene
Heptachlor [76-44-8], 1504, 1505, 1506, 1540–1545
Heptadecane [629-78-7], 1226–1227
n-Heptadecane [629-78-7], 1239
1,4-Heptadiene [5675-22-9], 1242–1243, 1253
1,6-Heptadiyne [2396-63-6], 1254–1255, 1256
1,1,1,2,2,3,3-Heptafluoropropane [2252-84-8], 1189
1,1,1,2,3,3,3-Heptafluoropropane [431-89-0], 1189
2-Heptanamine, see 2-Aminoheptane
Heptane [142-82-5], 1226–1227
Heptane-1, 1242–1243

n-Heptane *[142-82-5]*, 1223, 1236, 1326
trans-Heptene-3 *[14686-14-7]*, 1242–1243
1-Heptene *[592-76-7]*, 1249
2-Heptene *[592-77-8]*, 1249
3-Heptene *[592-78-9]*, 1249
cis-Heptene-2 *[6443-92-1]*, 1242–1243
Heptylamine, see 1-Aminoheptane
n-Heptylamine *[111-68-2]*, 1090
Hexachlorobenzene *[118-74-1]*, 1445–1446, 1477–1492
Hexachloropentadiene, 1505
n-Hexadecane *[544-76-3]*, 1239
Hexadecane *[544-76-3]*, 1226–1227
Hexadecene *[26952-14-7]*, 1242–1243
Hexadecylamine *[143-27-1]*, 1132
Hexadecyldimethylamine *[112-69-6]*, 1132
Hexadecylmethylenediamine *[929-94-2]*, 1146
1,3-Hexadiene *[592-48-3]*, 1242–1243 1253
1,4-Hexadiene *[592-45-0]*, 1253
1,5-Hexadiene *[592-42-7]*, 1242–1243, 1253
1,1,1,2,2,3-Hexafluoropropane *[677-56-5]*, 1189
1,1,1,2,3,3-Hexafluoropropane *[431-63-0]*, 1189
1,1,1,3,3,3-Hexafluoropropane *[690-39-1]*, 1189
Hexahydroaniline, see Cyclohexylamine
Hexamethylbenzene *[87-85-4]*, 1342
Hexamethylenediamine *[124-09-4]*, 1089, 1091, 1142, 1145–1146
Hexane *[110-54-3]*, 1226–1227
n-Hexane *[110-54-3]*, 1223, 1224, 1234
1,6-Hexanediamine, see Hexamethylenediamine
2,5-Hexanedione *[110-13-4]*, 1224, 1224 1225
1-Hexene *[592-41-6]*, 1249
2-Hexene *[592-43-8]*, 1249
cis-2-Hexene *[7688-21-3]*, 1242–1243
trans-2-Hexene *[4050-45-7]*, 1242–1243
n-Hexylamine *[111-26-2]*, 1090, 1126–1127
Hexylbenzene *[1077-16-3]*, 1351
1-Hexylene, see Hexene
Hydrogen cyanide *[74-90-8]*, 1092
Hydroquinone *[123-31-9]*, 1008, 1308, 1590–1592
4-Hydroxyacetanilide *[103-90-2]*, 1017
p-Hydroxyacetanilide *[103-90-2]*, 954
4-Hydroxyacetanilide sulfate *[10066-90-7]*, 1017
3-Hydroxyaniline, see m-Aminophenol
m-Hydroxyaniline, see m-Aminophenol
o-Hydroxyaniline, see o-Aminophenol
p-Hydroxyaniline, see p-Aminophenol
p-Hydroxybenzaldehyde *[123-08-0]*, 1339
4-Hydroxybenzanamine, see p-Aminophenol
p-Hydroxybenzoic acid *[99-96-7]*, 1339
2-Hydroxydiphenyl, see o-Phenylphenol
N-2-Hydroxyethylacetamide, see N-Acetylethanolamine

2-Hydroxyethylamine, see Monoethanolamine
2-N-Hydroxyethylamine, see Monoethanolamine
N-(Hydroxyethyl)diethylenetriamine *[1965-29-3]*, 1147
(2-Hydroxyethyl)dimethylamine, see Dimethylethanolamine
Hydroxylamine *[78-3-49-8]*, 950
2-Hydroxy-4-nitroaniline, see 2-Amino-5-nitrophenol
4-Hydroxy-3-nitroaniline, see 4-Amino-2-nitrophenol
p-Hydroxyphenyl, see Hydroquinone
3-Hydroxypropylmercapturic acid *[23127-40-0]*, 1134

IBA, see Isobutylamine
2,2'-Iminobisethanol, see Diethanolamine
1,1'-Iminodi-2-butanol, see Di-n-butanolamine
2,2'-Iminodiethanol, see Diethanolamine
2,2'-Iminodiethanol *[111-42-2]*, 1112
2-Iminodiethanol *[111-42-2]*, 1106
Iomex *[57285-10-6]*, 1393
Ionil, see Di-tert-butylmethylphenol
Iproniazid *[54-92-2]*, 1095
Isoamylamine *[107-85-7]*, 1090, 1126
Isobutane *[75-28-5]*, 1226–1227, 1230–1231
Isobutene *[115-11-7]*, 1242–1243, 1247–1248
Isobutylamine *[78-81-9]*, 1090, 1122–1123
Isobutylbenzene *[538-93-2]*, 1304–1305, 1350
Isobutylcarbylamine, see Isoamylamine
Isoflurane *[26675-46-7]*, 1207
Isoheptane *[31394-54-4]*, 1226–1227, 1236
Isoheptylamine, see 4-Aminoheptane
Isohexane-2, 1242–1243
Isohexane *[107-83-5]*, 1226–1227, 1235
Isohexene *[27236-46-0]*, 1249
Isohexylamine *[617-79-8]*, 1127
Isoniazid *[54-85-3]*, 1097
Isooctene *[11071-47-9]*, 1249–1250
Isopentane *[78-78-4]*, 1226–1227, 1232–1233
Isopentene *[26760-64-5]*, 1248
Isopentylamine, see Isoamylamine
Isopentyne *[598-23-2]*, 1254–1255, 1256
Isoprene *[78-79-5]*, 1242–1243, 1252–1253
Isopropanolamine *[78-96-6]*, 1112. See also Monoisopropanolamine
Isopropylamine *[75-31-0]*, 1089, 1090, 1105
4-Isopropylaminodiphenylamine *[101-72-4]*, 1008
Isopropylbenzene *[98-82-8]*, 1346–1347
Isopropyltoluene *[25155-15-1]*, 1349, 1350

Kelthane *[115-32-2]*, 1505
Kepone *[143-50-0]*, 1504, 1505, 1506, 1545–1550

Kerosene *[8008-20-6]*, 1394, 1397, 1404, 1410–1412, 1413

Laurylamine, see Dodecylamine
Ligroin, see Petroleum ether
Limonene *[138-86-3]*, 1276–1277, 1278–1279, 1282–1283
d-Limonene *[5989-27-5]*, 1268
Lindane *[58-89-9]*, 1503–1504, 1504, 1505, 1506, 1550–1557
Luteinizing hormone *[9002-67-9]*, 1520
Lycopene *[502-65-8]*, 1242–1243

MBOCA, see 4,4'-Methylenebis(2-chloroaniline)
MDA, see 4,4'-Methylenedianiline
MEA, see Ethylamine
p-Mentha-1,8-diene, see Limonene
Mercapturic acid *[616-91-9]*, 1451–1452, 1458
Mesitylene *[108-67-8]*, 1306. See also 1,3,5-Trimethylbenzene
Methanamine, see Methylamine
Methane *[74-82-8]*, 1223, 1226–1227, 1226–1228
Methanol *[67-56-1]*, 1226
Methoxychlor *[72-43-5]*, 1521
Methoxyflurane, 1207
N-Methydiethanolamine, see Monomethyldiethanolamine
Methyl n-butyl ketone *[591-78-6]*, 1224
Methyl parathion *[298-00-0]*, 948
Methyl Yellow, see 4-Dimethylaminoazobenzene
Methylacetylene, see Propyne
1-Methylallene, see Butadiene
Methylamine *[74-89-5]*, 1089, 1090, 1105, 1109, 1111, 1113
1-Methyl-2-aminobenzene, see o-Toluidine
2-Methylaminoethanol *[109-83-1]*, 1107, 1112. See also Monomethylethanolamine
Methylaniline, see N-Methylaniline
N-Methylaniline *[100-61-8]*, 985
2-Methylaniline, see o-Toluidine
o-Methylaniline, see o-Toluidine
p-Methylaniline, see p-Toluidine
Methylbenz(a)anthracene *[43178-22-9]*, 1387
4-Methylbenz(a)anthracene *[316-49-4]*, 1381
7-Methylbenz(a)anthracene *[2541-69-7]*, 1380
8-Methylbenz(a)anthracene *[2381-31-9]*, 1380
10-Methylbenz(a)anthracene *[2381-15-9]*, 1380
12-Methylbenz(a)anthracene *[2422-79-9]*, 1380
6-Methyl-1,2-benzanthracene *[316-14-3]*, 1379
7-Methyl-1,2-benzanthracene *[2541-69-7]*, 1379
9,10-Methyl-1,2-benzanthracene *[57-97-6]*, 1379
10-Methyl-1,2-benzanthracene *[2541-69-7]*, 1379
1-Methylbenzenamine, see o-Toluidine
3-Methylbenzenamine, see m-Toluidine
4-Methylbenzenamine, see p-Toluidine
N-Methylbenzenamine, see N-Methylaniline
2-Methyl-1,3-benzenediamine, see 2,6-Toluenediamine
2-Methyl-1,4-benzenediamine, see 2,5-Toluenediamine
3-Methyl-1,3-benzenediamine, see 2,5-Toluenediamine
4-Methyl-1,2-benzenediamine, see 3,4-Toluenediamine
Methylbenzoic acid *[25567-10-6]*, 1339
p-Methylbenzoic acid *[99-94-5]*, 1339
2-Methyl-3,4-benzophenanthrene, 1381
p-Methylbenzyl alcohol *[589-18-4]*, 1339
2-Methyl-1,3-butadiene, see Isoprene
2-Methylbutane *[78-78-4]*, 1232–1233
2-Methyl-2-butene, see 2-Isopentene
3-Methyl-1-butene, see 1-Isopentene
3-Methylbutylamine, see Isoamylamine
p-Methyl-tert-butylbenzene, see tert-Butyltoluene
3-Methylbutyne *[598-23-2]*, 1256
3-Methylcatechol *[488-17-5]*, 1339
4-Methylcatechol *[452-86-8]*, 1339
Methylcholanthrene *[56-49-5]*, 1383, 1379
3-Methylcholanthrene *[56-49-5]*, 985, 1338, 1383, 1392–1393
20-Methylcholanthrene, see 3-Methylcholanthrene
1-Methylchrysene *[3351-28-8]*, 1389
2-Methylchrysene *[3351-32-4]*, 1389
3-Methylchrysene *[3351-31-3]*, 1379, 1389
5-Methylchrysene *[3697-24-3]*, 1389
5-Methylcyclohexadiene-1,3 *[1489-57-2]*, 1276–1277
Methylcyclohexadiene *[30640-46-1]*, 1283
Methylcyclohexane *[108-87-2]*, 1268–1269, 1270–1271, 1272
Methylcyclopentadiene *[26519-91-5]*, 1283
Methylcyclopentane *[96-37-7]*, 1268–1269, 1273
4-Methyl-2,6-di-tert-butylphenol, see Di-tert-butylmethylphenol
Methyldiethanolamine *[105-59-9]*, 1112. See also N-Methyl-2,2'-iminodiethanol
1-Methyl-2,4-dinitrobenzene, see 2,4-Dinitrotoluene
1-Methyl-2,6-dinitrobenzene, see 2,6-Dinitrotoluene
4-Methyl-2,6-dinitrophenol, see 2,6-Dinitro-o-cresol
Methyldiphenyl *[644-08-6]*, 1365
Methylene dianiline, see 4,4'-Methylenedianiline
Methylenebis(o-chloroaniline), see 4,4'-Methylenebischloroaniline

## CHEMICAL INDEX

4,4'-Methylenebischloroaniline *[38483-48-6]*, 957–958, 961–962, 967
4,4'-Methylenebis(2-methylaniline) *[31291-65-3]*, 999
4,4'-Methylenedianiline *[101-77-9]*, 999, 1000. See also 4,4'-Methylenedianiline
N-Methylethanolamine, see Monomethylethanolamine
Methylethanolamine *[109-83-1]*, 1160. See also 2-Methylaminoethanol
Methylethylbenzene *[620-24-4]*, 1304–1305, 1345, 1346
2-Methyl-2-heptene, see Isooctene
1-Methylheptylamine, see 2-Octylamine
2-Methylhexane *[591-76-4]*, 1236
3-Methylhexane *[589-34-4]*, 1236
1-Methylhexylamine, see 2-Aminoheptane
2,2'-(Methylimino)diethanol *[105-59-9]*, 1107
N-Methyl-2,2'-iminodiethanol *[105-59-9]*, 1112
2,2'-(Methylimino)diethanol *[105-59-9]*, see Monomethyldiethanolamine
1-Methyl-4-isopropenylcyclohex-1-ene, see Limonene
N-Methylmethanamine, see Dimethylamine
Methylnaphthalene *[1321-94-4]*, 1372, 1377
1-Methylnaphthalene *[90-12-0]*, 1304–1305, 1372
3-Methylnaphthalene *[91-57-6]*, 1304–1305
2-Methylnaphthalene *[91-57-6]*, 1372
2-Methyl-5-nitroaniline, see 5-Nitro-*o*-toluidine
Methylnitrobenzene, see *o*-Nitrotoluene
2-Methylpentane *[107-83-5]*, 1235
3-Methylpentane *[96-14-0]*, 1226–1227, 1235
Methylphenol, see Cresol
2-Methyl-1,4-phenylenediamine, see 2,5-Toluenediamine
2-Methyl-1-phenylpropane, see Isobutylbenzene
2-Methyl-2-phenylpropane, see *tert*-Butylbenzene
2-Methylpropane *[75-28-5]*, 1230–1231
1-Methylpropylamine, see 2-Butylamine
2-Methylpropylamine, see Isobutylamine
3-Methylsalicyclic acid *[83-40-9]*, 1339
α-Methylstyrene *[98-83-9]*, 1363
β-Methylstyrene, see 2-Propenylbenzene
∞-Methylstyrene *[637-50-3]*, 1304–1305
*p*-Methylstyrene *[622-97-9]*, 1304–1305
Methyltrinitrobenzene, see 2,4,6-Trinitrotoluene
1-Methyl-2,4,6-trinitrobenzene, see 2,4,6-Trinitrotoluene
Mirex *[2385-85-5]*, 1504, 1506
MMA, see Methylamine
MNBA, see *n*-Butylamine
MNPA, see Propylamine
MOCA, see 4,4'-Methylenebis(2-chloroaniline)

Monoallylamine *[107-11-9]*, 1111. See also Allylamine
Monoamine oxidase *[9001-66-5]*, 1095
Mono-*sec*-butanolamine *[13552-21-1]*, 1157
Monobutylethanolamine *[111-75-1]*, 1162
Monochlorobenzene *[108-90-7]*, 1445–1446, 1447–1454
Monoethanolamine *[141-43-5]*, 1150–1153
Monoethylamine, see Ethylamine
Monoethyldiethanolamine *[139-87-7]*, 1164
Monoethylethanolamine *[110-73-6]*, 1159–1160
Monoisopropanolamine *[78-96-6]*, 1156
Monoisopropylethanolamine *[109-56-8]*, 1161–1162
Monomethylamine, see Methylamine
Monomethylaniline, see *N*-Methylaniline
Monomethyldiethanolamine *[105-59-9]*, 1164
Monomethylethanolamine *[109-83-1]*, 1157–1158

1-Naphthalamine, see α-Naphthylamine
Naphthalene *[91-94-1]*, 1304–1305, 1371, 1372, 1373–1377
1,5-Naphthalenediamine, see 1,5-Diaminonaphthalene
Naphthalidine, see α-Naphthylamine
1-Naphthylamine, see α-Naphthylamine
2-Naphthylamine *[91-59-8]*, 961, 1040
α-Naphthylamine *[134-32-7]*, 950, 1039–1042
β-Naphthylamine *[91-59-8]*, 950, 956, 959, 984, 1041–1042
1,5-Naphthylenediamine, see 1,5-Diaminonaphthalene
Narcylene, see Acetylene
Natural gas *[64741-47-5]*, 1394
NEA, see Triethylamine
Neoheptane *[58055-59-7]*, 1226–1227, 1236
Neohexane *[75-83-2]*, 1226–1227, 1235
Neononane, 1238
Neopentane *[463-82-1]*, 1226–1227, 1233
NG, see Nitroglycerin
Nitramine, see Tetryl
*m*-Nitraniline, see *m*-Nitroaniline
*p*-Nitraniline, see *p*-Nitroaniline
*o*-Nitraniline, see *m*-Nitroaniline
2,2',2''-Nitrilotriethanol *[102-71-6]*, 1107. See also Triethanolamine
1,1',1''-Nitrilotris-2-butanol, see Tri-*sec*-butanolamine
*m*-Nitroacetanilide *[122-28-1]*, 964
1-Nitro-2-aminobenzene, see *o*-Nitroaniline
2-Nitroaniline, see *o*-Nitroaniline
3-Nitroaniline, see *m*-Nitroaniline

m-Nitroaniline *[99-09-2]*, 955, 964, 988–989, 1019–1021
o-Nitroaniline *[88-74-4]*, 955, 966–967, 968, 987–988, 989–990
p-Nitroaniline *[100-01-6]*, 950, 955, 965, 989–990
2-Nitrobenzenamine, see o-Nitroaniline
3-Nitrobenzenamine, see o-Nitroaniline
4-Nitrobenzenamine, see p-Nitroaniline
Nitrobenzene *[98-95-3]*, 950, 952, 953–954, 963–964, 966, 1016–1018
p-Nitrobenzene *[100-25-4]*, 965, 1023
m-Nitrobenzoic acid *[121-92-6]*, 966
o-Nitrobenzoic acid *[552-16-9]*, 966
p-Nitrobenzoic acid *[62-23-7]*, 966
4-Nitrobenzoic acid *[62-23-7]*, 1054
Nitrobenzol, see Nitrobenzene
4-Nitrobenzyl alcohol *[619-73-8]*, 1054
4-Nitrobenzyl glucuronide *[89076-35-7]*, 1054
4-Nitrobenzyl sulfate *[89090-09-5]*, 1054
S-(4-Nitrobenzyl)glutathione *[6803-19-6]*, 1054
o-Nitrobiphenyl, see 2-Nitrobiphenyl
p-Nitrobiphenyl, see 4-Nitrobiphenyl
2-Nitrobiphenyl *[86-00-0]*, 1035
4-Nitrobiphenyl *[92-93-3]*, 957, 1035–1036
1-Nitro-2-chlorobenzene, see o-Chloronitrobenzene
2-Nitrochlorobenzene, see o-Chloronitrobenzene
p-Nitrochlorobenzene, see p-Chloronitrobenzene
3-Nitro-1,2-diaminobenzene, see 3-Nitro-1,2-phenylenediamine
4-Nitro-1,2-diaminobenzene, see 4-Nitro-1,2-phenylenediamine
4-Nitrodiphenyl *[92-93-3]*, 984
o-Nitrodiphenyl, see 2-Nitrobiphenyl
p-Nitrodiphenyl, see 4-Nitrobiphenyl
2-Nitrofluorine *[607-57-8]*, 957
Nitroglycerin *[55-63-0]*, 950, 1064–1067
Nitroglycerol, see Nitroglycerin
Nitroglycol, see Ethylene glycol dinitrate
5-Nitro-2-methylaniline, see 5-Nitro-o-toluidine
2-Nitronaphthalene *[581-89-5]*, 960
2-Nitrophenol, see o-Nitrophenol
2,4-Nitrophenol, 954
3-Nitrophenol *[554-84-7]*, 1017
4-Nitrophenol *[100-02-7]*, 975–976, 1017
m-Nitrophenol *[554-84-7]*, 954, 966, 967, 974–975
o-Nitrophenol *[88-75-5]*, 966, 967, 974
p-Nitrophenol *[100-02-7]*, 948, 954, 966, 967
4-Nitrophenyl glucuronide *[10344-94-2]*, 954
3-Nitrophenyl sulfate *[3233-64-5]*, 1017
4-Nitrophenyl sulfate *[1080-04-2]*, 954, 1017
m-Nitrophenyl sulfate *[3233-64-5]*, 954

p-Nitrophenyl sulfate *[1080-04-2]*, 954
p-Nitrophenylamine, see p-Nitroaniline
1-Nitro-2,4-phenylenediamine, see 2-Nitro-1,4-phenylenediamine
2-Nitro-1,4-phenylenediamine *[5307-14-2]*, 1009–1013
2-Nitro-p-phenylenediamine, see 2-Nitro-1,4-phenylenediamine
3-Nitro-1,2-phenylenediamine *[3694-52-8]*, 1010
3-Nitro-o-phenylenediamine, see 3-Nitro-1,2-phenylenediamine
4-Nitro-1,2-phenylenediamine *[99-56-9]*, 1010–1011
4-Nitro-o-phenylenediamine, see 4-Nitro-1,2-phenylenediamine
S-(p-Nitrophenyl)glutathione *[73962-43-3]*, 1025, 1026
Nitrophenylmethane, see o-Nitrotoluene
Nitrosobenzene *[586-96-9]*, 952
N-Nitrosodiethanolamine *[1116-54-7]*, 1155
Nitrosodiphenylamine *[86-30-6]*, 1013
2-Nitrosonaphthalene *[6610-08-8]*, 956
S-Nitrosothiol, 1065
p-Nitrothiophenol *[1849-36-1]*, 1026
2-Nitrotoluene, see o-Nitrotoluene
3-Nitrotoluene, see m-Nitrotoluene
4-Nitrotoluene, see p-Nitrotoluene
m-Nitrotoluene *[99-08-1]*, 967, 1052–1053
o-Nitrotoluene *[88-72-2]*, 967, 1051–1052
p-Nitrotoluene *[99-99-0]*, 967, 1053–1054
p-Nitro-o-toluidine, 952
5-Nitro-o-toluidine *[99-55-8]*, 1045–1046
Nitrotoluol, see o-Nitrotoluene
m-Nitrotoluol, 952
o-Nitrotoluol, 952
p-Nitrotoluol, 952
Nitroxanthic acid, see Picric acid
Nonadecane *[629-92-5]*, 1226–1227
n-Nonadecane *[629-92-5]*, 1239
Nonane *[111-84-2]*, 1226–1227
n-Nonane *[111-84-2]*, 1223, 1237–1238
Nonene *[27215-95-8]*, 1242–1243
NOP, see 4-Nitro-1,2-phenylenediamine
3-NT, see m-Nitrotoluene
2-NT, see o-Nitrotoluene
4-NT, see p-Nitrotoluene

Octadecanamine, see Octadecylamine
n-Octadecane *[593-45-3]*, 1239
Octadecane *[593-45-3]*, 1226–1227
Octadecene *[27070-58-2]*, 1242–1243
Octadecylamine *[124-30-1]*, 1090, 1132
Octadecyldimethylamine *[124-24-7]*, 1132
1,7-Octadiene *[3710-30-3]*, 1242–1243

Octafluorocyclobutane *[115-25-3]*, 1192
Octanamine, see 1-Octylamine
Octane *[111-65-9]*, 1226–1227
n-Octane *[111-65-9]*, 1223, 1236–1237, 1237
Octene *[25377-83-7]*, 1242–1243
1-Octene *[111-66-0]*, 1249–1250
2-Octene *[111-67-1]*, 1249–1250
Octylamine, see 1-Octylamine
1-Octylamine *[111-86-4]*, 1130
2-Octylamine *[693-16-3]*, 1130–1131
α-Octylene, see 1-Octene
β-Octylene, see 2-Octene
Oil of mirbane, see Nitrobenzene
ONCB, see o-Chloronitrobenzene
Orthamine, see o-Phenylenediamine
Orthoxenol, see o-Phenylphenol
4,4'-Oxydianiline, see 4-Aminophenyl ether

PAP, see p-Aminophenol
Paranox, see Di-*tert*-butylmethylphenol
PCNB, see Pentachloronitrobenzene
PCP, see Pentachlorophenol
Penchloral, see Pentachlorophenol
Penta, see Pentachlorophenol
Pentabromophenol *[608-71-9]*, 1616–1617
Pentachloroaniline *[527-20-8]*, 1032
Pentachlorobenzene *[608-93-5]*, 1447, 1474–1477, 1477, 1488, 1489
Pentachloronitrobenzene *[82-68-8]*, 1031, 1483
Pentachlorophenol *[87-86-5]*, 1032, 1477, 1487, 1489, 1605–1613
Pentachlorothiophenol *[133-49-3]*, 1487, 1488
Pentadecane *[629-62-9]*, 1226–1227
n-Pentadecane *[629-62-9]*, 1239
Pentafluorodimethyl ether *[3822-68-2]*, 1189
1,1,2,2,2-Pentafluoroethane *[354-33-6]*, 1189
1,1,3,3-Pentafluoropropane, 1188
1,1,1,2,2-Pentafluoropropane *[1814-88-6]*, 1189
1,1,2,2,3-Pentafluoropropane *[679-86-7]*, 1188
Pentamethylbenzene *[700-12-9]*, 1342
Pentamethyldiamine, see Pentamethylenediamine
Pentamethylene, see Cyclopentane
Pentamethylenediamine *[462-94-2]*, 1091, 1145
Pentane *[109-66-0]*, 1226–1227
n-Pentane *[109-66-0]*, 1223, 1231–1232
1,5-Pentanediamine, see Pentamethylenediamine
1-Pentene *[109-67-1]*, 1242–1243, 1248
2-Pentene *[109-68-2]*, 1248
cis-2-Pentene *[627-20-3]*, 1242–1243
trans-2-Pentene *[646-04-8]*, 1242–1243
Pentylamine, see n-Amylamine
Pentyl-1-pentamine *[2050-92-2]*, 1106
1-Pentyne *[627-19-0]*, 1254–1255, 1256

Perchlorobenzene, see Hexachlorobenzene
Perfluorobutene *[11070-66-9]*, 1192
Perthane *[77-47-4]*, 1505
Petroleum benzine, see Petroleum ether
Petroleum ether *[8032-32-4]*, 1396, 1398, 1407
Petroleum naphtha *[8030-30-6]*, 1394, 1397, 1399–1400, 1406–1407
Phenacetin *[62-44-2]*, 972
Phenanthrene *[85-01-80]*, 1285, 1379, 1380, 1386
Phenergan *[58-33-3]*, 1008
Phenobarbital *[50-06-6]*, 985
Phenol trinitrate, see Picric acid
Phenylacetylene *[536-74-3]*, 1304–1305, 1364
Phenylamine, see Aniline
Phenylaniline, see Diphenylamine
p-Phenylaniline, see 4-Aminobiphenyl
Phenylbenzene, see Diphenyl
N-Phenylbenzeneamine, see Diphenylamine
1-Phenylbutane, see n-Butylbenzene
2-Phenylbutane, see sec-Butylbenzene
Phenylbutene *[37338-09-3]*, 1363
1-Phenylbutene-2 *[1560-06-1]*, 1304–1305
1,2-Phenylenediamine, see o-Phenylenediamine
1,3-Phenylenediamine *[108-45-2]*, 1019–1021
1,4-Phenylenediamine, see p-Phenylenediamine
m-Phenylenediamine *[108-45-2]*, 952
o-Phenylenediamine *[95-54-5]*, 1004–1005
p-Phenylenediamine *[106-50-3]*, 955, 1007–1009
1,3-Phenylenediamine dihydrochloride, 1006
Phenylglyoxylic acid *[611-73-4]*, 1357
Phenylhydroxylamine *[100-65-2]*, 950, 952
Phenyl—naphthylamine *[135-88-6]*, 960
4-Phenylnitrobenzene, see 4-Nitrobiphenyl
p-Phenylnitrobenzene, see 4-Nitrobiphenyl
o-Phenylphenol *[90-43-7]*, 1617–1618
N-Phenyl-1,4-phenylenediamine, see p-Aminodiphenylamine
Phosphatidylethanolamine, 1152
Phosphine *[7803-51-2]*, 1254
Picric acid *[88-89-1]*, 980–981
Picronitric acid, see Picric acid
Picrylmethylnitramine, see Tetryl
Picrylnitromethylamine, see Tetryl
2-Pinene, see α-Pinene
α-Pinene *[80-56-8]*, 1276–1277, 1289
β-Pinene *[127-91-3]*, 1289
γ-Pinene *[5947-71-7]*, 1289
PNA, see p-Nitroaniline
PNB, see 4-Nitrobiphenyl
PNCB, see p-Chloronitrobenzene
Polytetrafluoroethylene *[9002-84-0]*, 1199
Porphyrin *[101-60-0]*, 1481, 1487, 1488, 1491
PPDA, see p-Phenylenediamine
Prilocaine *[721-50-6]*, 1044

Pristane *[1921-70-6]*, 1226–1227, 1239
Propadiene *[463-49-0]*, 1242–1243, 1250
Propane *[74-98-6]*, 1226–1227, 1228–1229
n-Propane *[74-98-6]*, 1223
1,2-Propanediamine *[78-90-0]*, 1091, 1107, 1144–1145
1,3-Propanediamine *[109-76-2]*, 1107, 1144–1145
1,2,3-Propanetriol trinitrate, see Nitroglycerin
Propene *[115-06-1]*, 1242–1243
1-Propenylbenzene *[637-50-3]*, 1362
2-Propenylbenzene *[300-57-2]*, 1362
Propylamine *[107-10-8]*, 1090, 1105
n-Propylamine *[107-10-8]*, 1089, 1109, 1119
Propylbenzene *[103-65-1]*, 1346
n-Propylbenzene *[103-65-1]*, 1304–1305, 1348
1-Propylbutylamine, see 4-Aminoheptane
Propylene *[115-07-1]*, 1229
Propylenediamine, see 1,2-Propanediamine
Propyne *[74-99-7]*, 1254–1255, 1256
Protoporphyrin *[553-12-8]*, 1324
Pseudocumene, see 1,2,4-Trimethylbenzene
Putrescine, see Tetramethylenediamine
Pyrene *[129-00-0]*, 1379, 1381, 1389–1390
Pyrenite, see Tetryl
Pyrocatechin, see Pyrocatechol
Pyrocatechol *[120-80-9]*, 1308, 1584–1586
Pyrogallol *[87-66-1]*, 1595–1597

QCB, see Pentachlorobenzene
Quinone *[106-51-4]*, 1593–1595
Quintochlorobenzene, see Pentachlorobenzene
Quintozene, see Pentachloronitrobenzene

Resorcin, see Resorcinol
Resorcinol *[108-46-3]*, 1586–1590

Santobrite, see Pentachlorophenol
Santophen 20, see Pentachlorophenol
Sodium bicarbonate *[144-55-8]*, 1581
Sodium dichromate *[10588-01-9]*, 950
Sodium nitrite *[7632-00-0]*, 950
Sodium pentachlorophenate *[131-52-2]*, 1605–1613
Sorbitol dehydrogenase *[9028-21-1]*, 1481
Squalene *[111-02-4]*, 1242–1243
Stilbene *[588-59-0]*, 1370
cis-Stilbene *[103-30-0]*, 1304–1305
trans-Stilbene *[645-49-8]*, 1304–1305
Styrene *[100-42-5]*, 1304–1305, 1351–1358, 1353, 1354–1355
Succinic dehydrogenase *[9002-02-2]*, 1481

TDA, see 2,4-Toluenediamine
TEPA, see Tetraethylenepentamine

Terphenyl *[26140-60-3]*, 1368–1369
m-Terphenyl *[92-06-8]*, 1304–1305, 1368–1369
o-Terphenyl *[84-15-1]*, 1304–1305, 1368–1369
p-Terphenyl *[92-94-1]*, 1304–1305, 1368–1369
Terraclor, see Pentachloronitrobenzene
TETA, see Triethylenetetramine
2,3,4,5-Tetrachloroaniline *[634-83-3]*, 1032
2,3,5,6-Tetrachloroaniline *[3481-20-7]*, 1032
1,2,3,4-Tetrachlorobenzene *[634-66-2]*, 1471–1474
1,2,3,5-Tetrachlorobenzene *[634-90-2]*, 1471–1474
1,2,4,5-Tetrachlorobenzene *[95-94-3]*, 1444, 1447, 1471–1474, 1477, 1488
Tetrachlorobenzene, mixed *[12408-10-5]*, 1471–1474
1,1,1,2-Tetrachloro-2,2-difluoroethane, 1187, 1195
1,1,2,2-Tetrachlorodifluoroethane *[76-12-0]*, 1187, 1195
Tetrachloroethane *[79-34-5]*, 1185
Tetrachlorohydroquinone *[87-87-6]*, 1487–1488
Tetrachloronitrobenzene *[117-18-0]*, 1032
2,3,4,5-Tetrachloronitrobenzene, 1032
2,3,5,6-Tetrachloronitrobenzene, 1032
Tetrachlorophenol *[25167-93-3]*, 1615
2,3,4,5-Tetrachlorophenol *[4901-51-3]*, 1477
2,3,4,6-Tetrachlorophenol *[58-90-2]*, 1487, 1614
2,3,5,6-Tetrachlorophenol *[935-95-5]*, 1477, 1487
Tetradecane *[629-59-4]*, 1226–1227
n-Tetradecane *[629-59-4]*, 1239
Tetradecylamine *[2016-42-4]*, 1132
Tetraethylenepentamine *[112-57-2]*, 1091, 1148–1149
1,2,2,2-Tetrafluoroethane *[811-97-2]*, 1189
Tetrafluoroethylene *[116-14-3]*, 1199
1,1,2,2-Tetrafluoropropane *[40723-63-5]*, 1189
Tetralin *[119-64-2]*, 1268, 1276–1277, 1278–1279, 1285, 1288
Tetralite, see Tetryl
Tetramethylbenzene *[25619-60-7]*, 1342
1,2,3,4-Tetramethylbenzene *[488-23-3]*, 1304–1305
1,2,3,5-Tetramethylbenzene *[527-53-7]*, 1304–1305
1,2,4,5-Tetramethylbenzene *[95-93-2]*, 1304–1305
Tetramethylenediamine *[110-60-1]*, 1091
1,4-Tetramethylenediamine *[110-60-1]*, 1142
Tetramethyl-p-phenylenediamine *[100-22-1]*, 1008
2,4,6-Tetranitroaniline, see Tetryl
Tetranitromethylaniline, see Tetryl
Tetryl *[479-45-8]*, 998

2,4,6-Tetryl, see Tetryl
Thalidomide [50-35-1], 1140
Thyroxine glucuronide [21462-56-6], 977
Tilicarex, see Pentachloronitrobenzene
TIPA, see Triisopropanolamine
TNBA, see Tri-n-butylamine
TNP, see Picric acid
TNT, see 2,4,6-Trinitrotoluene
Toluene [108-88-3], 1304–1305, 1326–1332, 1327, 1328
Toluene diisocyanate [26471-62-5], 1056
Toluenediamine, see 2,4-Toluenediamine
2,3-Toluenediamine [2687-25-4], 1048
2,4-Toluenediamine [95-80-7], 967, 1048–1050
2,5-Toluenediamine [95-70-5], 1050
2,6-Toluenediamine [823-40-5], 1049, 1050–1051
3,4-Toluenediamine [496-72-0], 965, 1051
3,5-Toluenediamine [108-71-4], 1051
m-Toluenediamine, see 2,4-Toluenediamine
o-Toluenediamine, see 3,4-Toluenediamine
p-Toluenediamine, see 2,5-Toluenediamine
Toluic acid [25567-10-6], 1338
p-Toluic acid [99-94-5], 1339
m-Toluidine [108-44-1], 1046–1047
o-Toluidine [95-53-4], 984, 1043
p-Toluidine [106-49-0], 1047–1048
Toxaphene [8001-35-2], 1504, 1505, 1506, 1521
Triallylamine [102-70-5], 1090, 1106, 1111, 1134–1135
Tri-n-amylamine [621-77-2], 1126
1,4,7-Triazaheptane, see Diethylenetriamine
2,4,6-Tribromophenol [118-79-6], 1616–1617
Tri-sec-butanolamine [2421-02-5], 1157
Tributylamine [102-89-9], 1105
Tri-n-butylamine [102-89-9], 1090, 1123–1124
1,2,3-Trichlorobenzene [87-61-6], 1468–1471
1,2,4-Trichlorobenzene [120-82-1], 1444, 1447, 1464, 1468–1471
1,3,5-Trichlorobenzene [108-70-3], 1447, 1468–1471, 1488
Trichlorofluoroethane [27154-33-2], 1189
Trichlorofluoromethane [75-69-4], 1179, 1187, 1189, 1192, 1193, 1195, 1196, 1197, 1199, 1201, 1202
2,3,5-Trichlorophenol [933-78-8], 1614
2,4,5-Trichlorophenol [95-95-4], 1487, 1614, 1615
2,4,6-Trichlorophenol [88-06-2], 1614, 1615
1,1,1-Trichlorotrifluoroethane [354-58-5], 1187
1,1,2-Trichlorotrifluoroethane [76-13-1], 1187, 1193, 1195, 1196, 1197
Tricresol, see Cresol
Tri-o-cresyl phosphate [78-30-8], 1273
Tridecane [629-50-5], 1226–1227

n-Tridecane [629-50-5], 1239
Tridecylamine [2869-34-3], 1132
Triethanolamine [102-71-6], 1112, 1154–1157.
    See also 2,2',2''-Nitrilotriethanol
Triethylamine [121-44-8], 1089, 1090, 1105, 1118
N-Triethylaminoethanol [302-61-4], 1160
Triethylenetetramine [112-24-3], 1091, 1107, 1148
Triethylenetriamine, 1110–1111
Triethylolamine, see Triethanolamine
1,1,1-Trifluorodimethyl ether, 1189
Trihydroxytriethylamine, see Triethanolamine
2,4,6-Triiodophenol [609-23-4], 1616–1617
Triisobutylamine [1116-40-1], 1124
Triisopropanolamine [122-20-3], 1107, 1112, 1156–1157
Trimethylamine [75-50-3], 1089, 1090, 1097–1098, 1115–1116
Trimethylaminoethane, see tert-Butylamine
1,2,3-Trimethylbenzene [526-73-8], 1304–1305, 1340, 1341
1,2,4-Trimethylbenzene [96-63-6], 1304–1305, 1340, 1341
1,3,5-Trimethylbenzene [108-67-8], 1304–1305, 1340, 1341
2,6,6-Trimethylbicyclo[3.1.1]hept-2-ene, see α-Pinene
Trimethylcarbinylamine, see tert-Butylamine
Trimethylene, see Cyclopropane
Trimethylenediamine [109-76-2], 1091
Trimethylhexane, 1238
2,2,5-Trimethylhexane [3522-94-9], 1226–1227
Trimethylnaphthalene [28652-77-9], 1378
2,2,4-Trimethylpentane [540-84-1], 1226–1227, 1237, 1268
2,3,4-Trimethylpentane [565-75-3], 1226–1227, 1237
Trimethylphenanthrene [30232-26-9], 1380
Trinitrin, see Nitroglycerin
sym-Trinitrobenzene, see Trinitrobenzene
Trinitrobenzene [99-35-4], 952, 1023–1024
1,3,5-Trinitrobenzene [99-35-4], 968
Trinitroglycerin, see Nitroglycerin
Trinitroglycerol, see Nitroglycerin
2,4,6-Trinitrophenol, see Picric acid
Trinitrophenol, see Picric acid
Trinitrophenylmethylnitramine, see Tetryl
Trinitrotoluene [118-96-7], 967
2,3,4-Trinitrotoluene, see 2,4,6-Trinitrotoluene
2,3,5-Trinitrotoluene, see 2,4,6-Trinitrotoluene
2,3,6-Trinitrotoluene, see 2,4,6-Trinitrotoluene
2,4,5-Trinitrotoluene, see 2,4,6-Trinitrotoluene
2,4,6-Trinitrotoluene [118-96-7], 1061–1063
3,4,5-Trinitrotoluene, see 2,4,6-Trinitrotoluene

α-2,4,6-Trinitrotoluene, *see* 2,4,6-Trinitrotoluene
*sym*-Trinitrotoluene, *see* 2,4,6-Trinitrotoluene
Trinitrotoluol, *see* 2,4,6-Trinitrotoluene
Tripentenamine, *see* Triallylamine
Tripentylamine, *see* Tri-*n*-amylamine
Triphenylamine *[603-34-9]*, 1015
Tripropylamine *[102-69-2]*, 1105, 1120–1121
Tris(hydroxyethyl)amine, *see* Triethanolamine
Tritisan, *see* Pentachloronitrobenzene
Trolamine, *see* Triethanolamine
Tropilidene, *see* Cycloheptatriene
Tuaminoheptane, *see* 2-Aminoheptane
Turpentine *[8006-64-2]*, 1276–1277, 1278–1279, 1286–1287

Undecane *[1120-21-4]*, 1226–1227
*n*-Undecane *[1120-21-4]*, 1239
Undecylamine *[7307-55-3]*, 1132

Varnish Makers' and Painters' Naphtha *[8032-32-4]*, 1396, 1400, 1408

Vinyl chloride *[75-01-4]*, 1351
Vinylcyclohex-1-ene *[2622-21-1]*, 1280
Vinylcyclohex-4-ene *[100-40-3]*, 1280
1-Vinylcyclohexene *[2622-21-1]*, 1276–1277, 1281–1282
4-Vinylcyclohexene *[100-40-3]*, 1251, 1276–1277, 1278–1279, 1281–1282
Vinylethylene, *see* Butadiene
Vinylstyrene, *see* Divinylbenzene
Vinyltoluene *[25013-15-4]*, 1358, 1359–1360
VM&P Naphtha, *see* Varnish Makers' and Painters' Naphtha

Xenylamine, *see* 4-Aminobiphenyl
*p*-Xenylamine, *see* 4-Aminobiphenyl
Xylene, mixed *[1330-20-7]*, 1302, 1304–1305, 1332–1339, 1333, 1334–1335
*m*-Xylene *[108-38-3]*, 1304–1305
*o*-Xylene *[95-47-6]*, 1304–1305
*p*-Xylene *[106-42-3]*, 1304–1305